Service Systems Management and Engineering

Service Systems Management and Engineering

Creating Strategic Differentiation and Operational Excellence

C. M. Chang
The State University of New York at Buffalo

JOHN WILEY & SONS, INC.

This book is printed on acid-free paper. ♾

Published by John Wiley & Sons, Inc., Hoboken, New Jersey

Published simultaneously in Canada

Library of Congress Cataloging-in-Publication Data:

Chang, Ching M.
Service systems management and engineering : creating strategic differentiation and operational excellence / Ching M. Chang.
p. cm.
Includes index.
ISBN 978-0-470-42332-5 (cloth)
1. Systems engineering. 2. Management information systems. 3. Service industries—Information technology—Management. I. Title.
TA168.C39 2010
658—dc22

2009045962

Printed in the United States of America

10 9 8 7 6 5 4 3 2 1

Dedicated to my loving family, wife Birdie Shiao-Ching,
son Andrew Liang Ping, son Nelson Liang An,
daughter-in-law Michele Ming Xiu, and grandson Spencer Bo-Jun.

Contents

Preface

INTRODUCTION

The U.S. government is forecasting significant growth rates for the U.S. service sectors over the next few decades. According to the U.S. Bureau of Labor Statistics, in 2008 service-providing industries made up 77.2 percent of total employment in the United States. This percentage is projected to increase to 78.96 percent by 2018, including the impact of the general economic recession of 2007 to 2009. It asserts that the overall growth of the service sectors is to continue its upward trend going forward.

Service systems engineering, which employs scientific and engineering principles to add value to service sectors, is a relatively new field, A specific case of service systems engineering is the efficient application of computer-IT-related technologies and the proper management of teams, projects, and processes involved in achieving productivity, time-to-market, convenience, and value-added benefits to customers.

Services require a higher degree of customization and are less amendable to mass production than products. NSF has initiated a large-scale research program to support new research activities in the emerging field of service systems engineering. It is quite likely that, in view of the significant manpower demand projected for the service sectors, schools will also start to participate in training future systems engineers for the service sectors.

The service company objectives, systems view of service enterprises, T-professionals, SSME-12 skills, and unique strategy of text design will be discussed next.

SERVICE COMPANY OBJECTIVES

Companies in the service sector need to pursue both strategic differentiation and operational excellence in order to enlarge and sustain long-term profitability.

Strategic differentiation is essential for companies to create and maintain market competitiveness in the form of differentiable service packages. Market competitiveness must be built on service innovations as well as on customer orientation. Service systems engineers need to become well versed in creating innovations that are both desired and required by customers. Specific thinking strategies are discussed in Chapter 12, "Innovations in Services," to promote the generation of new ideas and concepts in service environments. Besides training in creative thinking, students need to be exposed to marketing management, financial analysis, and cost accounting so that they can make strategic choices in order to realize long-term competitive advantages in the marketplace and attain sustainable profitability for their employers.

Operational excellence is also important, as service-providing employers need to develop, produce, distribute, and support their service packages at low cost, high efficiency, and competitive speed. Systems engineers need to learn methods to cut waste, streamline operations, and invigorate productivity of their operations. They need to be exposed to engineering management principles related to planning, organizing, leading, and controlling. Tools advanced in industrial engineering, such as Lean Six Sigma, value stream mapping, DMAIC methodology, 5 S, optimization, and others can be readily applied to add value to service organizations. In addition, service companies can also achieve cost savings and productivity enhancement by using software modules based on Service-Oriented Architecture (SOA) to perform its noncore activities, as well as utilizing Web-based enabler programs.

For systems engineers to contribute effectively to the important corporate goals of achieving strategic differentiation and operation excellence, they need to exercise their leadership skills in envisioning the future. They would benefit greatly from being exposed to the current "best practices" in the service industries. Many of these "best practices" will be introduced throughout the text. The motto to be instilled in graduates is "constantly searching out best practices and then superseding them." Successful systems engineers and leaders must also be able to "think globally and act locally."

THE SYSTEM-INTEGRATIVE VIEW

When educating future service leaders, we agree with Gosling and Mintzberg (2004) that we need to focus on wisdom, as well as knowledge, in order to emphasize the capacity to combine knowledge from different sources and apply this knowledge judiciously. Educators need to emphasize the systems view, including thoughtful reflection and exposure to alternative ways of thinking in order to augment their students' understanding of the world.

Systems approach—the study of and reactions to the interconnectivity and interactions between functionally related components of a complex enterprise system—applies well to today's service systems. According to Norton (2000), systems engineering should be a required course in every business school and executive program as today's business needs are met through a management framework.

By taking a system-integrative view, knowledge workers would be able to see a panoramic view of the forest; mountains beyond and blue sky above, instead of only the individual trees. These workers would recognize the components in a system, the roles played by them, the interactions between them, and the synergy produced that adds value above and beyond the sum of the individual parts.

This conceptual framework is supported by Section 1106 of the National Competitiveness Investment Act, which Congress passed on August 2, 2007 (U.S. Congress HR 2272-2007). In this section, *service science* is defined as:

> Curricula, training, and research programs that are designed to teach individuals to apply scientific, engineering, and management disciplines that integrate elements of computer science, operations research, industrial engineering, business strategy, management sciences, and social and legal sciences, in order to encourage innovation in how organizations create value for customers and shareholders that could not be achieved through such disciplines working in isolation.

Clearly embodied in this definition are the concepts of (1) systems engineering perspectives, (2) application orientation, instead of only focusing on academic research, (3) integrating multiple knowledge domains to bring about innovations, and (4) value creation emphasis to benefit customers and stakeholders.

From the perspective of the service enterprises, it is evident that they are indeed complex business systems, composed of a number of discrete and multidisciplinary components that interact with one another. Systems thinking is of critical importance (Boardman and Sauser 2008). Service leaders of tomorrow need to adopt a systems view about their enterprises, so that attention is paid to all functional units in order to assure success in the marketplace. Figure 1.7 illustrates some typical functional units in a service enterprise.

Science is a very important component in this business system. Science enables great contributions in the form of new types of services, new ways of delivering services, and new ways of supporting customers. Spohrer and Maglio (2010) offer a comprehensive discussion on several aspects of service science. However, science is just one of many contributors to the success of a service enterprise. Thus, we prefer to call SSME *Service Systems, Management and Engineering* (as opposed to Service Science, Management and Engineering) to recognize the important roles that a variety of other disciplines contribute to the success of this business system.

It is known that certain academic programs are said to have failed to impart useful skills, prepare effective leaders, instill norms of ethical behaviors, and lead graduates to good corporate jobs, because of their misplaced emphasis on academic rigor of scientific research (Schoemaker 2008), instead of focusing on the graduates' competency and understanding of the important drivers of business performance (Bennis and O'Toole 2005). We believe that educating future service leaders with a systems view, which is valid from both academic and relevance standpoints, would avoid falling into the same trap.

According to Chesbrough and Spohrer (2006), universities should develop curricula and advocate research that is focused on (1) business models, (2) productivity, (3) quality of life, and (4) competition and innovations through services. Universities should adopt a multidisciplinary approach involving disciplines such as business, engineering, information technology, computer science, and the social sciences. The research required to meet these goals must be mission oriented and application focused.

The rapidly changing environment of today's global economy calls for people capable of holistic thinking, balancing analysis and intuition, living with ambiguity, and practicing strategic flexibility. The skills and capabilities the students of today acquire need to be practical and relevant.

For SSME graduates, relevance is far more important than scientific rigor, so that they are prepared to deal with the messy ambiguities of the real world. This text emphasizes this practical relevance.

T-PROFESSIONALS

As suggested by Maglio et al. (2006), knowledge workers in services should have an in-depth understanding of a particular field (e.g., business, engineering, computer science, psychology, or anthropology), the vertical part of the T. They also need to have a fundamental understanding of service economy issues, core disciplines that interact with their core knowledge to solve world problems, and the ability to see applications across industries for their knowledge (the horizontal part of the T). In addition, they need to possess broad perspectives and collaborative skills to manage teams, projects, programs, and interact with other professionals. The specific objective of this text is to prepare service engineers to acquire the horizontal part of the T-personality.

The aim of this text is to enable service systems engineers to become much more valuable to service industries by emphasizing broadened perspectives and the knowledge and skills required to strengthen the horizontal part of the T. This text has the potential for educating service systems engineers of any technical background (e.g., industrial engineering, business consulting, information/computation, health science, and other fields). Furthermore, the proposed text could be used by professionals with technological specialty trainings to intensify their marketability and potential value to service enterprises.

Knowledge workers in service sectors must be ready to contribute in two specific ways: strategic differentiation and operational excellence. Both are important, as indicated by Chang (2007):

> *T-shaped professionals need to be capable of creating strategic differentiation and operational excellence for their service employers. Strategic differentiation emphasizes the creation of novel service packages that lead to increased sales revenues. It involves strategy formulation, marketing, design, innovation and supply chains management. Operational excellence, on the other hand, focuses on achieving short-term improvements in processes that leads to lowered cost of services sold. It emphasizes productivity, measurements, quality, operations, human resource management, engineering and computing. T-shape professionals are thus required to become familiar with the principles and methodologies of visioning the future and leading cross-functional teams to bring about breakthrough innovations needed in the marketplace, as well as, applying proven engineering technologies and other tools to achieve gains in productivity, efficiency, quality and cost.*

This text, along with a number of engineering/technological courses, provides the preparation needed to accomplish these goals.

SSME-12 SKILLS

A key strategy of this text is to address the skills and capabilities deemed essential to service system engineers. SSME-12 skills are composed of two groups of skills and capabilities. The first group was suggested by Sorby et al. (2006), comprising twenty-four specific skills important to B.S. engineers in the service industry. These twenty-four skills may be organized into the following six categories:

1. Management of service systems
2. Operations of service systems
3. Service processes
4. Business management

5. Analytical skills
6. Interpersonal skills

In a 2009 report, the Carnegie Foundation for the Advancement of Teaching emphasized critical thinking, problem solving, teamwork, and a multidisciplinary approach. Choudaha (2008) suggested a list of required skills and attitudes for master's degree graduates, based on an online Delphi study. This list includes integration, collaboration, adaptability, critical thinking, interpersonal competence, problem solving, system conceptualization, and diversity orientation. Some of these skills are similar to those identified by Sorby et al. (2006).

For service systems engineers at the master's degree level, I have added the following six groups of skills and capabilities to round out the twelve skill sets:

7. Knowledge management
8. Creativity and innovations in services
9. Financial and cost analysis and management
10. Marketing management
11. Ethics and integrity
12. Global orientation (thinking globally and acting locally)

These twelve categories of skills and capabilities constitute the SSME-12 skills, which, in turn, fully encompass the eleven attributes specified by National Academy of Engineering to be important for Engineers of 2020:

1. Strong analytical skills
2. Practical Ingenuity
3. Creativity
4. Communication
5. Business and management
6. Leadership
7. High ethical standards
8. Professionalism
9. Dynamism
10. Agility, resilience, and flexibility
11. Life-long learning

This text is focused on training students in the SSME-12 skills by using a set of systematically assembled and properly designed class examples, class problems, application notes, specific assignments, and supplemental readings. Many of these skills and capabilities will be illustrated in the text. Business cases and best practices are used throughout to further promote the intellectual exchanges between instructors and students. The objective is to make sure that the students will not only be able to think strategically, make decisions rationally, and execute effectively, but also keep up with innovations in learning. This new text is expected to be of general use to all service systems engineers, regardless of the specific service sectors they elect to enter.

UNIQUE STRATEGY OF TEXT DESIGN

The primary objective of this text is to ready readers to contribute toward the needs of service sector companies in strategic differentiation and operational excellence. The text is designed to prepare students to lead service systems teams, projects, technologies, or programs in a knowledge economy within competitive global environments.

The preparation of this text follows the proven *Three-Decker Leadership-Building Architecture* of our awarding winning text: C. M. Chang, *Engineering Management: Challenges in the New Millennium* (Upper Saddle River, NJ: Prentice Hall, 2005), which has been adopted by more than twenty U.S. universities and numerous international schools, as well as being translated into the Korean language.

The new chapters in this text include: "Introduction to SSME," "Innovations in Services," "Knowledge Management," and "Operational Excellence." Other chapters such as "Globalization," "Service Systems Engineers as Leader," "Ethics," and "Marketing Management" have been significantly revised to include service-based examples, problems, and business cases.

A large number of service-related business cases are contained in the text-end Appendix. These cases cover various service sectors and address diversified SSME issues. They represent a useful extension of the exercise problems listed at the end of each chapter.

By using a set of systematically assembled and properly designed class examples, class problems, application notes, specific assignments, and supplemental readings, selected skills and capabilities are incorporated in the relevant chapters.

This book is organized in three parts. Part I reviews the functions of engineering management. Chapters 2 through 5 are entitled "Planning," "Organizing," "Leading," and "Controlling." Part II covers the business management fundamentals, Chapters 6 through 8, which include "Cost Accounting and Control," "Financial Accounting and Management," and "Marketing Management." Part III addresses the service leadership in the new millennium and contains Chapters 9 through 14: "Service Systems Engineers as Managers and Leaders," "Ethics," "Knowledge Management," "Innovations in Services," "Operational Excellence," and "Globalization." The text-end Appendix includes more than thirty selected business cases. Figure 1.11 illustrates the current Three-Decker Leadership-Building Architecture, which forms the basis of this text design.

This book is suitable for use as a text for two 3-credit graduate courses in 15-week semesters, when selected business cases are added. It may also be used as text for one 3-credit graduate course in a 15-week semester in the absence of case studies, as well as for an elective course at the senior undergraduate level. It could be part of the required core courses in a typical master's degree program centered on service systems management and engineering (SSME), which needs to be effective yet flexible, as the service industry is composed of numerous divergent sectors, such as professional and business consulting, transportation, distribution and logistics, healthcare, IT services, financial services, entertainment, education, government, and others. A sample curriculum for such a master's degree program might include the following three-credit courses:

1. Required courses: SSME-1 and SSME-2 (based on this text).
2. Four more required courses form the core, and are selected from the following: (a) Six Sigma Quality in Services, (b) Supply Chain Management and Global Operations, (c) Services-Oriented Architecture and Web Services,

(d) Data Mining in Services, (e) Innovations in Services, (f) Project Management, (g) E-commerce Technology, (h) Technical Communications, and (i) others.

3. Three elective courses are to be selected from a broader list of service sector–specific courses in order to better customize the degree program to the needs of individual students: (a) Critical Issues in Healthcare, (b) Linear Programming and Optimization, (c) Process Simulation and Control, (d) Advanced Marketing, (e) Consumer Behavior, (f) Knowledge Management, (g) Operations Management, (h) Logistics, (i) Stochastic Methods, and (j) others.
4. One capstone project, which is to emphasize the application of SSME course knowledge and skills to add measurable value to a service organization.

This curriculum, designed in the spirit of the T-personality, not only enables graduates to understand the requirements necessary to be successful within any given service enterprise but also imparts the necessary technological background for the actual implementation of this learning.

In summary, this text aims to equip the future service systems engineers and leaders of tomorrow with the necessary knowledge and skills to achieve strategic differentiation and operational excellence in the growing service sector market.

REFERENCES

Bennis, Warren G., and James O'Toole. 2005. "How Business Schools Lost Their Way." *Harvard Business Review* (May).

Boardman, John and Brian Sauser. 2008. *Systems Thinking: Coping with 21st Century Problems.* Boca Raton, FL: CRC Press.

Choudaha, Rahul. 2008. "Competency-based Curriculum for a Master's Program in Service Science Management and Engineering: An Online Delphi Study." Doctoral Dissertation, University of Denver, Denver, Colorado.

Chesbrough H., and Jim Spohrer. 2006. "A Research Manifesto for Service Science." *Communications of the ACM* 49 (7): 35–41.

Chang, C. M. 2007. "Contributed Comments to IBM-Cambridge Symposium entitled "Succeeding through Service Innovation."" July 2007, University of Cambridge, Cambridge, England. URL: /www.ifm.eng.cam.ac.uk/ssme/comments/business.html#carl-chang

Gosling, Jonathan, and Henry Mintzberg. 2004. "The Education of Practicing Managers," *MIT Sloan Management Review* 45 (4): (Summer).

Heftly, Bill, and Wendy Murphy, eds. 2008. "Systems Engineering." In *Service Science Management and Engineering—Education for the 21st Century.* New York: Springer.

Maglio, P. P., S. Srnivasan, J. T. Kreulin, and J. Spohrer. 2006. "Service Systems, Service Scientists, SSME and Innovation." *Communications of the ACM* 49 (7): 81–85.

Norton, David P. 2000. "Is Management Finally Ready for the Systems Approach?" Harvard Business School Balance Scorecard Report, Article Reprint No. B0009E.

Schoemaker, Paul J. H. 2008. "The Future Challenges of Business: Rethinking Management Education." *California Management Review* 50 (3): (Spring).

Sorby, Sheryl A., Leonard J. Bohmann, Tom Drummer, Jim Frendewey, Dana Johnson, Kris Mattlia, John Sutherland, and Robert Warrington. 2008. *Defining a Curriculum for Service.* Houghton: Michigan Technological University.

Spohrer, Jim and P. P. Maglio. 2010. "Service Science: Toward a Smarter Planet," Chapter 1 in Salvendy, Gavriel and Waldemar Karwowski (Eds), *Introduction to Service Engineering*. John Wiley (January).

Acknowledgments

It is indeed a pleasure to acknowledge the invitation extended to me in 1987 by Dr. George Lee, then dean of the School of Engineering and Applied Sciences of the State University of New York and current SUNY Distinguished Professor, to design and teach two graduate courses on engineering management. The course notes developed at that time and updated regularly ever since have been built on a design model containing three parts: (1) engineering management fundamentals (planning, organizing, leading and controlling), (2) business management fundamentals (cost accounting, financial accounting and management, and marketing management) and (3) engineering leadership in the new millennium (engineers as leaders, Web-based tools, ethics, globalization, and challenges in the new millennium). With the energetic support of Dr. George Lee, as well as his successor, Dr. Mark Karwan, current Praxair Professor in Operations Research, I have been able to test-teach this "Three-Decker Leadership-Building Architecture" continuously since 1987. I am deeply indebted to both Dr. George Lee and Dr. Mark Karwan for their constant support and encouragement over the years.

I want also to express my appreciation for an excellent opportunity to serve as a member of a departmental committee, which was charged to design a new master's degree program in service systems engineering (SSE) during the period of November 2006 to August 2007. For a short period of time in 2007, I also served as the director of this new master's degree program. Because of these activities, I was highly motivated to scan, collect, and study a vast amount of service, engineering, and business literature, which subsequently enabled me to modify quite a few sections in the first two parts and new service-centered chapters to the third part of this Three-Decker Leadership-Building Architecture. These new chapters include Chapter 9, "Service Systems Engineers as Managers and Leaders," Chapter 11, "Knowledge Management," Chapter 12, "Innovations in Services," and Chapter 13, "Operational Excellence." During the ensuring semesters, I was able to test-teach all these new and modified chapters.

Various example problems in the current text came from the test problems and business cases used in two graduate courses I taught on engineering management for SSE. The *combinatorial, heuristic and normatively guided* method of producing innovative ideas was tested in about twenty team projects, each of which was conducted in multiple rounds with graduate students as team participants. During the same period of time, I directed all capstone projects of all students enrolled in this master's degree program in SSE. Many of these capstone projects, which were sponsored by industrial firms, contributed to the discussions contained in the current text. I appreciate the valuable inputs many of the SSE graduate students have made, which have enhanced the usefulness of this text enormously.

Finally, I wish to acknowledge the extraordinary support of the John Wiley team, led by Robert L. Argentieri, and assisted by Daniel Magers, Nancy Cintron, William/Cheryl Ferguson, Brian Roach, Jeffrey Faust, and Victor Aranjo. Their dedication and commitment have been invaluable toward the completion and success of this text. I also wish to express my appreciation for the able assistance offered by Andrea Strudensky.

C. M. Chang, Ph.D., MBA, PE

State University of New York at Buffalo
Buffalo, New York

Chapter 1

Introduction

1.1 INTRODUCTION

Services represent by far the largest contributor to the U.S. economy. Based on data published by U.S. Bureau of Labor Statistics, services and the total employment in the service sectors make up over 75 percent of the U.S. gross domestic product (GDP). As well, up until the year 2018, 96 percent of all 15.3 million new jobs are expected to come from the service sectors. Accordingly, the importance of services to the U.S. economy is clearly self-evident (Bartsch 2009). In fact, systems engineers are ranked at the very top of the list of "Best Jobs in America," with a 45% growth over a ten-year period, according to a Moneyline article (Anonymous 2009).

This chapter provides the definition and characteristics of services in contrast to products. Following this explanation, service sectors in the United States are introduced. A systems view of service enterprises is presented and the principles of service systems engineering are delineated. The skills and capabilities deemed essential to service systems engineers and leaders are then discussed, including how this text will help future graduates acquire the T-personality to meet the challenges of the new millennium. Conclusions are then presented.

1.2 SERVICES VERSUS PRODUCTS

Services are defined as "combinations of deeds, processes, and/or performances provided to customers in exchange relationships among organizations and individuals" (Zeithami et al. 2006). Services have seven key characteristics:

1. Provider and recipients are in direct face-to-face contact—based on the service roles, self-selected by the providers to prefer for such direct contact.
2. In service sectors, the merits of quality and productivity are not well defined (e.g., no physical parameters as existed in the goods sector)—raising issues related to whether cognitive science, organization, and engineering systems are more prominent in service delivery, productivity, and quality.
3. Although the physical assets depreciate over time and use in goods sector, key assets are generally reusable in the service sector. These service assets may actually increase in value. Examples are organization and human resources that derive from knowledge bases and skills realized in service interactions.

4. In the goods sector, equipment is usually newly designed and hence protected by intellectual rights. In services, equipment in application is often purchased and nonprotectable.
5. Services focus on knowledge-based understanding of technology and on how to use technology.
6. For service organizations, the keys to success are to adapt, utilize, and incorporate technological processes and equipment.
7. The right strategy of management of technology for services needs to take these factors into account.

Services are activities that cause a transformation of the state of an entity (e.g., a person, product, business, and region/nation) in a manner that is mutually shaped by its provider and the client. The transformation of the state of a person can be accomplished by services related to foods, healthcare, leisure, hospitality, travel, financial/investment advisement, banking, legal, education, entertainment, mail/package delivery, and others. The transformation of the state of a product is made possible by the design, operations, and maintenance services rendered. The transformation of the state of a business is the result of pursuing management consulting, outsourcing, e-procurement, marketing research, mergers and acquisitions, and others such corporate activities. The transformation of the state of a region/nation requires consulting advice and analysis related to regional/national economic advancement strategy, taxation policy, and other such macroscopic issues.

Services activities are becoming increasingly more diversified. Individual services are relatively simple, although they may require customization and a significant back-office support (e.g., database, knowledge management, analysis, forecasting, etc.) to assure quality and a timely delivery. Product services are also relatively straightforward, as product specifications, performance standards, quality control, installation guidelines, and maintenance procedures require good communication and understanding between providers and users. Business services are complex; some may involve intensive negotiation, work process alignment, quality assurance, team collaboration, and service coproduction. Regional and national services are even more complex, as they may affect policy, custom regulations, export permits, local business practices, logistics, distribution, and other such issues.

Services play an important role in an economy, as illustrated in Fig. 1.1 (Guile and Quinn 1988).

Services may also be classified into either front-stage or back-stage activities, depending on how close/remote the activities involved are to/from the customers. Front-stage activities are those in which provider and client interact directly. Customization leads to high value and high profit, whereas standardization tends to diminish profit margins. Back-stage activities do not directly involve customers and are mostly related to the efficient production of the services.

Services have a varying degree of front- and back-stage activities, which, in turn, have a varying degree of client interaction intensity. Figure 1.2 illustrates these specific characteristics of services. Services are also known to require different levels of labor intensity and degrees of customization, as depicted in Fig. 1.3. Table 1.1 illustrates a number of examples for the front- and back-stage activities involved in services.

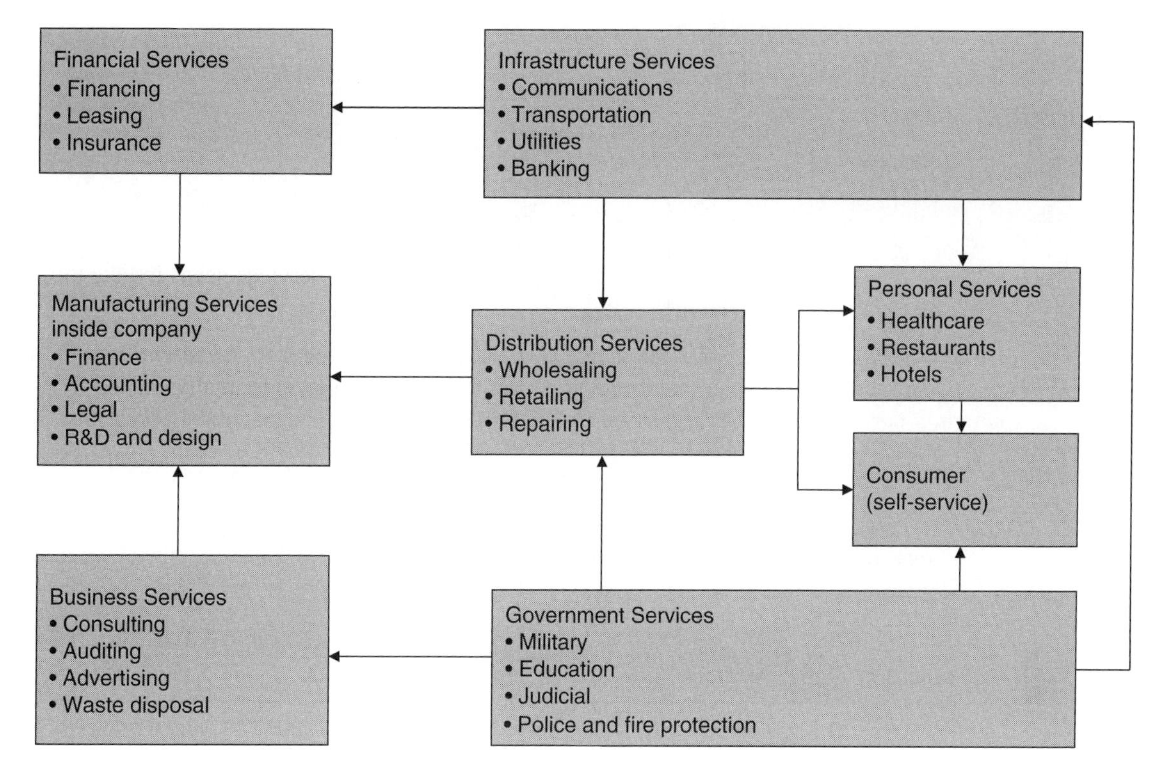

Figure 1.1 Roles of services in an economy. (Adapted from Guile & Guinn, 1988).

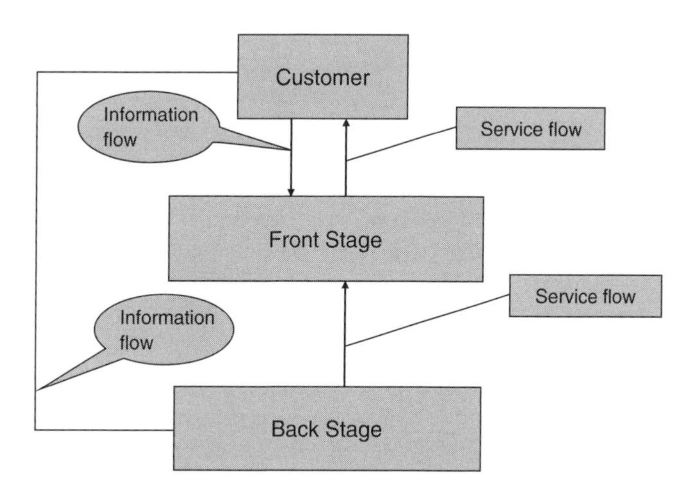

Figure 1.2 Service characteristics.

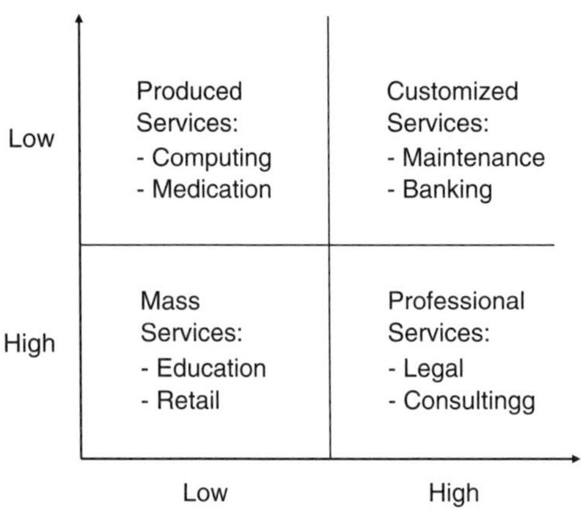

Figure 1.3 Service customization versus labor intensity. (Adapted from Fitzsimmerman et al. 2008)

Table 1.1 Front- and Back-Stage Service Activities

Number	Service Sectors	Front Stage	Back Stage
1	Health care	Working with patients	Setting up and maintaining facilities
2	Education	Delivering lectures	Setting up and maintaining educational facilities
3	Retail	Sales experience	Logistics
4	Professional	Assessment and consultation	Research, data analysis, interpretation, knowledge creation, insights preservation
5	Information	Presenting and delivering	Gathering, sensing, and organizing
6	Communications	Billing	Setting up infrastructure
7	Transportation	Transport experience	Maintaining the fleet
8	Utilities	Delivering, billing, and support	Setting up and maintaining infrastructure

Services may also be classified as high technology and low technology. Flipping hamburgers and sweeping floors are low-tech service activities, whereas conducting an e-market transaction and offering an engineering consultation service are high-tech activities. Technology-intensive services have at least five special features (Tien and Berg 2003):

1. *Information-driven.* The creation, management, and sharing of information is crucial to the design, production, and delivery of services.

2. *Customer-centric.* Customers are generally a co-producer of the services, as in the case of self-service. Customers require a certain degree of service adaptation or customization, and customers must be satisfied with the rendered services.
3. *E (electronics)-oriented.* Services are becoming more e-oriented. Thus, e-access, e-commerce, and e-customer management are crucial to e-services.
4. *Productivity-focused.* Both efficiency and effectiveness are important in the design, delivery, and support of services.
5. *Value-adding.* Services need to add value to the target clients. For profit-seeking service companies, the value so produced assures their profitability. For non-profit service entities, it reinforces the goodness of its policy.

Services differ from products in a major way. On the one hand, services involve intensive interactions with customers in the front-stage activities, with a varying degree in magnitude. On the other hand, products are mostly dominated by back-stage activities, which receive only small amounts of customer inputs. This contrast is illustrated in Fig. 1.4.

Nambisan (2001) also offers an excellent contrast between services and products, as depicted in Table 1.2, as related to (1) intellectual property rights, (2) complementaries to other offerings, (3) returns from economy of scale, (4) abstracting knowledge and integrating technology, and (5) connection with users.

Tidd and Bessant (2009) offer another useful comparison between six basic characteristics of products and services:

1. *Tangibility.* Products are more tangible than services.
2. *Perceptions.* Service quality is perceived based on physical evidence (the physical setting at which the service is offered), responsiveness (speed of service and willingness to help), competence (ability to perform the service dependably), assurance (knowledge and courtesy of staff and ability to convey trust and confidence), and empathy (provision of caring, and individual attention)
3. *Simultaneity.* Products are typically made in advance of consumption, whereas services are consumed mostly at the time of production. Simultaneity brings about potential of quality management problems related to the identification and correction of service errors as well as capacity-planning problems to match supply with demand.

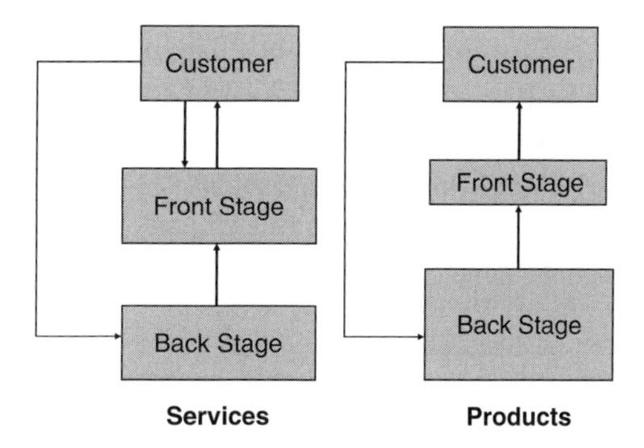

Figure 1.4 Services versus products.

Table 1.2 Services versus Products

Number	Key Issues	Products	Services
1	Intellectual property rights	Very important	Less important
2	Complementarities to other offerings	Very Important	Less Important
3	Returns from economy of scale	A fixed cost structure allows for higher marginal returns from scale.	A variable-cost structure makes increased returns from scale rare
4	Abstracting knowledge and integrating technology	Capture generic product knowledge so that the product can be used in a variety of contexts.	Knowing clients' idiosyncrasies is more important than knowledge abstraction
		Architecture-level technology integration is important for the smooth running of the end product.	Relying on data-interface technology integration; the primary emphasis is on development efficiency.
5	Connection with users	Long-term relationships; users are technologically sophisticated.	Project-driven relationships; users are technologically unsophisticated.

Source: Nambisan 2001

4. *Storage.* Capacity-management problems may arise due to an imbalance between supply and demand. Such problems may be mitigated by pricing (discounts at off-peak time to induce demand), adding temporary workers, and/or outsourcing.
5. *Customer contact.* Services demand high level of customer contact, some more (medical, business consulting) and some less (financial services, information).
6. *Location.* The proximity factor is more important for services than for products, making services more local and less competitive. Only about 10 percent of services in the developed economies are traded internationally.

For services to do well, a company must pay attention to the following:

- Control the variable costs—reusable software assets, knowledge management, process rigor, efficiency.
- Hire the right people for interacting with customers to excel in customization.
- Automate back-office work and outsource low-value activities to achieve speed and quality advantages.

As we have now discussed what services are, let us take a look at the service sectors in the United States.

1.3 SERVICE SECTORS

The U.S. economy consists of three major sectors: agriculture, manufacturing, and services. Over the years, significant changes have occurred in each of these sectors due to technological advancement, market expansion, customer preferences, and globalization.

According to U.S. Bureau of Labor Statistics (Bartsch 2009), the service-providing industry made up about 77.2 percent of total employment in the United States in 2008. This percentage is projected to increase to 78.96 percent by 2018. The total number of jobs is expected to increase by 15.3 million over the ten-year period from 2008 to 2018, and 95.6 percent of this increase will come from the service-providing sectors. The growth of U.S. service sectors is clearly astounding. Figure 1.5 illustrates this remarkable trend.

The service-providing industry in the United States is divided into thirteen sectors:

1. Professional and business services
2. Healthcare and social assistance
3. State and local government

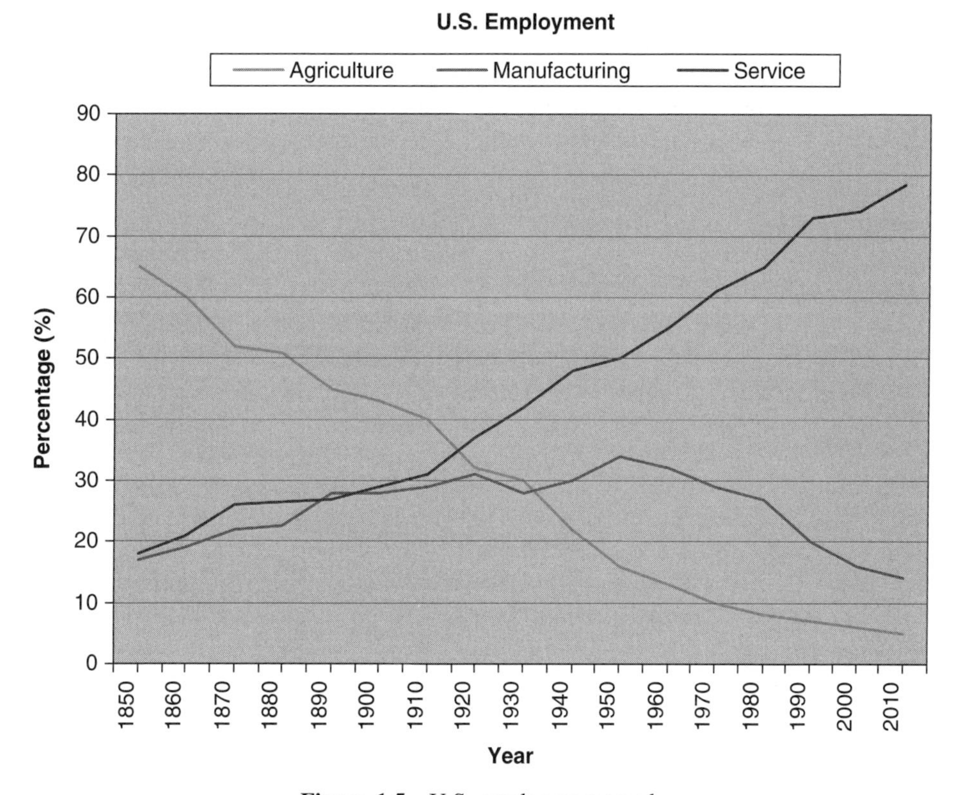

Figure 1.5 U.S. employment trend.

4. Leisure and hospitality
5. Other services
6. Educational services
7. Retail trade
8. Financial activities
9. Transportation and warehousing
10. Wholesale trade
11. Information
12. Federal government
13. Utilities

These are ranked in the order of relative job growth. The total job growth of all service sectors is projected to be 14,601,000. The percentage in Table 1.3 presents the fraction of this total contributed by each service sector. By 2018, a whopping 66.86 percent of all service-providing new jobs will come from the first three sectors alone. Figure 1.6 illustrates the projected percent change in U.S. employment by industry, from 2008 to 2018. This percentage is calculated by the net job change in a given service sector divided by the base job number for that sector in year 2008 (see Bartsch 2009).

It is also interesting to note that the projected total job growth in the United States for the same period is only 15,274,000, only slightly more than the total new jobs projected for service sectors. This chart clearly points out that job losses in other industries, such as manufacturing, agriculture, and mining, are projected to be quite substantial. Projected job growth and decline in occupations are shown in Table 1.A1 (Appendix 1.12.2).

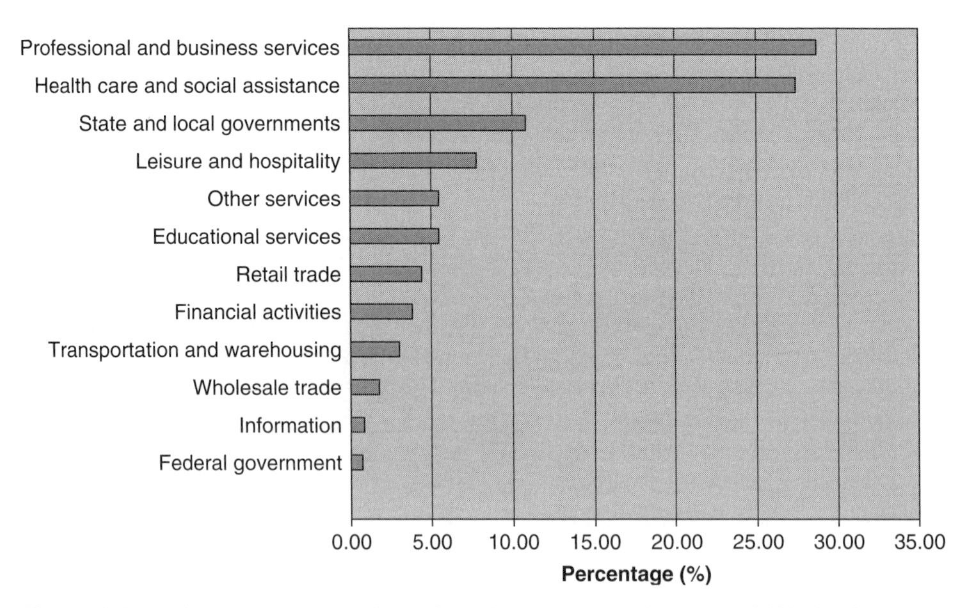

Figure 1.6 Projected percentage change in U.S. employment by sector, 2008–2018 (Bartsch 2009).

Table 1.3 Projected Job Growth by Service Sectors (Units: 1,000 jobs)

	Year 2008	Year 2018	Change	Percentage
Service-providing sectors	116,452	131,053	14,601	100%
Professional and business services	17,778	21,968	4,190	28.70%
Health care and social assistance	15,819	19,816	3,997	27.37%
State and local government	19,735	21,327	1,592	10.90%
Leisure and hospitality	13,459	14,601	1,142	7.82%
Other services	6,333	7,142	809	5.54%
Educational services	3,037	3,842	805	5.51%
Retail trade	15,356	16,010	654	4.48%
Financial activities	8,146	8,703	557	3.81%
Transportation and warehousing	4,505	4,950	445	3.05%
Wholesale trade	5,984	6,220	256	1.75%
Information	2,997	3,115	118	0.81%
Federal government	2,764	2,859	95	0.65%
Utilities	560	500	−60	−0.41%
Total U.S. job growth, all sectors	150,932	166,206	15,274	
Fraction of service jobs in total			95.60%	

Source: BLS, (2009) "Employment Projections 2008–18," (December 10).

Table 1.4 Global Demographics

Continents	% of Population Aged 60 or Older 2002	% of Population Aged 60 or Older 2050	Working Age Persons per Aged 65 or Older Person 2002	Working Age Persons per Aged 65 or Older Person 2050
Europe	20.00%	37.00%	3.90%	1.80%
North America	15.70%	27.10%	5.00%	2.80%
Asia	8.60%	22.90%	11.10%	3.90%
Latin America	7.90%	22.10%	11.00%	3.80%
Africa	5.00%	10.00%	16.80%	8.90%

Source: 2002 Population Data, United Nations

There are numerous reasons why the service sectors in the United States are expected to experience such rapid growth. A principal reason is the market demand of services due to a change in demographics. Table 1.4 indicates that from 2002 to 2050, the percentage of elderly people in all continents will drastically increase, and the ratio of working age persons to older people is expected to decrease accordingly.

Elderly people need services related to healthcare, hospitality, leisure, financial consultation, and investment, among others. Plentiful of these services will become increasingly computerized and automated, thus offering growth opportunities to computer-IT industries, which are mainly responsible for producing hardware platforms and software tools to facilitate the delivery of such services. The U.S. Congress has initiated programs that foster the digitalization of patient's medical records, so that healthcare services can be delivered faster and more accurately (Kinsbury 2008). The National Science Foundation has initiated a Service Enterprise Engineering program that sponsors research activities involving multidisciplinary engineering collaboration in healthcare.

Beside market demand for services, another major reason for the expected high growth in the service sectors is the continuous improvement of quality of life in all age groups. Companies in the service sectors strive to innovate new ways to make present-day services cheaper, faster, better, and more convenient to users.

Opportunities are also available due to the generally perceived need to invigorate the productivity in the service sectors, which typically lag behind that of the manufacturing sectors. Tools such as Lean Six Sigma, total quality management, automation, value stream mapping, and others that were advanced and perfected in the manufacturing industry are now increasingly being applied to the service industry.

Services are known to have made increasingly larger contributions to the national GDP than products. Apte et al. (2008) pointed out that the economy is also moving from predominantly material economy (noninformation) to an information economy. As a consequence, the economy may be decomposed into four subsectors:

1. Material products
2. Material services
3. Information products
4. Information services

Future service jobs are likely to become increasingly more information and knowledge intensive. Bloomberg (2010) addresses comprehensively the characteristics of work in a service economy.

As the need for highly skilled systems engineers in the service sectors becomes evident, both U.S. government and service sector leaders will need to start encouraging new educational programs at universities to prepare future service systems engineers to enter this emerging growth field.

1.4 SYSTEM-INTEGRATIVE VIEW OF SERVICE ENTERPRISES

Systems thinking is defined in Frank (2006) as the ability to see the whole picture and its relevant aspects (e.g., emergent properties, capabilities, behaviors, and functions), above and beyond its components, parts, and salient details. Engineering systems thinking is enabled by two components:

1. *Thinking skills*—interdisciplinary knowledge (having expertise in one and being knowledgeable in several others), ability to communicate with others in their fields and cognitive characteristics.
2. *Personal traits*—behavioral competence.

The cognitive characteristics involve seeing the big picture, having an understanding of the whole system and its environment, recognizing the interconnections between components, system synergy, analogies and parallelism between systems, and appreciating the system from multiple perspectives, such as economical, managerial, and social. One needs to be able to take a panoramic view to appreciate the forest, lakes, and the snow-capped mountains, rather than seeing only the individual trees.

For example, the United States has constantly reviewed its energy independence policy, as over 50 percent of the oil consumed must be imported from abroad to meet the domestic consumption. In practicing a system's thinking methodology, any emergent energy resource ought to be evaluated from the viewpoints of (1) technological feasibility, (2) economic viability, (3) environmental acceptability, and (4) resource conservational characteristics.

Boardman and Sauser (2008) offer a useful methodology to graphically diagram the interrelationship between the components of complex systems, such as UK Rail Network, Digital TV Business Model, Intel Community, USAF Combat Strategy, and others. Luczak and Gudergan (2010) point out the evolution of service engineering toward the design of integrative services.

Martin (2009) promotes integrative thinking as a key methodology for managers to make critical choices by processing multiple, sometimes conflicting, views affecting a given complex situation. Table 1.5 exhibits a modified four-stage approach of the integrative thinking process, as compared to conventional thinking.

According to Norton (2000) and Mott (2010), the systems approach focuses on evaluating and reacting to the interconnectivity and interactions between functionally related components of a complex enterprise system. This system-integrative approach applies well to today's complex service systems. It is the management framework that meets the current needs. Figure 1.7 illustrates the system-integrative view about a service enterprise, for which the following notations apply:

A. Enterprise offers service for sales to target customers and clients in exchange for payment.
B. Customer's purchases lead to enterprise profitability, which is the key purpose of the enterprise's existence.
C. Profitability is monitored and documented by financial management, which in turn feeds the information to enterprise management.
D. Business management commits resources to support various internal functions.

Table 1.5 Integrative Thinking

Stage	Conventional Thinking	Integrative Thinking	Paradigm
1. Determining salience	Focus only on obviously relevant features.	Seek less but potentially relevant factors.	Preserve as many different perspectives as possible.
2. Analyzing causality	Consider one-way linear relationships between variables in which more of A produces more of B.	Consider multidirectional and nonlinear relationships among variables.	Key is to discover correlations between many variables.
3. Envisioning the decision architecture	Break problems into pieces and work on them separately or sequentially.	See problems as a whole, examining how the parts fit together and how decisions affect one another.	Take a systems view.
4. Achieving resolution	Make either-or choices, settle for best available options.	Creatively resolve tensions among opposing ideas, generating innovative outcomes.	Combine features to define something new.

Adopted and modified from Martin (2009)

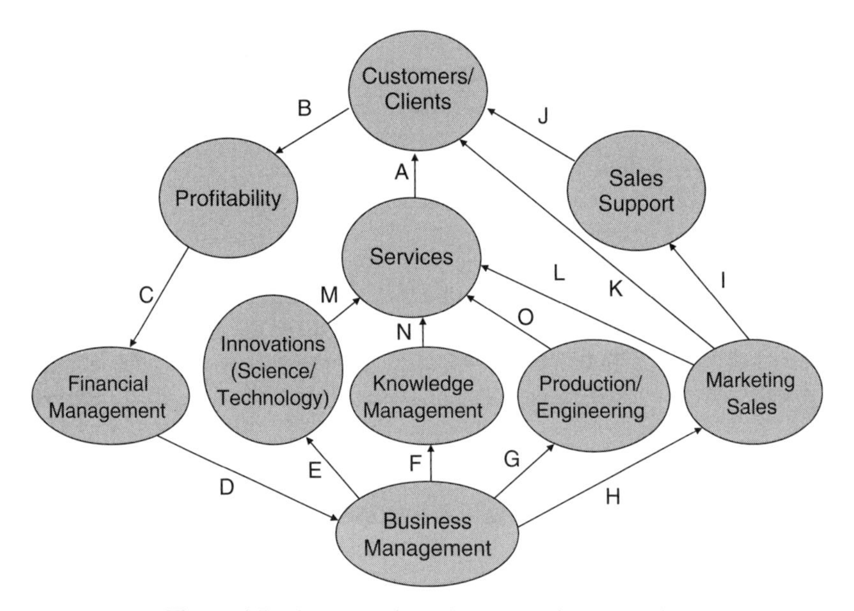

Figure 1.7 Systems view about a service enterprise.

E. Business management supports innovation. Here science and technology play an important role in advancing and delivering new services.
F. Business management supports knowledge management, which represents the key to creating competitive advantages in the marketplace by creating, updating and applying corporate knowledge, both the explicit and tacit kind.
G. Business management supports the production and engineering functions, including cost and quality control.
H. Business management supports the important marketing/sales functions.
I. The marketing/sales supports the sales support, which is critical to achieving customer satisfaction.
J. Sales support interacts with customers, and its activity must be managed carefully.
K. Marketing and sales generate service awareness in the marketplace, solicit feedback from customers, and interact with them to customize the service offering.
L. Marketing/sales suggest new services to be offered to improve competitiveness and profitability.
M. Innovations represent a key contributor to foster enterprise's competitiveness and service differentiation in the marketplace.
N. Knowledge management utilizes the firm's core competency in providing the competitive services to customers with superior knowledge contents.

O. Production/engineering takes care of the generation of services offered, and the control of their cost, quality, and reliability.

Taking the system-integrative view from a service enterprise, we can see various external components that exert significant impact. See Fig. 1.8, for which a different set of notations apply:

A. Governmental regulations have impact on the service sector involved, affecting opportunities and threats present in the marketplace.
B. Global competitions could present threats to the service acceptance in the marketplace.
C. Economic conditions (money supply, employment situation, consumer confidence, investment climate, etc.) have a direct impact on profitability.
D. Change in consumer preferences must be carefully monitored.
E. Global market of talents is critical to the supply of right knowledge workers.
F. Suppliers of capital (e.g., investors, bankers and funds).
G. Technology suppliers (e.g., IT, Web services, open innovations, etc.) have a profound impact on the service business.
H. Globalization (scale and scope) will influence all service sectors.

Viewed as systems, service enterprises may be considered as consisting of ten interacting components, requiring a systems thinking methodology to address its problems and opportunities (Fig. 1.7). This system-integrative approach must be actively

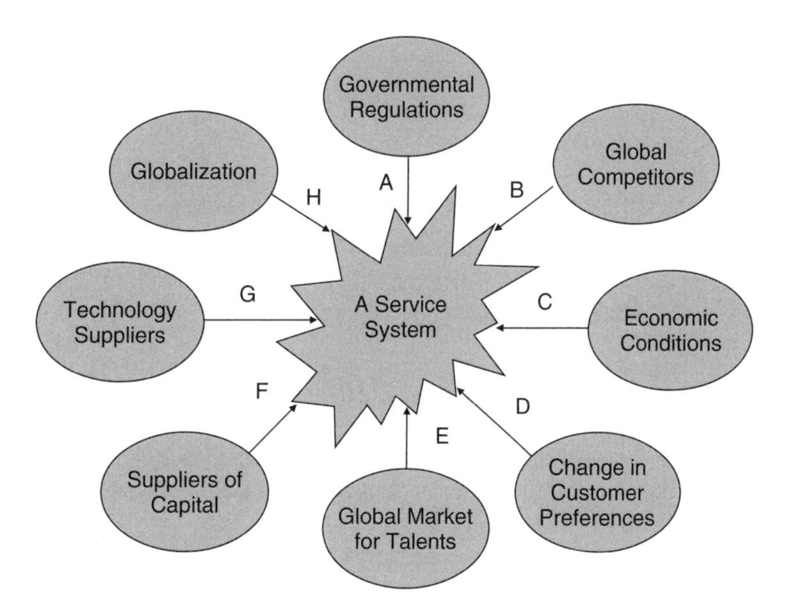

Figure 1.8 Systems view from a service enterprise.

nurtured by service systems engineers and leaders in order for them to maximize their contributions to their employers. Furthermore, they should recognize the following four system characteristics of a service enterprise:

1. A variety of different disciplines contribute to the overall success of the system. Service workers must be able to do their best to communicate and collaborate with others.
2. All functions will need to interact closely with one another and be properly coordinated via engineering management functions such as planning, organizing, leading, and controlling. The use of existing engineering tools such as Lean Six Sigma, value stream mapping, failure mode and effect analysis, and others will be essential for the service enterprise to achieve operational excellence.

   Productivity can also be improved by reorganizing the enterprise's operations and then employing service-oriented architecture (SOA)–based software modules to perform its noncore activities (Merrifield et al. 2008). SOA software modules are designed by following standard specifications so that countless different users can access them via the Internet to accomplish specific predefined outcome. Cost is reduced and productivity is improved by outsourcing these noncore activities to Web-based service vendors. Gartner, an IT research and consulting firm, predicts that by 2010, more than 80 percent of major companies with mission-critical activities will be implementing the SOA-based systems, compared to 50 percent in 2007.
3. Innovation remains the key driving force for a service enterprise to capture new opportunities (e.g., consumer preference, new technologies, special skills, and talents) in the marketplace, to overcome threats imposed from the outside, and to cultivate strategic differentiation and sustain long-term profitability. Innovation requires the active participation of all people in the organization.
4. The success of a service enterprise will need to be built on both strategic differentiation and operational excellence.

1.5 SERVICE SYSTEMS ENGINEERING

Service systems engineering is defined as a multidisciplinary engineering field that addresses a service system from the management, life-cycle, customer, and value-creation perspectives (Tien and Berg 2003). It is a relatively new field that employs scientific and engineering principles to add value to the clients of the service sectors companies. Table 1.6 summarizes the basic definitions and concepts related to service systems engineering.

The basic disciplines in support of service systems engineering were outlined in Tien and Berg (2003), which I have modified as displayed in Table 1.7. It is interesting to note that the disciplines of industrial engineering, operations research, and business management contribute by far the most useful tools for service systems engineering.

Table 1.6 Service Systems Engineering

#	Concepts	Definitions	Attributes
1	Service System	An assemblage of components having interactions with or interdependence interactions with or interdependence of each other	Types: physical or conceptual; open or closed; static or dynamic; business or social; elements components, attributes, relationships; level of abstraction: high (with more component) or low (with less components)
2	Engineering	Disciplines that apply scientific principles to specific projects or tasks to add value or to create knowledge.	Work of engineering involves definition synthesis, analysis, design, test and, evaluation.
3	Management	Plan, organize, lead and control	Feedback (evaluate performance against standards); control (communications, self-regulating, adaptation, optimization and/or management)
4	Life cycle	Elapse time progressing from the initial to final stages of a system's useful life.	Process involves needs assessment, design/development, production/ construction, utilization/support and phase-out/disposal
5	Customer	Recipient of the value package	Understand needs/requirements, manage expectations and assure satisfaction.
6	Value	Outcome of significance resulted from having spent the efforts	Profitability, market share, firm's reputation and industrial standing, better design, innovative products/ services, new knowledge, improved understanding of issues at hand, quality of life to constituents, well being of citizens

Adopted and Modified from [Tien and Berg (2003)]

Table 1.7 Selected Disciplines Regarded as Keys to Service Systems Engineering

Disciplines	Specific Methodologies and Technologies
Industrial engineering	Quality management, cost analysis, risk analysis simulation, human factors, cognitive ergonomics scheduling, manufacturing systems, lean Six Sigma project management, facility operations
Operations research	Optimization, linear programming, logistics Game theory
Mathematics	Probability, modeling
Statistics	Data mining, visualization
Computer science	Software programming, artificial intelligence data mining, service-oriented architecture
Bioengineering	Bioinformatics, informatics
Business management	Strategic planning, operational planning, supervision, financial analysis, marketing management, supply chain management

Source: Adopted from Tien and Berg (2003) with modifications.

Example 1.1 Several articles in the business literature proclaim the potentially large contributions that data mining could make to the service sectors. Present an example in which the application of data mining actually made a difference to a service company.

ANSWER 1.1 Data mining is an advanced numerical modeling technique that could be used to extract valuable insights from databases that record the transactional activities and consumption patterns of customers. A well-known example is the way Harrah's Entertainment applied data mining to improve its profitability (Loveman 2003).

Harrah's Entertainment operates twenty-six casinos in thirteen states. In 2003, it posted $4 billion total revenue. Its current CEO was a former Harvard Business School professor, who spearheaded a database-driven marketing and service-delivery strategies that drastically improved Harrah's financial performance.

For example, the company identified that 26 percent of its gamblers generated 82 percent of its revenue, the Pareto principle. These core customers are typically former teachers, doctors, bankers, and machinists, not the limousine-riding higher rollers. These are middle-aged and senior adults who have a discretional amount of time and income. They prefer free casino chips rather than rooms or dinners. Many of them visit a casino on the way home from work or on a weekend night out. The company also found that happy customers spend 24 percent more annually, whereas disappointed customers spend 10 percent less.

Such information allows the company to fine-tune its knowledge of customer segments using advanced numerical modeling techniques. By designing its marketing program to focus on these elements of customer service, the company secured customer loyalty and superior profitability.

1. Since the core customers are slot machine players, the company redesigned the floor plan to make it easier for customers to find the preferred slot machine designs (e.g., machines with proclaimed odds, the look of machines, etc.) as well as to benefit the company.
2. To encourage customer loyalty, the company introduces the "Total Rewards" card that allows customers to accumulate playing and other credits from any of the company's casinos. These credits form the bases of incentive rewards for the customers. The card system provides a very useful source of proprietary customer data for the company.
3. Train all staff to focus on speed and friendliness (smiling and addressing the customer by name). A detailed customer satisfaction survey is conducted. Bonuses are assigned to those sites that improved their customer satisfaction score by 3 percent or more per year.

   This group-based reward system promotes a self-managed correction at each site, as the weaker departments will be inspired by others to improve quickly in order to avoid dragging down the overall performance of the entire site. The key reason that such a bonus program worked well is because the reward depends on everyone's performance.
4. As speed of service (check in at reception, dining rooms, car parking, etc.) is critical to most customers, the company divides customers into three tiers: gold, platinum, and diamond, in the order of increasing level of service. Three service lines are offered so that customers can observe the perks others

are getting, thus becoming highly motivated to want to move to higher-tier groups. These tiers are typically defined based on the projected "lifetime worth" of the customers to the company.

5. Because customers spend more when they stay in casino hotels, the company offers free rooms to high-value gamblers.

Information technology and telecommunications technologies have aided the service sectors in its growth, access, speed and reduction of costs. These technologies enable real-time decision making from a system engineering approach. The productivity of the service sectors has increased significantly in recent years. High technology services are usually enabled by advanced information/communications technologies. Examples of such technologies are listed in Table 1.8. The enabling technologies therein are described below.

Collaborative software provides Web tools for employees and business partners to work together to make services better, faster, and cheaper. Business intelligence software extracts information from data for optimizing revenue-generating strategies, enhancing cost efficiency, and/or improving customer relations. Synchronization software enables the edits made to one copy to automatically propagate to all copies in the database. Autonomic computing performs self-monitoring and allocates storage resources dependent on demands imposed by data and information. Peer-to-peer networking permits serverless file sharing to promote collaboration. Distributed computing performs decentralized computing by aggregating the unused power of individual computers connected through a network. Extensible markup language (XML) is a metalanguage, which separates the structure and semantics of data from its presentation.

Table 1.8 Examples of Communication and Information Technologies Useful to Services

#	Service Characteristics	Enabling Technologies
1	**Information-driven**	
	Creation	Collaborative software, business intelligence software
	Management	Synchronization software, autonomic computing
	Sharing	Peer-to-peer networking, distributed computing, extensible markup language
2	**Customer-centric**	
	Co-production	Intranet, extranet, Internet
	Customization	Software agents, synchronization software, peer-to-peer networking
	Satisfaction	Software agents
3	**E-Oriented**	
	E-access	Wireless, Internet-on-a-chip
	E-commerce	E-procurement, e-fulfillment, e-supply chain, e-outsourcing, e-auction
	E-customer management	Customer relationship management software
4	**Productivity-Focused**	
	Efficiency and effectiveness	Enterprise resources planning software
5	**Value-adding**	
	Profitability	Financial ratio analysis software, Economic value added analysis

Adapted and modified from Tien and Berg (2003).

Intranet, extranet, and Internet are typical communications channels reserved for internal employees, external interchanges with business partners, and external interactions with customers. Software agents are smart software programs that are capable of processing a vast amount of factors linked by probabilities, causes, and effects in order to define decisions that would lead to customer-preferred outcomes.

Wireless enables high-bandwidth Internet access by cellular phones, laptops, and personal data assistants to create real-time connectivity from anywhere and at any time. Internet-on-a-chip contains protocols necessary for Internet connectivity, allowing the interactions between many sensing devices to facilitate maintenance and monitoring services. E-procurement, e-fulfillment, e-supply chain, e-outsourcing, and e-auction are all software technologies that make the respective processes fast and cost-effective. Customer relations management software helps to track customer activities, provide better customer services, and customize the enterprise's marketing campaigns.

Enterprise Resources Planning (ERP) software links the performance data of all departments of an enterprise and creates high quality reports to allow for fast decision making. Financial Ratio analysis compares the financial performance indices of an enterprise against those in industry in order to allow an external benchmarking. Economic Value Added analysis defines the real value added due to the operation of an enterprise by subtracting the cost of doing business from the net profit reported in its Income Statement. Such analysis is known to focus management attention to company activities which are important from both the short term and long term perspectives.

Example 1.2 Customer focus is something every service company is energetically talking about. However, it is easier said than done. How can customer focus be realized?

ANSWER 1.2 Based on a study of Royal Canadian Bank, Harrah's Entertainment, and Continental Airlines, Gulati and Oldroyd (2005) suggest that pursuing a customer-focus strategy is a journey that requires three preparatory steps:

1. Build a comprehensive database about the customers.
2. Analyze and interpret the data to gain insight into customer from past behavior.
3. Anticipate what customers would need in the future.

Information generated about customers is then made available to all customer-facing employees. The company induces a major shift in employee attitude to keep customer initiatives at the forefront by training their employees to acquire and practice customer-oriented skill sets. The company institutionalizes the customer-focus program by constantly monitoring and reviewing its companywide implementation. An employee reward system is set up to recognize those who deliver outstanding customer service.

1.6 SKILL SETS FOR SERVICE SYSTEMS PROFESSIONALS

Different services require skills at different quality levels, as shown in Table 1.9. New and innovative services are typically realized by people at service quality level ten.

Table 1.9 Skill Quality Levels

Level	Description
1	Able to follow role script in method when all resources are made available and there are no exceptions and supervisory function is active to validate each step before execution to avoid errors.
2	Able to follow role script in method when all resources are made available and there are no exceptions to be processed, with minimal supervision and corrections from project manager and other levels/roles using method.
3/4/5	Able to follow roles script in method when all resources are made available and there are few/several/many exceptions (nonstandard requirements)
6	Able to follow role script in method when not all resources are made available and there are many exceptions (nonstandard requirements)
7	Able to follow role script in method and improvise as required
8	Able to do all roles in method
9	Able to do all roles in all methods
10	Able to improvise and innovate new offerings

Companies in the service sector need to pursue strategic differentiation in the services they offer in order to sustain and extend market competitiveness and achieve long-term profitability. Market competitiveness must be built on service innovations such that the service packages offered to customers are constantly renewed and uniquely differentiable from the competition. Marketing management, financial analysis, and cost accounting are important tools to employ, when choices need to be made to achieve long-term strategic advantages.

Operational excellence is also important, as the service-providing employers need to minimize waste, streamline operations, and enhance productivity in order to maximize profitability. Engineering management principles related to planning, organizing, leading, and controlling will be needed to guide these operations. Quite a number of tools perfected in industrial engineering can be readily applied to pursue short-term results.

The importance of "Adoptive Innovators" for the service industry is advocated in Anomalous (2007). These *T-shape professionals* have the breath of cross-disciplinary knowledge and capabilities to interact with others in building and managing teams, programs and projects, corresponding to the horizontal bar of the T, while possessing in-depth knowledge and capabilities in specific technical domains, corresponding to the vertical bar of the T; see Fig. 1.9. Specifically, T-shape professionals have a service mindset and are versed in a large number of service disciplines.

1.6.1 Service Mindset

A service mindset consists of practicing a customer-focus paradigm in the creation, delivery, and servicing of value packages to meet customer's important needs, with a keen understanding that achieving customer satisfaction is the driving force for long-term corporate profitability. Adoptive innovators must adopt and practice this service mindset.

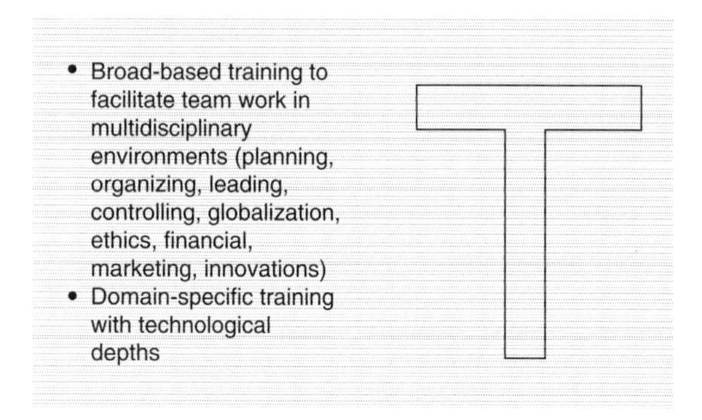

Figure 1.9 T Personality.

1.6.2 Service Disciplines

Anonymous (2007) outlines fourteen service disciplines that T-shape professionals need to master. I added a few more and put together a set of seven composite service disciplines, which are mutually exclusive and collectively exhaustive:

1. *Service vision and leadership.* Create vision, set directions, and make decisions that determine value in relation to new service offerings, business models, supply chains partnerships, market segments, and emerging technologies.
2. *Service creativity and innovation.* Invent new service offerings, experience, value proposition, or systems and pursue innovation programs to achieve market success.
3. *Service productivity and value creation.* Utilize management tools to monitor business activities (e.g., regulatory compliance, innovation, and financial data), create value through customer interactions, effectively utilize human resources, manage service knowledge portfolios, assess financial viabilities of new service offerings, improve all processes, control quality, and appraise performance.

   Management tools may include IT/computer-related tools (e.g., data mining, Monte Carlo simulation, analytical modeling, ERP, etc.) that support decision making, customer interactions, and service distribution and logistics, as well as industrial engineering tools (e.g., Lean Six Sigma, TQM, value stream mapping, FEMEA, 5 S, and Kanban).
4. *Service design and development.* Engage emerging technologies (e.g., Web-based collaboration and design tools) to invent new service systems as well as to improve existing ones.
5. *Service customer focus.* Apply marketing tools to understand customers' needs, foster customer collaborations, and create customer expectations.
6. *Service ethics and professionalisms.* Set high standards and implement corporate codes of conduct.
7. *Service globalization.* Assure methods of supply through tiered partnerships and exploit global networks of resources to capture new opportunities.

Not all of these service disciplines are equally important to the service enterprises. Table 1.10 points out that the majority of these composite service disciplines contribute

Table 1.10 Services Disciplines and Their Relative Roles in Companies

#	Composite Service Discipline	Strategic Differentiation	Operational Excellence
1	Vision and leadership	x	
2	Creativity and innovations	x	
3	Productivity and value creation		x
4	Design and development	x	
5	Customer focus	x	
6	Ethics and professionalism		x
7	Globalization	x	

directly to the creation of strategic differentiation (thus, sustainable competitive advantages of the service enterprises), while the remainder are useful to foster operational excellence.

These seven composite service disciplines are designed to impart broad-based knowledge and perspectives to graduates corresponding to the horizontal part of the T, whereas their in-depth technological training in selected specialty fields corresponding to the vertical part of the T.

For service systems engineers to be successful, they need to diligently acquire and practice skills and capabilities that would enable them to add value to the service industry. A group of twenty-four skills were recommended in Sorby et al. (2006) to be pertinent to B.S. service systems engineers. These twenty-four skills can be regrouped into six categories:

1. *Management of service systems.* These skills include scheduling, budgeting and management of information systems, leadership.
2. *Operations of service systems.* Engineers should be proficient in process evaluation and improvement, quality improvement, customer relationships, uncertainty management.
3. *Service processes.* These skills include performance measurement, flowcharting, work task breakdown.
4. *Business management.* Business skills include project costing, business planning, change management.
5. *Analytical skills.* These skills include problem solving, economic decision analysis, risk analysis, cost estimating, probability and statistics.
6. *Interpersonal skills.* Increasingly, service systems engineers are expected to excel in professional responsibility, verbal skills, technical writing, facilitating, and team building).

In a 2009 report, Carnegie Foundation for the Advancement of Teaching emphasized critical thinking, problem solving and teamwork, and a multidisciplinary approach. Choudaha (2008) suggested a number of required skills and attitudes for master's degree graduates, based on an online Delphi study. This list includes integration, collaboration, adaptability, critical thinking, interpersonal competence, problem solving, system conceptualization, and diversity orientation.

For service systems engineers at the master's degree level, I have added the following six categories of skills and capabilities to be covered in two 3-credit courses at the graduate level:

7. *Knowledge management.* Service systems engineers should be familiar with definitions, strategies, success factors, hurdles, and best practices in industry.
8. *Creativity and innovations in services.* These skills include creative thinking methods, success factors, value chain, best practices, and future of innovations.
9. *Financial and cost analysis and management.* Additional business skills include activity-based costing, cost estimation under uncertainty, T-account, financial statements, ratio analysis, balanced scorecards, and capital formation.
10. *Marketing management.* Market forecast, market segmentation, marketing mix—service, price, communication, and distribution—are important marketing tools.
11. *Ethics and integrity.* Service systems engineers must be held to high ethical standards. These include practicing ethics in workplace and clear knowledge of guidelines for making tough ethical decisions, corporate ethics programs, affirmation action, and workforce diversity, as well as global issues related to ethics.
12. *Global orientation.* Increasingly, engineers must be aware of emerging business trends and challenges with regard to globalization drivers, global opportunities, and global leadership qualities.

These twelve categories of skills and capabilities constitute the SSME-12 skills, which, in turn, are closely linked to the seven composite service disciplines. Service systems engineers are advised to focus their education on acquiring and practicing these SSME-12 skills, as a part of their basic education for the horizontal part of the T personality, so that they are in a position to contribute effectively to these seven composite service disciplines.

Example 1.3 Assume that you are a highly paid consultant to an ambitious university administrator who is starting a new master's degree program in service systems engineering. In three years' time, the administrator wants to make the degree program a great success in the eyes of students, industry professionals, and the university, Explain your planning advice to this administrator and what specifics you would like to see included in his plan.

ANSWER 1.3 For the degree program to be successful from the viewpoints of students, the university and industry professionals, the interests of all three groups of stakeholders must be sufficiently satisfied.

Students want courses that are reasonably tough and demanding, that challenge their thinking abilities, and that allow them to gain new perspectives and skills. They want to be exposed to ways they can continue to accumulate knowledge, receive good grades, have time to enjoy university life, network with a large number of new friends, and secure good jobs after graduation.

Universities like to see an expanding enrollment size, new research opportunities that could bring in grant money and scholarly recognition, novel courses that would set them apart from other schools, and collaborations with industry in the forms of internship, joint development, and industrial advisement.

Employers welcome flexible degree programs, which allow customization to their specific needs, high quality of service workers who can contribute without the need of excessive retraining, and opportunities of joint development program with universities to address their specific problems and needs.

The administrator should conduct a comprehensive survey to ascertain these interests, and then define specific action steps to address each of them. Together, the action plan comprising of all these action steps will move the program toward success in the interest of all three groups of stakeholders.

1.7 ROLES OF TECHNOLOGISTS VERSUS MANAGERS/LEADERS

Any services enterprises will need both technologists and managers/leaders, although their roles are different. As a technical contributor, the service engineer focuses primarily on the operational aspects of the work—what it takes to get a technical assignment accomplished and how the assignment can be done in the most efficient and speedy manners. The service engineer in a managerial position will focus on the strategic aspects of the work, such as what work should be done, why it should be done, who should be assigned to do the work, what resources should be used to do the work, and in what order of priority. Specifically, service managers get involved in the following steps:

- Setting goals for the group, department, or enterprise.
- Establishing priorities.
- Defining policies and procedures.
- Planning and implementing projects and programs to add value.
- Assigning responsibilities and delegating the commensurate authorities to others while maintaining control.
- Attaining useful results by working through people.
- Processing new information and handling multidisciplinary issues.
- Making tough decisions under uncertain conditions.
- Finding the proper solution quickly among several feasible alternatives.
- Doing things right the first time, with a sense of urgency.
- Coaching, teaching, and mentoring others.
- Dealing with people—handling conflicts, motivation, and performance correction.

Table 1.11, which is adopted and revised from Aucoin (2002), illustrates the fundamental differences between the work done by engineers and that performed by managers.

Table 1.11 Work Done by Service Engineers and Managers

Characteristics	Service Engineers	Managers
Focus	Technical/scientific tasks	People (talents, innovation, relationships); resources (capital, knowledge, process know-how); projects (tasks, procedure, policy)
Decision-making basis	Adequate technical information with great certainty	Fuzzy information under uncertainty (people's behavior, customer needs, market forecasts)
Involvement	Perform individual assignments	Direct work of others (planning, leading, organizing, controlling)
Work output	Quantitative, measurable	Qualitative, less measurable, except financial results when applicable
Effectiveness	Rely on technical expertise and personal dedication	Rely on interpersonal skills to get work done through people (motivation, delegation)
Dependency	Autonomous	Interdependent with others
Responsibility	Pursue one job at a time	Pursue multiple objectives concurrently
Creativity	Technology centered	People centered (conflict resolution, problem solving, political alliance, networks building)
Bottom line	"How" (operational)	"What" and "why" (strategic)
Concern	Will it work technically?	Will it add value (market share, financial, core technology, customer satisfaction)?

Adapted and revised from Aucoin, B. M. "From Engineer to Manager: Mastering the Transition." Artech House Publishers (2002).

In a 2002 report, the National Science Foundation described the employment situation of U.S. engineers and scientists. Out of a total of 2,343 engineers and scientists, 46.1 percent held management and administrative positions. Figure 1.10 suggests that this percentage varies only slightly with age. About every one of two engineers or scientists has taken on managerial or administrative responsibilities.

Service engineers who aspire to become managers are advised to fully understand these differences and the requirements associated with the management work. An individual should assess the compatibility of these implied requirements with his or her own personality, aptitude, value system, personal goals, preparations, and other factors, so that he or she is convinced that taking on managerial responsibilities will indeed lead to long-term happiness.

According to a survey reported by Badawy (1995), service engineers move into management because of one or more of the following reasons:

- Gaining financial rewards
- Exercising authority, responsibility, and leadership
- Acquiring power, influence, status, and prestige
- Receiving career advancement, achievement, and recognition
- Combating fear of technological obsolescence
- Responding to a random circumstance—an opportunity that is suddenly available

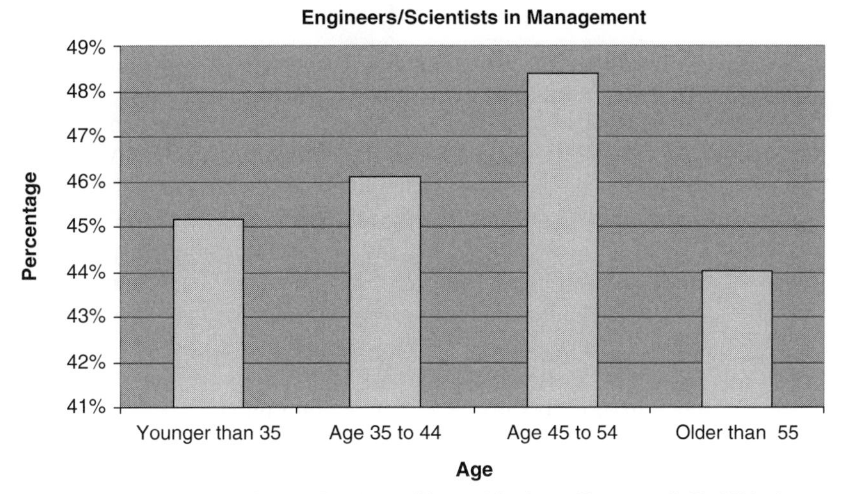

Figure 1.10 Engineers/scientists in management.

It is important that individual services engineers become aware of the advantages and disadvantages of being either technologists or managers/leaders, so that they make the best choices for themselves. Those who are not so sure of which way to go could consult resources such as those contained in Appendices 1.12.3 to 1.12.6.

1.8 PREPARATION OF SERVICE SYSTEMS ENGINEERS/LEADERS

1.8.1 Customer Focus

One way to determine how service systems engineers/leaders should prepare themselves is to first understand what a service enterprise aims to accomplish. Frei (2008) offers an excellent perspective in this respect and points out four things a service enterprise must get right:

1. *Understand what the customers are particularly looking for (convenience, friendliness, flexible choices, price, and others) and what the enterprise is willing to deliver.* It requires a strategic decision to optimize the benefit to cost ratio.
2. *Define ways to pay for the added work that is required to improve customer satisfaction.* For example, raise price, not change price but collect feedback for future use, and introduce more self-help capabilities for customer.
3. *Manage employees.* Recruit, select, train, monitor, and job redesign to deliver the needed service excellence.
4. *Manage customer behavior.* Design the service to foster this preferred behavior.

It is evident from this list that service systems engineers/leaders must learn to be customer focused. However, being customer focused is necessary but not sufficient for service systems engineers/leaders to be effectual in a service enterprise. There exist areas beyond customer focus that require training in order for systems engineers/leaders to be fully prepared for the market.

1.8.2 Three-Decker Leadership-Building Architecture

Future service systems engineers/leaders need acquire the useful T personality by having a broad set of skills for managing relationships, communicating and collaborating with others in multiple disciplines, and envisioning the future to define what are the best courses of action (the horizontal part of the T). As well, these engineers need to acquire a set of in-depth technological skills to enable them to properly implement these actions. They should strive to master the SSME-12 skills set at the proper skill quality levels, acquire a systems view of service enterprises, communicate with workers in multiple disciplines, practice the engineering management functions (planning, organization, leading, and controlling) to teams, projects, and programs, while making use of their understanding of the business fundamentals, and exert leadership in contributing to achieving strategic differentiation and operational excellent of their employers. For this purpose, this text is organized into three parts:

I: The Functions of Engineering Management (Planning, Organizing Leading, and Controlling).

II: Business Fundamentals for Service Systems Engineers and Leaders (Cost Accounting and Analysis, Financial Accounting and Analysis, Marketing Management).

III: SSME Leadership in the New Millennium (Systems Engineers as Managers and Leaders, Ethics, Knowledge Management, Innovations in Services, Operational Excellence, Globalization and Appendix, which contains thirty-plus business cases addressing SSME issues of relevance).

Figure 1.11 illustrates our "Three-Decker Leadership-Building Architecture," which is the design basis of this text.

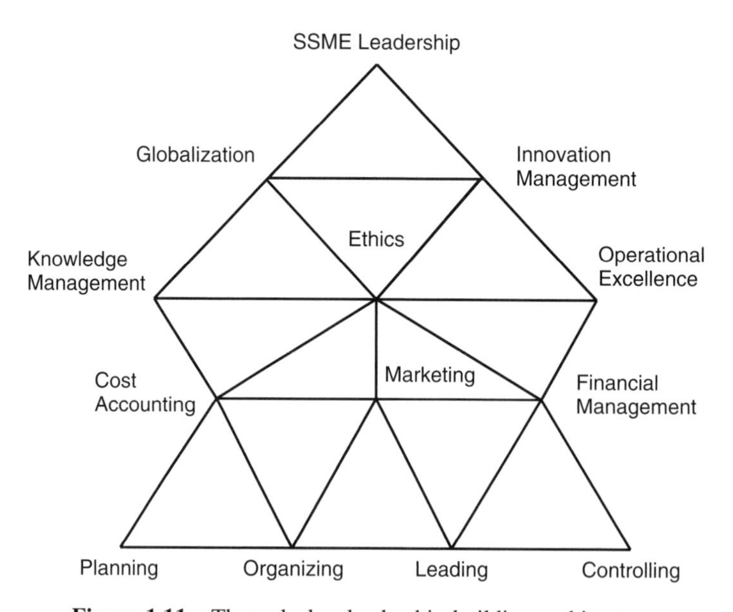

Figure 1.11 Three-decker leadership-building architecture.

1.9 CHALLENGES IN THE NEW MILLENNIUM

Globalization will continue to intensify the cross-border transfer of investment capital, technologies, talents, and other resources, as service companies seek newer markets, link with foreign business partners, and capture location-specific opportunities. Service systems engineers/leaders need to prepare themselves for this future by being capable of eight functions:

1. Thinking globally, acting locally.
2. Recognizing new local opportunities and mobilizing the required corporate and other resources to capture them effectively.
3. Engaging in open innovation to foster the creation of strategic differentiation.
4. Creating business partnerships and alliances on a global scale.
5. Managing global teams of members with diverse backgrounds (e.g., culture, business practices, language, and value) to pursue organizational objectives.
6. Resolving conflicts of planning, organizational, and personal types.
7. Implementing local and emerging technologies to add value.
8. Investing to master location-specific business factors (culture, language, business methodologies, governmental regulations, personal network, etc.).

The challenges faced by service systems engineers/leaders are indeed multidimensional.

Chapter 14, titled: "Globalization," contains a six-dimensional model of these new challenges; these six dimensions include inside, outside, present, future, local, and global.

1.10 CONCLUSIONS

Services are activities that involve intensive interactions with clients or customers in front-stage operations. Service enterprises are best reviewed and studied as systems, as they contain components that interact with one another and react as units to external threats and opportunities. A systems engineering approach is particularly suitable for studying, making decisions, and taking actions involving issues in services. A variety of established techniques in industrial engineering, management science, and other disciplines may be beneficially applied to improve both the competitiveness and productivity of services.

Because of the perceived high demand for services in the future, some of the service sectors in the United States are projected to grow steadily in the next ten to twenty years. These sectors offer the best job opportunities for those who exhibit the service-dominant and customer-oriented mindset, acquire and practice the right skills (e.g., SSME-12 skills) to add value, and display the T-personality to contribute to both the strategic differentiation and operation excellence of service enterprises.

This text provides three parts (engineering management functions, business fundamentals, and service leadership) designed to provide a broad foundation for preparing future service systems engineers/leaders to meet the challenges in the new millennium. The challenges have six dimensions of inside, outside, present, future, local and global.

1.11 REFERENCES

Anonymous. 2007. "Succeeding through Service Innovation." Cambridge Service Science. Management and Engineering Symposium (July 14–15, 2007). Cambridge, UK: Churchill College, University of Cambridge.

Anonymous. 2009. "Best Jobs in America: Money/Payscale.com's List of Great Careers," http://money.cnn.com/magazines/moneymag/bestjobs/2009/full_list/index.html.

Apte, Uday M., Uday S. Karmarkar, and Hiranya K.Nath. 2008. "Information Services in the U.S. Economy: Value, Jobs, and Management Implications." *California Management Review* 50 (3): (Spring).

Aucoin, B. M. 2002. *From Engineer to Manager: Mastering the Transition*. Artech House Publishers.

Badawy, M. K. 1995. *Developing Managerial Skills in Engineers and Scientists: Succeeding as a Technical Manager*, 2nd ed. New York: Von Nostrand Reinhold.

Bartsch, Kristina. 2009. "Employment Projections, 2008–2018," Monthly Labor Review, US Bureau of Labor Statistics (November) http://www.bls.gov/opub/mlr/mrlhome.htm

Bloomberg, J. 2010. "Work in Service Economy," Chapter 3 in Salvendy, Gavriel and Waldemar Karwowski (Eds), *Introduction to Service Engineering*. John Wiley (January).

Boardman, John, and Brian Sauser. 2008. *Systems Thinking: Coping with 21st Century Problems.* Boca Raton, FL: CRC Press.

Broder, D. 1992. "Clinton's Performance Won't Mimic Carter's." St, Louis Post, - Dispatch, (November17), editorial page.

Choudaha, Rahul. 2008. "Competency-based Curriculum for a Master's Program in Service Science Management and Engineering (SSME): An online Delphi Study." Doctoral Dissertation, University of Denver, Denver, Colorado.

Fitzsimmons, J. A., and M. J. Fitzsimmons. 2008. *Service Management: Operations, Strategy and Information Technologies*, 6th ed. New York: McGraw Hill, p. 23.

Frank, Moti. 2006. "Knowledge, Abilities, Cognitive Characteristics and Behavioral Competences of Engineers with High Capacity for Engineering Systems Thinking." *Systems Engineering* 9 (2).

Frei, Frances X. 2008. "The Four Things a Service Business Must Get Right." *Harvard Business Review* (April).

Gabarro, J. J., and J. P. Kotter. 2008. *Managing Your Boss* (Paperback). Harvard Business School Press (January 8).

Guile, Bruce R., and James Brian Quinn, eds. 1988. *Technology in Services: Policies for Growth, Trade and Employment.* Washington D.C.: National Academy Press, p. 214.

Gulati, Ranjay, and James B. Oldroyd. 2005. "The Quest for Customer Focus." *Harvard Business Review* (April).

Kinsbury, Katheleen 2008. "Medical Records Go digital: *Time* (March 17).

Luczak, H. and G. Gudergan. 2010. "The Evolution of Service Engineering – Toward the Implementation of Designing Integrative Solutions," Chapter 26 in Salvendy, Gavriel and Waldemar Karwowski (Eds), *Introduction to Service Engineering*. John Wiley (January).

Loveman, Gary. 2003. "Diamonds in the Data Mining." *Harvard Business Review*. (May).

Martin, Roger L. 2009. *The Opposable Mind: Winning Through Integrative Thinking.* Boston: Harvard Business School Press.

Merrifield, Ric. Jack Calhoun, and Dennis Stevens. 2008. "The Next Revolution in Productivity." *Harvard Business Review* (June).

Mott, M. R. 2010. "Applying the Methods of Systems Engineering to Service Engineering," Chapter 8 in Salvendy, Gavriel and Waldemar Karwowski (Eds), *Introduction to Service Engineering*. John Wiley (January).

Nambisan, Satish (2001). "Why Service Businesses are Not Product Businesses." *MIT Sloan Management* Review 42 (4): (Summer).

Norton, David P. (2000). "Is Management Finally Ready for the Systems Approach?" Harvard Business School Balance Scorecard Report. Article Reprint No. B0009E.

Tidd, J., and John Bessant (2009). *Managing Innovation: Integrating Technological, Market and Organizational Change*, 4th ed. Hoboken, NJ: John Wiley & Sons, Inc.

Tien, James M., and Daniel Berg (2003). "A Case for Service Systems Engineering." *Journal of Systems Science and Systems Engineering* 12 (1): 13–38.

Zeithami, V., M. J. Bitner, and D. D. Gremier. 2006. *Services Marketing: Integrating Customer Focus Across the Firm,* 4th ed. New York: McGraw-Hill.

1.12 APPENDICES

1.12.1 Definitions of General Service Terms

These are selected definitions commonly used in service systems engineering. They are adopted from Anomalous (2007).

- **Service**—Value co-creation by interactions between knowledgeable providers and their customers with unsatisfied needs.
- **Service system**—A configuration of resources (people, technology, organization, shared information) that co-creates value.
- **Service mindset**—A focus on the innovating interactions between customers and service providers to co-create value.
- **Adoptive innovators**—Capable of integrated systems thinking, broad-based communications in cross-disciplinary domains, while being deeply involved in some technology/engineering areas, the T-professionals.
- **Back-stage service activities**—Activities that do not involve direct interaction with the customer (e.g., information processing).
- **Customer service system**—A system that provides value propositions and searches for win-win co-creation opportunities.
- **Service-dominant logic**—This logic advocates that service involves co-creation interactions during the process of creating, proposing, and realizing value propositions.
- **Front-stage service activities**—Activities that involve direct interaction with customers.
- **Service computing**—The use of information technology (IT) to support customer-provider interactions.
- **Service design**—The application of design methods and tools towards the creation of new service systems and service activities with special emphasis on quality, satisfaction and experience.
- **Service economics**—The definition and measurement of service activities in an economy. Typical measures include productivity, quality, regulatory compliance and innovation.

- **Service engineering**—The application of technologies, methodologies and tools toward the advancement of new service offerings and the improvement of service systems.
- **Service experience and service outcome—**The customer's perceptions of the process and results of a service interaction or relationship. Customer's perceptions are based in large part on customer's expectations, which may change over time, causing some experience to be viewed as deteriorating when an objective measurement of results demonstrates otherwise.
- **Service innovation**—A combination of specific innovations in technology, business model, social-organization, and demand with the objective to improve existing service systems (incremental innovation), generate new value propositions (offerings) or build new service systems (radical innovations). Service innovations may also result from novel combinations of existing service elements.
- **Service mindset**—An orientation geared toward the innovation of customer-provider interactions, combined with interactive skills to enable teamwork across technical disciplines and business functions.
- **Value proposition**—A specific package of benefits and solutions that a service system intends to offer and deliver to others.

For additional definitions, reader should refer to Anomalous (2007).

1.12.2 Projected Growth and Decline of Occupations in the United States (2008–2018)

Bartsch (2009) provides a detailed projection of jobs in various occupations for the year 2018, whereas the actual job numbers for the year 2008 are used. The "Change Percentage" denotes the projected job change in percentage over the ten-year period of time from 2008 to 2018 in each occupation is calculated by dividing the "Total Change" by the base number of 2008 jobs. As illustrated in Table 1.A1, some occupations are projected to grow very rapidly in the United States, whereas others are forecast to decline drastically.

1.12.3 Are You Management Material?

Would you like to know more about yourself? Take an "Emotional Quotient Test" offered by Portfolio.com, at www.portfolio.com/infographics/2007/05/eq-quiz.

1.12.4 Ten Factors for Survival and Success in Corporate America

To be successful in corporate America, one needs to pay attention to the following common-sense success factors:

1. *Excellent performance.* Make sure that all assignments are performed well, as "You are only as good as your last performance." Pay attention to ensure that both the performance and its impact are properly recorded and made known to people in the organization who affect your career growth. Self-promote as needed.

Table 1.A1 Growing Occupations—Projected Employment from 2008 to 2018 (in thousand)

Occupational group	Employment 2008	Employment 2018	Change Number	Change Percent
Total, all occupations	150,932	166,206	15,274	10.1
Total Service-related Occupations	116,452	131,053	14,601	12.5
Management, business, and financial occupations	15,747	17,411	1,664	10.6
Professional and related occupations	31,054	36,280	5,227	16.8
Service occupations	29,576	33,645	4,069	13.8
Sales and related occupations	15,903	16,883	980	6.2
Office and administrative support occupations	24,101	25,943	1,842	7.6
Transportation and material moving occupations	9,826	10,217	391	4
Construction and extraction occupations	7,810	8,829	1,019	13
Installation, maintenance, and repair occupations	5,798	6,238	440	7.6
Production occupations	10,083	9,734	−349	−3.5
Farming, fishing, and forestry occupations	1,035	1,026	−9	−0.9

Source: BLS Employment Projections 2008–2018

2. *Personality.* Project a mature, easy-to-work-with, positive, reasonable, and flexible personality. How one acts and behaves is important.
3. *Communication skills.* Pay attention to skills related to asking, telling, listening, writing, and understanding.
4. *Technical skills and ability.* Keep one's own professional capabilities (e.g., analysis, design, integration, product development, tools application, etc.) current and marketable.
5. *Human relations skills.* Constantly review ways of interacting with people and make sure that you are creating and maintaining acceptable working relationships. Avoid being labeled as "not able to work well with other people."
6. *Significant work experience and assignments.* Seek diversified business and engineering exposure and high-impact assignments to build up your experience portfolio. Doing so will increase your ability to add value to the organization.
7. *Self-control.* Improve your ability to stay cool and withstand pressure and stress by, for example, taking courses in leadership training. According to a CNN report in 2001, a British military training camp was offering training services to business executives, subjecting the executives to a high-pressure artificial military environment to toughen them up for handling the real-world business environment.
8. *Personal appearance.* Follow the example of superiors to fit yourself into the corporate image. Dress for success.

9. *Ability to make tough business decisions.* Take careful chances when needed. Anyone can make the easy plays, but only great people make the tough plays.
10. *Health and energy level.* Take care of your health and maintain a high level of physical vitality.

1.12.5 Most Common Reasons for Career Failures

Some service engineers fail in their careers for one reason or another. Enumerated below are seven common reasons for career failure; these are relevant to technologists as well as managers.

1. *Poor interpersonal skills.* A lack of interpersonal skills is the single biggest reason for career failure. Few people are fired or invited to resign due to deficiencies in their technical capabilities. As a measure of social intelligence, interpersonal skills are important to achieve success in any organization. One needs to be sensitive to the feelings of others, able to listen and understand the subtext in communication, give and take criticism well, strive to build team support, and be emotionally stable.
2. *Wrong fit.* From time to time, a person may find it hard to adapt one's abilities, styles, personalities, and values to the culture and business practices of the workplace. The workplace may assume a cultural norm that is unfamiliar to some individuals. It is well known that rigidly layered corporations operate differently from dynamic partnerships or entrepreneurial start-ups. The individual's core value system, with priorities, profit motives, and social or environmental preferences may not be fully compatible with those of coworkers on the job. In addition, the chemistry among coworkers within a unit, department, or company could also be a source of conflict. Often, the management style of the superior is difficult for the individual to adapt to. In cases of such a wrong fit, the best strategy for the individual is to move on.
3. *Unable to take risks.* Lack of risk-taking abilities is a major stumbling block to the advancement of one's career. For fear of failure, some engineers stay in their current positions for too long and are not willing to accept promotions that require relocation within the company or to venture out for new positions outside of the company.

   Others feel comfortable with the technical work they do because they are able to control all of the key components of their work (e.g., data, facts, analysis, procedure, and equipment) and the quality of its outcome. Naturally, some of them may feel uneasy when requested to take on managerial responsibilities that involve (1) people who may react differently; (2) data that are often incomplete and inaccurate; (3) objectives that are usually multifaceted; and (4) decision-making steps primarily based on personal intuition and judgment. The inability to take calculated risks could lead to failure in one's career progression.
4. *Bad luck.* Sometimes, engineers get hurt by business circumstances that are beyond their control or expectation (e.g., mergers and acquisitions, corporate downsizing, change of market conditions, change of business strategies,

advancement of technologies, etc.). Career disruptions due to bad luck can happen to anyone. However, one should be able to recover quickly if one's record demonstrates that past achievements consistently produced value to employers, and such value-creation capabilities are widely marketable.

5. *Self-destructive behavior.* Certain engineers exhibit work habits or behavior patterns that are self-destructive. Examples of self-destructive behavior include working in secret, resistance to change, being excessively aggressive, having an uncooperative attitude, picking fights with people, becoming overly argumentative, being readily excitable about trivialities, and displaying a lack of perspective. Such behavior is clearly unwanted in any group environment.
6. *Lack of focus.* Some engineers pride themselves on being a jack-of-all-trades, getting busily involved in almost everything, but being good at nothing. Failing to focus on creating value to the employer is detrimental to one's own career.
7. *Workplace biases.* Under ideal conditions, all workplaces should be free of biases with respect to race, gender, age, national origin, religious beliefs, and other individual qualities. In reality, some workplaces are managed more effectively and progressively than others. Individual workers need to monitor the real situation at hand and take proactive steps to avoid being hurt by such biases. Engineers serve themselves well by constantly checking against these bias-based failures and proactively managing those over which they can exercise some control.

Certain articles address the factors that cause engineers to fail as managers. Engineers may be handicapped by a number of perceived shortcomings that do not allow them to become good managers. These perceived shortcomings are reviewed next, even though some of them represent a repetition of those already itemized:

A. Lack of Political Savvy. Engineers tend to be straightforward, honest, and open, and have strong views based on verifiable facts and data. According to Broder (1992), some engineers

1. Hate company politics. They tend to be technically intelligent, but sometimes politically amateurish.
2. Do not build a personal network.
3. Are uneasy trying to fit into an organizational culture because of strong beliefs, unique value systems, rigid principles, and inflexible attitudes.
4. Have an engineering mindset that is rational, efficient, and introspective. They may see things as either right or wrong, and may not be willing or able to accept different shades of gray. For holders of midlevel and top-level managerial positions, this mindset may confer disadvantages when attempting to resolve conflicts, handle disagreements, and foster alliances.

B. Uncomfortable with Ambiguous Situations. The technical training of engineers is based on equations, logic, experimental data, and mathematical analyses; this tends to make engineers see the world as orderly, certain, and black and white. The business environment in the real world causes some engineers to

1. Be uncomfortable with approximate or incomplete answers, since they have been trained well to recognize indeterminate problems and declare them unsolvable. They are not used to the idea of introducing additional assumptions and making such problems solvable. Some engineers

   a. Hate problems with inaccurate or unknown factors.

   b. Dislike planning with uncertainty (e.g., strategic planning).

2. Want to avoid using intuitive knowledge. They prefer cognitive knowledge, which is based on facts and data, and thus they lack the ability and willingness to make tough decisions by using intuition and gut feelings.

C. Tense Personality. Some engineers are too serious in their approach to professional life. They may be unable to say no and unable to request for help, allowing their personal ego and pride to get in the way. Afraid to be wrong, they may have the tendency to take mistakes personally.

D. Lack of Willingness to Take Risks. Some engineers are conservative in nature. They may have a low tolerance for uncertainties. Because of their tense personalities, they may not be comfortable taking risky opportunities to reach for higher levels of rewards. Some graduates with a master's degree in business administration (MBA) are said to be "often wrong, never in doubt"; they continue, in spite of often being wrong, to try new risky approaches until they reach their goals. Engineers are quite different in this respect. While having strong self-confidence in their own technical capabilities, many engineers do not like to take chances.

E. Tendency to Clinch on Technology. Some engineers do not feel comfortable leaving their fields of technological strengths when assuming managerial responsibilities. They tend to lean on technology as a safety net. From their viewpoints, technology is more readily controllable, and they are fearful of losing their sense of control. Some of them even have the uninformed notion that technology is the only thing worthy of respect, valuable, intellectually pure, and deserving of their efforts. Their perspectives are limited, causing them not to recognize that other functions that may be technically less intensive, such as customer service, marketing, procurement, production, and supply-chain management can also contribute equal, or in some cases, more value to the organization than engineering and technology.

F. Lack of Human Relations Skills. Because of their conservative nature, some engineers may be reactive in social settings and remain inflexible in dealing with a diversity of issues and people. Some of them may be readily argumentative and self-righteous when confronted with viewpoints radically different from their own. Over time, some of them may be perceived as suffering from a lack of human relations skills and the inability to become good team players.

G. Deficiency in Management Skills and Perception. One of the noted shortcomings of highly talented engineers is their lack of willingness to delegate. Some of them are not able to work through people and help others to succeed. They would prefer to ensure high quality by doing the projects or assignments themselves. A plentiful of them are unwilling to train subordinates for fear that one day the trainees may become technically more talented than themselves.

Other engineers may have the tendency to utilize self-imposed, ultrahigh standards in appraising employees. They have difficulty tolerating below-par performance by teammates or coworkers. Still other engineers have the tendency to over-manage and over-control their subordinates.

H. Not Cognitive of a Manager's Roles and Responsibilities. Some engineers are unexpectedly promoted to managerial ranks because they performed well as technologists. Due to a lack of preparation on their part, these newly promoted engineers have either limited or no understanding of what managers are supposed to do—that is, to add value by efficiently assigning resources to the right projects. They lack the preparation to do a manager's job. They are not aware of the fact that people problems require more time and attention than technical problems. Because of a lack of exposure to nontechnical, but equally important issues, they have not acquired the background required to deepen a well-rounded business sense.

I. Narrow Interests and Preparation. Some engineers are specialists in narrow technical fields. As a consequence, they have narrow technical viewpoints and limited vision and perspectives beyond technologies. They are not prepared to deal with accounting, marketing, production, finance, and other broad-based corporate issues outside of technology.

Numerous engineers may suffer, to varying extents, from some of the shortcomings just cited. Engineers who aspire to become managers should carefully examine their strengths and weaknesses and commit their efforts to making sure that all of these factors for failure are minimized over time.

1.12.6 How to Manage One's Superiors

Both engineers and engineering managers need to properly manage their respective superiors. On the one hand, the superior needs the active support of all employees to succeed, as most of the work is done by the subordinates. On the other hand, all of the subordinates need their superiors' support to move forward along their individual career paths.

The power of a superior should be taken seriously. One of the primary reasons for job turnover is personality conflict with the individual's own superior, not because of technical performance.

It is also of critical importance that one understands the corporate mindset. Whenever the organization appoints a group leader or manager, the following unwritten rules apply:

1. The organization knows that no one is perfect and that the appointee is no exception.
2. The appointee's strengths are valued more than the trouble caused by his or her weaknesses. Even if the appointee appears to be difficult for some subordinates to deal with, the organization counts on him or her to lead the group and add value. Unless the appointee clearly violates the stated rules, the organization will back the appointee most of the time.
3. To achieve the goals of the organization, the organization trusts the views and desires of the appointee over those of his or her subordinates.

The organization also expects employees to behave in certain ways. These include being attuned to the superior and not insisting that the superior adjust to the employees. Work closely to support the superior and help him or her to succeed. Avoid questioning the superior's judgment and decisions, as the superior typically has access to more and better information and data than the employees and may not be in a position to share such information or data freely.

In readying oneself to manage superiors effectively, it is useful for employees to form the following habits:

1. Understand the business and personal pressure the superior is under, his or her values and motivators (achievement, success, recognition, money, value systems, priorities, principles, and other factors), work style (peacekeeper, conflict lover, riser or setter, channel oriented), and personal style (optimistic, fighter).
2. Expect modest help, and request it only when you really need it. It is better to get help from your own networked coworkers and friends.
3. Be sensitive to the superior's work habits. Watch how he or she receives data and information and works on it. Learn his or her preferred mode of communications—face to face, phone, e-mails, or staff meetings, for instance.
4. Stay in touch with the superior, unless he or she does not want to be bothered regularly.
5. Present materials clearly and without complex details and jargon.
    - Emphasize the significance (the benefits and realizable impact) of your technical work to the group or company, not its technological complexity, sophistication, or elegance.
    - Use concise language to elucidate ideas and recommendations clearly.
6. Do not defend a cause unless it deserves it. Keep it in perspective. Do not complain when you do not get all that you asked for.
7. Exercise self-control. Manage your own overreactions or counterproductive behavior.

The following set of guidelines (Gabarro and Kotter 2008) for managing the superior–subordinate relationship is recommended:

1. Accept that your superior's support is important to you. Understand how important your support is to your superior.

2. Understand your own response to the superior's style and personality, and manage it. Respect the style and orientation of your superior to his or her work. Understand your response to your position in the hierarchy and how you feel about working within a structure.
3. Learn to take feedback objectively, not personally, and maintain your sense of self and your own uniqueness.
4. Push back when necessary, but only for business reasons and to maintain personal integrity; do not push for political gain or to embarrass the superior.
5. Learn the superior's goals, aspirations, frustrations, and weaknesses. Study and understand what the superior thinks is important. Study and be able to emulate the superior's communications style for the sake of being heard.
6. Be dependable; follow through on serious requests for information and work output.
7. Display respect to others and expect respect from others in all matters of business and on-the-job interpersonal interactions regarding time, resources, and alternative work styles.
8. Be honest and share all relevant data about the situations and concerns at hand.
9. Keep private any criticism and conflict that may arise between the two of you, and always work for a jointly satisfactory solution.
10. Be manageable by and available to those beneath you.

1.13 QUESTIONS

1.1 Tom Taylor, the sales manager, was told by his superior, Carl Bauer, to take an order from a new customer for a bunch of products. Both Tom and Carl knew that the products ordered would only partially meet the customer's requirements. But Carl insisted that the order was too valuable to lose. What should Tom do?

1.2 Nancy Bush, the plant manager, needs to decide whether to make or buy a component for the company's core product. She would like the advice of her production supervisors, since they must implement her decisions. However, she fears that the supervisors will be biased toward making the component in house, as they tend to favor retaining more work for their people. What should Nancy do?

1.3 Student A, in order to graduate on February 4, works hard to finish her master of engineering report by the due date of January 8. She is planning to return to her country immediately thereafter and get married. If she graduates on June 10, the next available graduation date, she will have to pay a tuition fee to keep her student status active for one more semester. That would be a substantial financial burden to her.

Her advisor, Professor B, is hesitant to accept the report as presented. The report includes a major marketing activity designed by Student A to promote a new service package of a local company. Because of logistics, this major marketing activity is scheduled on January 20. No customer feedback data, which are required to demonstrate the value brought about by the report, are available before January 8. Professor B cannot bend the rules to pass the report without these data.

Put your innovation hat on and recommend a way to resolve this conflict.

1.4 The engineering manager of Company A proposes to install an automated bar-code scanner costing $4,000. He estimates that he can save about 100 hours of labor time per month, as products can now be scanned much faster. He reasons that at the wage rate of $15 per hour, the benefit for using the automated bar-code scanner is $1,500 per month, and the scanner can be paid back in 2.67 months.

As the president of the company A, do you agree or disagree with the way the president computes the cost-benefit ratio? Why or why not?

1.5 The new millennium imposes a number of new challenges to business managers, who are different from engineering managers and technology managers. Name a few of such challenges.

1.6 In the literature, it is generally said that innovations in the service sectors are lagging behind those in the manufacturing sector. Explain why it might indeed be so.

PART I

The Functions of Engineering Management

Part I of this book addresses the basic functions of engineering management, such as planning (Chapter 2), organizing (Chapter 3), leading (Chapter 4), and controlling (Chapter 5). These functions provide service systems engineers and leaders with basic skills to manage themselves, staff, teams, projects, technologies, and global issues of importance.

Best practices are emphasized as pertinent standards for goal setting and performance measurement. Service systems engineers and leaders solve problems and minimize conflicts to achieve the company's objectives. They use Kepner-Tregoe method, among others, to make rational decisions and take lawful and ethical actions. They utilize Monte Carlo methods to assess projects involving risks and uncertainties. They engage emerging technologies, motivate a professional workforce of diversified backgrounds, develop new generations of services in a timely manner, and constantly surpass the best practices in industry.

The roles of service systems engineers and leaders in strategic planning, employee selection, team building, delegating, decision making, and managing creativity and innovations are explained. The development of managerial competencies is emphasized.

Chapter 2

Planning

2.1 INTRODUCTION

A major function of service management involves planning. In service companies it is the engineering manager who determines the best course of action for a given project. During the planning stages, a manager defines the course of action required to successfully delegate authority, decide on appropriate methods, schedule the optimal time and/or location, and choose the best resources. Louis Pasteur said, "Chance favors the prepared mind." The purpose of planning is to intensify the effectiveness and efficiency of the service enterprise by providing focus and direction (Coke 2002; Hamel and Prahalad 2003). Companies define important strategic and operational objectives and consider the composition of their portfolio and the specific attributes of their industry in order to maintain and strengthen core business that drive incremental growths, align management with short-term and long-term financial targets, and pursue growth beyond their core (Dye and Sibony 2006).

Planning is necessary due to rapid changes in technology (e.g., Web-based tools, enterprise resource-planning software, broadband communications options, automation, and mobile access), information (e.g., data, knowledge, open innovation sources, tacit interactions), environment (e.g., customers, global resources, competition, and marketplace), and organization (such as mergers, acquisitions, supply chain networks, alliances, and outsourcing). If service systems engineers and leaders fail to plan, then they plan to fail.

In this chapter, we will discuss the differences between strategic and operational planning, the planning roles of engineering managers, and the four specific planning activities every engineering manager needs to master.

2.2 NEW BUSINESS TECHNOLOGY TRENDS

Effective planning anticipates and takes into account new trends in business technology. Several emerging trends are transforming countless markets and businesses. These trends focus on relationships, resources, and information (Manyika et al. 2008):

- *Relationships*. Companies pursue open innovation by involving customers, suppliers, small specialty businesses, and independent contractors in the creation of new services. Doing so may lead to cost reduction, time to market advantage, work simplification related to intellectual properties rights, and internal R&D optimization. Web 2.0 technologies facilitate the involvement of consumers to act as innovators. For example, Threadless, an online clothing store, invites people to submit new designs for T shirts and allows the community at large to vote for their favorites. The top four to six designs are printed on shirts and sold in their store. The winners receive a combination of cash and store credit. In general, companies that involve customers in design, testing, marketing and the after-sales process have better insight into their customer's needs and behavior. Companies also outsource specific work to specialists, free agents, and talent networks (finance, IT, operations). For example, TopCoder builds a network of software developers for hire. This trend should gain momentum in sectors such as software, healthcare delivery, professional services, and real estate. The use of talent aggregators will become more frequent if companies can master the art of breaking down and recomposing jobs.

- *Extracting more value from interactions*. Companies gain value from the active interactions between employees. After having outsourced manufacturing and other rule-based activities, workers left behind are encouraged to engage in negotiations and conversations, knowledge transfer and creation, making judgment, and collaborating. Such tacit interactions may be facilitated by ways of wikis, virtual teams, and videoconferencing. Health care and banking are generating value by such interactions.
- *Increased degree of automation.* Companies gain productivity by pursuing automation of repetitive tasks and processes such as forecasting, supply chain management, enterprise resources planning, (ERP), customer relations management (CRM), and human resource management (HR).
- *Better utilization of corporate resources.* Companies use in-house information technologies and other resources more efficiently by leasing part of them to outside companies.
- *More science into management.* Companies engage such technologies as data mining and modeling to exploit a large amount of data in order to make smarter decisions and cultivate deeper insights into their customer's needs and behaviors. As the quality and quantity of data continue to grow, the use of these technologies will lead to strengthened corporate competitiveness. Decision alternatives are also more readily tested due to available technologies. For example, Amazon.com recommends other books to a specific customer based on his/her past search. Toll-road operators are segmenting drivers and charging them different prices based on the time of day and traffic conditions. Harrah's casinos study customer data to target promotions and drive exemplary customer service.
- *Creating new business from information.* New business opportunities may be invented by pulling information from a vast network of data sources (cross-field data analysis). For example, Zillow, a portal for real estate information, makes marketing data available to eliminate the dark reaches of the supply curve. The lack of transparency in real estate data was responsible for some agencies which thrived by keeping buyers and sellers partly in the dark.

- Identifying new opportunities. Planners should reflect and identify patterns that may next shape their markets and industries and consider whether there are opportunities to catalyze these changes and control their outcomes, instead of reacting to changing business patterns.

2.3 TYPES OF PLANNING

Managers engage in two types of planning at various levels in a company: strategic planning and operational planning. Both types of planning add value to the company.

2.3.1 Strategic Planning

Strategic planning sets the goals, purpose, and direction of a company. The top-level engineering managers (i.e., chief technology officer and vice president of engineering) are usually involved in strategic planning for the company (Elkin 2007; Nolan 2008).

Strategic planning focuses on identifying worthwhile future activities. Specifically, strategic planning assures that the company applies its resources—core competencies, skilled manpower resources, business relationships, and others—effectively to achieve its short- and long-term goals (Corbic 2000). It deals with questions such as the following:

- What are the company's mission, vision, and value system?
  - The mission statement of a company specifies why the company exists in the first place, what entities it serves, and what it will do to serve them.
  - The vision statement spells out the aspirations of the company with respect to its asset size, market position, business standing, ranking in industrial sectors, and other factors.
  - The value system is the externalization of five or six specific corporate values emphasized by the company. Some typical values favored by U.S. industrial companies include quality, innovation, social responsibility, stability, honesty, quality of life, and empowerment.
- What business should the company be in?
- Does the company need to change its product portfolio, market coverage, production system, or service capabilities? If so, why?
- What specific goals—profitability, market share, sales, technology leadership position, global penetration, etc.—should the company accomplish? By when should these goals be accomplished, with what investment, and by utilizing which core competencies?
- What business networks should the company pursue via supplier alliances, comarketing partnerships, production joint ventures, and other forms of collaboration?
- Which new services should the company offer?
- What core technologies should the company maintain, extend, acquire, or utilize?
- Which performance metrics are to be used for monitoring the company's progress?

The horizon of strategic planning is usually spread over five years, although it may be reviewed at more frequent intervals to adjust to changes in the marketplace. Strategic planning will be discussed in more detail in section 2.4.

2.3.2 Operational Planning

Managers at both middle levels (managers and directors) and lower levels (supervisors and group leaders) perform operational planning in order to define the specific tactics and action steps needed to accomplish the goals specified by top management (Gunther 2006). Managers and directors break down the company goals into short-term objectives. Supervisors and group leaders specify events and assignments that can be implemented with the least amount of resources within the shortest period of time. Operational planning ensures that the company applies its resources efficiently to achieve its stated goals. Questions considered in operational planning include the following:

- What is the most efficient way of accomplishing a project with known objectives?
- What is the best way to link up with three top suppliers in the marketplace for needed parts?
- What are the operational guidelines for performing specific work?

Operational planning involves a process of analysis by which a corporate goal or a set of corporate intentions is broken down into steps. These steps are then formalized for easy implementation. Operational planning focuses on the preservation and rearrangement of established categories (e.g., major strategies defined by upper management, existing products, and organizational structures). Operational planning is essentially a programming chore that is aimed at making various given strategies operational.

For production operations, tactical planning is composed of three planning levels: the sales and operation planning (S&OP), the master planning schedule (MPS) and material requirement planning (MRP). Literature is full of various models proposed to facilitate these planning tasks (Comelli et al. 2006). Some of these tools must be modified to become useful to services.

Operational planning is also called *platform-based planning* because it extrapolates future results from a well-understood, predictable platform of past experience. Results of such planning are predictable because they are based on solid knowledge rather than assumptions.

Compared with strategic planning, operational planning is easier for engineers to accomplish because past experience and examples are usually available as references.

2.4 STRATEGIC PLANNING

Strategic planning is important but difficult, because no one is prophetic enough to know what the future holds. In this section, strategic planning will be explored from different perspectives.

2.4.1 The Inexact Nature of Strategic Planning

Strategic planning requires an immense amount of strategic thinking (Aaker 2001; Schmetterer 2003). In turn, strategic thinking involves synthesis that likewise requires intuition and creativity. Strategic thinking brings about an integrated perspective of the enterprise, a foresight—albeit not too precisely articulated—of the company's direction that is built on insights from experience and hard data from market research. Strategic thinking is based on learning by people at various levels involved in conducting specific business activities. Strategic planning invents new categories rather than rearranging existing ones, and synthesizes experience to move the company in a new direction. Managers should inspire others to join in the journey and to shape the company's course, thus creating enthusiasm along the way. Broad participation is therefore strongly advisable.

Strategic planners use various kinds of inputs. Study after study has indicated that, in addition to hard data, the most productive managers rely on soft information—gossip, hearsay, and various other intangible scraps of information—to refine plans. A key part of strategic planning is to formulate a vision for the company as to what the company aspires to be. To formulate such a vision, the planners must be able to "see" (Corbic 2000). This is only likely when they are willing to get their hands dirty digging for ideas and extracting the strategic messages from them. Collecting these ideas as building blocks is instrumental to the development of useful strategic plans. Mintzberg (1994) said insightfully, "The big picture is painted with little strokes."

Once information becomes available, strategic thinkers comprehend it, synthesize it, and learn from it. They test ideas and verify the convergence of ideas before they define new strategies. Sometimes strategies must be left in flexible forms, such as broad visions, in order to adapt to a changing environment.

Strategic planning is also called *discovery-driven planning* (McGrath and MacMillian 1995). In situations involving the definition of the company's future direction, most planning inputs are based on assumptions about the future. Because the ratio of assumptions to facts is usually high, the success rate of the resulting plans is typically low. Therefore, strategic planning should be a continuous process and not a single assignment or event to be taken care of at well-defined milestone dates. It involves constant learning, acquisition, and interpretation of hard data and soft information, as well as staff discussions related to operational decision making, resource allocation, and performance management. Strategic planning requires the discipline of systematically identifying and validating key assumptions introduced in the planning. As more data and knowledge are discovered, more assumptions are validated to form an increasingly solid knowledge base for updating the planning. Strategic planners should engage a variety of participants at various levels to benefit from the relevant corporate expertise available.

The major difficulties of strategic planning can be traced back to three inherent characteristics of such planning:

1. *Prediction of the future.* Certain future events are more predictable than others; for example, seasonal variations of weather and election-year cycles. Other

predictions, such as the forecast of discontinuities—technological innovations, price increases, changes of governmental regulations affecting marketplace competition—are virtually impossible to predict accurately. In general, future is uncertain and varying rapidly and assumptions must be revised steadily.

2. *Applicable experience and insight.* Strategies cannot be detached from the subject involved. Planners must have in-depth knowledge and relevant hands-on experience of the subject at hand in order to set forth useful strategies.
3. *Random process of strategy making.* The strategy-making process cannot be formalized, as it is not a deductive but a synthesis process.

Strategic plans often fail due to one or more of the following seven reasons:

1. Not thinking strategically; for example, by limiting the strategy only to short-term needs and processes of the company.
2. Failure to identify critical success factors for the company.
3. Lack of firm and long-term commitment from company management (people, technologies, management attention, and focus).
4. Reluctance of senior management to accept responsibility for tough decisions; incompatible company culture in risk taking.
5. Not leaving enough flexibility in the plans, causing difficulties in adjusting to the changing environment.
6. Failure to properly communicate the plan and thus not securing support and management buy-in.
7. Difficulty to implement, as divisions do not always collaborate on futuristic stuff, while focusing on day-to-day operations.
8. Poor employee compensation scheme, which does not invigorate strategic planning and implementation.
9. Lack of integration of strategy with implementation.

2.4.2 Major Challenges in Strategic Planning

Generally speaking, there are two divergent views on strategic planning (Garvin and Schoemaker 2006), the view that holds promise and the one that signals reality:

1. "Most corporate planning is like a ritual rain dance. It has no effect on the weather that follows, but it makes those who engage in it feel they are in control" (Ackoff, Magidson and Addison 2006).
2. "The emphasis being placed on strategic planning today . . . . reflects the proposition that there are significant benefits to gain through an explicit process of formulating strategy, to insure that at least the policies (if not the actions of functional departments) are coordinated and directed at a common set of goals" (Porter 2008). There are six elements of an effectual strategic planning program:

a. Environmental scanning (customers, markets, industry, governmental regulations, etc.).

b. Competitive assessment (varies greatly, as all are reacting to new challenges).

c. Creative brainstorming (does not always yield the best new ideas)—a divergent process to encompass important strategic options.

d. Targeted goals setting (must be doable in time and within constraints).

e. Translation of goals to specific program and initiatives—convergent on selecting specific ideas for goals setting.

f. Effective and aligned implementation (collaboration, commitment).

But why is this list not sufficient to ensure an outcome? The list is not sufficient because many of these elements are difficult to implement successfully. It is especially difficult to bring forth a useful outcome utilizing environmental scanning and competitive assessment. Although poor in terms of producing outcomes, these six strategies provide a useful set of criteria for evaluating the strategic planning processes of any given company.

2.4.3 Methods Used to Plan Strategically

In adopting strategies, companies choose what to do and what not to do. Strategizing is about making choices. In general, managers use three methods to make choices (Gavetti and Rivkin 2005):

1. *Deduction*. Based on data, managers invoke general administrative and economic principles to a specific situation, weigh alternatives, and make rational choices (Tregoe and Kepner 2006). This method needs a lot of data, and is useful for mature and stable industries.
2. *Trial and error*. Experiment with several options and select one. This is good for ambiguous, novel, or complex situations in order to experiment and learn from the experience.
3. *Analogies*. Various strategic decisions are made by *analogies*. Decision makers often think back to a familiar situation, draw lessons from it, and apply those lessons to the current situation. However, making decisions by analogy must be done carefully.

Using analogies, managers pay attention to selected features of a problem, as opposed to considering every aspect. The case study method works toward building student's analogical thinking based on past experiences in order to enable the transfer of lessons from one industry to another.

To maximize the effectiveness of the analogy method in strategic planning, managers must be able to recognize similarities along dimensions that truly drive business performance, and not be fooled by superficial similarities between sources and targets.

The danger of focusing on superficial similarities is very real. Managers must be certain that deep similarities and surface resemblances are distinguishable.

Other uncertainties that impede analogical reasoning are emotional attachment to an early choice and halo effect—decision makers seek out supporting evidence and ignore conflicting data (e.g., confirmation biases).

There are four ways to avoid superficial similarity when practicing analogies:

1. *Recognize the analogy and identify its purpose.* Is there an analogy between the source and the target to allow the target to contribute to a desirable purpose?
2. *Understand the source (old case).* Source the environment, the solution that worked well in that environment, and the link between the environment and the winning strategy. Then, check if the relationship discovered here applies to the target environment.
3. *Assess similarities and differences between the source and target.* Are the similarities only superficial? Do the similarities exist along crucial dimensions (e.g., dimensions that drive economic performance)?
4. *Translate, decide, and adapt.* Decide whether the original strategy, properly translated, will work in the target industry.

The best strategy involves using all three methods (deduction, experimenting, and analogy).

Example 2.1 The method of analogy is important for cultivating strategic thinking. How can this method be applied systematically and in a rational manner in order to minimize its potential misuse?

ANSWER 2.1 The following rational method may be useful when applying the analogy methodology in strategic planning:

(a) Enumerate dimensions that characterize the source case and exemplify the past known case. The solution to the past case is to potentially apply the present case, the "target case." One such dimension is "linkage between solution and the achieved success in the past case." It is important that such linkages exist between cases in order to justify the use of the past case solution.
(b) Classify the dimensions as major or minor in describing the past case.
(c) Outline some key attributes of the past case in each dimension.
(d) Evaluate the present case in each of the listed dimensions in terms of relevant attributes.
(e) Compare the attributes of the present case with those of the past case.
(f) Assess the degree of similarly and difference between the two.
(g) Enter the result in the last column, "similar," "somewhat," or "no." Of course, it is crucial that the major dimensions are assessed correctly. If they are deemed similar, then the current case can be determined as possessing a "deep similarity" with the past case.
(h) Apply the solution to the past case as the strategic solution to the current case, only if a similarity is sufficiently established. See Table 2.1.

Table 2.1 Rational Method of Applying Analogy in Strategic Planning

Dimension	Major/minor	Source Case (Attributes)	Target Case (Attributes)	Is Target similar to Source?
1	Major	(1), (2)		Yes
2	Major	(3), (4)		Somewhat
3	Major	(5), (6)		No
4	Major	(7), (8)		Yes
5	Minor	(9), (10)		Maybe
6	Minor	(11), (12)		No

This method can be demonstrated using the examples of Intel and Toys R Us.

2.4.4 Technique to Gain Strategic Insights

Companies must consider the external environment, such as customers and competition, when they plan strategically. Urbany and Davis (2007) offer a special "Three Circles" technique to deepen strategic insights; see Fig. 2.1.

Let us draw three circles, one for customer's needs, one for a company's offerings, and one for a competitor's offerings. When a company or a competitor's offerings match with a customer's needs, these circles overlap. Specifically:

- **A:** Area representing our advantages. How big and sustainable are our advantages? Are they based on distinctive capabilities?
- **B:** Area designating points at parity. This area indicates that we are at par with our competitors. Are we delivering effectively in the area of parity?

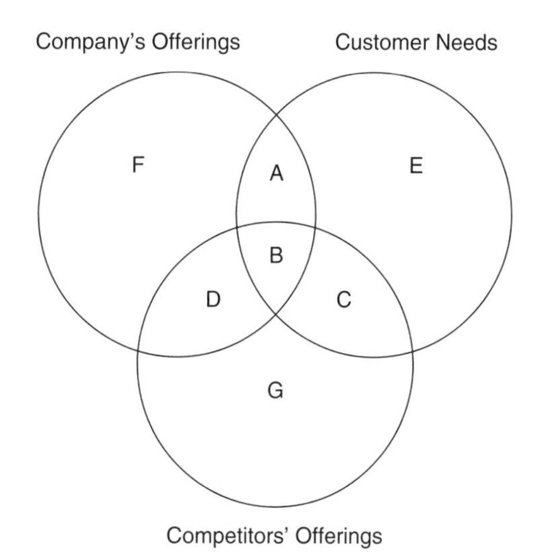

Figure 2.1 Strategic insights in three circles. (Adapted from Urbany and Davies, 2007)

C: Area depicting the competitor's strengths. How can we counter our competitor's advantages?

E: White space. This is the area for growth. Ask customers how our company's strengths can be made useful to them.

D, F, G: Areas denoting values produced by company and competitors that customers do not need.

A key planning question to ask is: How can we devise strategies that would increase area A while reducing D and F?

2.4.5 Techniques to Improve Strategic Planning

Dye and Sibony (2007) suggest specific techniques to augment strategic planning. These techniques are presented in five steps:

1. *Define the key long-term strategic issues first, before considering budget and operational issues.* Some best practices include the following:
   - Ask how a set of economic, social and business trends will affect specific divisions of a company and the best ways to capture the opportunities or counter the threads that these trends pose.
   - Define a roster of three to six priorities for the coming year to serve as basis for strategic deliberations among the responsible managers to achieve alignment between them. The results are then summarized for subsequent in-depth discussions.
   - Ask senior management to produce a group of ten strategic questions addressing the business of divisional units. Division leaders are to respond in six months in order to come up with answers. A meeting will be held to discuss the findings, propose solutions, and reach decisions.
   - Hold in-depth interviews of senior managers and selected corporate and business executives to generate a lineup of the most important strategic issues facing the company. The senior management team prioritizes this lineup and assigns managers to explore each issue and report back in four to six weeks. This is useful in companies where internal consensus building is imperative.
2. *Bring together the right people.* Key requirements are: (a) include the most knowledgeable and influential participants (CED, business unit leaders, who will implement outcomes), (b) find ways to simulate and challenge the participants thinking, (c) hold honest, open discussions about difficult issues, and (d) make the final decision. The interactions among participants play a key role. Forwarding the key issues and their supporting operational, financial, and other data prior to strategic meeting is useful. In order to be workable, the write-up should be about ten pages long and written in a concise manner.
3. *Adapt planning cycles to the needs of each business*. Allow strategic planning for every two to three years rather than annually, and rotate the units involved. If external circumstances justify its implementation, a planning review can be brought in at any time.

4. *Implement a strategic-performance-management system.* Having a strategy but no implementation does not bring about useful results. Strategies must be implemented effectively. Metrics to monitor implementation include (a) quality of available talents, (b) number of ideas and projects at each stage in development, (c) number and skill levels of people placed in important strategic projects, (d) financial and strategic goals (financial targets are "lagging" indicators), and (e) revenue from new products/service sales (lagging indicators). Selecting the right strategic metrics is important in order to allow mid-course corrections.
5. *Integrate human-resources systems into the strategic plan.* Tie the compensation of managers to the progress of new initiatives. The company needs to set up a compensation structure to account for short-term financial and operational targets, as well as longer-term, innovation-based growth targets. For example, allot 20 percent of managers' compensation toward the achievement of specific strategic goals, and link their success closely to bonus packages.

2.4.6 Integration of Strategic Planning with Implementation

According to Kaplan and Norton (2008), a comprehensive and integrated management system that explicitly links strategy formulation and planning with operational execution is needed to ensure positive outcomes. Successful implementation of these strategies is the key to drawing real value from any planning effort.

1. *Develop the strategy*. Use tools such as (a) mission, values and vision statements, external competitive, (b) economic and environmental analyses, (c) five forces and competitive positioning framework, (d) resource-based review of strategy, (e) blue ocean strategies, (f) scenario planning, (g) dynamic simulations and (h) war-gaming.
2. *Plan the strategy to define targets and strategic initiatives.* Use strategy maps, and balanced scorecards.
3. *Align the organization with the strategy*. Distribute cascading strategy maps and balanced scorecards to all organizational units (communicating with employees, lining their personal objectives and incentives to strategic objectives).
4. *Plan operations*. Use quality and process management, reengineering, process dashboard, realigning forecasts, activity-based costing, resource capacity planning, and dynamic budgeting.
5. *Execute.* Commit resources and devote management attention to implement the defined programs and initiatives.
6. *Monitor and learn about problems, barriers, and challenges*. Use management review meetings.
7. *Test and adapt the strategy.* Initiate a new cycle of integrated strategy planning and operation execution using data from internal operation, external environment and competition.

As illustrated in Fig. 2.2, these seven steps are keys for successfully integrating strategic planning with the required implementation.

Figure 2.2 Seven steps to integrate strategic planning with implementation. (Adapted from Kaplan & Noton, 2008)

2.5 PARTICIPANTS IN THE PLANNING PROCESS

2.5.1 Decision Makers

Top-level decision makers must be closely involved in planning process, to ensure that there is a firm commitment of corporate resources (e.g., talents, technologies, knowledge, funds, business relations and management attention) for the implementation of the planning outcome.

2.5.2 Workers with Knowledge

Those who have direct knowledge of the specific subject matters involved should take care of the planning.

In the 1960s, strategic planning was accorded emphasis and attention by the top management of U.S. corporations. Company after company set up high-level corporate planning departments made up of full-time planners who would devise business strategies. The approach failed to generate the expected business results. As outlined by Mintzberg (1994), one of the key weaknesses of this approach was that the strategic planners, while being superior analysts of hard business data, were outsiders insofar as the various specific business functions (marketing, production, engineering, and procurements) were concerned. What was not apparent at the time was the fact that planning new strategies for the future required both hard data and intuitive assumptions. The success of the decision to introduce assumptions, and the extent to which these assumptions could be validated, depended very much on the planner's hands-on management experience, intuitive know-how, and in-depth insight of the specific business activities involved. As such, numerous plans devised by these strategists were poor. Furthermore, business managers in operating departments did not wholeheartedly embrace the plans envisaged by these outsiders. Since the 1960s, innumerable companies have abolished their corporate strategic planning departments altogether and have delegated this important planning function to the business units themselves.

The moral of the story is that the most productive way of creating strategic plans for specific businesses or activities is to entrust such planning to those who are intimately involved with the particular businesses and activities. This paradigm is consistent with the empowerment doctrine whereby decisions are delegated downward to lower-level persons who have direct knowledge and in-depth understanding of the subject matters at hand (Barney 2002).

2.6 PLANNING ROLES OF ENGINEERING MANAGERS

Engineering managers at low levels will predominantly devise operational plans to achieve the short-term goals of the unit or department. As the engineering managers move up the corporate ladder, they are expected to participate increasingly in strategic planning, with emphasis placed on technology, product, and production planning. They may find it useful to follow the planning guidelines listed next in order to add value to the company.

2.6.1 Assist Their Own Superiors in Planning

It is important that engineering managers spend time and effort to actively assist their direct superiors in planning. These tasks may include: (1) analyzing hard data (industry, competition, and marketing), (2) offering alternative interpretations to data available, (3) raising insightful questions to challenge conventional assumptions, and (4) communicating the resulting outputs of planning—programs, schedules, and budgets—to help effectuate buy-in from others.

2.6.2 Ask for Support from Subordinates

In order to optimally benefit from the knowledge, expertise, and insights of staff, engineering managers are advised to engage their staff and other employees in the planning activities.

2.6.3 Develop Action Plans

It is quite self-evident that engineering managers need to perform four key planning tasks:

1. *Time management*. All managers need to plan and prioritize their personal daily activities (such as problem solving, staff meetings, job specification, progress monitoring, performance evaluation), according to the value each assignment may add, so that high-value tasks are completed before others. This is to maximize the value added by the daily activities (Lane and Wayser 2000).

   Also to be included in the daily to-do list are activities such as networking, continuing education, and scanning emerging technologies, that are deemed important for advancing one's own career.

2. *Projects and programs.* Engineering managers need to plan projects and programs assigned to them by upper management. In planning for projects and programs, engineering managers need to fully understand the applicable project objectives, the relevant performance metrics used to measure success, and the significance of the outcome of such projects and programs to the company. They should carefully select staff members with the relevant skills, expertise, and personality to participate in the project and seek their inputs regarding tasks, resource requirements, preferred methodologies, and task duration. They should then integrate all inputs to draft a project plan and distribute the plan among all participants to iteratively finalize the relevant details. These details include budgets, deliverables, and dates of completion. Managers must also secure authorization from upper management before initiating work related to the projects and programs.
3. *Corporate know-how.* The preservation of corporate know-how is of critical importance to the company for maintaining and enhancing its competitiveness in the marketplace. Corporate know-how comes in diverse types and forms. Certain documentable knowledge, such as patents, published memoranda, operational manuals, and trouble-shooting guides, is easy to retain. Others, such as insights related to procedures or perfected ways of designing specific products and services of the company, may require extra efforts to preserve. Managers should plan to systematically capture, preserve, and widely disseminate such know-how in order to maximize its use within the company.

   Certain other cognitive knowledge is typically retained mentally by the experts. Managers need to find workable ways to induce such experts to externalize these skills and insights for use by others in the company.

   Problem-solving expertise is yet another type of corporate know-how worth preserving. Typically, engineering managers are busy resolving conflicts that may arise from disagreements in assignment priorities, personality conflicts, customer complaints, interpretation of data, and other conflicts. Managers need to solve these problems promptly. They should be mentally prepared to jump from one job to another to handle such time-sensitive issues. What should be planned under these circumstances is the preservation of the learning experience garnered from each incident so that the company will become more efficient in solving similar problems in the future.
4. *Proactive tasks.* Engineering managers should plan to devote their efforts to proactively pursue certain other activities. These activities include:

   - Utilizing new technologies to simplify and augment the products and services of the company.
   - Looking for business partners to form supply chains.
   - Offering new or enhanced services to customers (e.g., self-service, an information-on-demand system, and a Web-based inquiry center).
   - Initiating new programs to promote healthy customer relationships.
   - Promoting new products with upgraded attributes (e.g., product customization to serve customers better, cheaper, and faster).

- Reengineering and simplifying specific operational processes to increase efficiency.
- Outsourcing specific jobs to augment cost effectiveness.

2.7 TOOLS FOR PLANNING

Engineering managers utilize a number of tools to prepare strategic plans. Some of these tools generate hard data, whereas others offer qualitative insights into specific subject areas (Kaufman 2000). The following are examples of some useful planning tools.

2.7.1 Market Research

Market research applies a number of tools to discover the preference of customers with respect to the company's products, services, marketing strategy, product prices, competitive strengths, and brand reputation in the marketplace. Specific tools include polling by questionnaires, product concept testing, focus groups, and pilot testing. The outputs of market research help assess the company's current marketing position and future growth opportunities in the marketplace.

2.7.2 SWOT Analysis

SWOT is the abbreviation for strength, weakness, opportunities, and threats. Each company has strengths and weaknesses in comparison to its competitors. The competitors offer products in direct competition with the products of the company. On the one hand, because of the company's strengths or core competencies, there may be opportunities offered in the marketplace that the company ought to exploit aggressively. On the other hand, because of the strengths of the competitors and the conditions in the marketplace, the company might be subjected to certain future threats. Such potential threats could be the result of technology advancement, business alliances, marketing partnerships, and other such step changes accomplished by the competition. New governmental regulations and policies may also affect the company's business in the future. A systematic monitoring of publications—patents, technical articles, news release, and financial reports—represents an initial step in conducting a competitive analysis. A well-performed SWOT analysis will bring to the fore the assessment of the company's current position.

As such, the SWOT analysis procreates a road map by which a company can make informed decisions about improving its core competencies to meet its current and future business and operational needs. The analysis answers questions such as the following: (1) What does the company have in place today? (2) In which direction is the company headed in the next three to five years? and (3) What is the company's process of managing changes?

2.7.3 Financial What-If Analysis and Modeling

Spreadsheets are useful in modeling the financial performance of an operation. Financial statements (such as an income statement, balance sheet, and funds flow statement—see section 7.3) are usually modeled in a spreadsheet program. *What-if* analyses are readily performed to discover the sensitivity of the company's financial performance relative to the changes of specific input variables. In addition, such analyses permit the verification of various assumptions incorporated into the financial models.

Most businesses have inherent uncertainties due to the unpredictable nature of the business climate, the liquidity of financial markets, certain governmental regulations and international trade policies, currency stability, customer preferences, competition in the marketplace, disruptions induced by new technologies, and other factors. What-if analyses and Monte Carlo simulations may be applied to assess the impact of some of these uncertainties on a company's business. (Monte Carlo simulations will be discussed in section 6.6.)

2.7.4 Scenario Planning

Scenario planning defines the major forces that may move a company in different directions, map out a small number of alternative futures (scenarios), specify narratives to elucidate these scenarios, and define options for managing within these future worlds (Garvin and Levesque 2006). Figure 2.3 displays the scenario-planning process involving a number of components.

The key focal issue depicts a significant upcoming decision or a strategic uncertainty. The driving forces are themes and trends that are most likely to influence the key focal issue in fundamental ways. Examples include social dynamics, economics, political affairs, and technology. Some of these driving forces are more important than others. The top two and most influential of these driving forces are critical uncertainties. Using critical uncertainties as axes, a 2 × 2 map, the scenario framework, is formulated. This 2 × 2 map defines four quadrants, the four futures, which will influence a company's future directions. Each of these four scenarios is then written up as a story. Planning participants then deliberate on the implications of these four stories with respect to the

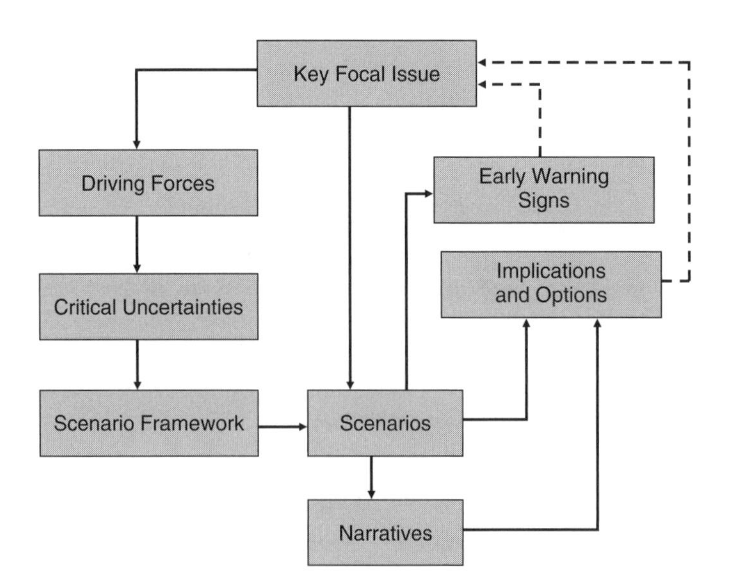

Figure 2.3 Scenario-planning process. (Adapted from Garvin and Levesque, 2006)

key focal issue at hand. This group's deliberations will likely result in the identification of the company's strengths and weaknesses, as well as define strategic options for future directions. Early warning signals are leading indicators that highlight the likely emergence of one scenario or another. These signals must be constantly monitored in order to initiate a strategic reevaluation.

Extension to Three- or More Dimensional Scenario Diagram. Why do we limit ourselves to only two critical uncertainties in the scenario-planning process? This is possibly because business people will typically use two-dimensional diagrams (2×2 maps), which are easily transposed onto paper to delineate product positioning, product classification, and business-related issues. Engineers are capable of envisioning three-dimensional diagrams. Thus, this concept could easily be extended to three key uncertainties, providing a total of eight ($=2^3$) scenarios for possible consideration when each future is affected by three uncertain driving forces. Three uncertainties are more realistic than two in evaluating the future of businesses. Molecular physicists are trained to define the "state" of molecules by seven variables—namely, three special coordinates, three velocity components, and one time variable. They typically think in seven-dimensional space and use a seven-dimensional distribution function to define the state of a molecule, as it is used in the Boltzmann equation in physics. Thus, one can easily envision a scenario defined by seven (not two) uncertainties, offering a total of 128 ($=2^7$) scenarios for consideration. Engineering managers should, try to use at least three key uncertainties for scenario planning. Table 2.2 exhibits eight futures influenced by three uncertainties.

As pointed out by Schoemaker (1995), scenario planning can capture a whole range of possibilities in rich details. It is a disciplined method for imagining possible futures. Scenario planning reduces a lot of data into a limited number of possible states. Each scenario tells a story about how various factors interact. Scenario planning is different from contingency planning, sensitivity analysis, and complex computer simulations, as described below.

1. *Contingency planning* examines only one uncertainty. What if we don't get the patent? Contingency planning starts with a base case and considers an exception to it (what-if analysis). Scenario planning, by contrast, explores the joint impact of *various uncertainties* that take place concurrently. It tries to determine the new state, as a result of the changing of several independent variables.

Table 2.2 Eight Scenarios Based on Three Critical Uncertainties

#	Uncertainty 1	Uncertainty 2	Uncertainty 3
1	High	Low	Low
2	High	High	Low
3	Low	Low	Low
4	Low	High	Low
5	High	Low	High
6	High	High	High
7	Low	Low	High
8	Low	High	High

2. *Sensitivity Analysis* examines the effect of a change of one variable while keeping all other variables constant. This approach is useful for monitoring small magnitudes in changes in order to determine new states.
3. *Complex computer simulations.* Scenarios can include elements that cannot be modeled, such as subjective judgments.

Whenever the future needs to be assessed, scenario planning is applicable. It applies generally to decision making under uncertainty. The methods of developing scenario planning include the following steps:

1. Define the scope.
2. Recognize the major stakeholders.
3. Point out basic trends.
4. Identify key uncertainties.
5. Construct initial scenario themes.
6. Check for consistency and plausibility.
7. Formulate learning scenarios.
8. Suggest research needs.
9. Develop qualitative models.
10. Evolve toward decision scenarios.

Scenarios should paint generically different futures rather than variations on one theme. The article (Schoemaker 1995) contains an example of scenario planning for an advertising agency.

Garvin and Schoemaker (2006) pointed out that "there is no right or wrong scenario." Scenario planning has a number of advantages and disadvantages. There are at least six advantages:

1. Scenario planning prepares managers for multiple, alternative futures.
2. It builds a common vocabulary.
3. Scenario planning fosters strategic conversations.
4. Managers are forced to articulate critical uncertainties.
5. Managers must consider the interactions between these uncertainties and the likely implications.
6. Scenario planning legitimizes divergent ways of thinking that are essential for success.

Disadvantages of scenario planning could include the following three:

1. Scenario planning might lead to no more than "pipe dreams and happy talk."
2. It might stipulate wishful thinking and lead to a desire for one particular future and outcome.
3. The absence of a tight link between scenario planning and formal goal setting and action planning could be a cause for concern.

2.7.5 Performance Benchmarks

Performance benchmarks are those that have been achieved by successful companies in the same industry in which the host company operates. When planning, it is important to define these benchmarks to measure corporate progress. Watson (2007) offers an excellent set of broad-based benchmarks:

- *Customer-related measures*—product defects, just-in-time delivery, life-cycle product cost, customer satisfaction score, order-processing efficiency, percent sales from new customers, service quality, time taken between orders and product delivery, etc.
- *Process-related measures*—time to market (i.e., the elapse of time from the initiation of product design and development to product delivery to the marketplace), quality standards, unit product cost, core competence development, labor hours per product, etc.
- *Financial measures*—gross margin, net income-to-sales ratio, current ratio, sales per employees, return on equity, sales growth rate, market share percentage, inventory turn ratio, etc.
- *Employee-related measures*—turnover ratio, employee satisfaction score, skill building and development expenses per employee, etc.
- *Competition-related measures*—market share, cost of innovation, acquisition cost per new customer, number of new products commercialized per year, and so on.

A large number of these quantitative metrics are available either from the financial statements of the companies in the same industry or from public sources such as the following:

Banks that offer loans to companies in a specific industry

Financial institutions that analyze and compare companies' performance on behalf of investors

Companies that offer credit ratings of companies seeking debt financing (Zagorsky 2003)

These metrics serve well as industrial benchmarks against which to assess the current status of a specific company and to define its new strategic direction.

Kaplan and Norton (2007) suggested the use of balanced scorecard as a strategic management system. The balanced scorecard revolutionized the performance metrics. Besides financial performance metrics, this system adds metrics to customers, internal business process, and learning and growth. Financial metrics are lagging indicators, whereas learning and growth and customers are leading indicators. Internal business processes are current indicators. These are good ways to present them, as relying on financial metrics alone will be too late to incorporate corrective means.

Example 2.2 Quality is usually defined differently by different people in a company. Explain why. Which quality definition is the correct one for the company to adopt?

ANSWER 2.2 Different people in a company may have different interests and perspectives in defining quality. Examples are presented here:

- *Production*—quality is reject rate and deviation from specifications (view of a production engineer).
- *Value based*—quality is defined in terms of price and costs (view of a marketing person).
- *User based*—Quality is the degree to which a product satisfies the customer's needs. Customers do not appreciate less or more quality than they need (view of a customer).
- *Product based*—Quality is related to the number of attributes offered by a product (view of a product designer).

For the company to succeed in the marketplace, quality is in the eyes of the beholder. The user-based definition is preferred.

2.7.6 Product Life-Cycle Analysis

Every product has a life cycle that moves typically through the stages of initiation, growth, market saturation, and decline. Engineering managers need to examine the life cycles of all products marketed by their employers. Doing so will guide them in introducing new products or product enhancements in a timely manner in order to sustain company profitability.

Tools for operational planning include project management tools: Microsoft® Office Project Standard or Microsoft® Enterprise Project Management Timeline, Critical Path Method (CPM), and Program Evaluation Review Technique (PERT). Other tools, such as action planning, design procedures, risk analysis, and operational guidelines, are also useful.

2.8 PLANNING ACTIVITIES

The activities of planning involve forecasting, action planning, issuing policies, and establishing procedures. Strategic planning requires forecasting, action planning, and issuing policies. Operational planning necessitates action planning, issuing policies, and establishing procedures. Some planning activities are proactive; others are reactive in nature.

2.8.1 Forecasting

The objective of forecasting is to estimate and predict future conditions and events. Forecasting activities center on assessing future conditions in technology, products, marketplace, and other factors affecting the business success of the company. The marketplace revolves around customers, competition, economy, global supply chains, human resources, capital, and facilities.

Forecasting helps to define potential obstacles and opportunities and establish the premises for the plan. It sets boundaries for possibilities to help focus on specific future conditions, defines worthwhile objectives, promotes intergroup coordination, provides basis for resources allocation (manpower, budget, facilities, and business relations), and induces innovation through forecasted needs.

Saffo (2007) offers six insightful rules for effective forecasting:

1. Define a cone of uncertainty.
2. Look for the S curve.
3. Embrace the things that don't fit.
4. Hold strong opinions weakly.
5. Look back twice as far as you look forward.
6. Know when not to make a forecast.

Examples are displayed to illustrate the use of these rules.

Forecasting may be implemented by using the following steps:

1. *Identify*—critical factors that have the most profound effects on the company's profitability.
2. *Determine*—the forecasting horizon as short term (one year), intermediate term (two to five years) or long term (five to ten years).
3. *Select*—forecasting methods such as:
   - *Mechanical projection.* The future is projected assuming essentially the same characteristics as in the past.
   - *Analytical projection.* The future is estimated based on an extrapolation of the past (trend analysis). Statistical tools such as linear or nonlinear regression, moving averages, exponential smoothing, time series, and others may be applied.
4. *Forecast*—future eventualities and their likelihood of occurrence.
5. *Prepare*—the forecast, as well as the pertinent database.
6. *Adjust*—forecasts regularly to incorporate pertinent changes related to assumptions and desirable results.
7. *Assure*—understanding and acceptance by all parties affected by the forecast.

Several observations are worth noting. Major economic events such as prices, wages, and raw materials tend to change gradually. The farther an event is projected into the future, the greater the probability of significant deviations between the forecast and reality. Certain future events tend to result from current and past occurrences, as long as there are no disruptive changes in technology or society such as wars, natural disasters, and major incidents. The future may be planned with detailed, factual knowledge of the present and the past under those conditions. It is important to screen ideas by using proper criteria that are consistent with the company's objectives, technical capabilities, financial viability, and marketplace compatibility. Useful inputs may be offered by customers, salespeople, production employees, service clerks, and others who possess intimate knowledge of specific subjects.

Engineering managers are likely to get involved primarily in technology forecasting. Technology forecasting is of critical importance to those companies whose products are composed of high-technology components. Companies must constantly examine, monitor, and utilize emerging technologies to magnify business performance.

For example, George Washington University (Halal, Kull, and Leffmann 1997) provides a forecast of emerging technologies in which major events in the fields of energy, environment, information technology, manufacturing, materials, transportation, and others are predicted on the basis of the opinions of a group of selected scholars. The latest forecast findings may be obtained from the *GW Forecast* Web site, which is on the Internet at www.gwforecast.gwu.edu.

Engineering managers need to understand the value that any of these emerging technologies may have on the products and services offered by their employers and then plan accordingly.

Another technological example is the speed of computing. Currently, the top speed is achieved by Japan, running at 40,960 gigaflops (GFlops). It is almost a certainty that the computing speed will continue to increase over time. The question for engineering managers is as follows: How can business benefit from such an advancement? Computing speed may be used advantageously in computationally intensive problems, whose solutions of finer granularity provide value to the business. Here are four examples that illustrate how a refined granularity can help:

1. Data-mining applications related to customer-relations management may be of direct benefit to business. Assuming that customer data (e.g., prices paid, items bought, purchase habits, and payment methods) are collected and available, a detailed analysis could lead to an in-depth understanding of customer behavior, not available heretofore, thus allowing companies to structure customized selling and marketing programs to achieve better customer satisfaction, strengthened brand loyalty, and improved company profitability.
2. Refined modeling of key components in turbomachinery (e.g., compressors, blowers, and pumps) by using computational fluids dynamics programs could raise aerodynamic performance and reduce energy consumption.
3. Plant operations groups typically have collected a significant amount of data, observations, and experience in maintaining and troubleshooting equipment and facilities. Such information is dispersed widely, cannot be easily reapplied, and remains useless. A data-mining application could help in getting the information organized and ferreting out valuable knowledge from the wasteful piles of raw data. Application of such knowledge could lead to productivity enhancement in plant operations.
4. To design better, safer, faster, and cheaper cars, sophisticated computer models could be devised to impact-test automobiles, instead of crash-test expensive vehicles at up to only 40 miles per hour of speed.

It is quite certain that engineering managers will be able to envisage plentiful other computationally intensive problems that can be processed to reap business benefits.

In general, forecasting the impact of new technologies on future businesses is difficult. For example, in the past, few companies understood the significance of the Internet to company operations and the marketplace. Questions like those enumerated below did not have clear answers:

1. What will be the impact of broadband technology (cables, optical networks) to communications?

2. How will nanotechnology affect engineering activities such as product design and equipment operation in the future?
3. Will the next wave of new products be smart appliances and intelligent devices?
4. What happens if processors get more powerful and intelligent devices get smaller and more mobile?
5. What about the molecular switching devices that Hewlett-Packard is said to be working on that could lead to computational devices about one million times smaller than those we have today?
6. What will be the impact of *pervasive computing* on consumer markets?
7. How quickly will personal computers (PCs) lose their market values, once alternative devices that allow customers to access the Internet, get and send messages, purchase goods and services, activate entertainment programs online, and control home appliances remotely become widely available?
8. How will the new technologies related to intrinsic and extrinsic smart materials, which exhibit sensing and other capabilities, affect the industrial product design in the future?

To forecast the impact of new technologies, engineering managers must be properly prepared. Some business researchers portend that people with broad perspectives and variable professional experience and exposure may have a better chance of accurately forecasting the impact of emerging technologies, market trends, and other future conditions that require "foresight." Teams whose members have diversified backgrounds in engineering, product design, manufacturing, marketing, service, and sales are said to be better equipped in handling technology forecasting that could benefit from the divergent experience and insights of composite teams.

Engineering managers whose backgrounds are broad based are likely to be more successful in technology forecasting if they are supported by teams that also have diverse experience.

Example 2.3 The U.S. economy is shaped by a number of factors. The war in Iraq and global anxiety rankle the business environment and influence employment and consumer spending. Correctly reading trends in the economy can make or break a business. Where can an engineering manager find data that could help predict the direction of the economy?

ANSWER 2.3 There are leading and trailing indicators for the economy. The 2001 U.S. recession differed from others in its cause, severity, and scope. According to a recent assessment (Stock and Watson 2003), many of the commonly used indicators did not forecast well. These indicators include stock prices, unemployment claims, housing starts, orders for new capital equipment, and consumer sentiment.

Gene Sperling, who served as director of President Clinton's Council of Economic Advisors, offers some advice for business leaders to get ahead of the competition by becoming their own economists (Sperling 2003). Specifically, he recommends that the following set of indicators be used: (1) CEO opinions, (2) temporary jobs, (3) consumer spending, (4) bank loans, (5) semiconductors, (6) commercial structures, and (7) housing markets.

> Consumer spending contributes to about 70 percent of the US Gross Domestic Products. The collapse of the US housing markets led to the crisis in the financial industry in 2007. Indeed, the housing markets, jobs, consumer spending and bank loans are the primarily factors responsible for the 2007–2009 recession in the U.S.

2.8.2 Action Planning

Another important activity related to planning is action planning, the process of establishing specific objectives, action steps, and a schedule and budget related to a predetermined program, activity, or project (Kerzner 2009). Action planning helps to focus on critical areas that need attention and action. The identification of critical areas enables the company management to pay attention primarily to planning for deviations that may arise—the principle of management by exception. Furthermore, action planning states specific results to be accomplished. Defining results to be accomplished requires the planner to make judicial selection and exercise judgment. In addition, action planning provides standards as milestones that facilitate control and clarify accountability for results. It also permits an effective delegation of responsibilities (who is responsible for what results), promote teamwork, and ensures an evaluation of the overall performance of the program, activity, or project on a continuous basis.

Action planning mandates engineering managers to take the following specific steps:

1. *Analyze critical needs*. Critical needs are those associated with staff development, staff maintenance, and staff deficiency, as well as those related to special assignments. Managers define these needs by reviewing standards related to position charters, duties, management expectations, and company goals. Short-term needs must be in balance with long-term needs.
2. *Define specific objectives*. Specific objectives need to be defined to satisfy the critical needs. The results statement (who will attain what desirable results by when) must be specific. Establishing objectives predetermines the results to be accomplished.
3. *Define standards*. Standards measure the attainment of the objectives. The standards should preferably be quantitative in terms of performance ratios, percentages, cost figures, resource parameters, and other factors in order to be measurable (Kaplan and Norton 2007).
4. *Define key action steps*. The definition of key action steps establishes the sequence and priority of steps required to attain objectives. Specifically, major steps are lined up in the order in which they are to be performed; this list includes the evaluation of uncertainties for the steps planned, the definition of contingency steps to ensure the expected results, and the specification of who is responsible for each step and who is accountable for achieving the target value associated with each step.

   Action steps must be reasonably implementable. After the expected results are defined, engineering managers should plan these steps with the active participation of involved workers to benefit from their creativity and expertise in the subject.

5. *Devise a schedule*. Scheduling establishes both a time sequence for action steps and the interrelationship among the steps, as some might be prerequisites for others. It is advisable to estimate the optimistic (earliest), the pessimistic (latest), and the most likely (most probable) dates of possible completion of each step. Doing so will permit a more realistic modeling of the project schedule.

   Sufficient scheduling flexibility should be included to account for contingency—more for projects related to new development and less for routine design and analysis work. *Contingency* refers to the slag and cost buffers introduced to account for undefinable, yet generally anticipated, deviations from the plan.

   The most important outcome of the scheduling effort is the definition of the project or program completion date. The engineering manager, as the leader, is accountable for completing the project or program on time.

6. *Develop a budget*. Budgeting allocates resources necessary to accomplish objectives. The planner determines the basic units (man-hours, man-weeks) to accomplish each task, estimates the total resources needed for the project, and adds a contingency to the total amount for potential deviation (e.g., 7 to 10 percent of the total budget, dependent on the customary percentages used in each industry) to arrive at a total budget.

   The budget estimate is typically the basis for seeking management approval for the project or program. The project leader is accountable for completing the project or program within the approved budget.

For complex projects that involve multiple participants (e.g., peer departments, external suppliers, and outsourcing service organizations) project management tools such as PERT or CPM may be applied. These tools deliver timelines, graphically diagram the tasks network to facilitate monitoring and control, and determine the tasks linked along the critical path. Consequently, the shortest time in which the project can be completed is determined. Managers are then reminded to monitor these critical-path tasks carefully in order to avoid project delays.

A well-designed project plan serves the purpose of promoting communication, monitoring progress, evaluating performance, and managing knowledge.

2.8.3 Issuing Policies

For companies to operate smoothly and consistently, corporate rules and regulations are used to prescribe acceptable practices. Company policies address important issues such as employee hiring and termination, Equal Employment Opportunity (EEO) policies, annual performance appraisals, savings plans, benefits, medical insurance, pension plans, sick leave, safety, contact with representatives of competitors, and other issues. At the departmental level, specific rules may be defined to regulate activities that are repetitive in nature, such as filing reports after each completed business trip, submitting monthly or quarterly progress reports to summarize achievements and to outline future work, attending scheduled staff meetings, publishing engineering or scientific articles, participating in professional and technical conferences, and other duties.

Managers may write policies to offer uniform answers to questions of common concern. In general, policies are continuing directives promulgated to address repetitive issues, assignmentss, and problems in an organization. Policies are useful for predeciding answers to basic repetitive questions, capturing the distilled experience of the organization, saving management time, and facilitating delegation. Issuing policies is a part of the manager's planning responsibilities.

To be practically workable, a policy must have certain common characteristics, including the following: (1) applies uniformly to the organization (or specific engineering unit) at large; (2) remains relatively permanent, unless and until repealed; (3) fosters the objectives of the company; (4) frees managers and employees to focus on important matters; (5) encourages productive teamwork by reducing disagreements, conflicts, and differences in interpretation; and (6) is issued by top management or authorized managers with perspective, balance, and objectivity.

2.8.4 Establishing Procedures

Companies perform various important activities such as product design, plant operation, project management, equipment installation, facility maintenance, manufacturing, system engineering, parts procurement, product delivery, customer service, and others. The specific methods by which these activities are performed serve as the valuable corporate know-how employees have learned to perfect. Over time, companies want to preserve these "tried-and-true" procedures in manuals.

Developing procedures is of critical importance to a company, not only because doing so will preserve the best way to perform repetitive work (to achieve high productivity), but also because doing so will accomplish the following: (1) provide the basis for method modernization, (2) ensure standardized action (such as quality control, resource saving, and work reproducibility), (3) simplify training, and (4) retain corporate memory, such as know-how, knowledge, heuristics, proven safety practices, and problem-solving techniques.

Establishing and preserving procedures is part of the planning responsibility of managers. If generated in suitable formats, such procedures could be widely applied within the company and among its business partners to garner competitive advantages in the marketplace. Techniques for developing procedures include the following:

- Concentrating on procedures for critical work that is in high demand, repetitive, and time consuming.
- Charting the work required—inputs, workflow, outputs, skills, and resources.
- Reviewing work characteristics carefully in order to decipher (a) why (is the work really necessary?), (b) what (results are to be obtained?), (c) when (is the best time to do it?), (d) where (is the best place—group, station, facility, or equipment—to do it?), and (e) who (is the person with the relevant training to do it?)
- Proposing procedures in the context of existing objectives, policies, and programs by keeping the procedures to a minimum (This will avoid restricting employee imagination and incentive, as well as assure consistent applications to minimize deviations.)
- Defining refinements to procedures and updating them regularly.
- Formulating the procedures in writing.
- Communicating with all affected parties to ensure understanding and acceptance.

2.9 SOME SPECIFIC ADVICE ON PLANNING

Good up-front planning is essential for any company to achieve its desired corporate objectives. Managers need to pay sufficient attention to planning activities in order to make sure that certain pivotal factors are sufficiently addressed in the strategic or operational plans they formulate.

2.9.1 Assumptions

Plans are typically built on both hard data and assumptions. Assumptions are usually based on extrapolations of past experience and intuitive projections into the future. It is important for managers to constantly seek and interpret additional resources and insights to verify their assumptions. This is to ascertain that the plans they introduce are built on an increasingly solid foundation.

2.9.2 People

Any plan is worthless unless its objectives are achieved through a successful implementation. Implementation requires dedicated people who are supportive of and ardent about the subject matter involved. Managers need to take into account the suitability of people, including their background, personality, training, mental flexibility, interpersonal skills, collaborative attitudes, adaptability, and emotional attachments to specific ways things are done.

Most plans contain activities related to a change of the current status. Unfortunately, most people resist change, particularly sudden change. Change may induce business instability, technology obsolescence, organizational restructuring, and other unwanted disruptions. People may be more amenable to gradual changes if such changes occur at rate they can understand and accept.

Managing change is a challenging assignment for managers. Managers need to recognize early that change is coming. They may want to delineate the change in detail and analyze the implications of the change as a way of preparing the staff and allowing them to become gradually accustomed to such a change.

By paying close attention to how changes are being communicated to the staff, managers may be able to minimize the resistance to change and gain support for the implementation of new plans. It is helpful for the managers to isolate and identify areas of threats and opportunity. If needed, they should employ contingency plans for handling threats, but focus on opportunities that will advance the company business.

2.9.3 Benefit versus Cost

When planning, managers need to be guided by the expected value that a given project or program may bring about. Low-value projects justify the commitment of low-level efforts, whereas high-value projects justify the allocation of high-level efforts. Efforts applied should be commensurate with the value added by the expected results. Otherwise, corporate resources may be wasted. The saying, "Things worth doing are worth doing well," is valid only to the extent justifiable by the expected value.

2.9.4 Small but Sure Steps

To be effective in planning, managers should (1) identify clearly the desired end results and the series of small steps required to reach them, (2) allow a timely control and mid-course correction, if needed, and (3) aim at attaining a series of small progressions (or continuous improvements) that are more acceptable in numerous old-style companies than one large achievement (or a step change) after a long period of time. However, some start-up companies with an entrepreneurial spirit may be able to exercise patience, take risks, and go for "blow-the-roof-off" breakthrough technologies and step-change products or services. Managers need to adjust accordingly.

2.9.5 Contingency Planning

As discussed before, strategic planning for the future entails considerable risks and uncertainties. Some of the changes in future conditions are unpredictable. Yet strategic planning for the future must be done today. Besides striving for acquiring hard data and soft information to continuously validate the assumptions introduced in the planning, managers should take an additional uncertainty-modulating step: Study exhaustively the sensitivity of various assumptions to the company business and incorporate contingency steps, including fallback positions, in order to minimize the adverse impact of questionable assumptions (Childs 2008).

2.9.6 Commitment

Managers need to secure company commitment before any plan can be implemented successfully. Company management must declare their intentions and their readiness to allocate resources needed to achieve the planned objectives. Without a firm company commitment, nothing of value will emerge from the planning efforts.

Example 2.4 Joe Engineer took a graduate school course at SUNY–Buffalo, where he learned the importance of planning. Joe knows that luck plays a big role in one's life. But he is convinced that proper planning will help him to have an orderly progression in his career. He thinks that it would be cool to become a CEO of a publicly owned, multinational company at the age of sixty and retire at sixty-five with a net worth of $5 million. He wants some guidance with career planning. How can you help him?

ANSWER 2.4 It is advisable for Joe Engineer to follow a number of planning steps, enumerated here:

I. Set Objectives and Specify Subgoals

Before starting the planning process, we need to introduce an important assumption. In order for Joe Engineer to be entrusted with a given management position in a publicly held major company, he needs to have acquired and successfully demonstrated certain business management capabilities beforehand. Obviously, this assumption may not be valid for small and medium sized companies that are privately held.

The CEO of a major company must be familiar with a lot of functional areas, such as (1) strategic management, (2) business management, (3) operational management, (4) project or program management, (5) engineering management, (6) production and manufacturing, (7) marketing management, (8) financial control, and (9) globalization. The future CEO must be able to demonstrate sufficiency in various skills:

- Public speaking and writing
- Business analysis and planning
- Public relations
- Problem solving and conflict resolution
- Interpersonal skills
- Negotiations
- Business relations development
- Other skills

Therefore, for Joe Engineer to qualify for the CEO job, he must have garnered useful management experience, possibly as a company president a few years back. Future capabilities are, by and large, based on past experience. Applying such a logic in a backward-chaining manner, Joe Engineer could readily establish a set of milestones in his plans:

Corporate president at 55
Division president at 50
Vice president at 45
Director at 40
Manager at 35
Supervisor at 30
Group leader at 25

II. Develop Action Plans

Joe should consider a forward chaining plan, which moves from the present to the future. As examples, the following plan illustrates the qualifications that should be built up when advancing from one stage to another:

A. Preparation (By a Certain Date)

1. Take steps to collect pertinent career development references and acquire perspectives.
2. Talk with experienced engineers to garner insights related to the costs and benefits of the targeted objectives. The advantages and disadvantages of being a manager are well known: power, prestige, and money versus travel, fifty- to sixty-hour workweeks, job pressure, office politics, balance between work and home, and related factors.
3. Understand one's own career objectives and the requirements to succeed. What are the "success factors" involved?
4. Be aware of one's own strengths and weaknesses, personality type, value system, and personal requirements for happiness.
5. Confirm desirable objectives of moving into the managerial career path.

B. Group Leader
 1. Get a master of engineering degree to demonstrate technical competence (by a certain date, say 1/20xx).
 2. Become well versed in engineering management concepts and practices (e.g., take courses or training).
 3. Practice good interpersonal skills by doing volunteer work.
 4. Network inside and outside the company (join technical societies, attend technical conferences, publish technical papers, etc.) and know some professional people well.

C. Supervisor
 1. Seek training on supervision and practice teamwork with dedication.
 2. Take advanced technical courses, if needed, to help become established as a technical leader.
 3. Broaden into marketing, production, and sales through business interactions.
 4. Function as a gatekeeper for technology.
 5. Demonstrate innovation.
 6. Continue networking and become known to countless others inside and outside the company.
 7. Attain recognizable technical achievements.
 8. Demonstrate managerial potential.
 9. Become known as a good problem solver.

D. Manager
 1. Exhibit prowess in strategic planning, operation, and all other engineering management skills.
 2. Showcase capabilities in interacting with sales, marketing, production, service, and customers.
 3. Demmonstrate success in initiating and implementing new technology projects that affect the business success of the company.
 4. Achieve organizationwide recognition.
 5. Form networks with important people at various levels.

E. Director
 1. Become widely known in one's own industry.
 2. Participate actively in industrial trade and technical groups.
 3. Demonstrate leadership in strategic planning affecting the company.
 4. Be recognized for operational efficiency.
 5. Make major contributions to direct the company's businesses.
 6. Master the technology-marketing interface.
 7. Guide the company in utilizing emerging technology to constantly strengthen competitiveness.
 8. Represent the company well to the press.
 9. Have real friends in high places.

F. Vice President

Joe Engineer is advised to fill in the remainder of this plan as an exercise.

1. Budget and Commitment

A. Invest the proper amount of resources (time, money, and efforts) to ready oneself for the next stages.

B. Make a firm commitment to carry out action steps specified in the plan.

2. Review and Update

Review the plan and make adjustments regularly to exercise proper control of this career path. Knowing what it takes to move to the next stage, and preparing oneself in time for that big opportunity ahead, set forth a good mantra for Joe Engineer to follow.

2.10 PLANNING IN THE HEALTHCARE INDUSTRY

According the Zuckerman (2006), the state of healthcare strategic planning is lagging behind that of other industries. The healthcare industry applied a set of ten best practices of strategic planning:

1. Establish a unique, far-reaching vision.
2. Attach critical issues.
3. Formulate focused clear strategies.
4. Differentiate from competition.
5. Achieve real results.
6. Organize preplanning.
7. Structure effective participation.
8. Think strategically.
9. Manage implementation.
10. Manage strategically.

Zuckermann offered another five state-of-the-art approaches to healthcare planning, borrowed from other industries:

11. Use knowledge management.
12. Advocate innovations and creativity in strategic approaches.
13. Emphasize bottom-up versus top-down strategic planning.
14. Use an evolving, flexible and continuously improving process.
15. Shift from static to dynamic strategic planning.

As pointed out by Ginter and Swayne (2006), these best practices stand in for some generally applicable guidelines, which are not specific to the four unique characteristics of the healthcare industry.

1. *Built-in characteristics.* Unrelated diversifications or vertical integration may not be appropriate for organizations with healthcare as their core. Their product development strategies take a very narrow focus, as most technologies are derived from outsiders (drug companies, research institutions, teaching hospitals,

equipment manufacturers, and robotic companies). Certain service practices in healthcare organizations cannot be cut to save cost—for example: emergency room services, OB/GYN services, or drug addiction treatment (mandated, mission required).

2. *Culture*. Physician managers maintain medical practices, as well as discharge their own administrative responsibilities. Strategic planning would be an add-on responsibility. (However, is it a good enough reason for not doing it?) Strategic planning must be streamlined. Where physicians act like independent contractors, excellence is usually built around physicians' reputations, and not that of the institution. Hospitals have semi-autonomous power centers, which may interfere with organizational strategic planning efforts. The organizational structure in hospitals can hinder the bottom-up planning and participation methods.
3. *Outside control on hospitals and physicians drive the entire process*. Their needs may be in conflict with other customers (patients, governments). Medicare, insurance companies and other third-party payers influence pricing and market power.
4. *Society has an impact on healthcare organizations*. Healthcare strategic planning is affected by the concept of healthcare as a right, access to care, quality of life or death, who pays for the cost of care, and other such social issues.

As the Obama administration is currently embarking on a major restructuring of the U.S. healthcare system, significant changes in healthcare planning and implementation are to be expected in the near future.

Example 2.5 Describe the top-five key lessons/insights you have learned/gained from this chapter, including justifications.

ANSWER 2.5 The most important capability for future leaders to nurture and cultivate is strategic planning, which requires vision for defining future directions. Forecasting technologies, business conditions, marketplace development, and customer's needs are critical but difficult to determine, as choices may have to be made based on intuition, judgment, and hunches, rather than hard data. Future leaders need to nurture strategic-planning skills through keen observations, deep reflection, and trial and error. Operational planning is more or less straightforward; it is essential for implementation, which includes 85 percent of the efforts required in order to achieve success in any endeavor. Planning activities are essential to both corporate and personal life. Policies and procedures are mostly administrative in nature; however, they need to be taken care of even though they are not being particularly exciting or rewarding.

2.11 CONCLUSIONS

Both strategic and operational planning are important, because the success of a company depends on creating new paths to the future as well as on implementing short-term operational plans.

Service systems engineers and leaders are expected to involve themselves in both strategic planning and operational planning. Both types of planning require forecasting and action planning. To regulate work, managers may also participate in issuing policies and establishing procedures. Among these planning activities, forecasting and strategic planning are difficult, as they involve making estimations of the future. The remaining planning activities related to policies and procedures are administrative or operational in nature. These activities are rather straightforward and should appear to be relatively easy to understand and implement. Doing extremely well in these administrative activities will not necessarily make a manager outstanding, but not doing well in them will project a negative image of the engineering manager.

To demonstrate managerial leadership, service systems engineers and leaders need to be proficient in technology forecasting and strategic planning. Technology forecasting involves the critical evaluation and adaptation of emerging technologies so that the company's products and services offered to the marketplace become better, cheaper, and faster to deliver. A primary opportunity for engineering managers to add value is to participate actively in creating technology projects affecting the company's future in major ways.

Efficiency portrays "doing thing right" with the associated metrics around cost, productivity, resources utilization and cycle time. Effectiveness mean "doing the right things," with corresponding metrics oriented to customer value, competitive positioning, and other such differentiation measures. To be world class, service enterprise must be both efficient and effective. Planning is an essential step to get there.

2.12 REFERENCES

Aaker, D. A. 2001. *Developing Business Strategy*, 6th ed. New York: John Wiley & Sons, Inc.

Ackoff, Russel, J. Magidson, and Herbert J. Addison. 2006. "*Idealized Design: How to Dissolve Tomorrow's Crisis Today.*" Wharton School Publishing (April 30).

Barney, J. 2002. *Gaining and Sustaining Competitive Advantages*. 3rd ed. Upper Saddle River, NJ: Prentice Hall.

Childs, D. R. 2008. Prepare for the Worst, Plan for the Best: Disaster Preparedness and Recovery for Small Businesses. 2nd ed. New York: John Wiley & Sons, Inc.

Coke, A. 2002. *Seven Steps to a Successful Plan*. New York: AMACOM.

Comelli, Michael, Michael Gourgand, and David Lemoine. 2006. "A Review of Tactical Planning Models." Proceedings of 2006 International Conference on Service Systems and Service management, Troyes, France (October).

Corbic, C. 2000. *Great Leaders See the Future First: Taking Your Organization to the Top in Five Revolutionary Steps*. New York: Dearborn Trade.

Dye, Renee, and Oliver Sibony. 2006. "Planning for Change." *McKinsey Quarterly* 3.

Dye, Renee, and Oliver Sibony. 2007. "How to Improve Strategic Planning." *McKinsey Quarterly* 3.

Elkin, P. 2007. *Mastering Business Planning and Strategy: The Power and Application of Strategic Thinking*, 2nd ed. London: Thorogood.

Gavetti, Giovanni, and Jan W. Rivkin. 2005. "How Strategists Really Think." *Harvard Business Review* 83 (4) (April).

Garvin, David A., and Paul J. H. Schoemaker. 2006. "Strategic Planning at United Parcel Service." Harvard Business School Teaching Note # 5-307-003 (July 13).

Garvin, David A., and Lynne C. Levesque. 2006. "A Note on Scenario Planning." Harvard Business School case # 9-306-003 (Rev. July 31, 2006).

Ginter, Peter M., and Linda E. Swayne. 2006. "Moving Toward Strategic Planning Unique to Healthcare." *Frontiers of Health Services Management* 23 (2): 33–37.

Gunther, N. J. 2006. *Guerrilla Capacity Planning: A Tactical Approach to Planning for Highly Scalable Applications and Services*. New York: Springer.

Halal, W. E., M. D. Kull, and A. Leffmann. 1997. "Emerging Technologies: What's Ahead for 2001–2030." *The Futurist* (November–December).

Hamel, G., and C. K. Prahalad. 2003. "Competing for the Future." *Harvard Business Review OnPoint Article* (September 1).

Kaufman, R. A. 2000. *Mega Planning: Practical Tools for Organizational Success*. New York: Sage Publications.

Kaplan, Robert S., and David P. Norton. 2007. "Using the Balanced Scorecard as a Strategic Management System." *Harvard Business Review* (July–August).

Kaplan, Robert S., and David P. Norton. 2008. "Integrating Strategy Planning and Operational Execution: A Six-Stage System." Harvard Business School Balanced Scoreboard Report. Book review report on Robert S. Kaplan and David P. Norton. "The Execution Premium." Harvard Business Publishing, June.

Kerzner, H. 2009. *Project Management: A Systems Approach to Planning, Scheduling and Controlling*, 10th ed. New York: John Wiley & Sons, Inc.

Lane, H. L., and C. Wayser. 2000. *Make Every Minute Count: 750 Tips and Strategies that Will Revolutionize How You Manage Your Time*. New York: Marlowe & Co.

Manyika, James M., Roger P. Roberts, and Kara L. Sprague. 2008. "Eight Business Technology Trends to Watch." *McKinsey Quarterly* 1.

McGrath, R. G., and Ian C. MacMillian. 1995. "Discovery-Driven Planning." *Harvard Business Review* (July–August).

Mintzberg, H. 1994. "The Fall and Rise of Strategic Planning." *Harvard Business Review* (January–February).

Nolan, T. N., Leonard D. Goodstein, and Jeanette Goldstein. 2008. *Applied Strategic Planning: An Introduction*, 2nd ed. Hoboken, NJ: Pfeiffer.

Porter, Michael. 2008. "The Five Competitive Forces that Shape Strategy." *Harvard Business Review* (January 1).

Saffo, Paul. 2007. "Six Rules for Effective Forecasting." *Harvard Business Review* 85 (7/8): (July/August).

Schmetterer, B. 2003. *Leap: A Revolution in Creating Business Strategy*. Hoboken, NJ: John Wiley & Sons, Inc.

Schoemaker, Paul J. H. 1995. "Scenario Planning: A Tool for Strategic Thinking." *Sloan Management Review* 36 (2): 25–40.

Sperling, G. 2003. "The Insider's Guide to Economic Forecasting." *Inc.* (August).

Stock, J. H., and M. W. Watson. 2003. "How Did Leading Indicator Forecasts Perform During the 2001 Recession?" *Economic Quarterly—Federal Reserve Bank of Richmond*, Summer 89 (3).

Tregoe, B. B., C. H. Kepner. 2006. *The New Rational Manager*. Princeton, NJ: Princeton Research Press.

Urbany, Joel E., and James H. Davis. 2007. "Strategic Insight in Three Circles," *Harvard Business Review* 85 (11).

Watson, G. H. 2007. *Strategic Benchmarking Reloaded with Six Sigma: Improving Your Company's Performance Using Global Best Practices*. Hoboken, NJ: John Wiley & Sons, Inc.

Zagorsky, J. L. 2003. *Business Information: Finding and Using Data in the Digital Age*. New York: McGraw-Hill.

Zuckerman, Alan M. 2006. Advancing the State of the Art in Healthcare Strategic Planning. *Frontiers of Health Service Management* 23 (2): 3–13.

2.13 QUESTIONS

2.1 Study the Case *"eBay (A) The Customer Marketplace," Harvard Business School Case* # 9-602-071 (Rev. June 15, 2005) and answer the following questions. The case materials may be purchased by contacting its publisher at www.custserve@hbsp.harvard.edu. or 1-800-545-7685.

A. What is this case about, and how do you describe the eBay business model?

B. The eBay experience: If you have experienced eBay auctions firsthand, recall your motivation for participating and evaluation of the experience. If you have not participated in an eBay auction before, search through www.ebay.com. How do you, as a new user, feel about the site content and navigation?

C. Company uniqueness: What is really unique about eBay service that enables it to earn high ROS? What is novel about eBay service transactions and operations?

D. Customer relationship management: How effective has the company been in managing customer behavior? What, if anything, would you recommend the company to do differently in this respect?

E. Corporate participation: Is corporate participation good or bad for eBay? What could eBay do to mitigate the conflicts between individual customers?

F. Lessons learned from this case: What are the lessons you have learned from having studied this case?

2.2 Strategic planning plays a very important role in any service organization. In this chapter, we learned that planning well for the future is a difficult job as it involves a number of activities such as scenario planning, directions setting, and plan implementing. Study the specific business case: David A. Garvin and Lynne C. Levesque (2006), "Strategic Planning at United Parcel Service," Harvard Business School Case # 9-306-002 (Rev. June 19)* and answer the following questions. The case materials may be purchased from the publisher at 1-800-545-7685) or www.custserv@hbsp.harvard.edu.

A. What are the advantages and disadvantages of scenario planning as a tool in support of the strategic planning process?

B. How does the scenario planning exercise at UPS for Horizon 2004, in comparison with that for Horizon 2017?

C. What are the major challenges of strategic planning?

D. What are the key stages in UPS's strategic planning process?

E. How do you evaluate UPS's Charter, the Centennial Plan, the Strategic Road Map and the Strategy Integration Plan?

F. What are the lessons you have learned after having studied this case?

2.3 Mission and value statements are indicative of the direction in which a company is headed. What are typically included in the statements of mission and values of well-known companies in the United States? Please comment.

2.4 What are included in the typical operational guidelines some industrial companies have developed? Please comment.

2.5 There are always risks (risks of failure) associated with the experimentation of a new manufacturing process or with the entry into a new global market. How should one decide whether to proceed with a risky venture? What is the proper level of risk for a company to take?

2.6 The marketing director needs to submit a strategic plan for entering a new market. She knows she needs long periods of uninterrupted time. She considers two options: (1) staying at home to do the plan or (2) delegating some parts of the plan to her subordinates. What are the factors the director needs to consider when she chooses the best way to come up with this plan?

2.7 In planning for a project, the critical path method (CPM) is a tool used widely in industry. Elucidate the basic concepts and techniques involved, illustrate by using an example, and review its advantages and disadvantages.

2.8 XYZ Company has been a one-product company focused on developing and marketing a package of innovative ERP software specialized for law firms and operated in computers running on a proprietary operating system software developed by the company. Customers must purchase both the hardware and software as a bundled package from XYZ Company. The company also provides around-the-clock services to ensure that the combined hardware and software system performs reliably, as lawyers are known to be typically disinterested in troubleshooting computer systems. This product-bundling strategy works out well for the company, and the sales revenue of XYZ increases dramatically during its first three years in business.

However, market intelligence suggests that new ERP software products are now being introduced by competitors. These new ERP software products are quite capable of performing all of the data-processing functions typically required by law firms. Furthermore, these new ERP software products can run on any computers using their existing operating systems, thus eliminating the need for customers to purchase dedicated computers. The president of XYZ Company recognizes the potential threat imposed by these new ERP software products. He wants to know the best counterstrategy he should plan and implement. Design and explain this counterstrategy.

2.9 Sandy Smith is about to graduate from UB with a master's degree in engineering and a GPA of about 3.8. She wants to find a good job that allows her to best utilize her strengths and capabilities. Her short-term goal is to become an operations manager in a manufacturing enterprise in ten years. Modeled after Example 2.4, how should she plan to achieve this specific goal?

2.10 On the eve of leaving her alma mater, Stacy Engineer remembers the encouraging words of the commencement speaker: "Graduation is the happy beginning of an exciting life ahead." She is, of course, excited about her new master of engineering degree that she received with honor. But she is also a bit confused about what to do now to make her new life exciting and filled with happiness. Apparently, what she needs is a roadmap into the future. How can you help her?

Chapter 3

Organizing

3.1 INTRODUCTION

Organization is another important function of service systems management and engineering. Organizing means arranging and relating work so that it can be done efficiently by the appropriate people (Galbraith 2008).

Corporate efficiency is usually achieved by appropriately delegating assignments, defining the correct management and arrangement of interrelated groups participating in projects, and addressing factors that influence outcomes such as time constraints, resource limitations, and business priorities. (O'Reilly and Pfeffer 2000).

Managers are empowered to design the organizational structure—the team, group, department, and so on—and to define the working relationships conducive for achieving the company's objectives. Doing so will:

- Ensure that important work gets done.
- Provide continuity.
- Form the basis for wage and salary administration.
- Aid delegation.
- Facilitate communication.
- Promote growth and diversification.
- Foster teamwork by minimizing personality conflicts and other problems.
- Stimulate creativity.

Although it is generally true that dedicated people can make any organization work, dedicated people in well-organized units can achieve outstanding results.

This chapter introduces the basic concept that organizational design needs to follow service strategy, compares several basic forms of organizational structures commonly employed in industry, and elucidates a number of possibilities to invigorate corporate performance through organizational practices. Illustrative examples are included for specific organizational structures that promote key improvements, such as: drive corporate profitability in service enterprises, intensify innovation, resolve conflicts at the interface between design and manufacturing, promote collaboration at the interface between research and development (R&D) and marketing, attain a balanced corporate strategy between short-term performance and long-term health, facilitate the all-important interactions between self-directed professionals, and foster employee motivation. Special emphasis is placed on teams composed of cross-functional members. The critical managerial jobs of assigning responsibilities while maintaining control and establishing work relationships are also addressed.

3.2 THE CONCEPT OF STRUCTURE FOLLOWING STRATEGY

Services consist of high-contact and low-contact activities, corresponding to front- and back-office work. Organizational design needs to follow the strategy chosen to position a company's services. There are four such service strategies to create and maintain competitiveness in the marketplace: (1) cost leadership, (2) cheap convenience, (3) dedicated services, and (4) premium services (Metters and Vargas 2000).

3.2.1 Cost Leadership

To compete on price, companies typically decouple work. They decompose a complex job into many simpler tasks, standardize the tasks to decrease quality variance, centralize selected tasks to spread overhead costs and even out variations in demand, and employ special labor and technologies to achieve scale economies. Examples of such service companies include McDonald's, Dunkin' Donuts, and Taco Bell, which all centralize food preparation for their retail stores.

Such service strategies have their disadvantages. These include: (1) diminished customerization and service quality, (2) poor dependability and accuracy, (3) less flexibility in service scope, (4) longer response time (although time of individual tasks is cut down, but the waiting time and hands-off between them are not, resulting in longer total process time), (5) diminished face-to-face contact with customers, and (6) work not motivating to back-office workers, who do not see results of their efforts.

3.2.2 Cheap Convenience

The focus of this service strategy is to offer convenience at a reasonably low cost. Convenience is accomplished by deploying a lot of service units, each having limited capabilities. To effectively implement such a *kiosk* strategy, companies need to (1) cross-train workers, (2) engage a number of employees to man service units, and (3) incorporate the back-office support work into these customer-facing service units. For example, Edward Jones, a stock brokerage firm, has opened 3,800 offices in the United States, each manned by a single broker offering a limited services line. 7-11 and Dollar General operate in the same way. Some of their offices remain open for 51 hours per week. These service strategies suffer from the difficulty of maintaining a uniform quality standard across the system, as well as limitations in their service contents.

3.2.3 Dedicated Service

Companies pursue this strategy by emphasizing increased flexibility and varieties in the services they offer, while making cost a secondary consideration. They are targeting their services to customers who require a high quality of services. To implement this strategy, companies need to take the following steps:

1. Select the customer-facing workers carefully with respect to personality and people-oriented work attitude.

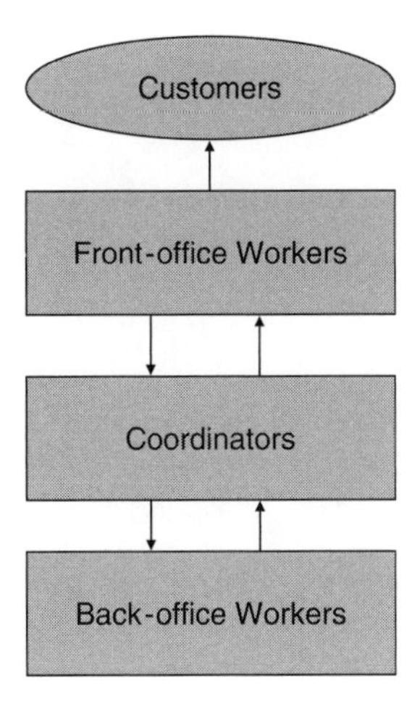

Figure 3.1 Coordinators between front- and back-office workers.

2. Centralize the back-office work on a regional basis.
3. Engage selected employee *coordinators* to resolve problems between the front- and back-office workers; see Fig. 3.1.
4. Reward the back-office worker based on work quality. The disadvantage of this service strategy is that it can elevate costs in order to maintain the implemented systems.

The types of companies that adopted this service strategy include real estate companies, stock brokerage firms, law firms, insurance companies, and hospitals (where surgeons are supported by nurses and other staff).

3.2.4 Premium Service

Companies pursuing this strategy want to maximize service customization and responsiveness to customers. They manage their organizations with a relationship orientation, rather than a transactional orientation. They offer a broad service line, select and train employees to acquire versatile skills, empower local decision making, gain insights from customers, inspire customer-facing employees to actively involve themselves in community affairs as a way of attracting new clients, and focus on strengthening a long-term and deep professional relationship with their customers. They decouple the back-office work from the front-office only if doing so yields a scale of economies. Business and professional consulting firms, such as McKinsey, Accenture, and Arthur D Little pursue this premium service strategy.

Figure 3.2 illustrates these four service strategies, each catering to a different customer base. As market conditions change, companies may opt to change their service strategies and their organizational design approaches to maximize their corporate effectiveness. Organizational design must be compatible with the service positioning strategy selected by the company to compete in the marketplace.

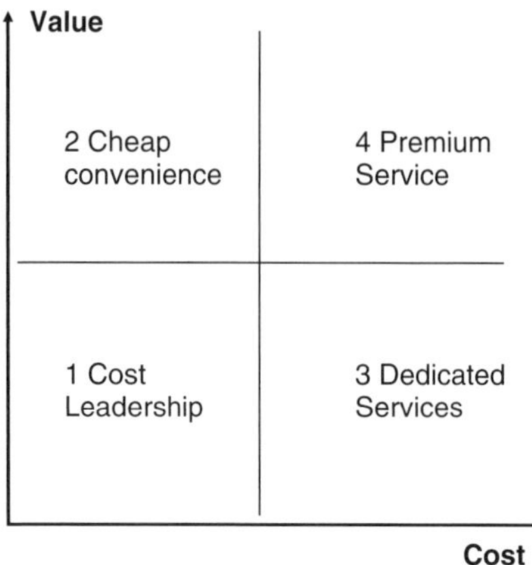

Figure 3.2 Options available for positioning services in the marketplace.

3.3 ACTIVITIES OF ORGANIZING

As a function of engineering management, organizing consists of several specific activities described here:

- *Organizing one's own workplace for productivity*. This includes the organization of one's own office, file systems, and daily routine so that work can be done efficiently (Fabrizio and Tapping 2006).
- *Developing organizational structure*. Identify and sort the work so that it can be done efficiently by qualified people in teams, taskforces, committees, departments, and other suitable arrangements (Dickinson and Jelphs 2008, Cox 2001).
- *Delegating*. Entrust responsibility and authority to others and the creation of accountability for results. Managers must learn to delegate effectively in order to achieve results by working through people. They distribute the workload while maintaining control to make the best use of available talent in the organization (Burns 2002).
- *Establishing working relationships*. Create conditions necessary for the mutually cooperative efforts of people. Managers must make commitments, set priorities, and provide needed resources (money, physical facilities, skills, and know-how) to foster teamwork and collaboration among people (Straus and Layton 2002).

All of these organizing activities exist for the purpose of achieving improved efficiency in performing work. Each of these activities is discussed below.

3.4 ORGANIZING ONE'S OWN WORKPLACE FOR PRODUCTIVITY

How well is the office of a typical engineering manager organized? A simple test is to inquire how much time it would take for the engineering manager to find a phone number, a piece of paper, or a file when his or her superior calls. Surveys point out that an average executive spends about five weeks per year looking for lost items (Von Hoffman 1998).

Engineering managers need to be organized with respect to time, paper, and space. A basic guideline recommended by efficiency experts is as follows: "The less you have, the less you have to sort through." Here are a few rules of thumb that are recommended for the engineering manager to become more efficient:

- *Use an online calendar that indicates time slots blocked out for important activities.* Such a calendar allows others—one's own secretary and peer managers—to schedule meetings conveniently. One should also prepare agendas before holding or attending meetings.
- *Maintain a to-do list.* Set priority to activities and separate urgent tasks from others by assigning most urgent tasks to list A, moderately urgent tasks to list B, and least urgent tasks to list C. Consult the lists regularly. If one is computer literate, then use electronic systems to generate and maintain such a list.
- *File papers based on access, or use a logical keyword system under which to find the document later.* The file system may be based on categories such as projects, persons, or deadlines. Keep a master copy of the file index nearby and update it often. This master index helps locate a file and safeguards against creating duplicate files. A document should be kept if:
  1. The information it contains cannot be easily found elsewhere. (How difficult would it be to obtain or reproduce it again?).
  2. The information it contains helps the engineering manager to reach a goal. (Does this piece of paper require action?).
  3. It has been consolidated as much as possible.
  4. It is up to date. (Is it recent enough to be useful?).
  5. It is really necessary to keep this document. (What is the worst thing that could happen if this document is unavailable?).

Most professional workers are said to use only about 20 percent of the paperwork they keep. The challenge is, of course, to decide which 80 percent can be thrown away. Question every piece of paper that crosses the desk. Use the wastebaskets frequently. Reserve a time slot during each day (e.g., after work, but before departing for home) to sort, file, and toss unneeded files. Make use of travel time (at the airport, on the plane, and in the hotel) to organize one's own files.

- *Implement a system for keeping track of names and phone numbers.* This might be a Rolodex for business cards, address book, or smartphone.
- *Cultivate the use of the phone. Prepare notes before placing calls* and make the calls brief (e.g., by standing up).

With practice, every engineering manager can get *his or her workspace and daily routine* organized for productivity (Hemphill 1999).

Example 3.1 * David Pope, engineering director, started out the day uptight. His young child had the flu the night before, and he had been up all night to help. Upon arrival at his office, David had to make urgent phone calls to approve a

* Condensed and adapted from Shannon, Robert E. 1990. *Engineering Management.* New York: John Wiley and Sons, Inc., p.32.

two-week overtime work plan due to a plant fire the night before and to plan for a product committee meeting the next day to counter environmental concerns about a wastewater treatment plant.

Then he spent thirty minutes reviewing the qualifications of new candidates and decided on one. He called for salary information and wanted to examine the offer before it was sent. He requested for further justification for the budget requested by industrial engineering for a minicomputer. Without reading it, he approved the research proposal from material engineering. He rejected an invitation to speak at a regional meeting of the American Society of Plant Engineering by giving an untrue reason.

David made a note for a United Way board meeting coming up soon. At 10:00 a.m., he met with two consultants for one hour and forty-five minutes on a formal wage and salary plan and then directed his administrative assistant to work out the details. He promised to inform all department heads and asked for cooperation.

As he walked back to his office after lunch, David noticed several engineers were still playing bridge after 1:30 p.m., and he planned to remind their department heads of this truancy from work.

As soon as he walked into his office after lunch, George Wallace, the general sales manager, called to complain about inadequate responses from engineering to field sales requests. David promised to look into it after receiving specific details. In return, he requested for Wallace's support at the product committee meeting the next day.

David gave a retirement plaque to Glen Sanford in his own office in the presence of the personnel director at 1:45 p.m. Furthermore, he approved the request of two engineers for a week of overtime to design a new, final quality-control station.

At 1:30 p.m., he was invited to attend a three-hour budget meeting at 2:30 p.m. called by the president. In the meeting, guidelines and a timetable for next year's budget requests were deliberated. For engineering, he was told there would be an increase of only 10 percent. He then arranged for a meeting with the president and the controller at 2:00 p.m. the next day to request for more money.

As he was about to leave for the day around 6:30 p.m., his wife called to say that his child is doing all right, but he has to go to the party of the executive vice president alone.

What do you see are David Pope's problems? How do you suggest improving his day?

ANSWER 3.1 David Pope had four major problems: (1) poor time management, resulting in the day being spent responding mostly to others; (2) lack of delegation; (3) inadequate utilization of administrative assistant; and (4) deficient guidelines for handling minor projects. David Pope could revise his day as follows:

1. Review the day's schedule in the morning and call in the administrative assistant to:

- Get background information on wage and salary plan for the 10:00 a.m. meeting with the two consultants. Prevent this initial meeting from dragging out to one hour and forty-five minutes.
- Request the personnel director to invite peers of Glen Sanford to attend the plaque-awarding ceremony in his own office.

- Collect information on the budget request for the minicomputer from industrial engineering.

2. Return all phone calls.
   - Authorize the two-week overtime work plan due to plant fires.
   - Send Jamieson (who wrote the report) to the product committee meeting to defend the wastewater treatment plant.
   - Approve Oscar Ford to use two engineers for one week to design a new, final quality-control station. Direct the administrative assistant to draft new guidelines for manpower allocation in minor projects.
3. Upon receiving notice for the 2:30 p.m. meeting called by the president, get the administrative assistant to start preparing the engineering positions on budget. Review these positions at 2:00 p.m. Call an urgent meeting for 2:15 p.m. with department heads to finalize the engineering position.
4. Present a plaque to Glen Sanford before a group of his peers, express appreciation for his services, and wish him well.
5. Become well-prepared to attend the 2:30 p.m. meeting. If more discussions are needed, request a follow-up meeting with the president. If additional budget preparations are required the next morning, leave a note to the administrative assistant.
6. Go to the party alone, and be happy.

3.5 DEVELOPING ORGANIZATIONAL STRUCTURE

The purpose of developing organizational structure is to help ensure that important work related to the key objectives of the unit or department is performed. By developing the right organizational structure for pursuing specific work, managers hope to eliminate or minimize the overlap and the duplication of responsibilities. Also, by logically grouping work according to positions in the organizational structure, managers will be in a better position to utilize available talents, advocate mutual support among workers, provide technological foci, and facilitate problem solving. Doing so will ensure that management, technical, and operating work are distinguishable so that people can be most efficient in performing such work (Wetlaufer 1999; Dean and Susman 1989; Chesbrough and Teece 2002). More recently, Karwowski et. al. (2010) discussed the customer-centered design of service organizations.

Numerous industrial organizations adopt one or more of the following structures.

3.5.1 Functional Organization

The functional structure is a very widely used organizational form in industry. Companies that favor this organizational form include (1) manufacturing operations, process industries and other organizations with limited service diversity or high relative stability of workflow, (2) start-up companies, (3) companies with narrow product range, simple marketing pattern, and few production sites, and (4) companies following the lead of their competitors.

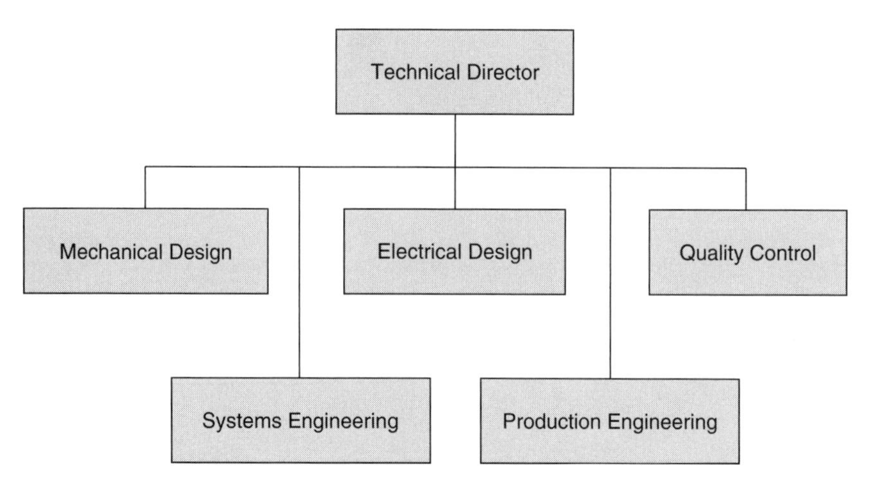

Figure 3.3 Functional organization.

Companies that prefer the functional structure establish specific departments responsible for manufacturing, finance, marketing, sales, engineering, design, operation, procurement, and other such functions (see Fig. 3.3).

The functional structure has certain advantages, as it (1) permits a hierarchy of skills to be acquired and maintained, (2) facilitates specialization in order to achieve high levels of excellence, (3) simplifies coordination as experts in various functional areas are logically grouped together, and (4) allows the use of current technologies and state-of-the-art equipment.

However, it also has some disadvantages, as it (1) reinforces excessive centralization, (2) delays decision making due to barriers brought about by the departmental silos, (3) compounds communication line loss, (4) restricts the development of managerial skills of employees, and (5) limits employee growth because of constrained exposure to professional experience outside of the departments.

3.5.2 Discipline-Based Organization

Universities, governmental laboratories, and some contract research firms are organized according to disciplines. These organizations contain departments for mechanical engineering, physics, business administration, and other specific disciplines so that specialists may focus on these disciplines in order to excel in research and other activities they pursue (see Fig. 3.4).

3.5.3 Product/Service/Region-Based Organization

Large companies may offer and market products/services of various types to different customers in geographically dispersed locations. More often than not, each of these products may require different production, sales, and business strategies to achieve success in the marketplace. Thus, some companies elect to organize themselves into a product-based structure (see Fig. 3.5). If the company is marketing products in various geographical regions, each demanding location-specific strategies to penetrate the local markets, and each implementing different product-customization strategies according

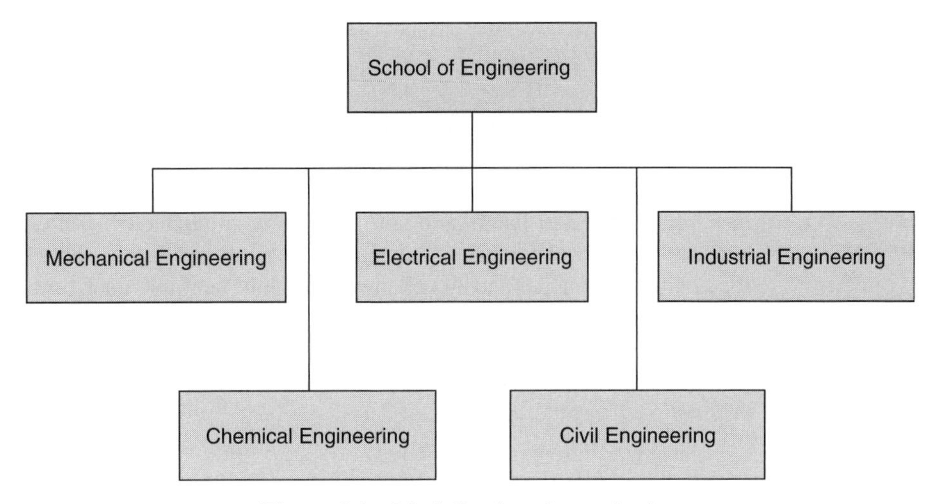

Figure 3.4 Discipline-based organization.

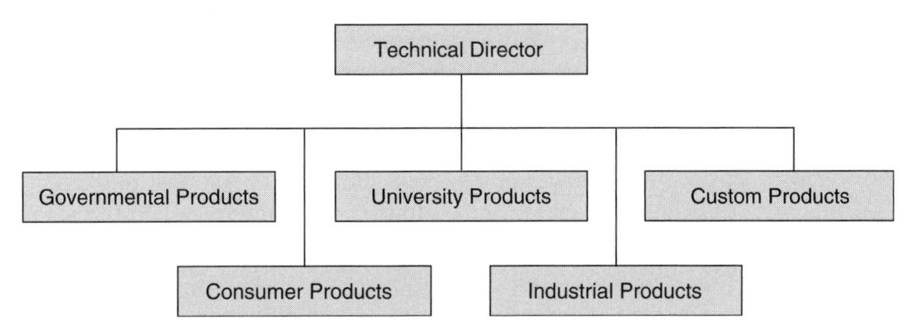

Figure 3.5 Product/service organization.

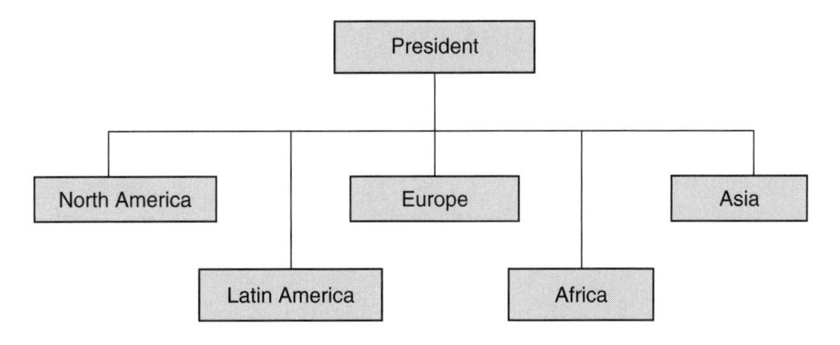

Figure 3.6 Region-based organization.

to local needs, a region-based structure may be preferred (see Fig. 3.6). In either of these cases, a product/service or regional manager could head up the activities with the overall profit-and-loss (P/L) responsibility for the product or region involved. This manager is further supported by the relevant experts in production, marketing, and other needed functional areas.

This type of product/service- or region-based organization enjoys the following advantages: It (1) focuses on end product/services or geographical regions for refined local adaptation, (2) facilitates companywide coordination, (3) fosters management development of employees, (4) provides for decentralization, and (5) opens ways for unlimited growth.

The disadvantages for such organizational structure are the following: (1) costs may be high due to layers and autonomous or duplicated facilities, (2) it may require added management talents, (3) specialists may easily become obsolete due to a lack of focus and dedication, and (4) changes are slow to implement because of the complex organizational bureaucracy.

The organizational structure types just described have one thing in common: They all have a hierarchical structure with a clearly defined chain of command (a structure originating from the military).

Service companies often employ the *multiunit design concept* to deliver services effectively (Garvin and Levesque 2008). As the economy becomes more service-intensive, more companies will take on the multiunit enterprise form. Multiunit organizations use several organizational design principles.

- Have branches, service centers, and stores to focus on implementation, alignment, and the standardization of service practices.
- Pursue some degree of local customization (such as in marketing, merchandizing, compensation, degree of local differentiation from global uniformity).
- Assign field units to concentrate on operational implementation, whereas headquarters takes care of strategic planning related to service positioning, advertisement, annual budget, and setting the performance targets.
- Define the responsibility of field managers, whose role is somewhere between being a general manager and a middle manager.

Typically, service companies introduce four levels of field managers: senior VP, regional VP, district manager, and store manager. The multiunit organizational design is characterized by five functions:

1. Allowing overlap in roles and responsibilities.
2. Using integrators at all levels.
3. Setting up information funnel and filters.
4. Appointing translators to convert strategies into actions.
5. Sharing responsibilities for talent development.

3.5.4 Matrix Organization

Some companies utilize the matrix organizational structure as a short-term arrangement for specific projects and activities involving both functional group employees and project managers (Gunn 2007).

Managers of functional groups supervise technically capable people who have valuable skills and know-how. Project managers are those entrusted by upper management with the responsibilities of accomplishing specific projects, such as capital projects, the design of new products to specifications, and the creation of business entry

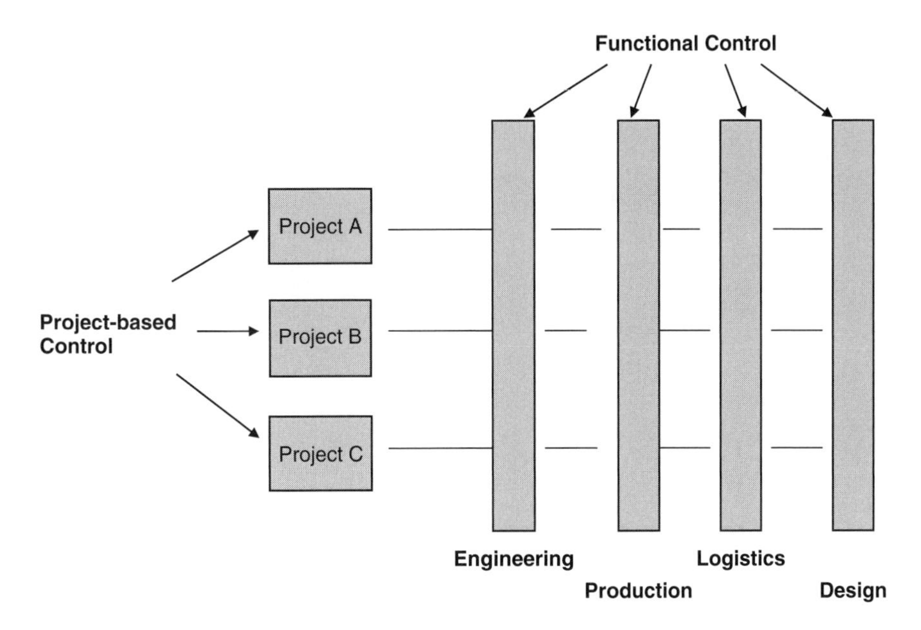

Figure 3.7 Matrix organization.

strategies. Project managers have resources—money, time, facilities, and management support—and they "borrow" employees from the functional groups to accomplish the work (see Fig. 3.7). When functional group employees complete their work for a given project, they usually return to their respective home groups to continue their original assignments.

The advantages of matrix organizations are that project managers focus on schedule and cost, whereas functional managers concentrate on quality and expertise. This arrangement offers a balance of workload, and it is excellent for participating employees to achieve wide exposure within the company by interacting with those outside of their special areas of expertise.

The main disadvantage is that this structure requires participating employees to report to two bosses (dual reporting), thus violating the *unity command* principle. When employees are assigned to work on several projects, they may be subjected to marching orders issued by several superiors. In practice, conflicts between the functional and project managers are frequent and severe, mostly with respect to task priority, manpower assignment, interests, quality versus urgency, performance appraisal, employee promotion, and other issues.

Matrix organization demands a delicate balance of power between functional and project managers. On the one hand, functional managers control manpower, particularly who works on what projects, when, and for how long, in addition to controlling knowledge and facilities. On the other hand, project managers have an approved spending budget and the support of upper management. A lack of a balance of power will occur when the functional managers have their own funds to support their own people, thus making them less dependent on project managers. An unbalance of power will also occur when project managers outsource some of the needed work that is not delivered by the functional managers. Under these circumstances, the matrix organization could break down.

Because of the aforementioned built-in conflicts, innumerable companies in industry are moving away from the matrix organizations in favor of teams.

Example 3.2 Once the functional manager and project manager agree on a project schedule, who is responsible for getting the work performed? Who is accountable for getting the work performed? Why the difference, if any?

ANSWER 3.2 Responsibility and accountability are two different management concepts.

In a matrix organization, the project manager delegates jobs to the functional manager, who, in turn, assigns specific activities to individual employees in his or her functional group. The functional manager remains responsible for getting the work performed, whereas the project manager is accountable for the results of the work that has been delegated to and done by the functional manager (or his or her people).

The project manager is accountable for achieving specific project objectives. He or she defines the pertinent jobs to be accomplished. If the jobs are defined improperly, causing the objectives to be impossible to attain, the project managers are accountable for such mistakes. The functional manager, by contrast, is responsible only for supplying the right people with the proper skills and dedication to accomplish the stated activities The functional manager is responsible for accomplishing the agreed-on tasks in an efficient and professional manner.

3.5.5 Team Organization

A team is composed of members who are "on loan" from their respective functional departments and are thus assigned to work full time for the team leader in tackling high-priority, short-duration tasks or projects (Katzenbach and Smith 2003). Since all team members report to the team leaders only, conflicts arising from dual reporting are eliminated (see Fig. 3.8). Proponents of the team structure believe "together everyone achieves more." A Japanese proverb says: "none of us are as smart as all of us." Examples of team organization include service development teams and special task forces.

3.5.6 Network Organization (Weblike Organization)

In response to rapid changes in customers' needs, advancements in technology, marketplace competition, and globalization, some companies have started pursuing certain new business paradigms that are based on thinking globally and acting locally (Ashkenas 2002).

One such paradigm is the inclusion of local suppliers' inherent technical skills and capabilities as a part of corporate strength.

Companies form alliances, build business networks, and establish supply chains with regional companies to manufacture, assemble, market, deliver, and service products for specific regional markets. (See Fig. 3.9.) At the nodes of such networks are

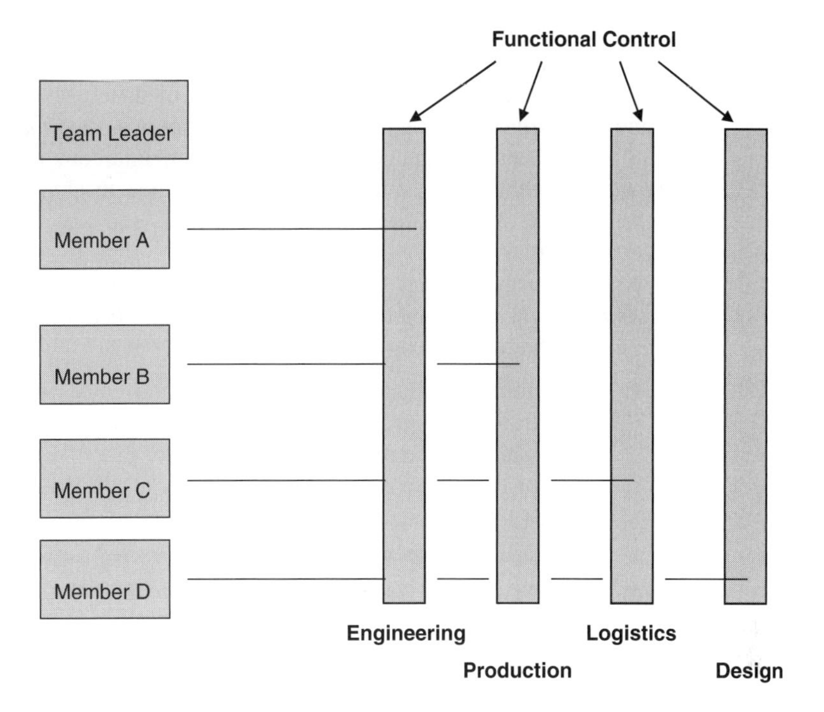

Figure 3.8 Team organization.

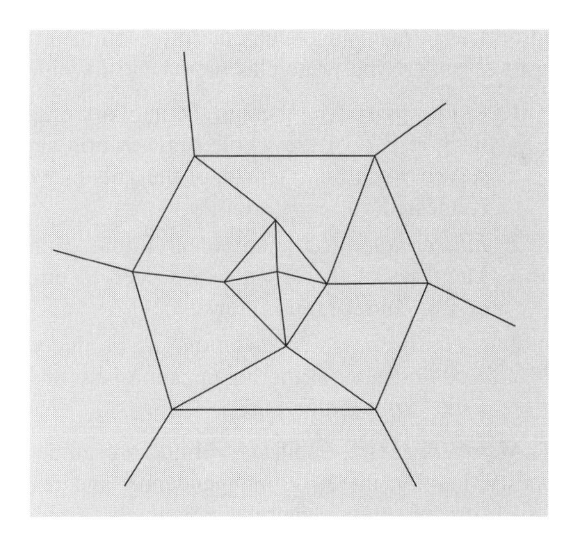

Figure 3.9 Network (weblike) organization.

knowledge workers who manage relationships with others (e.g., suppliers, customers, functional groups within an organization, and other such partners).

The number of such business network arrangements is expected to grow with time. Once formed, these networks must be properly maintained. Specific parts of these networks may be activated from time to time for business strategy development, product development, system design, quality control, logistics, customer service, and other such important projects.

Partners linked by these networks may be of different cultural and business backgrounds with divergent value systems and perspectives, and may be dispersed in various regions. Engineering managers serving on such intercompany network organizations may also be challenged by the expected resistance to change, difference in working habits, absence of motivation and control means, and slowness in consensus building and decision making.

Network organizations behave a step closer to organizations in chaos. According to *complexity theory* (sometimes also called *chaos theory*) these organizations exhibit three unique features (Kelly and Allison 1999):

1. All members are independent, flexible, and empowered. They behave responsibly, free of the traditional top-down command and control structures. The Newtonian cause–effect paradigm does not properly describe the organizational behavior that is influenced by the actions of all of its members.
2. Members tend to self-organize themselves by ways of intensive interactions between the members and to form self-directed network organizations.
3. The flexible organizational form fosters creativity and innovation of empowered members.

The complexity theory claims that organizations composed of members with personal autonomy will be better able to operate in economical, political, and sociocultural environments that are turbulent and rapidly varying (Dolan, Garcia, and Auerbach 2003; Hoogerwerf and Poorthuis 2002). Rigid objectives and instructions will no longer be appropriate in managing such enterprises in the emerging global economy. The following three underlying principles have been formulated for such organizations:

1. *Connectivity.* Members in the network organizations recognize themselves as an integral part of the whole organization and believe that their best interests are served when the interests of the greater whole are served. Members are closely connected with one another.
2. *Indeterminacy.* The turbulent future cannot be readily predicted or planned. Members of the organization need to empower themselves to act with confidence, courage, and integrity.
3. *Consciousness.* Organizations are products of collective thinking. The collective consciousness of the organization is defined by all of its members. Everyone is a vital contributor.

Members exercise self-control and are guided by shared essential values, such as honesty, loyalty, integrity, independence, and responsibility. Teamwork is a hallmark of such organizations, wherein individuals are increasingly autonomous, flexible, and dedicated. Product customization is a likely means for such organizations to achieve success in the marketplace.

In practice, research universities behave like semichaotic organizations. As a rule, tenured professors devote 50 percent of their time to academic research, 30 percent to teaching, and 20 percent to services. Each of them is completely independent as far as research is concerned. However, each of them receives instructions as to what courses to teach, which committees to serve on, and how many students to advise in a given year.

Although they are all empowered to conduct research with personal autonomy, some of them do more than others. It is not unusual to see school administrators dangling the distinguished professorships, achievement awards, and other inducements to entice educators to become more active in research.

Another instance of complexity theory in practice is the replacement of a well-known international bridge between the United States and Canada (Rienas et al 2007). It had been known for some time that the specific bridge under consideration needed to be replaced or expanded due to increased traffic, both commercial and residential. On the one hand, the Canadian provincial government acted decisively by having a truck-processing facility built on its side way ahead of schedule. On the other hand, various groups on the American side empowered themselves in the decision-making process. Involved are local citizens, politicians, special interest groups, members of Congress, state government officials, business groups, and others. These autonomous groups could not agree on the type of bridge to build (e.g., signature hanging bridge, double-decker, and another design). Some are in favor of speed of construction, whereas others favor a landmark design to add distinction to the region. Lawsuits were initiated. Everyone talked and no one listened. As a result of this complexity theory in action, no agreement was reached as to which bridge to build for a long time. This is the case wherein Bertrand Russell said "Too little liberty brings stagnation and too much brings chaos."*

Network organizations involving independent supply partners could be difficult to manage. It is not likely that much can be accomplished by leaving them alone to direct themselves, as suggested by the complexity theory. Engineering managers need to prepare themselves to effectively lead such network organizations.

Example 3.3 Company X manufactures automobile jacks, hubcaps, and a variety of fittings. These products are sold as replacement parts through chain auto-supply stores. The business of the company is growing, with production facilities located in rented buildings over various parts of the city. The production staff is expanding constantly. Now the president of the company wants to expand into the brass-fittings business. However, the president realizes that, after this newest expansion is accomplished, the company should consolidate to make its production operations more efficient.

Which organizational structure should the company adopt now, so that it can best accommodate its current needs of business expansion and also lay the foundation for anticipated consolidation thereafter? What information is needed to set forth such an organizational structure? What are the crucial variables that should be considered in the design of such an organization?

ANSWER 3.3 To expand the brass-fittings business, Company X should set up a multidisciplinary team initially. This team will be empowered to come up with a business plan to enter the brass-fittings business on the basis of market research and competitive analysis. The plan should include market-share position, time to market goals, marketing strategies, capital investment, and production or sourcing requirements. It may be useful to retain competent external business consultants for advisement.

Source: www.quotationspage.com/quote/29083.html.

The business plan defines the needs of personnel (capabilities, experience, and number), facilities, and other resources required for implementation. A product-centered department is set up to be responsible for the profit and loss of the brass-fittings business. Although the distribution of brass fittings may be readily handled by the existing organization in the company, special attention needs to be paid to production, marketing, engineering, and service of brass fittings.

Once the brass-fittings business is fully established, the production facilities of brass fittings could be integrated into the production organization of other company products in order to realize economies of scale advantages.

The organizational design must be flexible to effectively serve the purpose at hand. Crucial variables to be considered in the organizational design include the significance of the brass-fittings business to the company's overall performance (e.g., market size, competition, and profitability), extent of the resources required of the company to enter the brass-fittings business, and the management and technical talents available inside and outside of the company.

3.6 ENHANCING CORPORATE PERFORMANCE BY ORGANIZING—EXAMPLES

Organizing is an efficacious method to achieve the critically important objective of doing things right (or performing tasks in the most efficient ways). Management can put the right people together and keep the wrong people separated so that work efficiency and goal attainment can be greatly enhanced (Aguinis 2008). The following are several best practice examples of how company productivity and profitability could be raised by employing organizing strategies.

3.6.1 Organizing for Profitability—The Service Profit-Chain Model

For every service company there are five stakeholders—namely, customers, employees, suppliers, investors, and the community in which the company operates. Not all of them are equally important from the company's standpoint. In the new economy of services, frontline workers and customers need to be the center of management's concern. Customers are value-oriented, emphasizing results in relation to costs. As profits come from customers, the key concept of the *service profit-chain model* is to treat employees well so that they make customers happy, who in turn buy more, which leads to ameliorated corporate profitability, and allows suppliers and investors to be paid properly, as well as, avoids upsetting the communities in which they operate (Heskett et al. 2008). Fig. 3.10 illustrates this model.

In Fig. 3.10, Organizational strategies refer to preparatory activities and tasks related to workplace design, job design, employee selection and development, employee rewards and recognition, and the introduction of applicable tools for serving customers. Specifically, companies strive to take the following steps:

1. Hire employees with the right attitude toward jobs, coworkers, and companies, as people are the heart of service businesses.

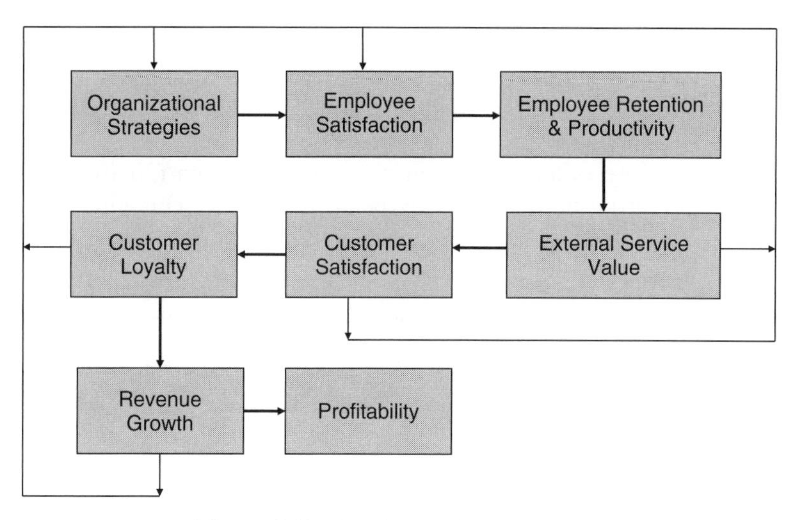

Figure 3.10 Service profit chain model.

2. Promote ability (tools, technology infrastructure) and empower service workers to achieve results for customers.
3. Communicate with customers and employees to get feedback and recognize good performance
4. Achieve employee satisfaction as related to jobs, training, pay, advancement, fairness, treating people with dignity, teamwork, and company's interest in employees' well being. Set up an 800 phone number to help solve employees' problems.
5. Link managers' compensation (say 20 percent) to customer satisfaction.

Customer satisfaction is the ultimate goal all service companies want to achieve. Customer satisfaction can be measured by a set of metrics using third-party interviews: These metrics include: (1) percentage of loyal customers with repeat businesses, (2) percentage of sales due to repeat customers, (3) funds spent to promote customer retention, and (4) company's understanding of the reasons for customer defect, and actions taken to mitigate. Service quality is a function of the gap between perceptions of active service received and customer's expectations beforehand (reliability, timeliness, empathy, authority of service delivery, and extent of evidence for service rendered). It is of critical importance to the company that service errors are quickly corrected. External service value refers to the value received by customers based on the extent in which the service designed and delivered meets their needs.

Customer loyalty results in retention, repeat business, and referrals. For service companies, the service profit-chain model serves as a useful organizational design concept because it emphasizes, among other stakeholders, the two most important ones: employees and customers. Companies that follow this best practice model include Banc One, Intuit, Southwest Airlines, Service Master, USSA, Taco Bell, and MCI, among others.

3.6.2 Organizing for High Performance by Using a Flexible Structure

Organizational structures are known to have impact on corporate performance. Some companies allow the structure to continuously evolve in order to adapt to changing opportunities in the marketplace.

According to Aufreiter, Lawyer, and Lun (2000), new-style companies achieved great business success by being relentlessly committed and by exercising discipline from the top, as well as practicing three organizational principles:

1. *Make it everyone's job to identify new opportunities.* Company culture must support such a principle by way of feedback loop and financial incentives.
2. *Decide quickly on project priority.* Speed and coordination are critically important in implementation. Use technology to support decision making wherever possible, and fill the gaps with fast, centralized, senior-level decision making.
3. *Hire people for specific roles such as marketing and technology support.* This is needed for implementing ideas to quickly cash in on new opportunities in the marketplace.

The following are some of the best examples of organizational practices in industry:

- *Starbucks Corporation* sells coffee and related products to consumers. It encourages all employees to suggest new ideas (using a one-page form) and promises a quick response from the senior executive team. The person suggesting a given idea is always invited to be a full-time member of the launching team.

  Leaders are critically important for any team to achieve success. Starbucks assembles teams headed by leaders with applicable marketing expertise. Starbucks hires two kinds of people: (1) integrators who are marketers with broad skills and who are capable of carrying out major responsibilities in coordinating the delivery of products and services to the market, and (2) specialists with unique capabilities and expertise for launching projects. All of them need to have an entrepreneurial mindset and be able to thrive in an uncertain and rapidly changing environment. For the "Store of the Future" project, the leader was hired from the outside. Another outsider was hired to spearhead its *Lunch Service Concept* project.

  Organizational flexibility is a hallmark of new-style companies. When in-house packaging and sales-channel management skills were deemed lacking, Starbucks elected to team up with another company, *Dreyer's Grand Ice Cream*, to launch a new ice cream product. Within four months, its coffee ice cream product became the top-selling brand in the industry. Frappuccino, a cold coffee drink, was also marketed jointly with *PepsiCo Inc.* in 1994. Within one year, its revenue accounted for 11 percent of the total sales of Starbucks in 1995.

  Starbucks uses a high-level steering committee for making rapid decisions based on two criteria: (1) the effect on company's revenue growth (the new ideas must have the potential of producing a minimum revenue of about $4 million per year) and (2) the impact on the complexity of the company's retail stores required for implementation.
- *First USA* is a financial services company. It reconfigures its structure dynamically and makes routine organizational changes. Once new market opportunities

worth pursuing are identified, the company determines what specialized skills are required and puts a suitable "dream team" together. The company maintains a pool of managers having special skills that enable them to launch new credit card products. Other people are added from internal resource pools or by hiring from the outside. Within a three-year period (1995–1998), First USA issued five times as many credit cards as another well-known financial company organized in traditional ways.

- *Dell* sells computers and other technology products directly to consumers and businesses alike. It focuses on identifying and capturing market opportunities. Managers are inspired to turn the needs of customers into products and services as quickly as possible. Business units are smaller in size, and more people are empowered to assume the profit and loss (P&L) responsibilities. Successful managers are rewarded accordingly.
- *Minnesota Mining and Minerals* (officially known as *3M*) is a major industrial company in the United States. It markets a diversified set of products. It is known to organize its employees to conceive new ideas by allowing them each to spend up to 15 percent of their work time on new initiatives of their own and by supporting such projects with grant money outside of the departmental budgets. To win the rights for funding future new projects, 3M managers must derive at least 30 percent of sales from ideas brought into being during the four preceding years.

When reviewing the merits of an organizational strategy, the basic criterion is, of course, its impact on business results. Aufreiter, Lawyer, and Lun (2000) studied the compound annual growth rate (CAGR) of several companies practicing flexible organizational strategies for the period of 1994–1998 and showed that the CAGR of Trilogy, First USA, Dell, Starbucks and The Home Depot are significantly higher than the average growth rates of the next three largest competitors in the same industry.

3.6.3 Organizing for Promoting Interactions of Self-Directed Professionals

In today's service organizations, professionals play an increasingly important role. The major duties they perform are to interact with their professional peers in order to produce and exchange knowledge, and to drive excellence in economies of scale and scope. Each professional is expected to have 5,000 bilateral relationships (Bryan et al., 2005). Such relationships are the key power bases (core competencies) of consulting professionals. Companies adopt several techniques to promote interactions: (1) organize line organizations according to service, functions, geographies or customers, while eliminating matrix or ad-hoc organizational structures, (2) segregate support functions from line organizations, (3) establish enterprisewide formal networks to promote collaborations, (4) separate managers who care about short-term operational results from professionals who devote efforts to long-term development work, allowing the professionals to "wonder in the woods" in trying out new initiatives, and (5) allow top-level managers to make decisions to balance the short- and long-term resource allocation problems. Self-directed professionals are best managed by setting aspirations (objectives, future state, and value propositions) and using performance metrics that motivate them to organize their work.

Once the interactions between self-directed professionals are established, companies will still need to organize toward building collaborative advantages (Hanson and Nohria 2004). Collaboration is manifested in sharing knowledge and joint development of new products and services. The interunit collaboration, which leverages dispersed resources, will be the key driver for competition beyond the economies of scale and scope. There exist four barriers to interunit collaboration:

1. *Lack of willingness to seek input and learn from others.* This is the not invented here syndrome. Some people routinely reject ideas, knowledge, and inventions developed elsewhere.
2. *Inability to seek and find expertise.* They do not know where the experts are. In this respect, the use of connectors, tenured employees who know where experts are, may help.
3. *Not willing to help*. These people tend toward "hoarding-of-expertise" due to conflicts of interest. These behaviors can be changed by performance evaluation policies and promotion criteria.
4. *Inability to work together and transfer tacit or specific knowledge.* In this instance, job rotation may help. Companies need to pay attention to the factors that undermine fruitful collaboration between units.

There are three preferred forms of organization that promote exchange of ideas among self-directed professionals: (1) off-line teams, (2) knowledge and talents marketplaces, and (3) networks. Off-line teams may be set up across the functional disciplines to stimulate communication. Talent marketplace allows professionals to change jobs so that they will naturally interact over time with different people within the organization. Knowledge marketplace advocates the exchange of knowledge and promotes the trading of "knowledge objects" between buyers and sellers. Networks need to be formalized in order to effectively promote knowledge exchange. Companies may designate a network owner, offer incentives to participating members, define knowledge territories, set standards and protocols, and provide a support infrastructure to augment the operations of such networks.

Networks determine which ideas become breakthroughs (Uzzi and Dunlap 2005); they deliver three unique advantages:

1. *Private information*. People make judgments based on both public and private information. Public information is easily available but it offers no competitive advantages. Private information is derivable from personal contacts, although its value depends on the trust that exists in the relationship.
2. *Access to diverse skill sets*. Linus Pauling attributes his innovative access to his diverse contacts. "The best way to have a good idea is to have a lot of ideas."*
Networks allow access to the multiple skills needed for today's complex world.
3. *Power*. Networks allow information brokers to link specialists with trustworthy and informative ties.

For networks to be useful, they need to be composed of people with diverse views and experiences. Professionals tend to connect with others based on either the *self-similarity* or *proximity* principle. The self-similarity principle states that professionals

Source: www.brainyquote.com/quotes/quotes/l/linuspauli163645.html.

tend to connect with others who share similar views. The proximity principle says that professionals are likely to form networks with those who work in nearby units or geographies. In relation to broad-based perspectives, networks that are formed on the self-similarity or proximity principles do not always provide the divergent background, experience, views and insights needed to be valuable. According to Uzzi and Dunlap (2005), the best approach is to use the *shared activities* principle, which involves looking for network partners in sports team, community service ventures, interdepartmental initiatives, voluntary associations, for-profit boards, cross-functional teams, and charitable foundations, and creating strong bonds of loyalty within these diverse groups.

3.6.4 Organizing for Innovation

Some companies are more focused than others on developing and sustaining corporate competitiveness by nurturing innovation. Innovation can be affected by company structure.

According to Chesbrough and Teece (2002), there are two types of inventions: autonomous and systemic inventions. On the one hand, *autonomous inventions* are those that can be pursued independently of other inventions. For example, inventions related to turbochargers could be pursued independently of inventions related to automobiles, to which turbochargers are applied for boosting power output. Similar relationships exist between filters and air conditioners, motors and compressors, and color-print films and analog cameras. *Systemic inventions*, on the other hand, are those that must be advanced in close coordination with others. Examples include the development of product design, supplier management and information technology for lean manufacturing, and films and cameras for instant photography. Because autonomous inventions may be pursued independent of other inventions, they are better suited for virtual organizations. Vertical organizations, by contrast, are more likely to succeed in pursuing systemic inventions that require close coordination and intensive information sharing.

Organizing a company's structure to be virtual means that all noncore functions are typically outsourced. These virtual companies form supply chain partnerships and business alliances to advance, manufacture, market, distribute, and support their offerings. They use incentives—signup bonuses, stock options—to attract highly trained independent inventors to generate breakthrough inventions. Because of the self-interests of all participants, the coordination between these partners can be difficult, rendering the resulting inventions of value only to some, but not to others.

Vertically organized companies are those that have rigid hierarchical organizational structures. They maintain control of all functions and typically have well-established procedures for settling conflicts and for planning all activities that promote innovations. Systemic inventions require information sharing and adjustments, which these vertically integrated companies can readily promote and safeguard. However, they cannot offer the high incentives that virtual companies use to attract independent inventors. As a consequence, they may not be able to access top-level talent for creating inventions.

Therefore, when organizing for innovations, one must choose between talents and control. The key is to select the right kind of organizational form to match the type of innovation (autonomous, systemic, or a mix of the two) the company needs. At one extreme, virtual companies are suitable for pursuing autonomous innovations. At the other extreme, vertical companies are excellent for pursuing systemic inventions. As the invention type changes gradually from purely autonomous to purely systemic, the company should consider intermediate forms of organizations such as alliances, joint ventures, and collaborations with autonomous divisions.

Nowadays, few companies can afford to develop all needed technologies internally. A mix of approaches is usually adopted. Some technologies are developed in-house to serve as the core part of the value chain. Other less critical ones are typically purchased outright or acquired through license, partnership, and alliance. Over the long run, however, key value-added advances will need to come from within.

3.6.5 Organizing for Performance at the Design-Manufacturing Interface

Conflicts are known to exist at the interface between product design and manufacturing. These conflicts cause frequent cost overruns and product introduction delays. In countless traditional companies, the product-design group signs off on the design, and then they "throw it over the wall." The group responsible for manufacturability takes over and reexamines the design for cost-effective mass production. Although product design may have focused on performance and aesthetics, manufacturing looks after production efficiency. Also contributing to these conflicts are other factors such as (1) funding periods for design and manufacturing that do not overlap, (2) differences in education between design and manufacturing staff, and (3) offices that are not at the same location.

Some of these difficulties may be removed by way of organizing. Organizational options for improving the design-manufacturing interface include the following (Dean and Susman 1989):

- *Institute a manufacturing sign-off*. Manufacturing has veto power over the final product design. Software programs (e.g., *Assembly Evaluation Method* by Hitachi and *Design for Assembly* by Bootheroyd Dewhurst) are available to calculate a producibility score for checking on manufacturability.
- *Appoint an integrator* who performs liaison work between design and manufacturing and offers a balanced view.
- *Form a cross-functional team* composed of members of design and manufacturing, with the final authority resting with the engineering department. The use of such cross-functional teams is known to have significant benefits, such as assuring compatibility between the design and manufacturing processes, saving time, simplifying the design process, and reducing design changes.
- *Combine the manufacturing process and product design* into one department.

In general, if the company's culture is conducive to absorbing organizational changes, then the organizational options of the team or combined department are to be preferred. However, if the products and manufacturing processes are fixed, then the organizational options of the sign-off or integrator tend to make more sense.

3.6.6 Organizing for Heightened Employee Motivation

As a rule, teams are temporary in nature because they are built for specific objectives and will be disbanded after their specific objectives have been achieved. Only in exceptional cases will teams be exhaustively utilized on a permanent basis to achieve business success. This is the case of AES Corporation.

Located in Arlington, Virginia, AES Corporation is the largest global power company, with sales at $16.07 billion and market capitalization of $9.48 billion (December 31, 2008). In 2008, it had a total electricity generation capacity of 43,000 MW and distribution business serving over 11 million people in twenty-nine countries. Seventy-five percent of its business was in contract generation. Fig. 3.11 illustrates its rapid revenue growth for the period 1990 to 2008.

The company has very few organizational layers and, except for a corporate accounting department, keeps no staff for functional specialties. It is organized into eleven regions, each headed up by a manager. Each region is further organized into five to twenty teams, and each team has five to twenty members. Teams are formed primarily for the combined functions of plant operation and maintenance.

Each team has no more than one of each kind of expert or specialist. As a result, everyone on the team becomes a well-rounded generalist. In-house qualification exams are held to ensure minimum expertise before job rotation requests of employees are approved. Each team owns what it does and is empowered to make decisions with commensurate authority to implement its decisions. The roles for company leaders are limited to advisors, guardians of the company principles, encouragers, and officers accountable to the outside world.

Employees are compensated according to the following formula: (1) 50 percent on financial performance and safety and environmental impact and (2) 50 percent on how well employees follow the four company values—fairness, integrity, social responsibility, and fun. The hiring practice of the company focuses on cultural fit first and technical skills second.

Company representatives attribute their business success at organizing the company in teams to the heightened level of employee motivation made possible by the team empowerment practice.

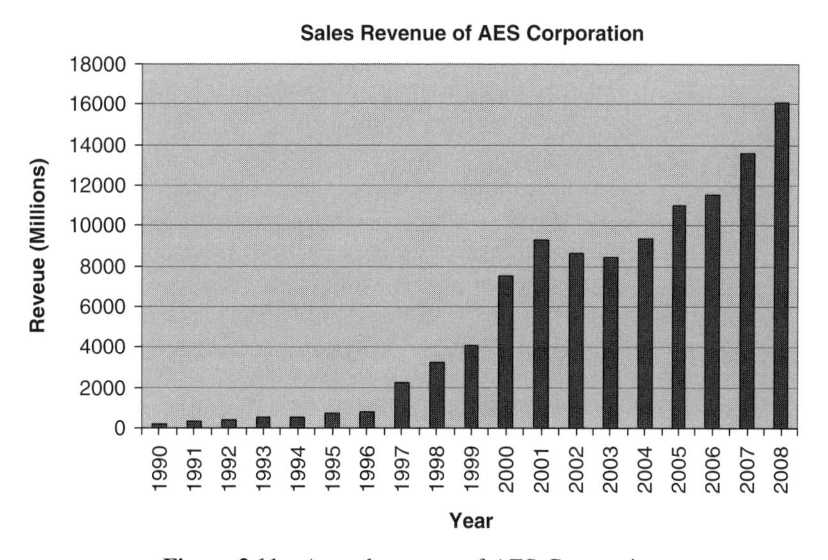

Figure 3.11 Annual revenue of AES Corporation.

It should be noted that the AES Corporation example may indeed apply well to other low-tech operations such as warehouses, distribution centers, supermarkets, hardware stores, and service centers wherein repeat common practices are the norm. Everything you would ever want to know about operating and maintaining a conventional power plant has already been sufficiently preserved in manuals; in-depth technical expertise and innovation are not required (Wetlaufer 1999).

3.6.7 Organizing for High-Tech Marketing

Some companies in industry are high tech, and others are not. *High tech* refers to products and services characterized by (1) their strong scientific–technical bases, (2) the possibility of being quickly obsolete because of new technologies, and (3) the capability to expand or revolutionize markets and demands when built on new emerging technologies. Examples of high-tech products include semiconductors, microcomputers, and robotics.

High-tech companies need to pay special attention to the interface between R&D and marketing. To achieve business success, a linkage between R&D and marketing must be established. When marketing high-tech products or services, companies typically follow two basic approaches: market driven and innovation driven.

1. *Market driven.* When pursuing the market-driven strategy, companies use marketing to define the needs of customers and direct R&D to provide the required innovations to satisfy such needs. Customer suggestions are typically good sources of new needs. In this case, marketing uses tools such as *concept testing*, *product prototyping*, and *pilot testing* to define the specific product or service features needed by customers. Marketing efforts precede the R&D efforts. The consequence of practicing a market-driven strategy is that there may be a possible delay in breakthrough innovations, preventing the company from a timely use of the opportunities offered in the marketplace.
2. *Innovation driven.* R&D employees take the lead in the innovation-driven approach by first making breakthrough inventions based on *preemptive needs* as perceived by researchers. Then the researchers consider pursuing the inventions to satisfy the real needs and wants of the customers. In such situations, marketing applies techniques such as *focus groups* to verify new product concepts and applications. There are risks associated with this approach, as identified customers' needs and wants may not be satisfied by the breakthrough inventions at hand.

A lack of coordination between R&D and marketing is known to be the cause of business failures in diverse companies. Furthermore, any new technology advancement by competition can change the company's market position instantly.

Organizing a workable interface between R&D and marketing is a way of avoiding the potential loss of market opportunities due to invention delay and the lack of compatibility of inventions to the actual market needs. Setting up a team of representatives of R&D and marketing to constantly monitor relevant activities and foster communications is a good organizing strategy (Viardot 2004).

3.6.8 Organizing for Balancing the Short- and Long-Term Corporate Needs

Companies must execute strategies in the present to take care of short-term performance needs and adapt effective modalities in the long term to ensure the overall health of their organization (Beinhocker 2006, Dobbs et al. 2005). In general, companies do not do well in balancing their short-term and long-term needs because of three distinct barriers to adaptability.

People—The Price of Experience. Most managers are not as perfectly rational as generally assumed. They frequently exhibit common biases, overoptimism, and errors in judgment. Typically, people acquire experience and learn new things following the if-then rules. We cast experience in such rule formats and build hierarchies from them. The hierarchies grow into categories, which interlink with rules and weightings. Over time, such mental models become increasingly difficult to change. As a result, some managers suffer from being rigid and adverse to novelty, causing incorrect responses to new and unfamiliar situations. This is a major reason why innumerable workers resist changes. The *hero-rogue* syndrome is when an executive succeeds in one business environment, raising their public profile and then immediately fails when they encounter a new business environment. This is primarily due to an inability to think adaptively.

Companies usually *place* experienced people in top positions who possess mental models suitable for execution but whose models also build an inertia in their ability to respond to new environmental changes.

Structure—The Risk of Complexity Catastrophes. Large organizations, like networks, are difficult to adapt. Complexity catastrophes refer to the interdependences of networks that generate conflicts, which, in turn, provide constraints and, consequently, make changes impossible. Established companies experience less freedom because of these interdependencies between their units. As a company's size increases, complexity grows, and consequently the degree of freedoms drops. Startups, by contrast, have a great degree of freedoms and as a result are able to adapt much more easily than established companies. Example: NCR Corp. versus IBM.

One way to counteract this discrepancy is to run larger companies as a conglomerate of small divisions, each independent of one another. However, even in this scenario, the scale advantages of large companies are no longer readily realizable.

Resources—The Path to Dependency. Resources are keys for implementing a chosen corporate strategy. Resources include physical assets, talents, knowledge, brands, reputation, and relationships. Resources constrain what opportunities can be exploited. Setting the right priority in the utilization of resources is the central issue to be decided on when balancing the short-term and long-term needs of a company.

Companies have four options to overcome the aforementioned barriers to adaptability:

1. Reduce hierarchy to prevent a small number of mental models from dominating the organization.
2. Increase autonomy to diminish the risks of complex catastrophes.

3. Foster diversity by embracing divergent mental models, resources and business plans.
4. Modify company culture by redefining norms for collaboration, innovation, and performance appraisal.

Example 3.4 It is self-evident that organizations need to focus on both short-term performance and long-term health and take a balanced approach to management (Dobbs et al. 2005). Short-term performance is needed to maintain cash flow and profitability, whereas long-term health is critical to continued competitiveness. Explain why there are only a few companies that are in a position to achieve the required balance in practice. What strategies are required in order to facilitate such a balance by way of organizational design?

ANSWER 3.4 The key reason for not being able to achieve a desirable balance between short-term and long-term objectives is that managers are under pressure to meet quarter-to-quarter earnings targets first, before considering allocating resources for R&D and marketing. Managers defend their own short-term performance in silos, arguing about the virtue of one metric as opposed to another, deflecting debate to other parts of the organization and setting up barriers to change.

Such a self-preservation attitude is understandable, and so is the resulting preference for short-term performance over the long-term health of the organization. There are six actions companies can take to mitigate this situation:

1. A company should set strategies to include initiatives for a set of different time horizons—a portfolio of initiatives, including short-term performance and options for the future (new products, services, markets, process, or value chains).
2. Introduce metrics for monitoring progress in new service development, customer satisfaction, government relations, and retention of corporate talents Examples of these metrics could include:
    - Financial
    - Operations (quality and consistency of key processes)
    - Organizational issues (depth of talents, ability to motivate and retain talents)
    - Service markets and position (quality and customer relationship)
    - Nature of relationships with external parties (suppliers, regulators, and nongovernmental organizations)
3. Build separate funds for long-term health efforts, independent of the budgets for short-term performance activities. Separate the managerial responsibility of those who take care of short-term performance from others who are in charge of long-term R&D and other competency building efforts. Compensate and reward these managers according to separate metrics.
4. Communicate actively to stock analysts and employees of company's strategic plans.
5. Demonstrate leadership in developing future leaders and reward long-term performance.
6. Exercise governance.

3.7 CROSS-FUNCTIONAL TEAMS

Cross-functional teams have become crucial to business success (Parker 2003). In general, teams are set up to (1) generate recommendations, such as a strategy to enter a specific regional market or solve a specific customer-related problem, (2) make or do things—for example, design products, advance new processes, or install new assembly lines, and (3) run things—for example, operate plants. The performance of a team is the sum of the performance by individual members plus the work product brought about by the members together. It is the work product delivered by the members together (the team synergism) that is responsible for the superiority of the overall team performance over the sum of the performances of the individuals (Payne 2001). Teams may fail if they are not led properly (Rees 2001, Huettner, Brown and James-Tanny 2007).

Organizing multifunctional teams is often the preferred choice to address complex coordination issues at interfaces, in addition to the interface between design and manufacturing discussed before. For example, in a typical functional organization, the development of new products follows a sequential process enumerated as follows:

1. *Marketing* conducts research to identify the customers' needs and defines product features, such as functionality, reliability, ease of repair, resale value, warranty, and so on.
2. *Design engineering* releases specifications, performs functional design, selects material, obtains vendor and supplier inputs, and conducts engineering analyses to incorporate these features into a product.
3. *Manufacturing* engineering reviews and simplifies the product design for manufacturability and reliability considerations.
4. *Service* organization further changes the design to facilitate serviceability.
5. *Production* is finally set up to define manufacturing techniques and to mass-produce the product.

Such sequential processes are known to be inherently ineffective with respect to coordination, information sharing, and decision making. In a concurrent engineering team, representatives of all of the functional groups just mentioned, plus those from procurement, finance, vendors and suppliers, product testing, and logistics, are included as members. All aspects related to product development are considered early on and concurrently. The goal is to bring forth an optimum product for the company within the shortest period of time and at the lowest possible cost, while satisfying all constraints and meeting all requirements.

All team members have the full support of their respective departments, functional units and home bases, so that the specific inputs they make on the team, at various stages of product development, are always the best possible inputs (Lencioni and Okabayashi 2008). There are three keys to the success of concurrent engineering teams:

1. Management commitment.
2. Ongoing communications that use advanced communication tools such as the intranet, e-mails, and electronic data interchange (EDI).
3. Teamwork training for all members.

The value of the concurrent engineering team concept is evident from the following statistics:

- Mercury Computer Systems, Inc., Lowell, Massachusetts—Concurrent engineering team whittled down the time to market of its add-on process boards for VME (Versa Module Eurocard) bus from 125 days to 90 days.
- Hewlett-Packard Company, Palo Alto, California—Cut the time to market of its 54,600 Oscilloscope by two-thirds.
- Toyota Motor Corporation, Tokyo, Japan—Concurrent engineering decreased its product cost by 61 percent.

In general, concurrent engineering delivered impressive benefits in the order of magnitude displayed in Table 3.1 (Loureiro and Curran 2007).

Engineering managers need to become proficient in leading and participating in concurrent engineering teams. Teams may be formed to address any important corporate activity. For teams to add value, team leaders must pay attention to team discipline, team learning, and factors affecting team effectiveness.

3.7.1 Team Discipline

For "blow-the-roof-off" performance, a team is often the vehicle of choice. But to excel, the team needs the right training and preparation, to make sure that all members listen well, respond constructively, support one another, share team values, and have a discipline (Katzenbach and Smith 2000).

A team is a small number of people (usually between two and twenty-five) who are committed to a common purpose and who possess complementary skills, a set of performance goals, and an approach for which they hold themselves mutually accountable.

Mutual accountability refers not only to the team leader, but also to the team members among themselves. Team members are said to have developed mutual accountability if and when they have reached the emotional state of "being in the boat together," based on commitment and mutual trust. Team leaders need to strive to organize the team so that team members hold each other accountable. Without mutual accountability, there can be no team of real value.

3.7.2 Team Learning

One decisive factor that affects the team's responsiveness is its learning capability (Michaelsen, Sweet, and Parmellee 2009). In corporate settings, teams need to learn new

Table 3.1 Benefits Derivable from Concurrent Engineering Teams

Activities	Percentage (%)
Reduction of time of product development	30–70
Shrinking the number of engineering changes	65–95
Decrease of time to market	20–90
Improvement of product quality	200–600

technologies (such as three-dimensional computer-aided design (CAD), visualization software, project management tools, videoconferencing, Web-based net-meeting tools, and others) or new processes (such as new ways of working and new relationships for collaborative work). How fast a team can learn will affect its overall performance in timely attaining the specific objectives at hand.

A learning team is one that is skilled at creating, acquiring, and transferring knowledge and at modifying its own behavior to reflect new knowledge and insights (Garvin 2003; Garvin 1997). The team needs to have systems and procedures in place to do the following:

- Solve problems systematically.
- Experiment with new approaches.
- Learn from own experience and past history.
- Learn from other's experience and best practices.
- Disseminate knowledge effectively throughout the team and organization.

Edmondson, Bohmer, and Piscano (2003) studied a large number of cardiovascular surgical teams. They believe that team learning may be speeded up if the team leaders possess both technical and managerial skills. Because learning has both technical and organizational aspects—status, communications—the organizational skills of the team leader affect the team learning. Factors affecting team learning include the following:

- *Team composition.* When selecting team members, team leaders should give preference to members' technical competence—retention of a mix of skills and expertise, ability to work with others, willingness to deal with ambiguous situations (risk takers), and self-confidence in making suggestions and proposing ideas while not inhibited by other members' ranking and corporate status. The most effective learning takes place during the process. Teams with a mix of expertise and experience tend to be able to draw on members' relevant past experience and expedite learning.
- *Team cultures*. Team leaders should build a team culture in which some experimentation is encouraged and failure is acceptable.
- *Leader's style.* Teams will learn better and faster if, as motivation, the team leaders frame the learning as a challenge for all team members.

Factors that do not affect team learning are said to include (1) educational background, (2) prior experience in practicing old technologies, (3) top-management commitment, (4) status of the team leader, and (5) reporting and auditing processes.

Example 3.5 Some people feel that working as a team, instead of allowing experts to invent more creative outcomes, actually results in watered-down compromises and bland solutions. They view teamwork as a series of exercises in "sharing ignorance." Do you agree or disagree, and why? What can be done to advance the technical qualities of the team outcomes?

ANSWER 3.5 The concern about the watered-downed outputs of teams is real. Team members of different background and expertise may indeed have different opinions, which often force the team members to compromise. It is quite true that sometimes the views of the domain experts on the team are not shared and accepted by others on the team, who do not and will not want to understand.

One obvious way to ensure the technical quality of the team results is to select people to lead who are technically qualified and able to render technical judgment. Another way is to bring in an impartial outside consultant to comment and advise on the relative technical merits of the options under consideration.

Team consensus is good to have, because it allows the team members to jointly own the team outcome. This ownership represents a strong motivation factor to team members who are then inspired to actively implement the team outcome. A technically superior team outcome adds little value if it is not implemented properly. It is certainly true that a lousy team outcome remains lousy, even if implemented well. Teams need both technologically superior outcome and proper implementation capability.

3.7.3 Guidelines for Building Collaborative Teams

Teams are a favored organizational form to address issues of importance, such as the acquisition of new businesses, the overhaul of Information technology systems, development of new services, and others. Several factors influence the working of teams (Gratton and Erickson 2007). The team size needs to be limited to approximately twenty members. Any size beyond twenty will tend to decrease the team's effectiveness. The distance of communication between members needs to be short. An increase in geographical dispersion tends to decrease collaboration between participants. Team members who perceive themselves as being alike tend to collaborate more easily and naturally. Other factors of importance include: nationality, age, educational level, tenure, association, and degree of acquaintance. However, lack of diversity in team membership will likely cause the team to be short of balanced perspectives, leading to a poor team outcome. Members with high educational levels tend to collaborate less.

Companies engage in a number of strategies to build collaborative teams. Company executives demonstrate their personal collaborative behavior by repeatedly visiting various groups. They nurture a culture in which the exchange of experience is viewed positively. Company executives support the building of social relationships among employees and provide facilities to promote interactions. They mentor, coach, and train employees to acquire the skills needed to collaborate (e.g., appreciating others, engaging in purposeful conversations, conflict resolution, and program management).

PricewaterhouseCoopers conducted training programs which include teamwork, emotional intelligence, networking, holding difficult conversations, coaching, corporate social responsibility, communicating the firm's strategy and shared value, recommended ways to influence others effectively, and methods to build healthy partnerships. They promote a sense of community by holding group activities such as cooking weekends, women's networking, and tennis coaching. They carefully select flexible workers to be team leaders. These are *ambidextrous* leaders who are both task oriented (action planning, project management) and relationship oriented (utilizing relationship to get things done, improving social relationships and networks). They select team members based on their past relationships in order to increase the chances for them to collaborate together. They specify the roles of individual team members in order to gain productive outcomes through collaborations.

In summary, companies pay attention to key success factors for building a collaborative team: (1) fostering relationship and trust among members, (2) setting personal examples of senior leader's cooperative roles, (3) adopting smart ways to form teams, (4) offering clarity of participant's roles, and (5) anticipating challenges and uncertainties.

3.8 DELEGATING

After a specific form of organization is established—unit, department, team, division, or regional group—the next step is for the engineering manager to delegate the proper responsibility and authority to the selected leaders and workers, and establish the upward-directed accountability needed to achieve the defined organizational objectives. Therefore, delegating is for the purpose of improving the engineering manager's overall efficiency by assigning responsibility and authority and by creating accountability (Burns 2002).

The benefits of delegating are plentiful. To engineering managers, as superiors, delegating is beneficial because it (1) upgrades the quality and quantity of work performed, (2) relieves the engineering manager for pursuing more important duties or gaining more time for management work, (3) makes the engineering manager knowledgeable of the employee's capabilities, (4) prepares the employee to step in for the engineering manager when needed and hence enabling the engineering manager to be absent from the job occasionally, (5) distributes the workload properly, (6) improves leadership qualities, (7) eases the engineering manager's job pressure, and (8) cuts costs through more efficient operating decisions.

Delegating is also beneficial to engineers as technical contributors because it (1) makes the job more satisfying, (2) provides encouragement, incentives, and recognition, (3) gains new skills and knowledge, (4) promotes self-confidence, (5) facilitates teamwork, (6) advocates growth and development, and (7) fosters initiative and competence.

It is important for engineering managers to keep in mind what should and should not be delegated. Problems and activities of the following kinds are to be delegated: (1) those that require exploration and recommendation for a decision, (2) those that are within the scope and capabilities of employees involved, (3) those that are needed to achieve company objectives, (4) those that promote the employee's development in technologies, business perspectives, and leadership skills, and (5) those that save the engineering manager's time if done properly by the employees.

For delegating, four guidelines, illustrated in Fig. 3.12, may be helpful.

1. What the engineering manager cannot do and the employee can do, the employee does.
2. What both the engineering manager and employee cannot do, the engineering manager does.
3. What both the engineering manager and the employee can do, the employee does.
4. What the engineering manager can do and the employee cannot, the engineering manager does.

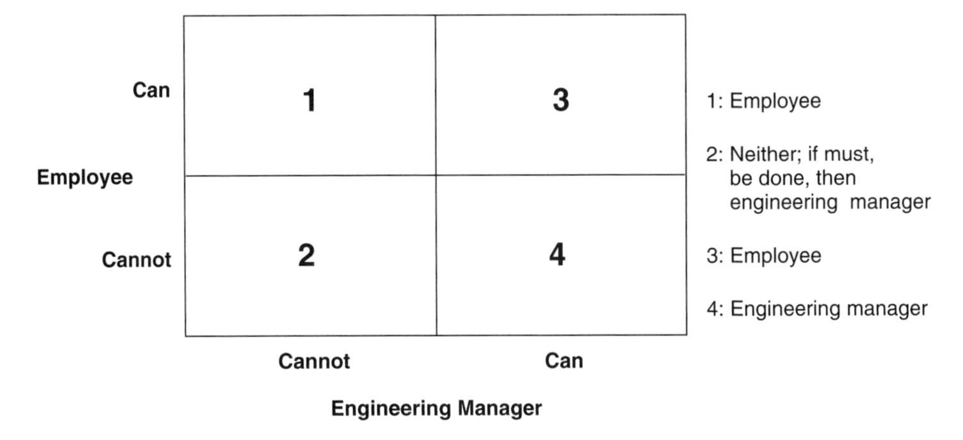

Figure 3.12 Delegation matrix.

Which problems or activities should the engineering manager not delegate? Such problems or activities include (1) planning—creating plans within larger plans and objectives, (2) resolving morale problems in the group or department, (3) reconciling differences and conflicts, (4) coaching and advising employees, (5) reviewing performance of employees, (6) completing assignments given to engineering managers by their superiors, and (7) pursuing other assignments only engineering managers themselves should handle (such as confidential committee assignments, "pet" projects, and activities without proper talents to delegate to).

Delegating requires skill and practice. The following are guidelines for efficacious delegation:

- Explain the importance of the assignment.
- Check on understanding and confidence.
- Give the employee leeway in their choice of method, unless the procedure has been specified and standardized before.
- Set a goal, timetable, or deadline. A short-term goal is better than a long-term goal.
- Be reasonable. Keep the goal within the employee's capabilities.
- Assign responsibilities that go with the job. Allow commensurate authority of decision making, and let employees accept responsibility for poor as well as good work. (The engineering managers remain accountable for the delegated assignment with respect to their own superiors.)
- Trust the employee.
- Give recognition for good work.
- Share the engineering manager's own worries. Let the employees know your concerns about the assignment; explain them openly and fully. Recognize difficulties in the assignment and solicit suggestions on how best to handle the assignment.
- Make it a project; let the assignment be a challenge.

- Do not rush in and take over. The employee could use more training if a lack of progress is apparent.
- Do not expect or want perfection.

Certain barriers to delegation do exist. Engineering managers need to beware of two key barriers:

1. *Psychological*. Engineering managers have concerns. If engineering managers let the employee do the work, they may fear their own technological obsolescence, while their employees shine. This fear may be particularly relevant to engineering managers who themselves are technically very strong.
2. *Organizational*. Unclear responsibility and relationship and confused understanding of line versus staff positions may hinder effectual delegation.

There are several more noteworthy observations:

- *Delegation* tends to be limited by the availability of workable controls. If there is no control, there should be no delegation. Engineering managers should delegate safely only to the extent that one can determine if the work can be correctly carried out and decisions can be made in the manner that they should be. Furthermore, engineering managers should make sure that the plans are sound and that controls are in effect.
- *Authority* must be commensurate with responsibility. Engineering managers should delegate enough authority to allow decision making by the employees related to the work.
- *Accountability* is demanded of employees who are obligated to their superiors for achieving the expected results. Accountability is achieved by properly discharging responsibility and using authority delegated. Effectiveness of and success in delegation depends on the willingness and ability of the employee to perform the work, make decisions, and achieve results.
- *Control* must be in place. Engineering managers should introduce midcourse corrections, if needed. Otherwise, delegation will lead to disaster. Setting up performance metrics and constantly monitoring performance will help.

Countless engineers fail to achieve success in managerial ranks, partly because they do not know how to delegate properly. Good delegation is a prerequisite for being a good manager.

3.9 ESTABLISHING WORKING RELATIONSHIPS

Another organizing activity to be performed by engineering managers is the establishment of the proper working relationships between employees and between units (Straus 2002). This is to ensure that people are working together well enough to achieve the company objectives. Specifically, the activity calls for role clarification and conflict resolution.

3.9.1 Role Clarification

In complex organizational settings, clarifying roles addresses the issues of authority and accountability. For a specific project involving personnel of multiple departments or business units, the need for defining roles of all participants is self-evident. Fig. 3.13 illustrates an example of role clarification.

Companies issue organizational charts that describe the roles and responsibilities of major business units. Employees may assume line roles, coordinating roles, and advisory roles:

- *Line roles.* Employees with line roles are those in profit centers with monopoly rights within the company to provide products and services to clients and customers. Examples of profit centers include business management, production, and sales. Profit centers are business units empowered to generate profits for the company. Managers of profit centers are accountable for offering quality products and services at competitive prices to ensure that the company makes profits. Profit centers define the services they might need. Managers of profit centers approve the annual budgets of cost centers that provide such needed services.
- *Coordinating roles.* Employees in some cost centers have monopoly rights for drafting and recommending constraints on the position duties of others. These constraints can take the form of approvals, policies, procedures, or planning objectives—legal, financial control, human resources, and so on. They are accountable for achieving higher-level organizational objectives such as

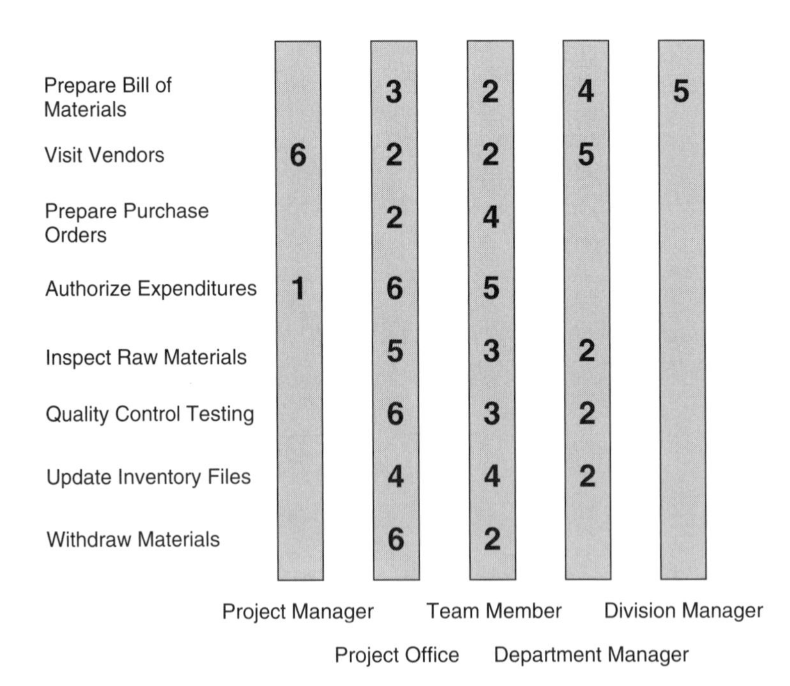

Figure 3.13 Roles assignment example.

consistency in work method, integration regarding external contacts, or cost efficiencies.

- *Advisory roles.* Employees in other cost centers provide services in support of the profit centers. Examples of such cost centers include R&D, maintenance, investors' relations, financial accounting, procurement, and others.

3.9.2 Conflict Resolution

In the real-world environment, there are conflicts of numerous types, such as the following: (1) technical, including design, analysis, and interpretation of test results, (2) operational, including procedures to perform specific tasks and assign responsibility, (3) emotional, such as treating bruised egos and hurt personal feelings, and (4) political, such as knowing who to consult and who has a say on specific projects or issues.

Engineering managers need to learn how to effectively resolve conflicts. Conflicts may be resolved by (1) *dominance*—dictating a solution, (2) *compromise*—engaging in negotiation based on a relative power base, and (3) *collaboration*—finding a win–win solution. The key requirement for conflict resolution is openness. By fostering mutual respect and trust, nurturing common interest to achieve project success, and focusing on commitment to task, most conflicts can be successfully resolved.

Example 3.6 Professional knowledge workers are those who earned B.S. degrees or higher in a specific engineering/scientific discipline. They constitute about 25 percent of the total workforce in the service industry. Their productivity has a significant impact on the well being of any service enterprise.

If you were the director of an engineering department employing a large number of professional knowledge workers, how would you go about managing these people, using planning, organizing and leading tools and practices, so that their productivity is at the highest levels possible?

ANSWER 3.6 Professional workers are a major source of core competencies, as ideas are the key competitive strengths of many companies. They should be managed in such a way that they have independence, low levels of supervision, variety and challenging job assignments, and opportunities to innovate in improving the company's processes or products.

Knowledge management is vital to a variety of companies, especially in relation to the creation, transfer, and utilization of tacit knowledge. Companies should offer assistance to enable employees to interact frequently with other professionals in teams, conferences, and seminars, to form and maintain networks with other professionals, as well as to form *communities of professionals*, in which to pursue technical subjects of interests and allow them to share their personal insights with others.

Besides offering them the proper recognition for positive results they bring forth, companies should motivate professional knowledge workers by providing opportunities to realize their high-level needs, such as peer recognition, wherein they become better recognized by their professional peers by way of technical publications and committee activities in professional societies, as well as self-actualization opportunities, wherein they could fully utilize their inherent talents.

In applying the principle of inclusion and participation, the director should also take the initiative to have personal dialogues with these professional workers from time to time and solicit their opinions regarding issues of long-term importance to the company.

3.10 INFORMAL ORGANIZATIONS

Aside from the formal organization of a business dedicated to streamlining labor, there are a number of informal organizations within every enterprise. Typical informal organizations are of the following types:

- *Social.* People form groups to pursue common interests, shared values, and beliefs—for example, beer clubs, bowling clubs, company outings, golf leagues, tennis groups, and so on.
- *Status.* People tend to be drawn to people well-known for their technical skills, abilities, special accomplishments, experience, tenure, charisma, interests, peer recognition, and acceptance, and to admire and respect such achievers for their status.
- *Group.* Coalitions form to advance shared interests. Fitness center on site, day-care centers, toastmasters groups, foreign-language study groups, bridge clubs, and so on are some examples.
- *Location.* Depending on the flow of vital information, people tend to migrate toward critical locations: work stations of executive assistants and secretaries, and water coolers.

To increase the success of an operation, engineering managers should inspire employees to participate in informal organizations. Informal organizations promote bonding between employees, increase job satisfaction, and have an overall positive effect on the company.

In recent years, social networking sites have emerged as external informal organizations. In today's environment, it is not what you know but who you know that counts; therefore, numerous professionals are making active use of social networking sites to connect with people outside of their organizations. Among a handful of social networking sites, four have become increasingly prominent:

1. *LinkedIn* (www.linkedin.com). LinkedIn offers a convenient way for professionals around the world to network. As of January 2010, it claims to have 55 million members from all Fortune 500 companies in more than 200 countries. Participating members are invited to offer profiles that summarize their professional expertise and accomplishments. These members are then encouraged to invite their trusted contacts to join LinkedIn. An individual's network consists of his connections, the connection's connections, and the people they know, thus linking the individual to a vast number of professionals and experts.

   Through such a network, an individual can (a) manage the information that is publicly available about himself, (b) find and be introduced to potential clients, service providers, and subject experts, (c) initiate and collaborate on projects (e.g., problem solving, data exchange, and brainstorming), (d) be found for business opportunities and partnerships, (e) discover inside connections that might be useful for new jobs, and (f) advertise job openings to find talented job seekers.

2. *Facebook* (www.facebook.com). Facebook offers a convenient platform by which individuals can keep up with friends, upload photos, share links and videos, and learn about people. As of January 2010, Facebook maintains 250 million members. 85 percent of college students use Facebook.
3. *LiveJournal* (www.livejournal.com). LiveJournal focuses on providing a user-friendly venue for creative individuals to share common interests, meet new friends, express themselves and socialize online. The communities established are different in each country. Currently, more than 16 million journals are in place on topics such as politics, entertainment, fashion, literature and design. LiveJournal's overall philosophy of creativity, open-source, diversity, and tolerance permeate all the journals on its site.
4. *Twitter* (www.twitter.com). Twitter is a real-time short messaging service that works over multiple networks and devices. In countries all around the world, people follow the sources most relevant to them and access information via Twitter as it happens—from breaking world news to updates from friends.

It should be of value to those professionals interested in networking socially to try out some of these sites.

3.11 CONCLUSIONS

Organizing is an important function of service systems management and engineering with a direct impact on the manager's ability to get work done efficiently. This function empowers a manager to choose the right organizational forms, be they teams, committees, task forces, functional or matrix arrangements, or other specific organizational structures. Managers assign the right-skilled and compatible people to work together, each one having clearly defined roles and responsibilities, along with commensurate authority. Managers assign responsibilities to employees so that work gets done and employees can acquire broadened experience. Managers allocate the right resources (such as skills, money, equipment, time, and technology) to accomplish the work efficiently.

Organizing is also important for enhancing the quality of work output. Flexible organizational structures allow companies to better respond to the changes of a dynamic marketplace. Certain organizational forms are superior to others in fostering creativity and inducing innovations. Multifunctional teams are known to be superior in handling conflicts at the interface between design and manufacturing or between R&D and marketing. Teams empowered by management to pursue specific assignments tend to be strongly motivating to the team members.

Engineering managers need to understand the power of organizing and use the function intelligently.

3.12 REFERENCES

Aguinis, H. 2008. *Performance Management*, 2nd ed. Upper Saddle River, NJ: Prentice Hall.

Ashkenas, R. 2002. *The Boundaryless Organization: Breaking the Chains of Organizational Structure*. San Francisco: Jossey-Bass.

Aufreiter, N. A., T. L. Lawyer, and C. D. Lun. 2000. "A New Way to Market." *The McKinsey Quarterly* 2.

Beinhocker, Eric. 2006. "The Adaptive Corporation." *McKinsey Quarterly* 2.

Bryan, Lowell, and Claudia Joyce. 2005. "The 21st Century Organization." *McKinsey Quarterly* 3.

Burns, R. 2004. *Making Delegation Happen: A Simple and Effective Guide to Implementing Successful Delegation*. St. Leonards, New South Wales, Australia: Allen & Unwin.

Chesbrough, H. W., and D. J. Teece. 2002. "Organizing for Innovation: When Is Virtual Virtuous?" *Harvard Business Review* (August).

Cox, T. 2001. *Creating the Multicultural Organization: A Strategy for Capturing the Power of Diversity*. San Francisco: Jossey-Bass.

Dean, J. W. Jr., and G. I. Susman. 1989. "Organizing for Manufacturable Design." *Harvard Business Review* (January–February).

Dickinson, H., and Kim Jelphs. 2008. Working in Teams. *Policy Press* (May 7).

Dobbs, Richard, Keith Leslie, and Lenny T. Mendonca. 2005. "Building the Healthy Corporation." *McKinsey Quarterly* 3.

Dolan, S. L., S. Garcia, and A. Auerbach. 2003. "Understanding and Managing Chaos in Organizations." *International Journal of Management* 20 (1): 23.

Edmondson, A., R. Bohmer, and G. Piscano. 2001. "Speeding Up Team Learning." *Harvard Business Review* (October).

Fabrizio, T., and Don Tapping. 2006. 5 S for the Office: Organizing the Workplace to Eliminate Waste. Productivity Press.

Galbraith, J. R. 2008. *Designing Matrix Organization that Actually Work*. San Francisco: Jossey-Bass.

Garvin, D. A. 1997. "What Makes for an Authentic Learning Organization: An Interview with David Garvin." *Harvard Business Review* (June 1).

Garvin, D. A. 2003. "Learning in Action: A Guide to Putting the Learning Organization to Work." *Harvard Business Review* (February 13).

Garvin, David A., and Lynne C. Levesque. 2008. "The Multiunit Enterprise." *Harvard Business Review* 86 (6): (June).

Gratton, Lynda, and Tamara J. Erickson. 2007. Eight Ways to Build Collaborative Teams. *Harvard Business Review* 85 (11): (November).

Gunn, R. A. 2007. *Matrix Management Success.* Infinity Publishing.

Hanson, Morten T., and Nitin Nohria. 2004. "How to Build Collaborative Advantage." *MIT Sloan Management Review* (Fall).

Heskett, James L., Thomas O. Jones, Gary W. Loveman, W. EarlSasser Jr., and Leonard A. Schlessinger. 2008. Putting the Service-Profit Chain to Work. *Harvard Business Review* (July–August).

Hoogerwerf, E. C., and A. M. Poothuis. 2002. "The Network Multilogue: A Chaos Approach to Organizational Design." *Journal of Organizational Change Management* 15 (4): 382.

Huettner, B. M., Katherine Brown, and Char James-Tanny. 2007. *Managing Virtual Teams*. Sudbury, MA: Jones and Bartlett Publishers.

Karwowski, W., G. Salvendy and T. Ahram. 2010. "Customer-Centered Design of Service Organizations," Chapter 9 in Salvendy, Gavriel and Waldemar Karwowski (Eds), *Introduction to Service Engineering,* John Wiley (January).

Katzenbach, J. R., and D. K. Smith. 2000. "The Discipline of Teams, HBR on Point Article." *Harvard Business Review* (July 14).

Katzenbach, J. R., and D. K. Smith. 2003. *The Wisdom of Teams: Creating the High Performance Organization*. San Francisco: Jossey-Bass.

Kelly, S., and M. A. Allison. 1999. *The Complexity Advantage*. New York: McGraw-Hill.

Lencioni, P. M., and K. Okabayashi. 2008. *The Five Dysfunctions of a Team, Manga Edition.* San Francisco: Jossey-Bass.

Loureiro, G., and Richard Gurran. 2007. Complex Systems Concurrent Engineering: Collaboration, Technology Innovation and Sustainability. New York: Springer.

Metters, Richard, and Vicente Vargas. 2000. "Organizing Work in Service Firms." *Business Horizons* (July–August).

Michaelsen, L. K., Michael Sweet, and Dean X. Parmelee. 2009. *Team-Based Learning: Small Group Learning's Next Big Step.* San Francisco: Jossey-Bass.

O'Reilly, C. A., and J. Pfeffer. 2000. *Hidden Value: How Great Companies Achieve Extraordinary Results with Ordinary People*. Cambridge, MA: Harvard Business School Press.

Parker, G. M. 2003. *Cross-Functional Teams: Working with Allies, Enemies, and Other Strangers*, 2nd ed. San Francisco: Jossey-Bass.

Payne, V. 2001. *The Team-Building Workshop: A Trainer's Guide*. New York: AMACOM.

Rees, F. 2001. *How to Lead Work Teams: Facilitation Skills*, 2nd ed. San Francisco: Jossey-Bass/Pheiffer.

Rienas, Ron, Alan E. Taylor, Daniel D'Angelo and Amy Jackson-Grove. 2007. "Peace Bridge Expansion Project: Draft Design Report and Draft Environmental Impact statement," Buffalo and Fort Erie Public Bridge Authority, Buffalo, New York.

Shannon, Robert E. 1990. Engineering Management. New York: John Wiley & Sons, Inc.

Straus, D., and Thomas C. Layton. 2002. *How to Make Collaboration Work: Powerful Ways to Build Consensus, Solve Problems and Make Decisions*. San Francisco: Berrett-Koehler.

Uzzi, Brian, and Shannon Dunlap. 2005. "How to Build Your Network." *Harvard Business Review* 83 (12): (December).

Viardot, E. 2004. *Successful Marketing Strategy for High-Tech Firms*, 3rd ed. Artech House.

Von Hoffman, C. 1998. "Getting Organized." *Harvard Management Update* (January).

Wetlaufer, S. 1999. "Organizing for Empowerment." *Harvard Business Review* (January–February).

3.13 APPENDIX—DEFINITIONS

Before reviewing the managerial function of organizing, it is useful to introduce a few definitions.

- **Span of control.** The span of control refers to the number of people supervised by a manager or supervisor. It may be small (a few people) or large (twenty to thirty people). The choice of a small or large span of control depends on workforce diversity, task volume, and complexity of work, as well as on the geographic dispersion of workers. Large span brings about lower costs and greater organizational efficiency, but it also leads to a lower intensity of control. The current trend is moving toward larger span of control, increasing from seven to twenty or more, due to:

- Reduced levels of middle management.
- Enhanced communication tools.
- Empowered knowledge workers, allowing decision making at lower levels by people with more applicable knowledge.
- Improved morale, productivity, and profitability made possible by less detailed supervision, particularly over professional workers.

- **Organization types**. The *line organization* (e.g., a profit center) performs activities directly related to the company's main goals. Examples include business management, product management, sales and marketing, product design and engineering, production, and customer services. The *staff organization* (e.g., a cost center) provides advice and comments in support of the line organization's work. Examples include research and development, financial and accounting, information services, procurement, legal affairs, public relations, and facility engineering.
- **Overlap and duplication of responsibility.** This refers to a situation where two or more people do the same work and make the same decisions. Such undesirable situations are to be avoided in any organization, as they represent sources of conflicts.
- **Specialization.** Specialization refers to the increased degree of skill concentration in narrow technical domains. Specialization of work leads to better efficiency. However, overspecialization may cause monotony, fatigue, disinterest, and inefficiency on the part of the worker.
- **Work arrangement.** Work needs to be arranged in a rational and logical manner. The logical arrangement of work promotes task accomplishments and increases personal satisfaction for more workers over a longer period of time.
- **Selected management terms.** *Authority* refers to the legal or rightful power of a person, by assignment or by being associated with a position, to command, act, or make decisions—this is the binding force of an organization. *Responsibility* is the duty to perform work assumed by a position holder in an efficient and professional manner. *Accountability* stands for an upward-directed obligation to secure the desired results of the assigned work.

3.14 QUESTIONS

3.1 Review the Harvard Business School case "Peterson Industries," # 9-396-182 (rev. February 12, 1996) and answer the following questions. The case materials may be purchased by contacting its publisher at 1-800-545-7685 or www.custserve@hbsp.harvard.edu.

A. What are the relationships among Peterson's strategy, the design of the organization, and the Rosegrant family's role in the management of the company?

B. How effective is Peterson's resources allocation process? What are its strengths and weaknesses?

C. In what ways is the resource allocation process helped or hindered by

1. The divisional performance measurement system?

2. The incentive compensation system?

D. What role does Louis Friedman play in Peterson's resource allocation process?

1. What are his goals?
2. What forces is he trying to keep in balance?
3. Why the ambiguity? Are the uncertainty and ambiguity necessary and desirable?

E. What is your assessment of Friedman's handling of Kells's two projects?

F. Would the resource allocation process be improved if Jenkins's two proposals were adopted? Why or why not?

1. What are the strengths and weaknesses of Jenkins's proposal?
2. Do you agree with Jenkins's view of the role and functioning of the resource allocation process? Why or why not?

G. What are the lessons you have learned from having studied this case?

3.2 A materials manager suspects that the quality of work within her department has been deteriorating. She wants to introduce a program of change to advance quality. What steps should she take?

3.3 The company has recently concluded a multimillion-dollar contract to supply products to a third-world country. The first elite group of engineers from that country has just completed a two-month training course on maintenance and operations. The company's training manager reports that the level of skill and knowledge of that country's engineers was so low that no amount of training would ever enable them to properly operate and maintain the products in question. "It might be better for that country to buy a less sophisticated product from our competitor," the training manager suggests. What should the company do?

3.4 Six months ago, the company hired an engineer for his expertise in hydraulic drives. The decision to hire him was based on a product development plan that projected a need for such expertise. Market conditions have suddenly changed in favor of more sophisticated electric drives. The new engineer turns out to be very good in his area of specialization, but it is difficult to retrain him for other assignments in the company. Should the company discharge this engineer?

3.5 The company has been making most of its sales to a few large customers. The company president wishes to broaden its customer base. To do so may require changes in the company culture, the product line strategy, marketing and sales programs, and the service organization. How should the president go about making the required changes?

3.6 The company is considering a plan to upgrade its current product line. The cost of upgrading is high. There is a small company that has advanced the technology required for this product upgrade. What strategy should the company follow if it wants to continue selling into its current market with the new, upgraded product?

3.7 As the company's sales are coming down unexpectedly, the president invites you to chair a task force with the objective of recommending solutions to correct the situation. Who do you want to be on this task force? How should the task force resolve this problem?

3.8 A loyal and high-volume customer has warned the company's marketing department that project X is extremely critical to their needs and that if this project is late, they may be forced to buy elsewhere. The project manager knows that the best estimates available to date from various in-house groups suggest that, at the current rate of progress, project X will be late by about six months. What should the project manager do?

3.9 Sally Lee, the engineering manager, delegates tasks as a good manager should. However, Mark Hayes, the engineering director, has the bad habit of calling up Sally unexpectedly to get detailed reports on various ongoing activities in Sally's department. Sally does not want to hold daily staff meetings in order to satisfy Mark's information needs because Sally

Table 3.A1 Distribution of Employees

Department	Number of Employees
A	3
B	7
C	4
D	6
E	9

is quite certain that asking her professional staff to stand by and make daily reports will definitely be counterproductive, as all of them are known to prefer independence. What should Sally do?

3.10 In an organization offering a dual-ladder career progression system, technically trained people may opt to progress along a technical ladder instead of the traditional managerial ladder. How does this work?

3.11 The organization chart of Company X reveals that different numbers of employees report to its five departments, as indicated in Table 3.A1. Why do the numbers differ?

3.12 Organizing is a powerful tool that managers should employ to achieve an optimum utilization of company resources (e.g., funds, people, equipment, know-how, relationships, and others) in achieving corporate objectives. Among a variety of value-adding activities managers regularly oversee, two are of particular importance: (1) interactions between self-directed professionals and (2) development of innovative products/services. Explain how managers may use the tool of "organizing" to promote and enhance these two activities.

3.13 What type of organizational structure is best suited for developing a new product that requires a high level of specialization in several functions and for which the time to market represents a critical factor?

Chapter 4

Leading

4.1 INTRODUCTION

Leading in service systems management and engineering refers to the function of an engineering manager that causes people to take effective action. After deciding what is worth doing, the engineering manager relies on communication and motivation to get people to act. By selecting people who are inclined to collaborate, the engineering manager influences people to take action. In addition, skills training and attitude development may also enable others to take action. In this chapter, five specific leading activities will be reviewed: deciding, communicating, motivating, selecting, and developing.

The true measure of the quality of an engineering manager's leadership is his or her demonstrated ability to guide and direct the efforts of others to attain organizational objectives (Bardaracco 2002). A leader has vision, sets a good personal example, is able to attract and retain good people, motivates people to use their abilities, and inculcates them to willingly do their best. Managers derive their authority from occupying higher positions within the organization. Leaders, by contrast, have the power of influence over people. Their power is attained by earning employees' respect and admiration (Friedman 2008; Zalenznik 1992). However, some leadership skills can be learned. Engineering managers with good leadership qualities are the most valuable ones to their employers.

In this chapter, specific methods for making decisions are also illustrated. These include the Kepner–Tregoe method, decision making by gut instinct, and decision making in teams. Furthermore, a few special topics on leadership are addressed, including (1) leading changes, (2) advice for new leaders, and (3) guidelines for superior leadership.

4.2 LEADING ACTIVITIES

The managerial function of leading includes performing five specific activities related to leading the engineering unit or department to achieve organizational objectives. These activities are outlined here (Kotter 2001):

1. *Deciding.* Arrive at conclusions and judgments with respect to priority, personnel, resources, policies, organizational structures, and strategic directions.
2. *Communicating.* Beget understanding by sharing information with others and talking, meeting, or writing to others.

3. *Motivating.* Inspire, encourage, or impel others to take required action and creating workplace conditions to ensure work satisfaction.
4. *Selecting people*. Choose the correct employees for positions in the organization or for specific team activities.
5. *Developing people.* Help employees improve their knowledge, attitudes, and skills.

Some of these tasks are relatively easy, and others are more difficult. Engineering managers need to practice them in order to become proficient over time in carrying out these activities. Each of these activities will be reviewed in detail in the following sections.

4.3 DECIDING

Making decisions is a key responsibility of service system engineers and leaders. The quality of decisions is the hallmark of excellent managers. The purpose of making decisions is to align the choices of project priorities, people, financial resources, technology, and relationships with the attainment of corporate objectives. As the engineering manager gains experience, the overall quality of his or her decisions is expected to increase.

Oftentimes, there is insufficient information available for guidance, or the future business or market conditions are very fluid and fuzzy. Under these circumstances, managers need to make spontaneous decisions based on intuition, gut instinct, and hunch. Otherwise, they make reasonable decisions based on systematic studies and logical analyses of available quantitative data. Engineering managers are typically quite proficient in handling this latter type of decision making, which follows the typical steps enumerated here:

1. Assessment of facts and evaluation of alternatives.
2. Use of full mental resources.
3. Emphasis on creative aspects of problem solving.
4. Consistency in thinking.
5. Minimization of the probability of errors.

Management decisions are usually difficult for the following reasons:

- They involve problems and issues that are ill defined, as they are wider in scope and affect more people than typical technical problems and issues.
- Needed data and information may be insufficient or excessive, and there may be no time available to collect or interpret the data.
- Available information is of poor quality, because it is based on guesswork, rumors, opinions, hunches, or hearsay.
- Decision making involves human behavior, which is not always predictable.
- The nature of problems and issues changes continuously.
- Consequences of management decisions depend on opinions available, and as such, the consequences are also changing.

- Rarely does a perfect solution exist for management problems, since all options involve compromise, whose validity changes over time.
- Decision making must consider implementation, which, in turn, depends on consensus and commitment of the affected people. Oftentimes, political considerations come into play, as well.
- A critical decision may involve multiple layers of management or peer departments, and thus it requires coordination.

Listed here are seven decision-making guidelines useful to engineering managers:

1. Acquire decision-making experience, for example, by recognizing patterns and rules and by studying management cases.
2. Prioritize problems for which decisions must be made. Do nothing with problems that are perceived to have minor significance and impact.
3. Follow a rational process to identify the problem and establish options to remove the root cause of the problem (Tregoe and Kepner 2006).
4. Involve others in the decision-making process, especially if the implementation of the pending decision affects them. Group decisions (see section 4.4.5) are superior from the standpoint of implementation. Such decisions, however, may take longer time to reach, and they usually denote compromises for all involved.
5. Make decisions based on available information and assumptions introduced. Check the validity of all assumptions and update the decisions accordingly. Account for necessary uncertainties and avoid becoming paralyzed by stress and uncertainty.
6. Delay making decisions until the last allowable moment, as the problems and available options may continue to change. Above all, meeting all deadlines with a decision is better than having none.
7. Avoid making decisions about issues that are not pertinent at the time or decisions that cannot be practically implemented (or both). Also, do not make a decision that ought to be made by others.

How can we judge the quality of a given decision? A simple way to find out is to raise the following three questions:

1. Did the decision achieve the stated purpose; did it correct or change the situation that caused the problem in the first place?
2. Is it feasible to implement the decision; is it meaningful with respect to the required resources and the anticipated potential value?
3. Does the decision generate noticeable adverse consequences or risks to the group or the company?

The decision at hand is regarded as good if the first two questions are answered by a yes and the last one by a no.

As a rule, managers are expected to make decisions. However, there are circumstances in which managers should delegate the decision-making authority to the staff or work alongside with them to come to a decision.

The following are problems or issues that should be handled by the managers only: (1) prioritizing activities and projects, office assignments, and work group composition, (2) handling personnel assignments, performance evaluations, and job action, (3) dispensing budget allocation, (4) applying administrative policies, procedures, and regulations, and (5) dealing with highly confidential business matters that are explicitly declared to be so by the top management (e.g., compensation, promotion, corporate strategies, and new marketing initiatives).

Managers should include the staff when making decisions on the following problems or issues: (1) considering staff needs for development (e.g., attending professional meetings, technical conferences, seminars, and training courses, as well as committing to study programs at universities), (2) explaining policies and procedures, involving staff interactions with other departments, and (3) determining team membership (e.g., considering personality fit, skills compatibility, working relationships, balancing workload, etc.).

Managers should delegate the decision-making authority to staff members for the following matters: (1) techniques to accomplish assigned activities or projects, (2) options to continuously update current operations and work processes, and (3) social events involving staff participation, such as group picnics, golf outings, and Christmas parties.

As the saying goes, "Practice makes perfect," service systems engineers and leaders should seek opportunities to constantly acquire experience regarding when, where, and what decisions are to be made. How decisions can be made is deliberated in the next section.

4.3.1 Decision Makers

Companies deal with strategic and operation issues such as which markets to enter or exit, which businesses to buy or sell, where to allocate capital and talent, how to ferret out service innovation, what is the best way to position brands, and how to manage channel partners. Businesses lose ground if decisions cannot be made quickly and effectively and are not executed in a timely and consistent manner.

Decisions may get stalled at the interface of the following four bottlenecks (Rogers and Blenko 2006):

1. Global versus local (e.g., marketing—who decides on pricing and advertising; brand building—how much to localize; product development—how to balance customization and economies of scale).
2. Headquarters versus business units (e.g., major capital investment).
3. Function versus function (e.g., balance between product development and marketing).
4. Inside versus outside partners (related to outsourcing, joint ventures, strategic alliances and franchising; internal disagreements on strategic directions and external misalignment on operational issues.

It is advisable to assign roles and responsibilities clearly. Member in a project team need to know who should make recommendations, who needs to be in agreement, who

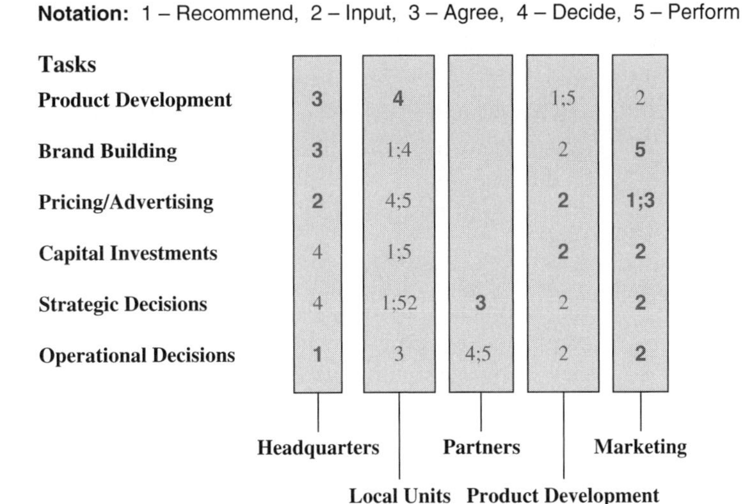

Figure 4.1 Roles and responsibilities.

should have inputs, who has the ultimate authority for executing the decision, and who is accountable for following through. Figure 4.1 illustrates such a division of labor.

4.3.2 Decision-Making Styles

Decision-making style is defined by two components: how information is used and how options are created (Brousseau et al. 2006). *Maximizers* study data exhaustively until they have found the best answer. *Satisficers* find key facts, introduce hypotheses, test them as they go, and get ready to act. *Single-focus managers* take one course of action to pursue. *Multifocus managers* generate options and pursue multiple courses. There are four common styles:

1. *Decisive*—little information, one course of action.
2. *Flexible*—little information, many options.
3. *Hierarchic*—lots of data, one course of action.
4. *Integrative*—lots of data, numerous options.

Decision-making style changes according to managerial authority. Managers change styles when going from first-line supervisors to managers, directors, VPs, and senior execs. When moving up, managers' style moves away from decisive to become more flexible. Senior executives are predominantly of the flexible and integrative types.

4.3.3 Framework for Decision Making

Managers need to make a variety of decisions and provide responses in their daily work. Snowden and Boone (2007) proposed a useful framework for guiding the decision-making process, from simple to chaotic. Such a framework helps to categorize the issues at hand and thus facilitates the right decision making approach:

1. *Simple.* This is the domain of *known knowns,* applicable to repeatable work processes and procedures and eliminates deviations based on the past best practices. Need to be alert for new changes that alter the context, making some of the self-evident steps no longer appropriate as solutions. Beware that best practices are past practices, and they may no longer be sufficient for the new context.
2. *Complicated.* Experts provide multiple right answers, Leaders sense, analyze, and respond. They define options and make a choice. The analysis is usually dominated by experts, who may be biased, causing a rejection of another's views, thus forgoing useful opportunities. Holding brainstorming sessions involving more people will help. Another potential difficulty is *analysis paralysis,* where a group of experts cannot agree on any one answer. This is the domain of *known unknowns.* There is usually one right answer to this situation.
3. *Complex.* This is the domain of emergence. Business conditions are complex, because of concurrent changes in many of their parts (some of them are unpredictable and in flux). Leaders need to probe, sense, and respond and allow a solution to emerge from the situation at hand. It should be sense, probe (experiment), and then respond. It requires patience. The command and control style will not work here. This is the domain of *unknown unknowns.*
4. *Chaotic.* This is the domain of rapid response. It is pointless to search for the right answers. The domain of *unknowables*, for example, includes the events of September 11, 2001. Fight the fires to stabilize the situation and avoid causing it to become complex. Do not ask for inputs, just communicate directives. Leaders must move away from this command and control style once the situational context changes.

Chaos can be a useful impetus for innovative development to capture new opportunities. Leaders need to change their style to fit the context at hand and prepare their organizations for diverse contexts. They need to know when to share power and when to wield it alone, when to look to the wisdom of the group and when to take their own counsel. A deep understanding of context, the ability to embrace complexity and paradox, and a flexible attitude are required for leaders who want to make things happen in a time of increasing complexity.

4.3.4 Rational Decision-Making Processes

A rational decision-making process is generally useful in facilitating decision making for numerous problems or issues in engineering when an adequate amount of information is available. It consists of a set of logical steps outlined as follows:

A. *Assess the apparent problem based on observed symptoms.*

B. *Collect the relevant facts.* Usually, not all facts are available due to resource, cost, or time constraints. Facts must be related to five decision-making factors:

   1. *Situation* (what, how)—the sequence of events leading to the problem and its conditions.

2. *People* (who)—personalities, preferences, personal needs, and egos.
3. *Place* (where)—significance of location.
4. *Time* (when)—pressure to bring forth an immediate solution.
5. *Cause* (why)—why the problem originally occurred, and why it occurred in one situation but not in another.

Past experience demonstrates that there are several good sources for identifying the relevant facts related to the problems at hand. (See Table 4.1.)

C. *Define the real problem.* As the most important initial step, the real problem at hand must be defined. It will be helpful to pose the following three questions:
   1. What is the deviation between actual performance and the expected norm?
   2. What are the desired measurable results in a problem-solving situation?
   3. What embodies success based on well-defined metrics and the proper method of measurements?

The answers to these questions are likely to point to the root cause of the problem.

D. *Develop alternatives to solve problems.* Once the root causes of the problem at hand are defined, it is then useful to come up with options to address them. Engineering managers should do the following:
   1. Freely invite creative suggestions from people who have direct knowledge of the problem at hand.
   2. Brainstorm in group settings (without criticisms or comments so as not to deter imaginative suggestions).
   3. Take into account both short-term and long-term impacts.

E. *Select the optimal solution.* Engineering managers need to choose among the options to address the root causes. A useful rational tool for making such a choice is introduced in section 4.3.5. Engineering managers must ensure that the chosen option produces minimum adverse consequences to the company or unit. They also need to plan for contingencies and make midcourse corrections, if required, to secure the greatest probability of achieving the desired results. They should also avoid committing to a final choice prematurely until its implementation becomes feasible.

Table 4.1 Sources for Facts Related to Problems

Problem Categories	Sources of Useful Facts
Equipment	Plant operations personnel
Technical	Engineers with direct working knowledge
Customer inquiry	Salespeople
Customer complaints	Service and sales personnel
Materials and parts	Delivery and inspection personnel
Product quality	Production staff
Customer preference	Marketing personnel
Market competitiveness	Marketing personnel

A decision is nothing but the choice among several available options to solve a problem or address an issue. If there is only one option available, then no decision is needed.

F. *Set a course of action to implement the decision.* Once a decision is made, engineering managers should devise an applicable action plan to implement the decision. The manager must consider such details as (1) policies that limit possible action, (2) programs (the sequence of action steps), (3) schedules (dates and milestones), (4) procedures (the action steps carried out in an orderly manner), and (5) budgets and expenses for equipment and manpower. Decisions that are not efficaciously implemented are useless.

4.3.5 Kepner–Tregoe Decision Analysis Tool

The Kepner–Tregoe method is a renowned analysis tool available to support decision making (Tregoe and Kepner 2006). It prescribes the following steps (see Table 4.2) to arrive at a rational decision:

1. Define a set of decision criteria needed for making the decision. The necessary criteria are those that must be met. For example, all entry-level engineering applicants must have undergraduate degrees in engineering to be considered for employment. Some hiring companies may define a grade point average (e.g., 3.5) as the cutoff academic performance level below which an applicant would not be considered. The sufficiency criteria are those that are not necessary, but are good to have. For hiring entry-level engineers, companies may specify these to be summer work experience, internship activities, project work, leadership positions held in student organizations, and others.
2. Rank-order the sufficiency criteria by assigning weight factors ranging from 10 (as the most preferable) to 1 (as the least preferable).
3. Evaluate all options against each of the identified as necessary decision criteria. For example, the options that meet the necessary criteria may be designated with words "go."
4. Remove from further consideration those options that fail the necessary criteria.
5. Rank all remaining options relatively, with respect to specific sufficiency criteria. Assign a relative score of 10 to the most satisfactory and 1 to the least satisfactory option.
6. Repeat the scoring process for each of the remaining sufficiency criteria.

Table 4.2 Kepner-Tregoe Decision Analysis Method

Criteria	Weight Factor	Option A	Option B	Option C
Criteria 1	R	Go	Go	Go
Criteria 2	10	4	8	10
Criteria 3	5	6	10	7
Criteria 4	8	10	6	8
Total Weighted Score		150	178	199

7. Compute a weighted score for each option by multiplying its relative score for a specific sufficiency criterion with its corresponding weight factor. Add up the weighted scores for all sufficiency criteria to obtain the overall weighted score for this option. Repeat the computation for each of the remaining options.
8. Compare the overall weighted scores and choose the option with the highest overall weighted score.

The Kepner–Tregoe method forces decision makers to externalize all necessary criteria and to assign weight factors to all sufficiency criteria before making decisions. The chosen criteria must constitute a *mutually exclusive and collectively exhaustive* set of criteria for the decision at hand. By ranking options against each of the defined criteria, all options are properly evaluated in a rational, equitable, and comprehensive manner (Ragsdale 2007).

The Kepner–Tregoe method is particularly useful in a team environment where members may needlessly argue for specific options without externalizing their decision criteria and the relative ranking they have assigned to the criteria. Oftentimes, the advocates for a specific option make implicit assumptions that remain hidden and unknown to others on the team. In addition, personal biases may influence the relative scores assigned to the options when these options are evaluated against a given decision criteria. Experience has demonstrated that the personal biases tend to become minimized when the relative scores are polled from all teammates during a meeting.

The Kepner–Tregoe method is also practically useful for decision making on an individual basis. Some engineering managers tend to emotionally overemphasize certain decision criteria and downplay the importance of others. Again, having all decision criteria and their respective weight factors explicitly delineated will facilitate a rational decision.

Example 4.1*[*] Bill Pickens, manager of the test division, called John Riley, the group head of mechanical testing, into his office and told him that there was a new opening for a manager of product development position in the company. For John, it would be a promotion to a higher managerial rank with an appropriate increase in salary. However, the new position is temporary, in that it may be eliminated in a year. Although Bill hates to lose a very valuable worker like John, he wants to let John himself make the decision. The product development division has specifically requested that this opening be recommended to John. After having given it some thought, John decided to take the new position.

The next day, Bill Pickens and John Riley sit down together again to name a group head successor. Among the three section heads in the group, Dodd is the most experienced. However, Dodd is quiet and does not communicate well. He may have difficulty in selling testing services to others. Yeager is competent, but has made hasty decisions that have been very costly to the group. Bennett is ambitious and aggressive, but has poor interpersonal skills. They concluded that none could be immediately promoted to take over. Finally, they agreed to rotate the acting head

[*]Condensed and adapted from Robert E. Shannon, 1990. *Engineering Management*. New York: John Wiley & Sons, Inc., p. 203.

job among the three, to test out each of them, since there is an outside chance that John may come back to his old position after one year.

Shortly thereafter, Bill Pickens was promoted out. John decided not to return to take Bill's position. Terry Smith was brought in to take over Bill Pickens's job as test division manager. However, before Bill Pickens left, he hinted to Yeager that Yeager would likely get the job, based on the results of the trial periods.

Terry found significant rivalry and ill feeling between the three section heads. The group had low morale and poor productivity. Under such circumstances, Terry decided to appoint a new employee, Dennis Brown, to the mechanical testing head position instead of one of the three.

Did Terry make the right decision? Apparently, the job rotation idea failed. What would have been the right way for Bill Pickens to handle this problem?

ANSWER 4.1 The decision made by Terry was not the right one. The reasons are as follows:

1. The personnel situation was brought into being by Bill Pickens's and John Riley's inability to make a staffing decision by choosing the best one among the three and minimizing the impact of the new head's shortcomings. Rotating the acting head job brought about chaos due to infighting. Experienced managers should have anticipated this.
2. Terry's decision negated an implied management promise that one of the three would be promoted after a one-year trial period. This broken promise could be the basis of a future lawsuit.
3. The appointment of a new employee, without consultation with and concurrence by the three section chiefs, reflects a lack of sophistication on the part of Terry. It raises the fairness issue and generates an employee loyalty problem. The three section chiefs are not likely to be motivated to work with the new person.
4. It is not known if Dennis Brown has the necessary technical and managerial skills to be more successful than any one of the three tried entities.
5. The likely results are as follows:
   - Lost management credibility due to broken promises and a lack of personnel staffing capabilities. (Riley neglected to groom a successor by correcting the perceived shortcomings of his chosen successor during the last five years.)
   - Management is perceived to be lacking fairness in decision making. (This will result in lower group morale and decreased employee loyalty. Employee turnover may be increased.)

The job rotation idea is very poor. It was selected only because Pickens and Riley were not able to make good staffing decisions. They were looking for a perfect person, and overlooked the possibility that most of the identified shortcomings could be easily compensated for or corrected.

What Pickens and Riley should have done was to make a hard choice in the beginning, either bringing someone in from the outside or promoting one of the

Table 4.3 Making a Personnel Decision

Criteria	Weight Factor	Dodd	Yeager	Bennett
Minimum technical experience	R	Go	Go	Go
Experience	10	10	8	6
Communications skills	8	5	10	10
Decision-making abilities	6	10	5	10
Human relations skills	6	10	10	5
Total Weighted Score		260	250	230

Table 4.4 Making a Refined Choice

	Rank without improving Shortcoming	Probability of correcting Shortcoming	Adjusted Total Score
Dodd	260	80	292 ($260 + 0.8 \times 8 \times 5$)
Yeager	250	90	277 ($250 + 0.9 \times 6 \times 5$)
Bennett	230	60	248 ($230 + 0.6 \times 6 \times 5$)

three employees. Assuming that no suitable outside candidates were available, then the Kepner–Tregoe decision analysis should have been used to come up with a choice, as depicted in Table 4.3.

At the first glance, Dodd appears to be the winner. However, as the candidates weighted scores are rather close, the refinement step presented in Table 4.4 may be taken.

The adjustments are made based on the expected values of improving the relative score of the identified weakness from 5 to 10. The adjusted total score represents the final ranking of these three individuals, after each is allowed to minimize his weaknesses. Dodd remains the winner in this case.

Example 4.2 Due to global competition, the company faces a tough time in the marketplace, and so it must scale down its workforce. The board has advanced the following options for the employees:

1. *Quit voluntarily*. This could be attractive to several bright young engineers whom the company does not want to lose.
2. *Last in and first out*. This could result in the loss of young and more versatile operators.
3. *Early retirement of those within ten years of their normal retirement age*. This could cause a loss of engineers with valuable product knowledge.
4. *Reverse ranking in performance records*. This could lead to unfair selection, as the Uniform Performance Appraisal System has been operating in the company only for the last few years.

Table 4.5 Choosing the Method of Downsizing

Criteria	Weight Factor	Option 1	Option 2	Option 3	Option 4	Option 5 (1&4)	Option 6 (4&1)
Not to lose knowledge and experience	10	5	10	3	8	6	7
Easy for affected employees to find jobs	8	10	10	5	8	9.5	8.5
Easy for company to find replacements	8	5	10	3	8	6	7
Easy for company to avoid legal problems	10	10	9	8	5	8	6
Total Weighted Score		**270**	**350**	**174**	**258**	**264**	**254**

What methods, or combination of methods, should the company use to reduce employment?

ANSWER 4.2 The company should use the Kepner–Tregoe method and assign weight factors to all the criteria. The relative score as displayed in Table 4.5 should be assigned to evaluate all options.

Based on the results obtained from Table 4.5, the method of last in and first out (Option 2) should be chosen to reduce employment.

4.3.6 Additional Support Tools for Decision Making

A number of spreadsheet-based tools are commercially available to engineering managers to support the decision-making process (Nagal 1993, Donker 2007, and Figueira, Greco and Ehrgott 2004). The following tools, among others, are widely used in the industry:

- Forecasting (exponential smoothing, time series, and neural network computing)
- Regression analysis (single variable and multivariable)
- Risk analysis and project management
- What-if solver
- Simulation modeling (Monte Carlo)
- Decision trees
- Optimization (linear programming and integer and dynamic programming)
- Artificial intelligence and pattern recognition tools
- Expert or knowledge-based systems

Engineering managers should familiarize themselves with all of these decision support tools so that they can employ the right tools under the right circumstances (Ragsdale 2004).

Example 4.3 A certain company makes two products: P_1 and P_2. Each P_1 requires 5 kg of material M and 3 kg of material N. Each P_2 requires 3 kg of M and 3 kg of N. In the warehouse, there are 350 kg of M and 270 kg of N available. The profit

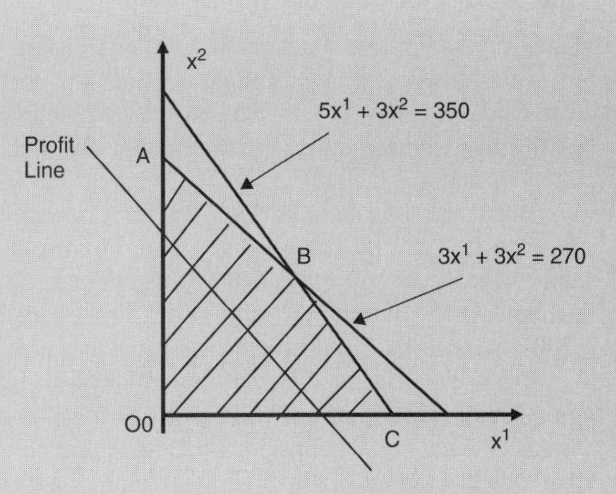

Figure 4.2 Linear programming problem.

is \$50 for each P_1 and \$40 for each P_2. What mix of P_1 and P_2 products should the company make and sell in order to maximize its total profit?

ANSWER 4.3 This is an optimization problem to be solved by linear programming:

$$\begin{array}{lll} \text{Maximum Profit} = & 50 \times x_1 + 40 \times x_2 & \{\text{Objective function}\} \\ \text{Subject to} & 5x_1 + 3x_2 <= 350 & \{\text{Constraints}\} \\ & 3x_1 + 3x_2 <= 270 & \\ & x_1 >= 0 & \\ & x_2 >= 0 & \end{array}$$

Graphically, the solution method can be displayed as presented in Fig. 4.2. Because of the four constraints, the solution space is bound by the area $OABCO$: O (0,0), A (90,0), B (40,50), and C (70,0).

As the profit line ($50 \times x_1 + 40 \times x_2 =$ Profit) moves to the right, the profit is maximized when it passes through the point B (40,50), the right-most location it can take while still satisfying the stated constraints. Thus, the maximum profit is (\$50 × 40) + (\$40 × 50), or \$4,000, and numbers of products P_1 and P_2 to make and sell are 40 and 50, respectively.

The simplex method is programmed to handle linear optimization programs with n independent variables.

Suppose we want to know more about the impact of material constraints on the company's profit. Let us assume that material N can be increased by 30 units, subject to the new constraint:

$$3x_1 + 3x_2 \leq 300$$

This new constraint, represented as a straight line $3x_1 + 3x_2 = 300$, intersects the straight line $5x_1 + 3x_2 = 350$ at the point D (25,75), which is not exhibited in Fig. 4.2. The new optimum solution is then D, producing a new total profit of (\$50 × 25) + (\$40 × 75), or \$4,250. This new profit is \$250 over the previous total, because of added material N, which has a shadow price of \$250/30, or \$8.33.

4.3.7 Decision Making by Gut Instinct

Since, up to middle managerial levels, decision making is mostly quantitative, the tools described in the preceding section are useful. However, at upper and senior managerial levels, problems and issues get much more complex and ambiguous. When such circumstances defy systematic analyses, decision making is typically based on intuition and gut instinct.

Bob Lutz, president of the Chrysler Corporation, was reported to have made the decision in 1988 by pure instinct while driving alone along a country road, to build and market a new sports car, the Dodge Viper. This car later turned out to be a great success. Using interviews with several other managers as source material, Hayashi (2001) studied their intuitive decision-making processes.

What is gut instinct? According to Hayashi, our minds process information all the time. Our left brains process conscious, rational, and logical thoughts, whereas our right brains take care of subconscious, intuitive, and emotional thoughts. Some people claim that they can tap into right-brain thinking by jogging, day dreaming, listening to music, or using other meditative techniques. Others have reported that they get innovative ideas while taking long showers or placing themselves in unfamiliar situations.

The theory of intuitive thinking claims that accumulated past experience enables some people to bundle information so that they can easily store and retrieve it. Experts further claim that such information is retrieved from memory by the observation of patterns. Professional judgment can often be reduced to patterns and rules. Accordingly, all other things being equal, people with varied and diverse backgrounds are likely to be more capable of thinking intuitively and learning faster because they recognize more patterns. When using gut instinct, people essentially draw on rules and patterns that reside in their memories. Diverse backgrounds facilitate cross-indexing, allowing one to see similar patterns in disparate fields.

According to Matzler et al. (2007), intuitive decision making is really one's ability to recognize patterns at a fast speed. Because complex decision making actuates a process in which experience, knowledge, and emotions are linked, a person's intuitive decision-making capability may be enhanced by five factors:

1. *Experience*. Extensive experience produces more patterns. "Gut instincts" refer to recognizable patterns known in the past. Studies in Psychology have found that one needs 10 years of domain-specific experience to cultivate the gut-feeling needed to make good instinctive decisions.
2. *Network*. Personal networks are helpful to offer feedback to intuitive decisions.
3. *Emotional intelligence*. Emotion is known to precede cognition. 90 percent of the differences between top-performing and average-performing senior executives can be explained by emotional intelligence (being able to recognize and interpret one's own emotions).
4. *Tolerance.* Willingness to tolerate mistakes and take risks.
5. *Curiosity.* A prerequisite for discovering new opportunities. An opportunity-oriented way to think and act allows the acquisition of new experience that strengthens the foundation of intuition.

However, instincts can, at times, be wrong. People who make decisions intuitively are advised to secure constant feedback in order to minimize the impact of incorrect decisions and to learn from these decisions. Over time, the process of learning on the basis of feedbacks has the potential of improving the patterns and rules stored in peoples' memory, thus enabling them to make better intuitive decisions in the future.

In order to update and modify their own patterns and rules, engineering managers at low- or middle-managerial levels should carefully observe how top-level leaders make important decisions and reverse-engineer such decision-making processes.

4.3.8 Decision Making in Teams

Typically, individual engineering managers make decisions by using one or more of the methods just described (analytical, rational, or intuitive). However, engineering managers may also elect to make decisions by using the inputs generated by teams. Under such circumstances, additional factors come into play, such as personality clashes, conflicts of interest, and coalitions or alliances among the team members, which affect the resulting decisions. Team leaders need to pay special attention to a set of additional guidelines that inform better decision making in group settings.

Garvin and Roberto (2001) advanced the idea that the group decision-making process must be managed properly to consider social and organizational aspects. Doing so will secure the needed support for implementation, which ultimately determines the success of any decision. Three factors are important for the team leader to take into account when managing a group decision-making process: conflict, consideration, and closure.

Group decision making requires a set of leadership talents somewhat different from those demanded in other situations. These include (1) active solicitation of divergent viewpoints, (2) acceptance of ambiguity, (3) the wisdom to end a debate, (4) the ability to convince people of the merits of the decision made, and (5) the ability to maintain balance to embrace divergence and unity—divergence in opinion during the debate and the required unity of participants needed to implement the decision.

Engineering managers are encouraged to follow the preceding guidelines when managing team decision-making processes.

4.3.9 Hidden Traps in Decision Making

The quality of decision depends on a number of factors. These factors include (1) decision criteria not being independent and all-inclusive, (2) weight factors not appropriately assessed, (3) not clearly defining the alternatives and options, (4) not collecting the right information, and (5) costs and benefits not accurately weighed. According to Hammond et al. (2006), there are six hidden traps in decision making, which, if not avoided, can result in poorly made decisions.

1. *Anchoring trap.* Minds are influenced by the initial ideas, impressions, estimates or data, on which people tend to anchor subsequent thoughts. To avoid this trap, managers should view a problem from different perspectives, establishing their own opinions about a given situation before consulting others. They should be open-minded and not anchor an adviser, consultant or network, when soliciting information and counsel. Managers should first think through their own position, before engaging in negotiations.

2. *Status quo trap.* In order to avoid damaging their own egos, people tend to perpetuate and preserve the status quo, as opposed to breaking with convention. To avoid this trap, managers should check objectives and understand the value of the status quo. They should develop alternatives and assess their advantages and disadvantages in order to avoid exaggerating the switch cost of moving away from the status quo.
3. *Sunk-cost trap*. People tend to protect a flawed decision, to avoid admitting a mistake. This trap will be further enforced if penalties for making bad decisions are high in the company. To avoid this trap, managers should solicit independent views from people not involved in creating the sunk-cost events and accept the fact that good decisions may lead to bad consequences. They should make sure that all subordinates are free of sunk-cost biases, and reward decisions of good quality, as opposed to just the consequences of the outcomes.
4. *Confirming—evidence trap*. Seeking out information that favors your own instincts and your own viewpoints. This is the *selective listening* bias that assigns too much weight to supporting evidences and not enough to conflicting ones. To avoid this trap, managers should check all evidence with equal rigor, and consult with independent-minded people who challenge their positions. Managers should also establish their own counterarguments.
5. *Framing trap*. How a question is framed has a significant impact on the recipient's response. A frame can establish the status quo, introduce an anchor, highlight the sunk cost or lead to confirming-evidence. To avoid this trap, managers should try to reframe the question in neutral and redundant ways to avoid distortions and challenge themselves with the same questions but framed differently.
6. *Estimating and forecasting trap.* Some people tend to be overconfident in assessing the probability of uncertain events, others are overcautious. The estimations of others may be overly influenced by dramatic events due to the strong impressions left on their memory. To avoid this trap, managers should consider the likely upper and lower bounds of estimates, exam the validity of these bounds, provide honest inputs and test them to assess impact. Managers should exam all assumptions to make sure that they are not grounded on impressions.

It is essential that decisions are made in a timely manner. Furthermore, actions must be taken to follow through with the decision. Feldman and Spratt (2001) offer an enlightening riddle:

> "Five fogs are sitting on a log. Four decide to jump off. How many are left?"
>
> "Why, one frog of course," or "Five frogs." Both are wrong.

Even though four frogs decided to jump off, nothing is said about their having actually done so. Deciding to do something and actually doing so are two different things. Decision is not equal to action. To add value, a decision must be followed by an effectual implementation.

4.4 COMMUNICATING

The purpose of communicating is to promote understanding and acceptance of facts, impressions, and the feelings being communicated.

When communicating, engineering managers must have a clear purpose in mind and ensure that their message is understood and retained. A proper form of communication needs to be selected, such as one-on-one meetings, phone conversations, written memos, staff meetings, e-mail, videoconferences, Web postings, or net meetings. It is advisable for engineering managers to keep channels of communication open. Managers should be straightforward and honest, respect confidential information, welcome suggestions, anticipate resistance to changes, and dispel fears by disclosing full information (Clampitt 2001).

According to Sosa et al. (2007), operational problems such as cost overruns, schedule slippages, and quality problems often result from a failure to provide timely information or resources. Ford and Bridgestone Firestone had failed to communicate properly with each other regarding the tires of Ford Explorer. Airbus miscommunicated with respect to the design of the electrical harnesses of A380 plane fuselage.

There are four key actions to achieve efficacious communication: asking, telling, listening, and understanding. These actions will be addressed next.

4.4.1 Asking

Service systems engineers and leaders should proactively request information as opposed to waiting for information to be given to them. A lack of information can prevent understanding. Open-ended questions—those that cannot be answered by yes or no—should be raised to gain new knowledge. The quality of the questions represents a gauge of the questioner's background, education, and depth of understanding of the issues involved. Voltaire said, "Judge a man by his questions rather than his answers." Good insights come from asking good questions. Great creative thinkers ask more questions than others. One needs to dig deeper and deeper until magic insights reveal themselves. Asking questions is to unpeel the layers of packaging to get to the real heart of the matter, by repeating why, how, where, when, who, what, what if and what else.

4.4.2 Telling

Telling means transmitting information (verbally or in written form, or both) to keep employees informed about matters of concern, inform managers about problems and pertinent developments (e.g., to avoid surprises that would trigger spontaneous and low-quality decisions), and to pass information on to peers. Knowing how to produce efficacious written communications, such as proposals, executive summaries, and important e-mails, is more crucial than ever in today's fast-moving and global business environment (Anonymous 2006).

Engineering managers need to exercise judgment in terms of how much information to disclose, as too much can lead to overload and confusion and too little can cause employee mistrust and poor productivity. A typical industry rule of thumb is that information is dispersed on *a need to know basis*. Managers will share information freely if it is required for performing specific work or has an impact on the individual's work environment.

4.4.3 Listening

Engineering managers need to work on their listening skills to beget their understanding of both words (spoken and written) and any possible subtext. They should maintain their concentration by exercising self-discipline and rigorous control of their own urge to talk and interrupt. All of us are given two ears and one mouth; we need to use them proportionally. Woodrow Wilson said: "The ear of the leader ring with the voices of the people."*

Some managers may have skewed perceptions about their openness to unfavorable and challenging news. These managers overestimate their personal openness to receiving difficult messages (unwittingly signal that they don't want to hear bad news) and also underestimate the extent to which the power difference discourages subordinates from speaking their minds—they sensor themselves (Barwise and Meehan 2008). Conducting 360-degree surveys may help to uncover such hidden misperceptions.

4.4.4 Writing

Written communications need to be *concise* (using the least number of words to express the maximum number of concepts), *logical* (allowing easy comprehension), and *pertinent* (focusing on impact to the business purpose at hand). Check the writing advice offered by Felden (1964) with respect to readability, correctness, appropriateness, and thought. HBS Press (2003) offers seven principles of good writing:

1. Have a clear purpose.
2. Be audience focused.
3. State key message clearly.
4. Stay on topic.
5. Observe economy of words.
6. Use simple sentence.
7. Consider the right delivery strategy.

Consult the books by Strunk Jr. and White (2000) and Hacker (2000), with respect to style.

4.4.5 Understanding

The ultimate goal of communication is to promote understanding—to hear with the head and to feel with the heart. Engineering managers need to recognize shared meaning (emotional and logical) and to assess the degree of sincerity by observing body language, intonation, and facial expression.

Four communication barriers exist and should be taken into account by engineering managers:

1. *Interpretations of words and terms*. Words are symbols or semantic labels applied to things or concepts. The same words may have different meanings to different people.
2. *Selective seeing*. Some people have the tendency to see only what they want to see and remain blind to other information unfavorable to the position they take.

**Source*: http://quotations.about.com/cs/inspirationquotes/a/Leadership1.htm.

3. *Selective listening*. Some people hear only what they want to hear by screening out information that may seem threatening to them, thus hobbling their ability to appreciate different perspectives and points of view. Others in conflicts may want to understand only that which allows them to pursue their own self-interest.
4. *Emotional barriers*. All people have emotions. Engineering managers need to appreciate the fact that people's feelings are as important as their intellectual knowledge. Sometimes people's attitudes and feelings may be so strong that they impair their understanding of what is being conveyed. Generally, personal biases will distort the understanding of what is being communicated.

The barriers just cited may cause the communications process to fail in creating the desired degree of understanding. Experience has shown that appeals to emotion tend to be understood and accepted much more readily than appeals to reason, analysis, or cold logic.

To communicate efficaciously, engineering managers are advised to pay attention to the following guidelines:

- *Know what to say and say what is meant.* Engineering managers should focus on key messages when communicating. Avoid noise or meaningless sounds, pointless statements, and inconclusive remarks often used by people to impress others, but not to express themselves. Examples of such noise include, "The answer is definitely a maybe," and, "It is not probable, but still possible."
- *Understand the audience.* Engineering managers should tailor the communication to the receiver's frame of reference—their beliefs, concerns from the job, background and training, attitudes, experience, and vocabulary.
- *Secure attention.* Engineering managers should try to appeal to the receiver's interests; anticipate and overcome emotional objections (fear, distrust, and suspicion); talk in the receiver's terms; and lead from the present to the future, the familiar to the unknown, and the agreeable to the disagreeable.
- *Achieve understanding.* A useful communication technique is to start with agreements and the statement of facts (not conclusions), use simple words (not ponderous, confusing, or abstract terms), and communicate in bursts (avoiding information overflow and knowledge digestion problems).
- *Ensure retention.* The *rule of four* states the following:
  1. Before trying to get an idea across, tell your receivers what you are going to say.
  2. Say what you have to say.
  3. Tell them what you said.
  4. Get them to tell you what they have understood. Obviously, engineering managers must practice such a rule tactfully when the receiver happens to be the company president instead of a young intern engineer who may have just recently started to work.
- *Receive feedback.* Engineering managers need to proactively pose questions and learn to listen in order to get feedback about what was communicated.
- *Get actions to promote communications.* Engineering managers should have the receivers take action on the just-completed communication as a way to secure

its impact. This could be in the form of a commitment by the receivers to take specific steps by agreed-on dates.

Creating understanding is what communication is all about. Engineering managers need to practice posting insightful questions, conveying messages clearly, and listening attentively so that understanding is accomplished at each and every communication endeavor.

Example 4.4 The company decided to move its engineering center to another location, since it was running out of space. The new location was to be modern and had been planned as a showpiece for the company. Management felt certain that the employees would welcome the move. Negotiations were started with several local governmental authorities for suitable accommodations.

To keep the workforce fully informed, it was agreed that the employees would be told that a move was to be made, but that as yet no site had been chosen.

This communication let to wide speculations among the engineers as to the new site, and various rumors circulated. Some engineers with families decided to look for alternative employment elsewhere, fearing that the new location would not be within commuting distance. Morale fell and productivity suffered.

Negotiations took longer than anticipated, and no suitable location had been found after six months. By then, morale was so low that the company decided to abandon its relocation plan altogether. To overcome the space problem, the company split the engineering group by putting a smaller team into another factory site nearby.

What went wrong? How would you have handled this case differently?

ANSWER 4.4 "To communicate or not to communicate"—that was the question. A well-intended but premature relocation announcement induced anxiety in the minds of affected engineers. A lack of progress in site negotiations compounded these anxieties, causing low morale and decreased productivity, leading to an eventual abandonment of the plan.

It would have been better for the company management to keep the plan secret initially, negotiate for and decide on a specific site, and then have the company president announce the relocation plan in a town meeting. The announcement should have included the following:

- The location of the new site, with emphasis on the advantages in transportation, health care, weather, and historical, cultural, and recreational attractions.
- A request for the support of all engineers in making the relocation as smooth as possible. The purpose of the relocation is to provide a better facility for everyone. The company is investing x million dollars to support this move, which will allow for further expansion in the near future.
- The date by which relocation is to be completed.
- A delineation of the company's plans to fund all relocation costs and offer assistance in selling and buying homes, if the relocation is more than 100 miles away. The company will also assist the affected spouses to find jobs at the new site.
- The description of a human resources desk that will be set up to answer specific questions.

Example 4.5 Your department is going to institute a major change. Some members have expressed that they believe the change may be needed. However, in the past, members of the department have tended to resist changes that they did not initiate. The department as a whole has a good performance record. Discuss the advantages of the following alternatives:

1. Permit the members of the department to determine if the change is needed.
2. Let the group make recommendations, but see that your objectives are adhered to.
3. After the group deliberation, adjust the goals, if possible, and monitor performance to see that the change is followed.

Which do you think is most appropriate? Are there other strategies that are more appropriate?

ANSWER 4.5 People resist changes unless they are convinced that the contemplated changes are necessary. They tolerate changes better if the changes are introduced gradually and they have had some say in making the changes.

Out of the three alternatives given, the second is the most appropriate. Management must state clearly the objectives of the planned changes and how the attainment of objectives is to be measured. The members of the department should be allowed to participate in deciding how to achieve the stated objectives.

Company management must set the goals, which are not negotiable. Members of the department are not to be empowered to decide if changes are needed or not. Although the department has a good performance record, change may still be needed for achieving significantly better performance in view of the competition in the marketplace. Thus, the first alternative is not appropriate.

The third alternative is also not appropriate because the goals of the department should not be adjusted on the basis of what members of the department would like to do. Performance monitoring is a valid approach to ensure that the stated objectives are met.

In general, staff participation is useful to ensure an active implementation of the decision made. Thus, a combination of alternatives (1) and (2) may be proper. Management specifies the objectives of changes and the ways the attainment of these objectives are to be measured. Members of the department are asked to make recommendations regarding the best ways to implement the changes. The actions taken by the department are to be monitored constantly to make sure that the stated objectives are met, even if gradually.

4.5 MOTIVATING

The engineering manager secures results by motivating people. Examples of motivators include opportunities to do challenging, interesting, and important work, leadership, position power, prestige, and compensation. Ralph Waldo Emerson said: "Nothing great was ever accomplished without enthusiasm."*

Source: Barlett, John. 2002. "*Barlett's Familiar Quotations*," 17th eds. Little Brown & company, Boston. Page 456.

To motivate is to apply a force that excites and drives an individual to act in preferred ways. In general, emphasis is given to motivational forces that cause the individuals to willingly apply their best efforts.

It is advisable for engineering managers to accept differences in personal preferences, values, and standards, and not try to change people. According to one theory, personality traits are usually firmed up at the early ages of four or five by environmental conditions. Managers should also recognize that every employee has inherent drives to fulfill their own needs, such as self-actualization, recognition, ego, self-esteem, group association, and money.

4.5.1 Methods of Motivating

In general, engineering managers have three key methods of motivation at their disposal:

1. *Inspire.* Infuse a spirit of willingness into people to perform most effectually by way of their own personality and leadership qualities, personal examples, and work completed.
2. *Encourage.* Stimulate people to do what has to be done through praise, approval, and help.
3. *Impel.* Force and incite action by any necessary means, including compulsion, coercion, fear, and, if required, punishments.

The first methods are well suited to motivate professionals, and the last category is not. Being assigned challenging work is a useful motivator for professionals.

4.5.2 Specific Techniques to Enhance Motivation

Engineering managers may implement the following techniques to inspire and encourage professionals to act:

- *Participation.* Invite employees to take part in setting objectives and making decisions. Doing so will ensure emotional ownership and the application of specialized knowledge. Participative management is known to have a positive motivational impact on employees.
- *Communication.* Set clear standards, relate the importance of the work, keep expectations reasonable, and give answers to suggestions offered by employees.
- *Recognition.* Give credit where it is due, as sincere praise tends to promote further commitment. Fair appraisals induce employee loyalty and trust.
- *Delegate authority*. Trust the employees and do not overcontrol them. Achievers will seek additional responsibilities; security seekers will not. Delegate what to do and leave how to do it to the individuals. Delegate technically doable work only to those who want it. Empowerment is known to be an effective motivator to employees (Bowen and Lawler 1995).
- *Reciprocate interest.* Show interest in the desired results to motivate employees to achieve these results.

4.5.3 McGregor's Theory of Worker Motivation

Confucius says, "Reciprocity is the foundation of human relations." One very forceful method of employee motivation is indeed to offer help needed by the employees, who will surely be inclined to reciprocate. The key is then to define such needs.

McGregor's theory of worker motivation (McGregor 1967) is built on Maslow's need hierarchy model (Smith 2008). According to Maslow, a person's needs may be grouped into hierarchical levels, as follows:

1. *Physiological needs*—hunger, thirst, and need for clothing and shelter.
2. *Safety*—protection from threats and danger.
3. Social—giving and receiving affection, group membership, and acceptance by peers.
4. *Esteem*—ego and self-confidence to achieve recognition.
5. *Self-actualization*—continued self-development and realization of one's own potential.

A satisfied need no longer dominates the individual's behavior, and the next higher-level need takes over. But a higher level need only arises when lower ones are already satisfied. The central premise of the Maslow hierarchy model is that an unsatisfied need acts as a motivator. Accordingly, a need-based motivation strategy suggests that engineering managers should learn to understand the specific needs of their professionals at any given time and find ways to help satisfy these needs.

Experience has shown that the motivation strategies presented here can be helpful in motivating professionals who typically have high-level needs related to self-actualization and esteem:

- Present a variety of work assignments perceived to be desirable and that offer the opportunity for personal growth.
- Offer work that has a broad enough scope for the employee to cultivate self-expression and individual creativity.
- Manage with minimum supervision and control, as professionals favor independence and individuality. Professionals tend to prefer having the freedom to make their own decisions and choose their own work methods for achieving the stated objectives.
- Provide work that fully utilizes the individual's professional experience, skills, and knowledge.
- Assign work that enables the employee to receive credit and peers' recognition. Examples include teamwork, publication of technical articles, patents, company awards, and activities in professional and technical societies.

By contrast, pay and benefits have only a minor impact, as physiological needs do not serve as a motivator for most professionals. Because the higher-level needs are never completely satisfied, engineering managers have ample opportunities to motivate professionals to act with their best abilities in achieving the corporate objectives.

Example 4.6 Company X recently installed an incentive system in the production department. Each person receives incentive payments (in addition to hourly wages) for any work done beyond the work standards established for each job. After one month, the production manager noted that there was only a meager increase of 4.5 percent in production. How would you comment on this result?

ANSWER 4.6 The incentive program appears to have failed in realizing the projected benefit. This could be due to one of three key reasons:

1. The incentive offered may be too small relative to the base hourly wage that these workers have been earning all along.
2. Management may have made the mistake of not having consulted with workers to understand their specific hierarchy of needs. Additional pay may not be a strong motivator to them in comparison to other nonmonetary factors, such as peer recognition, self-expression, social acceptance among peers, and others. It is known that team participation (such as a quality circle) has been a strong motivating factor for countless production workers in the automobile and other industries.
3. If the production workers are unionized, the union leadership may have played a role in discouraging workers to compete against each other for pennies.

It would be worthwhile for the production manager to set the target of a desirable productivity improvement at, for example 10 percent, over the next 12 to 18 months. Then, the production managers should form a team that is empowered to suggest recommendations regarding the specific ways to achieve the stated improvement goals in productivity. The team should be made up of workers on the plant floor, union leader, production engineers, and others who have direct knowledge of the production process involved. By having participated in such teamwork, workers on the plant floor become part owners of the resulting action plan. The resulting plan is more likely to be successfully implemented.

4.6 SELECTING ENGINEERING EMPLOYEES

On the one hand, the long-term success of an engineering organization depends on the employees' abilities and the effectual use of these abilities. On the other hand, job satisfaction is known to have a profound impact on employees' willingness to apply their skills to the best of their abilities.

Through employee selection, engineering managers have some control over employees' abilities, their willingness to apply their best efforts, and their job satisfaction. Engineers who are likely to be productive and happy workers in corporate settings are those who are firmly dedicated to their assignments have excellent interpersonal skills, the team player mentality, sound basic training, and the capability to learn new things quickly.

4.6.1 Selection Process

Typically, the employee selection process includes the following steps:

1. *Define needs.* Specify the needs of the new positions by taking into account the immediate requirements and long-term growth demands of the organization.

2. *Specify jobs.* Compose a job description for each of the open positions to define the roles and responsibilities of the position holders, the position grade levels, and the minimum qualifications of the ideal candidates (i.e., levels of basic training and work experience).
3. *Acquire applicants.* Publicize job openings in newspapers, professional publications, company Web sites, employment agencies, and Internet job sites to solicit candidates.
4. *Review and prescreen.* Select applicants by matching personal objectives with company goals. Check documents and references carefully.

   For entry-level candidates who are recent college graduates with no professional experience, diverse industrial employers place a significant amount of weight on the grade point average (GPA). Some companies have even specified a minimum GPA level as a prescreen criteria.

   This overemphasis on the GPA is probably due partially to an ignorance of better, more objective criteria than the GPA in assessing the mastery of basic course subjects. This overemphasis is also due to the notion that the GPA is a composite reflection of the level of personal responsibility demonstrated by an individual in doing his or her principal job of learning during the college years.
5. *Conduct interviews.* Each applicant may be interviewed by several managers. The basis of assessment is typically *studying the past to predict the future*. The quality of past work is a very good predictor for the future, as people are known not to change significantly for the better overnight. Here are a few useful guidelines:

   - Query about the candidate's capabilities pertaining to the new position.
   - Listen carefully to what the candidate says during the interview. Avoid spending too much time selling the job opportunity to the candidate.
   - Prompt the candidate to characterize his or her last job. Be cautious of candidates who speak negatively about their past employers.

     For example, a recent college graduate may complain loudly about his or her research-centered alma mater's negligence in undergraduate teaching and use it as a reason for the individual's poor GPA records. This individual may be likely to behave as a blame-shifting, finger-pointing, and irresponsible individual in a professional environment.
   - Suggest that candidates tell you something negative that you should know about them. Look for honesty. Determine if the candidate is aware of his or her personal flaws, and what active steps have been taken to correct them.
   - Urge the candidates to explain what they would do if they got in over their heads at work. An employee who turns to a colleague is a team player. An employee who turns to a supervisor behaves like a child. An employee who isolates himself when in trouble can be extremely damaging to the business.
   - Encourage candidates to elucidate their aspirations. Raise the question, "Where do you see yourself in five years?" or "What are your future goals?" Companies need employees who can grow and evolve over time.

A candidate who does not have goals or ambitions may resist learning new skills or taking on additional responsibilities.

6. *Decide on job candidates.* Match the candidate's personality, technical capabilities, work ethics, values, and other qualities with those of the company.

Generally speaking, the selection process just described, which is widely practiced in industry, has not always yielded desirable results for employers. Typically, four to five managers may interview an engineering applicant during a one-day site visit. For employers, the easy part is to assess the candidate's technical capabilities, as such capabilities are readily supported by documents (e.g., academic records, internship reports, thesis, publications, and reference letters from professors and company executives). The more difficult part is the assessment of the individual's soft skills, as presented in the next subsection.

4.6.2 Soft Skills

Engineers' future success in a company is strongly affected by their *soft skills* in team work, interpersonal relationships, leadership quality, collaborative attitude, mental flexibility, and adaptability. These soft skills are linked to the engineers' personality traits, psychological profiles, value systems, and deep-rooted beliefs. However, companies generally do not require candidates to undergo specific psychological tests, and most interviewers are not trained to assess candidates for soft skills.

Part of the difficulty in assessing the soft skills of engineers is brought about by the engineers themselves. Nowadays, most engineers, armed with the knowledge of interviewing guides, know quite well how to "talk the talk and walk the walk" in interviews. They have polished responses to almost any type of questions in interviews and are thus proficient in displaying the characteristics they believe a variety of employers are looking for.

Results in literature have indicated over and over again that most professionals who failed in industry—those who have been laid off or voluntarily quit due to personal dissatisfaction—were deficient in soft skills, not in technical capabilities. Future engineering managers need to learn more about how to assess the soft skills of candidates.

Some companies have devoted significant efforts to addressing this issue. Shown here are industrial practices that delineate what two progressive companies—Mazda Motor Manufacturing Corporation (U.S.A.), Flat Rock, Michigan, and Diamond Star Motors Corporation, Normal, Illinois—have done to assess the soft skills of their candidates and the selection criteria they used when selecting these blue-collar workers (Hampton 1988):

- *Interpersonal skills*—ability to get along with people.
- *Aptitude for teamwork*—team dedication and participation, focus on the impact on the team and company instead of individual performance.
- *Flexibility*—learn several jobs, change shifts, and work overtime.
- *Drive to improve continuously*—make and take constructive criticism.

The basic strategy followed by these companies is to *pick the best employees and train them well*. It is noteworthy that "best" is defined by the soft skills of the candidates, not by their hard (technical) skills. These companies select 1300 candidates out of 10,000 applicants at a cost of $13,000 per person, using a multiphase process involving tests, exercises, and role playing in group activities.

4.6.3 Character

The general public has found renewed interest in business ethics sparked by the questionable practices of companies such as Enron, Global Crossing, Adelphia Cable, Arthur Andersen, and others.

Chapter 10 provides detailed delineations of various ethical issues. However, it is important to note here that it serves companies better in the long run to hire employees with character and then train them to acquire the requisite technical skills to become productive.

Guthridge et al. (2008) stressed the importance of making talent a strategic priority for all companies. Although companies say that employees are their biggest source of competitive advantage, the reality is that nowadays, few of them are prepared to meet the challenges of finding, motivating and retaining capable workers. The talent question is complicated by the following realities:

- Candidates for the engineering and general-management positions in emerging markets have wide variations in suitability. Poor English skills, dubious education qualifications, and culture issues (lack of teamwork experience, reluctance to take initiative or assume leadership roles) are among the barriers to suitability.
- Generation Y (born after 1980) demand more flexibility, meaningful jobs, professional freedom, higher rewards, and better work–life balance. They tend to switch jobs every two to three years. They represent 12 percent of U.S. workforce. They are substantially harder to manage.
- Global companies need people who are willing to work abroad, hire local talents with an international mindset, while understanding local ways of doing business and local customers—the expanding middle class.
- Knowledge workers make up 35 percent of total U.S. workers. They contribute more profit than other employees. They need minimum oversight and use, produce, and share information. Companies are not uniformly extracting good value from these knowledge workers.
- Companies overwhelmingly focus on short-term operations and business problems, neglecting the long-term talent sourcing and career development functions.

Some companies suffer from a vicious cycle of inefficiency: They lack talented employees, thus blocking their growth and creating pressure on management. This added pressure diverts management's attention to short-term issues and sacrifices the long-range health of the company.

Further, Guthridge et al. (2008) suggest that companies should address the needs of talents at all levels of the organization, rather than only at the top levels. They recommend that companies cultivate different value propositions for a number of needed talents to fit different groups of candidates (e.g., opportunities for real decision making, career development, housing, educational benefits and learning and flexible work practices). They also propose that companies strengthen the role of human resources department in terms of acquiring business expertise, so that HR can play a more significant role in shaping corporate talents.

Example 4.7 The company has set up a new organization, which is chartered to spearhead new and innovative services whose importance has been recognized. Because of your personal achievement and stature within the company, you have been empowered by the company CEO to select a person to head up this new organization.

The human resources department has been very helpful in advertising and bringing in new candidates. There are three finalists and all of them appear to be qualified for the job at hand insofar as they possess basic intelligence, maturity, and drive to excellence.

Candidate A has a Ph.D. degree in engineering, several U.S. patents to his credit, and about five years of working experience in industry as a technologist. He is very much self-assured in his field of learning, which is directly related to the services of interest. He thinks rationally but has a certain degree of closed-mindedness. His knowledge of the marketplace and industry is somewhat limited. He sets very high standards for himself and has a low tolerance for mediocre people who do not perform up to the expected levels. His former employer views him as a high achiever. He gets along reasonably well with people. He has no experience managing a group or department.

Candidate B has a M.S. degree in engineering and about ten years of management experience. His major strength is his easy-going manner in dealing with people, and as a consequence, he was able to score great accomplishments by working through people. He communicates well. His technical capabilities are only average and he is not known for his innovative capabilities. Even though he is not a risk taker, he is a well-respected and reliable doer who has been able to get things done in time and within budget. As a team player, he is well liked by people in his former group.

Candidate C also has a M.S. degree in engineering and has about seven years of industrial experience as a technologist. She is hard working and eager to learn constantly. Her technical skills are above average and she has good knowledge about the products of the company. She reads widely and has acquired a broad-based understanding of the market, industry, and technologies. Her communications skills are excellent, as demonstrated by the intelligent questions she raised and insightful discussions she offered at staff meetings. She demonstrates strong capabilities for strategic thinking and is also rather open-minded. She is mindful of the company's long-term goals beyond the day-to-day operations. She likes to take calculated risks when venturing into new uncertain projects. So far she has no experience in managing a group or department.

Which candidate will you choose, if you must choose one from these three? How can you substantiate your decision?

ANSWER 4.7 The very first step to take is to define the decision criteria for selecting the new head of the organization. These criteria must be mutually exclusive and collectively exhaustive, as shown in the table that follows.

Then we need to assign the weight factors to reflect their relative importance. Finally, we will evaluate all candidates against each criterion separately. The weighted score defines the top choice.

Number	Decision Criteria	Weight	A	B	C
1	Domain technical knowledge	9	10	5	8
2	Personal innovativeness and creativity	10	10	5	8
3	Team management competencies	8	5	10	5
4	Get along well with people	8	5	10	8
5	Industrial and market understanding	6	5	8	10
6	Risk-taking mindset	8	8	5	10
7	Communications	5	5	10	8
			389	393	436

Based on the chosen criteria and the weight factors assigned, Candidate C is to be recommended. In reality, there is no perfect candidate for any specific job. Company management is thus well advised to offer some team management training to Candidate C and advise her to practice creative thinking methods.

4.7 DEVELOPING PEOPLE

Developing employees is another important activity of the managerial leader. The objective of developing employees is to shape their knowledge, attitudes, and skills to intensify their contributions to the company and to foster their personal growth. Knowledge is the cognizance of facts, truths, and other information. Attitudes are habitual personal dispositions toward people, things, situations, and information. Skills are the abilities to perform specialized work with recognized competence.

In well-organized companies, managers are evaluated on the basis of several performance metrics, including how they have taken care of the development needs of their employees. To be successful, employees must demonstrate initiative in seeking to continuously improve their own knowledge, attitudes, and skills. Examples of what can be done practically are presented in the next subsections.

4.7.1 Employees

There are several ways in which engineering managers may help develop employees. Employees may be prompted to follow the personal examples of continuous improvement in knowledge, attitude, and skills set by engineering managers. Engineering managers may coach inexperienced employees on the job by demonstrating preferred ways of performing specific assignments. In addition, engineering managers could enrich employees' work experience by institutionalizing job rotation. If the company budget and policy so allow, the engineering managers may send specific employees to attend professional meetings, technical conferences, training seminars, and study programs at universities. Furthermore, team assignments may be used to permit a better utilization of the employees' talents and expertise to other critical projects, while offering the employees an opportunity to become known to a larger circle within the company.

In training employees, engineering managers need to emphasize employee participation, as the goal is to achieve the employee's objectives while simultaneously attaining the company's objectives. Employees should be appraised with respect to their present performance in determining what steps might be needed to qualify them to make greater contributions in the future. If the employee's current performance is deemed to be inadequate, managers need to be positive and forward looking in helping the individual recognize the need for self-improvement. By setting a personal example of continuous improvement, the manager is likely to positively motivate the individual to seek further development.

4.7.2 Successors

Besides training employees, engineering managers are also expected, as a part of their managerial duties, to find suitable candidates within their organizations to succeed themselves one day. This is consistent with career planning programs that some industrial companies are actively implementing in order to promote leaders from within, discourage turnover, and maintain continuity.

4.8 SPECIAL TOPICS ON LEADING

Corporate change needs strong leadership. Leaders promoted into new positions need special strategies to succeed. Successful leaders share certain common attributes. The next subsection elucidates the special topics related to the managerial function of leading.

4.8.1 Trust

Surveys reported by Hurley (2006) suggest that about 50 percent of company employees do not trust their leaders. A high-trust work environment is known to be fun, supportive, productive, and comfortable, whereas a low-trust environment is characterized by being stressful, threatening, divisive, unproductive, and tense.

Let us define a *truster* as the person who makes the decision to trust others or not, whereas a *trustee* is the person who is asking for trust. Managers and employees may both be trustees and trusters when they interact professionally. Everyone needs to recognize that trust is the result of a rational decision-making process, and a mental calculation, conducted by all trusters, based on a trust model that contains the following ten factors:

1. *Risk tolerance.* The lower the risk tolerance level of the truster, the harder it is for this person to trust anyone. The Japanese are known to be lower risk takers than U.S. citizens. Spending time to explain options and uncertainties involved and offering some sort of safety net may help
2. *Level of adjustment.* The level of adjustment refers to the degree in which the truster has adjusted to the environment This factor affects the amount of time needed by the truster to build trust. A higher level of adjustment leads people to have more self-confidence and highly self-confident people tend to

trust others sooner. Trustees need to be patient. It can be useful to recognize the achievements of the truster and correct his/her failures through coaching in order to build confidence.

3. *Relative power.* A trustee at a higher power position will easily trust others because he can sanction the trustee who violates his trust. In such as case, it may be useful for the trustee to provide a list of choices and goals outlining organizational interests, so that the truster could find more comfort in becoming engaged in the relationship.
4. *Security*. The higher the security stakes, the less likely people are to trust. Today, job security is less secure, so people trust others less. Trustees need to find ways to temper the uncertainties inherent in this situation.
5. *Number of similarities*. People trust others who are similar. Examples of similarity include: strong work ethics, membership in a defined group (church, work unit, professional association, gender, shared personality traits, etc.). Companies can augment the mutual trust of employees by emphasizing a unifying culture that is focused on candor, integrity and fair processes. Using the word "we" instead of "I" may help.
6. *Alignment of interests.* The truster usually weights the question: "How likely is this trustee to serve my interests?" An alignment of interest promotes mutual trust. A good company leader will turn critical success factors for the company into a common interest that is shared by employees in order to promote mutual trust. Misaligned interests exacerbate doubt and suspicion. A transparent decision making process leads to higher levels of organizational trust (e.g., Kepner–Tregoe method).
7. *Benevolent concern*. People are usually self-centered. Demonstrating benevolent concerns for employees (new experience, skills, advancement preparation, etc.) will engender not only trust, bust also loyalty and commitment. Engaging in fair process will help.
8. *Capability.* The trustee must be cognizant of similarities, aligned interests and benevolent concern in order to generate trust. Managers assess subordinate's capabilities before deciding to trust or delegate authority to them.
9. *Predictability and integrity*. The degree to which the trustee's action can be predicted has a direct impact on trust (doing what one says one will do). People who overpromise and underdeliver are not trustworthy. Demonstrating consistency and rationality in behavior will help.
10. *Communication.* Open and honest communication tends to support the decision to trust. Increasing the frequency and candor of communications will help.

When creating and promoting personal networks or establishing new organizational units, some of these factors may be used in the selection process. All ten factors may be influenced by the actions of trustees. This is where trustees need to focus their efforts on winning over the trusters.

4.8.2 Leading Changes

In twenty-first century corporate America, internal changes occur frequently and often in reaction to changes in the external environment. Changes are typically forced on companies by the market entry of new competitors, the declining market-share position of the company, the emergence of new technologies threatening the company's products or services, financial performance that is worse than expected (e.g., measurements of gross margin, earnings, and other indices), and other factors (Capodagli and Jackson 2001; Murphy and Murphy 2002; Hesselbein and Shrader 2008).

Changes result in modified ways that the company conducts business. Changes are difficult to introduce because people like to stay in their comfort zones. After changes are introduced, it is important for leaders to see to it that the changes are sustained beyond the transformational period. Corporate changes demand strong leadership. According to Kotter (2007), there are two reasons that countless transformation efforts fail:

1. *Large-scale corporate transformation takes time.* In one specific example, the maximum number of changes in a corporation was reached in the fifth year of transformation. (See Fig. 4.3.) Leaders must be patient in marshalling corporate resources to push forward.
2. *Corporate transformation must follow a process of eight consecutive steps to succeed*. Every one of these eight steps is critical, as failure in any one will affect the overall transformation performance.

The process of eight consecutive steps delineates essentially the success factors for leading a transformational change:

1. *Establish a sense of urgency.* Leaders must examine the market and competitive realities and identify the crises, potential crises, or major opportunities available in the global market. The goal is to convince at least 75 percent of the

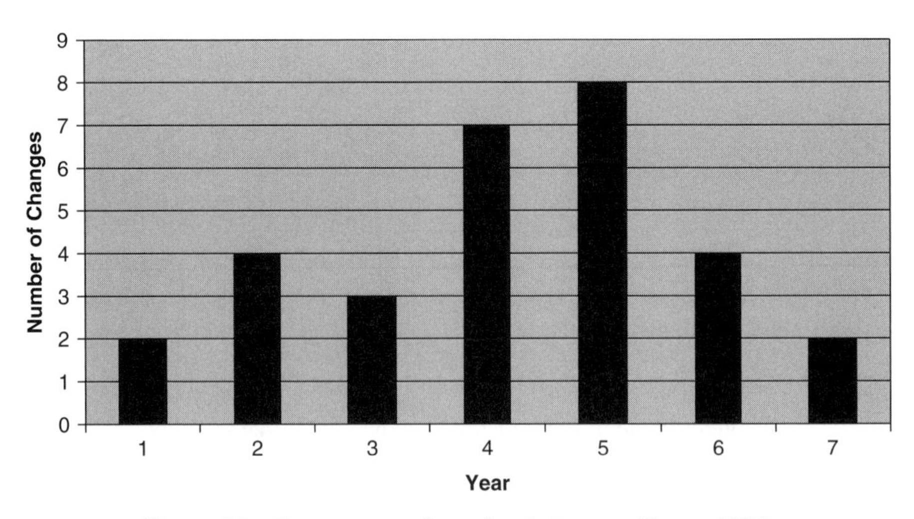

Figure 4.3 Corporate transformational changes. (Kotter, 2007)

management that remaining in the status quo is more dangerous to the health of the corporation than launching in a new corporate direction.

2. *Form a powerful guiding coalition.* Major renewal programs often start with one or two people. But a leadership coalition must grow over time. In addition to the top leaders, there should be another five to fifty people committed to renewal. The group must be powerful in term of titles, reputations, and relationships. The renewal process will move forward only when there are enough leaders in the senior ranks.

   Leaders should inspire the group to work together as a team. The specific goal is to secure shared commitment to change by the top management and by the most influential people. A line position holder must guide the coalition.

3. *Create a vision.* A coherent and sensible vision is needed to help direct the effort to change. Leaders need to cultivate strategies to achieve that vision. The vision should be easy to communicate to stockholders, employees, and customers. Ideally, it should be explainable to an audience within five minutes and achieve their understanding and acceptance.

4. *Communicate the vision.* Transformation needs a lot of people to make it happen. Employees need to be persuaded and motivated to help make the changes. The vision needs to be repeated whenever there are opportunities—through a newsletter, review meetings, training seminars, company picnics, and other means. Leaders should use every channel possible to communicate the new vision and strategies, teach new behavior by the example of the guiding coalition, and "walk the talk," as communication occurs in both words and deeds.

5. *Empower others to act on the vision.* Leaders need to do away with obstacles to change, modify systems or structures that seriously undermine the vision, and promote risk taking and nontraditional ideas, activities, and actions.

   Examples of obstacles to remove include (1) structure (narrow job categories), (2) compensation and appraisal systems, and (3) managers who refuse to change.

6. *Plan for producing short-term wins.* Leaders need to plan for visible performance betterment, achieve these improvements, recognize and reward employees involved in the progressions, and achieve at least some success within the first one to two years. Otherwise, the renewal effort may lose momentum.

7. *Consolidate improvements and procreate still more change.* Leaders need to use increased credibility to change systems, structures, and policies that do not fit the vision; hire, promote, or train employees who can implement the vision; and reinvigorate the process with new projects, themes, and change agents.

   Leaders should resist declaring victory too early to avoid killing the momentum.

8. *Institutionalize new approaches.* Leaders should articulate the connections between new behaviors and corporate success, establish the means to ensure leadership development and succession, and anchor the changes in company culture (values, behaviors, and social norms) so that the changes continue into the next generation of top management.

The eight-step process just described is recommended for top-level engineering leaders who are planning to initiate and implement major corporate transformational changes. With minor modifications, this process applies also to midlevel engineering managers who may be called on to change the performance of a division or a department.

4.8.3 Advice for Newly Promoted Leaders

If a new engineering manager is hired from the outside to take over a department or division, he or she might need to follow a special strategy during the transition period (e.g., the first six months on the job) in order to be productive. This is because going into an unfamiliar situation is akin to sailing in dense fog and only having visibility for a short distance.

Watkins (2001) points out that new leaders often make a number of common mistakes. These mistakes include being isolated, having "the answer," not strengthening the team, attempting too much, trusting the wrong people, and setting unrealistic expectations. In order for an engineering manager to become productive in a new organization, Watkins offers specific advice consisting of seven rules: leverage time before entry, organize to learn, secure early wins, lay a foundation for major advancements, envisage a personal vision, build winning coalitions, and manage oneself.

For newly promoted engineering managers, the leadership strategy just outlined serves as a useful guiding light during the initial six-month period of sailing through dense fog. Although the steps recommended here cover all important aspects of a new leadership job—technical, cultural, political, and personal—individual engineering managers may need to further customize these steps to fit their personal style, organizational needs, and the people involved in a given situation.

4.9 CONCLUSIONS

Leading is another key function of engineering management. It encompasses the specific managerial activities of making decisions and selecting, developing, motivating, and communicating with people. Carrying out these specific activities well will transform an engineer into a strong engineering manager.

Decision making plays an important role in the career life of an engineering manager. In the engineering community, the rational decision-making method is regarded as a standard. Engineering managers need to become familiar with a number of other decision support tools so that the right tool can be fittingly applied to specific circumstances.

This chapter offered guidelines for engineering managers to become better prepared for the special cases of (1) building trust, (2) introducing major corporate changes, and (3) working as a new leader in an engineering management environment.

Engineering managers are advised to practice various guidelines associated with leadership whenever they find opportunities to do so.

4.10 REFERENCES

Anonymous. 2006. *Written Communications that Inform and Influence the Results-Driven Manager Series* (paperback). Cambridge, MA: Harvard Business School Press.

Bardaracco, J. J. 2002. *Leading Quietly: An Unorthodox Guide to Doing the Right Thing*. Cambridge, MA: Harvard Business School Press.

Barwise, Patrick, and Sean Meehan (2008), "So You Think You're a Good Listener." *Harvard Business Review* 86 (4): (April).

Bowen, David E. and Edward E. Lawler III, (1995), "Empowering Service Employees," *Sloan Management Review* 36 (4): (Summer).

Brousseau, Kenneth R., Michael J. Driver, Gary Hourihan, and Rikard Larsson. 2006. "The Seasoned Executive's Decision-Making Style," *Harvard Business Review* 84 (2): (February).

Capodagli, B., and L. Jackson. 2001. *Leading at the Speed of Change: Using New Economy Rules to Transform Old Economy Companies*. New York: McGraw-Hill.

Clampitt, P. G. 2009. Communicating for Managerial Effectiveness: Problems, Strategies, Solutions, 4th ed. Thousand Oaks, CA: Sage Publications.

Donker, L. 2007. *Strategic Decision-Making for Excellence*. Bloomington, IN: Author House.

Felden, L. 1964. "What Do You Mean I Can't Write?" *Harvard Business Review* (May –June).

Feldman, M. L., and Spratt, M. F. 2001. *Five Frogs on a Log: A CEO's Field Guide*. New York: John Wiley & Sons, Inc.

Figueira, J., Salvatore Greco, and Matthias Ehrgott (eds.). 2004. *Multiple Criteria Decision Analysis: State of the Art Surveys.* New York: Springer.

Friedman, S. D. 2008. *Total Leadership: Be a Better Leader, Have a Richer Life.* Cambridge, MA: Harvard Business School Press.

Garvin, D. A., and M. A. Roberto. 2001. "What You Don't Know about Making Decisions." *Harvard Business Review* (September).

Guthridge, Matthew, Asmus B. Komm, and Emily Lawson. 2008. "Making Talent a Strategic Priority." *McKinsey Quarterly* (January).

Hacker, D. 2009. A Pocket Style Manual with 2009 MLA Update, 5th ed. Boston: Bedford/St. Martin's.

Hammond, John S., Ralph L. Keeney, and Howard Raiffa. 2006. "The Hidden Traps in Decision Making." *Harvard Business Review* 84 (1): (January).

Hampton, W. J. 1988. "How Does Japan Inc. Pick Its American Workers?" *Business Week* (October 3).

Hayashi, A. H. 2001. "When to Trust Your Gut." *Harvard Business Review* (February).

HBS Press. 2003. "Good Writing: It Begins with Principles." Chapter 1 in *Business Communications.* Cambridge, MA: Harvard Business School Press.

Hesselbein, F., and Alan Shrader. 2008. *Leader to Leader: Enduring Insights on Leadership.* San Francisco: Jossey-Bass.

Hurley, Robert F., 2006. "The Decision to Trust." *Harvard Business Review* 84 (9) (September).

Kotter, J. P. 2001. "What Leaders Really Do." *Harvard Business Review* (HBR Classic) (December 1)

Kotter, J. P. 2007. "Leading Change: Why Transformation Efforts Fail." *Harvard Business Review* (HBR Classic) (January 1).

Matzler, Kurt, Franz Bailom, and Todd A. Mooradian. 2007. "Intuitive Decision Making." *MIT Sloan Management Review* 49 (1) (Fall).

McGregor, D. 1967. *The Professional Manager*. New York: McGraw-Hill.

Murphy, E. C., and M. A. Murphy. 2002. *Leading on the Edge of Chaos: The 10 Critical Elements for Success in Volatile Times.* Upper Saddle River, NJ: Prentice Hall.

Nagal, S. S. 1993. *Computer-Aided Decision Analysis: Theory and Applications*. New York: Quorum Books.

Ragsdale, C. 2007. *Spreadsheet Modeling and Decision Analysis: A Practical Introduction to Management Science,* 5th ed. Stamford, CA: South-Western College Publishing.

Rogers, Paul, Maria Blenko. 2006. "Who Has the D?" *Harvard Business Review* 84 Issue (1): (January).

Shannon, R. E. 1980. *Engineering Management*. New York: John Wiley & Sons, Inc.

Smith, C. P. (ed.). 2008. *Motivation and Personality: Handbook of Thematic Content Analysis.* London: Cambridge University Press.

Snowden, David J., and Mary E. Boone. 2007. "A Leader's Framework for Decision Making." Harvard Business Review 85 (11) (November).

Sosa, Manual E., Steven D. Eppinger, and Craig M. Rowies. 2007. "Are Your Engineers Talking to One Another When They Should?" *Harvard Business Review* 85 (11) (November).

Strunk, Jr., W. and E. B. White 2008. *The Elements of Style,* 50th Anniversary Edition, White Plains, NY: Longman (October 23).

Tregoe. B. B., and C. H. Kepner. 2006. *The New Rational Manager*. Paperback Princeton, NJ: Princeton Research Press.

Watkins, M. 2001. "Seven Rules for New Leaders." *Harvard Business School Notes*, (revised) No. 800-288, June.

Zaleznik, A. 2004. "Managers and Leaders: Are They Different?" *Harvard Business Review* (HBR Classic) (January 1).

4.11 APPENDICES

4.11.1 Factors Affecting One's Influence on People

This section deliberates on the various factors known to affect a person's influence on people. It is advisable that the engineering manager pay attention to them.

A. *Credibility*. A person's credibility is based on the following six attributes:

1. *Composure (ways to handle oneself).* Degree of poise, stability, and patience; skills to handle a crisis situation effectually; humor under stress; self-confidence; and ability for public speaking are important attributes. Composure is the most important factor impacting credibility in the short term.
2. *Character* Integrity and honesty (not lying, not cheating, attempting to do things above board and maintaining high-moral standards), cooperative spirit, and professional behavior—return all phones calls and respond to all mail; keep promises, have an open and forthright attitude; and be fair in all situations. Character is the most important factor affecting credibility in the long run (integrity and honesty in professional versus private matters).

Table 4.A1 Competence Factors on Influence Exerted Upward (Accuracy +/- 10%)

	First-Line Supervisor (percent)	Mid-Manager (percent)	Executive (percent)
Technical	70	30	5
Managerial	25	40	25
Visionary	5	30	70

3. *Competence*. Technical (e.g., job-specific skills, experience, and training), managerial (e.g., planning, organizing, leading, and controlling), and visionary (e.g., capability to envision the future with strategic thinking). Table 4.A.1 displays the competence factors that exert an influence on superiors.
4. *Courage*. Commitment to principles; the willingness to stand up for beliefs, challenge others, and admit mistakes; and the ability to make tough decisions under uncertain conditions and accept responsibility for the consequences.
5. *Conviction (beliefs)*. Commit to the vision, demonstrate passion, and display confidence in the direction being pursued.
6. *Care for people*. Know people (family, aspirations, current and future needs, favored learning modes, upward mobility, etc.), treat people with respect and dignity (listening to understand), comment only on issues and situations and not on the person, and be a team player.

B. *Personal power (independent of position power)*. Personal power is affected by the following three factors:
   1. *Personal attributes*—physical appearance and size, drive, dedication, and personal values.
   2. *Knowledge*—common sense, historical perspective, political knowledge needed by others (how to get things done through which doors, by what means, and with whom—otherwise known as tricks of the trade.
   3. *Relationships*—business connections and power by association.

C. *Variable leadership style*. Leadership style needs to be varied in accordance with the circumstances involved. In general, there are four situations that each requires a different style of leadership. (See discussion in section 9.5.1).

The influence exerted on people by an engineering manager is affected by his or her credibility, personal power, and leadership style. Engineering managers need to do the right thing at the right time and place, to the right people, for the right people, or with the right people.

4.11.2 Motivation of Mission-Critical People

In the competitive world today, all companies struggle to attract and retain innovative knowledge workers who are critical to the mission of their operations. To be successful, companies in general, and engineering managers in particular, need to tailor specific motivation strategies to the needs of individuals. In general, most knowledge workers have the following types of needs.

1. *Need for power (40 percent)*—setting goals and offering positive recognition to allow one to stand out and be unique.
2. *Need for affiliation (40 percent)*—focusing on mission, vision, and the difference the individual can make in teams.
3. *Need for self-achievement (20 percent)*—offering task variety, learning, development, and growth opportunities.

Nortel Networks and Cisco Systems, Inc. experienced high turnover (about 40 percent) in their information technology sector. Surveys point out that people with mission-critical jobs left because of three specific deficiencies:

1. *The work itself*—meaningfulness, relevancy, learning opportunity, enjoyability, variability, and so on.
2. *Appreciation*—thanks, recognition; being told that their jobs were mission critical.
3. *Money*—just compensation.

The preceding list of needs and the types of deficiencies causing knowledge workers to want to leave are consistent with the Maslow need hierarchy model reviewed in section 4.5.3. It is obvious that unsatisfied needs will strip employees of motivation if they stay unsatisfied for long. Under such circumstances, knowledge workers are likely to migrate to places where they can satisfy these unmet needs.*

4.12 QUESTIONS

4.1 Study the data given in Harvard Business School Case "GM Powertrain" #9-698-008 (Rev. April 2000) and answer the following questions. The case materials may be purchased by contacting its publisher at 1-800-545-7685 or www.custserve@hbsp.harvard.edu.

A. What is this case about?

B. Joe Hinrichs's objectives—What are his objectives as a plant manager? What happens if these objectives are not met? What are the conflicting pressures faced by a plant manager like Joe?

C. Joe Hinrichs's specific change processes—What are the major changes introduced by Joe Hinrichs? Is there anything unique or novel about his method of introducing changes? Are these changes significant for the plant? Can these changes be sustained?

D. How would you assess Joe Hinrichs's leadership capabilities?

E. What is to be done with respect to the 1,500-ton press? What are the options available? How should Joe Hinrichs make his decision?

F. Lessons learned—What are the lessons you have learned after having studied this case?

4.2 As advised by the company president, the sales department received a set of specific recommendations provided by an outside management firm to reorganize for maximum effectiveness the sales manager believes that a few of the sales staff may disagree with

---

**Sources:* P. McKenna and D. H. Maister. *First among Equals: How to Manage a Group of Professionals*. New York: Free Press, 2002; J. M. Ivancevich and T. N. Duening. *Managing Einstein: Leading Edge High-Tech Workers in the Digital Age*. New York: McGraw-Hill, 2002; D. M. Soat. *Managing Engineers and Technical Employers: How to Attract, Motivate and Retain Excellent People*." Norwood, MA: ArTech House, 1996.

the recommended changes. The sales manager herself is also not fully convinced of the merits of all of the recommendations, but she wants to implement them, at least in part. How should she proceed?

4.3 The engineering director of the company is called on to send one engineer abroad to assist in the installation of equipment. There are three qualified candidates, each working for a different manager under the director. The director knows that all three engineers will want to go, but their superiors will oppose any of them going for fear of losing time in completing their own critical projects. How should the director make the choice?

4.4 The marketing department needed to submit a proposal to a global customer, and it called a review meeting the next morning. By the time Bill Taylor, the design manager, was informed in the late afternoon, all of his design staff had left for the day and there was no one available. Bill Taylor decided to work on the proposal himself through the night so that he could talk with his design staff the next morning, one hour before the marketing review meeting.

All of the staff agreed with the proposed design except Henry King, a senior staff member who I recognized as the most experienced and best designer in the group. His objections were that the current design was too complex and that it would take another week to modify on the design to ensure its functional performance.

In order to pacify him, Bill Taylor invited Henry King to come along to the marketing review meeting so that Henry King would feel the pressure that marketing was exerting on design. Unexpectedly, Henry King stood up at the marketing review meeting and reiterated all of his design objections, causing tremendous embarrassment to Bill Taylor and his superior, Stanley Clark, the design director. Bill Taylor became furious. What should Bill Taylor and Stanley Clark do?

4.5 Jerry Lucas is the division director. As branch chief, Bob Sanford reports to Jerry Lucas. Bob Sanford has four section chiefs reporting to him.

Bob Sanford is technically competent, with extensive experience in solid rocket propulsion; he is also regarded as the best expert in this field. He is highly dedicated to his work, but inexperienced in managing technical people, as he has been on the job for only two years. Bob Sanford handles his subordinates quite roughly. He reverses his section chiefs' decisions without prior consultation with them. He demands that no information or data be transmitted to persons outside the group without his knowledge and concurrence. He also bypasses his section chiefs to go to people and inspires them to come to him directly with problems. Rumors have it that he places spies or informants within the group. As expected, he delegates no decision-making authority to his section chiefs and regards all of his section chiefs as technically incompetent. He brings into being an atmosphere of fear and suspicion, with low group morale.

Bob Sanford does not report to Jerry Lucas candidly on project progress and on difficulties he encounters. He does not understand his own responsibility of building teamwork, enhancing group morale, and creating employee satisfaction while achieving the goals of his group. He is lacking the skills and willingness to resolve conflicts within the group.

Finally, the section chiefs as a group go in to see Jerry Lucas and complain about the lack of authority and the oppressive atmosphere in the section. What should Jerry Lucas do?*

4.6 The board of directors receives a proposal from a business partner to jointly set up an assembly plant in a developing country. This new plant will assemble final products with

*Condensed and adapted from Robert E. Shannon, 1990 *Engineering Management*, New York: John Wiley & Sons, Inc., p. 227.

key components made by the company. Financial terms are attractive and the future marketing outlook is bright. There is just one problem. The developing country is not a democracy, has a poor human rights record, neglects to protect its own environment, and does not safeguard workers' rights. An investment placed by the company would boost this country's economy and thus the political position of its current dictator. Should the company accept the proposal, and why?

4.7 What are some important characteristics of effective leaders? Which of these characteristics are more difficult for engineers to acquire?

4.8 The plant manager noticed a need to lessen the amount of waste materials, which occurred in the production process. A task force was set up, composed of the plant manager herself and two of her supervisors, to examine the problem. They met for three months and regularly published the task force objectives and findings on the plant bulletin board.

The plant manager found, to her surprise, that the workers on the shop floor exhibited limited interest in the task force and ignored the bulletin board entirely. At the end of the three-month period, the task force came up with several excellent recommendations, which require changes in work practices. Most of the workers implemented the recommended changes very reluctantly, and some even secretly worked to sabotage the new practices. Eventually, all recommendations were withdrawn.

What went wrong? How should the plant manager have handled this case?

4.9 The project was running late and the section manager thought that it was time for a pep talk with his staff. He realized that he was considered to be somewhat autocratic by his staff, but this time he thought that he would impress on them that he was really one of the members of the team and that they would work together as one in order to succeed.

The section manager thought he made quite a good speech. He pointed out that the project was running late and that, if they failed, the customer could cancel the contract. He explained further that, as manager, he was responsible for the success of the project, and so everyone would be equally to blame for the failure of the project.

Unexpectedly, a group of staff came in to see him a few days later to clarify whether they were all under threat of unemployment should it turn out in the future that they were indeed late and the contract was cancelled by the customer.

What went wrong? What would you have done differently?

4.10 A regional sales manager suspected that one of her customers was having financial troubles. However, she was reluctant to mention it to her superior because she felt that she could be wrong. She kept quiet for several months, continuing to take large orders from this customer and hoping that the customer could recover from their troubles. Eventually, the customer went bankrupt and defaulted on the payment of several large bills. What went wrong? What would you have done differently?

4.11 Company X selects someone who is weak technically, but very strong in group-process skills to lead a team in advancing a new engineering product. Would such a person be successful as a team leader? What can be done to ensure that the engineering product advanced by the team will be satisfactory from the technical standpoint?

4.12 Conflicts between technologists and managers may arise when the technical professionals with the skills to make a decisions have to deal with a manager, who has the right to decide. Why do such conflicts often exist in organizations wherein everyone works toward the same common goal?

4.13 Company X makes the decision to substitute aluminum for steel in a component of its product. What factors probably have contributed to this decision? At what managerial level would this decision most likely have been made?

4.14 As the department head you urgently need to find an experienced person to fill a vacancy. The work involves close cooperation and coordination with others inside and outside of the department. Candidate A has exactly the experience required, but appears to be very unsociable. Candidate B has experience in a related job and seems to have a pleasant personality, though not an extrovert. Candidate C has business experience in a different industry and is extremely sociable. All three candidates have scored sufficiently high on intelligence tests to quality for the job in terms of general ability. Which candidate would you choose, and why?

4.15 Joe Engineer has just graduated from the University at Buffalo. He earned a 3.8 GPA for his Master's of Engineering degree. Before he finished all academic work, he sent out numerous job applications and received three specific job offers, A, B. and C, All of them require him to make a decision for acceptance or rejection within one week.

Company A has an annual sales revenue of $5 billion dollars and is located in New York City. The job of "Engineer" pays $95,000 a year, plus full benefits (e.g., 401 (K), health insurance, 4 weeks vacations per year, educations assistance, and relocation assistance). The city living is, of course, exciting and fun, but very expensive. The company has a structured training program for new employees to become familiar with its operations. His future boss is friendly and acts professionally. Joe believes that he can get along well with him. The work is in line with his basic technical training. As it is typically the case with big companies, other employees in the department are all quite smart and the internal competition among co-workers is relatively strong. He envisions that he may be able to get a promotion to the next level in five years time. New York City is huge and there are at least five competitors to Company A offering similar products/services in the city. Another potential benefit of working in New York City is that Joe may be able to meet a lot of interesting young people and find a future spouse.

Company B is a $500 million dollar mid-size company located in Rochester, New York. The job of "Engineer" pays $80,000 a year plus full benefits, which are similar to that of Company A. Rochester is a mid-size city with some cultural and entertainment activities. Its cost of living is reasonable. Rochester is close to the Finger Lakes region, a well-known recreational area. The next large city is Toronto, which is about two hours away by car. The future boss is quite enthusiastic about Joe's employment at the company and is eager to welcome him. The work is of a technical nature, but offers some managerial training opportunities that Joe likes. Joe believes that he could be promoted in the next three years and allowed to assume a higher level of responsibilities thereafter. The city has only a few large companies, such as Xerox, and Kodak, but no company that competes directly against Company B. Joe thinks that it may be a bit more difficult, if not absolutely impossible, to meet a lot of interesting young people in Rochester.

Company C is a $50 million dollar small company located in Buffalo, New York, where Joe did some summer work during his school years. The job of "Engineer" pays $70,000 a year and offers some benefits. The benefits are not as good as in either Company A or B. Buffalo is slightly larger than Rochester, but is still way behind New York City in terms of culture and entertainment activities. Like Rochester, Buffalo suffers from a declining industrial base. There have been no new companies relocating into the Buffalo or the Rochester areas in recent years. Company C has no competitors operating in Buffalo and nearby regions. Joe knows his future boss because of his earlier summer work at the site. The work is quite exciting, as the future boss views Joe as one of the bright new stars and shows a significant willingness to personally train Joe for higher-level roles and responsibilities. Being in a small company, Joe understands that he needs to face up the challenges of getting involved quickly in several disciplines beyond the principal one that forms the basis of his Master's Degree. Because Buffalo is also a regional city, Joe believes that it could be hard for him to meet a lot of interesting young people and subsequently find a mate.

Which job would you elect to accept? Explain the detailed decision making methodologies and reasons you used to arrive at your decision.

4.16 Preparation of the company product that was promised to a major customer is running late, and there is intense pressure on the production team to deliver the product. The director of production is eventually told by the company president to deliver, "or else." The director therefore decides to ship the product, even though it had not gone through all of its testing procedures. Members on the production team become upset due to their uncertainty about the functionality and reliability of the shipped product. The director, however, insists that "We will just have to take that chance."

As the director of production, how would you have acted differently?

Chapter 5

Controlling

5.1 INTRODUCTION

The function of controlling in service systems management and engineering refers to activities taken on by an engineering manager to assess and regulate work in progress, evaluate results for the purpose of securing and maintaining maximum productivity, maintain and upgrade critical talents, and reduce and prevent unacceptable performance.

Although the bulk of the controlling activities appear to be primarily of administrative and operational nature, controlling has strategic importance. To efficiently implement any assignment project, program, or plan, managerial control is crucial. Any forward-looking strategic plan becomes useless if its implementation is poor. Furthermore, without adequate control, managerial delegation is ineffective, rendering the managerial leadership of questionable quality. The function of controlling also contributes to corporate renewal by pruning the dead wood, if needed.

Engineering managers exercise control by carrying out the specific jobs of (1) setting standards, (2) measuring performance, (3) evaluating performance, and (4) controlling performance. In addition, this chapter addresses the manager's control of time, personnel, business relationships, projects, and company knowledge.

5.2 SETTING PERFORMANCE STANDARDS

To set standards is to specify criteria by which work and results are measured and evaluated. Setting performance standards explicitly defines the expected results. It is important for both the company and its employees to distinguish between performance grades of "outstanding," "better than average," "average," "below average," and "unacceptable" and to understand how performance is measured. Here, "average" is defined as the generally expected performance level for the job at hand by persons with adequate training and dedication.

Setting standards provides specific guidelines for exercising authority and making decisions. Standards represent a yardstick for measuring the performance of employees and units (such as cost centers and profit centers). Proper standards facilitate employee self-evaluation, self-control, and self-advancement.

Standards are typically established in the form of how many (number of units), how good (quality, acceptance), how well (user acceptance), and how soon (timing), as imposed by the company management, the customers, or the marketplace. It is worth noting that current trends of setting standards emphasize the inclusion of customer viewpoints.

There are technical, historical, market, planning, safety, and equal employment opportunity (EEO) standards. Technical standards specify metrics related to quality, quantity, mean time between failure (MTBF), maintenance requirements, and so on. Historical standards are based primarily on past records. Market standards are those related to competition, sales, return on investment (ROI), earnings expected by securities analysts, and other factors. Planning standards mostly relates to the strategic and operational planning needs of the company. They address topics such as objectives, programs, schedules, budgets, and policies. Safety standards refer to the metrics related to the safe operation of the company's facilities. Government programs such as the U.S. Occupational Safety and Health Administration (OSHA) promulgate some of these safety standards. EEO standards are specific practices related to affirmative action for the purpose of achieving equal employment opportunity for all.

Effective standards are characterized as being company sponsored, measurable, comparable, reasonable, and indicative of the expectation of work performance on an objective basis in order to promote worker growth. They are also identified as being considerate of human factors by soliciting the workers' inputs and securing their understanding and acceptance.

Indeed, as pointed out by Hammer (2007), a variety of companies make mistakes in setting performance standards. Some metrics are set that will make everyone involved look good. Others focus on local but not global performance and on component but not system performance. Yet others are chosen to be unrelated to customers' needs. Still others are selected without consideration of human factors. Performance standards must be selected to focus on the end-to-end business processes and the cross-organizational sequence of activities that add customers values.

Example 5.1 The engineering director asks her managers if they have any nominations for promotions from within their respective departments. The maximum number of promotions allowed for the entire division is two. The nominations must be selective and only for people whose performance has been outstanding. One manager thought his whole team had been outstanding, so he recommended all ten for promotion. He reasoned, "It is better for the morale of the team that they know that I support all of them fully. If the director now promotes one of the ten or none at all, then they will not feel so bad knowing that at least I have thought them all worthy of being promoted." What should the director do?

ANSWER 5.1 The director should call the manager in and reprimand him for (1) not following the instructions given of being selective in nominating candidates for promotion, (2) neglecting the delegated responsibility of evaluating the performance of his staff fairly and objectively, and (3) wanting to pass on his own responsibility to his director so as to be the "good guy."

She should order the manager to repeat the nomination process and come up with a rank-ordered list of outstanding performers for possible promotion within two days. If he refuses to do so, this will be registered as one incidence of disobedience. Repeated offenses of this type will lead to immediate discharge.

There are quite a few barriers that prevent good standards from being readily established. Standards may be too subjective, as technically strong engineers tend to set unrealistically high standards for themselves and others. On the one hand, if standards are set too high, workers may become demoralized and fearful of not measuring up. On the other hand, if standards are set too low—not challenging enough for the workers—management will lose the respect of employees. Standards may be confusing and may not clearly indicate the criteria for excellence. Standards may be qualitative and vague and thus subject to different interpretations by different people. Standards may also be set without proper consideration for the constraints related to resources and implementation.

A useful way for companies to set proper standards is the practice of benchmarking, discussed in the next section.

5.3 BENCHMARKING

Benchmarking is a method of defining performance standards in relation to a set of internal and external references. There are two types of benchmarking:

1. *Internal benchmarking* uses references internal to a company to set performance standards, as illustrated by the following example:

   Company X has achieved a productivity of $150,000 per employee in 2003 and is determined to continue improving performance on a yearly basis. The company president sets a new performance standard for the year 2004 at $165,000 per employee, a level 10 percent higher than the previous year.

   Internal benchmarking is convenient to utilize, as it offers a reasonable, short-term performance assessment. However, it may bring about a false sense of corporate well-being, as the resulting performance standards, in the absence of an external reference, may be deficient in the long run. That is, the 10 percent productivity increment may be inadequate, or even outright dangerous to the long-term health of the company, if the industrial average productivity gain has been 12 percent per year in the past, and some competitors of Company X are aiming at 15 percent for the coming year.

2. *External benchmarking* uses references external to the company to set performance standards. The following examples illustrate its applications:

   - *Financial ratios.* Assuming that the gross margins of many service-oriented public companies are in the range of 35 to 40 percent for the last several years, as published regularly by Wall Street securities firms, then setting the gross margin target for a specific company at the 40 percent level for 2004 is a performance standard defined by external benchmarking. Setting the gross margin at 40 percent will ensure that the company will be among the best performers in the industry, provided that the standards are met.
   - *Performance metrics.* A large number of performance metrics are published in business literature to evaluate production, product delivery, quality control, time to market, customer service, reliability, customer problem solving, and others. Using these known metrics as references to set performance standards makes explicit the relative competitive position of a company.
   - *Best practices.* Another set of important external benchmarks is called *best practices* (Brown and Heller 2003; Rodier 2000). These are work processes or procedures perfected and upgraded by various companies over the years to address specific problems and issues in engineering, production, marketing, strategic management, business development, and other areas of corporate

governance. These are extremely valuable "tried-and-true" methods to achieve useful results and add value to the companies. Those companies too undisciplined to consistently implement best practices in carrying out their corporate activities may discover that they have become less and less competitive over time.

Best practices can be readily copied by others. Thus, any advantages so accomplished will be of short-term nature, unless these practices are further upgraded or superseded. Best practices which companies acquire typically from the outside involving such processes as project management, strategic planning by team and talent development (Gratton and Ghoshal 2005).

- *Critical success factors.* Serving well as external benchmarks are critical success factors, the necessary and sufficient conditions for achieving success in specific business, engineering, production, and marketing domains. These factors are derived from the successes and failures experienced by companies in various industries. Having a clear understanding of these factors allows a company to make strategic choices to move in new directions, initiate new products to gain advantages, and capture opportunities in new markets by optimally employing corporate strengths while minimizing any exposure of weaknesses.
- *Target pricing.* In recent years, Japanese companies have successfully applied the technique of target pricing, another external benchmarking method, to achieve success in the marketplace. First, the company conducts a survey to determine the current prices of products that are in direct competition with the new products that the company is planning to market. Using these prices as references, the company sets a target selling price (such as 80 percent of the competitive product prices) for its new product. Then, the company deducts the required gross margin for the company to sustain itself while meeting the shareholders' expectations. The resulting dollar amount is the *cost of goods sold* target (the cost target) for the new product. The cost target is then imposed on the production and engineering departments as performance standards for advancing the new product. Funds are available for developing the new product only if the cost target can be met. This is tantamount to an *innovation under duress* model of management control. The resulting new product is ensured a competitive edge in the marketplace.

Traditionally, bringing a new product to the marketplace follows a sequential process. First, the product is designed by engineers on the basis of inputs from marketing. Then, it is redesigned by service engineers for serviceability. Afterward, it is redesigned again by production engineers for manufacturability. Finally, it is mass produced, and its final product cost is accurately estimated. The company management adds the gross margin to set the product prices. At each step, engineers tend to introduce contingencies, or "cushions," to ensure that the work is done properly within their respective units. However, oftentimes this sequential process leads to products that are not cost competitive in the marketplace.

In fact, the basic concept of target pricing can be applied to innumerable other corporate activities to ensure that the company remains competitive. This is particularly pertinent to the current business environment, which is becoming increasingly globalized.

- *Balanced scorecard.* In recent years, business researchers have noted the biases many companies have against monitoring corporate performance primarily by using financial metrics only on activities related to the past. (See section 7.4.) Progressive management needs to devise a balanced set of performance metrics to properly monitor other company activities that also have a profound impact on its future success.

  Examples of such forward-looking performance metrics include the establishment of (1) the percentage of company business generated by products introduced in the last five years, (2) the percentage or amount of corporate funds spent on projects initiated in the last five years, (3) the number of patent disclosures, patent applications, and patent awards in a given year, (4) the number of new supply chain partners engaged in the last five years, (5) the percentage of sales realized from new customers acquired during the past year, and (6) the fraction of product cost arising from new technologies adopted in the last five years.

It is self-evident that setting adequate, forward-looking standards based on both internal and external benchmarking helps guide the company to success in the future.

5.3.1 Sample Benchmarking Metrics

Many sets of performance metrics are available from business literature. Table 5.1 contains selected samples of metrics used in various business domains.

5.3.2 Limitations of Benchmarking

Benchmarking is useful but has certain limitations. For example, some reference data might not be available, and in such cases, estimates must be made. This might cause the value of such benchmarks to be less robust. Benchmarking metrics are always based on past performance, and they do not predict the future. Neither can they be used to predict new competition. However, even with these limitations, past-oriented benchmarks are still valuable. As Confucius says, "Studying the past will lead to an understanding of the future."

Example 5.2 Product quality has a number of dimensions. Which dimensions of product quality have the most impact on the product's success in the marketplace?

ANSWER 5.2 According to Garvin (1984), product quality has the following eight dimensions:

1. Performance (operational characteristics)
2. Features
3. Reliability
4. Conformance to design specification
5. Durability
6. Serviceability
7. Aesthetics (look and feel)
8. Perceived quality (affected by brand name and company reputation)

Which of these quality dimensions affect the product's success in the marketplace will depend on the type of products involved:

A. Automobiles—(7), (8), (1), (6)
B. Consumer products—(8), (1), (6), (3)
C. Industrial gases—(1), (3), (6), (5)
D. Office supplies—(1), (8), (7)
E. Home appliances—(3), (6), (1), (5)

Reference: D. A. Garvin. 1984. "What Does 'Product Quality' Really Mean?" *Sloan Management Review* 26 (1) (Fall): 25–43.

Table 5.1 Sample Performance Metrics

Domains	Metrics
Financial	ROI (return on investment)
	ROA (return on assets)
	ROS (return on sales)
	Debt–to–equity ratio
	Number of inventory turns each year
	Number of units produced per employee
	Number of units produced per hour
	Sales per employee
	Profit per unit of production
	Break-even volume
Nonfinancial	Average number of defects detected by customers in the first month of ownership
	Average number of defects detected during manufacturing and repair
	Hours lost to production due to unscheduled maintenance
	Work in progress in the plant
	Number of machines per worker
	Length of time to change a machine or introduce a new operation
	Number of job classifications in the plant
	Amount of materials made obsolete by model changes
	Average energy consumption per unit of production
	Rate of absenteeism in the workforce
	Number of months required to introduce a new product model
	Number of engineering change requests during a new product development program
	Number of units produced prior to a model change (batch production)
	Time metrics (response time, lead time, uptime, downtime, etc.)
Product-related	Parts count
	Number of material types used
	Material utilization in each component
	Assembly process used in production
	Service quality—field repair versus field replacement
	Failure-mode effect analysis
	Quality of product as experienced by customers
	Long-term durability of product
	Fraction of sales to repeat customers
	Company responsiveness to service requests

5.4 TALENT MANAGEMENT

Nowadays, service companies compete primarily on talents. Self-directed professionals contribute in cultivating innovative service concepts, inventing new business models, forming new alliances and expanding other such value added activities. Managers need to know how to acquire and retain critical professionals, as well as train and upgrade employees, who need new skills to serve the company.

5.4.1 Preserving Talents

Professionals join companies for rational reasons, such as better compensations, benefits, and career opportunities. But they stay and work hard for emotional reasons—namely, that there is a sense of connection between the company's mission and their work, sufficient support and resources, and clear alignment with the company's culture and values. Lawler et al. (2008) pointed out that there are various reasons why those professionals, who appear to be seemingly happy, would leave their organizations voluntarily. Some of the "Push" factors, which drive them out, include (1) poor support from marketing and sales, (2) no career advancement opportunities, (3) insufficient coaching to new employees, and (4) the feeling that they are not valued or listened to. Managers can take a number of steps to prevent the talent drain. These include:

- Talking often to people to sense the situations (devoting 40 percent of time to people to make sure that they are happy)—Microsoft provides a weekly blog to address questions raised by people during the "listening tours" and solicits for suggestions. Companies respond to those suggestions to bring about the open atmosphere conducive to keeping talents.
- Encourage and safeguard employees' complaints. In line with what the company can afford to offer, find out employees needs and satisfy them.
- Emphasize talent management as a major responsibility of top managers.
- Hire a third party to do surveys and conduct exit interviews.
- Read the senior managers' body language and read their facial expressions, when communicating with them.
- Solicit feedback from employees and act on the information.

5.4.2 Measuring Performance

From time to time, managers need to appraise the performance of employees. After the performance standards are set, the next step of management control is to measure performance. Performance measurement refers to the recording and reporting of work done and the results attained. Engineering managers take the following steps to measure performance: (1) collect, store, analyze, and report information systematically, (2) compare the performance against established standards, and (3) issue reports such as data, results, and forecasts to document results.

Engineering managers can use techniques of time study, work sampling, and performance rating, among others (Chang 2005), for measuring the performance of routine work, which represents a part of the service value chain. The performance of professional workers needs to be measured with respect to the contributions made toward the attainment of the company's short- and long-term objectives. All measurements must be factual and accurate in order to be justified as valid basis for performance evaluation.

5.4.3 Evaluating Performance

To evaluate performance is to appraise work in progress, assess jobs completed, and provide feedback. Engineering managers take the following steps to evaluate employee performance: (1) establish limits of tolerance, (2) note variations (deviation within the tolerance limits) and exceptions (deviation outside of the tolerance limits), and (3) provide recognition for good performance and give timely, proper credit, if justifiable. Voltaire said: "Appreciation is a wonderful thing: It makes what is excellent in others belong to us as well."* Paying attention to deviations encourages employee self-appraisal, fosters initiative, and enhances managerial efficiency.

A rating method often used in industry is to rank an employee in one of five categories: (1) outstanding, (2) above average, (3) average, (4) poor, and (5) failure. Category ranking is based on performance metrics that are specific to the individual. More often than not, in order to indicate the importance of these performance metrics to the company, top management will assign and publicize the weight factors associated with all performance metrics. A weighted score, similar to that calculated by the Kepner–Tregoe method discussed in section 4.3.5, is then determined for each employee. The weighted score and a written statement are then submitted to superiors at the next management level for review. Once the approval of higher management is secured, these evaluation results become the official basis for salary administration and promotional considerations in the future.

It is quite obvious that this type of rating system has basic weaknesses. Some managers may suffer from a *halo effect* and assign an employee the same rating for all performance metrics. Others may be handicapped by a *recent effect* in that they are predominantly influenced by the recent events. By nature, some managers are more lenient than others, resulting in different interpretations of *outstanding performance.* Furthermore, the competitive nature of getting one's own employees promoted sooner tends to cause inflated ratings. To exert some control over this potentially chaotic situation, a few industrial companies are known to have advised department managers to steer the overall evaluation results toward a Gaussian type distribution, which places the majority of employees in the "average" group, and only about 5 percent and 15 percent, respectively, in the excellent and above average groups.

At the annual appraisal time, the individual employees will be notified of the approved evaluation results. Feedback from the individual employees is solicited and documented. If the individual employee disagrees with some or all parts of the evaluation, his or her written comments are incorporated into the official evaluation files, which are reviewed again by superiors at the next management level. If deemed proper, the approved evaluation results may be modified. The needs for the individual's future development are also discussed and noted. Specific goals of development are then agreed upon as a gauge for monitoring the individual employees' progress at the next annual appraisal time.

---

* *Source*: www.quotationspage.com/quote/1186.html.

The level of average performance is defined as the level of performance that can be generally expected of a person with adequate training and dedication at a given position. To be rated as excellent or above average, one must perform extraordinarily well and deliver an unexpected positive outcome which has a recognizable impact on the company's objectives. Poor performance is usually associated with work that is not meeting the acceptable quality standards, exceeding the approved budget, not completing the work on time, and/or suffering from problems related to communications, personality conflicts, work devotion, interpersonal skills and other deficiencies.

Should the performance of an employee be rated as poor or failure, engineering managers must initiate action to correct such performance in a timely manner.

5.4.4 Correcting Performance

To correct performance is to rectify and improve work done and results obtained. The performance evaluation may show that the quality of the employee's work is below expectation. Engineering managers must understand that there are reasons for performance deficiencies. Some employees may not know that their performance is deficient because of a lack of known performance standards or feedback. Others may not possess the required technical capabilities to perform the assignments at hand. Still others lack devotion to the profession or do not possess a good work ethic and personal initiative.

Engineering managers should correct an employee's mistakes by focusing on future progress and growth. They should take short-term action to overcome variances, such as getting assistance from outside consultants or hiring temporary workers. They should also consider long-term management action to avoid repeating the noted deficiencies by improving training, modifying procedures and policies, transferring employees, or recommending dismissals.

When correcting performance, engineering managers should also offer negative feedback without attacking the employee's self-esteem. Feedback must be focused on results and outcome—not the person—and directed toward the future, not the past. Engineering managers need to avoid upsetting employees or harboring punitive motives. Managers should demonstrate a helpful and sincere attitude and pose no threat to their workers.

Generally speaking, it is not good to make mistakes with the fundamentals of engineering. It is acceptable to fail in new and risky development projects, but making the same mistakes again and again is viewed negatively by management.

5.5 MEANS OF CONTROL

Engineering managers have a number of tools available to exercise control. They can perform personal inspections, review progress, and define any variance to plans. This is the strategy of *management by exception*.

Managers may set priorities with respect to job assignment, resource deployment, and technology application. They may also exercise control by managing resources.

5.6 GENERAL COMMENTS

Engineering managers must constantly define which tasks should be performed, and have employees performing these tasks correctly. The principle of critical few (the Pareto principle) says that, as a rule, 20 percent of factors may affect 80 percent of results. (See Fig. 5.1.) The key is, of course, for the engineering manager to define these critical few, allocate resources to pursue them, and achieve the desirable outcome.

Control should be focused on where action takes place. In general, self-control imposed by the persons involved is the most useful type. However, by and large, people also resent control, and extensive control may lead to loss of motivation. Therefore, engineering managers must manage both the positive and negative exceptions. With information available and mechanisms in place, the preferred type of control is flexible and coordinated.

5.7 CONTROL OF MANAGEMENT TIME

Time is a valuable and limited resource for everyone, including managers. Management activities have several common characteristics: Important tasks often arrive at unpredictable times, trivial tasks often take up a disproportionately large amount of time, and interruptions are common to a manager's schedule.

Engineering managers may waste a lot of time if the roles and responsibilities of employees are not clearly defined. Oftentimes, a lack of self-discipline—procrastination, confused priorities, lack of personal drive, or lack of planning—wastes time. Other engineering managers suffer from a lack of effective delegation; for example, there may be a delegation of responsibilities without the commensurate authority, or the application of too little or too much control. Still other managers waste time because of poor communication related to policies, procedures, meetings, and other subjects (Anonymous 2003).

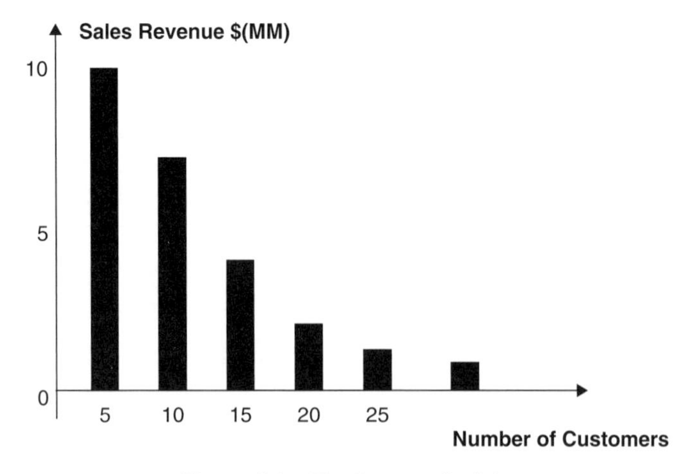

Figure 5.1 The Pareto principle.

The following techniques might help the engineering managers avoid wasting time:

- Set goals for the day, the week, the month, and the year.
- Prioritize tasks to be done, beginning with the most important tasks (A tasks), then the less important ones (B tasks), leaving the least important ones (C tasks) untouched if time does not allow. Make sure that the tasks are relevant and they indeed add value. Reserve blocks of time to pursue A tasks.
- Plan each task beforehand and group some of them together.
- Minimize interruptions by keeping the office door closed for a specific time period and asking the secretary to hold all phone calls.
- Make use of waiting time at the airport, on the train, in the doctor's office, and so on.
- Keep reports and memoranda short and to the point. Some managers in industry have the habit of browsing over only the first page of any report and stopping if the information is uninteresting, of secondary importance, or irrelevant to their current needs.

Enumerated in Table 5.2 are some time-saving tips that are adopted from several published sources: Cottringer (2003), Adair (2009) and Anonymous (2003B). Engineering managers might find them useful.

Table 5.2 Time-saving Tips for Engineering Managers

1. Set goals	Write specific, measurable outcomes that you want to achieve in the next week, month, year, and five years. Consider your work, relationships, play, and well-being. Progress from goals to plan to work.
2. Use a master "to-do" list	Categorize all "to-do" ideas according to which goal each serves. Estimate all others.
3. Get the big picture	Plan your priorities so that you work foremost on whatever gives the biggest payoff and potential.
4. Cluster common tasks	Do similar tasks in the same time block (e.g., a batch of letters, several phone calls, etc.)
5. Create systems	Keep tools, forms, checklists, and information handy and organized for repetitive tasks.
6. Establish place habit	Keep everything in its predetermined place.
7. Delineate time blocks	Schedule blocks of uninterruptible time (2–4 hours) to work on projects requiring concentration. Assure colleagues of availability otherwise.
8. Design your environment	Make your setting conducive to concentration (e.g., sit with your back to traffic passing your office, and screen calls).
9. Cut meeting time	Use proven meeting time savers (e.g., go to other's offices for meetings, do stand-up meetings, set an agenda and follow through rigorously).
10. Lessen panic	Handle what worries you the most. Ask yourself, "Will this matter seem urgent 10 years from now?"
11. Take the one-minute test	Periodically take a minute to ponder: "Am I doing this in the best way to meet my goals, serve others, and take care of myself?"

Condensed and adopted from Cottringer (2003), Casavant (2003), and Anonymous (2003B).

Example 5.3 The customer service manager is a busy person. He rushes from one problem to another without actually taking time to complete any job and solve any problem properly.

What control problem does the customer service manager have? What can he do to enhance his job effectiveness?

ANSWER 5.3 The customer service manager has a time-control problem. He reacts to problems and does not discharge his job responsibilities effectively. He can do several things to rectify the situation:

- Organize the customer service department into groups (e.g., repairs, parts supply, warranty, problem solving by phone, etc.) and know the capabilities of his support staff for these groups of services.
- Set up the call center operation to automatically channel customer calls to the respective service groups.
- Delegate the responsibilities of providing customer services to the leaders of these groups, requiring them to follow through on each and every one of the customer problems and keeping good records of all services rendered.
- Refer specific problems (e.g., nasty customers) that the group leaders cannot resolve, to the customer service manager to handle personally.
- Assign the analysis of service records to someone who could invoke statistics and define trends suggested by these data (e.g., parts with highest failure rate, nature of frequent complaints, average time spent on problem solving, number of calls from customers unhappy with company's services, etc.).
- Establish a system to solicit customer feedback on service and suggestions to upgrade.
- Set up metrics to measure service quality (e.g., number of complaints per week, average service time spent per customer, cost per service call, etc.), making use of "best practices."
- Monitor progress and seek ways to constantly augment the service operation.

5.8 CONTROL OF PERSONNEL

Managerial control is exercised primarily for the purposes of maximizing company productivity and minimizing potential damages arising from ethics, laws, safety, and health issues. For highly skilled personnel, less control is more effective and acceptable. Excess control induces undesirable reactions and brings forth adverse effects. This is best illustrated by the supervision curve displayed in Fig. 5.2.

To manage creative people, or those who are able to deliver new and useful results, managers need to set targets, monitor the employees periodically, apply a low level of supervision, and maintain a collaborative and creativity-inducing work environment (McKenna and Maister 2002; Ivancevich and Duening 2001).

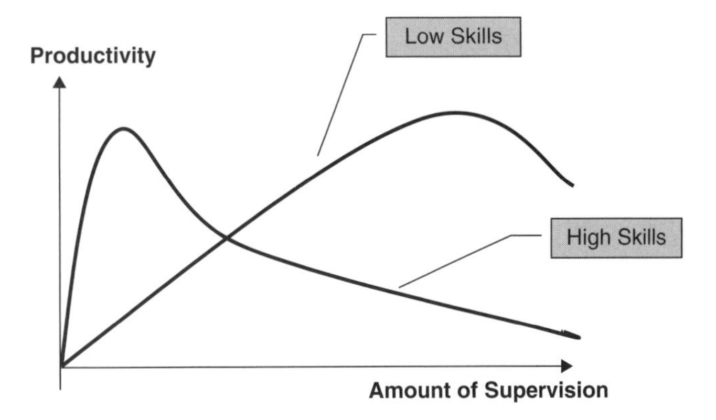

Figure 5.2 Supervision curve.

Example 5.4 Mary Stevenson, the shop manager, works well with all of the staff members. She regards Mike Denver, who has the longest tenure and most extensive experience in the group, as the second in command for the day-to-day operation. The shop is modernizing its operation with the use of computers. Mary Stevenson and her boss, Craig Martin, decide to bring in a young computer specialist, Janet Carter, from the outside.

To make sure that the shop modernization process moves forward, Mary Stevenson spends a lot of time with Janet Carter. Mike Denver sees less and less of Mary Stevenson, although Mary still depends on Mike for the day-to-day operation. Mike resents being shut out from the work done by Mary and Janet. Mike does not complain, but after six months, he tenders his resignation and goes to work for a competitor. Mary Stevenson is shattered. She deeply regrets this major loss to the shop.

What went wrong? What was not controlled? What would you do differently?

ANSWER 5.4 The key issue at hand is that Mike perceived a loss of trust and respect from Mary. Mike also incorrectly perceived an approaching obsolescence in the shop's current technology. Both of these misperceptions caused Mike to have serious doubts about his future standing in the shop. Mary did not anticipate these perceptions and misunderstandings and did nothing to correct them in a timely manner.

Mary Stevenson should have done the following:

- Bring Mike Carter into the loop of hiring a computer specialist to broaden the skill sets of the shop. Allow Mike to participate in the selection and interview process.
- Announce in a staff meeting that Mary needs to spend time to bring Janet along initially and that Mike will actively assist in taking care of the day-to-day operations in the shop.
- Get Mike involved in the work done by Janet, so that Mike, as the second in command, is kept up to date with this new type of computer work planned for

the shop. Should Mike need to take over the management of the shop at some future point in time, he would have been given time to familiarize himself with the computer-related operation. Grooming a candidate in the shop for possible succession in the future should have been Mary's responsibility anyway.

- Maintain a balance of management attention given to both computer-related and other chores in the shop.

5.9 CONTROL OF BUSINESS RELATIONSHIPS

As industrial markets become increasingly global, business relationships (defined by whom the company knows and how well) represent an increasingly important competitive factor in the marketplace (Kuglin with Hook 2006; Carrig and Wright 2006).

It is highly advisable that engineering managers acquire the habit of proactively forming, maintaining, and controlling new business relationships for the benefit of their employers and themselves.

Contacts may be established with noncompetitors. At proper occasions, engineering managers should be accustomed to introduce themselves to others with a five-second *commercial*—a brief self-description of their key areas of expertise. Making note of others' professional specialty areas and following up with periodical exchanges will nurture the relationships. Over time, such a network of professional contacts may become a very powerful business asset to engineering managers and their employers.

5.10 CONTROL OF PROJECTS

Engineering managers exercise control over projects when serving as project leaders (Evans 2003; Oliver 2006; Katz and Thompson 2003). Tools for project control include PERT, CPM charts, or suitable computer or Internet-based project-management software. (See section 13.4.4.) Project control focuses on several key issues, as indicated in Table 5.3.

Zwikael and Globerson (2008) compared the quality of project management in services with that of other industrial sectors. Their survey indicates that the project management in services is poor in technical performance and customer satisfaction, two of the key metrics defining quality (see Table 5.4). They recommend that in order for project management in service to improve outcomes, companies need to emphasize the planning, assurance and control aspects of their projects:

- *Quality planning*—cost benefit analysis, benchmarking, cost of quality
- *Quality assurance*—Quality audit, process analysis, cost benefit analysis
- *Quality control*—cause-effect diagram, control charts, flow charting, histogram, Pareto charts, run charts, scatter diagram and inspection

To manage and control projects, engineering managers need skills related to organization and planning, people management, problem solving, communication, and change management. Successful managers have an optimistic outlook, drive, and energy. They possess a broad technical knowledge and are goal oriented and customer focused.

Table 5.3 Project Control Issues

1. Cost control	Monitor the actual versus projected percentage of cost expenditures and take proper actions to minimize deviations.
2. Schedule control	Monitor the actual versus projected percentage of completion time and take the proper actions to minimize deviations.
3. Critical path activities	These are activities without time slacks, which must be managed with extra care to avoid schedule delay.
4. Task deviation from plan	Delays arise from slow equipment delivery or installation equipment damage in transportation; construction delay due to labor action, weather, utilities, and other causes; and changes of personnel.
5. Collaboration	Securing collaboration among team members is a key success factor for any project.
6. Conflict resolution	Resolving instantly all conflicts and problems among team members will assure a smooth progress toward achieving the project goals.

Table 5.4 Comparison of Project Quality

Industry Type	Services	Construction and Engineering	Software and communications	Production and Maintenance
# Questionnaires	79	49	132	15
Technical performance (1–10 scale)	7.8	8.3	8.1	8.1
Customer satisfaction (1–10 scale)	7.9	8.4	8.2	8.2
Schedule overrun (%)	35	17	30	42
Cost overrun (%)	22	16	24	44

Source: Zwikael and Globerson (2008).

5.11 CONTROL OF QUALITY

To achieve success in the marketplace, companies focus on product and service quality as two of several attributes deemed important to customers. For some companies, to plan and implement quality control programs represents a major engineering management undertaking.

5.11.1 Early Programs on Quality

A number of years ago, Deming (2000) promoted the concept of product quality in the United States and got few followers. He went to Japan and was enthusiastically welcomed there. Since then, a number of quality-control practices have been advanced by the Japanese, such as quality circles, Kaizen, Kanban, just in time (JIT), lean production, Taguchi, Ishikawa, and the 5S campaign. Kanban means looking up to the board in order to adjust to a constantly varying production schedule. The 5S campaign includes (1) *Seiri*—arrangement, (2) *Seiton*—tidying up, (3) *Seisou*—cleaning, (4) *Seiketsu*—cleanliness, and (5) *Shukan*—customizing.

Kaizen means change for the better, or continuous sustainable progressiveness. It includes a number of quality practices such as customer orientation, total quality control, robotics, quality circles, suggestion system, quality betterment, JIT, zero defects, small group activities, productivity furtherance, and corporate labor-management relationships. Kaizen begins and ends with people. An involved leadership guides workers to strive for lower cost, higher quality, and faster delivery of goods and services to customers (Martin and Osterling 2007; Laraia, Moody, and Hall 1999). The elimination of all nonvalue-adding activities is a key emphasis in the Kaizen approach. Table 5.5 presents numerous sample Kaizen steps taken by various manufacturers. Many of these steps are based on common sense.

These quality concepts are logical, reasonable, practical, and, above all, obvious. In fact, there is really nothing new or novel in them. Plentiful of these quality concepts were subsequently "reimported" back to the United States with Japanese labels. Thereafter, American managers started paying attention to quality. As of today, several American automotive companies are still in the mode of catching up with their Japanese competitors.

Knowing what to do is useful, but it is not enough. Pronouncing a few quality terms in Japanese will impress no one. Practicing the quality concept meticulously is the only key to success in quality enhancement and control. To achieve useful results in quality, management commitment is essential. Management commitment is reflected in company value, vision, resources assignment, customer focus, long-term strategic orientation, and rewards systems. In addition, worker dedication—drive to excellence, attention to details, and continuous betterment—must be assured.

Table 5.5 Sample Kaizen Steps

#	Sample Kaizen Steps
1.	Get the most useful ideas to upgrade operations from the workers involved.
2.	Conduct time studies and observe the actual work activities to evaluate productivity.
3.	Use questionnaires to collect data and small-group discussions to encourage worker participation.
4.	Workers should put everything they need next to them to minimize time (JIT and lean production).
5.	Use a checklist that is constantly revised to inspect and study the shop floor activities.
6.	Combining several process steps to lessen efforts required saves resources.
7.	Incorporate prefabricated component modules to cut cost.
8.	Use lighter and more versatile manufacturing equipment to whittle away manpower and utility costs.
9.	Use flexible welding jigs and general purpose pallets to cut welding costs.
10.	Use electric-driven robots and adopt a new server gut welder to increase spot welding speed.
11.	Use general-purpose carriers with common pickup points to handle all car models.
12.	Combine primer and top coats in a single operation to gain 15 percent in speed and cut down energy consumption based on the use of a resin (a polyacetal), which would arise from the primer coat to the surface and separate the primer from the top coats.
13.	Decrease the number of seat sets and utilize more common components in design.
14.	Use a combination of automatic, semiautomatic, and small lot stations to make products of different models and volumes.

It is interesting to note that, during the past twenty years, foreign automakers opened a total of seventeen factories in the United States to manufacture cars and using American workers. Their superior product quality output can only be attributed to superior management practices, as there is no difference in culture, language, or value systems between American workers employed by General Motors and Ford versus those working in U.S. factories for Toyota, Nissan, and other foreign carmakers. Maynard (2004) pointed out that General Motors had 60 percent of the American car market in 1960 and only about half of that in 2003. Its market share dropped further to 22.1 percent in 2008, whereas that of Toyota increased to 16.4 percent. Toyota is predicted to overtake General Motors in becoming the largest car company in the world in a not too distinct future. Maynard further pointed out that there is a difference in the background of lead managers involved. General Motors and Ford have been headed by financial professionals, whereas Toyota, Volkswagen, BMW, and Mercedes are led by engineers who are "passionately interested in everything to do with cars." This could be a factor affecting the varying extent of management commitment devoted to product quality.

J.D. Power Associates conducts the Initial Quality Study annually to register the number of problems encountered by customers per hundred vehicles within the first ninety days of car ownership. Those nameplates with lower numbers of problems than the industrial average in a given year are perceived to be of better quality than the others. This industrial average number of car problems has continuously declined over the years, as shown in Fig. 5.3. As an externally imposed and steadily tightening standard, it exerts a remarkable pressure on car makers to get better quality continuously.

5.11.2 FMEA

Another case in point is the implementation of the quality program. Ford Motors Company is credited for having spearheaded the well-known "failure mode and effect analysis" (FMEA) method in the U.S. automotive industry around 1977. FMEA may be applied to design, process, service, and other engineering or business activities that may go wrong. Murphy's law says, "If something can go wrong, it will." FMEA is a proactive program intended to catch all possible failures modes before they actually

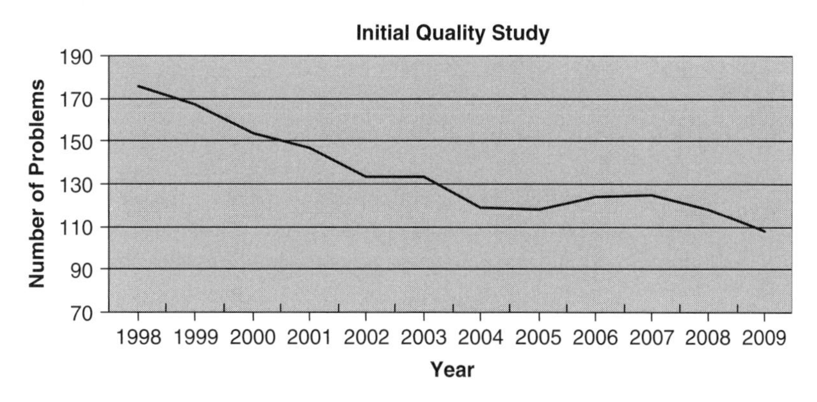

Figure 5.3 Industrial average problems with new cars.

Table 5.6 FMEA Worksheet

A.	Step number
B.	Process description
C.	Potential failure mode
D.	Potential effects of failure
E.	Severity
F.	Potential causes of failure
G.	Occurrence
H.	Current process control and detection
I.	Detection
J.	Risk response number (RPN)
K.	Recommended actions
L.	Responsibility and target completion date
M.	Actions taken
N.	New severity
O.	New occurrence
P.	New detection

occur. It is solidly based on the understanding that the correction of potential failures before they actually occur will be much less costly than the required remedial actions after the fact.

The FMEA concept is rather straightforward. Table 5.6 includes the headings of an FMEA worksheet. For a given step number, the process description, potential failure mode, and potential effects of the failure in question are to be entered. Then, the severity factor (Row E) is assessed, using a number between 1 and 10 (10 being the most severe), as to impact on customers. Row F registers the potential root cause for the failure. Row G defines the probability of its occurrence, again using a number between 1 and 10, with 10 being the most likely to occur. Row H specifies the current control and detection practices. The detection factor (Row I) is assigned a value between 1 and 10, with 10 being the most difficult to detect.

The factor RPN (risk response number, Row J) is the product of three numbers, namely severity (Row E), occurrence (Row G), and detection (Row I). Remedial or preventive actions (Row K) are to be taken according to the priority order based on the RPN. Row L tabulates the person responsible for taking the recommended actions. Row M documents what actions were in fact taken.

Rows N, O, and P contain the results of reevaluation of the failure mode in question after remedial and preventive actions have been taken. The new RPN number is expected to be significantly lower than its corresponding preaction number, and it documents the corrective impact on the failure mode. Such documents may be used to satisfy the customer's requirements and serve as guidelines to drive continuous improvement.

Practicing FMEA systematically should lead to a continuous reduction of the effects of failure modes in product design, product manufacturing, service, and other engineering or business applications. Several U.S. automotive companies and their tiered suppliers have applied FMEA with varying degrees of success. Although Ford initiated the use of FMEA back in 1977, it is still behind other industrial leaders today, in terms of problems per hundred vehicles, according to the 2004 study by J.D. Power

Associates. This drives home the point that knowledge of quality is useful, but actual implementation is the key in achieving quality performance. Successful implementation requires management commitment and worker dedication. Four American automotive nameplates (Cadillac, Buick, Mercury, and Chevrolet) scored above the industrial average, compared with eleven foreign nameplates, as shown in the 2004 study of J.D. Power Associates. Technically, FMEA can be successfully implemented. The management challenge is how to achieve better quality performance consistently for all nameplates or products.

Like Kaizen and FMEA, total quality management (TQM) is also a very powerful program addressing the issues related to product quality and organizational productivity (Peratec 2009; Goetsch and Davis 2003; Evans 2004; Mohanty and Lakhe 2006). In a typical academic course at the graduate level, TQM covers the concepts of (1) customer satisfaction, (2) empowerment of employees in problem solving, (3) continuous improvement, and (4) management excellence by creating and implementing corporate visions. To achieve TQM success, management commitment and worker dedication are also required.

5.11.3 Six Sigma

Six Sigma is a quality management and control tool initiated by Motorola in the 1980s. Typical failure rate of common processes is four sigma, about 6,000 defects per million. By definition, six sigma level of quality means no more than 3.4 defects per million. A large number of companies, such as General Electric, AlliedSignal, Bombardier, Seagate Technologies and others, had applied Six Sigma successfully to high volume and standardized production operations, focusing on eliminating wastes and non–value-added steps, and thereby reducing cost significantly. Six Sigma consists of five specific steps (DMAIC):

1. *Define*. Define process, customer requirements, and key process output variables.
2. *Measure*. Develop and evaluate measurements, take performance data related to current process.
3. *Analyze*. Analyze date to prioritize key input variables and to identify waste.
4. *Improve*. Introduce improvements and pilot new processes.
5. *Control*. Finalize control systems and verify long-term capability.

A Six Sigma black belt is said to be able to undertake five to six projects a year, each saving about $175,000.

There have been some divergent discussions in the research community regarding the suitability of Six Sigma to service industries. Contrary to products, which involves high volume and standardized operation procedures with very little customer interactions, services are characterized by the following:

- Intangibility and heterogeneity of service outputs—containing non-physical elements such as courtesy, responsiveness and accessibility.
- Consumption and delivery take place simultaneously.
- High degrees of labor intensity and customer interactions.

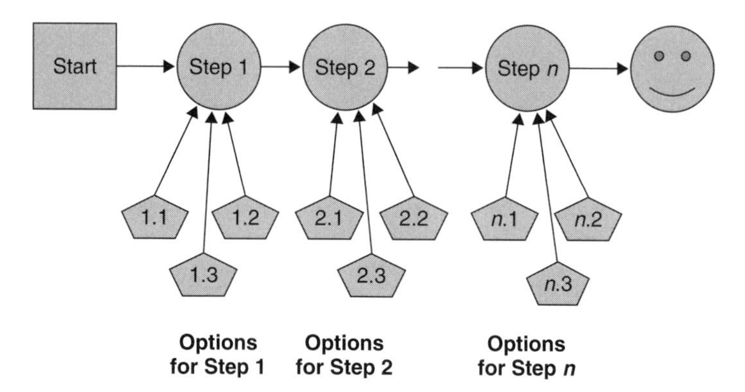

Figure 5.4 Applying systems principles to improve service quality.

Prajogo (2005) argues that while certain tools used for manufacturing operations may not be directly applicable to services, the basic principles remain useful to both products and services. Six Sigma can be useful to some of the repetitive and standard processes within the service industry, such as (1) accounts receivable, (2) sales, (3) R&D, (4) finance, (5) information systems, (6) legal, (7) marketing, (8) public affairs, and (9) human resources. Therefore, it is important that service companies apply Six Sigma to the right candidate processes in order to derive value (Bioloa 2002).

An example of applying the principles of Six Sigma to services in order to enhance service customization and customer satisfaction is illustrated in Fig. 5.4.

Here, the steps involved—*measure, analyze,* and *improve*—can be focused on coming up with technically feasible options for each value chain step so that options can be delivered in the most efficient manner, thus achieving a low-cost and speedy delivery of customized services.

When a service value chain is broken down into a set of sequential and partially parallel steps, each of these steps may be implemented by a number of predetermined options, thus allowing a high degree of customization while achieving an acceptable quality and speed. New innovative options may be continuously produced for any of these steps. Companies need to preserve these options, in order to promote competitiveness by satisfying customer's individual needs to a greater degree than otherwise possible.

The use of a number of such feasible options will provide services which are aligned to a higher degree with the needs of individual customers while the traditional Six Sigma process strives to define the best option for each step to achieve the goal of lowest waste and highest productivity, the suggested approach maximizes the degree in which customerization can be achieved.

In the implementation of each option, consideration should be given to relevant soft issues, such as empowerment, people management, courtesy, and others, so that all options are exercised appropriately to assure proper procedure. To further improve the outcomes from these options, tools such as Statistic Process Control (SPC) and others may be evoked.

Quality control is an important function in which engineering managers play a key role. They need to be actively and persistently involved with workers to invoke common sense in eliminating wastes, speeding delivery, simplifying processes, paying attention to the gritty details of practice, and improving the way work gets done continuously.

Example 5.5 Hospital emergency rooms are highly dynamic service environments, in which many key variables change in unpredictable ways, such as patient flow and types of ailments. In addition, this service delivery system is constrained by space, equipment, and medical staff. The outcome of this system has life and death consequences. The basic industrial and systems engineering discipline covers several known quality assessment and control concepts. In what ways could these industrial and systems engineering concepts be applied to assess the service delivery process in hospital emergency rooms and to upgrade its service quality?

ANSWER 5.5 There are four operating parameters that affect the service quality of a hospital emergency room:

1. Patient care—prompt treatment of urgent patients and timely treatment of nonurgent patients by doctors; nurse practitioners can treat nonurgent ailments.
2. Wait time for nonurgent patients, who might otherwise complain loudly.
3. Staff attentiveness to patients and work pace.
4. Reasonable operating costs.

Incoming patients are categorized by a triage nurse to fit one of the four categories: Category 1—immediate attention; Category 2—treatment within ten minutes, Categories 3—treatment within thirty minutes, and Category 4—nonurgent. The emergency room beds are assigned in accordance with the patient's category:

Resuscitation beds—Category 1, or 2.
Observation Beds—Category 2, 3, maybe 4.
Cubicle Beds—Category 4.
Spare rooms—Category 4, if needed.

Such a dynamic service delivery system is best modeled using a number of industrial engineering tools, such as block flow diagram, production floor layout, time study, and others:

1. Use block flow diagram to categorize and link the service activities needed so that a stable service demand can be handled with sufficient productivity, see Figs. 5.5A and 5.5B.*
2. Consider layout of service stations in order to minimize interferences between patient flows and their respective service activities.
3. Define bottlenecks in the block flow diagram, if the load becomes dynamic. Most likely, the emergency room operation breaks down because of one of the three bottlenecks (1) shortage of doctors, (2) shortage of nurses, and (3) shortage of beds. Establish pools of reserve resources that can be called on, using suitable overtime and other arrangements. Temporary bed systems may be used as one possibility.

*Stephen M. Grisanti (2008), "MFS Emergency Department Patient Throughput and Operational Efficiency Initiative (ED Length of Stay)," Master's Project under the Direction of Dr. C. M. Chang, Department of Industrial and Systems Engineering, University at Buffalo, Buffalo, New York.

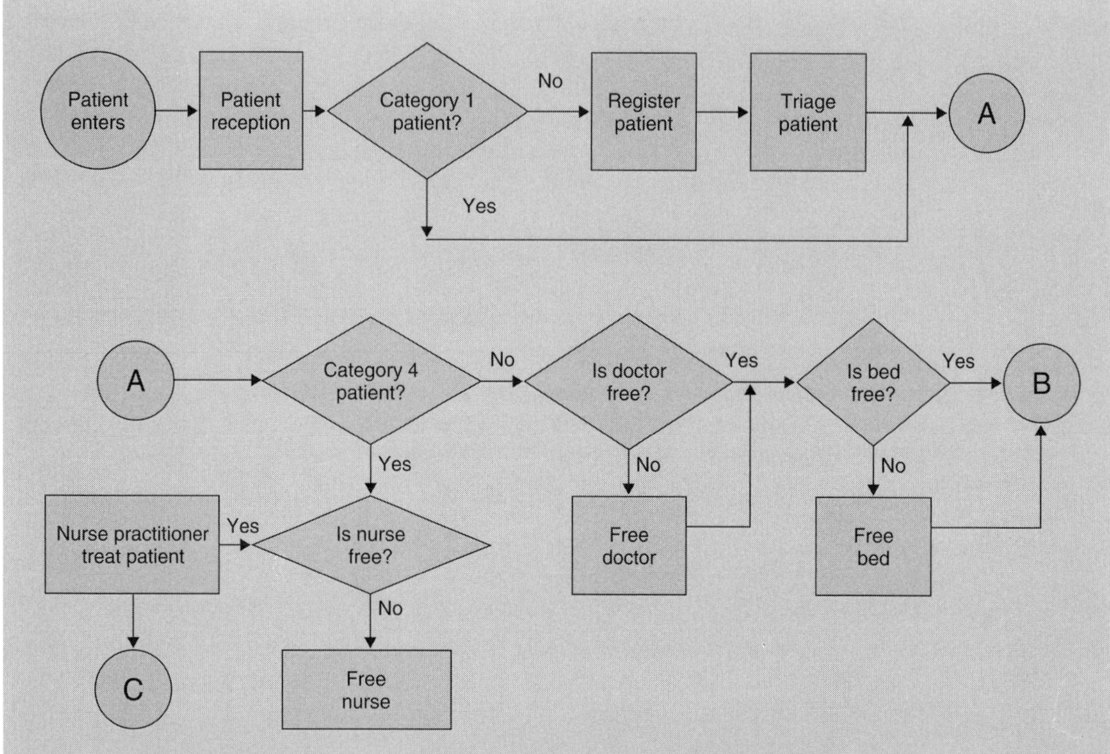

Figure 5.5A Block flow diagram for hospital emergency room operation.

4. Communications infrastructure must be in place so that the patient flow and service load can be constantly monitored and the required actions taken instantly (such as calling in reserve resources, specialists not available on staff at the time, communication with patient's primary physician).
5. Understand the best practice performance of leading emergency room services (such as patient wait time for being registered, checked, and discharge), so that the system can be further iteratively revised to reach the best practices service quality.
6. Collect operational information at various stations so that assessing the needed reserve resources can be done automatically. Utilize value stream analysis to further streamline and upgrade the work process and enhance operational efficiency.
7. Useful operational practices should include being alert in making sure that all patients in the waiting room are taken care of. People have died in emergency rooms after having waited for thirty-four hours without being noticed by medical staff (Kaufman, 2007). Time studies should be conducted to define the variations in patient arrival time, admission rates, and the times taken to register, triage, treat, discharge, as well as any other steps needed for preparing analysis towards creating an optimized operational strategy under the constraints of space and medical staff.

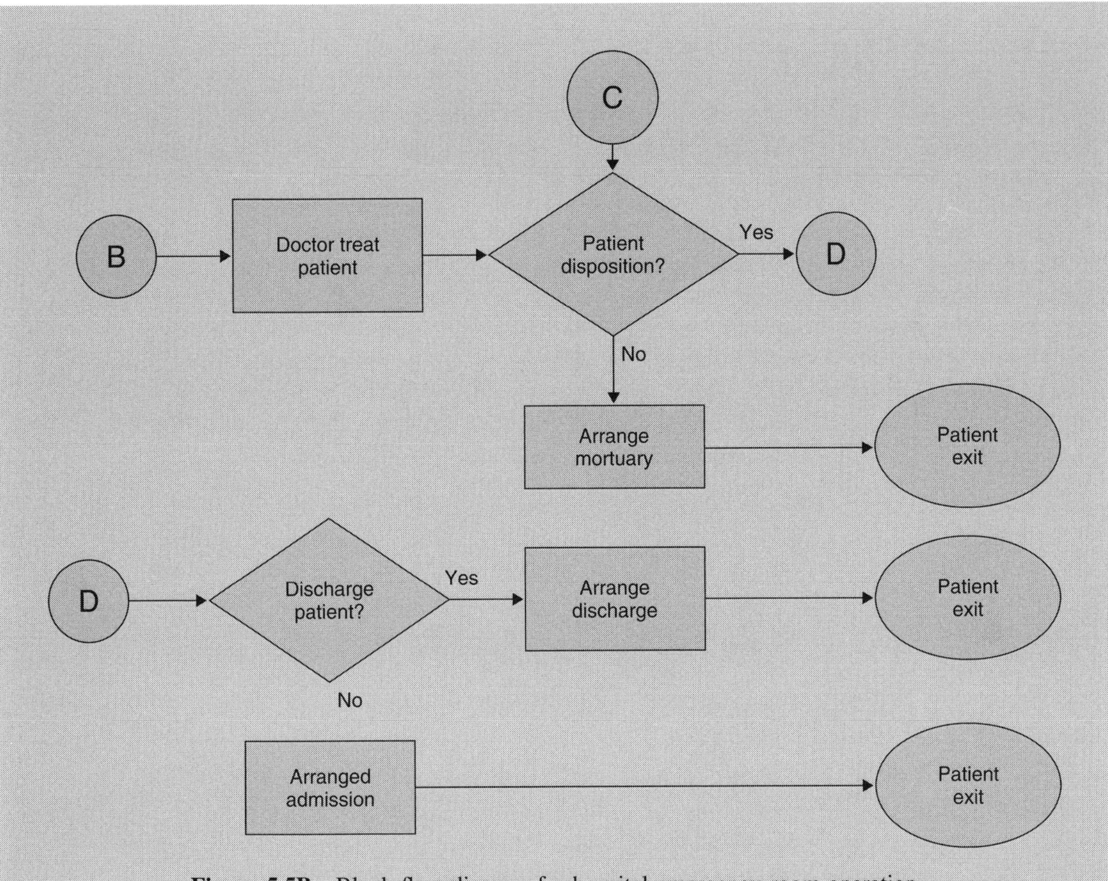

Figure 5.5B Block flow diagram for hospital emergency room operation.

Example 5.6 What is the generic problem-solving approach applicable to solve most engineering and management problems?

ANSWER 5.6 The generic problem-solving approach may look like the following:

1. Perform a situation analysis (e.g., assess strengths, weakness, opportunities, and threats).
2. Formulate a statement of the problem.
3. Define performance standards that are observable, measurable, and relevant to the goal.
4. Generate alternative solutions to the problem.
5. Evaluate these alternatives in terms of their consequences to the organization.
6. Select the best alternative solution with the most value and least adverse effect on the organization.
7. Implement a pilot test of the proposed solution and revise as indicated from practical experience.

Table 5.7 Xerox Problem-Solving Process

Step	Questions to be Answered	Expansion/ Divergence	Contraction/ Convergence	What Is Needed to go to the Next Step
1. Identify and select the problems.	What do we want to change?	Multiple problems for consideration.	One problem statement, one "desired state" agreed upon.	Identification of the gap and description of the "desired state" in observable terms.
2. Analyze problems.	What's preventing us from reaching the "desired state"?	Multiple potential causes identified.	Key cause(s) identified and verified.	Key cause(s) documented and ranked.
3. Generate potential solutions.	How could we make the change?	Multiple ideas on how to solve the problem.	Potential solutions clarified.	Solution list.
4. Select and plan the solution.	What's the best way to do it?	Multiple criteria for evaluating potential solutions.	Criteria to use for evaluating solutions agreed upon.	Plan for making and monitoring the change. Measurement criteria to evaluate solution effectiveness.
5. Implement the solution.	Are we following the plan?		Implementation of agreed-upon contingency plans (if necessary).	Solution in place.
6. Evaluate the solution.	How well did it work?		Effectiveness of solution agreed upon. Continuing problems (if any) identified.	Verification that the problem is solved, or agreement to address continuing problems.

8. Implement the solution.
9. Evaluate the outcome.
10. Revise the process as necessary.

Note that there is no one universal solution to all management problems but that the correct solution will depend on the unique needs of the situation.

The problem-solving process practiced by Xerox Corporation is quite representative of the approaches taken by numerous companies in industry (Garvin, Edmondson and Gino 2008). The Xerox problem-solving process contains the details shown in Table 5.7.

5.12 CONTROL OF KNOWLEDGE

Knowledge refers to the sum total of corporate intellectual properties that are composed of (1) patents, (2) proprietary know-how, (3) technical expertise, (4) design procedures,

Table 5.8 Knowledge Management

1. Experimentation	Put systems and processes in place to facilitate the search and test of new knowledge.
2. Benchmarking	Learn from one's own experience, and best practices of others in industry, by reflection and analysis.
3. Preservation of knowledge	Set policies concerning the preparation of reports, design procedures, and data books. Practice knowledge acquisition tools to preserve valuable heuristic knowledge.
4. Dissemination of knowledge	Rotate experts to different locations or jobs so that knowledge may be shared with and learned by others. Make knowledge or data available electronically companywide.

(5) empirical problem-solving heuristics, (6) process operational insights, and others (Gautschi 1999). Some of these knowledge chunks are documentable. Others typically reside in the employees' heads.

Engineering managers are responsible for producing, preserving, safeguarding, and applying corporate engineering and technology knowledge. They also need to draft policies to facilitate the control of knowledge. Some examples of knowledge management strategies are listed in Table 5.8. Additional discussions on knowledge management are offered in Chapter 11.

One of the major problems in knowledge management is that innumerable knowledge chunks are dispersed throughout various documents within the company. Data-mining software products represent new tools to help extract and group together related information from diverse sources for wide dissemination and effective utilization by others. Another major problem in knowledge management is that most experts do not like to share their knowledge with others, rendering the use of knowledge acquisition tools somewhat ineffective.

Knowledge control is particularly important from the viewpoint of countering industrial espionage. Special policies regulating employee contact with competitors at neutral sites (such as professional meetings, university environments, and industrial seminars) must be defined to safeguard company knowledge.

Preserved or acquired knowledge adds little value to the company if there is no accompanying refinement in the way work gets done. Engineering managers also need to focus on effectively transferring the knowledge gained to cause a modification of the company's behavior that reflects the new knowledge and insights. Doing so will make the company a true learning organization that steadily enriches itself.

5.13 CONCLUSIONS

Control is another important function of engineering management, which focuses mostly on the administrative and operational aspects of the job. Controlling is essential to the implementation of any project or program activities by specifying performance metrics, monitoring program for initiating corrective steps if necessary, and assuring the accomplishment of useful outcomes. Particular attention should be paid to external benchmarking when setting standards in order to avoid causing a company to lose its long-term competitive strengths. Delegation without proper control will result in wasting company resources (e.g., time, manpower, technology and management attention).

Control is routinely applied to team members, knowledge, business relationships, and the allocation of resources, as they all contribute directly to the specific project objectives at hand. Care must be taken in dealing with professionals who generally prefer to be guided by specifying objectives and then being empowered to select the right courses to achieving the project objectives within the given time and budget constraints. On the one hand, extreme tight control destroys employee initiatives. On the other hand, too loose a control will most likely lead to missing the project objectives. Exercising the right level of control, dependent on people, nature of project, urgency and impact and the risks entailed, is of critical importance.

As a general rule, setting clear and understandable standards and performance goals and insisting on good planning efforts will encourage employees to exercise self-directed control, while promoting participation, involvement and employee motivation.

5.14 REFERENCES

Adair, J. 2009. *Effective Time Management: How to Save Time and Spend it Wisely,* rev. ed. New York: Pan Books.

Anonymous. 2003. "Key Tips for Effective Time Management." *Logistics and Transport Focus* (January–February).

Bioloa, Jim. 2002. "Six Sigma Meets the Service Economy," *Harvard Management Update* (November).

Brown, Tand Robert Heller (eds). 2003. Best Practice: Ideas and Insights From the World's Foremost Business Thinkers. New York: Basic Books.

Carrig, K., and Patrick Wright. 2006. *Building Profit Through Building People: Making Your Workforce the Strongest Link in the Value-Profit Chain.* Alexandria, VA: Society for Human Resource Management.

Chang, C. M. 2005. Engineering Management: *Challenges in the New Millennium*. Upper Saddle River, NJ: Prentice Hall.

Cottringer, W. 2003. "How to Save 12 Hours a Day." *SuperVision* (March).

Deming, W. E. 2000. *The New Economics: For Industry, Government, Education*. Cambridge, MA: MIT Press.

Evans, P. M. 2003. *Controlling People: How to Recognize, Understand, and Deal with People Who Try to Control You*. Avon, MA: Adams Media Corp.

Evans, J. R. 2004. *Total Quality Management, Organization and Strategy,* 4th ed. Cincinnati, OH: South-Western College Publishing.

Garvin, D. A. 1984. "What Does 'Product Quality' Really Mean?" *Sloan Management Review* 26 (1): (Fall).

Garvin, D. A., Amy C. Edmondson, and Francesca Gino. 2008. "Is Yours a Learning Organization." Harvard Business Review (March 1)

Gautschi, T. 1999. "Knowledge as Advantage," *Design News* 54 (August 2): 156.

Goetsch, D. L., and Stanley B. Davis. 2003. *Quality Management: Introduction to Total Quality Management for Production, Processing and Services*, 4th ed. Upper Saddle River, NJ: Prentice Hall.

Gratton Lynda, and Sumantra Ghoshal. (2005). "Beyond Best Practices." *MIT Sloan Management Review* (Spring).

Hammer, Michael. 2007. "The Seven Deadly Sins of Performance Measurement and How to Avoid Them." *MIT Sloan Management Review* 48 (3).

Ivancevich, J. M., and T. N. Duening. 2001. *Managing Einsteins: Leading High-Tech Workers in the Digital Age*. New York: McGraw-Hill.

Katz, E., A. Light, and W. Thompson. 2003. *Controlling Technology: Contemporary Issues*, 2nd ed. New York: Prometheus Books.

Kaufman, Hatti, "Emergency Room Tragedy at Los Angles King-Harbor Hospital," CBS News Video (June 14, 2007), www.cbsnews.com/video/watch/?id=2926948n.

Kuglin, F. A., with J. Hook. 2006. *Building, Leading and Managing Strategic Alliances: How to Work Effectively and Profitably with Partner Companies*. New York: AMACOM.

Laraia, A., P. E. Moody, and R. W. Hall. 1999. *The Kaizen Blitz: Accelerating Breakthroughs in Productivity and Performance*. New York: John Wiley & Sons, Inc.

Martin, K., and Mike Osterling. 2007. The Kaizen Event Planner: Achieving Rapid Improvement in Office, Service and Technical Environments. London: Productivity Press.

Maynard, M. 2004. *The End of Detroit: How the Big Three Lost Their Grip on the American Car Market*. New York: Broadway Press.

McKenna, P., and D. H. Maister. 2002. *First among Equals: How to Manage a Group of Professionals*. New York: Free Press.

Mohanty, R. P. and R. R. Lakhe. 2006. Handbook of Total Quality Management. Mumbai, India: Jaico Publishing House.

Oliver, L. 2006. *The Cost Management Toolbox*. New York: AMACOM.

Prajogo, Daniel I. 2005. "The Comparative Analysis of TQM Practices and Quality Performance Between Manufacturing and Service Firms." *International Journal of Service Industry Management* 16 (3): 217–228.

Peratec LTD. 2009. *Total Quality Management* (Paperback). New York: Springer.

Rodier, M. M. 2000. "A Quest for Best Practices." *IIE Solutions* 32 (February): 36.

Zwikael, Ofer, and Shlomo Globerson. 2008. "Quality Management: A Key Process in the Service Industries." *The Service Industries Journal* 27 (8): 1007–1020.

5.15 QUESTIONS

5.1 Study the case "Growth at Stein, Bodello & Associates, Inc.," Babson College # BAB080 (Revised April 23, 2004) and answer the following questions. The case materials may be purchased by contacting its publisher at www.custserve@hbsp.harvard.edu or at 1-800-545-7685.

A. What is this case about?

B. What are the key issues at hand? What are the root causes of difficulties encountered in resolving these issues? Why were the memos sent by TAHC not effective in getting these issues resolved?

C. Why does Stein see differently those issues of concern to TAHC? What are the sources of his difficulties in appreciating and understanding the views of TAHC?

D. What are the options left for the TAHC to overcome these difficulties? What are the advantages and disadvantages of these options? Which ones are like to succeed?

E. What are the lessons you have learned from having studied this case?

5.2 A company decides to offer an average annual raise of 8 percent, although the current inflation rate is 10 percent. Each engineering manager decides on the best way to distribute the salary increases to his or her staff. However, if everyone gets an increase of 8 percent, then there will be no differentiation between strong and weak performers for the previous year. What should you do as an engineering manager?

5.3 A key engineer in the department handed in her resignation notice; her reason for leaving was that she was offered a much higher salary from a competitor. The manager recommends to the director that the company match the competitor's offer, even though this would allow the engineer to earn above the maximum for her grade. "We can always give her smaller increases in subsequent years to bring her salary back into line," says the manager. What should the director do?

5.4 Motivation in the assembly shop is high. However, the shop manager notices that, although the daily production is above average for the shop, it drops down to a low during the first hour after the lunch break. It is further diagnosed that the operators tend to continue socializing until well after the lunch break.

The shop manager changes the lunch break and staggers it over a two-hour period so that the operators cannot go to lunch together. To his surprise, motivation begins to fall and productivity drops dramatically.

What do you think is the problem? How do you advise the shop manager to fix this problem?

5.5 Bill Carter is an excellent hardware designer, but he wants to move into management to broaden his experience. His manager is supportive and encourages Bill to go to evening classes on management. Bill works hard for two years and graduates first in his class. Soon there is a management opening in the procurement department. Bill applies for it, but an outside applicant eventually fills the position. Bill is turned down because he does not have sufficient experience in the procurement functions, a stated key requirement for the job.

Bill protests, "But I have better technical knowledge of components than all the procurement engineers put together, and I learned about procurement in my management courses. The only way I will get experience is to work on the job."

Do you think Bill should have been given the job? How would you have handled this situation differently?

5.6 The company president has noted a constant increase of reports passed on to her, a number of them through the mail, some come through the company's Intranet. Most of these reports remain unread, although the company president when traveling for business does find time to browse through some of the reports while on the airplane. It has clearly become difficult for her to keep track of all projects due to information overload. However, she does not want to abandon her personal objective of being constantly informed, and she does not want to query her vice presidents for summaries of major developments.

What are the alternatives available to the company president?

5.7 The production department is undergoing an upgrade of its automation program. It has a conflict that needs to be controlled and managed. The line supervisor wants to standardize machines supplied by an American vendor, as his people will eventually use them. On the other hand, the automation team leader thinks that the Swiss-made machines would result in greater productivity. Being a specialist in automation, she was brought into the department to find ways of significantly improving productivity.

The department head does not know what to do, as these two experts frequently fight in staff meetings. He regrets that he has not kept up with the automation technology to enable him to arbitrate and decide on the best way.

What should the department head do now?

5.8 Two junior members of the production department unexpectedly come in to see you, the production director, to complain that their manager, who reports to you, commits discrimination, practices favoritism, and misuses company facilities for possible personal gain.

How would you handle this complaint?

5.9 At present, the company is running at full capacity in advancing a new product for a major customer. The sales director has unexpectedly secured a small, but highly profitable, order, which requires some low-level development work and a minor change to the current production process.

Should the company accept the small order? If so, how should the company satisfy the small order?

5.10 Some foreign countries, particularly those in the early stages of industrial development, are known to illegally copy product designs and technologies that originated from developed countries.

What are the best ways for small businesses to protect their technical know-how in foreign countries?

5.11 John Elrod founded the Elrod Manufacturing Company 50 years ago. Vernon Scott is the vice president of plant engineering, reporting to George Elrod, who took over as the company president from his father, John five years ago. Vernon Scott and John Elrod have been good friends for many years. The company's products include automotive parts (such as gears, axles, and transmissions), metal stampings, and sheet-metal subassemblies.

Also reporting to George Elrod are six plant managers. Each plant has its sales, engineering, manufacturing, warehouse, and other functions. The plant managers are responsible for the profitability of the individual plants. The total employment of the company is 12,000 people.

The company has a standing policy on capital expenditure: Expenditures below $5,000 are to be approved by plant managers, those between $5,000 and $50,000 by Vernon Scott, and those above $50,000 by the executive committee.

Vernon Scott favors expenditures for machinery and equipment directly related to manufacturing, but not for maintenance, facility expansion, and improvements that are not related to manufacturing. Over the years, plant managers have become unhappy with Scott's refusal of expenditures for nonmanufacturing equipment, such as computers. Forced to keep their plants profitable, they cannot help, but to bypass him by breaking down the nonmanufacturing projects into numerous small components, each below $5,000.

Eventually, Vernon Scott finds out about the piecemeal purchase of a $27,000 computer by Paul Nelson, a very capable plant manager, whose plant has become the most profitable in the company. Scott demands that George Elrod fire Nelson, citing insubordination, cheating, and dishonesty.

What would you do if you were George Elrod?*

5.12 The manufacturing manager of Company X had installed a wage incentive program. She happily reported six months later that the system was a success because production was up and unit costs were down. A quality control manager said, however, that the percentage of

---

*Condensed and adapted from Robert E. Shannon, Engineering Management, New York: John Wiley & Sons, Inc. 1980, p. 296.

rejects had increased markedly and that this was creating a backlog of rework requirements. An industrial engineering manager reported that the expenditures for industrial engineering studies in the department were up by 50 percent. An industrial relations manager said that arbitration fees resulting from incentive grievances had tripled.

Discuss your observations in this case. What should be done to correct the situation in this production department?

5.13 A number of years ago, ISO Standards 9000 series was introduced to promote work quality by standardizing engineering design, testing, production, and other procedures. How many ISO standards are there, and how well have these standards been accepted in the United States?

PART II

Business Fundamental for Service Systems Engineers and Leaders

Part II covers the fundamentals of business management, including cost accounting (Chapter 6), financial accounting and management (Chapter 7), and marketing management (Chapter 8). This part is to enable service systems engineers and leaders to facilitate their interactions with peer groups and units and to acquire a broadened perspective of the company business and its stakeholders.

Part II also prepares service systems engineers and leaders to make decisions related to cost, finance, services, and capital budgets. Discounted case flow and internal rate of return analyses are also reviewed. These discussions are of critical importance, as decisions made during the service design phase typically define a major portion of the final costs of services offered. Activity-based costing (ABC) is presented to define indirect costs related to services and economic value added (EVA), which determines the real profitability of a service enterprise above and beyond its cost of capital deployed, is addressed.

Also discussed are capital formation through equity and debt financing, resource allocation concepts based on adjusted present value (APV) for assets in place, and option pricing for capital investment opportunities. By understanding the project evaluation criteria and the tools of financial analyses, service systems engineers and leaders will be in a better position to secure project approvals. A critical step to developing technological projects is the acquisition and incorporation of customer feedback. For SSME professionals to lead, a major challenge is the initiation, development, and implementation of major technological projects that contribute to the long-term profitability of the company.

The important roles and responsibilities of marketing in any profit-seeking service enterprise are introduced, along with the contributions expected of service systems engineers and leaders to support the marketing efforts. Many progressive enterprises are increasingly concentrating on customer relationship management to grow their businesses. Such a customer orientation is expected to continue to serve as a key driving force for service design, project management, process improvement, productivity enhancement, customer services, and many other service-centered activities.

Chapter 6

Cost Accounting and Control

6.1 INTRODUCTION

Costing is the assessment of the value of resources consumed in the generation of a service offered for sales. Cost accounting and control are very important management functions in both profit-seeking and nonprofit organizations.

A profit-seeking service organization strives to maximize its financial gains (for example, sales revenue minus costs) for its owners. These gains can be sustained over time only if all stakeholders of the firm (e.g., stockholders, customers, employees, suppliers, business partners, and the community in which the firm operates) are reasonably satisfied. A nonprofit organization (e.g., the United Way, the Ford Foundation, government agencies, educational institutions, church organizations, etc.) seeks to maximize its service value to its respective target recipients while minimizing operations costs.

This chapter covers the basics of cost accounting. The discussions focus on the costing of services, although all cost accounting concepts are equally applicable to the costing of products (Colander 2007; Horgren, Foster, Datar, and Rajan 2008). First, some commonly utilized accounting terms are introduced. Then, the cost analysis of a single period versus multiple periods is explained, leading to topics such as the time value of money and compound interest formulas. The costing of products follows, including the estimation of direct costs absorbed into the company's inventory. The complex problem of assigning indirect costs to products is illustrated by the conventional method of using overhead rates, as well as by the more sophisticated method of activity-based costing.

Estimation of costs with uncertainties is then presented. The Monte Carlo simulation is introduced as an effective method to account for cost uncertainties. Its superiority over the conventional estimation method, which uses deterministic data, is clearly demonstrated through examples of the output distribution functions of the Monte Carlo simulation. Finally, inventory accounting is addressed to arrive at the all-important cost of goods sold (CGS).

It is important for all systems engineers and managers to become well-versed in cost accounting. They need to know how to estimate service/product costs and manage overhead costs. Doing so will allow them to assess service costs and initiate steps to further reduce them.

6.2 SERVICE/PRODUCT COSTING

Service or product costing is one of the key responsibilities of engineering managers, as the proper computation of the cost of products or services is of paramount importance to a service- or product-centered company. Service and product costing are similar in that both encompass the direct costs (e.g., raw materials, labor) and indirect costs (e.g., general supports, maintenance, overhead and others), which are incurred during the process of generating the service or product involved. Direct costs are those that vary directly with the volume/quantity of the service or product manufactured; these are relatively easy to properly account for. Indirect costs, by contrast, are somewhat difficult to allocate because of the complex and varied nature of these costs and their nonobvious relationships to the cost objects at hand. Cost objects are the targets (e.g., services, customers, products, customers) for which costing is to be performed.

6.2.1 Traditional Method of Allocating Indirect Costs

The traditional practice of general ledger costing involves estimating all overhead costs for the upcoming year in a single cost pool (e.g., factory overhead, utilities, safety program, training, and salaries of foremen and compensations of factory managers). This total is then divided by the estimated number of labor hours to be worked. The result is an hourly overhead rate. For each product, the required labor hours are estimated. The total overhead cost for the product is then equal to the respective labor hours required to fabricate them, multiplied by the hourly overhead rate.

According to Granof et al. (2000) traditional costing systems accumulate costs into facilitywide or departmental cost pools. The costs in each cost pool are heterogeneous—of many different processes—and are generally not caused by a single method of resources utilization. The systems allocate costs to products/services using volume-based allocation base (e.g., units, direct labor hours, machine hours, or revenue dollars). The resulting unit costs for product or services may be overestimated or underestimated in this manner. Specifically, there will be a cost distortion of overcharging overhead to high-volume products and charging too little to low-volume products.

For example, let us assume that an enterprise has $800,000 in overhead (e.g., salary of manager, benefits, and other general charges), 2,000 direct hours per employee per year, and 20 employees. The overhead rate is then $20 an hour ((800,000/(2,000 × 20)). The major deficiency of this method of allocating indirect costs is that it does not reflect the true relationship between the indirect costs and the cost object (such as products or services produced). Therefore, the allocation is often improper, as diverse types of overhead costs are lumped together, making an in-depth analysis impossible (Cooper and Kaplan 1988). A better method of allocating indirect costs is activity-based costing, which is introduced in the next section.

6.2.2 Activity-Based Costing (ABC)

Activity-based costing (ABC) is a cost accounting technique by which indirect and administrative support costs are traced to activities and processes and then to the cost objects (e.g., services, products, or customers). ABC logic is that resources generate costs, activities utilize resources, and cost objects (products, services, customers) consume activities.

ABC is built on the notion that an organization has to perform certain activities in order to generate products and services. These activities cost money. The cost of each of these activities is only measured by and assigned to those products or services requiring identifiable activities and using appropriate assignment bases (called cost drivers). The results of ABC analyses offer an accurate picture of the real cost of each product or service, including the cost of serving customers. Nonactivity costs (such as direct materials, direct labor, or direct outside services) do not need to be included because these costs are readily attributable to the specific product or service considered.

ABC is most useful for companies with diverse products, service centers, channels, and customers, and for those companies whose overhead costs take up a large percentage of their overall product and service costs (Cokins 2001; Hicks 1999; O'Guin 1991).

In the service economy, direct manufacturing labor is no longer the overriding factor of production and the distinction between production and service departments has become decidedly blurred. The overall costs of products/services are more influenced by research, materials handling, procurement, equipment maintenance, quality control, and customer service requirements than by direct labor. ABC systems accumulate costs into activity cost pools, in accordance with the groups of major activities or business processes. The costs in each cost pool are largely caused uniformly by a single factor—the cost driver. ABC systems allocate costs to specific cost objects (products, services, customers, etc.) from the cost pools using these applicable cost drivers as the allocation bases. The cost information so provided is more accurate (Granof et al 2000).

According to Atkinson et al. (2001), ABC is particularly useful to service companies because:

- Most costs are indirect and appear to be fixed. Variable costs tend to be small and frequently near zero.
- Most costs are capacity-related costs. These costs are based on the amount required, rather than the amount used.
- Most costs are customer-specific rather that customer-independent.

For service companies, it is highly desirable to define the differential profitability of individual customers by implementing ABC.

All engineering managers should learn to practice ABC, because the traditional method of allocating overhead uses only high-level information about costs, and the general ledger system does not provide information related to time and resources spent on assignments and activities. In contrast, a well-practiced ABC method offers specific insights that include (1) a clearer picture for management as to what generates profits and losses for the company, (2) the ability to track operating profits for specific cost objects (such as customers, orders, and products), (3) the ability to determine whether a service center is efficient or deficient, and (4) the possibility to externalize the relative profitability among products and customers.

Even a company with overall profitability might lose money on certain products, orders, and customers in the absence of costing information created by ABC. According to the published best practices of some industrial pioneers (such as Honeywell Inc. and Coca-Cola®) on the use of activity-based costing, simpler ABC models deliver better results.

ABC has become increasingly popular with industrial companies, partly because it is useful for organizations of any size and does not require a massive effort to implement, and partly because of increased processing capabilities of personal computers (PCs), reduced prices for ABC software products, and increased competition forcing companies to achieve a better understanding of their own product costs. There are several ABC products on the market. Examples include Oros®, EasyABC Plus®, EasyABC Quick®, Metify ABM, and others.

6.2.3 Survey of ABC Uses in Companies

In July 2005, SAS conducted a survey of ABC uses among 529 companies in various industries, of which 56 percent were in services, 24 percent in manufacturing, and 20 percent in others. Of these companies, 42 percent had sales revenues at or below $100 million, and 26 percent were above $500 million.

The overall results showed that 35 percent of these companies were using ABC, 20 percent were engaged in piloting ABC, and another 32 percent were considering ABC. Only 10 percent of these companies had never considered ABC, and 2 percent were no longer using ABC.

The use of ABC increased with company size. ABC is being used or being considered for use by 71 percent of large companies, 58 percent of midsize companies and 42 percent of small business. Within the service industry, 46 percent of financial services and 58 percent of communications companies use ABC. The primary use of ABC by all industries is for costing and cost control. About 80 percent of all ABC programs were initiated by people in the finance department.

This survey demonstrated that ABC is important to the service industries and useful for future service systems engineers to know and master.

6.2.4 Sequential Steps to the Implementation of ABC

It is advisable to form a cross-functional team when implementing the ABC method of allocating indirect costs. The team should determine the cost objects. Examples include costs to serve customer; costs to purchase, carry, and process products; costs to order, receive, sell, and deliver products; and costs to perform other activities.

The team then needs to define activities that depict homogenous groups of work (such as accounting, machining, forging, and design) that lead to the cost objects.

Next to be determined are cost drivers. These are the agents that cause costs to be incurred in the activities. Cost drivers are factors that directly affect the cost of a given cost object. Examples of cost drivers are exhibited in Table 6.1.

Figure 6.1 shows a generic block-flow diagram for implementing ABC. It contains ten major steps:

1. The first very important step is to define an ABC project, whose objective aligns fully with business needs. The value of an ABC project, which would provide more structural knowledge regarding the cost objectives involved, must be convincing to leaders in the organization; otherwise, corporate commitment, management priority and availability of resources will likely to be in doubt. ABC is best to offer such cost structural information, when large overhead costs are shared among cost objectives (such as products, services, or customers groups).

Table 6.1 Cost Drivers

Activity	Cost Driver
Loading	Tons
Driving	Miles
Invoice processing	Number of invoice
Machining	Machining hours
Material movement	Weight
Production	Number of products

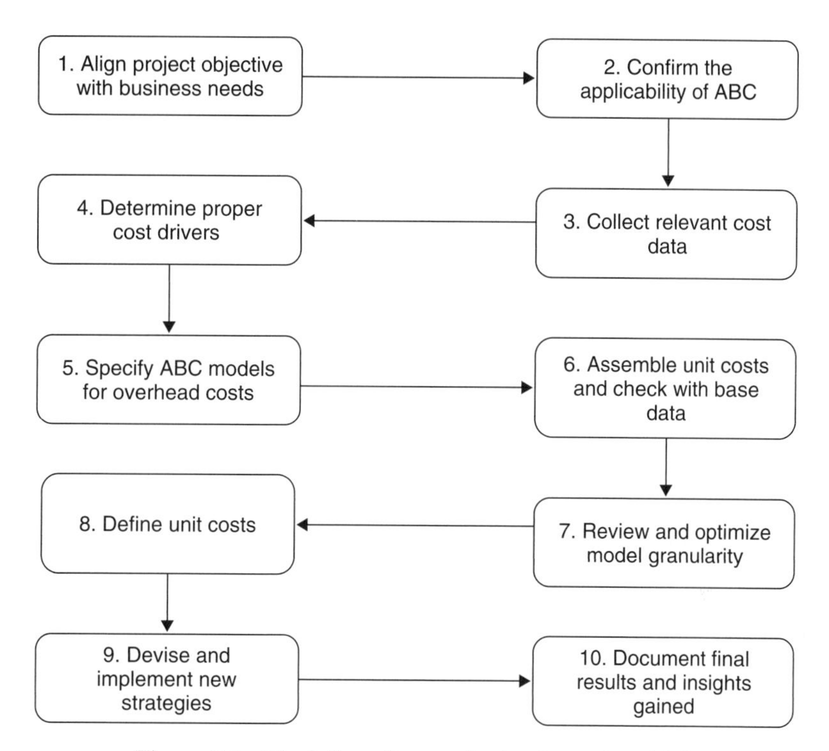

Figure 6.1 Block-flow diagram for implementing ABC.

2. Make sure that ABC is indeed the right method to accomplish the defined project objective.
3. Collect the relevant data from sources such as from general ledger, time sheets, procurement records, and other data sources, particularly those related to activities that are adding value to the cost objectives. Interviews with applicable managers may be needed to help identify additional details that may be useful in categorizing the data.
4. Define cost drivers for the various resources-consuming activities. Sources for this information could be industrial best practices, engineering literature, and/or rational and logical assumptions.

5. Create an ABC cost model for allocating overhead costs to the specific cost objectives at hand, including all resource-consuming activities that add value to the cost activities at hand.
6. The unit cost of the specific cost objective is then assembled. This result needs to be double checked to make sure that the total overhead cost agrees with the base data contained in the general ledger. If there are deviations, the ABC cost model must be readjusted to eliminate the differences.
7. The model is further reviewed to see if the ABC model's granularity is good enough for the purpose at hand. Past experience suggests that simple ABC models should be used in the beginning in order to deliver results and demonstrate value quickly. Detailed ABC models with a high level of sophistication should be used only if difference in outcome justifies the additional efforts required.
8. Define the final unit costs and determine its impact.
9. Devise new strategies to take advantage of the ABC results. Make sure that these strategies are implemented effectively to realize added profitability. Example of such new strategies may include promoting one product/service more than the other because of differences in gross margins, changing the customer support service strategy in favor of customers who are more valuable than others, modifying pricing strategies because of the cost differences between products/services, and others.
10. Document results and preserve insights gained so that lessons are shared with others in the future.

6.2.5 Practical Tips for Performing ABC

When a company initiates the process of creating an ABC system, it is highly recommended that the company start with a small group (pilot group) of well-informed and cross-functional workers. The team should interview other workers about what they do in their jobs. The team members should be cognizant of the potential fears of job restructuring that some employees may harbor as a result of the ABC studies.

The team should start with the "worst" department so that immediate success may be used to get faster "buy-in" from top management. The key for ABC success is to use "close-enough" data. The team should keep the level of information manageable by avoiding being bogged down with minute details. However, an ABC system that is too broad and general will not be useful. The team may have to try out ABC cost models of different granularities on small scales. For companies attempting to employ ABC cost models for the first time, useful outputs can be expected in six to twelve months.

Section 6.3.6 describes an updated ABC methodology, the time-based ABC, which is said to be able to handle complex and extended costing systems while keeping the expenses under control.

6.3 APPLICATION OF ABC IN VARIOUS SERVICE SECTORS

Activity-based costing may be applied to various businesses. In this section, its uses are described in banking and financial services, healthcare, governments, IT and library services and software development. Its improved version, the time-driven ABS, is also discussed.

6.3.1 ABC in Banking and Financial Services

Buckeye National Bank (Bamber and Hughes 2001) services both retail and business customers. The services include paying checks, providing teller services and responding to customers' service calls. All of these services consume labor-intensive activities. The resources involved are employees (salary and benefits), part-time workers, and those related to the operation of service call centers. The bank's traditional costing system suggests that it is more profitable for the bank to pursue additional retail customers and that business customers bring losses. As a result, the bank's retail customers grow, while the business customers remain stable. Yet the bank's profit is trending downward. Bank management becomes puzzled as to the reasons why.

Cost data were made available from the banks' traditional accounting system. Tables 6.2 and 6.3 display the data.

The solution obtained based on the traditional costing system is depicted in Table 6.4. It indicates that the annual profit of the retail account is $8.10, whereas that of the business account records a loss of $11.30. The logic of pursuing more retail customers appears to be correct.

An ABC pilot study was initiated to define (1) percent of time each employee spends on the aforementioned three activities, and (2) costs associated with toll-free phone lines, depreciation of equipment used for paying checks and providing teller services. The cost drivers were identified to be (1) number of checks processed for

Table 6.2 General Ledger Cost Data of Buckeye Bank

Buckeye National Bank	
Salaries of check-processing personnel	$700,000
Depreciation of equipment used in check processing	$440,000
Teller salaries	$1,000,000
Depreciation of equipment used in teller operations	$200,000
Salaries of call center personnel	$450,000
Tool-free phone line plus depreciation of related equipment	$60,000
Total costs	$2,850,000
Total profit	$650,000

Table 6.3 Additional Cost Data of Buckeye National Bank

	Retail	Business
$ value of checks processed	$9,500,000	$85,500,000
Checks processed	570,000	2,280,000
Teller transactions	160,000	40,000
Number of customer calls	95,000	5,000
Annual profit (interests) per account	$10	$40

Table 6.4 Traditional Cost Solutions

Traditional System	Retail	Business
$ value of checks processed	9,500,000	85,500,000
Cost per $ processed	$0.03	$0.03
Total cost	$285,000	$2,565,000
Cost per account	$1.90	$51.30
Annual profit per account	$8.10	($11.30)

paying check activity, (2) number of teller transactions for activity related to providing teller service, and (3) number of calls received for activity related to responding to customer inquiries.

Table 6.5 summarizes the unit cost of the three activities. Table 6.6 illustrates the cost assignments and the per-account cost. In fact, the retail accounts are shown to lose money, indicating that the bank policy of pursuing retail customers was based on an incorrect costing data.

6.3.2 ABC in Healthcare

The traditional cost accounting system tends to overestimate the unit cost of high-volume services and underestimate the cost of low-volume services. When the indirect costs are large, often the case in healthcare, the cost of services may be seriously misrepresented by the traditional cost accounting system (Walters et al. 2003).

The MaxSalud Institute for High Quality Healthcare in Chicalyo, Peru, is a private, nonprofit organization funded by USAID to provide health services to a low- and middle-income population of about 20,000 through two clinics and a central management support unit. This ABC application included (1) the description of all departments,

Table 6.5 Unit Costs of Three Principal Activities

ABC Solutions		
Cost per check processed	$0.40	(700K + 440K)/2850
Cost for teller transaction	$6.00	(1000K + 200K)/200K
Customer inquiries	$5.10	(450K + 60k)/100k

Table 6.6 ABC Solution

ABC Cost Assignment	Retail	Business
Paying checks (0.4*570k)	$228,000	912,000
Teller transactions (6 × 160)	960,000	240,000
Call centers (5.10 × 95,000)	484,500	25,500
Total	1,672,500	1,177,500
Per-account cost	$11.15	$23.55
Net Profit per Account	−1.15	16.45

services, and activities by department, (2) staff estimates of time spent on each activity and unproductive time, (3) estimated cost of all activities by each department using wage and other data, (4) trace activities and costs within and across departments to services provided, and (5) estimated service volumes from records and calculate unit costs (cost/volume).

This ABC study identified 107 distinct activities at MaxSalud, including training and meetings. ABC derived unit costs that were generally higher than previous estimates and much higher than fees charged. Among others, the ABC study discovered that the primary activity of delivering a baby accounts for only 23 percent of the total unit cost, with 42 percent from secondary activities (i.e., admission, general services, and others), and 35 percent from overhead (i.e., the central management support unit). The study also revealed information on activities associated with unproductive time, such as repeating lab tests.

The study concluded that ABC is potentially very valuable to MaxSalud to set policy, manage expenditures, and even raise funds, but it requires reliable data systems for costs and service statistics, management attention, staff support, and technical assistance.

Several useful guidelines were suggested when utilizing ABC in developing countries:

- ABC requires complementary accounting systems that provide reasonably accurate costs organized by cost category and department.
- ABC requires accurate information on the volume of services provided.
- Access to and strong cooperation from personnel is important.
- Technical assistance and guidance on the ABC methodology might be necessary initially.
- To derive long-term benefits, cost data trending is essential. Data trending requires continued efforts of costing analyses, which, in turn, require management commitment.

6.3.3 ABC in Governments

State and local governments have a common goal: to provide services to the public at the lowest possible cost. Many governments, however, are not perceived to be particularly efficient in realizing this goal. Most governments do not have a clear idea of all the costs associated with their own in-house operations. They typically underestimate the true costs of in-house operations by as much as 30 percent, typically because they omit many indirect costs when determining the total costs of performing any function (Buttross and Schmelzle 2008).

Traditionally, state and local governments have practiced cost control by simply aggregating costs for the units within a governmental body and comparing the total costs of those units from period to period. Overhead costs that applied to multiple units were normally allocated to the units based on arbitrary measurements that are common to all units, such as square feet occupied or direct labor costs. Volume-based measurements are used predominantly in traditional systems to determine overhead rates and to assign overhead costs to products and services (Briner, Alford and Noble 2003).

The Texas Department of Agriculture (TDA) operates six livestock facilities for inspecting animals before exporting them to Mexico. The inspection is necessary in order to be in compliance with Mexico's health regulations. The Texas Department of Agriculture charges fees for the inspections. When trucks carrying livestock arrive at the TDA export pens, the pen manager checks the driver's document before authorizing workers to unload the truck. The unloaded animals are placed in pens to rest. They are then moved to an inspection area and then inspected by veterinarians. Animals that do not pass inspection will be reloaded, fees are collected, and export to Mexico is denied. Those animals that do pass inspection are immediately reloaded or returned to the initial pens to wait for a truck. Once the truck has been reloaded and cleared for export, required fees are collected, the document is returned to the driver, and the truck is sealed for departure to Mexico. Four steps are taken to apply ABC to this case:

1. Identify costs and resources.
2. Identify the direct and indirect costs of activities.
3. Assign costs to activities.
4. Calculate unit costs.

There are altogether nine distinct activities. Costs are grouped into four primary categories and one indirect-cost category. Five cost pools are created. The total costs of each pool were expressed on a per-unit basis according to the appropriate driver for the particular activity. For details of its quantitative analyses, see Briner, Alford, and Noble (2003).

6.3.4 ABC in IT and Library Services

Under a contract by the Australian Department of Education, Training and Youth Affairs, Ernst & Young developed the activity-based costing methodology in 1998 for university environments [Ernst & Young, 2000]. The key concepts of this methodology include:

- *Resources*—economic elements directed to the performance of activities (e.g., salaries, consumables, telecommunications, computers, etc.).
- *Activities*—a unit of work performed within an organization (e.g., providing IT support).
- *Cost objects*—any entity of managerial interest for which a separate cost measurement is desired (e.g., customer, product, service, process, faculties, courses student types or modes of delivery).
- *Cost drivers*—a measure of the quantity of resources consumed by an activity (e.g., headcount of FTEs (full-time equivalents), square meters, number of computers, number of staff/students, etc.).

This work was further extended to the costing of information technology (IT) and library services (Information and Education Services Division 2001). It illustrates that ABC can be applied to inform university management of the actual costs of the IT and library services they provide, how these costs might vary with changes in demand from client groups, the cost of different student types enrolled, or the unit cost of providing particular services in comparison to other universities or providers.

The ABC model of Ernst & Young contains three modules based on the use of the Easy/ABC Plus Software: (1) Resource Module, (2) Activity Module, and (3) Cost Object Module. Within each of these modules, costs are grouped together within (1) centers—containing groups of accounts that represent a common cost, process or cost object, such as a department or program, (2) accounts—containing costs of resources (e.g., general ledger costs), activities and cost objects, and (3) cost elements—specific costs for accounts. It includes a detailed list of activity drivers for IT and library service activities.

The activities in libraries include material acquisitions, cataloging, loan processing, and the shelving of library materials. Activity cost pools are the accumulations of all overhead costs involved in the processing of each activity cost driver. The usual steps of utilizing ABC include:

1. Identify key activities and relevant cost drivers. Cost drivers are those that caused the occurrence of each activity (the cause-and-effect relationships).
2. Allocate staff time to activities.
3. Allocate staff salaries and other costs to activity cost pools.
4. Define cost per cost driver.

Table 6.7 lists the typical activities and the corresponding cost drivers when applying ABC to library services.

Table 6.7 Cost Drivers for Activities in Libraries

Number	ABC Application in Library Cost Activity	Cost Drivers
1	Item loans	Number of loans
2	Item returns	Number of book returns
3	Item renewals	Number of renewals
4	Items recalls	Number of recalls
5	East loans	Number of easy loans
6	Overdue books	Number of overdue books
7	Closed reserve—setup	Number of reserve Items
8	Serials maintenance	Number of serial titles
9	Interlibrary loans—requester	Number of items requested
10	Interlibrary loans—suppliers	Number of items supplied
11	Invoicing	Number of invoices
12	Intercampus loans	Number of items supplied
13	Film and video	Number of film and video loans
14	Shelving	Number of items shelved
15	Equipment maintenance and depreciation	Equipment use
16	Reference desk	Number of inquiries
17	Faculty work—businesses	Number of EFTSU—business
18	Faculty work—health	Number of EFTSU—health
19	Faculty work—education	Number of EFTSU—education

EFTSU = Equivalent full-time student units

6.3.5 ABC in Software Development

Traditional software development follows a "waterfall" approach by performing needs analysis, coding software, testing, documentation, training, and implementation. Component-based software development, by contrast, aspires to create reusable codes on a large scale. The overhead costs for developing component-based codes are expected to be significant due to the maintenance of reuse infrastructure and the development, locating, evaluation, and adapting of components. Table 6.8 lists some of the activities that are to be cost-estimated and allocated to the proper cost objects in such an environment using the ABC method (Fichman and Kemerer 2002).

In Table 6.9, the corresponding cost drivers are identified to implement the ABC costing system. Here, the basic steps of implementing ABC are quite similar to those in other service sectors.

6.3.6 Time-Driven ABC

Kaplan and Anderson (2004) identified a number of barriers to the implementation of ABC. For example, if the cost system is complex or of expanded scope, it would demand large efforts (interviewing staff, prepare data, maintain database) beyond what some companies would be willing to spend (rising cost and time demand). In addition, it was noted that frequent employee surveys of details activities, and use of resources might upset some, causing them to be unwilling to collaborate (employee irritation). Furthermore, the estimates offered by employees are not always accurate (e.g., not including break, slack time, and other idle and unused time, etc.).

As indicated in Fig. 6.2, the selection of the right granularity (degree of details and scope) for the ABC costing system is a key factor for its successful implementation. It needs to be refined enough to deliver accurate results, yet not too detailed in modeling for the system to become overly expensive. ABC can be time-consuming and complex, with costs potentially outweighing benefits.

A new time-driven ABC methodology, in which the numbers of individual activities are reduced without losing the complex nature of the operations involved, was suggested. The basic approach is to combine a number of activities into departmental processes and use managerial estimates, rather than employee surveys, to amass key data. Specifically, it recommends that managers directly estimate the resource demands imposed by each transaction or cost object (product or service) rather than assign resource cost first to activities and then to the cost objects. The capacity of most resources is measured in terms of time availability. For each group of resources, managers estimate only two parameters: (1) the cost per time unit (minute, hour, day or month) of supplying resource capacity, and (2) the unit times (minute, hour, day or month) of consuming the resource capacity by the cost objects. The practical resources capacity is recommended to be at about 80 percent of its theoretically available level. The cost-driver rates are then readily calculated, thus allowing unit times to be estimated even for complex, specialized transactions. The objective is to be approximately right, rather than overly precise. This approach eliminates the need for extensive surveys that annoy employees and reduces the high costs associated with data collection, maintenance, and processing.

Table 6.8 Software Development Activities to be Cost-estimated Using ABC

	Activity Center	Activity	Description
A	Reuse infrastructure creation.	1. Develop reuse infrastructure	Develop reuse policies, tools, processes and measures.
B	Reuse infrastructure maintenance and reuse marketing.	2. Maintain reuse infrastructure.	Maintain reuse policies, tools, processes, and measures.
		3. Communicate existence of components.	Advertise the reused components.
C	Reuse administration.	4. Administer reuse measurement accounting and incentives.	Measure ongoing levels of reuse on projects. Allocate costs of reuse. Reward individuals and/or teams for reuse achieved.
D	Reuse production.	5. Analyze reuse components.	Investigate opportunities for acquiring for developing components.
		6. Develop or acquire reuse components.	Develop components to be reusable, acquire and/or generalize components previously developed.
		7. Certify components.	Certify components for reusability.
		8. Document, classify, and store components.	Document, classify and stored offered components.
E	Reuse consumption.	9. Search for components.	Search libraries for components (or assist the search process).
		10. Retrieve, understand and evaluate components.	Understand and evaluate components found in libraries.
		11. Adapt and integrate components.	Modify (if necessary) and integrate components.
F	Reuse maintenance.	12. Maintain reusable components.	Correct and extend components in the library.
		13. Update reusable components.	Correct and extend components in systems.

Source: Fichman and Kemerer 2002

Table 6.9 Cost Drivers in Developing Reusable Software

Activity	Casual Cost Driver	Performance Measure
1. Develop reuse infrastructure.	Reuse policy and expected volume.	Breath and quality of infrastructure.
2. Maintain reuse infrastructure.	Size/complexity of infrastructure	Breath and quality of infrastructure
3. Communicate existence of components.	Size of reuse libraries Size of developer community	Number of communications
4. Administer reuse measurement accounting and incentives.	Reuse policy Volume of reuse	Number of measurements
5. Analyze reuse components.	Reuse policy	Number of components evaluated
6. Develop or acquire reuse components.	Reuse opportunity	Number and quality of reusable components acquired
7. Certify components.	Offered components	Number of components certified
8. Document, classify and store components.	Certified components	Number of components stored
9. Search for components.	Project policy Reuse potential of the domain Size/quality of reuse libraries	Number of searches Number of promising components found
10. Retrieve, understand and evaluate components.	Promising components found	Number of components evaluated and number of components accepted
11. Adapt and integrate components.	Retrieved components selected for use	Number of attempted integrations; number of successful integrations
12. Maintain reusable components.	Volume of reuse	Number of components maintained
13. Update reusable components.	Volume of reuse	Number of components updated

Use is made of *time equation,* which computes the activity time of a process, comprising groups of potential activities. An example of such time equation is as follows:

> Inside sales order entry process time = 5 + (3 × number of line items) + 15 [if new customers] + 10 [if expedited order].

The data collection process is simplified while still preserving the complex nature of the process involved. If the capacity of a resource (e.g., warehouse, memory storage) is not measured by time, then managers would calculate the resource per the correctly assigned unit (cost per cubic meter or cost per megabyte). The authors claim that the

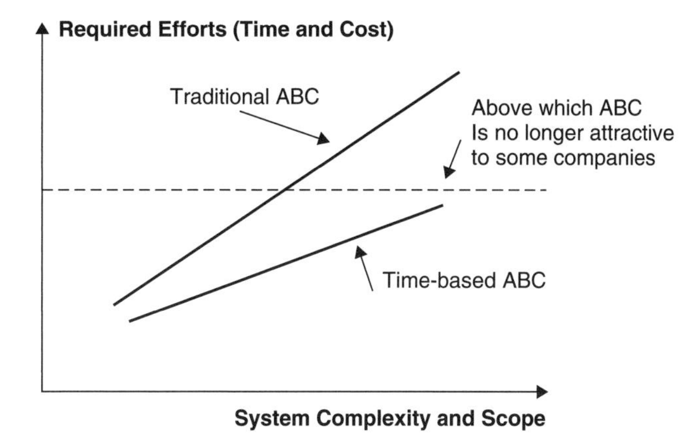

Figure 6.2 Complexity and scope of costing system versus required development efforts. (Source: Kaplan and Anderson)

resulting data precision, which is diminished somewhat due to the use of estimates, is still adequate for the purpose at hand and claim to have successfully applied this new method to over 1,000 companies. The "Compton Financial" case is described in Kaplan and Anderson (2008). The application of time-driven ABC to library services is described in Pernot et al. (2007).

6.4 APPLICATION OF ABC IN THE MANUFACTURING SECTOR

Product costing in a product-centered company requires the computation of costs related to direct material (DM), direct labor (DL), and factory overhead (FO) for the following operations:

- Raw materials (Stores)
- Work in progress (WIP)
- Finished goods (FG)

Additional Definitions:

Prime cost = DM + DL

Conversion cost = DL + FO

Indirect manufacturing overhead (FO), also called factory burden, includes all costs other than DM and DL.

Typically, T-accounts are set up for Stores operations, WIP operations, and FG operations.

A T-account is a tool used commonly by accountants to record transactions. As a convention, an increase in assets (e.g., cash, accounts receivable, inventories, land, machines, etc.) is debited to the left side of the T-account. A decrease in assets is credited to the right side of the T-account. The opposite is true for liabilities and owners' equities. For additional discussions on the use of T-accounts, see Appendix 7.10.1. Here's a schematic diagram that illustrates product costing:

Stores		WIP		FG	
Debit	Credit	Debit	Credit	Debit	Credit
(1)	(2)	(3)	(4)	(5)	(6)

Explanations:

1. Purchasing raw materials (an increase in asset).
2. Putting materials into a production process (a reduction in stores assets).
3. Production is initiated, adding value to raw materials, while consuming labor, materials, utilities, and other resources (assets are increased).
4. Production is complete and units shipped out to finished goods operations (a reduction of WIP assets).
5. Receiving of finished goods in storage (an increase of FG assets).
6. Finished goods are shipped out for sale (a reduction of FG assets).

Note that in the schematic diagram (a reduction of WIP assets), materials flow from left to right, whereas cash (financial resource) flows from right to left.

Here's a specific example of product costing:

Stores		WIP		FG	
Debit	Credit	Debit	Credit	Debit	Credit
Beg. 75	3 (b)	Beg. 22	440(f)	Beg. 50	
198(a)	187(c)	125(d)		440(f)	430(g)
		147(e)			
		187(c)			
End. 83(h)		End. 41(h)		End. 60(h)	

a. Purchased material for $198.00.
b. Received credit for having returned the material purchased.
c. Direct material actually shipped to WIP and used in the accounting period.
d. Direct labor used and cost assigned.
e. Factory overhead used and cost assigned.
f. Product completed and transferred out (cost of goods manufactured).
g. Finished goods shipped out to customer, receiving the cost of goods sold (CGS) as credit.
h. The sum of ending balances in Stores, WIP, and FG, representing inventory at the end of the accounting period.

In computing the costs of goods sold, the inclusion of materials and labor costs is rather straightforward, as these are direct costs and quite easy to track. The difficulty in product costing is the inclusion of indirect costs—namely, the factory overhead, as in the specific example just illustrated. For manufacturing operations in which the factory overhead makes up only a small fraction of the total product cost, the precision with which to allocate the indirect costs is irrelevant. However, in other circumstances, the allocation of indirect costs must be precise, especially when the factory overhead becomes a major portion of the product cost and the plant facility generates multiple products.

6.4.1 Specific Example of Applying ABC to XYZ Manufacturing Company

XYZ is a small manufacturing company with $10 million in annual sales. It makes components for the automotive industry, and the key processes involved are forging and machining. The product-related operating activities are as follows:

1. Buy steel bars from outside vendors.
2. Test steel bars upon delivery and moving them to storage.
3. Send the bars to the forging area when needed for an order, at which point they are sandblasted and cut to desired lengths. Since most of the bars are large, they are then moved in bins that hold twenty to twenty-five pieces.
4. Size the bars and moving them to a forging operation station where they are shaped. The bars are then moved to the in-process storage. In some cases, a steel bar may need to be forged up to three times.
5. Transfer the bars for each forging procedure from in-process storage to the forging areas and then back to the in-process storage.
6. Move the steel bars after the final forging from the in-process storage to the machining area where they are finished. The bars are then sent to finished-goods storage.
7. Sort, pack, and load the bars is done in the shipping area and onto trucks for delivery to customers.

Before using ABC, the company applied the traditional costing method that included the following three steps:

1. Assign manufacturing costs to products by using a plantwide costing rate on the basis of direct labor. The setup costs are included in the manufacturing overhead.
2. Assign the nonmanufacturing costs to products via a general and administrative (G&A) rate that is calculated as a percentage of the total cost.
3. Define the direct labor rate and the G&A rate on the basis of the actual results obtained for the preceding year.

The deficiencies of the traditional method are obvious. The traditional method is used because management does not know better methods.

When implementing ABC, the company did not buy any ABC-specific software. Instead, it used a standard spreadsheet program. Specifically, the company considered five factors:

1. *Setup costs.* Management assigned equipment setup costs only to the steel bars in a given equipment process.
2. *Forging costs.* Depending on the weight of the steel bar involved, one or two operators may operate the forging press. Before forging, each steel bar must be induction heated, with the heating cost being dependent on the mass of the bar involved. Thus, the forging cost consists of three parts:

(a) Press-operating costs on the basis of press hours.
(b) Production labor cost on the basis of labor hours.
(c) Induction-heating costs on the basis of the steel bar's weight.

3. *Machining costs.* The machining centers do not require full-time operators. Once the machines are set up, workers load and unload parts for multiple centers. On average, one machine-worker hour is required for every two and a half machine hours. Thus, the costs of machine-shop workers are treated as the indirect costs assigned to products on the basis of machine hours.
4. *Material movement costs.* Depending on the size of the bar, the bin size, and the required forging and machining steps, the material movement cost could vary significantly from one bar to another. Thus, the material movement cost is assigned to each bar on the cost-per-move basis.
5. *Raw material procurement and order processing costs.* These are readily traceable on the basis of records on hand.

The ABC cost model for the XYZ Manufacturing Company is illustrated in Table 6.10. The final results of ABC implementation are impressive. The company's sales tripled and its profit increased fivefold after the introduction of ABC. Much of this improvement came from a more profitable mix of contracts generated by a pricing and quoting process that more closely reflects the actual cost structure of the

Table 6.10 ABC Model for a Manufacturing Company

	Forging Press Hour Cost	Machine Hour Cost	Induction Heating Cost	Material Movement Cost
Directly Attributable Costs	Depreciation Utilities Manufacturing supplies Outside repairs	Depreciation Utilities Manufacturing supplies Outside repairs Straight-line wages Fringe benefits Payroll taxes Overtime premium Shift premium	Depreciation Utilities Manufacturing supplies Outside repairs	Depreciation Utilities Manufacturing supplies Outside repairs Straight-line wages Fringe benefits Payroll taxes Overtime premium Equipment leases
Distributions	Maintenance Buildings and grounds Manufacturing engineering Commodity overhead	Maintenance Buildings and grounds Manufacturing engineering Commodity overhead Supervision	Maintenance Buildings and grounds Manufacturing engineering Commodity overhead	Maintenance Buildings and grounds Human resources Supervision
Total	Total costs	Total costs	Total costs	Total costs
Rate	$ per press hour	$ per machine hour	$ per heating weight	$ per move

company. Particularly useful are the isolation and measurement of material movement costs that result in operational changes for increased efficiency.

6.5 TARGET COSTING

If the products or services offered by companies are new and without competition, companies can easily price them on the basis of "cost plus," which means defining product prices by adding an acceptable gross margin on top of their product costs. On the other hand, for products and services offered to a competitive marketplace, prices might need to be constantly reduced due to external market forces (i.e., competition, economy, technologies, industrial status, etc.). In such cases, companies are able to maintain their gross margins only if they are in a position to adjust their costs downward according to the externally imposed price reductions. Target costing is a useful management tool to assist companies in such circumstances (Swenson, Ansari, and Kim 2003). The principles of target cost are as follows:

- *Price-let costing*. Market prices are used to determine allowable—or target—costs. Target costs are calculated using a formula similar to the following: Market price − Required profit margin = Target cost.
- *Focus on customers*. Customer requirements for quality, cost, and time are simultaneously incorporated in product and process decisions and guide cost analysis. The value (to the customers) of any features and functionality built into the product must be greater than the cost of providing those features and functionality.
- *Focus on design*. Cost control is emphasized at the product and process design stage. Therefore, engineering changes must occur before production designs, resulting in lower costs and reduced "time-to-market" for new products.
- *Cross-functional involvement*. Cross-functional product and process teams are responsible for the entire product from initial concept through final production—brainstorming session to generate ideas to cut costs.
- *Value-chain involvement*. All members of the value chain (e.g., suppliers, distributors, service providers, and customers) are included in the target-costing process, as well as any cost saving opportunities offered by suppliers or based on customer inputs.
- *A life-cycle orientation*. Total life-cycle costs are minimized for both the producer and the customer. Life-cycle costs include purchase price, operating costs, maintenance and distribution (disposal) costs.

Companies use target costing to establish concrete and highly visible cost targets for their new products. The gaps between the target cost and cost projections for the new product are to be closed using basic value engineering tools.

6.6 RISK ANALYSIS AND COST ESTIMATION UNDER UNCERTAINTY

When estimating product costs, some costs are well-defined and firm, while others are not. Similarly, some projects in engineering based on past experience are risk-free, while others are not. Risks are defined as a measure of the potential variability of an outcome (e.g., cost or schedule) from its expected value. Risks must be properly accounted for in projects (Zio 2007; Bedford and Cook 2001).

6.6.1 Representation of Risks

Risks can be graphically represented by a probability distribution function. Three cases are examined next.

Case 1 refers to an investment in U.S. Treasury bills. The yield of 10-year U.S. Treasury bills has varied in the range of 8 percent (1990) to below 4 percent (2009). But once an investor buys the Treasury bills from the U.S. government, the yield is locked in until maturity. Such an investment is guaranteed by the assets of the U.S. government. This return is graphically represented by a vertical line at a fixed return rate (6 percent assumed in this example) with 100 percent probability. (See Fig. 6.3, Case 1.)

Case 2 is an investment in a blue-chip corporate stock with a most likely return of 8 percent. Due to market conditions being usually unpredictable, the return of such an investment has some measure of risk, as depicted by the bell-shaped curve centered on 8 percent in Fig. 6.3. The return may vary from 4 percent (minimum) to 12 percent (maximum). Risks are measured by the standard deviation of this probability distribution curve (e.g., sigma 2). (See Fig. 6.3, Case 2.)

The bell-shaped curve is mathematically represented by the normal probability density function:

$$F(x) = \frac{1}{\sigma_x \sqrt{2\pi}} e^{-\frac{1}{2}\left(\frac{x-\mu}{\sigma_n}\right)^2} \tag{6.1}$$

where

σ = standard deviation
μ = mean
π = 3.14159

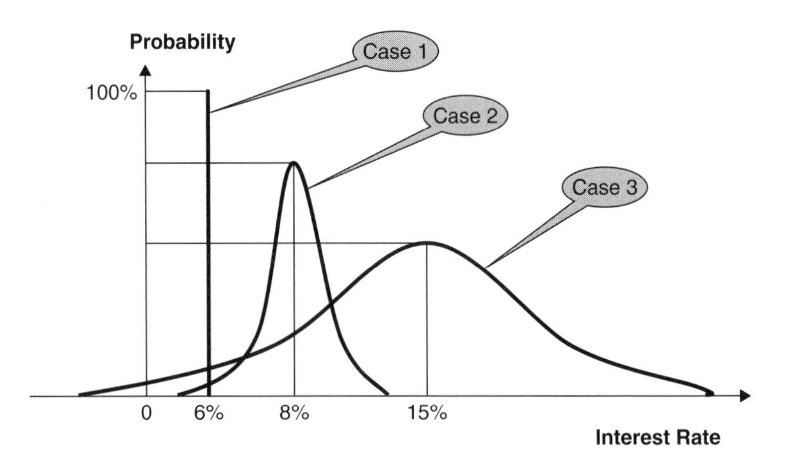

Figure 6.3 Representation of risks.

The area underneath the curve is normalized to be 1.

Case 3 is the return of an investment in real estate, centered, for example, at 15 percent. Because this investment requires tax payment, maintenance costs, and other expenditures, its minimum return can be negative. Its upside potential might be very large, however, if commercial developments and property-zoning results become favorable. This case is designated by a bell-shaped probability curve having a large standard deviation (sigma 3 being larger than sigma 2) and with its most likely return centered at 15 percent. Furthermore, the probability of achieving its most likely return of 15 percent is now only 50 percent. (See Fig. 6.3.)

Risky events can be represented mathematically by the normal probability density function, which is defined by two parameters, standard deviation and mean. Besides the normal, several other probability density functions (such as triangular, Poisson, and beta) can also be used to represent risky costs. (See Section 6.6.3.)

6.6.2 Project Cost Estimation by Simulation

Recent literature outlines the advancements of PC-based techniques that estimate project costs under uncertainty (Mun 2006; Fishman 2005; Rubinstein and Kroese 2007). The key elements of these PC-based techniques consist of the following steps:

1. Construct a *cost model* for the projects at hand with a spreadsheet program (e.g., Excel®, Lotus 123™) (Smith 2007). The spreadsheet program takes care of the required computation of the cost model, such as addition, subtraction, multiplication, and division. The numerical values entered in the spreadsheet cells are typically deterministic, each having a well-defined and fixed magnitude. The cost model encompasses all cost components and computes the total project cost.
2. Make a *three-point estimate* for each of the component costs, composed of the minimum, the most likely, and the maximum values. This is to account for the perceived-cost uncertainty. Past experience may serve as a guide in the selection of these values.

   Select a probability distribution function (e.g., triangular, normal, beta, or other distribution functions) to represent the three-point estimate of the component cost. Repeat this step for all other cost components of the project.
3. Activate *risk analysis software* to replace the deterministic values contained in the spreadsheet cells by the probability distribution functions chosen to represent the corresponding three-point estimates.

   Currently, commercial PC-based software products are becoming readily available. Examples:

   - CrystalBall® for Excel
   - @Risk and BestFit programs (Palisade Corporation) for Lotus 123 spreadsheet

     The BestFit program can be used to define a suitable probability distribution function on the basis of numerical three-point estimates or other user-specified input data.

The activated risk analysis software automatically converts all input probability density functions to their corresponding cumulative distribution functions. The technical fundamentals related to this conversion are illustrated in Appendix 6.11.10.

Simulations may also be performed using Microsoft Excel, but without the add-on software products identified above (Zarc 2009).

4. Conduct a *Monte Carlo simulation* to compute the total project cost. Upon activation of risk analysis software, a random number is first generated between 0 and 1. This random number designates a trial probability value (e.g., P1). Using this random number, the specific cost value is read from the cumulative distribution of the cost component C1, which represents a random input variable (e.g., C11, see Fig. 6.4). A second random number is generated (P2), which is then used to define the cost components C2 (e.g., C21). A third random number is generated to define the cost C31of a third cost component. This process is continued until the costs of all cost components are defined. The total project cost (e.g., TPC1) is then calculated with the spreadsheet program that contains the cost model. This is one outcome of the random output variable TPC.

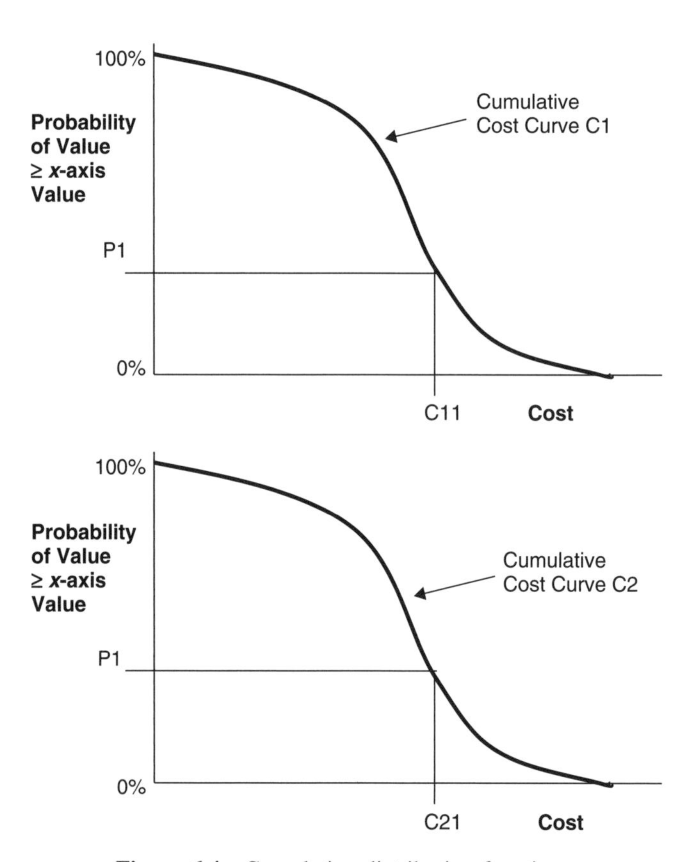

Figure 6.4 Cumulative distribution functions.

The sampling process is repeated thousands of times to create a distribution of the total project costs (e.g., TPC1, TPC2, etc.). These output results are then statistically grouped into bins (with zero to maximum value) to define a cumulative distribution. The resulting cumulative distribution for the total project cost TPC may then be converted back to its corresponding probability density function.

The total project cost so generated has a set of minimum, most likely and maximum values.

5. Interpret the *total project cost* represented in a cumulative distribution to arrive at the following typical results:

- There is an 80 percent probability that the total project cost will not exceed $\$D$.
- The minimum, most likely, and maximum total project costs are $\$A$, $\$B$, and $\$C$.
- The standard deviation of the total project cost is x, or the overall measure of the project risk.

Information of this type is extremely useful for decision making. It is particularly true in situations where multiple projects are being evaluated for investment purposes.

There is an additional benefit realized by using the just-described cost estimation by simulation. Because of the risk-pooling effect due to risk sharing among all input cost components, the total project cost is expected to have a lower overall risk than the risk levels of its individual components. Various studies (for example, Canada, Sullivan, and White 2004) have confirmed that the total project cost computed by simulations requires a smaller contingency cost for a given risk level than that computed by the traditional method using deterministic values.

Other important applications involving risk analyses include (1) project schedule and (2) portfolio optimization.

The value of risk analyses is typically to make explicit the uncertainties of input variables, to promote more reasoned estimating procedures, to allow more comprehensive analyses—or the simultaneous variation of all input variables involved—and to measure the variability of output variables. It is believed that a decision maker can make better decisions with a fuller understanding of the risk-based implications of the decision.

The use of risk analysis in the business and engineering environments is expected to become increasingly widespread in the years to come. Engineering managers are advised to become familiar with advanced tools for risk analyses.

6.6.3 Examples of Input Distribution Functions

In engineering cost estimation, several distribution functions are often used as inputs. Figure 6.5 displays the triangular probability density function. This is the easiest function to utilize, as the three-point estimates may be directly incorporated into this representation.

Figure 6.6 illustrates the normal probability function. Figure 6.7 displays the beta probability density function. Figure 6.8 depicts the Poisson probability function.

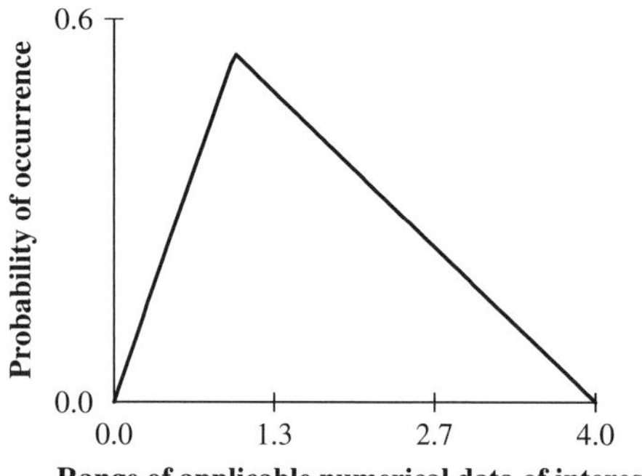

Figure 6.5 Triangular distribution function.

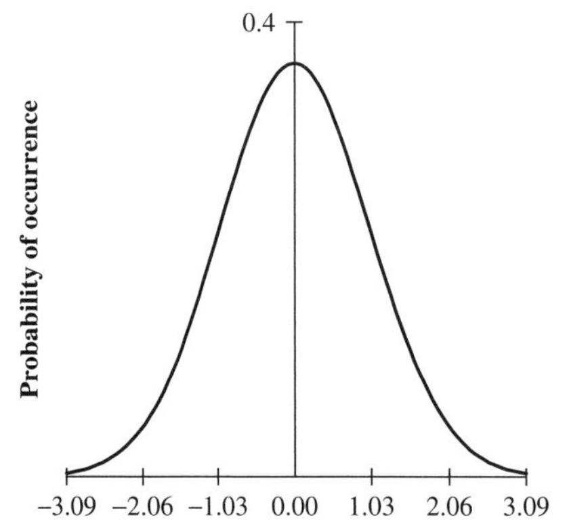

Figure 6.6 Normal distribution function.

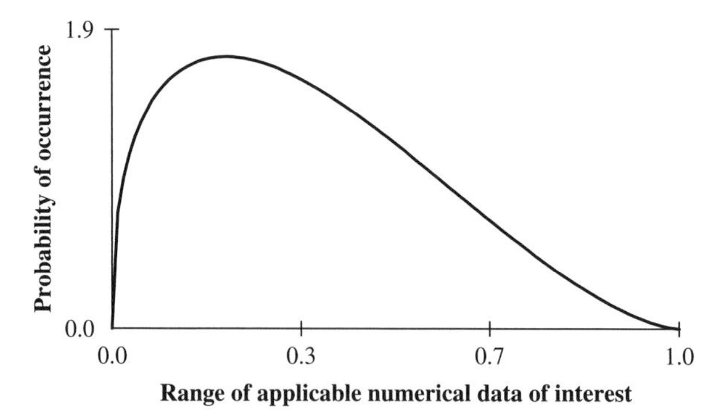

Figure 6.7 Beta distribution function.

6.6.4 Application—Cost Estimation of a Risky Capital Project

As an example, the cost estimation for a turnkey capital project is illustrated in Table 6.11. Project managers define the base (e.g., the most likely) estimates, as well as the lower and upper bounds for each cost item in the estimate. Doing so will force

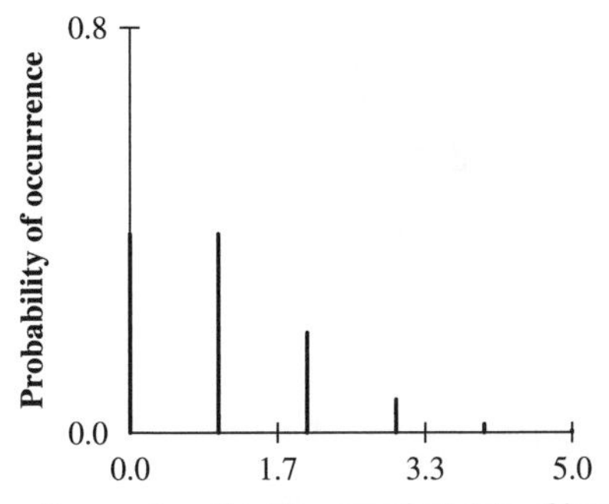

Figure 6.8 Poisson distribution function.

Table 6.11 Cost Model of a Capital Project

Number	Cost category	Base ($K)	MIN	MAX	Minimum ($)	Most likely ($)
1XXX	Cold box	$748.00	−5%	5%	$710.60	$748.00
2XXX	Rotating equipment	$742.00	−3%	3%	$719.74	$742.00
3XXX	Process equipment	$658.00	−2%	5%	$644.84	$658.00
4XXX	Electrical equipment	$194.00	−5%	3%	$184.30	$194.00
5XXX	Instrumentation	$295.00	−2%	10%	$289.10	$295.00
6XXX	Piping mat/specials	$121.00	−2%	10%	$118.58	$121.00
711X	Civil construction	$284.00	−2%	5%	$278.32	$284.00
712X	Mechanical construction	$390.00	−1%	20%	$386.10	$390.00
713X	Electrical construction	$85.00	−2%	10%	$83.30	$85.00
71?X	Other contracts	$83.00	−5%	10%	$78.85	$83.00
716X	Purchased enclosures	$48.00	−5%	12%	$45.60	$48.00
717X	Fabrication	$179.00	−5%	4%	$170.05	$179.00
7890	Freight	$80.00	−5%	15%	$76.00	$80.00
84X0	Field support	$188.00	−5%	7%	$178.60	$188.00
85XX	Startup	$60.00	−10%	30%	$54.00	$60.00
81X0	Product line design	$516.00	−1%	15%	$510.84	$516.00
8150	Project execution	$333.00	−5%	20%	$316.35	$333.00
	Total neat	$5,004.00				
	Contingency	$131.90				
	GRAND TOTAL	**$5,135.90**				

them to externalize the reasons for any variance and require them to think hard about the contingency plans for each.

The output total cost is depicted by the probability density and cumulative distribution functions. (See Figs. 6.9 and 6.10.) From the output cumulative distribution, the following results are readily obtained:

- The most likely total project cost is $5,136,000, which, of course, only echoes the input data.

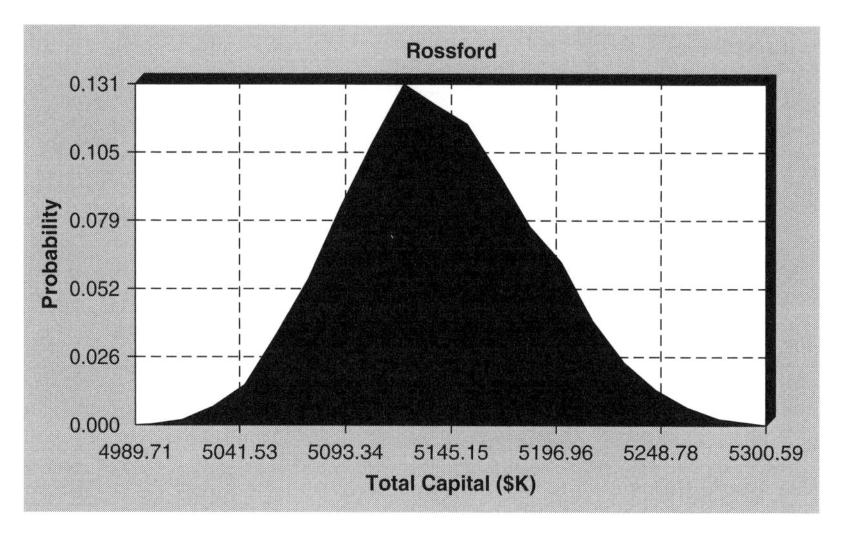

Figure 6.9 Total capital cost represented in a probability density function.

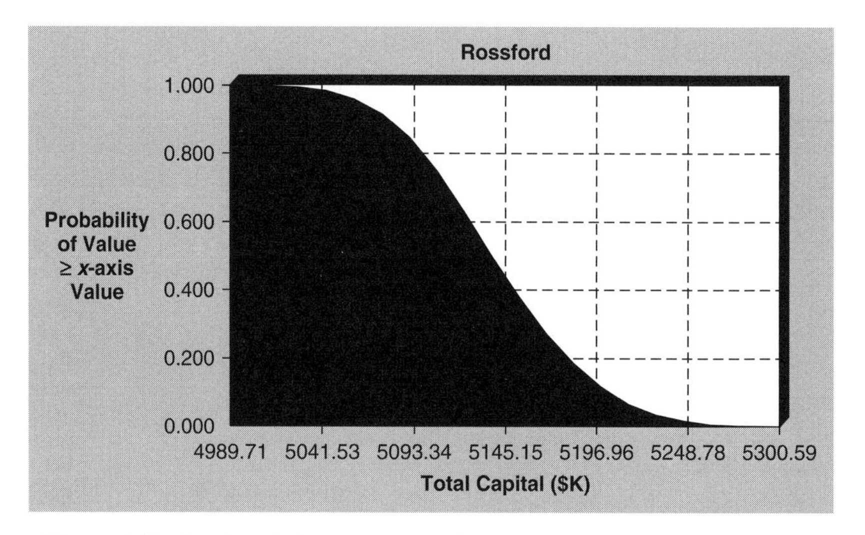

Figure 6.10 Total capital cost presented in a cumulative distribution function.

- There is an 80 percent probability that the project cost will exceed $5,100,000.
- There is a 20 percent probability that the project cost will exceed $5,170,000.
- The maximum project cost is $5,250,000.
- The minimum project cost is $4,989.710.

Note that the information offered by items B, C, D, and E is new. In a traditional, deterministic project-cost estimate, the cost figures for items A, D, and E would be the same, and there would be no information of the kind offered by items B and C. When choosing among various projects that may have similar outcomes in the most likely project costs, information offered by items B and C is especially critical for differentiating projects by their inherent risks.

For construction managers, the estimation of contingency is of critical importance. Instead of assigning specific contingencies as a percentage to each cost category items, the simulation calculates the contingency of the construction project. Probabilistic models have been used in the past to define construction project contingency (Touran 2003).

In addition, project control and tracking of coefficient of variation can be readily conducted. (See Figs. 6.11 and 6.12.) The coefficient of variation is defined as $CV = 100(\sigma/\mu)$wherein σ is the standard deviation and μ is the mean of the total project cost distribution function.

6.6.5 Other Techniques to Account for Risks

Several other techniques are also routinely applied in industry to assess and manage the risks associated with projects (Bierman 2006; Sullivan, Wicks and Koelling 2008).

A. *Sensitivity analysis.* Because of possible variation of specific input parameters, what-if analyses are typically performed to assess the sensitivity of the project cost and time to completion.

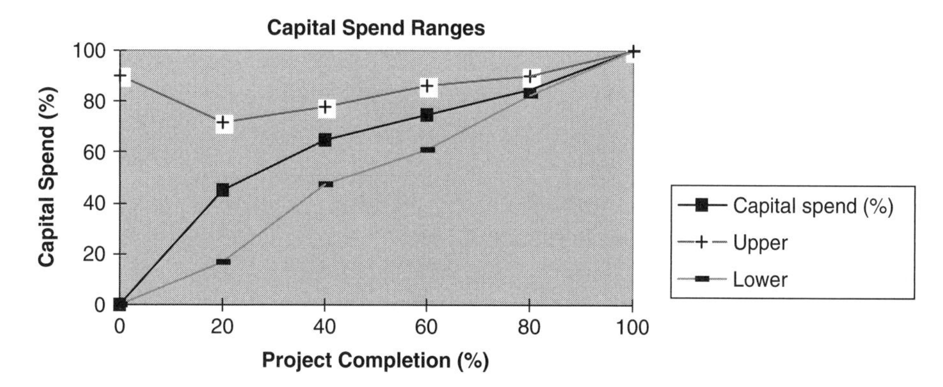

Figure 6.11 Project control and tracking.

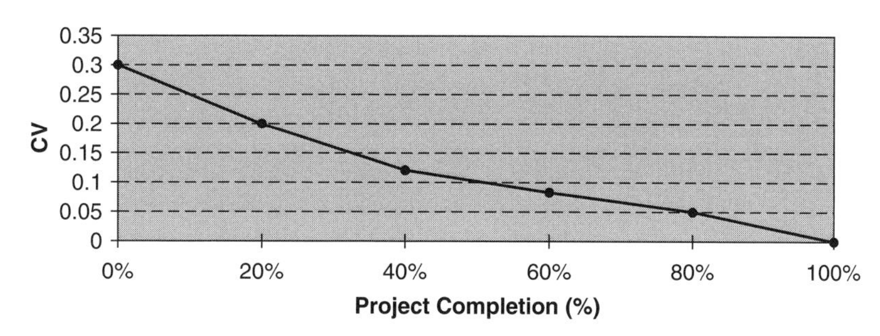

Figure 6.12 Coefficient of variation.

B. *Contingency cost estimation.* The cost of a risky project can be estimated by adding a contingency cost to each task (typically 5 to 7 percent of the task cost) to cover the risk involved.

C. *Decision trees.* Decision trees are used to evaluate sequential decisions and decide on alternatives on the basis of the expected values of probabilistic outcomes.

D. *Diversification.* With this risk-management technique, several risky projects can be engaged at the same time to spread the risks by means of diversification—combining high-risk, high-return projects with low-risk, low-return projects—to achieve a reasonable overall return on investment.

E. *Fuzzy logic systems.* These systems of reasoning are based on fuzzy sets (Nguyen and Walker 2005; Ross 2004). A fuzzy set defines the range of values for a given concept as well as the degree of membership. A membership of 1 indicates full membership, whereas 0 defines exclusion. The change of membership from 0 to 1 is gradual. For example, a fuzzy expert system employs rules such as the following:
If the temperature T is high (A_i) and the difference in temperature is small (B_i), then close the valve V slightly (S_i), wherein A_i, B_i and S_i are fuzzy sets. Fuzzy logic systems have been applied to assess project risks (Tanada and Niimura 2007).

Example 6.1 The company needs to decide either to develop a new product with an investment of $400,000 or to upgrade an existing product by spending $200,000, as illustrated by decision point 1 in Figure 6.13.

- If the new product strategy is pursued, then there is a 60 percent probability that the product will be in high demand, in which case the company will make $200,000 next year. Concurrently, there is a 40 percent chance for low demand, which will result in a loss of $100,000 for the company next year. (See Node A in the Decision Tree Diagram in Figure 6.13.) If the new product enjoys a high demand, then there is an 80 percent chance that the product will make $1,000,000 and a 20 percent chance that it will make only $50,000 in the year after the next. (See Node a in Figure 6.13.) If the new product meets a low demand, then there is a 30 percent chance for a revenue of $500,000 and a 70 percent chance of suffering a loss of $500,000 in the year after next. (See Node b in Figure 6.13.)
- If the company follows the strategy of improving the existing product, then there is a 60 percent chance for high demand, leading to revenue of $100,000; and a 40 percent chance of low demand, to yield zero revenue in the first year. (See Node B in Figure 6.13.) If the demand is high at the end of the first year, the company needs to make a second decision (decision point 2) whether to expand the product line. The expansion option will require an investment of $100,000, whereas the option of no expansion costs nothing. If the company expands the ungraded product line, then there is an 80 percent chance that it will reap revenue of $800,000 and a 20 percent chance of $100,000 revenue in the year after the next. (See Node c in Figure 6.13.) If the company elects

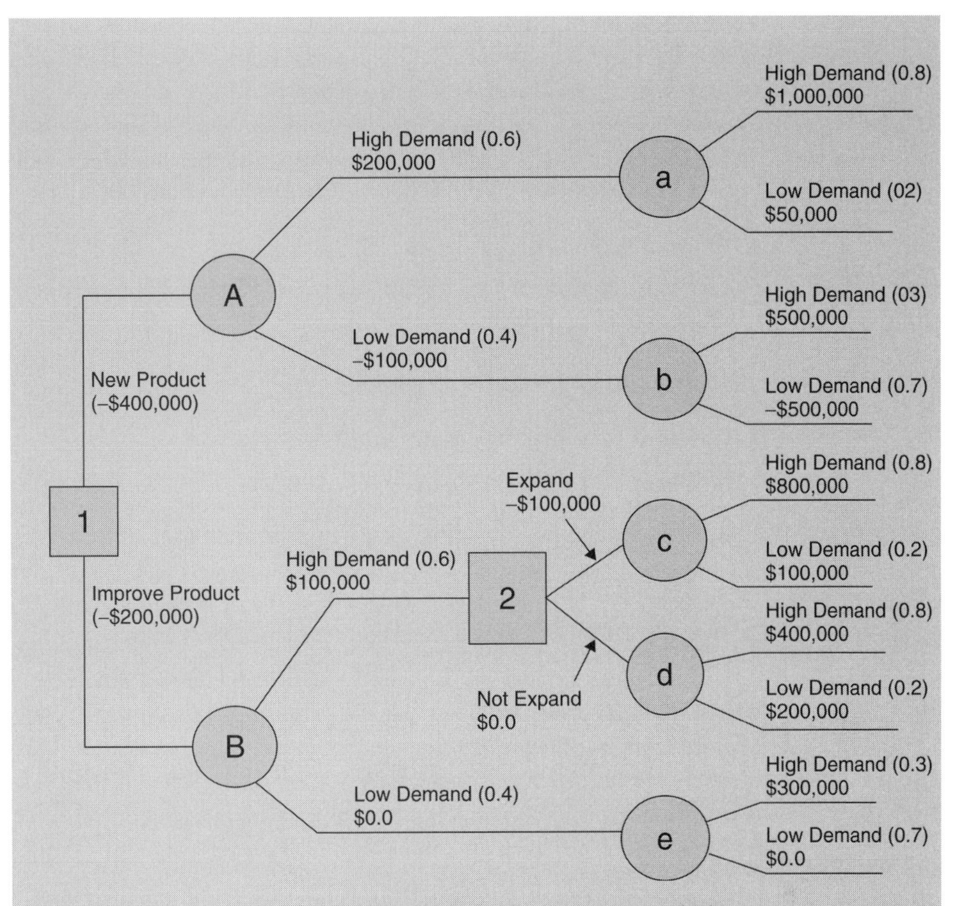

Figure 6.13 Decision tree analysis.

not to expand at the end of the first year, then there is an 80 percent chance that it will realize revenue of $400,000 and a 20 percent chance of $200,000 in revenue. (See Node d in Figure 6.13.) Should the upgraded product of the company see a low demand (at a probability of 40 percent) and get zero revenue in the first year, then there is a 30 percent chance that the product can generate revenue of $300,000 and a 70 percent chance of zero revenue. (See Node e in Figure 6.13.)

The interest rate is 10 percent. Determine which decisions at decision points 1 (product strategy) and 2 (expansion strategy) that the company should make.

ANSWER 6.1 This problem can be solved by using the decision tree method, which works from right to left, from future (the year after next) to present.

Decision Point 2:

At Node c, the expansion option has a total expected return of:

$$ER(c) = 0.8\ (\$800{,}000) + 0.2\ (\$100{,}000) = \$660{,}000$$

At Node d, the no-expansion option has a total expected return of:

$$ER(d) = 0.8\ (\$400{,}000) + 0.2\ (\$200{,}000) = \$360{,}000$$

The present value of expansion is:

$$PV(Expansion) = -\$100{,}000 + ER(c)/1.1 = \$500{,}000$$

And the present value of no expansion is:

$$PV(No\ expansion) = 0 + ER(d)/1.1 = \$327{,}272$$

Based on these present values, the decision should favor expansion.

Decision Point 1:

At Node a, the total expected return is:

$$ER(a) = 0.8\ (\$1{,}000{,}000) + 0.2\ (\$50{,}000) = \$810{,}000$$

At Node b, the corresponding total expected return is:

$$ER(b) = 0.3\ (\$500{,}000) + 0.7\ (-\$500{,}000) = -\$200{,}000$$

Thus, the present value for the High demand case is:

$$PV(High\ Demand) = \$200{,}000 + ER\ (a)/(1.1) = \$936{,}363$$

And the Present value for the low demand case is:

$$PV(Low\ Demand) = -\$100{,}000 + ER\ (b)/1.1 = -\$281{,}818$$

Thus, the present value for the New Product strategy is:

$$\begin{aligned} &PV(New\ Product) \\ &= -\$400{,}000 + [0.6\ PV(High\ Demand) + 0.4\ PV(Low\ Demand)]/1.1 \\ &= \$8{,}264 \end{aligned}$$

On the other hand, we have for the product improvement strategy:

$$PV(High\ Demand) = \$100{,}000 + PV(Expansion) = \$600{,}000$$

Note that the no expansion option is abandon.
The present value of low demand is:

$$\begin{aligned} PV(Low\ Demand) &= 0 + ER(e)/1.1 = 0 + [0.3(\$300{,}000) + 0.7(0)]/1.1 \\ &= \$81{,}818 \end{aligned}$$

Thus, the present value for the product improvement strategy is:

$$\begin{aligned} PV(Product\ Improvement) &= -\$200{,}000 + [0.6(\$600{,}000) + 0.4\ (\$81{,}818)]/1.1 \\ &= \$157{,}024 \end{aligned}$$

Since the present value for improving the product is larger than the present value for new product, the choice should be in favor of product improvement.
Even if the company decides to forgo expansion at decision point 2, the present value for product improvement is:

$$PV(Product\ Improvement\ with\ no\ expansion) = -200{,}000 + [0.6(\$427{,}272)$$

$$+ 0.4 (\$81,818)]/1.1$$
$$= \$62,809$$

which is still larger than that for new product development. The value of expansion is thus:

$$\text{Value (Expansion)} = \$157,024 - 62,809 = \$94,215.$$

6.7 MANAGEMENT OF OVERHEAD COSTS

Some of the overhead costs are related to finance, human resources, IT, and legal functions. Overhead costs are known to grow directly with gross margins, and they often grow faster than revenues. Several best practices are outlined in the literature to reduce such overhead costs (Nimocks 2005):

A. Aligning the overhead capacities with the company's strategic needs:

1. Streamline the planning process for marketing.

2. Use enterprise resource planning tools.

3. Outsource IT support.

4. Automat/centralize cash incentive preparation.

B. Reducing the needs for these overhead functions:

5. Change reporting from monthly to quarterly.

6. Change payroll from weekly to biweekly.

7. Eliminate nonstrategic technology initiatives.

8. Replace quarterly internal strategy meeting with monthly conference calls.

9. Eliminate statistical analysis group.

10. Reduce the frequency of long-term trend analysis.

C. Optimizing the delivery of these overhead activities:

11. Standardize reporting format and eliminate redundant ones.

12. Enforce purchasing standards.

13. Set travel and expense standards.

14. Use self-service for HT benefits.

15. Use temporary contract personnel.

16. Consolidate advertisement agency work and eliminate high-cost outliers.

Since overhead costs denote a large percentage of the overall cost of doing business, it is important for engineering managers and systems engineers to constantly look for ways to reduce these costs.

6.8 MISCELLANEOUS TOPICS

This section discusses several miscellaneous topics, including economic quantity of ordering, simple cost-based decision models, and project evaluation criteria.

6.8.1 Economic Quantity of Ordering

Ordering of parts, materials, and other supply items directly affects product costs. The ordering process must take into account the quantity needed, purchase price, order-processing fees, shipping costs, and the time value of money. Engineering managers may need to know how to arrive at the *economic quantity of ordering* in order to minimize the total cost of ordering. The next example illustrates this concept.

Example 6.2 A manufacturing company buys 6,000 steel bars a year at a fixed price of $18 each. It costs the company $85 to process and place each order. Assuming 10 percent interest compounded annually, what is the most economic quantity to order at one time?

ANSWER 6.2

Let N = Number of orders placed in a year

Let C = Total cost of ordering at year-end

$$N = 1\text{: } C = 6000(18)(1 + 0.1) + 85$$

$$N = 2\text{: } C = 3000(18)(1 + 0.1) + 85 + 3000(18)(1 + 0.1/2) + 85$$

$$N = 3\text{: } C = 2000(18)(1 + 0.1) + 85 + 2000(18)(1 + 0.1(2/3)) + 85 + 2000(18)(1 + 0.1(1/3)) + 85$$

Hence:

$$N = N\text{: } C = 6000(18)(1/N)[N + 0.1(1/N)(N + (N - 1) + (N - 2) + \ldots + 1)] + 85$$

$$N = 6000[1 + 0.1(1 + N)/(2N)] + 85N$$

To find the minimum of C, we perform a differentiation of the last equation and setting the derivative to zero:

$$\mathrm{d}C/\mathrm{d}N = 0 = 5400[1/N - (N + 1)/N^2] + 85$$

$$N = 7.98 = 8$$

$$6000/8 = 750$$

The economic quantity to order is 750 units, every 6.5 weeks, and 8 times a year.

6.8.2 Simple Cost-Based Decision Models

Engineering managers need to make regular choices among alternatives. In some cases, such choices can be made based on costs, as illustrated by two examples shown below.

Comparison of Alternatives. When faced with the option of buying one of several sets of capital equipment with similar functional characteristics, engineering managers can use the following annual cost formula to identify which has the lowest total annual cost.

The annual cost for a long-lived asset is defined as the sum of its depreciation charge, interest charge for the capital tied down by the purchase, and its annual operational expenses—that is:

$$\mathrm{AC} = \frac{(P - L) \times i}{(1 + i)^N - 1} + P \times i + \mathrm{AE} \tag{6.2}$$

where:

P = Initial investment ($)
L = Salvage value ($)
N = Useful life of a long-lived asset (year)
i = Interest rate (%)
AE = Annual expenses ($) – taxes, supplies, insurance repairs, utilities, etc.
AC = Annual cost ($)

The capital equipment with the lowest AC is preferable. The derivation of the Annual Cost Computation Equations is presented in Appendix 6.11.8.

Example 6.3 Your company has averaged 15 percent growth a year for the past seven years, and now you need additional warehouse space for purchased material as well as finished goods. Two types of construction have been under consideration: conventional and air-supported fabric. (The data are available in Table 6.12.) Which is the better economical choice?

ANSWER 6.3

Conventional:

$$AC_1 = (200{,}000 - 40{,}000) \times 0.08/[1.08^{40} - 1] + 200{,}000^{*}0.08$$
$$+ (1500 + 700 + 1.5 \times 200,000/100) = \$21,818$$

For Air-Supported:

$$AC_2 = (35{,}000 - 3000)^{*}0.08/[1.08^{40} - 1] + 35000 \times 0.08$$
$$+ (5000 + 5500 + 1.5 \times 35000/100) = \$16,833$$

Choice: Air-supported system.

Table 6.12 Warehouse Options

Type	Conventional	Air-Supported
First cost ($)	200,000	35,000
Life (years)	40	8
Annual maintenance ($)	1,500	5,000
Power & fuel ($)	700	5,500
Annual taxes ($)	1.5/100	1.5/100
Salvage value ($)	40,000	3,000
Interest rate	0.08	0.08

Example 6.4 A semiautomatic machine is quoted at \$15,000, while an advanced machine is quoted at \$25,000. The salvage value of these machines is assumed to be zero. A four-man party can make 500 parts a day with the semiautomatic machine, by using a machinist at \$200 a day, a maintenance worker at \$150 a day, a parts laborer at \$100 a day, and a warehouse clerk at \$80 a day.

A six-man party can produce 770 parts a day with the advanced machine, using a machinist at \$200 a day, an assistant machinist at \$180 a day, a maintenance worker at \$150 a day, two parts laborers at \$100 a day each, and a warehouse clerk at \$80 a day.

Material cost is \$10 a part. The factory overhead is 50 percent of the direct labor cost, only when parts are being made. Maintenance expense for the semiautomatic machine is \$250 a year, and for the advanced machine is \$500 a year. The estimated life of the semiautomatic machine is 20 years, and 15 for the advanced machine. The cost of money is 8 percent a year.

How many parts must be made each year to justify the deployment of the advanced machine?

ANSWER 6.4

Define $x =$ Number of parts produced per year.

$y =$ Number of working days to produce parts. For the semiautomatic machine, $y = \frac{x}{500}$; for the advanced machine, $y = \frac{x}{770}$. (See Table 6.13)

The annual cost of operating the semiautomatic machine is given by:

$$\begin{aligned} AC_{SA} &= (15{,}000 \times 0.08)/[1.08^{20} - 1] + 15000 \\ &\quad \times 0.08 + 250 + 10 \times x + 530(x/500) \times 1.5 \\ &= 1777.78 + 11.59x \end{aligned}$$

The annual cost of operating the advanced machine is given by:

$$\begin{aligned} AC_{A} &= (25{,}000 \times 0.08)/[1.08^{15} - 1] + 25000 \times 0.08 + 500 + 10 \\ &\quad \times x + 810(x/770) \times 1.5 \\ &= 3420.74 + 11.57792x \end{aligned}$$

Setting $AC_{SA} = AC_{A}$, we have:

Table 6.13 Comparison of Two Machines

	SemiAutomatic	Advanced
First cost (\$)	15,000	25,000
Daily production	500	850
Daily wage (\$)	530	810
Annual maintenance (\$)	250	500
Life (years)	20	15
Material cost (\$/part)	10	10
Factory overhead as percent	50%	50%
Interest	0.08	0.08

$$1775.78 + 11.59\text{ x} = 3420.74 + 11.57792x$$

$$x = 136,172$$

The advanced machine is justifiable if the production exceeds 136,172 parts each year.

Replacement Evaluation. Engineering managers are sometimes faced with the decision of whether to replace an existing facility with a brand-new one. Again, this replacement decision can be made by identifying the option with the lowest annual cost.

In this analysis, the existing facility is treated as if it is new, in that its residual equipment life and its residual book value (initial capital investment minus accumulated depreciation) are equivalent, respectively, to the useful product life and the capital investment cost of new equipment. Thus:

$$\text{AC}_o = \frac{i \times \text{BV}(t) - L_o}{(1+i)^{(N_o - t)} - 1} + \text{BV}(t) \times i + \text{AE}_o \tag{6.3}$$

$$\text{AC} = \frac{i \times [P - L]}{(1+i)^N - 1} + P \times i + \text{AE}. \tag{6.4}$$

P_o = Original investment cost ($)
N_o = Original estimate of useful life (year)
AE_o = Annual expenses of using the original equipment ($)
L_o = Salvage value at the end of the equipment's original useful life ($)
AC_o = Annual cost for using the existing equipment ($)
t = Present age of equipment (year)
$N_o - t$ = Remainder life (year)
$BV(t)$ = Book value of existing equipment at the end of the t(th) year ($)
P = Initial investment of the replacement equipment ($)
AC = Annual cost of using the replacement equipment ($)
L = Salvage value of the replacement equipment at the end of its useful life (year)
N = Useful life of the replacement equipment (year)
AE = Annual expenses for using the replacement equipment ($)

If $AC_o > AC$, use the replacement equipment to save costs.

Example 6.5 A compressor air-supply station was built eighteen years ago at the main shaft entrance to a coal mine at a cost of $2.6 million. The station was equipped with steam-driven air compressors that have an annual operating expense of $360,000. The salvage value at the estimated twenty-five-year life of the station is $130,000. It can be sold now for $800,000.

A proposal has been made to replace the station with electrically driven compressors that would be installed underground near the working face of the mine for a cost of $2.8 million. The new compressor station would have a life of 30 years and a salvage value of 10 percent. Its annual operating cost would be two-thirds of the steam-driven station. Annual taxes and insurance are 2.5 percent of the first cost of either station. The interest rate is 8 percent. Is there a financial justification to replace the steam station?

ANSWER 6.5 Table 6.14 summarizes the data of these two compressors.

A. Assuming approximate method (straight-line depreciation):

$$\text{Steam:}\quad AC = \left(\frac{800{,}000 - 130{,}000}{7}\right) + (800{,}000 - 130{,}000) \times \frac{0.08 \times 8}{2 \times 7}$$
$$+ 130{,}000 \times 0.08 + 425{,}000 = \$561{,}742.86$$

$$\text{Electric:}\quad AC = \frac{2{,}800{,}000 - 280{,}000}{30}$$
$$+ (2{,}800{,}000 - 280{,}000) \times \frac{0.08 \times 31}{2 \times 30}$$
$$+ (280{,}000 \times 0.08) + 310{,}000 = 520{,}560$$

Choice: Replace the old steam unit.

B. Assume the exact method for calculating depreciation (sinking fund):

$$\text{Steam:}\quad AC = (800{,}000 - 130{,}000) \times \frac{0.08}{1.08^{7} - 1} + 800{,}000 \times 0.08$$
$$+ 425{,}000 = 564{,}088.50$$

$$\text{Electric :}\quad AC = \frac{(2{,}800{,}000 - 280{,}000) \times 0.08}{1.08^{30} - 1} + 2{,}800{,}000 \times 0.08$$
$$+ 310{,}000 = 556{,}245.13$$
$$\times 0.08 + 312{,}500 = 558{,}745.13$$

Table 6.14 Comparison of Two Compressor Drives

	STEAM	ELECTRIC
Original investment ($)	2,600,000	2,800,000
Life (years)	25	30
Present age (years)	18	0
Remaining life (years)	7	30
Present salvage value ($)	800,000	
Final salvage value ($)	130,000	280,000
Annual expense ($)	360,000 + 65,000	240,000 + 70,000

6.8.3 Project Evaluation Criteria

Engineering managers are often required to make choices among capital projects that could deliver benefits but also consume resources on an annual basis over a number of periods. Several standard methods are used in industry to evaluate such projects. These include *net present value*, *internal rate of return*, *payback*, and *profitability index*.

Net Present Value.

$$\text{NPV} = -P + \sum_{m=1}^{n} \frac{\text{NCIF}(m)}{(1+i)^m} + \frac{\text{CR}}{(1+i)^n} \tag{6.5}$$

$$m = 1 \text{ to } n$$

NPV = Net Present Value (dollars)
P = Present investment made to initiate a project activity (dollars)
$NCIF_m$ = Net cash in-flow (dollars) in the period m, which represents revenues earned minus costs incurred, or $(R_m - C_m)$(dollars)
i = Cost of capital rate
N = Number of interest period (year)
CR = Capital recovery (dollars), which is the amount regained at the end of the project through resale or other methods of dispositions

Note that the first term on the right-hand side is the capital outlay for the project, or an outflow of value (cash). The second term on the right-hand side is the sum of discounted net cash in-flow earned over the years. The third term on the right-hand side is the discounted capital recovery of the project.

One major weakness of the NPV equation is that all benefits derived from a project must be expressed in dollar equivalents—within NCIF_m—in order to be included. Nonmonetary benefits, such as enhanced corporate image, expanded market share, and others, cannot be represented.

For the special case of $NCIF_m = CF =$ constant

$$\text{NPV} = -P + \text{CF}\frac{(1+i)^n - 1}{i \times (1+i)^n} + \frac{\text{CR}}{(1+i)^n} \tag{6.6}$$

Projects with the largest *NPV* values are preferable, as *NPV* stands for the net total value added (before tax) to the firm by the project at hand. Note that *NPV* may be determined only if the project's net cash inflow $NCIF_m$ is known. These *NPV* equations can, if applicable, be readily expanded to include corporate tax rate and investment tax credits. (See Appendix 6.11.9.)

Internal Rate of Return (IRR). Rate of return is generally defined as the earnings realized by a project in percentage of its principal capital.

The *internal rate of return* (IRR) is the average rate of return (usually annual) realized by a project in which the total net cash inflow is exactly balanced with its total net cash outflow, resulting in zero NPV value at the end of its project life cycle. In other words, this is the rate realizable when reinvestment of the project earnings is made at the same rate until maturity.

IRR is determined by the following equations:

$$0 = -P + \sum_{m=1}^{n} \frac{\text{NCIF}(m)}{(1+\text{IRR})^m} + \frac{\text{CR}}{(1+\text{IRR})^n} \tag{6.7}$$

$m = 1$ to n

For $\text{NCIF}_m = CF =$ constant

$$0 = -P + \text{CF}\frac{(1+\text{IRR})^n - 1}{\text{IRR} \times (1+\text{IRR})^n} + \frac{\text{CR}}{(1+\text{IRR})^n} \tag{6.8}$$

The IRR values (before tax) of acceptable projects must be greater than the firm's cost of capital. Projects with high IRR are to be preferred.

Payback Period. The payback period is defined as the number of years that the original capital investment for the project will take to be paid back by its annual earnings, or:

$$PB = P/CF \tag{6.9}$$

Where:
P = Capital investment
CF = Annual cash flow realized by the project

Cost reduction projects with small payback periods (e.g., less than two years) are preferable.

Profitability Index. Profitability index is defined by the ratio:

$$\text{PI} = \frac{\text{Present value of all future benefits}}{\text{Initial investment}} \tag{6.10}$$

Projects with large PI values are preferable.

Example 6.6 Your company is currently pursuing three cost-reduction projects at the same time.

- Project A requires an investment of $10 million. It is expected to yield a cost savings of $30 million in the first year and another $10 million in the second year.
- Project B demands an investment of $5 million. It is expected to realize a cost savings of $5 million in the first year and another $20 million in the second year.
- Project C needs an investment of $5 million. It is expected to bring about a cost savings of $5 million in the first year and another $15 million in the second year.

Table 6.15 Summary of Results

Project	TIME => 0	1	2	NPV	IRR (%)	PB	PI
A	−10	30	10	25.5	230	0.5	3.55
B	−5	5	20	16	156	0.4	4.22
C	−5	5	15	12	130	0.5	3.39

After the second year, there will be no receivable benefit or capital recovery from any of these projects. The cost of capital (interest rate) is 10 percent.

Determine the ranking of these projects on the basis of the evaluation criteria of NPV, IRR, PB, and PI.

ANSWER 6.6

P = Present investment
$n = 2$
CF = Variable
$CR = 0$
$i = 10\%$

NPV:

A. $NPV = -10 + 30/(1.1) + 10/(1.1)^2 = 25.537$
B. $NPV = -5 + 5/1.1 + 20/(1.1)^2 = 16.074$
C. $NPV = -5 + 5/1.1 + 15/(1.1)^2 = 11.942$

IRR:

A. $0 = -10 + 30/(1 + r) + 10/(1 + r)^2; r = 2.3$
B. $0 = -5 + 5/(1 + r) + 20/(1 + r)^2; r = 1.56$
C. $0 = -5 + 5/(1 + r) + 15/(1 + r)^2; r = 1.3$

PB:

A. $PB = 10/[(30 + 10)/2] = 10/20 = 0.5$
B. $PB = 5/[(5+20)/2] = 5/12.5 = 0.4$
C. $PB = 5/[(5 + 15)/2] = 5/10 = 0.5$

PI:

A. $PI = [30/1.1 + 10/(1.1)^2]/10 = 35.537/10 = 3.553$
B. $PI = [5/1.1 + 20/(1.1)^2]/5 = 21.074/5 = 4.214$
C. $PI = [5/1.1 + 15/(1.1)^2]/5 = 16.942/5 = 3.3884$

These results are summarized in Table 6.15.

6.9 CONCLUSIONS

This chapter reviews basic cost accounting issues related to product and service costing. Product costs have direct and indirect cost components. Although direct costs are relatively easy to assess, the indirect costs that account for overhead charges might need to be properly assessed by using tools such as activity-based costing.

Cost data might apply for a single period or for multiple periods. In the case of multiple periods, the time dependency of cost data must be considered. For time-dependent data, the concept of the time value of money and the compound interest formulas are to be applied.

Depreciation accounting affects the facility costs that are part of the indirect costs of products. Different depreciation methods will lead to more or less indirect costs for the products. Finally, the inventory costs are affected by the sequence in which the time-dependent product costs are selected.

Cost data might be uncertain because of factors related to economy, market condition, political stability, labor movement, and others. For uncertain cost data, risk analysis might be needed. The Monte Carlo simulation is an efficacious method to conduct risk analyses. Several other methods are also available to account for cost uncertainties.

Engineering managers need to become well-versed in cost accounting.

6.10 REFERENCES

Atkinson, A. A., R. D. Banker, R. S. Kaplan, and S. M. Young. 2001. "Management Accounting" Chapter 5, in *Activity Based Costing Management Systems*, 3rd ed. Upper Saddle River, NJ: Prentice Hall.

Bamber, Linda Smith, and K. E. Hughes II. 2001. "Activity Based Costing in the Service Sector—The Buckeye National Bank." *Issues in Accounting Education* 16 (3).

Bedford, T., and R. Cook. 2001. *Probabilistic Risk Analysis: Foundations and Methods*. New York: Cambridge University Press.

Bierman, H. Jr. 2006. *The Capital Budgeting Decision: Economic Analysis of Investment Projects*, 9th ed. London: Routledge.

Briner, Russel F., Mark Alford, and Jo Anne Noble. 2003. "Activity Based Costing for State and Local Governments." *Management Accounting Quarterly* (March 22).

Bragg, S. M. 2002. *Accounting Reference Desktop*. New York: John Wiley & Sons, Inc.

Buttross, Thomas, and George Schmelzle. 2008. "Activity Based Costing in the Public Sector," in Evan M. Bergman and Jack Rabin, "Encyclopedia of Public Administration and Public Policy," 2nd ed. Boca Raton, FL: CRC Press.

Canada, J. R., W. G. Sullivan, and J. A. White. 2004. *Capital Investment Analysis for Engineering and Management*, 3rd ed. Upper Saddle River, NJ: Prentice Hall.

Cokins, G. 2006. Activity-Based Cost Management in Government, 2nd ed. Management Concepts, Vienna, Virginia.

Colander, D. C. 2007. *Microeconomics*, 7th ed. New York: Irwin/McGraw-Hill.

Cooper, R., and R. Kaplan. 1988. "How Cost Accounting Distorts Product Costs." *Management Accounting* (April).

Ernst & Young. 2000. "A Study to Develop a Costing Methodology for the Austrian Higher Education." Sector: Final Report, DETYA (Department of Education, Training and Youth Affairs). Canberra.

Fichman, Robert G. and Chris F. Kemerer (2002), "Activity Base Costing for Component-Based Software Development," *Information Technology and Management*, Vol. 3, pp. 137–160.

Fishman, G. S. 2005. First Course in Monte Carlo Simulation, international edition. Brooks/Cole. Florence, Kentucky.

Granof, Michale H., David E. Platt, and Igor Vaysman 2000. "Using Activity Based Costing to Manage More Effectively," Grant Report, The PriceWaterhouseCoopers Endowment for the Business of Government.

Horngren, C. T., George Foster, Srikant M. Datar, and Madhav Rajan 2008. *Cost Accounting: A Managerial Emphasis*, 13th ed. Upper Saddle River, NJ: Prentice Hall.

Information and Education Services Division, The University of Newcastle. 2001. "Activity Based Costing—A study to Develop a Costing Methodology for Library and Information Technology Activities for the Australian Higher Education Sector", DETYA (Department of Education, Training and Youth Affairs), Canberra.

Kaplan, Robert S. 2006. *Activity Based Costing and Capacity*, rev. ed. Harvard Business School Notes #105059 (March 20).

Kaplan S. Robert, and Steven R. Anderson. 2007. "Time-Driven Activity-Based Costing: A Simpler and More Powerful Path to Higher Profits," Chapter 10 in *Compton Financial*. Boston: Harvard Business Press.

Mun, J. 2006. *Modeling Risks: Applying Monte Carlo Simulation, Real Options Analysis, Forecasting, And Optimization Techniques*. Hoboken, NJ: John Wiley & Sons, Inc.

Nimocks, Suzanne P. 2005. "Managing Overhead Costs." *McKinsey Quarterly* (2).

Nguyen, H. T., and E. A. Walker. 2005. *A First Course in Fuzzy Logic*, 3rd ed. Boca Raton, FL: Chapman and Hall/CRC.

Pernot, Elli, Filip Roadhooft, and Alexandra Abbeele 2007. "Time-Driven Activity-Based Costing for Inter-Library Services: A Case Study in a University." *Journal of Academic Librarianship* (33): 551–560: (September).

Rapier, D. M. 1996. "Standard Costs and Variance Analysis." *Harvard Business School Note*, No. 9-196-121, January.

Ross, T. 2004. *Fuzzy Logic with Engineering Applications*, 2nd ed. John Wiley & Sons, Inc.

Rubinstein, R. Y., and Dirk P. Kroese. 2007. *Simulation and the Monte Carlo Method*, 2nd ed. Hoboken, NJ: Wiley-Interscience.

Smith, G. N. 2007. *Excel Applications for Accounting Principles*, 3rd ed. Cincinnati, OH: South-Western College Publishing.

SAS. 2005. "Activity Based Costing: How ABC Is Used in the Organization." White paper published bySAS Institute, Cary, North Carolina (September). http://www.sas.com/offices/europe/switzerland/romandie/pdf/actualites/abm_survey_result.pdf.

Sullivan, W., Elin M. Wicks, and C. Patrick Koelling. 2008. *Engineering Economy*, 14th ed. Upper Saddle River, NJ: Prentice Hall.

Swenson, Dan, Shahid Ansari, Jan Bell, and Il-Woon Kim. 2003. "Best Practices in Target Costing." *Management Accounting Quarterly* (Winter).

Tanaka, K., and T. Niimura. 2007. *An Introduction to Fuzzy Logic for Practical Applications*. New York: Springer.

Touran, A. 2003. "Calculation of Contingency in Construction Projects." *IEEE Transactions on Engineering Management* 50 (2): May.

Waters, Hugh, Hany Abdalla, Diana Santillan, and Paul Richardson 2003. "Application of Activity Based Costing in a Peruvian NGO Healthcare System." *Operations Research Results* 1 (3, Revised).

Published for the U.S. Agency for International Development (USAID) by the Quality Assurance Project (QAP), Bethesda, Maryland.

Zarc, Greg. 2009. "Monte Carlo Simulation in Excel without Using Add-ins," Ivey Business Case #909E04, Version (A), Harvard Business School Press (April 20).

Zio, E. 2007. *An Introduction to the Basics of Reliability and Risk Analysis*. Singapore: World Scientific Publishing Company.

6.11 APPENDICES

6.11.1 Basic Terms in Cost Accounting

Engineering managers need to become familiar with the standard vocabulary used by cost accountants or cost engineers, as costs are important bases for corporate performance evaluation, product cost estimation, profitability analysis, and managerial decision making. While the cost accounting systems used by various firms do not need to strictly follow the generally accepted accounting principles (GAPP) adopted by the financial accounting profession, engineering managers are still advised to understand the meaning of various accounting terms in order to ensure that their cost-based decisions are made properly. The following is a general set of accounting terms used by many firms (Bragg 2002; Rapier 1996):

- **Cost center**. An organizational unit responsible for controlling costs related to its functional objectives (e.g., R&D, procurement, operations, engineering, design, and marketing).
- **Inventory costs**. The total sum of product costs, which are composed of the direct costs and indirect costs related to the manufacturing of the products involved.
- **Direct costs**. Materials and labor costs associated with the manufacturing of a product.
- **Indirect costs.** All overhead costs (e.g., rent, procurement, depreciation, supervision, supplies, power, and others) indirectly associated with the production of products involved.
- **Fixed costs**. Costs that do not strictly vary with the volume of products involved, such as the general manager's salary, rent for the facility, machine depreciation charges, and local taxes.
- **Variable costs**. Costs that vary in proportion to the volume of products involved, including, for example, material, labor, and utilities.
- **Step function costs**. Costs that would experience a step change when a specific volume range is exceeded; for example, the factory rent that might change stepwise if new floor space must be added because of the increased production volume.
- **Contribution margin**. The product price minus unit variable cost; the economic value contributed by selling one unit of product to defray the fixed cost already committed for the current production facility.
- **Cost pool**. An organizational unit where costs incurred by its activities performed for specific products (or other cost targets) are accumulated for subsequent assignments.
- **Cost drivers**. Bases used to allocate indirect costs to products. Products drive the consumption of resources and the utilization of resources incurs costs. Examples

include floor space, head counts, number of transactions, number of employees, labor hours, machine hours, number of setups, and material weight.

- **Cost objects**. Targets to allocate indirect costs, such as products and services sold by the firm.
- **Budget**. A quantitative expression in dollar value of a project or a plan of action. Examples include production budget, product design budget, engineering budget, R&D budget, sales budget, marketing budget, and advertising budget. Typically, budgets span a specific period of time (e.g., a month, a quarter, or a year).
- **Standard costs**. Direct and indirect costs budgeted for products. The standard costs are defined by using estimations or historical costs.
- **Variance**. The difference between standard costs and actual costs. Such variance could be the result of price variation, quantity change, technology advancement, and other factors. Conventionally, actual quantities are used when computing price variation to easily assess the procurement performance. On the other hand, the quantity-based variance is computed by using standard costs for an easy assessment of the production performance.
- **Current costs**. Costs for the total efforts (e.g., physical efforts, raw materials, and service fees) that must be spent in order to carry out an activity or implement a plan. Current costs form a key basis for managerial decision making.
- **Opportunity costs**. The benefit of the second-best alternative that must be forgone because of a commitment made to the first alternative. For example, an engineering manager who quits a job paying $100,000 a year to pursue a three-semester MBA degree at a university incurs an opportunity cost at graduation of $150,000 plus an out-of-pocket cost of $90,000 for tuition fees. Opportunity costs are included in managerial decision making, but are not included in a cost accounting system.
- **Sunk costs**. Costs that have already been spent or incurred. Such costs are typically included in all cost accounting systems, but they are not considered in any management decision-making for the future.

6.11.2 Cost Analysis

Managers perform variance analyses and study the reasons for the deviation of actual costs from standard costs. They issue periodic and systematic reports of their findings and take proper actions to enhance the efficiency and effectiveness of the organizational units (Sullivan, Wicks, and Koelling 2008).

The two major factors affecting cost analysis are time and accuracy. For management decisions, cost analyses can be performed for a single time period or for multiple periods. Cost data can vary or can be uncertain.

A. Single-Period Analysis. Single-period analysis applies primarily to a short period of time during which the costs involved remain essentially constant. The gross profit equation for a given product line is given by the following equation:

$$\begin{aligned}\text{Gross Profit} &= \text{Revenue} - \text{Costs}\\ \text{GP} &= P \times N - (\text{FC} + \text{VC} \times N)\end{aligned} \tag{6.11}$$

Where:
P = Product price (dollars/unit)
N = Number of products sold during the period
FC = Fixed costs (dollars)
VC = Variable costs (dollars/unit)
GP = Gross profit (dollars)

For the case of break-even (i.e., GP = 0), the break-even product quantity is given by:

$$n^* = \mathrm{FC}/(P - \mathrm{VC}) \tag{6.12}$$

The value $(P - VC)$ is defined as the contribution margin of the product. Selling each additional unit of a product generates a contribution in the amount $(P - VC)$ to defray the fixed cost (FC) that has been committed to the production line.

Organizational performance can be readily assessed as the number of cost items involved is limited. One needs to make sure that the values of these cost items are valid, although from time to time, the validity of such values might be tough to verify precisely, because of joint production activities and other cost-sharing systems involved.

B. Multiple-Period Analyses. The cost analyses over a longer period of time (e.g., multiple periods) are much more difficult to calculate for two reasons. First, costs might change predictably over time due to inflation, investment return, cost of capital, and other reasons. Second, future events are unpredictable (e.g., natural disasters, labor unrest, political instability, war against terrorism, spread of disease, investment climate, etc.) (Horngren, Foster, Datar and Rajan 2008.) The change of costs over time needs to be addressed by using concepts such as net present value (NPV) and internal rate of return (IRR). These concepts are built on the fundamentals of the time value of money, compound interest, and the cost of capital. These topics are introduced in Appendix 6.11.3. Depreciation accounting, an important part of the indirect costs of products, is included in Appendix 6.11.4.

In dealing with the uncertainties of future costs, risks must be included in product cost analysis. Risk analysis is elucidated in detail in section 6.6.

6.11.3 Time Value of Money and Compound Interest Equations

The time value of money refers to the notion that the value of money changes with time. This is because money at hand can lose value (purchasing power) if not invested properly. Money at hand can earn income through investment. A dollar that is to be received at a future date is not worth as much as a dollar that is on hand at the present. Thus, two equal dollar amounts at different points in time do not have equal value (purchasing power).

Before introducing basic compound interest equations useful for multi-period cost analyses, a few definitions are reviewed next:

- **Interest**, It denotes a fraction of the principal designated as a reward (interest income) to its owner for having given up the right to use the principal. It may also be a charge (interest payment) to the borrower for having received the right to use the principal during a given interest period.

- **Compound interest**, When the interest income earned in one interest period is added to the principal, the principal becomes larger for the next period. The enlarged principal earns additional interest under such circumstances. The interest is said to have been compounding.
- **Nominal interest rate**, The interest rate quoted by banks or other lenders on an annual basis, also called the annual percentage rate (APR).
- **Effective interest rate**, The interest rate in effect for a given interest period (e.g., one month). For example, if the nominal interest rate for a bank loan is 12 percent, then its effective interest rate for each month is 1 percent (12%/12).
- **Nominal dollar**, The actual dollar value at a given point in time.
- **Constant dollar**, The dollar value that has a constant purchasing power with respect to a given base year (e.g., the reference year 2005); the value is adjusted for inflation.
- **Consumer price index**, The index tracked by the U.S. Department of Commerce to indicate the price change for a basket of consumer products.

To introduce the compound interest formulas for multiple period cost analyses, the following notations are used (See Fig. 6.A1):

P = Present value ($), the value of a project, loan or financial activity at the present time
F = Future value ($), the value of a project, loan or financial activity at a future point in time
i = Effective interest rate for a given period during which time the interest is to be compounded (e.g., 1% per month)
A = Annuity ($), a series of payments made or received at the end of each interest period
n = Number of interest periods

When employing these compound interest formulas, the following guidelines should be kept in mind:

- The periods must be consecutively and sequentially linked with the end of one as the beginning of the next.
- Complex problems can be broken down into time segments so that the equations can be correctly applied to each of the segments.

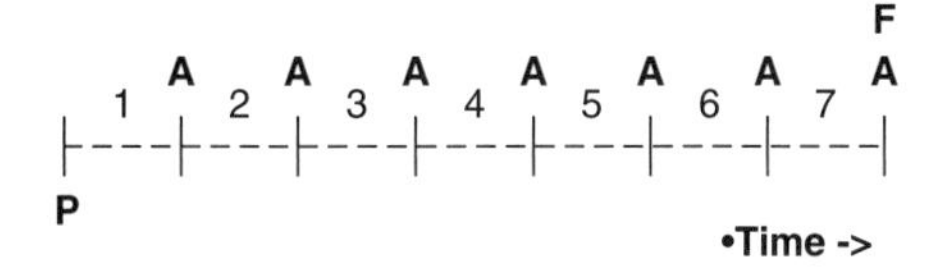

Figure 6.A1 Timeline convention.

A. Single Payment Compound Amount Factor.

$$F = P \times (1+i)^n$$
$$\frac{F}{P} = (1+i)^n = \left(\frac{F}{P}, i, n\right) \tag{6.13}$$

Equation 6.13 defines the total value of an investment P, with periodical returns added to the principals to earn more money at the end of n periods. Its derivation is illustrated in Appendix 6.11.6.

Example 6.7 Mr. Jones invests $5,000 at 8.6 percent interest compounded semi-annually. What will be the approximate value of his investment at the end of 10 years?

ANSWER 6.7 $F = P(1+i)^n = 5000(1+0.086/2)^{20} = \$11{,}605.29$

B. Present Worth Factor.

$$P = \frac{F}{(1+i)^n}$$
$$\frac{P}{F} = (1+i)^{-n} = \left(\frac{P}{F}, i, n\right) \tag{6.14}$$

Equation 6.14 defines the present value of a sum that will be available in the future. The factor $\frac{1}{(1+i)^n}$ is also called the discount factor.

C. Uniform Series Compound Amount Factor.

$$F = A \times \frac{[(1+i)^n - 1]}{i}$$
$$\frac{F}{A} = \frac{(1+i)^n - 1}{i} = \left(\frac{F}{A}, i, n\right) \tag{6.15}$$

Equation 6.15 determines the total future value of an account (e.g., retirement, college education, etc.) at the end of n periods, if a known annuity A is deposited into the account at the end of every period. Its derivation is presented in Appendix 6.11.7.

D. Uniform Series Sinking Fund Factor.

$$A = F \times \frac{i}{(1+i)^n - 1}$$
$$\frac{A}{F} = \frac{i}{(1+i)^n - 1} = \left(\frac{A}{F}, i, n\right) \tag{6.16}$$

Equation 6.16 calculates the amount of the required annuity (e.g., a series of period-end payments) that must be periodically deposited into an account in order to reach a desired total future sum F at the end of n periods.

E. Uniform Series Capital Recovery Factor.

$$A = P\frac{i \times (1+i)^n}{(1+i)^n - 1}$$
$$\frac{A}{P} = \frac{i \times (1+i)^n}{(1+i)^n - 1} = \left(\frac{A}{P}, i, n\right) \tag{6.17}$$

Equation 6.17 defines the amount of periodical withdrawal that can be made over n periods from an account worth P at the present time, such that the account will be completely depleted at the end of n periods.

Example 6.8 Mr. Jones wishes to establish a fund for his newborn child's college education. The fund pays \$60,000 on the child's eighteenth, nineteenth, twentieth, and twenty-first birthdays. The fund will be set up by the deposit of a fixed sum on the child's first through seventeenth birthdays. The fund earns 6 percent annual interest. What is the required annual deposit?

ANSWER 6.8 The future sum of a series of annual deposits is:

$$F1 = A1[(1.060^{17} - 1]/0.06 = 28.21288A1$$

Annual withdrawal when the child enters college is:

$$A2 = P2 \times 0.06(1.06)^4/[(1.06)^4 - 1] = 0.2885915P2$$
$$A2 = 60{,}000$$
$$P2 = F1$$

Answer: A1 = \$7369.2 (the required annual deposit).

F. Uniform Series Present Worth Factor.

$$P = A\frac{(1+i)^n - 1}{i \times (1+i)^n}$$
$$\frac{P}{A} = \frac{(1+i)^n - 1}{i \times (1+i)^n} = \left(\frac{P}{A}, i, n\right) \tag{6.18}$$

Equation 6.18 determines the total present value of an account to which an annuity A is deposited at the end of each period. For example, if A is the periodical maintenance costs for capital equipment, then this equation calculates the present value of all maintenance costs over its product life of n periods.

Example 6.9 The annual maintenance on the parking lot is \$5,000. What expenditure would be justified for resurfacing if no maintenance is required for the first 5 years, \$2,000 a year for the next 10 years, and \$5,000 a year thereafter? Assume that the cost of money is 6 percent.

ANSWER 6.9 Since the annual maintenance cost is the same after 15 years, the effect of resurfacing applies to the first 15 years only. The total present value of the "doing nothing" option is:

$$P1 = A[(1+i)^n - 1]/[i(1+i)^n] = 5000[1.06^{15} - 1]/[0.06(1.06)^{15}]$$
$$= 48{,}561.25$$

The total present value of the resurfacing option is:

$$P2 = P + 2000[1.06^{10-}1]/[0.06 \times 1.06^{10}]/(1.06)^5$$
$$= P + 10{,}999.77$$

Setting $P1 = P2$:
P = \$37,561.47 (the maximum amount for resurfacing the parking lot).

Example 6.10 You need a new high-speed computer system and find one on sale for \$3,900 cash, or \$500 down and \$200 monthly payments for 2 years. What nominal annual rate of return does this time-payment plan represent?

ANSWER 6.10 Loan amount = \$3,900 − 500= \$3,400 = P
N = 24 months
A = \$200

$$P = A[(1+i)^N - 1]/[i(1+i)^N]$$
$$3400 = 200[(1+i)^{24} - 1]/[i(1+i)^{24}]$$

This is an implicit equation for the single unknown i. It can be solved by trial and error: Input a trial value of i, and compute residual Δ = RHS − LHS. The i value that produces zero Δ (delta) is the answer.

Trial value of i	Δ (delta)
0.06	4.496
0.03	0.0645
0.025	−0.884985−0.884985
0.029	−0.1194−0.1194
0.02965	+0.0004288+0.0004288

Answer: Monthly rate = 2.965 percent, and the nominal annual rate is 35.58 percent.

For all multiple period problems, the timeline convention is regarded as standard (see Fig. 6.A1).

6.11.4 Depreciation Accounting

In calculating indirect costs associated with production facility, equipment, and other tangible assets related to production, depreciation charges must be included. Depreciation is a cost-allocation procedure whereby the cost of a long-lived asset is recognized in each accounting period over the asset's useful life in proportion to its benefit brought forth over the same period. This procedure is undertaken in a reasonable and orderly fashion. Specifically, the acquisition cost of an asset can be considered as the price paid for a series of future benefits. As the asset is partly used up in each accounting period, a corresponding portion of the original investment in the asset is treated as the cost incurred for the partial benefit delivered.

The U.S. Internal Revenue Service accepts three depreciation accounting methods. They are discussed next, using the following notations:

P = Initial investment (\$) at the present time
N = Useful life of a long-lived asset measured in years (e.g., $N = 25$ for buildings, $N = 15$ for equipment, $N = 5$ for automobiles, $N = 3$ for computers, etc.)
$D(m)$ = Depreciation charge (\$) in the asset's m(th) year
L = Salvage value (\$) recoverable at the end of the equipment's useful life
$AD(m)$ = Accumulated depreciation (\$), which is the total amount of depreciation charges accumulated at the end of the m(th) year
$BV(m)$ = Book Value (\$) of an asset in its m(th) year; $BV(m) = P - AD(m)$
$P - L$ = Depreciable base (\$)
$r(m)$ = Depreciation rate, a fraction of the depreciable base to be depreciated per year

A. Straight Line. By this depreciation method, an equal portion of depreciation base ($P - L$) is designated as the depreciation charge for each period of the assets' estimated useful life:

$$D(m) = (P - L)/N = \text{Constant} \tag{6.19}$$

$$BV(m) = P - m \times (P - L)/N; \ m = 1, 2, 3$$

$$r(m) = 1/N = \text{constant}$$

$$AD(m) = m \times (P - L)/N$$

More than 91 percent of publicly traded companies in the United States use this straight-line depreciation method.

B. Declining Balance. The depreciation charge is set to equal to the net book value (e.g., acquisition cost minus accumulated depreciation) at the beginning of each period (e.g., year) multiplied by a fixed percentage. If this percentage is two times the straight-line depreciation percentage, then it is called a double-declining balance method:

$$D(m) = P \times r \times (1 - r)(m - 1) \tag{6.20}$$

$$BV(m) = P \times (1 - r)^m$$

$$r(m) = \text{constant};\ r = 2/N(\text{double-declining balance method})$$

$$AD_{(m)} = P \times [1 - (1 - r)^m]$$

Note that the salvage value is not subtracted from the acquisition cost. To make sure that the total accumulated depreciation does not exceed the depreciation base $(P - L)$, the depreciation charge of the very last period (e.g., year) must be manually adjusted.

C. Units of Production Method. The depreciation charge is assumed to be proportional to the service performed (e.g., units produced, hours consumed, etc.). Companies that are involved with natural resources (e.g., oil and gas exploration) use units of production method to depreciate their production assets. Software companies also use this method to depreciate their capitalized software development costs.

Example 6.11 The company plans to change its depreciation accounting from the straight-line method to the double-declining method on a class of assets that have a first cost (acquisition cost) of $80,000, an expected life of six years, and no salvage value. If the company's tax rate is 50 percent, what is the present value of this change, assuming 10 percent interest compounded annually?

ANSWER 6.11

$$P = 80{,}000;\ N = 6, t = 0.5;\ L = 0$$

$$F = (1 - t)\Delta;\ P = F/(1 + i)^n;\ i = 10\%$$

Table 6.A1 displays the present values of the differences between these two depreciation charges for the assets' expected life of 6 years.

Answer = $2,191.72

Table 6.A1 Calculation of Difference due to Depreciation Methods

Year	SL	Double-Declining (r = 2 (1/6) = 0.333333)	Delta	F = (1 − t) * Delta	Present Worth
1	13333.33	26666.67	13333.34	6666.67	6060.61
2	13333.33	17778.66	4445.33	2222.66	1836.91
3	13333.33	11851.87	−1481.46	−740.73	−556.52
4	13333.33	7901.25	−5432.08	−2716.04	−1855.09
5	13333.33	5267.5	−8065.83	−4032.92	−2504.12
6	13333.33	10534.05	−2799.28	−1399.64	−790.06
			TOTAL		**2191.72**

Example 6.12 A new delivery truck costs $40,000 and is to be operated approximately the same amount each year. If annual maintenance costs are $1,000 the first year and increase $1,000 each succeeding year and, if the truck trade-in value is $24,000 the first year and decreases uniformly by $3,000 each year thereafter, at the end of which year will the yearly costs of ownership and maintenance be a minimum?

ANSWER 6.12 The average annual ownership cost is calculated as presented in Table 6.A2.

The numbers in the sixth column are calculated by dividing the numbers in the fifth column by the ownership duration in years. Answer: fifth year.

Table 6.A2 Annual Ownership Cost computation

Year	Maintenance Cost (Annual)	Trade-in Value	Accumulated Depreciation	Total Accum. Cost of Ownership	Average Annual Cost Over the Ownership Period
1	1,000	24,000	16000	17,000	17,000
2	2,000	21,000	19,000	22,000	11,000
3	3,000	18,000	22,000	28,000	9,333
4	4,000	15,000	25,000	35,000	8,750
5	5,000	12,000	28,000	43,000	8,600
6	6,000	9,000	31,000	52,000	8.667

Example 6.13 Ceramic hot-gas filters provide 2,400 hours of service life. Three of the processes in a refinery are each equipped with one of these filter sets. Each filter set has an initial cost of $1,200. When production is scheduled, each process runs 24 hours a day.

In the first quarter of the year, process A did not start operating until the beginning of the fourth week. Process B terminated at the end of the 10th week. Process C was on stream for the entire period. What depreciation charge should be allocated for ceramic filters during the first quarter?

ANSWER 6.13 The proper method of calculating the depreciation charge is on the basis of usage, as illustrated in Table 6.A3.

6.11.5 Inventory Accounting

After the direct and indirect costs are estimated, the product costs can be defined. When products are transferred from work-in-progress operations to a finished-goods warehouse, they become inventory. Inventory can be managed by one of two methods: first in and first out (FIFO) and last in and first out (LIFO). The FIFO method specifies that inventory that enters the warehouse first will leave the warehouse first. By the LIFO method, the inventory that enters the warehouse last is shipped out first.

According to the time value of money concept, these two inventory operational methods might render different costs of goods sold (CGS). The inventory accounting

Table 6.A3 Depreciation Based on Usage

	A	B	C
Cost ($)	1,200	1,200	1,200
Weeks in operation	10	10	13
Hours in operation	1,680	1,680	2,184
% of useful life	70%	70%	91%
Depreciation charge ($)	840	840	1,092

Total = $2,772 (Zero salvage value assumed)

takes into account such possible change of product cost over time, due, for example, to inflation. In general, companies utilize one of the following three inventory accounting methods:

1. FIFO (first in and first out)
2. LIFO (last in and first out)
3. Weighted average

LIFO is most useful during periods of high inflation, as it results in less reportable earnings with lower taxes paid; LIFO is not useful, however, when prices for raw materials decrease. LIFO also provides lower inventory value, thus understating the value of the inventory in the balance sheet. Finally, LIFO is a more conservative accounting technique than FIFO. Note that LIFO is prohibited by law in some countries, such as the United Kingdom, France, and Australia. As a product of creative accounting, FIFO defines an inventory value more closely matched with its market value. It tends to make the income statement look better than it really is. In periods when the business climate experiences stagnation or recession, innumerable companies frequently switch from LIFO to FIFO. The weighted average method denotes a compromise between the two. Table 6.A4 is an illustration of the use of FIFO and LIFO accounting techniques.

Assume that a manufacturing company has five units of products in inventory and each has a product cost of $100. Furthermore, the company makes five more units at $200 each in one period and then another five units at $300 each in a later period. During these periods, the company sells ten units to customers. Determine the average cost of goods sold on the basis of both FIFO and LIFO and assess its impact on the company's net income.

The impact of inventory accounting on net income is quite direct, as illustrated in Table 6.A5, an abbreviated income statement, wherein CGS is costs of goods sold, GS&A is general, sales, and administration expenses, and EBIT is earnings before interest and taxes. Upon switching from FIFO to LIFO inventory accounting, the tax liabilities are calculated to have been reduced from $2.4 million to $2 million.

6.11.6 Derivation of Single Payment Compound Amount Factor

$$F_1 = P + P \times i = P(1+i) \text{at the end of first year}$$

$$F_2 = F_1 + F_1 \times i = P(1+i)^2 \text{(end of second year)}$$

$$F_n = P(1+i)^n \text{(End of the nth year)}$$

Table 6.A4 Inventory Accounting

	FIFO	LIFO	Average	
(1) Beginning inventory				
5 * $100	$500	$500	$500	Withdrawal of 10 units
				=================
(2) Purchase and value added				5 * 100
5 * $200	$1,000	$1,000	$1,000	5 * 200
				5 * 300
5 * $300	$1,500	$1,500	$1,500	=================
(3) Ending inventory				
5 *	$1,500	$500	$1,000	
(4) Cost of goods sold	$1,500	$2,500	$2,000	

Table 6.A5 Effect of Inventory Accounting on Net Income

	FIFO (dollars)	LIFO (dollars)	Weighted average (dollars)
Sales	10,000	10,000	10,000
CGS	1,500	2,500	2,000
Gross margin	8,500	7,500	8,000
GS&A	2,000	2,000	2,000
EBIT	6,500	5,500	6,000
Interest	500	500	500
Taxable amount	6,000	5,000	5,500
Tax (40%)	2,400	2,000	2,200
Net income	3,600	3,000	3,300

$$\text{Hence:} F = P(1+i)^n (\text{This is Equation 6.13})$$

$$\text{And:} P = F(1+i)^{-n} (\text{Equation 6.14})$$

6.11.7 Derivation of Uniform Series Compound Amount Factor

$$F_1 = A$$

$$F_2 = F_1 + F_1 \times i + A = A + A \times i + A = A(1+i) + A$$

$$F_3 = F_2 + F_2 \times i + A = F_2(1+i) + A = A(1+i)^2 + A(1+i) + A$$

....

......

$$F_n = A(1+i)^{(n-1)} + A(1+i)^{(n-2)} + \ldots. + A(1+i) + A$$

$$F_{(n+1)} = A(1+i)^n + A(1+i)^{(n-1)} + \ldots + A$$

Form a difference between these two series:

$$F_{(n+1)} - F_n = A(1+i)^n$$

By definition:

$$F_{(n+1)} = F_n(1+i) + A$$

$$\text{Hence:} F_n(1+i) + A - F_n = A(1+i)^n$$

$$F_n \times i = A[(1+i)^{n-1}]$$

$$F_n = F = A[(1+i)^{n-1}]/i \text{ (This is Equation 6.15)}$$

Other factors such as equations 6.16, (6.17), and (6.18) are to be derived by substitution.

6.11.8 Derivation of Annual Cost Computation Equations

There are two methods to compute the annual cost: the exact method and the approximate method.

A.Exact Method (Depreciation based on sinking fund method).

$$\text{AC} = \frac{(P-L)i}{(1+i)^n - 1} + Pi + \text{AE} \tag{6.21}$$

The first term on the right-hand side is the annual cost (based on sinking found depreciation method) for the $(P - L)$ amount. The second term is the annual interest charge for the investment capital P. The last term is the annual expense.

B.Approximate Method (Depreciation based on straight-line method).

$$\text{AC} = \frac{(P-L)}{n} + (P-L)i\frac{(n+1)}{2n} + Li + \text{AE} \tag{6.22}$$

The first term on the right is depreciation charge based on straight-line method. The second two terms on the right are the *average annual interest charge*, which is an opportunity cost (lost interest income for having made the investment). This average annual interest charge may be derived as follows:

The interest charge for each year is:

Year	Formula
1	$= Pi = (P-L)i + Li$
2	$= \left[(P-L) - \frac{(P-L)}{n}\right]i + Li$
3	$= \left[(P-L) - \frac{2(P-L)}{n}\right]i + Li$
⋮	⋮
n	$= \left[(P-L) - \frac{n-1}{n}(P-L)\right]i + Li$

The sum of the total interest charge from year 1 to year n is:

$$\begin{aligned} \text{Sum} &= i(P-L)1 + (1-1/n) + (1-2/n) + \ldots. + (1-(n-1)/n + nLi \\ &= i(P-L)n - [1+2+3+4+\ldots.+(n-1)]/n + nLi \\ &= i(P-L)n - [[(n-1)+1)](n-1)/2n + nLi \\ &= i(P-L)(n+1)/2 + nLi \end{aligned}$$

Thus, the average annual interest charge is (Eq. 6.12):

$$AAIC = \text{Sum}/n = i(P-L)(n+1)/2n + Li$$

Because of this last averaging step, this method is called an *approximate method*.

6.11.9 Net Present Value Including Corporate Tax, Depreciation, and Investment

Tax Credit.

$$\begin{aligned} \text{NPV} = &-P(1-\text{ITC}) + (1-t)A\frac{[(1+i)^n - 1]}{[i(1+i)^n]} \\ &+ \text{DB}t\ \text{DF} + \text{SV}\frac{(1-t)}{(1+i)^n} \end{aligned} \tag{6.23}$$

Where:

- NPV = Net present value
- A = Net uniform annual income ($= R - AOC - Tax - M - X$)
- R = Revenue
- AOC = Annual operating cost
- Tax = Tax paid for property (local, county)
- M = Maintenance cost
- X = Other annual expenses
- P = Present investment
- t = Federal corporate tax rate (= 0.46 for large corporation)
- i = Interest rate (or WACC)
- n = Number of interest period (i.e., years)
- ITC = Investment tax credit (e.g., $ITC = 0.08$ or 0.1)
- DB = Depreciable base
- DB = P(1—0.5*ITC), if ITC = 0.1
- DB = P, if $ITC = 0.08$
- DF = Depreciation discount factor
- DF = Sum $DP(m)/(1+i)^m$, $m = 1$ to n
- $DP(m)$ = Depreciation percentage in period m (e.g., for straight-line $DP(m) = 1/n$ constant)
- SV = Salvage value (= L), representing the capital recovery at the end of n periods

The following is a numerical example:

P = 10,000; A = 4,600; ITC = 0.1; n = 5; t = 0.35; SV = 0;

$DP = P\ (1 - 0.5 \times 0.1) = 0.95\ P$

IRS Form 4562: 5-year ACRS schedule

$DP(1) = 0.15$; $DP(2) = 0.22$; $DP(3) = 0.21$;

$DP(4) = 0.21$; $DP(5) = 0.21$

$DF = \sum \text{DP(m)}/(1 + \text{i})^m$, $m = 1$ to 5

Results: NPV = 5,570 (for i = 0.08)

IRR = 0.2768 (for NPV = 0)

6.11.10 Conversion of a Probability Density Function to Its Cumulative Distribution Function

The process of converting a probability density function to its cumulative distribution function is straightforward and unique. Figure 6.A2 displays a triangular probability density function for the cost of the component C1. The vertical axis depicts probability and the horizontal axis represents cost. The triangular probability density function is the easiest one to apply when a three-point estimate for a risky input variable is known.

The component C1 is assumed to have a minimum cost of \$30 (point A), a maximum cost of \$80 (point Z), and a most likely cost of \$50 (point N). The area underneath the triangular probability density function is normalized to 1; this condition prescribes that the y coordinate for the point N is 0.04 based on the calculation of $1 = 0.04 \times 0.5 \times (80 - 30)$.

Let us insert a vertical cost line through x. With this cost line in place, we define the shaded area PNZXP as A_x, which is under the probability density function, but bound by the vertical cost line that passes through x on the left. We form a ratio of

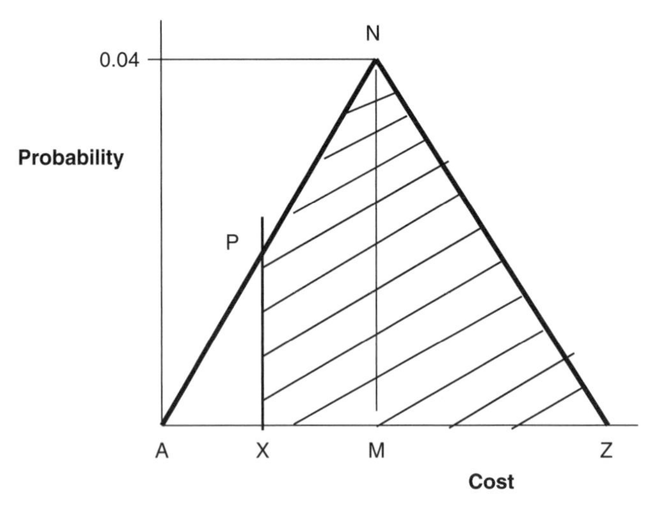

Figure 6.A2 Conversion of a triangular probability density function to its cumulative descending function.

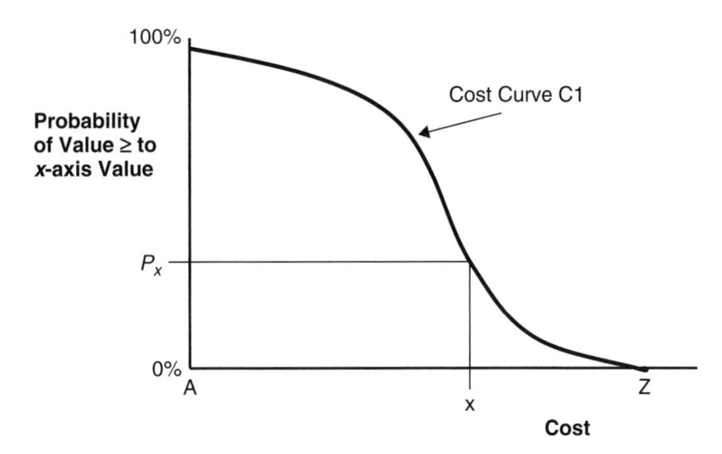

Figure 6.A3 Cumulative descending distribution function of a cost component.

A_x to the total area APNZXA underneath the same probability density function. This ratio is designated as P_x. The value of P_x varies from 0 to 1, as x moves from Z to M and then to A. P_x is the probability for the cost of this component to be equal to or in excess of x. The pair of P_x and x denotes a point on a cumulative distribution chart.

For another cost value, y, this process is repeated. A new pair of P_y and y defines another point in the descending cumulative distribution chart. After many repetitions, a descending cumulative curve is generated that resembles the one depicted in Fig. 6.A3. The vertical axis is the probability for the component cost to equal or exceed the value shown on the x-axis. The x-axis spans the minimum value of A on the left to the maximum value of Z on the right.

6.12 QUESTIONS

6.1 Study the case "Kanthal (A)," Harvard Business School Case # 9-190-002 (Rev. April 27, 2001) and answer the following questions. The case materials can be bought by contacting its publisher at www.custserve@hbsp.harvard.edu *or at* 1-800-545-7685.

A. What was the Kanthal president, Ridderstrale, attempting to accomplish with the Account Management System? Are these sensible goals?

B. Why did Ridderstrale feel that the previous cost system was inadequate for the new strategy? What causes a customer to be a "hidden loss" customer?

C. How does the new Kanthal 90 Accounting Management system work? What new features does it offer? What are its limitations that might restrict its effectiveness?

D. Consider a product line whose products generate a 50 percent gross margin (after subtracting volume-related manufacturing and administrative expenses from prices). The cost for handling an individual customer order is SEK 750, and the extra cost to handle a production order for nonstocked items is SEK 2,250.

D.1 Compare the net operating profits of two orders, both for SEK 2,000. One order is for stocked items and the other is for nonstocked items.

D.2 Compare the operating profits and profit margins of two customers, A and B. Both customers buy SEK 160,000 worth of goods during the year. A's sales came from three orders, for three different nonstocked items. B's sales came from 28 orders, of which 6 were for stocked items and 22 for nonstocked items.

E. What should Ridderstrale do about the two large unprofitable customers revealed by the account management systems?
F. What are the lessons you have learned from having studied this case?

6.2 The company's warehouse has been busy taking in and shipping out vendor-supplied automotive parts. Table 6.A6 illustrates the warehouse's activities in eight consecutive periods, during which time the price of the parts has steadily increased.

A. Determine the total LIFO prices for each stock withdrawal in periods 3, 4, 6, and 8.
B. Repeat the same price computation by using the FIFO technique.

Answer 6.2
The stock withdrawal prices are follows:

	LIFO	FIFO
Period 3	21,600	18,600
Period 4	11,400	12,000
Period 6	23,000	24,800
Period 8	11,200	11,000

6.3 A dam is being considered on a river that periodically overflows. Each time the river overflows, it causes about $600,000 in damage. The project horizon is 40 years. A 10 percent interest rate is being used.

Three different designs are available, each with different costs and storage capacities. (See Table 6.A7.) The U.S. Weather Service has provided a statistical analysis of annual rainfall in the area draining into the river. (See Table 6.A8.)

Table 6.A6 FIFO and LIFO Computation

Period	Units In	Unit Price Paid	Units Out
		(dollars)	
1	150	100	
2	250	120	
3			180
4			100
5	100	130	
6			200
7	100	140	
8			80

Table 6.A7 Design Options

Design Alternatives	Cost (dollars)	Maximum Storage Capacity
A	500,000	1 unit
B	625,000	1.5 units
C	900,000	2.0 units

Table 6.A8 Annual Rainfall and Probability

Units Annual Rainfall	Probability
0	0.1
0.1 to 0.5	0.6
0.6 to 1.0	0.15
1.1 to 1.5	0.1
1.6 to 2.0	0.04
2.0 or more	0.01

Assume that the dam requires no annual maintenance, has zero salvage value at the end of its forty-year life, and is essentially empty at the start of each annual rainfall season. Which design alternative would you choose?

Answer 6.3

The alternative B should be chosen.

6.4 The NPV equation (Eq. 6.24) is described as follows:

$$\mathrm{NPV} = -\mathrm{P} + \sum_{m=1}^{n} \frac{\mathrm{C}(m)}{(1+i)^m} + \frac{\mathrm{CR}}{(1+i)^n} \tag{6.24}$$

$$m = 1 \text{ to } n \tag{6.24}$$

The NPV equation is important for evaluating project-based investments. It is also a basic equation for defining the concept of "value addition," having a broad philosophical implication of what engineers do. Explain.

6.5 A manufacturing company makes three products, A, B, and C. The fixed factory overhead is \$60,000, consisting of \$10,000 for material handling, material waste, and procurement; \$30,000 for rent and utilities; and \$20,000 for safety and canteen costs. Other costs are presented in Table 6.A9.

A. Determine the product cost for products A, B, and C, using the activity based costing (ABC) method.

B. If products A, B, and C are sold at \$400, \$350, and \$150 a unit, respectively, what is the gross profit for each product?

C. What is the company's monthly gross profit if all units produced are sold?

Answer 6.5

A. The product costs for A, B, and C are \$167.25, \$134.58, and \$83.16, respectively.

B. The gross profits for A, B, and C are \$223.75, \$215.42, and \$66.84, respectively.

Table 6.A9 Product Options

	Product A	Product B	Product C
Number of units produced per month	250	400	900
Total material costs per month (dollars)	5,000	8,000	4,000
Labor hours per unit	4	3.5	1.5
Labor rate per unit (dollars per hour)	25	20	30
Machine hour per unit (hour)	1	1	3

C. The company's monthly profit is $204,511.50.

6.6 A company makes and sells three technology products: A, B, and C. It has a production plant with 17,000 square feet of floor area, consisting of machine setup (2,000 square feet), machining operation (9,000 square feet), assembly (4,000 square feet), and inspection, packaging, and shipping activities (2,000 square feet). The total annual expenditure for the plant is $200,000 for depreciation, $700,000 for utilities, $20,000 for phone and travel services, $150,000 for manufacturing supports, $200,000 for procurement, and $150,000 for supervision. The labor hours and material costs required to manufacture the products are displayed in Table 6.A10. The labor charges are $25 an hour for machine setup, $35 an hour for machining operation, $30 an hour for assembly, and $20 an hour for inspection, packing, and shipping.

The company plans to sell product A at $5,000 a unit, product B at $4,500 a unit, and product C at $4,100 a unit. All products manufactured during the year are assumed to be sold successfully. Apply the activity-based costing technique to determine the product cost and individual gross margin for each product.

Table 6.A10 Production Hours for Three Products

	A	B	C
Machine setup (hours)	2	3	4
Machine operation (hours)	16	12	8
Assembly (hours)	4	3	2
Inspection/packing/shipment (hours)	2	2	2
Raw materials/unit of product (dollars)	950	430	640
Purchased components/unit of product (dollars)	100	80	90
Outsourced service/unit of product (dollars)	20	30	40
Number of units produced per year	700	900	550

Answer 6.6

The product costs for A, B, and C are $2,679, $1,787, and $1,747, respectively. The individual gross margins for A, B, and C are $2,321, $2,713, and $2,553, respectively.

6.7 You are considering a good-looking Toyota hybrid car priced at $28,000 or an elegant GM luxury car at $24,000. The fuel efficiency is rated at 50 miles per gallon for the Toyota and 25 miles per gallon for the GM. The annual maintenance cost for both cars is about 0.5 percent of the car price. The gasoline in the local market is selling at $2 a gallon. The cars are to be driven about 10,000 miles a year. You plan to keep your car for five years. At the end of the fifth year, the resale values of the Toyota and the GM are about 40 percent and 30 percent, respectively, of their original prices. The interest rate is 6 percent.

Which car is the better choice from the standpoint of costs?

Answer 6.7

Toyota is the better choice from the standpoint of costs.

6.8 Company X manufactures automotive door panels that can be made of either sheet metal or plastic sheet molding (glass-fiber-reinforced polymer). Sheet metal bends well to the high-volume stamping process and has a low material cost. Plastic sheet molding meets the required strength and corrosion resistance and has a lower weight. The plastic forming process involves a chemical reaction and has a slower cycle time. Table 6.A11 summarizes the cost components for each.

Table 6.A11 Materials Options

Description	Plastic	Sheet Metal
Material cost (dollars per panel)	5	2
Direct labor cost (dollars per hour)	40	40
Factory overhead (dollars per year)	500,000	400,000
Maintenance expenses (dollars per year)	100,000	80,000
Machinery investment (dollars)	3 million	25 million
Tooling investment (dollars)	1 million	4 million
Equipment life (years)	10	15
Cycle time (minutes per panel)	2	0.1
Interest rate (percent)	6	6

Assuming that the machinery and tooling have no salvage value at the end of their respective equipment lives, what is the annual production volume that would make the plastic panel more economical?

Answer 6.8

For production volume up to 544,322 panels a year, the plastic panels are more economical.

6.9 Company X produces two products, A and B, respectively. Table 6.A12 summarizes the cost structures of these two products over a three-month period.

The company's manufacturing operation is limited to 30,000 machine hours available over a three-month period. Furthermore, because of a previous sales commitment, the company must make at least 1,000 units of Product B. Determine the maximum profit that the company can achieve in a three-month period.

Answer 6.9

The maximum profit is $296,900.

6.10 Buffalo Best Company markets three products for sales. Product A, Product B and Product C. Its production plant, which is located in the city of Buffalo, occupies 20,000 square feet of space. The use of this space is carefully planned as follows:

1. Product Assembly: 5,000 square feet
2. Machine setup: 2,500 square feet
3. Machining operation: 10,000 square feet
4. Inspection, packaging and shipping: 2,500 square feet

This plant has an expenditure of:

A. $200,000 for supervision
B. $250,000 for procurement

Table 6.A12 Product Options

	Product A	Product B
Selling price (dollars per unit)	10	12
Variable cost (dollars per unit)	5	10
Fixed costs (dollars)	600	2,000
Machining time (hours per unit)	0.5	0.25

Table 6.A13 Buffalo Best Company Operations Data

Number	Requirements (per unit of products)	A	B	C
1	Machine setup (hours)	2	4	5
2	Machine operation (hours)	20	15	12
3	Assembly (hours)	5	4	3
4	Inspection/packing/shipment (hours)	3	3	3
5	Raw material ($)	1000	500	700
6	Purchased parts needed ($)	150	120	130
7	Purchased services ($)	30	40	50
8	Number of units produced per year (-)	800	1000	700

C. $250,000 for depreciation

D. $750,000 for utilities

E. $25,000 for phones and travel expenses

F. $175,000 for manufacturing supports

As shown in Table 6.A13 each of the three products that this company makes requires different labor hours and material costs as follows:

The labor charges of the company are listed below:

Number	Type of Labor Hours	$/hour
1	Machine set up	$40.00
2	Machine operation	$50.00
3	Assembly	$35.00
4	Inspection/packing/shipping	$30.00

As shown in Table 6.A14 the company will sell the products at the following prices:

All products made by Buffalo Best Company are assumed to have successful sales rates.

A. Invoke the Activity Based Costing method to determine the unit cost and the individual gross margin for each of Product A, B, and C. Elucidate the detailed computations.

B. Which product has the highest gross margin percentage?

6.11 Monte Carlo simulations is a mathematical tool often used in engineering and business to solve complex problems involving uncertainties or risks.

A. Explain the technological foundation of this tool.

B. How can it be applied? Discuss the inputs to and outputs from Monte Carlo simulations applications.

Table 6.A14 Unit Prices

Number	Products	Price/unit
1	Product A	$6,000
2	Product B	$5,000
3	Product C	$4,000

C. What specific benefits can be derived from employing this tool, in comparison with those that can be readily obtained from employing a deterministic model in cost estimation, such as using an Excel spreadsheet program?

6.12 Employee A is about to retire. Based on his long tenure with the company, he is entitled to use his unused sick leave to pay for health insurance upon retirement. His current sick leave benefit is estimated to be worth $3,156 annually. The health insurance premium is estimated to be $2,532 a year for family coverage and $588 for single coverage, and these rates are projected to increase by 3 percent a year into the future. He has two options. Option one is to take the full amount of $3,156 now, which will continue during his lifetime. Afterward, his spouse will need to pay for the single coverage premium out of her own pocket. He will receive no refund from the company, even though his sick leave benefit ($3,156) exceeds the family coverage premium ($2,532) initially. Option two is to take only 70 percent of the full amount ($2,209) during his lifetime, and this benefit is guaranteed to continue beyond his death to cover the health insurance premium for the spouse, should the spouse survive him. Currently, Employee A is 73 years old and has a life expectancy of 11 more years. His spouse is 70, and her life expectancy is 18 years. Assume the cost of money is 4 percent. Which option is better for Employee A and his spouse?

Answer 6.12
Option #1 is to be preferred.

6.13 The company is evaluating two specific proposals to market a new product. The current interest rate is 10 percent. Proposal A calls for setting up an in-house manufacturing shop to make the product, requiring an investment of $500,000. The expected profits for the first to fifth years are $150,000, $200,000, $250,000, $150,000, and $100,000, respectively. Proposal B suggests that the manufacturing operation be outsourced by contracting an outside shop, requiring a front-end payment of $300,000. The expected profits for the first to fifth years are $50,000, $150,000, $200,000, $300,000, and $200,000, respectively. The expected profits would be lower in earlier years due to third-party markup.

Which proposal should the company accept?

Answer 6.13
Proposal B.

Chapter 7

Financial Accounting and Management for Service Systems Engineers

7.1 INTRODUCTION

Financial accounting and analysis serve the important corporate functions of reporting and evaluating the financial health of a firm (Kimmel, Waygandt and Kieso 2009). Financial statements are prepared by certified management accountants (CMAs) and certified public accountants (CPAs), according to the generally accepted accounting principles (GAAP) in a conservative, material, and consistent manner. These financial documents provide (1) internal reporting to corporate insiders for planning and controlling routine operations and for decisions on capital investments and (2) external reporting to shareholders and potential investors in financial markets (Sydsaeter and Hammond 2008).

All financial statements are designed to be relevant, reliable, comparable, and consistent. Financial accounting treats owners (shareholders) and corporations as separate entities. Owners of corporations are liable only to the extent of their committed investments. On the one hand, owners enjoy a flexible tenure and participation and, as investors, may buy or sell stocks of the company at any time. On the other hand, corporations are legal entities, fully responsible for their liabilities up to the limits of their total assets. Corporations are assumed to be going concerns and in operation forever, unless they cease to exist by declaring bankruptcy or being acquired by others.

Managerial finance focuses on the sources and uses of funds in a company. Capital budgeting is the key responsibility of the company's top management. Capital budgeting decisions take into account the need to expand production facilities, advance new products, create new supply chains, acquire new technologies to complement the company's own core competencies, penetrate into new regional markets, and other needs. Besides defining what is worth doing, top management must also find the investment capital needed to execute the chosen short-term and long-term strategies (Ross, Westerfield and Jordan 2007; Besley and Brigham 2007).

Investment capital may be raised from either internal sources (retained earnings) or external sources (equity and debt financing). Equity and debt financing has advantages and disadvantages. Topics relevant to capital formation include the weighted average cost of capital (WACC), capital structure, level of leverage and risks, and impact of leverage on the company's financial performance (Brigham and Ehrhardt 2007; Brealey, Myers and Alen 2008).

Resource allocation drives the company's overall performance. Allocation decisions address three major ways of utilizing investment capital: (1) assets in place—building new facilities and designing new products, (2) marketing and R&D efforts, and (3) strategic partnerships—acquisitions and joint ventures. In deciding on resource allocation, company management needs to know what a specific endeavor is worth. Then, different ways of estimating the value of companies, projects, and opportunities need to be established.

To be covered in this chapter are (1) language and concepts, (2) T-accounts, (3) financial statements, (4) performance ratios and analysis, (5) balanced scorecards or tools to monitor and promote corporate productivity, (6) the valuation of assets in place by the discounted cash flow methods—both the WACC and applied present value (APV) versions, and R&D and marketing opportunities by the simple option method. No discussion about the valuation of strategic partnerships is included, as engineering managers are not often involved in the financing aspects of acquisitions and joint ventures.

Engineering managers need to know how to read financial statements; monitor the firm's activity, performance, profitability, and market position; and assess the long-term health of a firm. They also need to understand the language, principles, and practices of financial management (Droms 1998). Doing so will allow them to initiate proper projects (e.g., plant expansion, new product and technology development, new technology acquisition, strategic alliances, etc.) at the right time to add value to their employers as well as to participate in capital budgeting decision making that involves major projects in which engineering managers provide significant inputs.

7.2 FINANCIAL ACCOUNTING PRINCIPLES

As practiced in the United States, all financial statements are formulated for a specific accounting period. A typical accounting period is three months, as the U.S. Securities and Exchange Commission prescribes that all publicly traded companies file Form 10-Q reports every quarter. All companies also need to publish their Form 10-K reports annually. The financial statements must adhere to the basic principles of accounting, discussed in the following subsections (Berman and Knight 2008, Miller 2005).

7.2.1 Accrual Principle

Accounting statements include both cash and credit transactions. Revenue is recognized when it is earned. For example, a manufacturing enterprise will recognize revenues as soon as products are shipped to the customer and an invoice is sent, irrespective of any credit payment already received or yet to be collected. Sports teams are known to sell season tickets ahead of the games for cash and then recognize the applicable revenue only after each game is played. According to the accrual principle of accounting, companies recognize revenues when earned, with the assumption that the collection of this revenue from approved credit accounts and the delivery of the promised products or services are both reasonably assured.

Similarly, the accrual principle specifies that costs and expenses are established when incurred, even before actual payments are made.

7.2.2 Matching

Expenses are recognized by matching them with the revenue generated in a given accounting period. For example, the cost of goods sold (CGS) is recognized as an expense only after products are sold and revenue is recognized. Before the products are sold, CGS stays as inventory—a part of the corporate current assets—even though costs for materials, labor, and factory overhead have already been spent for these unsold products.

7.2.3 Dual Aspects

The assets of a company are always equal to the claims against it (i.e., assets equal to claims). The claims originate from both creditors and owners. Each transaction has a dual effect in that it induces two entries in order to maintain a balance between assets and claims.

7.2.4 Full Disclosure Principle

All relevant information is disclosed to the users of the company's financial reports. Extensive footnotes contained in the annual reports of numerous publicly traded companies are testimonials for such disclosure practices.

7.2.5 Conservatism

Assets are to be recorded at the lowest value consistent with objectivity (e.g., book values are often lower than market values). While profits are not recorded until recognized, losses are recorded as soon as they become known. Inventories are valued at the lower of the cost or market value.

7.2.6 Going Concern

As stated earlier, it is assumed that the company's business will go on forever. This assumption justifies the current practice of using historical data (e.g., the original acquisition costs) and a reasonable method of depreciation (e.g., straight line) by which the book value of corporate tangible assets is defined. Otherwise, liquidation accounting must be applied to define the corporate asset value by using current market prices (Berman, Knight, and Case 2008A).

7.3 KEY FINANCIAL STATEMENTS

Companies in the United States use three financial statements: income statement, balance sheet, and funds flow statement (Berman and Knight 2008, Ittelson 2009).

7.3.1 Income Statement

The income statement is an accounting report that matches sales revenue with pertinent expenses that have been incurred (cost of goods sold, tax, interest, depreciation charges, salaries and wages, administrative expenses, R&D, etc.). Sometimes it is also called the *profit or loss statement*, *earnings statement*, or *operating and revenue statement*. The income statement contains the following key entries:

1. *Sales revenue* is the total revenue realized by the firm during an accounting period. Sales revenue is recognized when earned, for example, by having goods shipped and invoices issued.
2. *Cost of goods sold (CGS)* is the cost of goods that have been actually sold during an accounting period. In a manufacturing company, CGS is calculated as the opening inventory at the beginning of an accounting period, plus labor costs, material costs, and manufacturing overhead incurred during the period, and minus the closing inventory at the end of the period.
3. *Gross margin* is the sales revenue minus the cost of goods sold.
4. *Expenses* are those expenditures chargeable against sales revenue during an accounting period. Examples include general, selling, and administrative expenses; depreciation charges; R&D; advertising; interest payments for bonds; employee retirement benefit payments; and local taxes.
5. *Depreciation* is a process by which the cost of a fixed, long-lived asset is converted into expenses over its useful life. This noncash expenditure is to be claimed in proportion to the value it has produced during an accounting period. (See Appendix 6.11.4)
6. *EBIT* is the earnings before interests and taxes.
7. *Net income (earnings or NOPAT)* is the excess of sales revenue over all expenses (e.g., CGS, all items under point 4, and corporate tax) in an accounting period. Sometimes it is also called profit, earnings, or net operating profit after tax (NOPAT).
8. *Dividend* is the amount per share paid out to stockholders in an accounting period.
9. *Earnings per share* are the net income of a firm during an accounting period (e.g., a year), minus dividends on preferred stock, divided by the number of common shares outstanding.
10. *Costs* can be defined as follows: While all costs are also expenditures, not all costs are expenses, which are chargeable against revenues in a given accounting period. For example, direct and indirect costs contained in the products preserved as inventory are not recognized immediately as expenses. When products in inventory are sold, then the respective cost of goods sold (CGS) is recognized as expenses in the income statement, along with other expenses.
11. *Cash flow* is the sum of net income earned plus the depreciation charge claimed in a given accounting period.

    An income statement displays the firm's activity. An example is given in Table 7.1 for the XYZ Company. In general, sales revenue is referred to as the *top-line* and net income as the *bottom-line* figures. These line items are

Table 7.1 Example of XYZ Income Statement (Millions of dollars)

	Year 2001	Year 2002
Sales (net) revenue	8,380.30	8,724.70
cost of goods sold	6,181.20	6,728.80
Gross margin	2,199.10	1,995.90
General, selling, and administrative expenses	320.7	318.8
Pensions, benefits, R&D, insurance, and others	494.6	538.7
State, local, and miscellaneous taxes	180.1	197.1
Depreciation	297.2	308.6
EBIT	906.5	632.7
Interest and other costs related to debts	82.9	114.4
Corporate tax	(32.05 percent) 264.00	(20.84 percent) 108.00
Net income (NOPAT)	559.6	410.3
Common stock dividend	151.6	172.8
Retained earnings	408	237.5

examined closely by financial analysts, as are the line items of gross margin and EBIT. (For a detailed analysis and interpretation of income statement entries, see section 7.4.2 on ratio analysis.)

Engineering managers deploy company resources to make contributions to the financial success of their employers. The impacts of their engineering activities are registered in several line items contained in the income statement:

1. *Sales.* Engineering managers increase sales through well-designed products that satisfy the needs of customers. They introduce innovative products that address the needs of new customers in new markets. They refine products that are easy to serve and maintain, thus promoting market acceptance of the company's products. They also bring into being supply chains to increase the speed of product introduction and the extent of product customization in the marketplace.
2. *Cost of goods sold.* Engineering managers decrease product costs by innovative design, engineering, manufacturing, and quality control.
3. *R&D.* Engineering managers advance and utilize new technologies to enhance product features and to foster the rapid development of new global products.

Example 7.1 The Advanced Technologies company has had quite a successful year. At the end of the current fiscal year, its assets, liabilities, revenues, and expenses are exhibited in Table 7.2.

Determine the net income of the company for the current year.

ANSWER 7.1 To determine the company's net income, we need to create the income statement for the company. As presented in Table 7.3, only selected items of Table 7.2 are to be included in the company's income statement.

Table 7.2 Records of Financial Entries

#	Items	Thousands of dollars
1	Accounts payable	3,740
2	Accounts receivable	7,550
3	Advertising expense	3,340
4	Administrative expense	5,500
5	Building (net)	36,300
6	Cash	6,320
7	Cost of goods sold	31,000
8	Depreciation expense—building	960
9	Depreciation expense—equipment	1,310
10	Equipment (net)	14,640
11	Inventory	11,000
12	Insurance expense	840
13	Interest expense	2,100
14	Land	2,100
15	Long-term loans outstanding	42,000
16	Miscellaneous expense	1,480
17	R&D	5,200
18	Salaries payable	170
19	Sales revenue	60,300
20	Supplies expense	1,820
21	Taxes expense	2,630
22	Taxes payable	610
23	Utilities expense	2,070

Table 7.3 Income Statement of Advanced Technologies

	Thousands of dollars
Sales revenue	60,300
Cost of goods sold	31,000
Gross margin	29,300
Administrative expense	5,500
Advertising expense	3,340
Supplies expense	1,820
Utilities expense	2,070
Miscellaneous expense	1,480
Insurance expense	840
Depreciation—building	960
Depreciation—equipment	1,310
R&D	5,200
Operating income	6,780
Interest expense	2,100
Taxable income	4,680
Taxes expense	2,630
Net income	2,050

7.3.2 Balance Sheet

The balance sheet is an accounting report that presents the assets owned by a company and the ways in which these assets are financed through liabilities and owners' equity. Equation 7.1 depicts the balance between assets (A) and claims consisting of liabilities (L) and owners' equities (OE).

$$A = L + OE \tag{7.1}$$

Liabilities are creditors' (such as banks, bondholders, and suppliers) claims against the company. Owners' equity stands for the claims of owners (shareholders) against the company (Hawkins and Cohen 2002). The following key entries are included in a balance sheet:

- *Assets* are items of value having a measurable worth. They are resources of economic value possessed by the company. There are three classes of assets: current, fixed, and all others.
- *Current assets* are convertible to cash within twelve months. Examples include, in a descending order of liquidity, cash, marketable securities, accounts receivable, inventory, and prepaid expenses.
- *Cash* is money on hand or in checks, and is the most liquid form of assets.
- *Accounts receivable* is the category of revenue recognized prior to payment collection. It is money owed to the firm, usually by its customers or debtors, as a result of a credit transaction.
- *Inventory* designates stock of goods yet to be sold that is valued at cost, including direct materials, direct labor, and manufacturing overhead. It may consist of stores, work in progress, and finished goods inventories (See Appendix 6.11.5). Inventory is included in the balance sheet as a current asset.
- When finished goods are shipped and invoiced to customers in an accounting period, the cost of goods sold (CGS) is then recognized in the income statement as an expense.
- *Prepaid expenses* are paid before receiving the expected benefit (e.g., rent, journal subscription fee, or season's tickets). It is a current asset.
- *Fixed assets* are tangible assets of long, useful life (more than twelve months) such as land, buildings, machines, and equipment. (Improvement costs are added to the fixed asset value. Repair and maintenance costs incurred in a given accounting period are expensed in the income statement.)
- *Other assets* are valuable assets that are neither current nor fixed. Examples include patents, leases, franchises, copyrights, and goodwill. Amortization accounting applies to these assets in a similar manner, as depreciation is applied to fixed assets. (See section 6.11.4)
- *Goodwill* is a company's reputation and brand-name recognition. It is recognized as an asset only if it has been purchased for a measurable monetary value, such as in conjunction with a merger or acquisition transaction.
- *Accumulated depreciation* is the sum of all annual depreciation charges taken from the date at which the fixed asset is first deployed up to the present.

- *Net fixed assets* are the net value of the firm's tangible assets—original acquisition cost minus accumulated depreciation. (This net fixed asset value may deviate considerably from its market value or replacement cost in a given accounting period. The conservatism principle prescribes that the net fixed asset is carried on the balance sheet even if it is lower than its current market value. Otherwise, a loss entry must be added to the balance sheet to adjust the fixed asset value downward, should it become higher than its current market value.)
- *Liabilities* are obligations that are to be discharged by the company in the future. They denote claims of creditors (e.g., banks, bondholders, and suppliers) against the firm's assets. Sometimes, it is also called *debt*.
- *Current liability* describes amounts due for payment within twelve months. Examples include accounts payable, short-term bank loans, interest payments, payable tax, insurance premiums, deferred income, and accrued expenses. Accounts payable is always listed first within the category of current liabilities, with others to follow in no specific order.
- *Accounts payable* is an expense recognized before payment. It is an obligation to pay a creditor or supplier as a result of a credit transaction, usually within a period of one to three months.
- *Deferred income* is income received in advance of being earned and recognized (i.e., payment received before shipment and invoicing of goods). In the balance sheet, it is included as a current liability.
- *Deferred income tax* is the amount of tax due to be paid in the future, usually within twelve months.
- *Long-term liability* is defined as the amounts due to be paid in more than twelve months. Examples include corporate bonds, mortgage loans, long-term loans, lines of credit, long-term leases, and contracts.
- *Bonds* are long-term debt certificates secured by the assets of a company or a government. Bonds issued by a publicly held company are corporate bonds, and those issued by the U.S. federal government are Treasury bonds. In case of defaults, bondholders have the legal right to seize the assets that have been placed as collateral for the bonds in question from the company for recovery.
- *Debentures* are unsecured bonds issued by the firm.
- *Convertible bonds* are those issued by a company and allowed to be converted into common stocks according to a set of specifications.
- *Owners' equity* is the shareholders' original investment plus accumulated retained earnings. It stands for the residual value of the corporation owned by the shareholders after having deducted all liabilities from company assets; also called *net worth*.
- *Stock* is a certificate of ownership of shares in a company. Preferred stocks have a fixed rate of dividend that must be paid before dividends are distributed to holders of common stocks.
- *Capital surplus* is the premium price per share above the par value of the stock. It includes the increase in the owners' equity above and beyond the difference between assets and liabilities reported in the company's balance sheet.

- *Retained earnings* are the accumulated earnings retained by the company for the purpose of reinvestment and not to be paid out as dividends.
- *Book value* is defined as the tangible assets (such as fixed assets) minus liabilities and the equity of preferred stocks. It is the share value of common stocks carried in the books.
- *Stock price* is the market value of a firm's stock. It is influenced by the book value, earning per share, anticipated future earnings, perceived management quality, and environmental factors present in the marketplace.

The contributions of engineering managers affect only one line item in the balance sheet, namely, inventories. Inventories may be scaled down by employing superior production technologies, product design, and best practices of supply chain management.

The organization of entries in a balance sheet follows the specific convention:

- Assets are listed before liabilities, which are then followed by owners' equity.
- Current assets and liabilities are enumerated ahead of noncurrent assets and liabilities, respectively.
- Liquid assets are listed before all other assets with less liquidity.
- The listing of current liabilities follows no specific order, except that accounts payable must always be listed first in this section.

Table 7.4 illustrates a sample balance sheet of XYZ Company.

Example 7.2 Using the data given in Table 7.2, construct the balance sheet of Advanced Technologies and determine the owners' equity at the end of the current fiscal year.

ANSWER 7.2 The owners' equity is $29,120,000 at year-end. (See Table 7.5.)

7.3.3 Funds Flow Statement

The funds flow statement compares the firm's activities in two consecutive accounting periods from the standpoint of funds. It is an accounting report that elucidates the major sources and uses of funds of the firm. It is sometimes also called *statement of changes in financial position* or the *statement of sources and uses of funds*.

The principle behind the funds flow analysis is rather simple. An increase in assets signifies a use of funds, such as buying a plant facility by paying cash or using credit. A decrease in assets indicates a source of funds, such as selling used equipment to receive cash for use in the future. An increase in liabilities produces a source of funds, such as borrowing money from a bank so that more cash is available for other purposes. A decrease of liabilities yields a use of funds, such as paying down a bank loan by using money from the company's cash reservoir. Table 7.6 presents an example of the funds flow statement of XYZ Company.

The funds flow statement depicted in Table 7.6 is generated by applying the following procedure to the balance sheet of XYZ company (Table 7.4):

A. Increase in Plants and Equipment

1. Increase in fixed assets (11897.7 − 11070.4) = 827.3.
2. Increase in long-term receivable and other investments (735.2 − 574.8) = 160.4.

Table 7.4 Example of XYZ Balance Sheet (Million of dollars)

	Year 2001	Year 2002
Assets		
Cash	231.00	245.70
Marketable securities	450.80	314.90
Accounts receivable	807.10	843.50
Inventories	1,170.70	1,387.10
Total current assets	**2,659.60**	**2,791.20**
Fixed assets	11,070.40	11,897.70
Accumulated depreciation	6,410.70	6,618.50
Net fixed assets	**4,659.70**	**5,279.20**
Long-term receivables and other investments	574.80	735.20
Prepaid expenses	260.90	362.30
Total long-term assets	5,495.40	6,376.70
Total Assets	**8,155.00**	**9,167.90**
Liabilities		
Accounts payable	571.20	622.80
Notes payable	65.30	144.50
Accrued taxes	346.30	275.00
Payroll and benefits payable	433.70	544.30
Long-term debt due within a year	30.40	50.80
Total current liabilities	**1,446.90**	**1,637.40**
Long-term debt	1,542.50	1,959.90
Deferred tax on income	288.40	405.30
Deferred credits	27.00	36.30
Total long-term liabilities	**1,857.90**	**2,401.50**
Total Liabilities	**3,304.80**	**4,038.90**
Common stock ($1.00 par value)	81.40	82.20
Capital surplus	1,549.10	1,589.60
Accumulated retained earnings	3,219.70	3,457.20
Total Owners' Equity	**4,850.20**	**5,129.00**
Total Liabilities and Owners' Equity	**8,155.00**	**9,167.90**

3. As details are missing, we introduce the following reasonable assumption:
   (a) Long-term receivables = 59.6
   (b) Increase in other investment = 100.8
4. Total increase in fixed asset investment (827.3 + 100.8) = 928.1.

B. Increase in long-term debt
   1. Increase in long-term debt (1959.9 – 1542.5) = 417.4.
   2. Increase of long-term debt due within one year (50.8 – 30.4) = 20.4.
   3. Total (417.4 + 20.4) = 437.8

C. Increase in Common Stock and Capital
   1. Increase in common stocks (82.2 – 81.4) = 0.8.
   2. Increase in capital surplus (1589.6 – 1549.1) = 40.5.

Most other line items in the statement are directly verifiable.

Table 7.5 Balance Sheet of Advanced Technologies

	Thousands of dollars
Cash	6,320
Accounts receivable	7,550
Inventory	11,000
Total current assets	24,870
Land	2,100
Equipment (net) ($14,640 – 1,310)	13,330
Building (net) ($36,300 – 960)	35,340
Total assets	75,640
Accounts payable	3,740
Taxes payable	610
Salaries payable	170
Long-term loans outstanding	42,000
Total liabilities	46,520
Owners' equity	29,120
Total liabilities and owners' equity	75,640

Table 7.6 Example of XYZ Funds Flow Statement (Millions of dollars)

	2001–2002	Percentage
Sources		
Increase in long-term debt	437.4	26.50
Net income	410.3	24.86
Depreciation*	308.6	18.7
Decrease in marketable securities	135.9	8.23
Increase in deferred taxes on income	116.9	7.08
Increase in payroll and benefits payable	110.6	6.70
Increase in notes payable	79.2	5.00
Increase in accounts payable	51.6	3.13
Total Source of Funds	1,650.50	100.00
Uses		
Increase in fixed assets and other investments	928.1	56.23
Increase in inventories	216.4	13.11
Dividend paid	172.8	10.47
Increase in prepaid expenses	101.4	6.14
Increase in long term receivables	59.6	3.61
Decrease in accrued taxes	71.3	4.32
Increase in capital surplus	40.5	2.45
Increase in accounts receivable	36.4	2.21
Increase in cash	14.7	0.90
Increase in deferred credits	9.3	0.60
Total Uses of Funds	1,650.50	100.00

*Depreciation is a noncash expenditure that must be added back here to denote a source of funds available to the firm.

7.3.4 Linkage between Statements

The three financial statements described previously are linked to one another. The net profit in the income statement is linked with the retained earning in the balance sheet. The inventory account in the balance sheet is linked with the sales revenue in the income statement. The accumulated depreciation in the balance sheet is linked with the annual depreciation charge included in the income statement. Because the depreciation charge taken in a given period affects the net profit of the company during the same period, it is thus indirectly linked to the retained earning account in the balance sheet as well.

The linkage between the funds flow statement and the other two financial statements is self-evident, as all data in the funds flow statement are derived from changes in various line items in the other two statements.

7.3.5 Recognition of Key Accounting Entries

This section offers additional notes on the recognition of several key accounting entries—assets, liabilities, revenues, and expenses—according to GAAP practiced in the United States. Other countries may have slightly different rules governing the report of these items.

Assets. *Assets* are the resources under company control. They have economic value and can be used to produce future benefits. Assets recognition is based on two principles: historical cost and conservatism (Healy and Choudhary 2001A). All assets are reported by using historical cost—that is, the initial capital investment value at some time in the past. The book value of a given asset is defined as its initial acquisition cost minus the accumulated depreciation. Should the asset's current market value drop below its book value, the shortfall must be reported as an expense. If its market value exceeds its book value, however, the surplus is not reported in the company's balance sheet. This is to ensure that the asset value included in the balance sheet always denotes its lower bound. Thus, the balance sheet may understate the true value of the company's assets.

Asset reporting must address the two issues of asset ownership and the certainty of its future economic benefits. If neither the ownership nor the future benefits are clearly established, an asset cannot be recognized. For example, companies routinely invest in employee training in the hope that doing so will lead to increased productivity at a future point in time. Since the completed training is really owned by the employees, and employees may leave at any time they wish, companies do not have real ownership of the training results. Thus, employee training is regarded as an expense and not an asset. When companies acquire plant facilities to make products for sales in the marketplace, its future benefits are more or less certain. Plant facilities are thus reported as assets. When companies apply resources to expand R&D and advertising, the future benefits of these investments are neither certain nor measurable. R&D and advertising are thus recognized as expenses and not assets.

GAAP accounting rules in the United States contain one exception: Generally speaking, software development costs are to be reported as expenses as they are incurred. However, once the company management becomes confident that the software development efforts can be completed and the resulting software product will be used as intended, all costs incurred from that point on are to be reported as assets.

There are three key ambiguities in the accounting rules practiced in the United States:

1. *Buying versus developing.* If Company A acquires Company B by paying a purchase price that exceeds Company B's net asset value, then this excess value is called goodwill. Goodwill includes the intangible assets of Company B such as its brand name, trademarks, patents, R&D portfolio, and employee skills. After the merger, the surviving company has part of its R&D (from the original company A) recognized as expenses, and part of the R&D (acquired from company B) as assets.
2. *Valuing intangible assets.* "If you can't kick a resource, it really isn't an asset." This saying is typically the justification used by companies to rapidly write off intangible assets from their balance sheets. Oftentimes, goodwill is significantly overvalued in a merger or acquisition transaction due to potential conflicts of interests among the parties involved. Writing off intangible assets distorts the true value of the assets reported in the company's balance sheet.
3. *Market value.* U.S. accounting rules prescribe that marketable securities (e.g., bonds) are to be reported at their fair market values only if they are not to be held to maturity. Thus, at any given time, the real asset value of a company is distorted by not reporting the current true market values of these assets in balance sheets.

Liabilities. *Liabilities* are obligations to be satisfied by transferring assets or providing services to another entity (e.g., banks, suppliers, and customers).

A liability is recorded when an obligation has been incurred and the amount and timing of this obligation can be measured with a reasonable amount of certainty (Healy and Choudhary 2001B). For multiple-year commitments, the recordable obligation is the present value of expected future commitments wherein the discount rate is the prevailing rate when the obligation was first established. (See Section 6.11.3.)

Revenues. *Revenues* recognition must satisfy two conditions: Revenue is earned when (1) all or substantially all of the goods or services are delivered to the customers and (2) it is likely that the collection of cash or receivables will be successful. Generally speaking, the timing of product or service delivery may not be the same as that for payment collection (Healy 2001).

For magazine subscriptions, insurance policies, and service contracts, customers usually pay in advance. In these cases, payments received ahead of the service delivery dates are kept in a *deferred revenue* account. Only after the pertinent service is delivered will the applicable payments be credited to the revenue T-account during the accounting period. A detailed discussion on T-accounts is offered in Appendix 7.10.1. For products sold on credit, companies recognize the revenue as soon as the products are shipped out, and invoices are issued to customers ahead of the payment collection.

In the case of construction projects, which usually stretch out over a number of accounting periods, revenues are recognized in T-accounts by using the *percentage completion* method and are recognized in proportion to the expenses incurred in the project.

For products sold with money-back guarantees, companies recognize revenue at the time the product is delivered. At the end of an accounting period, management makes an estimate of the cost of returns (a liability) to adjust the revenue figure.

Expenses. *Expenses* are economic resources that either have been consumed or have declined in value during an accounting period. Expenses are typically recorded in the form of a reduction of asset value (e.g., cash) or by a creation of liability (e.g., accounts payable).

There are three types of expenses: (1) consumed resources having a cause-and-effect relationship with revenue generated during the same accounting period (e.g., cost of goods sold); (2) other resources consumed during the same accounting period, but having no cause-and-effect relationship with revenue (such as R&D expenses, advertising expenses, depreciation charges, local taxes, pension expenses, and other general administrative expenses); (3) reduction of expected benefits of company assets generated by past investments (e.g., the write-down of production facilities and equipment that is no longer of value due to the noncompetitiveness of products or technological obsolescence) (Healy and Choudhary 2001C).

7.3.6 Caution in Reading Financial Statements

In the United States, companies follow the rules set by the Financial Accounting Standards Board (FASB), an industrial panel, in preparing financial statements. Even so, these financial statements are not created equal; they need to be studied carefully because of the following built-in variations:

- *Depreciation accounting base.* Some companies may use straight-line method, whereas others may choose to use the double-declining method, as both are allowable.
- *Inventory accounting method.* Some companies use First in and First out (FIFO) methods, whereas others may choose to use Last in and First out (LIFO). In time periods with high price volatility, the inventory value will be affected by the method chosen.
- *Cost of capital.* Dependent on the debt to equity ratio and cost of raising equity and debt, the weighted average of cost of capital will be different from one company to another.
- *Difference between book and market values.* The book values of company's fixed assets are calculated by subtracting the accumulated depreciation from their initial acquisition prices. These values may be quite different from the assets' current market values, which, in turn, are dependent on the supply and demand in the marketplace and the overall economy.
- *Long-term liabilities reportable.* The current FASB rules do not require companies to report long-term liabilities associated with pension, healthcare and other such obligations in balance sheets, although these liabilities are typically disclosed in footnotes and other such obscure places within the companies' annual reports. Some companies with a heavily unionized workforce could therefore

create the illusion of having a higher net worth than otherwise before unsophisticated investors. In March 2006, FASB announced the intention of initiating new regulations to improve the transparency of this disclosure. It is thus important to keep such long-term liabilities in mind when reading financial statements.

Example 7.3 Superior Technologies sells a product at the unit price of $100. The unit cost of the product is $60. Annual sales have averaged 1 million units, and its annual selling expense has been $7 million. Market research has determined that, if the selling price of the company product is decreased to $90, there will be a 35 percent increase in the number of units sold. The engineering department estimates that, if the production volume is increased by 35 percent, it will reduce the unit product cost by 10 percent due to the scale of economies. To market the 35 percent increase in sales volume, the company's selling expense will need to increase by about 50 percent.

The company's current warehouse facilities are sufficiently large to accommodate the possible increase of 35 percent in sales volume without requiring new investment. Furthermore, regardless of the product price, the company is obliged to pay an annual loan interest of $2 million. Its corporate tax rate is 45 percent. It maintains an R&D department, whose operation is independent of the sales units, at an annual cost of $5 million. Its administrative expense is $15 million, which is also independent of the sales activities. In addition, the company incurs a pretax depreciation charge of $2 million.

Determine if the reduction of product price would increase or decrease the net income of the company, and by how much.

ANSWER 7.3 The reduction of product price will cause the company's net income to increase to $7.75 million from $4.95 million (see details in Table 7.7).

Table 7.7 Net Income Due to Increased Sales

	Current Operation (dollars)	Operation with Increased Unit Sales (dollars)
Units of product sold	1,000,000	1,350,000
Product price	100	90
Unit product cost	60	54
Sales revenue	100,000,000	121,500,000
Cost of goods sold	60,000,000	72,900,000
Gross margin	40,000,000	48,600,000
Selling expense	7,000,000	10,500,000
Administrative expense	15,000,000	15,000,000
R&D	5,000,000	5,000,000
Depreciation	2,000,000	2,000,000
EBIT	11,000,000	16,100,000
Interest	2,000,000	2,000,000
Taxable income	9,000,000	14,100,000
Tax (45 percent)	4,050,000	6,345,000
Net income	4,950,000	7,755,000

Example 7.4 Dell assembles, sells and services personal computers. The company markets Pentium models (75 percent) and 485 models (25 percent) directly to its customers and builds computers after receiving a customer's order. The company was growing rapidly at a Compound Annual Growth Rate (CAGR) of 57.5 percent during the period of 1991 to 1995. In 1996, Dell's competitors had days sales inventory (DSI) of 73 for Compaq, 48 for IBM, and 54 for Apple. DSI is the number of days inventory covers sales. It is defined as inventory times 360 divided by cost of sales. Dell's inventory was about 10.14 percent of its cost of sales (COS), whereas Compaq's inventory to COS ratio was 20.3 percent. During this time period, computer components experienced a drastic price reduction of about 30 percent per year, as new models were rapidly marketed. Dell's income statement and balance sheets for 1996 are exhibited in Tables 7.8 and 7.9 (Ruback and Sesia, 2003).

What are the advantages and disadvantages of Dell's working capital over Compaq?

ANSWER 7.4 The build-to-order model enables Dell to (1) have small working capital requirements, (2) benefit from reductions in component prices, (3) introduce new products quickly, and (4) finance growth by using working capital and profitability. Low inventory requires small working capital to be tied down by inventory and reduces the cost of financing working capital.

$$\text{Dell's DSI} = 360 \times \text{Inventory/COS} = 360 \times 429/4{,}229 = 36.5 \text{ days.}$$

In comparing Dell with Compaq, the extra inventory held at Compaq is worth an additional $4{,}229(73 - 36.5)/360 = \$429$ million. This amount of $429 million in working capital was avoided by Dell because of its low inventory. At an assumed annual bank interest rate of 8 percent, this leads to an interest avoidance of $34.3 million.

Another advantage of Dell's low inventory strategy is its lower exposures to the reduction in component prices, which dropped 30 percent a year due to introduction of new technologies. As Dell's inventory is only 10.14 percent of its COS, its exposure is thus limited only to 3.04 percent ($= 0.3 \times 10.14\%$) of COS, or $128.65 million, compared to Compaq's 6.09 percent ($= 0.3 \times 20.3\%$) of COS, or $257.5 million. The avoidance of about $128.65 million inventory write-down was to Dell's

Table 7.8 Income Statement of Dell

Income Statement ($ Million)	1996
Sales	5296
Cost of sales	4229
Gross margin	1067
Operating expense	690
Operating income	377
Financial and other income	6
Income tax	111
Extraordinary loss	0
Net profit	272

Table 7.9 Balance Sheet of Dell

	Year Ended 1/28/96
Current assets	
Cash	55
Short-term investments	591
Accounts receivables, net	729
Inventories	429
Others	156
Total Current Assets	1957
Property, plant and equipment, net	179
Other	12
Total Assets	**2148**
Current liabilities	
Accounts payables	466
Accrued and other liabilities	473
Total current liabilities	939
Long-term debt	113
Other liabilities	123
Total Liabilities	**1175**
Equity	
Preferred stock	6
Common stock	430
Retained earnings	570
Others	($33)
Total Equity	**973**
Total Equity and Liabilities	**2148**

favor. Besides not having to discount the outdated components, Dell did not have to market both the new and the old components, thus avoiding cannibalization.

One disadvantage of maintaining low inventory is that Dell could have shortage of components in stock, which could lead to lost sales from time to time.

7.4 FUNDAMENTALS OF FINANCIAL ANALYSIS

The purpose of conducting financial analyses is to assess the effectiveness of the company's management in achieving the objectives set forth by the company's board of directors with respect to a number of critically important business factors (Subramanyam and Wild 2008). Such factors include the following:

- *Liquidity*—the availability of current assets to satisfy the firm's operational requirements.
- *Activity*—the efficiency of resource utilization.
- *Profitability*—the extent of the firm's financial success.
- *Capitalization*—the makeup of the company's assets and its utilization of financial leverage.

- *Stock value*—the market price of the company's stock.
- *Market capitalization*—the product of stock price and the number of outstanding stocks.

Typically, the corporate objectives of diverse companies are growth, profitability, and return on investment (ROI).

The *growth* objective suggests that companies keep product prices low, increase marketing expenses, run plants at full capacity, take loans to keep inventory high, and strive for a larger market share and a more dominant market position. The *profitability* objective dictates that companies set prices high to maximize profits, run plants at a capacity that minimizes costs (production and maintenance), and use debt when called for. The *ROI* objective is achieved by operating the company to maximize its financial return with respect to the firm's investment (e.g., the "milk-the-cash-cow" strategy. See section 8.5.1).

Financial analyses focus on studying period-to-period changes in key financial data and on comparing performance ratios with the applicable industrial standards.

7.4.1 Performance Ratios

In this section, we elect to use a specific system of calculating performance ratios—grouping together the ratios for liquidity, activities, profitability, capitalization, and stock value (Berman, Knight, and Case 2008). In order to understand this system, we will first define each item involved.

Liquidity. *Liquidity* is the firm's capability to satisfy its current liabilities, such as buying materials, paying wages and salaries, paying interests on long-term debt, and other necessary expenditures. Without liquidity, there can be no activity.

Working capital is defined as current assets minus current liabilities. The changes in working capital over several periods provide an indication of the company's reserve strength to weather financial adversities.

Current ratio is the ratio of current assets to current liabilities. Current assets are frequently considered the major reservoir of funds for meeting current obligations. This ratio provides an indication of the company's ability to finance its operations over the next twelve months. A current ratio above 1.0 indicates a margin of safety that allows for a possible shrinkage of value in current assets such as inventories and accounts receivables. However, having a current ratio in excess of 2.0 or 3.0 may indicate a poor cash management practice.

Quick ratio is the ratio of quick asset to current liabilities. *Quick asset* is defined as cash plus marketable securities and accounts receivable. This ratio is more severe than the current ratio in that it excludes the value of inventory whose liquid value may not be certain. Thus, the quick ratio indicates the company's ability to meet its financial obligations over the next twelve months without the use of inventory that may take time to unload. It is sometimes also called the *acid test* or *liquidity ratio*.

Activity. *Activity* is the changes in sales and inventory. Successful activity leads to profitability.

Collection period ratio is the accounts receivable divided by average daily sales as measured in days. The average daily sales are the total annual sales divided by

360 days. This ratio measures the managerial effectiveness of the credit department in collecting receivables and the quality of accounts receivable.

Inventory turnover ratio is the cost of goods sold divided by the average inventory. It expresses the number of times during a year that the average inventory is recouped or turned over through the company's sales activities. The higher the turnover, the more efficient the company's inventory management performance will be, provided that there has been no shortage of inventory producing a loss of sales and failure to satisfy customers' needs.

Asset turnover ratio of net sales to total assets indicates the ability of the company's management to utilize total assets to generate sales.

Working capital turnover ratio is net sales to working capital. Working capital is defined as average current assets minus average current liabilities. It indicates the company's ability to efficiently utilize working capital to generate sales.

Sales to employee ratio is the company's net sales revenue divided by its average number of employees working during an accounting period. It measures the company's ability to effectually utilize human resources.

Profitability. To be profitable is the objective of most companies. Without liquidity and activity, there can be no *profitability*. If the company is profitable, it can readily obtain the required liquidity to keep its operations continuing.

Gross margin to sales ratio measures the company's profitability on the basis of sales. Gross margin is defined as sales minus cost of goods sold. Gross margin percentage is defined as the gross margin divided by sales.

Net income to sales ratio indicates the company's overall operational efficiency (e.g., procurement, cost control, current assets deployment, and utilization of financial leverage) in creating profitability based on sales. This ratio is also known as ROS, which stands for return on sales.

Net income to owners' equity ratio measures profitability from the shareholders' viewpoint. It is also known as ROE, which stands for return on equity. This very common measure points out the earning power of the ownership investment in the company.

Net income to total asset ratio is net income divided by total assets. It measures the management's ability to effectively utilize company assets in generating profits. It is known as ROA, which stands for return on assets.

Return on invested capital is the ratio of net income divided by capital. The capital of a company is the sum of its long-term liabilities plus owners' equity.

Capitalization. The sum of the company's long-term assets and owners' equity is defined as the total *capital* deployed by company management to pursue business opportunities. Several ratios are in use to check this capital deployment effectiveness.

Interest coverage ratio (EBIT divided by the interest expense) calculates the number of times the company's EBIT covers the required interest payment for the long-term debt—an indication of the company's ability to remain solvent in the near future.

Long-term debt to capitalization ratio is the ratio of long-term debt to the sum of long-term debt plus owners' equity—the total permanent investment in a company, indicating the percentage of long-term debt in the company's capital structure, excluding current liabilities. It is a measure of the company's financial leverage. Keeping this ratio small (hence, large owners' equity percentage) may not always be the smart choice, as the company will forgo the use of low-cost debt with tax deductible interest payments to enhance profitability.

Debt to equity ratio is the ratio of total liability to owners' equity. It also measures the company's financial independence and the relative stake of shareholders (insiders) and bondholders (outsiders). A low ratio suggests that the company is financially secure as far as the owners are concerned. A high ratio indicates that the firm may have difficulty borrowing money in the future.

Stock Value. This is the market price or value of the company's stock as defined by the financial markets. The company's management is obliged to pursue proper business strategies in order to steadily raise their *stock value*.

Earnings per share is the ratio of net income minus preferred stock dividends divided by the number of common stocks outstanding.

Price to earning ratio is the ratio of the market price of common share to earning per share.

Market to book ratio is the ratio of market price of stock to the book value per share. More precisely, the total book value of a company is defined as the total assets minus intangible assets, minus total liability, and minus the equity of preferred stocks. The book value per share is then the total book value divided by the number of outstanding common shares.

Dividend payout ratio is the ratio of dividends per share divided by earnings per share. It designates the percentage of annual earnings paid out as dividends to shareholders. The portion that is not paid out goes into the retained earnings account on the balance sheet.

7.4.2 Ratio Analysis

Ratios are useful tools of financial analysis (Bruns 1996; Troy 2008; Healy and Cohen 2000). Sometimes ratios of significant financial data are more meaningful than the raw data themselves. They also provide an instant picture of the financial condition, operation, and profitability of a company, provided that the trends and deviations reflected by the ratios are interpreted properly.

Ratio analyses are subject to two constraints: past performance and various accounting methods:

1. *Past performance* is not a sure basis for projecting the company's condition in the future.
2. *Various accounting methods* employed by different companies may result in different financial figures (e.g., inventory accounting, depreciation, etc.), rendering a comparison that is not always meaningful between companies in the same industry. Typically, accountants try to reconcile financial statements before conducting comparative analyses. Examples of adjustments that are frequently made include:

   - Adjusting LIFO inventories to a FIFO basis (Appendix 6.11.5.)
   - Changing the write-off periods of intangible assets, such as goodwill, patents, and trademarks (Appendix 6.11.4)

- Adding potential contingency liabilities if lawsuits are pending (Section 6.11.4).
- Reevaluating assets to reflect current market values (Section 7.3.5).
- Changing debt obligations to reflect current market interest rates.
- Restating reserves or charges for bad debts, warranties, and product returns.
- Reclassifying operating leases as capital leases.

When performing ratio analyses, it is advisable to follow this set of five generally recommended guidelines:

1. Focus on a limited number of significant ratios.
2. Collect data over a number of past periods to identify the prevalent trends.
3. Present results in graphic or tabular form according to standards (e.g., industrial averages).
4. Concentrate on all major variations from the standards.
5. Investigate the causes of these variations by cross-checking with other ratios and raw financial data.

An example of finding the causes of changes noted in ratios is the performance of Dell and Compaq as studied by Healy and Cohen (2000). The return on equity ratios of Dell and Compaq in the years 1995, 1996, and 1997 were compared. Table 7.10 displays the results.

Obviously, the results raised the question by numerous financial analysts of why the differences were so big. Additional analyses revealed interesting details. The following DuPont equation is pertinent (Bodie et al 2004):

$$\text{ROE} = \text{Net income/Equity} = (\text{Net income/Sales}) \times (\text{Sales/Assets}) \times (\text{Assets/Equity}) \quad (7.2)$$

Table 7.11 presents the ratios calculated for these two companies during the three-year period.

It became clear that the operational efficiency of these two companies was comparable. However, Dell had a much higher rate of efficiency in asset utilization and took a much more aggressive stand with respect to financial leverage than Compaq. Specifically, Dell was able to (1) outsource most of its manufacturing activities; (2) keep a very low inventory based on its direct sales model (which allows Dell to receive customer orders from the Internet prior to initiating the needed manufacturing operation); (3) collect accounts receivables fast, thus reducing the need for self-financed working capital; (4) practice a high-leverage financing strategy by raising debt, which, in turn, produced high sales for the company. In contrast, Compaq was operating much more conservatively than Dell during the same three-year period.

Table 7.10 Return on Equity

Company	1997 (percent)	1996 (percent)	1995 (percent)
Dell	90	58	33
Compaq	22	22	22

Table 7.11 Ratios for Dell and Compaq

	1997 (percent)	1996 (percent)	1995 (percent)
Dell			
Net income/sales	7.70	6.70	5.10
Sales/assets	3.40	3.02	2.83
Assets/equity	3.46	2.89	2.30
Compaq			
Net income/sales	7.50	6.60	5.40
Sales/assets	1.82	1.99	2.38
Assets/equity	1.61	1.69	1.69

In financial literature, many of the ratios just defined are being systematically collected and published for U.S. companies in various industries by investment services companies. Sources of information on ratios and other financial measures are typically reported regularly and made available for use by the public from such publications, including the following:

- Value Line® Investment Survey
- Standard & Poor's Industrial Surveys
- Moody's Investors Services
- Specific investment letters and publications (e.g., The Zweig Forecasts)

Engineering managers should acquire a habit of reading and considering such reports.

Other commercial sources are accessible through the Internet. For the 500 stocks comprising Standard & Poor's index, five specific ratios—the current ratio, long-term debt to capital, net income to sales, return on assets, and return on equity—are published in a widely available special guide for ten-year, consecutive periods (Standard & Poor's 2009).

Example 7.5 For the years 2001–2002, the financial statements of XYZ Company are given in Tables 7.12 and 7.13. Define the performance ratios and compare them with industrial standards.

ANSWER 7.5 The 2001 performance ratios of XYZ Corporation are displayed here:

1. **Liquidity**

   (a) Current ratio = CA/CL = 3:1

   From the creditor's standpoint, this ratio should be as high as possible. On the other hand, prudent management will want to avoid the excessive buildup of idling cash or inventories (or both).

   (b) Quick ratio = Quick asset/CL = 0.93:1

   A result far below 1:1 can be a warning sign.

   (c) Interest coverage ratio = 7.833

Table 7.12 XYZ Balance Sheet (thousands of dollars)

	2002	2001
Assets		
Cash	18,500	17,000
Marketable securities	0	5,000
Accounts receivable	39,500	28,500
Inventories	98,000	113,000
Total current assets	156,000	163,500
Plant and equipment (net)	275,000	290,000
Other assets	3,000	8,000
Total assets	434,000	461,500
Liabilities		
Accounts payable	34,500	18,000
Notes payable	20,000	25,000
Accrued expanses	18,500	11,500
Total current liabilities	73,000	54,500
Mortgage payable	20,000	30,000
Common stock	200,000	200,000
Earned surplus	141,000	177,000
Total liabilities and equities	434,000	461,500

Table 7.13 XYZ Income Statement (thousands of dollars)

	2002	2001
Sales	330,000	395,000
Cost of sales*	265,000	280,000
Gross profit	65,000	115,000
Selling and administrative	95,000	88,000
Other expenses	4,000	3,500
Interest	2,000	3,000
Profit before taxes	(36,000)	20,500
Federal income tax	0	10,000
Net income (loss)**	(36,000)	10,500

*Includes depreciation of $15,500 in 2001 and $15,000 in 2002
**No dividends paid in 2002

The earning before interest and taxes (EBIT) of the firm could pay 7.833 times the interest and other costs associated with the long-term debts. This ratio is good.

2. Debt versus Equity

(d) Long-term debt to capitalization ratio = 4.44 percent.
This debt level is prudent for firms in this industry.

(e) Total debt to owners' equity = 22.4 percent

Total debt = CL + Long-term debt

OE = Common stock plus capital surplus plus accumulated retained earnings

(f) Total debt to total asset ratio = 18.3 percent

3. Activity

(g) Sales to asset ratio = 0.86

(h) Ending inventory to sales ratio = 28.6 percent

(i) CGS/average inventory = 2.65 times

Average inventory = The average of ending inventory of two consecutive years (e.g., 2001 and 2002)

4. Profitability

(j) Net income to owners' equity ratio = 2.8 percent

(k) Net income to sales ratio = 2.66 percent

(l) Gross margin to sales ratio = 29.1 percent

(m) EBIT to total asset ratio = 5.1 percent

(n) Net income to total asset ratio = 2.3 percent

(o) EBIT to sales ratio = 5.9 percent

Selected investment services companies track various financial measures and report them regularly for use by the public. Engineering managers should become accustomed in reading and considering such reports.

7.4.3 Economic Value Added

Developed by Stern Stewart & Company in 1989, economic value added (EVA) is a superior valuation method for asset-intensive companies or projects. EVA is defined as the after-tax-adjusted net operating income of a company or unit minus the total cost of capital spent during the same accounting period. It is also equal to the return on capital minus the cost of capital, or the economic value above and beyond the cost of capital (Ferris, Ferris, Treadwell and Desai 2006; Tully 1993). Sometimes EVA is also called *economic profit*. In equation form, it is defined as:

$$\text{EVA} = \text{NOPAT} - \text{WACC} \times (\text{Capital deployed}) \tag{7.3}$$

where

NOPAT = net operating profit after tax (net income)

WACC = weighted average of cost of capital (equity and debt) employed in producing the earnings (section 7.6.3)

Capital deployed = Total assets – Current liabilities

If EVA is positive, the company or unit is said to have added positive shareholder value. If EVA is negative, the company or unit is said to have diminished shareholder value.

EVA may also be applied to a single project by calculating the after-tax cash flows generated by the project minus the cost of capital spent for the project. For example, the NPV equation can be modified as follows:

$$\text{NPV} = -P + \left[\sum_{t=1}^{N} \frac{C_t - P \times \text{WACC}}{(1 + \text{WACC})^t} \right] + \frac{\text{SV}}{(1 + \text{WACC})^N} \tag{7.4}$$

P = investment capital (dollars) for the project
C_t = net after - tax cash inflow (dollars) to be produced by the project in year (t)
$P \times \text{WACC}$ = cost of capital (dollars) spent during the year (t)
WACC = weighted average cost of capital (percent) in effect
SV = salvage value of the project at the end of N years
N = number of years

The major advantage of EVA over return on capital (ROC) is that it may encourage managers to undertake desirable investments and activities that will increase the value of the firm. The next example presents the difference between these two methods.

ABC Company has established that its WACC is 10 percent, and its ROC standard for investment purposes is 14 percent. Management is considering a new capital investment that is expected to earn a return of 12 percent. This new investment is attractive according to the EVA criterion, as 12 percent is larger than 10 percent. However, this new investment is a poor choice if evaluated with the ROC criterion, because 12 percent is less than 14 percent. Thus, by using the ROC criterion to evaluate investments, the company may lose the opportunity to create shareholder value.

A ten-year study has shown that there is a general correlation between EVA and stock returns of many companies. However, the correlation between EVA and wealth creation (in the form of stock price increase) is weak. Among users of EVA, the following firms are known leaders in industry: AT&T, Eastman Chemical, Coca-Cola, Eli Lilly, Wal-Mart, and the U.S. Postal Service.

Engineering managers should learn to apply EVA in order to strengthen their financial accounting skills.

7.4.4 Creation of Shareholder Value

Shareholders are primarily interested in the *total return to stockholders* (TRS), which consists of the yield of dividend paid out by the company plus the rate of the stock's long-term (e.g., ten years) appreciation potential (Rappaport 2006). The amount of annual dividend paid out is a function of the net income earned by the company in a given year. Countless established companies strive to continuously pay dividends in order to appease investors. Being able to pay dividend quarter after quarter regularly requires that the company focuses on its short-term operations.

The long-term appreciation rate of the company's stock, on the other hand, depends on the company's long-term investment strategies. The stock price is an indication of the market's assessment of the company's future expected cash flows. These future expected cash flows are to be created by projects, which add value to the company over a long period of time. The finance function is particularly qualified to offer advice in increasing spending in value-creating activities (such as research & development, design, marketing, advertising, etc.), in systematically reviewing and retaining only assets that maximize value creation, and in pursuing proper acquisition and divestiture strategies in order to maximize this expected value.

Example 7.6 The Global Business Services (GBS) organization within Procter & Gamble (P&G) provides services in finance, accounting, employee services, customer logistics, purchasing and IT to other divisions of the company. Over the years, P&G has prided itself on its inclusive company culture and progressive employment practices (e.g., job guarantees, length of employment, salary/benefit policies favorable to employees). In an attempt to maximize shareholder value, P&G management periodically reviews their strategies of deploying company resources. Since GBS is not one of the company's core competencies, there has not been enough management attention, and consequently resources devoted to the continuous improvement of GBS (Delong et al. 2005). Eventually, questions have been raised regarding the service quality and costs of the GBS organization to P&G. Finally, a management committee has recommended that P&G consider one of the four options:

1. Spin off GBS to be independent company that continues to provide services to P&G.
2. Outsource all services to one outside company and transfer all employees over.
3. Outsource services to a number of separate providers in order to improve quality.
4. Continue GBS in-house.

Which one is the right option for P&G to take in this case?

ANSWER 7.6 The best choice is to make use of Kepner–Tregoe method (see section 4.3.5) by first specifying a number of mutually exclusive and collectively exhaustive decision criteria, assigning weight to each criteria to specify its relative importance, ranking all options against each of these criteria, and finally computing the weighted scores, which indicate a relative ranking of the options under consideration, as illustrated in Table 7.14.

Option 1 is the preferred choice.

Table 7.14 Application of Kepner–Tregoe Method

Criteria	Weight	Option 1	Option 2	Option 3	Option 4
Compatibility with P&G policy and tradition	8	3	8	6	10
Service cost	9	10	6	8	3
Service quality	10	6	8	10	3
Value to affected P&G employees	8	8	6	3	10
P&G control	8	8	6	3	10
Total Weighted Score		302	294	268	297

7.5 BALANCED SCORECARD

Financial ratios are developed by accountants, who naturally emphasize the companies' financial performance of direct interest to shareholders. Nonfinancial ratios are limited in number and restricted in scope. Examples include accounts receivable, collection, inventory, utilization of fixed assets, and working capital.

All financial ratios are determined on the basis of past performance data; they are "trailing" indicators, and as such, they cannot foretell the future performance of a company. Because financial ratios are oriented to the short term, usually from one quarter to another, company management is inadvertently forced to overemphasize short-term financial results, oftentimes neglecting the company's long-term growth. The narrow focus of these financial ratios makes them no longer completely relevant to today's business environment, in which customer satisfaction, employee innovation, and continuous betterment of business processes are key elements of company competitiveness in the marketplace.

These basic deficiencies in financial ratios are well recognized in industry. Attempts have been made in the past to modify these ratios as corporate measurement metrics. Kaplan and Norton (2007) suggest that corporate measurement metrics are to be defined to cover four areas:

1. *Financial*—shareholder value.
2. *Customers*—time, quality, performance and service, and cost.
3. *Internal business processes*—core competencies and responsiveness to customer needs.
4. *Innovation and corporate learning*—value added to the customer, new products, and continuous refinement.

The significance of the balanced scorecard lies in its balanced focus on both short-term actions as well as long-term corporate growth. What you measure is what you get. Kaplan and Norton (2007) advocate the use of a total of fifteen to twenty metrics to cover these four areas to guide the company as it moves forward.

As an illustrative example, the balanced scorecard metrics for a manufacturing company may contain the following:

- *Financial*—cash flow, quarterly sales growth and operational income, increased market shares, and return on equity.
- *Customer*—percentage of sales from new products, percentage of sales from proprietary products, on-time delivery as defined by customers, share of key account's purchase, ranking by key accounts, and number of collaborative engineering efforts with customers.
- *Internal business process*—manufacturing capabilities versus competition, manufacturing excellence (cycle time, unit cost, and yield), design engineering efficiency, and new product introduction schedule versus plan.
- *Innovation and learning*—time to develop next-generation technology, speed to learn new manufacturing processes, percentage of products that equal 80 percent of sales, and new product introduction versus competition.

In general, balanced scorecard metrics for a given company must be built up according to its corporate strategy and vision, using a top-down approach. Doing so will ensure that performance metrics at lower management levels are properly aligned with the overall corporate goals. A unique strength of balanced scorecard metrics is that they link the company's long-term strategy with its short-term actions. These metrics contain forward-looking elements at the same time that they balance the internal and external measures. The creation of such metrics provides clarification, consensus, and focus on the desired corporate outcome.

According to Kaplan and Norton (2007), United Parcel Service has achieved an increase of 30 to 40 percent in profitability with balanced scorecard metrics. Mobil Oil's North American Marketing and Refining Division raised its standing from last to first in its industry after having implemented a balanced scorecard. Catucci (2003) recommends that managers, when implementing a balanced scorecard, take personal ownership, nurture a core group of champions, educate team members, keep the program simple, be ruthless about implementation, integrate the scorecard into their own leadership systems, orchestrate the dynamics of scorecard meetings, communicate the scorecard widely, resist the urge for perfection, and look beyond the numbers to achieve cultural transformation of the company.

A widespread use of broad-based metrics, such as those suggested by the balanced scorecard, is likely to shift the attention of corporate management from a focus primarily on financial performance to other areas of equal importance, such as customers, internal business processes, and innovation and learning. Contributions by engineering managers made in these nonfinancial areas will likely become readily and more favorably recognized in the future.

7.6 CAPITAL FORMATION

Capital formation refers to activities undertaken by a company to raise capital for short-term and long-term investment purposes.

In general, the net income earned by most companies, plus their internal resources of retained earnings, are not sufficient to finance all investments needed to achieve both their short-term and long-term corporate objectives. Even if these internal resources are sufficient, some companies still engage in external financing due to good reasons related to tax and strategic management flexibility. Various companies pursue some external resources, such as equity or debt financing (or both).

7.6.1 Equity Financing

Equity financing is the raising of capital by issuing company stocks. Stocks are certificates of company ownership, which typically carry a par value of one dollar. The price of the same stock in the market (e.g., the New York Stock Exchange) may fluctuate in time as a consequence of economic conditions, industrial trends, political stability, and other factors unrelated to company performance. Shareholders are those who own stocks. They have the residual claims to what is left of the firm's assets after the firm has satisfied other high-priority claimants (e.g., bondholders, bankruptcy lawyers, and unpaid employees). Basically, there are two kinds of stocks: *common stocks* and *preferred stocks* (whose dividends have a priority over that of common stocks).

Companies may issue new stocks to raise capital. The process requires the approval of the company's board of directors (which mirrors the interests of all shareholders) because such a move may result in the dilution of company ownership. In addition, a public offering needs to be registered with the U.S. Securities and Exchange Commission and organized by an underwriting firm. The underwriting firm helps set the offering price, prepares the proper advertisements, and assists in placing all unsold issues to ensure a successful completion of the equity financing process. Companies issuing new stocks will incur an issuing expense.

The capital received by issuing stocks has a cost with the following components: the cost of equity capital, which includes the stock issuing fees, the dividends to be paid in future years, and the capital gains through stock price appreciation expected by investors in the future. Typically, this cost of equity is set to equal the return of an equity calculated by using the capital asset pricing model (CAPM).

This well-known capital asset pricing model (CAPM) defines the return of an equity as (see Fig. 7.1)

$$R = R_f + \beta(R_m - R_f) \tag{7.5}$$

where

R_f = Risk-free rate (e.g., 6.0 percent of 10-year U.S. Treasury bills).

R_m = Expected return of a market portfolio, a group of stocks acting for the behavior of the entire market (e.g., S&P 500 Index). R_m depends on conditions not related to the individual stocks. The typical value for R_m is in the range of 15 percent, based on long-term U.S. market statistics.

β = Relative volatility of a stock in comparison to that of a market portfolio, which by definition has a β of 1.0. The Standard & Poor's 500 stocks serve as a proxy for the overall market. If the beta of a stock is 1.5, then its price will change by 1.5 percent for every 1.0 percent price change of the market portfolio in the same direction. An issue with a β of 0.5 tends to move 50 percent less. A stock with a negative β value tends to move in a direction opposite to that of the overall market. Stocks with large β values are more volatile and hence more risky. Beta values are published in the financial literature for most stocks that are publicly held.

$R_m - R_f$ = Market risk premium.

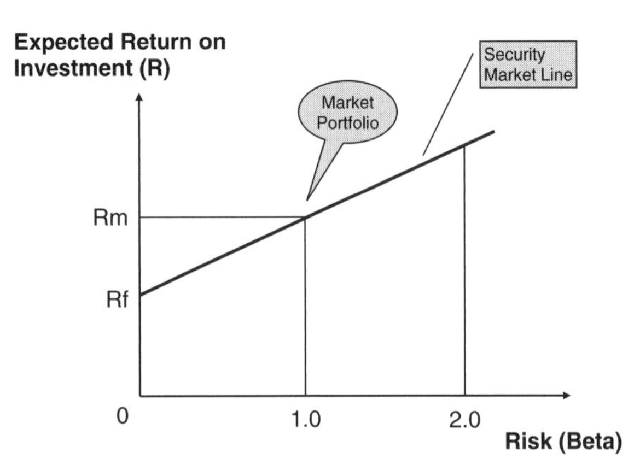

Figure 7.1 Security market line.

Since the value of β is based on historical data, numerous financial analysts view it as a major deficiency of the CAPM model. Another deficiency is its lack of timing constraints. The same cost is assumed to be valid for capital projects of a two-year or ten-year duration. In practice, when evaluating a specific capital investment project, some companies adjust the value of β manually in order to arrive at a pertinent cost of equity capital.

Recently, McNulty, Yeh, Schulze, and Lubatkin (2002) proposed the market-derived capital pricing method (MCPM) to calculate the cost of capital. According to this method, the cost of equity capital consists of two parts: the risk-free premium and the equity return risk premium. The risk-free premium is set to equal the yield of corporate bonds currently traded in the bond market. MCPM uses a *put option* to secure the capital gain in stock price within a fixed-duration period (e.g., five years); this is needed to ensure that the expected return for investors is greater than the dividend yield. The annualized cost of the put option is then divided by the company's current stock price to arrive at an *equity return risk premium*. The sum of corporate bond yield and the equity return risk premium then becomes the cost of the company's equity capital for the fixed-duration period involved. The McNulty et al. article presents a specific numerical example in which the cost of equity for General Electric was calculated. Based on a comparison of many sets of company data, McNulty and his coauthors further claim that MCPM yields more realistic costs of equity than CAPM.

7.6.2 Debt Financing

Debt is the liabilities incurred by the company to make contractual payments (e.g., interest payments) under specified terms. Debt depicts a fixed prior claim against the company's assets and poses a financial risk to the company. Debt financing by issuing industrial bonds or taking loans from financial institutions produces creditors. Debts are usually secured by a certain part of the company's assets. Creditors have the legal power to enforce payments and thus potentially drive companies into bankruptcy. When a company declares bankruptcy, it must satisfy the claims of creditors in a specific order: (1) secured debts (bonds or loans), (2) lawyers' fees, (3) unpaid wages, and (4) stockholders. Note that bankruptcy lawyers have a payment priority ahead of the hard-working employees and risk-averse shareholders.

Companies seeking debt financing through the issuance of bonds also need to engage an underwriting firm and follow a specific set of steps. There are several types of bonds:

- Corporate bonds
- Mortgage bonds
- Convertible bonds (convertible to stocks at a fixed price by a given date)
- Debentures (unsecured bonds)

The length of debts may be short term (less than one year's duration), intermediate (one to seven years), or long term (more than seven years). There is also a fee associated with this type of public offering.

Companies seeking debt financing through loans will typically negotiate for terms and conditions directly with the lending financial institutions involved. The interest rate charged is usually the prevalent prime rate plus a surcharge rate. The prime rate is published by the Federal Reserve Board on a regular basis as a means to control the liquidity of the financial markets. The surcharge rate varies with the company's credit rating, which, in turn, depends on the company's past financial performance and future business prospects. The higher is the company's credit rating, the lower is this surcharge rate. Hence, better performing companies may borrow money at cheaper rates. This reflects the lower risks involved for the lenders.

The cost of debt capital includes the issuing fee, bond rate, or interest rate to be paid in future periods, the opportunity cost associated with the diminished company growth opportunity, and other costs.

The opportunity cost related to reduced growth opportunity results from the fact that highly leveraged companies can no longer be as aggressive in pursuing new growth business opportunities. The obligatory interest burdens tend to temper the company's otherwise bold investment strategies. These burdens also constrain the company's investment flexibility and thus cause the company to lose the potential benefits that they could have otherwise realized from such new opportunities. Examples of such new opportunities include the development of new products, engagement in leading-edge R&D, entrance into new global markets, and creation of new and innovative supply chain partnerships.

The other costs could result from one or more of the following: (1) suboptimal operational policies that aim at the lower end of a range of sales forecasts, (2) vulnerability of the company to attacks by competitors, (3) the company's inability to access additional debt capital, if needed, and (4) the cost of bankruptcy.

Example 7.7 Innovative Products is a company that has enjoyed a high growth rate in recent years. Its growth has been largely financed by the retained earnings that belong to the common stockholders (see Table 7.15).

For the last three years, the company has earned an average net income of \$75,000, after having paid an annual interest of \$8,000 and the annual taxes of \$50,000. The company's tax rate is 40 percent.

Company management is considering the strategy of raising \$750,000 to double its production volume. Of this amount, \$500,000 would be used to (1) build an addition to the current office building, (2) purchase new IT equipment, and (3) install an advanced ERP software system. The remaining amount would be needed for working capital to add inventories and to enhance marketing and sales activities.

Market research suggests that, in spite of a doubling of the company's sales volume, the product price can be kept at the current level. The EBIT is projected to be \$275,000.

Two specific financing options are to be evaluated closely:

1. Sell enough additional stock at \$30 per share to raise \$750,000.
2. Sell 20-year bonds at 5 percent interest, totaling \$500,000.

Determine which financing option is to be favored from the standpoint of earning per share.

ANSWER 7.7 Earning per share is defined as the company's net income divided by the number of outstanding common stocks. To compare the earning-per-share data for these two financing options, the income statement must be constructed as in Table 7.16.

Table 7.15 Balance Sheet of Innovative Products (2002)

	Year 2002 (dollars)
Cash	40,000
Accounts receivable	160,000
Notes receivable	70,000
Inventories	260,000
Prepaid items	16,000
Total Current Assets	546,000
Land	25,000
Building (net)	165,000
Equipment (net)	350,000
Total Assets	1,086,000
Accounts payable	87,000
Notes payable	80,000
Taxes payable	60,500
Total Current Liabilities	227,500
Long-term loan (due 2007, 5 percent interest)	100,000
Common stocks	400,000
Retained earnings	358,000
Total Liabilities and Owners' Equities	1,086,000

Table 7.16 Income Statement of Innovative Products

	Present Operation No Financing (dollars)	Option 1 Equity Financing (dollars)	Option 2 Equity & Debit Financing (dollars)
EBIT (earnings before interest and tax)	133,000	275,000	275,000
Interest	8,000	8,000	25,000
Taxable income	125,000	267,000	250,000
Taxes (40 percent)	50,000	106,800	100,000
Net income	75,000	160,800	150,000
Outstanding shares of stocks	400,000	425,000	410,000
EPS (earning per share)	0.1875	0.3769	0.3658

The following explanations may be helpful:

1. From the known values of net income ($75,000), tax rate (40 percent), and interest payment ($8,000) of the present operation, the EBIT is calculated to be $133,000.
2. For the present operation, the number of outstanding common stock is 400,000, because the common stock (usually at the par value of $1) is described as $400,000 in the company's balance sheet (see Table 7.15).
3. For Option 1 and Option 2, the EBIT is known to be $275,000.

4. For Option 1, the interest charge remains at $8,000, but the number of outstanding stock is increased to 425,000.
5. For Option 2, the interest charge is increased by $25,000 annually due to the new loan. However, the old loan is being paid off, reducing the annual interest payment by $8,000. Thus, the net interest payment is $25,000 per year.
6. For option (2), the number of outstanding stock is increased to $410,000.

Based on this analysis, the earning per share of Option 1 is larger than those of the present operation and Option 2. Equity financing is to be preferred.

7.6.3 Weighted Averaged Cost of Capital (WACC)

The weighted average cost of capital (WACC) is a very important cost figure for any company. It is defined as

$$\text{WACC} = K_e\left(\frac{E}{V}\right) + K_d\,(1-t)\left(\frac{D}{V}\right) \tag{7.6}$$

where

D = debt (long – term loans, corporate bonds, etc.) (dollars)
E = equity (stocks) (dollars)
t = corporate tax rate (percent)
$V = E + D\,(dollars)$
K_e = cost of equity capital (e.g., = 0.15 to 0.18)
K_d = cost of debt capital [e.g., 0.08 = yield to maturity (YTM) rate for bonds, plus cost associated with lost growth opportunity]

In general, an increase in leverage (e.g., adding more debts) reduces the firm's WACC. This is due to the tax-deductibility of interest payments associated with the debt. For a large number of U.S. companies, WACC is typically in the range of 8 to 16 percent.

When defining WACC for global projects, additional costs must be taken into account to manage the location-specific risks (Desai 2008). Let us take a look at how the AES Corporation defines the WACC for such projects. The AES Corporation is a leading independent supplier of electricity across thirty-three countries with total revenue of $33 billion. The company defines applicable cost of capital for each power project to recognize the unique opportunity at hand and the specific risks. The following equation is proposed by Desai and Schlinger (2006) to take into account these additional location-specific risks:

$$WACC = K_e(E/V) + K_d(1-t)(D/V) + \text{Idiosyncratic risk measure} \tag{7.7}$$

The new element added to the right hand side is the *idiosyncratic Risk Measure,* which accounts for additional costs in dealing with location-specific risk factors. Several risk categories are included in this assessment, such as (1) operational, (2) counterparty, (3) regulatory, (4) construction, (5) commodity, (6) currency, and (7) legal. Weights are assigned to each risk category. Each project is then assessed with respect to these seven risk categories, in one of the four levels: 0, 1, 2, or 3. The weighted averaged *risk score*

Table 7.17 Calculation of the "Idiosyncratic Risk Measure"

Risk Category	Weight	Project 1	Project 2	Project 3	Project 3	Project 5	Project 6	Project 7
Operational	3.50%	3	2	2	1	0	0	0
Counterparty	7.00%	3	0	2	1	1	1	2
Regulatory	10.50%	3	2	2	3	1	0	1
Construction	14.50%	3	3	0	0	0	2	1
Commodity	18.00%	3	1	3	2	2	1	0
Currency	21.50%	3	2	0	3	1	0	2
Legal	25.00%	3	2	2	3	0	1	1
Risk Score		3	1.8	1.5	2.2	0.8	0.8	1.1
Idiosyncratic Risk Measure		**15.0%**	**9.0%**	**7.5%**	**11.0%**	**4.0%**	**4.0%**	**5.5%**

is then calculated for each project. The idiosyncratic risk measure is then computed by multiplying the risk score by 5 percent, which is assumed to be the incremental cost of capital for being exposed to each risk level. Table 7.17 illustrates the AES example:

The *idiosyncratic risk measure* so obtained is to be added to the right hand side of the WACC equation.

The inclusion of this specific example here is intended to illustrate one specific method by which location-specific risk factors can be accounted for and its cost impact considered, even though the actual choices of risk categories and weight distributions by Dasai and Schillinger (2006) for AES may not be applicable to other situations.

7.6.4 Effect of Financial Leverage

Financial leverage denotes the use of debts in financing corporate projects. A measure of financial leverage is given by the leverage ratio D/V. The company is said to be highly leveraged if its leverage ratio is more than 0.5.

Leverage ratio is known to have an impact on both the variability of reportable earning per share and the return on equity values. This is illustrated by the following example.

Assume that the total assets of a company are \$1,000. These assets may be financed 100 percent by equity (Case A) or by a combination of 40 percent equity and 60 percent debt (Case B). It is further assumed that under normal circumstances the company's EBIT is \$240. The EBIT value may be reduced to \$60 under bad economic conditions, but it could be increased to \$400 under good conditions. The corporate tax rate is assumed to be 50 percent. There is no interest payment in Case A. However, \$48 must be paid in Case B as an interest charge, which is tax deductible. The net incomes in these cases are different. The earning per share and return on equity data reported out by the company varies accordingly (see Table 7.18).

Under normal economic conditions, the earning per share is 1.20 in Case A (no debt) and 2.40 in Case B (with debt). Thus, companies engaged in debt financing will be able to report higher earning per share data than others that carry no debt, assuming everything else being equal.

In the absence of leverage (no debt, Case A), the earning per share varies from 0.3 to 2.0 and the return on equity from 3.0 to 20 (both from 25 to 167 percent). When

Table 7.18 Impact of Leverage on EPS and ROE

	Case A			Case B		
	Bad	Normal	Good	Bad	Normal	Good
Assets		1000			1000	
Debt		0			600	
Equity		1000			400	
Leverage ratio (%)		0			60	
EBIT	60	240	400	60	240	400
Interest (8 percent)	0	0	0	48	48	48
Taxable income	60	240	400	12	192	352
Tax (50 percent)	30	120	200	6	96	176
Net income	30	120	200	6	96	176
Number of shares	100	100	100	40	40	40
EPS	0.30	1.20	2.00	0.15	2.40	4.40
ROE (percent)	3.0	12.0	20.0	1.5	24.0	44.0

Notes: (1) When ROE exceeds interest (cost of debt), leverage has a favorable impact on EPS and vice versa. (2) Financial leverage increases the variability of EPS positively as well as negatively.

the company engages in debt financing, these same ratios vary more widely from 0.15 to 4.40 and from 1.5 to 44.0 (both from 6.25 to 183 percent), respectively. Financial leverage compounds the variability of companies' financial performance.

7.6.5 Optimum Leverage

The correct selection of the amount of leverage (i.e., debts) is of critical importance to the economic health of a firm. Incurring an excessive amount of debt (high leverage) will push the company to generate cash for meeting the interest payments, skimp on quality, keep inventory low, invest little in maintenance and capital expenditures, and make short-term-oriented decisions, since the company is not able to capitalize on future-growth opportunities. By contrast, if the debt level is too low (low leverage), the firm does not feel any real pressure to be as efficient as possible (e.g., management will tend to waste resources, tolerate excessive scrap, commit capital expenditures loosely, and initiate R&D indiscriminately) and will suffer from having a higher cost of capital than otherwise because they did not take advantage of the tax deductibility of interest payments on debt.

Most firms attempt to assume an optimum amount of leverage. This is because, as the leverage exceeds the optimum level, the valuation of the firm experiences a reduction due to (1) the cost of bankruptcy, (2) agency costs (e.g., lawyers, courts, and others), and (3) preemptive costs (i.e., loss of growth opportunities). Figure 7.2 illustrates such a leverage optimum, where V_u is the value of the firm with no leverage.

Example 7.8 Sampa Video is a small company that offers videocassette rental services in Boston area with thirty outlets (Andrade 2003). The company's background is described in Table 7.19.

Table 7.20 includes the projections of sales and other financial data for the five years indicated. The annual growth rate from the sixth year on is assumed to be 5 percent.

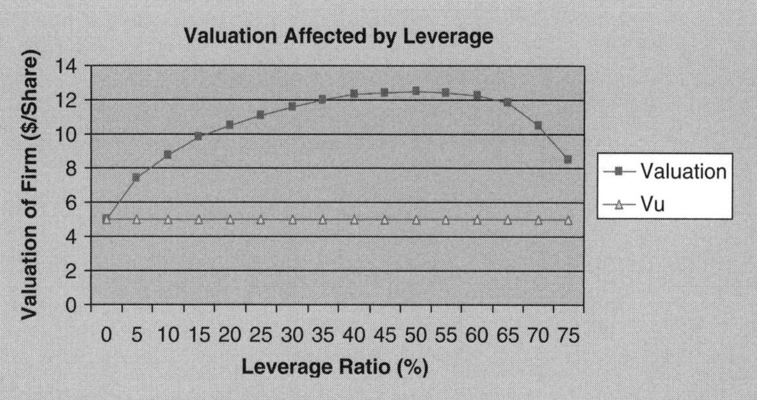

Figure 7.2 Optimum leverage.

Table 7.19 Sampa Video Company Background

Sampa Video Case (HBS Case # 9-201-094, October 7, 2003) (1000$)	
Sales (FY2000)	22,500
EBITDA (a)	2,500
Depreciation	1100
Operating profit	1,400
Net income	660
Risk-free rate	5.00%
Project cost of debt (R_d)	6.80%
Market risk premium	7.20%
Marginal corporate tax rate	40%
Project debt beta (b)	25.00%
Asset beta	1.50%

(a) Earning Before Interest, Taxes Depreciation and Amortization
(b) From Kramer.com and Cityretrieve.com

1. Assuming the firm was entirely equity financed, what is the value of the project? What are the discount rates? What are the annual projected free cash flows?
2. Assuming the firm raises $750 thousand of debt to fund the project and keeps the level of debt constant in perpetuity, what is the project's value using the adjusted present value (APV) approach?
3. Assuming the firm maintains a constant 45 percent debt to market value ratio in perpetuity, what is the project value using the weighted average cost of capital (WACC) approach?

ANSWER 7.8

1. The free cash flow of the project is defined as earnings before interest and tax (EBIAT), plus depreciation (which is a noncash expenditure and thus should be added back), minus capital expenditures and minus change in working

Table 7.20 Sales Projections and Financial Data

	Year 1	Year 2	Year 3	Year 4	Year 5	Terminal Value
EBIAT	−12	81	201	339	495	5% annual growth
Depreciation	200	225	250	275	300	
Capital expenditures	300	300	300	300	300	
Change in working capital	0	0	0	0	0	

capital (which symbolizes an increase in the use of cash). The expected return on assets is calculated using the CAPM model. It is equal to 15.8 percent, which is the risk free rate of 5 percent, plus a risk premium that is equal to an asset beta (1.5) times the market risk premium (7.2 percent).

Using this discount rate of 15.8 percent, the present value of this free cash flow stream is $2,728,500. Since the initial investment is $1,500,000, the net present value of the firm is $1,228,500. Table 7.21 shows the results.

2. When the firm borrows $750,000 in perpetuity to fund this project, the present value of the expected tax shield equals the expected interest tax shields discounted at the appropriate cost of capital. The cost of capital is 6.8 percent. Since the debt will be impacted forever, the value of the perpetual tax shield is:

$$\text{Value of tax shield} = \$750{,}000 \times 0.40 \times 6.8\%/6.8\% = \$300{,}000.$$

Thus, the NPV value of the project using the APV approach is equal $1,528.500(= $1228,500 + $300,000).

Table 7.21 Free Cash Flows and Net Present Value of Project (1000$)

	Formula	Year 1	Year 2	Year 3	Year 4	Year 5	Terminal Value
EBIAT	[A]	−12	81	201	339	495	
Depreciation	[B]	200	225	250	275	300	
Capital expenditures	[C]	300	300	300	300	300	
Change in working capital	[D]	0	0	0	0	0	
Free cash flow	[E] = [A] + [B] − [C] − [D]	−112	6	151	314	495	4812.5(a)
Discount rate		15.80%	15.80%	15.80%	15.80%	15.80%	
Discount factor		0.864	0.746	0.644	0.556	0.48	0.48
Present value		−96.7	4.5	97.2	174.6	237.7	2311.1(a)
Present value of cash flow		2,738.50					
Less: Initial investment		1,500					
Project's net present value		1,228.50					

(a) To calculate the sum all future values from Year 6 on, use the year 5 value multiplied by the factor $(1-r)/r$, wherein $r = [1-(1+\text{Growth rate})/(1+\text{Discount rate})]$. For growth rate = 0.05 and discount rate = 0.158, $r = 0.0932642$. See derivation in Table 7.30.

3. Weight average cost of capital (WACC) of a firm is defined as:

$$\text{WACC} = (1 - t)K_d(D/V) + K_e(E/V)$$

Wherein t = tax rate; K_d = return on debt; (D/V) is debt to total capital ratio, K_e = return on equity, and (E/V) is equity to total capital ratio. At a debt to value ratio of 45 percent, the tax impact is expected to be more favorable to the company's WACC than in the case of no leverage (15.8 percent). The results are displayed in Table 7.22.

Using the WACC as the discount rate, the project value is $1,527.720, about $299,200 more than the case with no leverage (see Table 7.23).

Table 7.22 Firm's WACC with a Debt to Value Ratio of 45 Percent

Asset beta	[A]	1.5
Risk-free rate	[B]	5.0%
Market risk premium	[C]	7.2%
Debt beta	[E]	0.25
Debt percentage	[F]	45%
Debt return	[G] = [B] + [F]* [C]	8.24%
Debt beta contribution	[H] = [E]*[F]	11.25%
Equity percentage	[I] = 1 − [F]	55%
Equity beta	[j] = ([A] − [H])/[I]	2.523
Equity return	[K] = [B] + [J]*[C]	23.17%
Equity beta contribution	[L] = [J]* [I]	1.3877
Tax rate	[M]	40%
WACC	[N] = (1 − [M])*[F]* [G] + [I]*[K]	14.97%

Table 7.23 NPV of Project with 45 Percent of Debt to Value Ratio

	Formula	Year 1	Year 2	Year 3	Year 4	Year 5	Terminal Value
EBIAT	[A]	−12	81	201	339	495	
Depreciation	[B]	200	225	250	275	300	
Capital expenditures	[C]	300	300	300	300	300	
Change in working capital	[D]	0	0	0	0	0	
Free cash flow	[E] = [A]+[B] − [C] −[D]	−112	6	151	314	495	5213.14 (a)
Discount rate		14.97%	14.97%	14.97%	14.97%	14.97%	
Discount factor		0.8698	0.7565	0.658	0.5724	0.4978	0.4978
Present value		−97.42	4.539	99.358	179.7336	246.411	2595.1(a)
Present value of cash flow		3,027.72					
Less: Initial investment		1,500					
Project's net present value		1,527.72					

(a) To calculate the sum all future values from Year 6 on, use the year 5 value multiplied by the factor (1 − r)/r, wherein r = [1 − (1 + Growth rate)/(1 + Discount rate)]. For growth rate = 0.05 and discount rate = 0.1497, r = 0.086718. See derivation in text on Table 7.30.

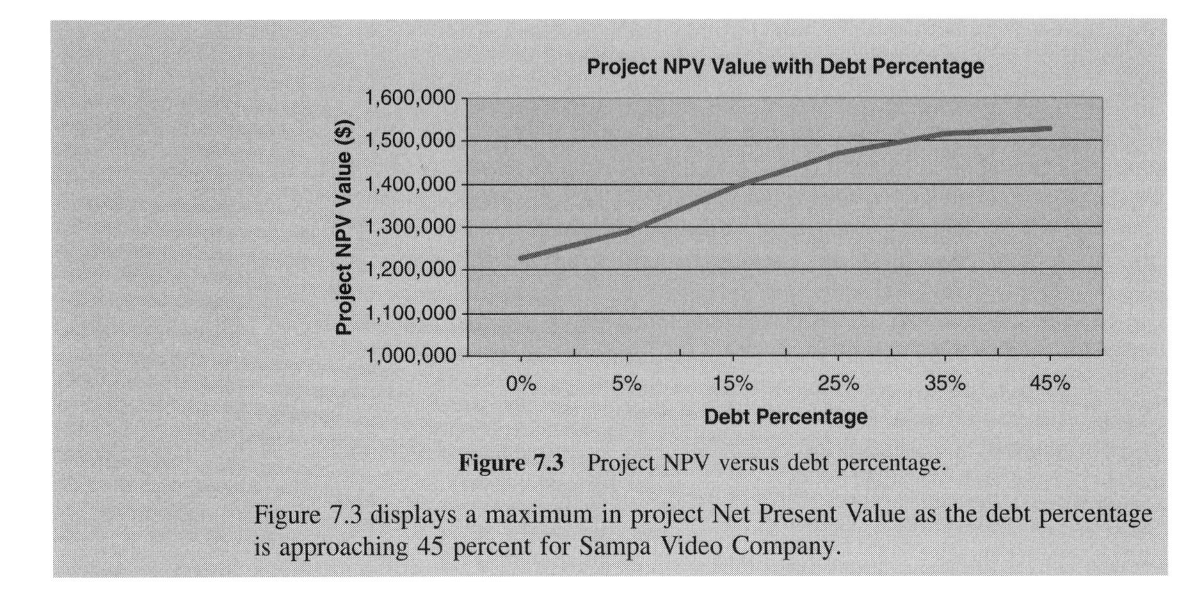

Figure 7.3 Project NPV versus debt percentage.

Figure 7.3 displays a maximum in project Net Present Value as the debt percentage is approaching 45 percent for Sampa Video Company.

7.7 CAPITAL ASSETS VALUATION

Financial management deals with three general types of capital assets valuation problems: assets in place (operations), opportunities (R&D and marketing), and acquisitions or joint ventures.

Capital budgeting problems related to assets in place are those that deliver a predictable string of cash flows in the immediate future. Examples include building a new plant facility, designing new products, and entering a new regional market. Sometimes these problems are grouped under the heading of operations, as investment in operations usually leads to immediate cash flows. Problems related to opportunities arise from decisions that do not generate an immediate flow of cash but preserve a likelihood that future gains may be realized. Examples include R&D and marketing efforts. The third type of problems is related to acquisition, joint venture, formation of supply chains, and others, all of which may require the company to participate in equity investment and to share future equity cash flows with its business partners.

Generally speaking, each of these types of valuation problem is best handled by different valuation methods. Luehrman (1997A) suggests a number of recommended methods, as shown in Table 7.24.

7.7.1 Operations—Assets in Place

There are several evaluation methods currently in use to assess capital projects in the investment category of operations.

Discount Cash Flow (Based on WACC). Since 1980 or so, most companies have been using the discounted cash flow (DCF) method to determine the net present value

Table 7.24 Recommended Capital Assets Valuation Methods

Valuation Problems	Recommended Methods	Alternative Methods
1. Assets in place (operations)	Adjusted present value (APV)	Multiples of sales, cash flows, EBIT, or book value; DCF (based on WACC), Monte Carlo simulations
2. Opportunities (R&D, marketing)	Simple option theory	Decision tree, complex option pricing, simulations
3. Equity claims	Equity cash flow	Multiples of net income; P/E ratios; DCF (based on WACC minus debt), simulations

of an operation with assets in place and WACC to specify the discount factor:

$$\text{NPV} = -P + \sum_{t=1}^{N} \frac{C_t}{(1+\text{WACC})^t} + \frac{\text{SV}}{(1+\text{WACC})^N} \tag{7.8}$$

Here

NPV = net present value (dollars)
P = initial capital investment (dollars)
Ct = cash flows = future net benefits (dollars)
SV = salvage value = capital gain (dollars)
N = total number of periods (year)
WACC = weighted average cost of capital (percent)

The net present value (NPV) is equal to the present value of all future net benefits (e.g., income minus relevant costs), plus discounted capital gain, if any, and minus the initial investment capital. It represents the net financial value added to a firm by a given capital investment. Projects with large positive NPV values are favored. This method is sometimes called the *DCF (based on WACC) analysis*, as it is based on the use of the weighted cost of capital as the all-important discount factor.

Companies accept capital projects if the NPV is greater than zero, meaning that an initiation of such projects leads to net positive value added to the companies.

Internal Rate of Return (IRR). A popular variation to DCF (based on WACC) is the internal rate of return (IRR). When applying Equation 7.8 to evaluate projects, IRR is the discount rate that is realizable when the present values of all discounted cash flows balance the initial investment (NPV equals zero). IRR represents the reinvestment rate, which is held constant over the duration of project. For example, assuming no salvage value, the IRR of a $10,000 investment that yields revenue of $5,000 per year for three years is 23.35 percent. Even though the concept is logical and its computation remains straightforward, IRR does not always symbolize the true annual return on investment (Kelleher and MacCormack, 2008). This is especially the case when the interim cash flows produced by the investment at hand can only be reinvested at a rate lower then IRR. For most projects, these interim cash flows may be realistically reinvested at WACC, the company's averaged cost of capital. If WACC is lower than IRR, then the computed IRR overestimates the true return on investment. Thus, the use of IRR could lead to wrong investment decisions and budget distortions, if the reinvestment rate of the interim cash flow is lower than the computed IRR.

In general, companies specify a hurdle rate that must be met or exceeded by the IRRs of all acceptable capital projects. By adjusting the hurdle rate according to conditions related to market, economy, and environment for a specific period involved, companies exercise control over the capital investment criteria. The hurdle rate is typically three to four times of WACC in value.

Adjusted Present Value (APV). In recent years, objections have been increasingly raised in financial literature opposing to the use of DCF (based on WACC) analysis for capital budgeting purposes. One of the principal objections is that WACC only insufficiently captures the tax-shield effects of debt financing in the real-world environment. As defined in Section 7.6.3, the tax-shield effects are incorporated within WACC in a single term displayed as the following equation:

$$\mathrm{WACC} = K_e \left(\frac{E}{V}\right) + K_d\,(1-t)\left(\frac{D}{V}\right) \tag{7.9}$$

Other organizations are involved in much more complex debt financing situations than the one depicted by this equation, which is limited to the special case of having a constant debt-to-equity ratio.

As a refinement to the DCF (based on WACC) method, the *adjusted present value* (APV) method is proposed. This new method segregates evaluation on the basis of equity from evaluation on the basis of debt financing and then recombines the results. Specifically, the *real cash flow* is first estimated and then discounted on the basis of the cost of equity capital only; that is,

$$\mathrm{NPV}_1 = -P + \left[\sum_{t=1}^{N} \frac{\mathrm{RC}_t}{(1+K_e)^t}\right] + \frac{\mathrm{SV}}{(1+K_e)^N} \tag{7.10}$$

where

K_e = cost of equity capital

RC_t = real cash flow (dollars) for period t

Then the *side effects* are estimated on the basis of such factors as issuing cost, tax shields, subsidized financing, hedge, and cost of financing distress:

NPV_2 = Cash flow due to interest tax shield is discounted by using cost of debt capital (Kd) only

NPV_3 = Sum of net present values due to all other side effects

The final net present value is then the sum of these individual items:

$$NPV = NPV_1 + NPV_2 + NPV_3 \tag{7.11}$$

By itemizing all side effects, the APV method provides additional information that is valuable to management. Because the tax shield is calculated separately, the debt-to-equity ratio need no longer be kept constant for all projects within a company, as it was so presumed in the DCF (based on WACC) method.

Table 7.25 Components of APV Results

NPV_1	(Baseline):	\$157 million
NPV_2	(Interest tax shield):	\$102 million
NPV_3		
	Higher growth:	\$ 34 million
	Asset sales:	\$ 16 million
	Networking capital improvement:	\$ 16 million
	Margin enhancement:	\$ 21 million
Total		\$346 million

Luehrman (1997B) works out a numerical example of using both DCF (based on WACC) and APV in estimating the value of ACME Filters. The results show that*:

$$NPV\ (DCF) = 417.1 \text{ million}$$

$$NPV\ (APV) = 346.0 \text{ million}$$

Not only is the NPV based on APV smaller in value, but it also contains additional useful information not unbundled by the DCF (based on WACC) method, see Table 7.25.

See Luehrman (1997B) for a detailed discussion on how each of these side effects is to be incorporated into the evaluation.

Multipliers. Another method to estimate the proper capital investment in a project is to use numerical multipliers that are based on historical data. Specifically, average multipliers are defined for use in conjunction with commonly available financial data such as sales, book value, earning before interests and taxes (EBIT), and cash flow.

In general, the financial data of countless publicly held U.S. companies in various industries is widely available in literature, including Standard & Poor's Industry surveys and Value Line Industrial surveys. Sales figures of numerous companies are readily obtained, and their relationship with company assets is typically recorded as the *asset turn ratios*. (See Section 7.4.1.) The reciprocal of this ratio is a multiplier that, when used together with the known sales figure, provides a rough estimate of the company's asset value.

For the XYZ Company described in Tables 7.12 and 7.13, the sales to total asset ratio for the year 2001 is $0.856\,(= 395{,}000/461{,}500)$. The reciprocal of this number is 1.168, which is the sales multiplier to determine assets. If a sufficient number of other companies in the same industry are surveyed, the resulting average industry-based multiplier can be used to generate a preliminary estimate of the assets employed to produce known sales revenue.

To determine the capital investment value of a plant expansion, new product development, and other products, the company's existing total sales to total asset ratio may serve as a good yardstick to ascertain a reasonable capital investment level, but only if the project outcome in terms of future sales can be estimated.

*Luehrman, T.A. 1997B. "Using APV: A Better Tool for Valuing Operations," Harvard Business Review, May-June.

How much debt should the company incur to finance a specific project? The *debt to asset ratio* is linked to the debt to equity ratio. (See Section 7.4.1.) For the XYZ Company described in Tables 7.12 and 7.13, the debt to asset ratio for 2001 is 0.08. Again, if an industrial average is found for this multiplier, it can serve as a useful tool to estimate a reasonable debt level for a project on the basis of its known asset value.

Similarly, industrial average multipliers may be found for cash flow. Cash flow for a given accounting period t is defined as

$$C_t = X_t - I_t \tag{7.12}$$

Where
X_t = accounting income (not including financial charges and depreciation)
I_t = investment

Specifically, cash flow is related to a number of financial accounting entries in the income statement, as follows:

Let	
Revenue	R
CGS	C
Gross margin	$R - C$
SGA	SGA
Depreciation	Dep
EBIT	$R - C$ − SGA − Dep
Interest	Int
Tax (tax rate T)	T ($R - C$ − SGA − Dep − Int)
NI	$(1 - T)$ ($R - C$ − SGA − Dep − Int)

Then

$$\text{CASH} = (1 - T)\,(\text{EBIT}) + \text{Dep} \tag{7.13}$$

Note that financial charges (e.g., interest payment) are usually ignored in the cash flow computation. Depreciation, a noncash expense, is added back. The following numerical example offers additional clarification, see Table 7.26.

$$X_t = 200{,}000 - 80{,}000 - 40{,}000 - 10{,}000 - 5{,}000 - 12{,}000 = \$53{,}000$$

$$\text{Dep} = \$20{,}000$$

$$C_t = X_t + \text{Dep} = 53{,}000 + 20{,}000 = \$73{,}000$$

For the XYZ Company in 2001, the cash flow is $10.5 million, calculated from data contained in Table 7.13.

Yet another multiplier is related to EBIT. This multiplier is estimated to be 14.49 for the XYZ Company in 2001, based on the EBIT to asset ratio of 0.051 (=23,500/461,500) evident in Tables 7.12 and 7.13.

The use of some of these multipliers in combination is likely to generate figures that can serve as an acceptable ballpark estimate of the required investment for a new project.

Table 7.26 Sample Data Set

Sales	$200,000
Manufacturing costs (including a depreciation of $20,000)	80,000
Sales and administrative expenses	40,000
Equipment service charges	10,000
Decrease in contribution of existing product	5,000
Increase in accounts receivable	15,000
Increase in inventory	20,000
Increase in current liability	30,000
Income tax	12,000
Interest paid for bonds used to finance projects	18,000

E. Monte Carlo Simulations. Monte Carlo simulations refer to a sampling technique that processes input data presented in the form of distribution functions. All input variables are simultaneously varied within each of their respective ranges, as defined by their distribution functions. The mathematical operations (e.g., addition, subtraction, multiplication, and division) of a given cost model are readily specified in spreadsheet programs (e.g., Excel or Lotus 123). (See Smith et al. 2002.) Upon activating a suitable risk-analysis software program, Monte Carlo simulations produce one or more outputs that are also in the form of distribution functions (Mun 2006). (See section 6.6)

In evaluating operations, Monte Carlo simulations may be usefully applied to the DCF (based on WACC) method. All future net cash flows and discount rates are modeled by distribution functions (e.g., triangular or Gaussion), as they are indeed expected to vary within ranges. The DCF (based on WACC) equation is readily modeled in a spreadsheet program. As the sole output, the net present value will also vary within a lower and upper bound. The following results will be helpful to decision makers of capital budgeting:

- The maximum probability at which the net present value is projected to be negative.
- The probability at which the net present value is projected to exceed a given value (e.g., $10 million).
- The standard deviation of the net present value output.
- The minimum net present value.
- The maximum net present value.

Monte Carlo simulations are also applicable to the calculation of internal rate of return (IRR) for the evaluation of operations.

7.7.2 Opportunities—Real Options

The second category of valuation problems is related to opportunities such as R&D and marketing projects. These problems do not lend themselves to DCF or APV analyses, both of which require the estimates of projected future cash flows. If there is no cash flow, there can be no net positive present value. Financial analysts and researchers in the literature recommend that the *European simple call option* method be used to price these opportunities.

Option is a common tool frequently used for trading securities in the financial markets. There are *call* and *put* options. A *call option* provides its holder with the right, but not the obligation, to buy 100 shares of an underlying company (e.g., IBM, Eastman Kodak, or General Electric) by a certain expiration date (typically three months from the present) at a specific price (*strike* or *exercise* price). The holder pays a fee, or premium, to buy the call option, which he or she may exercise on any business day up to and including the contract expiration date. Investors who predict that the stock price of a given company is going to rise will want to buy a call option to earn the difference between the higher stock price anticipated in the future and the strike price fixed by the option.

A *put option* gives its holder the right, but again not the obligation, to sell 100 shares of stocks within a period of time at a predetermined strike price. Investors who predict that the stock price of a given company is going to decline in the future will want to buy a put option to cash in on the difference between the lower stock price anticipated in the future and the strike price fixed in the option. The premium for an option is dependent on five factors: (1) current stock price, (2) strike price, (3) length of option contract, (4) stock price volatility, and (5) current opportunity cost (e.g., bank interest).

For the purpose of evaluating capital project opportunities, the European simple call option is more appropriate in that the call option can be exercised only on the expiration date specified in the contract and no sooner. Table 7.27 presents the equivalence between financial calls and real calls.

Companies with new technologies, product development ideas, defensive positions in fast-growing markets, and access to new markets have valuable opportunities to explore. When dealing with opportunities, three possible scenarios exist: (1) to invest immediately, (2) to reject the opportunity right away, and (3) to preserve the option of investing in the opportunity at a later time (Luerhman 2009).

This problem may be studied by using the Black–Scholes option-pricing model (BSOPM) (Black and Scholes 1973). The Black–Scholes option-pricing model is defined as

$$C = V\,[N\,(d)] - e^{-rT} X \left[N\left(d - sigma\,\sqrt{T}\right)\right] \tag{7.14}$$

$$d = \frac{\ln\left(\frac{V}{X}\right) + T\left(r + \frac{sigma^2}{2}\right)}{sigma\,\sqrt{T}} \tag{7.15}$$

where

C = Option price
V = Current value of the underlying asset
X = The exercise price of the option

Table 7.27 Equivalence of Call Options

Financial Calls	Real Calls
Current stock price	Underlying asset value
Length of contract	Length of time to invest
Volatility	Volatility of future project value
Strike price	Capital investment for the project
Current bank interest	Cost of equity capital

Table 7.28 Values of Cumulative Normal Distribution Function

d	N(d)	d	N(d)	d	N(d)
−3.00	0.0013	−1.00	0.1587	1.00	0.8413
−2.90	0.0019	−0.90	0.1841	1.10	0.8643
−2.80	0.0026	−0.80	0.2119	1.20	0.8849
−2.70	0.0035	−0.70	0.2420	1.30	0.9032
−2.60	0.0047	−0.60	0.2743	1.40	0.9192
−2.50	0.0062	−0.50	0.3085	1.50	0.9332
−2.40	0.0082	−0.40	0.3446	1.60	0.9452
−2.30	0.0107	−0.30	0.3821	1.70	0.9554
−2.20	0.0139	−0.20	0.4207	1.80	0.9641
−2.10	0.0179	−0.10	0.4602	1.90	0.9726
−2.00	0.0228	0.00	0.5000	2.00	0.9772
−1.90	0.0287	0.10	0.5398	2.10	0.9821
−1.80	0.0359	0.20	0.5793	2.20	0.9861
−1.70	0.0446	0.30	0.6179	2.30	0.9893
−1.60	0.0548	0.40	0.6554	2.40	0.9918
−1.50	0.0668	0.50	0.6915	2.50	0.9938
−1.40	0.0808	0.60	0.7257	2.60	0.9953
−1.30	0.0968	0.70	0.7580	2.70	0.9965
−1.20	0.1151	0.80	0.7881	2.80	0.9974
−1.10	0.1357	0.90	0.8159	2.90	0.9981

$Sigma$ = Annual standard deviation of the returns on the underlying asset
r = The annual risk-free rate
$N(d)$ = Cumulative standard normal distribution function evaluated at d
$\ln(x)$ = Natural log function of x
T = Time to expiration of the option (years)

Table 7.28 exhibits the representative data of the $N(d)$ function.

Example 7.9 If $1 million is invested immediately, there will be a loss of $100,000 due to the current economic condition and marketing environment. However, if the company waits for two years, things may be different. What should the company do, assuming that the current risk-free rate is 7 percent and the project volatility is 0.3?

ANSWER 7.9 The problem may be studied by using BSOPM. Let us define the following equivalents:

$$\text{Sigma} = 0.30$$

$$r = 0.07$$

$$X = \$1{,}000{,}000$$

$$V = \$900{,}000$$

$$T = 2$$

From equation (7.15),

$$d_1 = \frac{\left\{\ln(900{,}000/1{,}000{,}000) + 2\left(0.07 + \frac{0.3^2}{2}\right)\right\}}{0.3\sqrt{2}} = 0.2938$$

$$d_2 = d_1 - \Sigma\sqrt{2} = 0.2938 - 0.3 \times 1.41456 = -0.1306$$

From Table 7.28, we get by interpolation:

$$N(0.2938) = 0.6155$$

$$N(-0.1356) = 0.4462$$

From Equation 7.14, the option price becomes

$$C = 900{,}000 \times 0.6155 - \exp(-0.07 \times 2) \times 1{,}000{,}000 \times 0.4462$$
$$= \$166{,}042$$

Now the company has the alternative of investing $166,042 to preserve investment opportunities for two years, in addition to deciding for or against it right away. If this option is preserved, additional information that will cause a change in the underlying asset value may become available during the ensuing two years. The company may still decide not to invest in two years, but preserving the option to invest is still a valuable alternative.

7.7.3 Acquisition and Joint Ventures

When considering companies as candidates for acquisition, the evaluation of the assets of the target companies becomes critically important. A number of methods are practiced to assess the value of a company (Lacey and Chambers 2007; Titman and Martin 2007; Koller, Goedhard and Wessels 2005).

The value of a company is defined by its equity and debt, that is,

$$V = E + D \tag{7.16}$$

where
V = Firm's value in the market
E = Equity (stocks)
D = Debt outstanding (i.e., bonds, loans, etc.)

The company management continues to maximize the candidate firm's value for its shareholders. Two factors may affect this value maximization attempt: (1) *Takeover bids* (when acquiring firms are enticed to pay a higher than normal premium to absorb the acquisition candidates) tend to raise the stock price; and (2) *stock options* (rights to buy stock at a fixed price awarded to company's management personnel and new hires) tend to dilute the shareholder value.

Example 7.10 XYZ Company is considering the acquisition of Target Company, a smaller competitor in the same industry. The income statement of Target Company is displayed in Table 7.29. As a stand-alone company, its sales, cost of goods sold, depreciation, selling, and administrative expenses are all projected to increase by 4 percent per year.

To maintain the projected sales growth of Target Company, XYZ Company must make additional working capital investments (see Table 7.29).

1. Assuming a 10 percent discount rate, what is the maximum price XYZ Company should be willing to pay for this acquisition if it is to be run as a stand-alone subsidiary of XYZ Company?
2. XYZ Company could also integrate Target Company into its existing corporate IT operations. Web-based customer services, inventory management, order processing, and other activities can be readily added to lessen the required working capital by 50 percent from its stand-alone values. There is, however, an increased IT service charge of $1 million for the first year, and this charge increases by 4 percent per year. Again, assuming a 10 percent discount rate, what is the maximum price XYZ should pay for Target Company under the integration scenario?

Table 7.29 Income Statement of Target Company (1999–Future)

	(dollars in thousands)					Growth to Infinity (percent)
	1999	2000	2001	2002	2003	
Sales	61,000	63,440	65,978	68,617	71,361	4
Cost of goods sold	29,890	31,086	32,329	33,622	34,967	4
Depreciation	4,000	4,160	4,326	4,499	4,679	4
Selling and administrative	21,010	21,850	22,724	23,633	24,579	4
IT services	0	0	0	0	0	4
EBIT	6,100	6,344	6,598	6,862	7,136	4
Tax (35 percent)	2,135	2,220	2,309	2,402	2,498	4
EBIAT	3,965	4,124	4,289	4,460	4,638	4
Cash flow	7,965	8,284	8,615	8,960	9,318	4
Additional investments						
Working capital	8,000	8,320	8,653	8,999	9,359	4

ANSWER 7.10 For estimating the present values of cash flows for the period 2004 to infinity, the computations illustrated in Table 7.30 are needed. In Table 7.30, the letter *A* denotes a constant but unknown cash flow defined at the beginning of year 2003. I will show how this is actually done:

The present value of cash flow (at the beginning of year 1999) in Table 7.30 consists of the following parts:

1. $7965/1.1$
2. $8284/1.1^2$
3. $8615/1.1^3$
4. $8960/1.1^4$
5. $9318/1.1^5$

Table 7.30 Present Value of Cash Flows in 2004 (Growth Rate = 4% and Discount Rate = 10%)

	2003	2004	2005	2006	Infinity
Cash flow	A	A(1.04)	$A(1.04)^2$	$A(1.04)^3$	
Present value (2004)		A(1.04)/(1.10)	$A(1.04)^2/(1.10)^2$	$A(1-.4)^3/(1.10)^3$	
		$A(1-r)$	$A(1-r)^2$	$A(1-r)^3$	
PV of all future cash flow at start of 1999		$A[(1-r)/r]/(1.10)^5$			
r	0.05454545				
$(1-r)/r$	17.3333				

6. The present value of the cash flow amounts of 2004 to infinity (see equation in Table 7.30).

$$A = 9318$$

$$\text{Formula} = A[(1-r)/r)]/1.1^5$$

$$\text{Present value} = 9318 \times 17.33333/1.1^5 = 100{,}286.05$$

The sum of these six items is 32465.31 + 100,285.05 = 132,750.45. This is correctly exhibited in Table 7.31.

Table 7.31 Income Statement of Target Company (Standalone Subsidiary, Thousands of Dollars)

	1999	2000	2001	2002	2003	Growth to Infinity (percent)
Sales	61,000	63,440	65,978	68,617	71,361	4
Cost of goods sold	29,890	31,086	32,329	33,622	34,967	4
Depreciation	4,000	4,160	4,326	4,499	4,679	4
Selling and administrative	21,010	21,850	22,724	23,633	24,579	4
IT services	0	0	0	0	0	4
EBIT	6,100	6,344	6,598	6,862	7,136	4
Tax (35 percent)	2,135	2,220	2,309	2,402	2,498	4
EBIAT	3,965	4,124	4,289	4,460	4,638	4
Cash flow	7,965	8,284	8,615	8,960	9,318	4
PV (cash flows)	132,750					4
Additional investments						
Working capital	8,000	8,320	8,653	8,999	9,359	4
PV (WC Charge)	$13,333					
(A) NPV of target company	($119,417) (Not to acquire)					

Note: The closed-form equation $(1-r)/r$ may be derived as follows:

$$C = (1-r) + (1-r)^2 + (1-r)^3 + (1-r)^4 + \cdots\cdots\cdots + (1-r)^n$$

$$C(1-r) = (1-r)^2 + (1-r)^3 + (1-r)^4 + \cdots\cdots + (1-r)^n + (1-r)^{(n+1)}$$

$C - C(1-r) = (1-r) - (1-r)^{(n+1)}$

$C = (1-r)/r$ as n approaching infinity and $(1 - r)$ assumes a value of less than one.

The same factor $(1 - r)/r$ is applicable for the estimation of the present values of working capital for the period of 2004 to infinity. The spreadsheet represented by Table 7.31 enumerates the results:

The NPV of Target Company, as a stand-alone operation, is negative, not justifying its potential acquisition by XYZ Company.

However, if XYZ integrates Target Company into its IT operations, then the value of this target company is significantly improved, as displayed in Table 7.32.

Under the integration scenario, Target Company is worth about $116 million. Any price below this figure will denote a net gain for XYZ Company.

Common Stock Valuation Model (Dividend Valuation Model). The stock price of the acquisition candidate depends on the dividend stream it is able to generate. The *dividend valuation model* offers an estimate of the acquisition candidate's stock price as the sum of the present values of its future dividends:

$$P_o = \sum_{t=1}^{\infty} \frac{D_t}{(1+r)^t} \tag{7.17}$$

Here

P_o = equity price
D_t = dividend payout (DPS) per period.
$r = K_e$ = Cost of equity capital incurred by the firm
Upper limit = infinite (going concern)

Table 7.32 Income Statement of Target Company (Integrated Operations, Thousands of Dollars)

	1999	2000	2001	2002	2003	Growth to Infinity (Percent)
Sales	61,000	63,440	65,978	68,617	71,361	4
Cost of goods sold	29,890	31,086	32,329	33,622	34,967	4
Depreciation	4,000	4,160	4,326	4,499	4,679	4
Selling and administrative	21,010	21,850	22,724	23,633	24,579	4
IT services	1,000	1,040	1,082	1,125	1,170	4
EBIT	5,100	5,304	5,516	5,737	5,966	4
Tax (35 percent)	1,785	1,856	1,931	2,008	2,088	4
EBIAT	3,315	3,448	3,586	3,729	3,878	4
Cash flow	7,315	7,608	7,912	8,228	8,558	4
PV (cash flows)	121,934					4
Additional investments						
Working capital (50%)	4,000	4,160	4,326	4,499	4,679	4
PV (WC charge)	$6,667					
(B) NPV of integrated company	$115 268 (To bid for no more than $116 million)					

Two special cases may be considered:

1. *Constant dividend*
   If $D_t = D_o =$ constant, then

$$P_o = \frac{D_o}{r} = \text{capitalized value of dividend} \tag{7.18}$$

   The stock price P_o is equivalent to a single present value that is capable of producing a stream of constant dividends of value D_o indefinitely. (The derivation of this closed-form solution is included in Appendix 7.10.3.)

2. *Finite holding period*
   For investors who hold a given stock for only N periods and plan to sell the stock at the price P_n thereafter, the stock price may be calculated as

$$P_o = \sum_{t=1}^{N} \frac{D_t}{(1+r)^t} + \frac{P_n}{(1+r)^N} \tag{7.19}$$

where
$r =$ Effective rate of return required by the market of the firm's stock equals the cost of equity capital incurred by the firm (K_e)
$P_n =$ Investment recovery at the end of N periods

Once the stock price is known, the total value of the acquisition candidates is then equal to the stock price multiplied by the outstanding number of its stocks.

Dividend Growth Model. If the target company is capable of paying out dividends that grow at an annual rate of g percent, then its stock price is calculated as the capitalized dividend value at a rate equal to the cost of capital minus the dividend growth rate:

$$P_o = \frac{D_1}{(K_e - g)} \tag{7.20}$$

In this equation, whose derivation is given in Appendix 7.10.4,
$G =$ Growth rate of dividend per share (percent)
$K_e =$ Cost of equity capital (Note: $K_e > g$)
$D_1 =$ Dividend for the next period (dollars/share)

The growth rate of annual dividend is calculated by the equation

$$g = (1 - b)ROE \tag{7.21}$$

where
$ROE =$ Return on equity – Net income/Average equity capital
$b =$ Payout ratio of dividend = Dividend payout/Net income
$1 - b =$ Retained dividend ratio = Dividend retained/Net income

In Equation 7.21, the assumption is made that the company management is capable of utilizing the retained earning to foster dividend growth in the current year at the same rate as that of ROE, the return on equity accomplished by the company in the previous fiscal year.

Modified Earning Model. The stock price of a company that reinvests retained earnings to generate dividend growth is delineated by the equation:

$$P_o = \frac{\text{EPS}_1}{K_e} + \text{PVGO} \tag{7.22}$$

where

EPS_1 = Earning per share in the next period
K_e = Cost of equity capital
$PVGO$ = Present value of growth opportunities

Now, EPS_1/K_e is the capitalized value of EPS, and PVGO is the net present value of all returns (on the per-share basis) generated by having reinvested the retained earnings at the rate of ROE. Specifically,

$$\text{PVGO} = \frac{\text{EPS}_1\,(1-b)\,(\text{ROE} - K_e)}{K_e\,(K_e - g)} \tag{7.23}$$

Typically, growth stocks have large PVGO values that arise from the reinvestment of continuously increased earnings at the rate of

$$g = (1-b)ROE \tag{7.24}$$

The derivation of Equation 7.23 is given in Appendix 7.10.5.

Example 7.11 The company's stock is selling now at $50 per share. Its expected dividend next year is $2.00 per share. A 20 percent annual growth of dividend is anticipated for the next three years. From the fourth year on, its dividend growth rate will be reduced to only 6 percent annually. What is the expected long-term rate of return (R) from buying this stock at $50?

ANSWER 7.11 On the basis of the dividend model, we would formulate the following equation for evaluating the overall rate of return:

$$P_o = \frac{\text{Div}_1}{(1+R)} + \frac{\text{Div}_2}{(1+R)^2} + \frac{\text{Div}_3}{(1+R)^3} + \frac{\text{Div}_4}{(1+R)^4} + \frac{\text{Div}_5}{(1+R)^5\,(R-g)}$$

$$50 = \frac{2}{(1+R)} + \frac{2\,(1.2)}{(1+R)^2} + \frac{2\,(1.2)^2}{(1+R)^3} + \frac{2\,(1.2)^3}{(1+R)^4} + \frac{2\,(1.2)^3\,(1.06)}{(1+R)^5\,(R-0.06)}$$

By trial and error, $R = 0.11153$. The long-term rate of return of this stock is 11.153 percent.

Example 7.12 The company expects total sales revenue of $20 million and a total cost, including tax, of $15 million in the forthcoming year. During the subsequent five years (e.g., year 2 to year 6), the revenues and costs will increase 25 percent per year, and all profits will be reinvested back into the business. Thereafter, the company's growth will be decreased to only 5 percent per year and the company will need to reinvest only 40 percent of its profits.

If the company is offered $75 million in cash by a major competitor, is this a fair acquisition price, assuming the opportunity cost of capital is 12 percent?

ANSWER 7.12 Stock price is the present value of expected future dividend. On the basis of this model, we have

$$P_o = \frac{\text{Div}_1}{(1+R)} + \frac{\text{Div}_2}{(1+R)^2} + \frac{\text{Div}_3}{(1+R)^3} + \frac{\text{Div}_4}{(1+R)^4} + \frac{\text{Div}_5}{(1+R)^5} + \frac{\text{Div}_6}{(1+R)^6} + \frac{\text{Div}_7}{(1+R)^7\,(R-g)}$$

$$\text{Div}_1 = \text{Div}_2 = \text{Div}_3 = \text{Div}_4 = \text{Div}_5 = \text{Div}_6 = 0.0$$

$$\text{Div}_7 = (20-15)(1.25)^5(1.05)(1-0.4) = \$9.613 \text{ millon}$$

For the sixth year, $g = (1-b)$ ROE; $0.25 = (1-0)$ ROE

ROE = 0.25

For the seventh year,

$$g = (1-b)\ \text{ROE} = (1-0.6)\,0.25 = 0.1$$

$$P_o = \frac{9.613}{(1+0.12)^7\,(0.12-0.1)} = \$217.4 \text{ million}$$

The offered acquisition price is much too low, and the offer should be rejected.

Equity Cash Flows. For the valuation problems related to joint ventures and partnerships, ownership claims and equity cash flows must be evaluated. Financial specialists are typically involved in evaluating such problems by using sophisticated computer programs and models. For detailed discussions of this type of evaluations, see Luehrman (1997A). Other advanced corporate finance issues are discussed by Ogden, Jen, and O'Conner (2003).

Capital budgeting is a managerial responsibility related to the company's capital investments. Capital budgeting decisions are made on the basis of certain rational decision criteria. Engineering managers are advised to acquire the necessary background knowledge in order to become effective contributors to such decision-making processes.

7.8 CONCLUSIONS

This chapter introduces the basic accrual principle of financial accounting, presents the practice of T-accounting, and discusses the workings of income statements, balance sheets, and funds flow statements, with explanations of all applicable accounting terms. Also pointed out are the ways in which contributions by engineering managers are recorded in these financial statements.

Ratio analysis uses the financial data contained in these statements to assess companies' financial health. Economic value added (EVA) is described as an upgraded method of reporting the true financial value created by a company, unit, or a specific project.

The shortcomings of ratio analysis are outlined and a broad-based system of measurement metrics (the balanced scorecard) is illustrated. An adoption of this broad-based metrics system by corporate America will likely shift corporate management's attention from being predominantly focused on short-term financial results to a balanced emphasis on corporate long-term growths. This is possible with the use of metrics that address important competitive factors such as customer satisfaction, continuous improvement of internal business processes, and innovation for growth. As such broad-based measurement metrics become widespread, the critical contributions made by engineering managers will be increasingly recognized and rewarded.

This chapter also reviews the basic elements of financial management: raising and utilizing investment capital. Equity and debt financing are the two most common ways of obtaining financing. There are costs involved in each: the cost of equity capital and the cost of debt capital.

There are different types of capital projects in which a company might invest: operations (assets in place), opportunities (R&D and marketing), acquisitions, and joint ventures.

Different evaluation methods are used for these capital projects. For example, DCF (based on WACC), IRR, APV, multipliers, and Monte Carlo simulations are useful for evaluating operations. Option pricing is suitable for evaluating opportunities for which there are no predictable cash flows. Acquisitions and joint ventures are advanced financial topics, the evaluation of which should be deferred to knowledgeable financial specialists on the subject.

Engineering managers are advised to become well prepared to actively participate in the company's capital budgeting process by becoming familiar with the sources and costs of capital as well as the evaluation criteria adopted for capital budgeting. Doing so will enable them to constantly bring forth and screen useful projects and valuable opportunities (including risks assessment) and to be in a position to initiate winning capital project proposals on a timely basis.

7.9 REFERENCES

Andrade, Gregory. 2003. "Sampa video, Inc." Harvard Business School Case #9-201-094.

Berman, K., and Joe Knight 2008. *Financial Intelligence for Entrepreneurs: What You Really Need to Know About the Numbers*. Cambridge, MA: Harvard Business School Press.

Berman, K., Joe Knight, and John Case 2008. "The Power of Ratio: Learning What the Numbers Are Really Telling You." # 6561 BC. Harvard Business School Press.

Berman, K., Joe Knight, and John Case 2008A. "Understanding Balance Sheet Basics: Financial Intelligence for Entrappers." Harvard Business School Note #6551BC. Harvard Business School Press.

Besley, Scott, and E. F. Brigham 2007. *Essentials of Managerial Finance*, 14th ed. Cincinnati, OH: South-Western College Publishing.

Black F., and M. Scholes. 1973. "The Pricing of Options and Corporate Liabilities." *Journal of Political Economy* 81 (May–June): 637–654.

Bodie, Zane, Alex Kane and Alan J. Marcus. 2004. Essentials of Investments, 5th ed. McGraw-Hill Irwin, 458–459.

Brealey, R. A., S. C. Myers, and Franklin Alen 2008. *Principles of Corporate Finance*, 9th ed. New York: McGraw-Hill Higher Education.

Brigham, E. F., and M. C. Ehrhardt 2007. *Financial Management: Theory and Practice*, 12th ed. Cincinnati, OH: South-Western College Publishing.

Catucci, B. 2003. "Ten Lessons for Implementing the Balanced Scorecard." *Harvard Business School Balanced Scorecard Report*, # B0301E. January.

Delong, Thomas J., Warren Brackin, Alx Cabanas, Phil Shellhammer and David L. 2005. "Procter & Gamble: Global Business Services," Harvard Business School Case #9-404-124.

Desai, Mihir A. 2008. "The Finance Function in a Global Corporation," *Harvard Business Review* 86 (7/8).

Dasai, Mihir, and Doug Schillinger 2006. "Globalizing the Cost of Capital and Capital Budgeting at AES," Harvard Business School Case # 9-204-109 (Rev. October 23).

Droms, W. G. 2003. *Finance and Accounting for Nonfinancial Managers: All the Basics You Need to Know*. 5th ed. New York: Basic Books.

Ferri, F., W. P. Ferris, Steve Treadwell, and Mihir A. Desai. 2006. Understanding Economic Value Added (Revised). Harvard Business School Note #206016. Harvard Business School Press.

Healy, P. M. 2001. "Revenue Recognition." *Harvard Business School Note*, No. 9-101-017, February.

Healy, P. M., and P. Choudhary. 2001A. "Asset Reporting." *Harvard Business School Note*, No. 9-101-014, January.

Healy, P. M., and P. Choudhary. 2001B. "Liabilities Reporting." *Harvard Business School Note*, No. 9-101-016, November.

Healy, P. M., and P. Choudhary. 2001C. "Expenses Recognition." *Harvard Business School Note*, No. 9-101-015, February.

Healy, P. M., and J. Cohen. 2000. "Financial Statement and Ratio Analysis." *Harvard Business School Note*, No. 9-101-029, September.

Ittelson, T. 2009. *Financial Statements: A Step-by-Step Guide to Understanding and Creating Financial Reports*. (revised). Franklin Lakes, NJ: Career Press.

Kaplan, R. S., and David Norton 2007. Using the Balanced Scorecard as a Strategic Management System, HBR Classic # R0707M. Harvard Business Review.

Kelleher, J. C., and J. J. MacCormack. 2008. "Internal Rate of Return: A Cautionary Tale," *McKinsey Quarterly* (Special Edition): 70.

Kimmel, P. D., J. J. Weygandt, and Donald E. Kieso. 2009. *Financial Accounting: Tools for Business Decision Making*, 5th ed. Hoboken, NJ: John Wiley & Sons, Inc.

Koller, T., M. Goedhard, and David Wessels 2005. *Valuation: Measurement and Managing the Value of Companies*, 4th ed. Hoboken, NJ: John Wiley & Sons, Inc.

Lacey, Nelson J., and Donald R. Chambers. 2007. Modern Corporate Finance: Theory and Practice, 5th ed. Plymouth, MI: Hayden-McNeil Publishing.

Luehrman, T. A. 1997A. "What's It Worth? A General Manager's Guide to Valuation." *Harvard Business Review* (May–June).

Luerhman, T. A. 1997B. "Using APV: A Better Tool for Valuing Operations." *Harvard Business Review* (May–June).

Luerhman, T. A. 2009. "Real Options Exercise." Revised Edition. Harvard Business School Exercises # 208045. Harvard Business School Press.

McNulty, J. J., T. D. Yeh, W. S. Schulze, and M. H. Lubatkin. 2002. "What Is Your Real Cost of Capital?" *Harvard Business Review* (October).

Miller, G. S. 2005. "Financial Reporting, Tax Reporting and The Role of Deferred Taxes." Harvard Business School Note # 106026. Harvard Business School Press.

Mun, J. 2006. *Modeling Risks: Applying Monte Carlo Simulations, Real Options Analysis, Forecasting and Optimization Techniques*. Hoboken, NJ: John Wiley & Sons, Inc.

Ogden, J. P., F. Jen, and P. F. O'Conner. 2003. *Advanced Corporate Finance: Policies and Strategies*. Upper Saddle River, NJ: Prentice Hall.

Rappaport, Alfred. 2006. "10 Ways to Create Shareholder Value." *Harvard Business Review* 84 (9) (September).

Ross, S., R. Westerfield, and Bradford Jordan. 2007. *Fundamentals of Corporate Finance, Alternate Value*, 8th ed. New York: McGraw-Hill/Irwin.

Ruback, R., and A. Sesia. 2003. "Dell's Working Capital (Revised)." Harvard Business School Case # 201- 029.

Smith, Katherine T., L Murphy Smith, and Lawrence C. Smith. 2002. *Microsoft for Accounting: Managerial and Cost*. Upper Saddle River, NJ: Prentice Hall.

Standard & Poor's. 2009. *Standard & Poor's 500 Guide 2009* PB (Paperback). New York: McGraw-Hill.

Subramanyam, K. R. and John Wild 2008. *Financial Statement Analysis*, 10th ed. New York: McGraw-Hill/Irwin.

Sydsaeter, K., and P. Hammond. 2008. *Essential Mathematics for Economic Analysis*, 3rd ed. Upper Saddle River, NJ: Prentice Hall.

Titman, S., and John D. Martin. 2007. *Valuation: The Art and Science of Corporate Investment Decisions*. University Education Edition. White Plains, NY: Addison-Wesley.

Troy, L. 2008. *Almanac of Business and Industrial financial Ratios (2009)*. 40th ed. CCH, Inc.

Tully, S. 1993. "The Real Key to Creating Wealth." *Fortune* (September 30): 38.

7.10 APPENDICES

7.10.1 T-Accounts

Accountants use T-accounts as tools to register transactions in preparation for creating financial statements. T-accounts may be set up for any items that are assets, liabilities, equities, or other temporary holding entries. A T-account, as displayed in Fig. 7.A1, looks like the letter "T." On the left side of the "T," debits are recorded, and on the right side credits are recorded. This type of entry is also known as double-entry bookkeeping. Figure 7.A2 presents some additional details of such T-accounts.

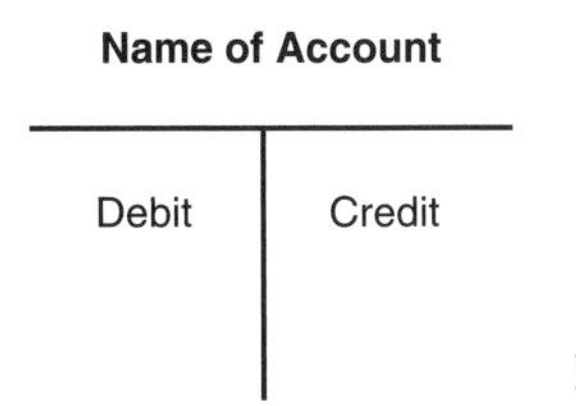

Figure 7.A1 T-account.

STORES		WIP		FG	
Debit	Credit	Debit	Credit	Debit	Credit
(1)	(2)	(3)	(4)	(5)	(6)

Explanations:
(1) Purchasing raw materials.
(2) Put materials into production process.
(3) Initate production and add value.
(4) Complete production.
(5) Receive finished goods in storage
(6) Ship finished goods for sale.

Figure 7.A2 T-Accounts with additional details.

Following the double-entry bookkeeping practice, every transaction affects at least two accounts. This is to ensure that a balance is continuously maintained between both the assets of, and the claims against, the company.

The company's assets include cash, accounts receivable, inventory, land, machines, plant facilities, marketable securities, and other resources of value. Liabilities include accounts payable, accrued expenses, long-term debts, and other claims creditors have against the company. Owners' equities include stocks, surplus, retained earnings, and other claims of the owners against the company. Equation 7.A1 depicts the balance between assets (A) and claims consisting of liabilities (L) and owners' equities (OE):

$$A = L + OE \tag{7.A1}$$

The convention of T-accounts is as follows: to increase the amount of an asset, debit the account; conversely, to decrease an asset amount, credit the account. All liabilities and owners' equities accounts are treated in the opposite way; that is, to increase a liability or equity amount, credit the account; and to lessen a liability or equity amount, debit the account.

For accounts that do not fall directly into one of these three categories (i.e., A = Assets, L = Liabilities, OE = Owners' equity), we need to first define their relationships to either A, L, or OE and then treat them accordingly. Revenues, expenses, and dividends are such examples.

Revenues raise the net income of the company. The resulting net income goes into the retained earning account for the owners. Thus, an increase of revenues needs to be credited to its T-account. The company's expenses are generally deducted from its revenues in order to arrive at its net income. An increase in expenses results in a reduction of net income and consequently a reduction of owners' equities. Therefore, an increase of expenses needs to be debited to its T-account. Similarly, an increase in dividends paid to shareholders whittles down the residual net income amount, which is then added to the retained earnings account of the owners. Thus, an increase of dividends needs to be debited to its T-account. Table 7.A1 summarizes the ways in which increases in the indicated assets, liabilities, owners' equities, or other amounts should be treated in their respective T-accounts.

Table 7.A1 T-Account Convention for an Increase in Selected Account Items

Accounting Items	Debit	Credit
Assets	x	
Cash, accounts receivables, inventory, land machines, marketable securities, etc.		
Liabilities		x
Accounts payables, accrued expenses, long term debts, etc.		
Owners' Equities		x
Stocks, capital surplus, retained earnings		
Revenue		x
Expenses	x	
Dividend	x	

Source: David F. Hawkins and Jacob Cohen, "The Mechanics of Financial Accounting." *Harvard Business School Note*, No. 9-101-119, June 27, 2001.

For engineers and engineering managers who are familiar with equations, the rule that follows may represent a convenient way of keeping them better oriented with the T-account convention. Rearranging the basic accounting equation (Equation 7A.1), we get:

$$\text{LHS} = \text{A} - \text{L} - \text{OE} = 0 \tag{7.A2}$$

where LHS stands for "left-hand side" of the equal sign. Note that, for each financial transaction, there are two account entries. The account entry that causes the LHS to increase temporarily should be debited to its respective T-account. Examples include increases in all assets and decreases in all liabilities and owners' equities. The account entry, which leads to a temporary reduction of the LHS, should be credited to its respective T-account. Equation (7A.2) remains valid after both entries of the financial transaction are entered.

Accountants use *T-accounts* to collect raw financial data that they check and recheck for validity and reliability. Then they make sure that the data are relevant to the accounting period at hand and consistent with past practices. Finally, to ensure comparability, accountants regroup them into known line items typically included in financial statements.

Example 7.13 Study the following accounts, which contain several transactions keyed together with letters shown in Table 7.A2:

Explain the nature of each transaction with the dollar amount involved.

ANSWER 7.13

a. Convert capital of $9,000, add $6,000 to the cash account, and purchase books worth $3,000 for the law library.
b. Pay the prepaid rent of $2,500 in cash.
c. Pay office supplies of $150 in cash.

Table 7.A2 Various T-Accounts

Cash		Office Equipment		Capital	
(a) 6,000	(b) 2,500	(d) 8,500			(a) 9,000
(e) 1,300	(c) 150				
	(f) 3,500				
	(g) 160				
Office Supplies		**Law Library**		**Legal Fees Earned**	
(c) 150		(a) 3,000			(e) 1,300
(d) 125					
Prepaid Rent		**Accounts Payable**		**Utilities Expenses**	
(b) 2,500		(f) 3,500	(d) 8,625	(g) 160	

d. Purchase office equipment (\$8,500) and office supplies (\$125) by credit, creating an account payable of \$8,625.
e. Receive the \$1,300 legal fees earned in cash.
f. Pay accounts payable of \$3,500 in cash.
g. Pay utilities expenses of \$160 in cash.

7.10.2 Risks

In general, the outcome (i.e., earnings) of any investment has a degree of inherent uncertainty; large in some and small in others. Investment risk is defined as the measure of potential variability of earning from its expected value. It is usually modeled mathematically by the standard deviation of the outcome when the outcome is expressed in the form of a probability density distribution function (e.g., Gaussian distribution function); see Fig. 6.3 and Section 6.6. The rate of return on risky security can be modeled as the risk-free rate plus a risk premium:

$$r = R_f + R_p \tag{7.A3}$$

The risk-free rate (e.g., R_f equals some constant rate such as 6.0 percent for ten-year U.S. Treasury bills) is the return for compensation of opportunity cost without uncertainties. The risk premium is the additional return needed to compensate for the added risks the investors undertake.

Figure 7.A3 illustrates several investment examples and displays the general risk–reward correlation between expected annual return and the associated risk of the investment in question. An investor may realize only 6 percent from the risk-free U.S. Treasury Bills, but a whopping 25 percent from highly precarious junk bonds.

A. Expected Value. *Expected value* is the return of an outcome multiplied by its probability of occurrence. For a portfolio containing N independent investments, the total expected return is the sum of the products of individual returns and its respective probabilities of occurrence. The total expected return is the average return weighted by

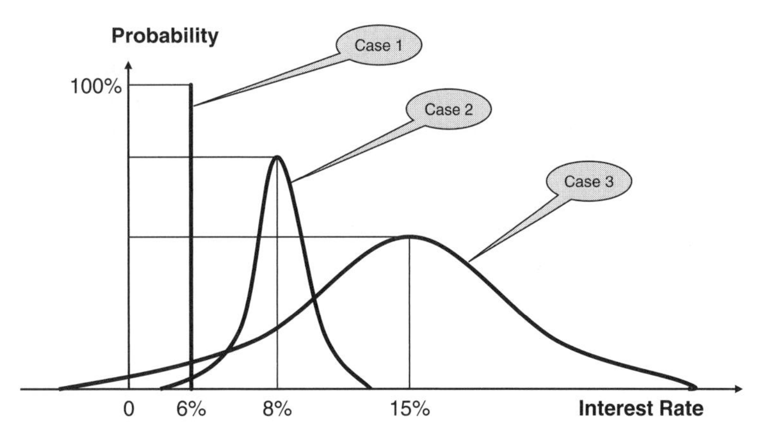

Figure 7.A3 Risk curves.

probability factors:

$$\text{EV} = \left[\sum P_i \times R_i;\, i = 1 \text{ to } N\right] \tag{7.A4}$$

R_i = Return on investment i
P_i = Single-valued probability of occurrence for i
$\left[\sum P_i;\, i = 1 \text{ to } N\right] = 1.$

B. Risk Aversion. *Risk aversion* delineates the generally expected unwillingness on the part of investors to assume risks without realizing the commensurate benefits. Most investors are unwilling to assume risks unless there are incremental benefits (the risk premium) that compensate them for bearing the added risks involved. Some audacious investors are more willing than others to take on additional risks in order to realize the added benefits.

Table 7.A3 exhibits an example in which the behavior of risk-averse investors can be studied. Investment A has a nominal value of $100 when the economy is in a

Table 7.A3 Behavior of Investors

State of Economy	Investment A (dollars)	Investment B (dollars)	Probability
Recession	90	0	0.333
Normal	100	100	0.333
Boom	110	200	0.333
Expected payoff value	100	100	
Range	90–110	0–200	
Amount at risk	–10	–100	

Note: Conservative persons would choose Investment A, whereas risk-preferring persons would choose Investment B. Risk-neutral persons would not have a preference.

normal state. This investment is projected to be valued at \$90 if the economy goes into a recession in the near future. On the other hand, its value may increase to \$110 if the economy booms. Investment B also has a nominal value of \$100 when the economy is normal, but its value decreases to zero in a recession and increases to \$200 in a booming economy.

If we further assume that there is an equal probability of 33.33 percent that the state of the future economy will be either normal, in recession, or in a booming state, then the expected pay-off value of these two investments is identical (e.g., \$100); however, their amounts at risk are different: –\$10 for Investment A and –\$100 for Investment B.

Among these two investment options, a risk-averse investor will choose Investment A for its lower downside risk; he or she will not choose Investment B because there is no gain in return for the added risks. A risk-preferring investor will choose Investment B for its reward potential of doubling the money if the economy booms in the future.

7.10.3 Derivation of an Infinite Series

Let

$$A = \sum \frac{1}{(1+r)^m} \qquad m = 1, 2, \ldots\ldots \infty \tag{7.A5}$$

$$B = \sum \frac{1}{(1+r)^{(m+1)}} \qquad m = 1, 2, 3, \ldots \infty \tag{7.A6}$$

Then

$$\mathrm{B} - A = \frac{1}{(1+r)^{(m+1)}} \tag{7.A7}$$

Furthermore, from Equations 7.A5 and 7.A6:

$$\mathrm{B}\,(1+r)^{(m+1)} - A\,(1+r)^m = (1+r)^m \tag{7.A8}$$

Substituting Equation (8.A8) into Equation (8.A7), we have

$$A\,(1+r)^{(m+1)} + 1 - A\,(1+r)^m = (1+r)^m \tag{7.A9}$$

or

$$A\,(1+r) + \frac{1}{(1+r)^m} - A = 1$$

$$A = \frac{1 - \dfrac{1}{(1+r)^m}}{r} \tag{7.A10}$$

As m approaches infinite, we obtain

$$A = \frac{1}{r} \tag{7.A11}$$

This relation is used in section 7.7.3 of this chapter.

7.10.4 Derivation of the Dividend Growth Model

$$P_o = \sum_{t=1}^{\infty} \frac{D_t}{(1+r)^t} \tag{7.A12}$$

where

$$D_1 = \text{Next year's dividend}$$

$$D_2 = D_1\,(1+g)$$

$$D_3 = D_2\,(1+g) = D_1\,(1+g)^2$$

$$D_4 = D_1\,(1+g)_3$$

$$\vdots$$

$$P_o = \frac{D_1}{(1+r)} + \frac{D_2}{(1+r)^2} + \frac{D_3}{(1+r)^3} + \cdots$$

$$= \frac{D_1}{1+r}\left[1 + \frac{1+g}{1+r} + \left(\frac{1+g}{1+r}\right)^2 + \left(\frac{1+g}{1+r}\right)^2 + \cdots\right]$$

$$= \frac{D_1}{1+r}\left(\frac{1+r}{r-g}\right) \tag{7.A13}$$

$$P_o = \frac{D_1}{r-g} \tag{7.A14}$$

7.10.5 Derivation of Present Value of Growth Opportunity (PVGO)

The net present value of reinvested retained earning is as follows:

$$\text{NPV} = -\,(1-b)\,\text{EPS} + (1-b)\,\text{EPS} \times \frac{\text{ROE}}{k_e} \tag{7.A15}$$

where the first term on the right-hand side (RHS) of the equal sign is the reinvested retained earning, with b being the payout ratio. The second term on the RHS describes the capitalized return, assuming at the same effective rate as ROE. By substitution, we have

$$\text{PVGO} = \frac{\text{NPV}}{\left[K_e - g\right]} = \text{EPS}\,(1-b)\,\frac{\text{ROE} - K_e}{K_e\,(K_e - g)} \tag{7.A16}$$

(See section 7.7.3.)

7.11 QUESTIONS

7.1 Study the case "Whirlpool Europe," Harvard Business School Case # 9-202-017 (Rev: November 22, 2002) and answer the following questions. The case materials may be purchased by contacting its publisher at www.custserve@hbsp.harvard.edu or at 1-800-545-7685.

A. What is this case about?

B. Cash flows are to be generated by several sources, if an investment in EPR system is made. There are four waves of implementation described in the case over the period of 1999–2007. Calculate the cash flows for all waves due to

   1. Improved inventory (One-time, nontaxable)

   2. Additional sales due to better product availability (continuous, taxable)

   3. Improved margin (continuous, taxable)

C. Analyze the interactions between these three sources of cash flows. What is your finding?

D. Calculate the present value of the combined after-tax cash flows from these three sources for the period of 1999 to 2007.

E. When deciding on the proposed ERP investment, should value be included for possible cash flows that occur beyond 2007? What does it depend on?

F. Would you recommend the ERP investment? What is your major concern?

G. What are the lessons you have learned from having studied this case?

7.2 For the years 2001–2002, the financial statements of XYZ Company are given in Tables 7.12 and 7.13. Prepare a funds flow statement that delineates the sources and applications of funds for the XYZ Company during 2001 and 2002.

Tables 7.12 and 7.13. Prepare a funds flow statement that delineates the sources and applications of funds for the XYZ Company during 2001 and 2002.

Answer 7.2

The total sources is $63,500.

7.3 The XYZ Company supplies products to a number of original equipment manufacturers (OEMs). It employs 5,000 mostly unionized workers and generates about $2.2 billion in revenue annually. In 2002, it won an "Excellence Award" from one of its major client companies for having supplied an OEM product at the quality level of ten defects per million in three consecutive years, thereby exceeding the specific target of 15 defects per million set by this client. However, the corporate parent of the XYZ Company was less than happy with its financial performance. It declared this subsidiary a "troubled operation" in 2002, giving notice that if its business performance did not get better within a certain timeframe, the XYZ Company would be closed down or divested. The specific financial targets set by the corporate parent consisted of (1) net income (NOPAT) to sales ratio at 5 percent, (2) earning before interest and tax (EBIT) to sales ratio at 10 percent, and (3) return on assets at 22.5 percent.

Needless to say, management worked diligently to find ways to improve the company's business performance. One strategy was to induce 700 early retirements from the unionized workforce by November 2002. As of April 2002, only 480 workers signed up for this heavily promoted early retirement program that qualified them each to receive a $35,000 cash incentive. Besides encouraging workers to retire earlier, what other strategies do you think the XYZ Company should pursue to achieve all three of the noted financial targets?

7.4 The company is considering the introduction of a new product that is expected to reach sales of $10 million in its first full year and $13 million of sales in the second and third years. Thereafter, annual sales are expected to decline to two-thirds of peak annual sales in the fourth year, and one-third of peak sales in the fifth year. No more sales are expected after the fifth year.

The cost of goods sold is about 60 percent of the sales revenues in each year. The sales general and administrative (SG&A) expenses are about 23.5 percent of the sales revenue. Tax on profits is to be paid at a 40 percent rate.

A capital investment of $0.5 million is needed to acquire production equipment. No salvage value is expected at the end of its five-year useful life. This investment is to be fully depreciated on a straight-line basis over five years.

In addition, working capital is needed to support the expected sales in an amount equal to 27 percent of the sales revenue. This working capital investment must be made at the beginning of each year to build up the needed inventory and implement the planned sales program.

Furthermore, during the first year of sales activity, a one-time product introductory expense of $200,000 is incurred. Approximately $1.0 million had already been spent promoting and test marketing the new product.

A. Formulate a multiyear income statement to estimate the cash flows throughout its five-year life cycle.
B. Assuming a 20 percent discount rate, what is the new product's net present value?
C. Should the company introduce the new product?

Answer 7.4
The net present value is −8.411 million and the product should not be introduced.

7.5 The XYZ Corporation has current liabilities of $130,000 with a current ratio of 2.5:1. Explain whether the individual transactions specified next increase or decrease the current ratio or the amount of working capital and by how much in each case. Treat each item separately.

1. Purchase is made of $10,000 worth of merchandise on account.
2. The company collects $5,000 in accounts receivable.
3. Repayment is planned of note payable, which is due in current period, with $15,000 cash from bank account.
4. The acquisition of a machine priced at $40,000 is paid for with $10,000 cash, and the lump-sum balance is due in 18 months.
5. The company conducts a sale of machinery for $10,000. Accumulated depreciation is $50,000, and its original cost is $80,000.
6. The company pays dividends of $10,000 in cash and $10,000 in stock.
7. Wages are paid to the extent of $15,000. Of this amount, $3,000 had been exhibited on the balance sheet as accrued (due).
8. The company borrows $30,000 for one year. Proceeds are used to increase the bank account by $10,000 to pay off accounts due to the supplier ($15,000) and to acquire the right to patents ($5,000).
9. The company writes down inventories by $7,000 and organization expenses by $5,000.
10. The company sells $25,000 worth (cost) of merchandise from stock to customers who pay in 30 days. Company has a gross margin of 40 percent.

Answer 7.5
The working capital is as follows: (1) unchanged, (2) unchanged, (3) unchanged, (4) $185,000, (5) $205,000, (6) $185,000, (7) $190,000, (8) $188,000, (9) $183,000, and (10) $211,667.

Table 7.A4 Income Statement of Superior Technologies

	Year 2002 (Thousands of dollars)	Year 2001 (Thousands of dollars)
Net sales	193,213	91,954
Cost of goods and services	128,434	60,776
Gross profit	64,779	31,178
Selling, general, and administrative expenses	26,369	13,844
Employee profit sharing and retirement	9,831	4,167
Total overhead	36,200	18,011
Operating profit before interest and tax	28,579	13,167
Other income	956	273
EBIT	29,535	13,440
Interest paid	680	505
Taxable income	28,855	12,935
Income tax	14,712	6,934
Net income	14,143	6,001

7.6 The income statement and balance sheet of Superior Technologies are exhibited in Tables 7.A4 and 7.A5. Conduct a ratio analysis and observe major trends and deviations from the available historical company information and industry data. (See also Table 7.A6.)

Answer 7.6
Selected answers for 2002 are as follows:
Days receivables = 54.1 days
Total long-term debt to net worth = 0.77
Total asset turnover = 1.8x
Gross profit (margin) = 33.5 percent

7.7 For the current fiscal year that starts on January 1, 2002, Buffalo Best Company projects its financial performance as shown in Table 7.A7.

In fiscal year 2002, the products of the company are priced at $10,000 each. The company president predicts that, because of the generally anticipated recovery of the U.S. economy, there will be a 4-percent-per-year increase in each of the next four years (i.e., 2003 to 2006) in (a) price of the company's products and (b) number of products sold. However, the company's administrative and selling expenses, as well as the product cost per unit, are projected to increase by 7 percent per year. Other expenses, such as R&D, interests, depreciation, and corporate tax rate, will remain unchanged. The anticipated increase in product cost is primarily due to an increase in materials costs and a decrease in labor productivity because of high turnover rate, poor employee supervision, inadequate workforce compensation policy, cumbersome production processes, and inferior employee training. If nothing is done, the company expects its net income to decrease by as much as 50 percent around 2005 to 2006.

At a management meeting on February 12, 2002, the vice president of engineering proposed to hire a consulting firm in the second half of 2002 to develop a customized program

Table 7.A5 Balance Sheet of Superior Technologies

	Year 2002 (Thousands of dollars)	Year 2001 (Thousands of dollars)
Cash	17,856	10,841
Receivables (net)	29,053	19,350
Inventories	23,282	6,869
Prepaid expenses	664	315
Payments received on government contracts	−6,013	−405
Total Current Assets	64,842	36,970
Property, plant, and equipment	60,806	26,773
Accumulated depreciation	20,083	10,281
Net property account	40,723	17,492
Patents, etc. (net)	0	249
Other assets	429	80
Total Assets	105,994	53,791
Accounts payable	10,368	5,337
Accrued wages, pensions, taxes, etc.	21,309	10,696
Other current liabilities	5,589	2,867
Total Current Liabilities	37,266	18,900
Long-term debt	29,935	9,250
Preferred stock	3,265	0
Common stock	3,915	3,257
Paid in surplus	8,205	6,228
Retained earnings	23,408	16,156
Total Liabilities and Net Worth	105,994	53,791

for improving labor productivity. She noted that three options are currently available at different investment prices and projected CGS reduction levels:

Program	Investment	CGS Reduction Percentage
A	$ 700,000	1.0%
B	$1,100,000	2.0%
C	$1,200,000	3.0%

The investment for any one of these options will be a one-time lump sum that can be expensed as an increase in overhead in 2002. The benefits of the consultation program will be realized in a projected reduction of product cost per unit (i.e., the originally anticipated product cost before the management consultation program) by a percentage (1, 2, or 3 percent, respectively) per year for the next four years (i.e., 2003 to 2006).

The company president welcomes the idea of management consultation as a possible way to enhance workforce productivity. Assuming that the company's cost of money is 10 percent, he wants to know which management consultation program he should approve.

Answer 7.7

Program C is to be preferred.

Table 7.A6 Selected Information of Superior Technologies

	2002	2001
(Depreciation and amortization:	$8,135	$4, 924)
(Preferred dividends	102	0)
(Other dividends	185	0)

Past Ratios

	Fill in 2002	2001	2000	1999	1998	1997	1996
Current ratio (ratio)		2	2	2.2	2.5	2	1.9
Acid test		1.6	1.2	1.3	1.7	1.3	1
Total debt to total assets (percent)		52.3	47.9	41.2	30.4	46.0	53.3
Long-term debt as percentage of capitalization		26.5	26.3	16.8	8.4	8.3	25.4
Total debt to net worth (ratio)		1.1	0.9	0.7	0.4	0.8	1.1
Days' receivables (days)		75.7	66.3	82.2	96.7	81.9	65.5
Ending inventory turnover (sales)		13.4	7.8	7	7.6	8.1	5.7
Ending inventory turnover (cost of sales)		8.8	5.5	5.1	5.5	5.7	4.1
Net property turnover		5.6	4.4	4.8	4	5.6	6.9
Total assets turnover		1.7	1.8	1.7	1.5	1.6	1.8
Net profit to total assets (percent)		11.2	10.0	8.6	8.1	7.9	8.5
Net profit to net worth (percent)		23.4	19.2	14.6	11.6	14.7	18.3
Net profit to net sales (percent)		6.5	5.6	5.1	5.5	4.9	4.7
Gross profit		34.0	29.5	26.5	27.4	29.3	29.2
		Industry Information					
Current ratio (ratio)		2.3	2	2.1	2.1	2.2	1.6
Acid test		1.2	1	1	1	1.1	0.6
Total debt to total assets (percent)		40.0	52.8	46.1	47.2	42.6	56.1
Long-term debt as percentage of capitalization		12.3	16.7	17.1	14.4	11.3	7.8
Total debt to net worth (ratio)		0.7	1.1	0.8	1.1	0.7	1.3
Days' receivables (days)		44	45	42	42	39	33
Ending inventory turnover (sales)		6.2	5.3	4.8	5.5	5.4	4.7
Ending inventory turnover (cost of sales)		4.8	4.1	3.8	4.2	4.4	3.8
Net property turnover		8.7	10.1	7.1	8.4	9	12.2
Total assets turnover		2	1.9	1.7	1.9	2	2.2
Net profit to total assets		6.1	5.2	4.8	5.6	4.4	5
Net profit to net worth (percent)		10.2	11.0	9.0	10.6	7.7	11.4
Net profit to net sales (percent)		3.1	2.7	2.8	2.9	2.3	2.3
Gross profit (percent)		22.8	21.7	20.6	23.2	18.6	19.9
Number of companies reported:		60	42	55	65	73	60

Note: Data derived from several literature sources, 1996–2000.

Table 7.A7 Income Statement of Buffalo Best

	Year 2002 (Thousands of dollars)
Sales	20,000
Cost of goods sold	13,000
Gross margin	7,000
Administrative and selling expenses	4,000
R&D	500
Depreciation	1,000
Earning before interest and tax (EBIT)	1,500
Interest	100
Taxable income	1,400
Tax (30 percent)	420
Net operating profit after tax (NOPAT)	980

7.8 The company's vice president of marketing proposes a new program to significantly increase the product sales by 250,000 units per year throughout the 1998 to 2004 period. Specifically, it is suggested that the company takes the following actions:

1. Spend $2.5 million over the period of 1998 to 2000 as promotional expenditures—for example, spend $1.0 million each in the years 1998 and 1999, and $0.5 million in the year 2000.
2. Make a one-time investment of $1.4 million in plants and equipment needed at the beginning of 1998 to generate these additional products. No new warehouse capability is needed. This investment is to be depreciated on a straight-line basis over the seven-year period. There will be no salvage values for these plants and equipment in 2005.

It is further assumed that the product unit cost is $8.00 in 1998, and it is estimated to increase by 3 percent per year. The product unit price is $20 in 1998, and it is estimated to change as manifested in the following table:

Items	1998	1999	2000	2001	2002	2003	2004
Unit Price	$20.00	$20.60	$21.00	$21.15	$21.25	$21.25	$21.00

The SG&A expenditure is estimated at $1.25 million in 1998, and it will increase by 3 percent per year during the six-year period. A corporate tax of 40 percent must be paid for any marginal income. There is an interest charge during this period, and the company's weighted average cost of capital (WACC) is 8 percent.

If the company's hurdle rate for this type of investment is 25 percent, would you recommend that the marketing initiative be approved?

Answer 7.8
No, the marketing initiative should not be approved.

7.9 One of the company's technology patents is about to expire, inducing a likely rush of product entries from the competition. Currently, the product line is projected to have stagnant sales of 1 million units for the next seven years from 1998 to 2005. The unit price is expected to decrease slightly by 1 percent per year during the same period. The profitability of this product is estimated to be $1,166,000 in 1998, as illustrated in Table 7.A8.

Table 7.A8 Simplified Income Statement

	1998 (dollars)
Sales	20,000.00
CGS	10,000.00
Depreciation	1,057.00
SG&A	7,000.00
EBIT	1,943.00
Tax at 40 percent	770.00
EBIAT	1166.00

Both the cost of goods sold (CGS) and SG&A expenses are expected to increase by 3 percent per year from 1998 to 2004. The depreciation charge is estimated to remain constant at $1 million per year, and there is no salvage value for the equipment at the end of 2004. If the product is continued as planned, the company can recover a working capital of $3.9 million at the end of 2004 from sales of residual inventory and collection of accounts receivable after having discharged all applicable short-term liabilities.

If the management decides to discontinue this product line at the end of 1997, then it can sell the fixed assets related to the product line (having a book value of $7 million) for about $3 million, and the loss of $4 million would be tax deductible. Furthermore, the company can recover the working capital (inventory plus accounts receivable, minus accounts payable and other expenses) worth about $3.9 million at the end of 1997.

Assuming that the appropriate discount rate is 12 percent, would you recommend that this product line be discontinued at the end of 1997 or be continued through 2004?

What is the next best alternative open to the company, besides either shutting it down immediately at the end of 1997 or continuing to run it until 2004?

Answer 7.9

The best option is to shut down the production line at the end of 1997. The next-best alternative is to discontinue the product line at the end of 2001.

7.10 Define the economic value added (EVA) of XYZ Company. (See income statement and balance sheet of Question 7.2 for the years 2001 to 2002.) Assume that the WACC is 12.35 percent for both years. Discuss the EVA results and contrast them with the EVA results for Superior.

Technologies in 2001 and 2002 (see Question 7.6) for which WACC is also assumed to be 12.35 percent.

Answer 7.10

The EVA of XYZ is −39,764,500 in 2001 and −80,583,500 in 2002. The EVA of Superior Technologies is $5,655,092 in 2002.

7.11 The 2000–2001 income statement and balance sheet for Buffalo Best are presented in Tables 7.A9 and 7.A10. The WACC for Buffalo Best is 10 percent for both years.

Review and comment on the performance of the company based on the following:

A. Liquidity, activity, and profitability
B. Uses and sources of funds
C. Value creation based on EVA analysis

Table 7.A9 Income Statement of Buffalo Best

	Year 2001 (Thousands of dollars)	Year 2000 (Thousand of dollars)
Sales	18,000	17,000
Cost of goods sold	11,000	10,500
Gross margin	7,000	6,500
Administrative and selling expenses	3,500	3,200
R&D	500	500
Depreciation	1,000	1,000
Earning before interest and tax (EBIT)	2,000	1,800
Interest	100	100
Taxable income	1,900	1,700
Tax (30 percent)	570	510
Net profit after tax (NOPAT)	1,330	1,190
Dividends	330	190
Retained earnings	1,000	1,000

Table 7.A10 Balance Sheet of Buffalo Best

	Year 2001 (dollars)	Year 2000 (dollars)
Assets		
Cash and securities	5,000	6,000
Accounts receivable	15,000	10,000
Inventory	10,000	7,300
Net fixed assets (Investment minus accumulated depreciation)	50,000	51,000
Other	1,000	1,200
Total assets	81,000	75,500
Liabilities		
Accounts payable	20,000	15,000
Other short-term liabilities	26,000	24,000
Long-term liabilities	1,000	1,500
Total liabilities	47,000	40,500
Equities at par value	1,000	1,000
Capital surplus	12,000	14,000
Retained earnings	21,000	20,000
Total liabilities and owners' equities	81,000	75,500

Answer 7.11

The sources total is \$10.53 million. The EVA is −\$2.17 million in 2001 and −\$2.46 million in 2002.

7.12 Company X manufactures technology products. It plans to expand its manufacturing operations. Based on past data, management anticipates the first project year of the as-yet-to-be-expanded operations to match the data in Table 7.A11.

A. Compute the working capital requirement during this project year.

B. Determine the taxable income during this project year.

Table 7.A11 Selected Data of Company X

	(dollars)
Sales	1,500,000
Manufacturing costs	
Direct materials	150,000
Direct labor	200,000
Overhead	100,000
Depreciation	200,000
Operating expenses	150,000
Equipment purchase	400,000
Borrowing to finance equipment	200,000
Increase in inventories	100,000
Decrease in accounts receivable	20,000
Increase in wages payable	30,000
Decrease in notes payable	40,000
Income taxes	272,000
Interest payment on financing	20,000

C. Calculate the net income during this project year.
D. Define the net cash flow from this project during the first year.

Answer 7.12
(a) \$290,000; (b) \$680,000; (c) \$408,000; (d) \$608,000.

7.13 The company has 10,000 shares of common stock outstanding, and the current price of the stock is \$100 per share. The company has no debt. The vice president of engineering discovers an opportunity to invest in a new technology project that yields positive cash flows with a present value of \$210,000. The total initial capital that is required for investing and developing this project is only \$110,000. It is proposed that capital be raised by issuing new equity. All potential purchasers of common stock will be fully aware of the new project's value and cost, and are willing to pay "fair value" for the new shares of the company.

A. What is the net present value of this project?
B. How many shares of common stock must be issued, and at what price, to raise the required capital, assuming the costs of underwriting these new shares are negligible?
C. What is the effect, if any, of this new project on the value of the stock of existing shareholders?

Answer 7.13
A. \$100,000; B. 1111.11 shares at \$99 per share; C. slight dilution.

7.14 XYZ Company is financed by debt (50 percent), preferred stocks (20 percent) and common equity (30 percent). Its common stock price is \$43 per share. It pays a dividend of \$3.00 and has a growth rate of –2 percent. Its annual preferred stock dividend is \$82 per \$1,000 share with a flotation cost of 7.5 percent per share. The interest for long-term debt is 11 percent. Its corporate tax rate is 30 percent. What is the company's weighted average cost of capital (WACC)?

Answer 7.14
WACC = 8.483%

7.15 XYZ Company receives a contract from one of its major customers. The contract calls for 20,000 hours of work billed at $75 per hour to be completed within one year. The normal billable work for each engineer is 2,000 hours per year. Thus, the contract requires the involvement of ten full-time engineers. At the present time, the company has only six full-time engineers who may be able to work on this contract. The engineers' wages average $40 per hour. If the engineers are to work overtime, their overtime pay is time and a half (i.e., $60 per hour).

A. What would be the contribution margin if 10 engineers were available with no need to work overtime?
B. What would be the contribution margin if all 6 engineers are used full time and the deficiency is made up through 8000 hours of overtime?
C. What would be the contribution margin if the company hires two more full-time engineers and makes up the rest with overtime?
D. If the cost of recruiting and training each new engineer is $15,000, what would be the contribution margin on hiring two new engineers, after factoring in recruiting and training costs?

Answer 7.15
A. 0.4667; B. 0.36; C. 0.4133; D. 0.3933.

7.16 XYZ Company plans to install a new production line consisting of several precision machines costing a total of $800,000. The installation of these machines requires another $150,000. The products made by the machines are projected to deliver a net income after tax of $400,000 per year for the next ten years. The useful life of each machine is estimated to be ten years. At the end of ten years, these machines have a salvage value of $20,000. Compute the cash flows generated by this new production line.

Answer 7.16
The annual cash flow is $493,000.

7.17 Using the data from Question 7.6, conduct an evaluation of Superior Technologies' common stock in 2002, using the three different methods specified, as follows:

A. The market value of the company's net property has risen and it is now about two times the value reported in the balance sheet. Calculate the stock price by using its *net asset value* as a basis.
B. Assuming that the company's cost of equity (K_e) is 16 percent, determine the stock price by applying the *dividend growth model*.
C. Price-to-earning (P/E) ratio reflects the general sentiment of the securities market toward a specific company or the industry in general. Assuming that the average P/E ratio is about 10 for the industry, of which the company is a member, define the stock price, using the *earning model*.

Answer 7.17
A. $19.47/share; B. $25.61/share; C. 24.29/share.

7.18 Company A has been performing reasonably quite well over the last several years. Table 7.A12 exhibits an abbreviated set of its financial data for the years 1998 to 2003.

A. Analyze the company's dividend-payout ratio to common stockholders and comment on the suitability of this dividend policy.

Table 7.A12 Financial Data of Company A (Millions of dollars)

	2003	2002	2001	2000	1999	1998
Net income	41	33	34	35	27	22
Preferred dividends	3	3	3	1	0	0
Common dividends	18	18	18	18	13	12
Total assets	492	455	417	403	280	258
Current liabilities	68	57	75	68	51	43
Long-term liabilities	113	114	75	83	57	57
Preferred stock ($100 par)	82	82	82	82	0	0
Common stocks ($6.25 par)	45	45	45	45	38	38
Capital surplus	6	0	0	0	34	34
Retained earnings	177	157	140	126	100	86

B. Discuss the debt-financing policy of this company over the years. In your opinion, does the long-term debt of the company represent a percentage too high (aggressive) or too low (conservative) in its capital structure? Why?

C. Do you regard the percentage of common stock equity in the company's capital structure as adequate, and why?

D. For 2004, the company needs an influx of $30 million to finance business expansion. Which financing option should the company pursue? Why?

Answer 7.18
The option of debt financing is recommended.

7.19 Company B is currently financed by common stock equity. It is considering two alternative ways of financing in order to increase the return on common equity. Table 7.A13 displays the two options under consideration, along with the base case.

The company's capitalization and EBIT (earning before interest and tax) remain constant at $30 million and $3 million, respectively. The composition of the capitalization changes from 100 percent common stock equity (base case) to a mix of common and preferred stocks (Option A) and to a mix of common stock and long-term debt (Option B). The corporate tax rate is 40 percent. Dividends of preferred stocks are paid at a 5 percent rate. The interest charge for the long-term debt is 4 percent.

1. Compute the rate of return on common stocks equity for the three cases. Explain why these numbers change from one case to another.
2. Compute the rate of return on capitalization for the three cases.

Table 7.A13 Financing Options of Company B

	Base Case (dollars)	Option A (dollars)	Option B (dollars)
Capitalization	30,000,000	30,000,000	30,000,000
Common stock equity	30,000,000	6,000,000	6,000,000
Preferred stock equity	0	24,000,000	0
Long-term debt	0	0	24,000,000
EBIT (operating income)	3,000,000	3,000,000	3,000,000

3. Among the three cases presented, which financing option is to be preferred by the company, and why?

Answer 7.19

None of the stated options is to be recommended. Instead, the company should pursue Option C, which limits the company's debt-to-capitalization ratio to no more than 50 percent, resulting in a return to common stock equity of 9.6 percent.

7.20 Company X is planning to introduce a new service that is expected to reach sales of $10 million in its first full year, and $13 million of sales in the second and third year. Thereafter, annual sales are expected to decline to two-thirds of peak annual sales in the fourth year, and one third of peak sales in the fifth year. No more sales are expected after the fifth year.

The cost of goods sold (CGS) is about 60 percent of the sales revenues in each year. The sales, general and administrative (SG&A) expenses are about 23.5 percent of the sales revenue. Tax on profits is to be paid at a 40 percent corporate rate.

A capital investment of one-half of a million dollars is needed to acquire the service delivery equipment. No salvage value is expected at the end of its five-year useful life. This investment is to be fully depreciated on a straight-line basis over five years.

In addition, working capital is needed to support the expected sales in an amount equal to 27 percent of the sales revenue of a given year. This working capital investment must be made at the beginning of each year to build up the needed sales support elements to implement the planned sales program.

Furthermore, during the first year of sales activity, a one-time service introductory expense of $200,000 is incurred. Approximately $1.0 million had already been spent promoting and test marketing the new service, before the sales start.

A. Formulate a multiyear income statement to estimate the cash flows throughout its five-year life cycle.
B. Assuming a 15 percent discount rate, what is the new service's net present value?
C. Should the company introduce the new service?

Answer 7.20

The new service's net present value is $2.08 million. The company should introduce the new service.

7.21 **Buffalo Best Company** has been performing reasonably well over the last several years. Table 7.A14 exhibits an abbreviated set of its financial data for the years 1998 to 2003.

For 2004, the company needs an influx of $60 million to finance business expansion. Which financing option should the company pursue? Why?

7.22 Buffalo Best is currently financed by common stock equity. The company is considering a couple of alternative ways of financing in order to increase the return on common equity. Table 7.A15 presents the two options under consideration, along with the base case.

The company's capitalization and EBIT (earning before interest and tax) remain constant at $30,000,000, and $3,000,000, respectively. The composition of the capitalization changes from 100 percent common stock equity (base case), to a mix of common and preferred stocks (Option A) and to a mix of common stock and long term debt (Option B).

Table 7.A14 Buffalo Best Company Financial Data (Millions of Dollars)

Number	Description	2003	2002	2001	2000	1999	1998
1	Net income	41	33	34	35	27	22
2	Preferred dividends	3	3	3	1	0	0
3	Common dividends	18	18	18	13	13	12
4	Total assets	492	455	417	403	280	258
5	Current liabilities	68	57	75	68	51	43
6	Long-term liabilities	113	114	75	83	57	57
7	Preferred stocks ($100 par)	82	82	82	82	0	0
8	Common stocks ($6.25 par)	45	45	45	45	38	38
9	Capital surplus	6	0	0	0	34	34
10	Retained earnings	177	157	140	126	100	86

Table 7.A15 Financing Options of Buffalo Best

	Base Case	Option A	Option B
Capitalization	$30,000,000	$30,000,000	$30,000,000
Common stock equity	30,000,000	6,000,000	6,000,000
Preferred stock equity	0	24,000,000	0
Long-term debt	0	0	24,000,000
EBIT (Operating Income)	$3,000,000	$3,000,000	$3,000,000

The corporate tax rate is 50 percent. Dividends of preferred stocks are paid at a 6 percent rate. The interest charge for the long-term debt is 8 percent.

A. Compute the rate of return on common stocks equity for the three cases. Explain why these numbers change from one case to another?

B. Compute the rate of return on capitalization for the three cases.

7.23 As the vice president of engineering of the Best Company in Buffalo, you need to make a decision regarding how a new product is to be manufactured. You have been offered two specific proposals.

Proposal A is to set up an assembly operation in house and to outsource the production of all subassemblies and parts to supply chain partners. This proposal would need a front-end investment of $2,000,000 for the assembly operations, an investment of $300,000 for the product design and development efforts, and another $100,000 for managing and coordinating the supply chain partners. The projected net profits for the products manufactured by this method are $0, $300,000, $600,000, $900,000, $1,200,000, and $600,000, in the first, second, third, fourth, fifth, and sixth year, respectively. There is no salvage value of the assembly equipment at the end of sixth year, at which time the sales of this product will be terminated. Interest is at 5.0 percent.

Proposal B is to build a production facility to manufacture all subassemblies and assemble the products in house. This proposal would need a front-end investment of

$3,000,000, which includes facility, equipment, engineering and all other required efforts. The projected net profits for the products manufactured by this method are $200,000, $400,000, $800,000, $1,200,000, $1,000,000, and $600,000 for the first, second, third, fourth, fifth, and sixth year, respectively. There is a salvage value of $400,000 of the facility at the end of sixth year. Interest rate is also at 5 percent.

Which proposal should you accept, and why?

7.24 Michele Brown started up a new consulting firm to offer design and testing services to industries. During the first three months of its operation, the firm has recorded a number of transactions, shown as follows:

A. Michele Brown initiated the business by investing $35,000 in cash; $1,200 for office equipment; and $23,500 for instrumentation.

B. Land for an office site was purchased for $17,500; $5,000 paid in cash and a promissory note signed for the balance.

C. A prefabricated building was purchased for $7,000 cash and moved onto the land for use as an office.

D. The premium of $800 was paid on an insurance policy.

E. A consulting project for Central Construction was completed and $900 was collected.

F. Additional test equipment costing $10,000 was purchased for $2,000 in cash, and a note payable for the balance was signed.

G. A consulting project was completed on credit for Eastern Manufacturing for $1,200.

H. Office supplies were purchased on credit for $300.

I. A bill was received for $250 and recorded as an account payable for rent on a special machine used for the Western Technologies project.

J. A project for Superior Design was completed on credit for $1,000.

K. Cash was received from Eastern Manufacturing for the consulting project they received on credit.

L. The wage of engineers, $5,000, was paid.

M. Office supplies purchased earlier (for $300) were paid for.

N. Cash of $125 was paid for repairs to test equipment.

O. The company president, Larry Brown, wrote a $65 check on the bank account of the consulting firm to pay for repairs to his personal automobile, which is not used for business purposes.

P. The wage of the office workers, $3,000, was paid.

Q. The maintenance expenses of the instruments, $300, were paid.

Open the following T-accounts: Cash, Accounts Receivable, Prepaid Insurance, Test Equipment, Building, Land, Notes Payable, Accounts Payable, Office Supplies, Larry Brown—Capital, Larry Brown—Withdrawals, Sales Revenue, Instrument Maintenance Expense, Test Equipment Rental Expense, Office Equipment, Instrumentation and Test Equipment Repairs. Record the transactions by entering debits and credits directly into the accounts. Use the transactions letters to identify each debit and credit entry. What is the cash account balance at the end of this three-month period?

Answer 7.24

The balance of the cash account is $13,575.

Chapter 8

Marketing Management for Service Systems Engineers

8.1 INTRODUCTION

Companies have essentially two major activities: marketing and innovation. Marketing is the whole business of the enterprise seen from the viewpoint of customers. The purpose of marketing is to provide services that meet the needs and wants of customers. Marketing impacts the top line (i.e., sales revenue) of any enterprise. Innovation strengthens the enterprise's competitive marketing position to sustain profitability by way of timely application of unique technologies and other core competencies (Kotler, Jain, and Maescincee 2002; Kotler 1999; Kotler 1997).

As the markets further proliferate due to expansions of customer segments, services, geographies and distribution channels, the marketing today is more complex, more costly and less effective than before (French and Knudsen 2007).

Service marketing is the general subject of a number of relatively new textbooks (e.g., Lovelock and Wirtz, 2006, Palmer 2007, Baron et. al. 2009, Zeithmal et. al. 2005, and Lovelock et. al. 2008). Some others cover certain specific topics related to service marketing, such as customer loyalty (Griffin 2002) and relations marketing (Harwood et. al. 2008). Still others address the marketing details in specific service sectors, such as professional services (Kotler et.al. 2002, Waugh 2001), architectural services (Jones 2006), financial services (Harrison 2006, Ennew and Waite 2006), legal services (Monk and Moyers, 2008), telecommunications services (Strouse 2004), hospitality marketing (Reid and Bohanic 2005), and healthcare marketing (Thomas 2008).

The objective of this chapter is to introduce various basic marketing concepts related to services (Paley 2001) and to prepare services engineers and engineering managers to interact effectively with marketing and sales personnel in service companies. Such concepts and applications include marketing management processes in profit-seeking organizations, identification of opportunities and threats facing an organization, and marketing tasks.

Services engineers are known to be technologically innovative. By mastering marketing techniques as well, services engineers can make significant contributions toward the success of the company and, as a result, gain entry into higher leadership positions. Service engineers and leaders are encouraged to continue studying additional current references on specific issues in marketing management.

8.2 THE FUNCTION OF MARKETING

Marketing and sales are critically important to profit-seeking companies because they strive to ensure satisfaction in the exchange of values between the producers and consumers of services, as exhibited in Fig. 8.1. Vargo and Lusch (2010) offer a useful service perspective of marketing.

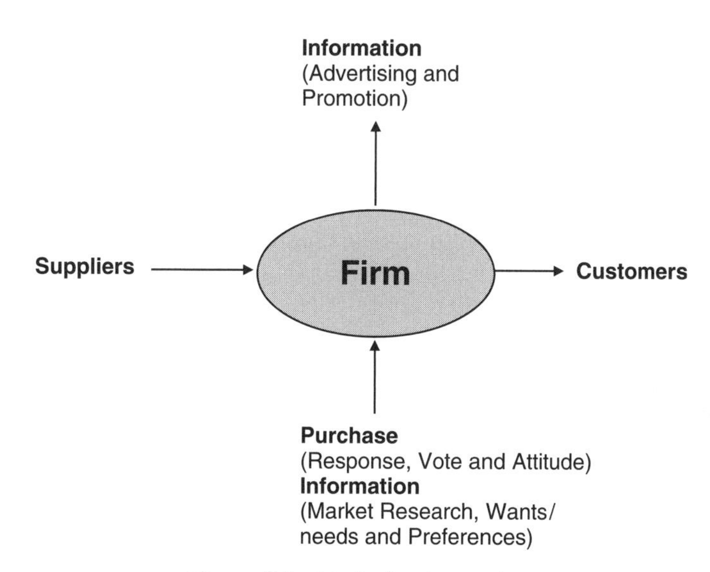

Figure 8.1 Marketing interaction.

8.2.1 Sales versus Marketing

Sales is a process by which producers attempt to motivate target customers to buy the available services. The mentality behind sales, as illustrated by Fig. 8.2, is that "someone out there will need this." At the time that Ford Motor Company was the dominant carmaker in Detroit, Henry Ford was famously quoted as saying about the Model T: "You can have any color you want, as long as it is black." Sales does not take the customers' concerns into account.

In contrast, companies with a marketing orientation offer something customers want by seeking feedback from the marketplace, adjusting the service offerings, and increasing the value to consumers. (See Fig. 8.3.)

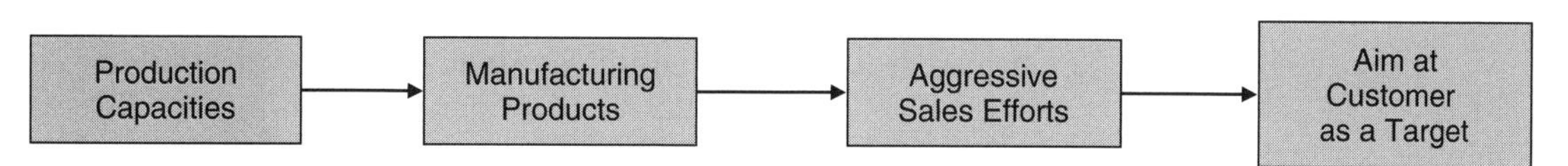

Figure 8.2 Sales orientation.

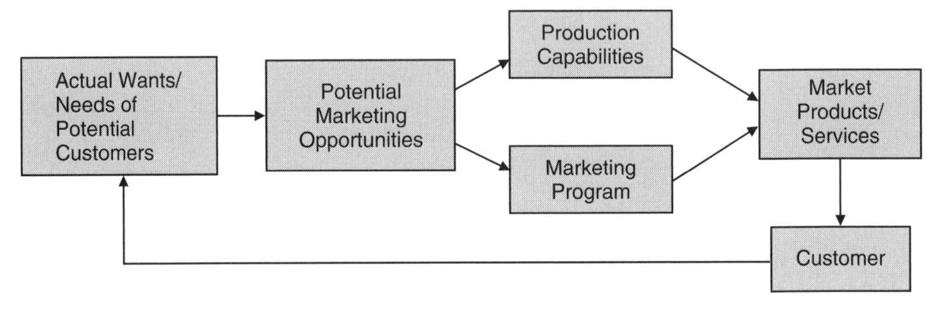

Figure 8.3 Marketing orientation.

In the pursuit of marketing strategies, companies solicit intelligence, financial data, and customers' responses to constantly reassess the market. They evaluate such factors as changing needs, competition, cost effectiveness, service substitution, and maturity of services. A long-term orientation ensures that benefits for both producers and customers will be sustained. Sales strategies are only a part of marketing.

8.2.2 The Marketing Process

The marketing efforts of companies are typically focused on four specific aspects (Kotabe and Helsen 2007):

1. *Customer focus.* The purpose of a profit-seeking business is to understand customers' needs, deliver value to customers, and offer services to ensure that customers are satisfied. In other words, the customer comes first.
2. *Competitor focus.* Companies seek advantages relative to competitors, monitor competitive behaviors, and respond to the strategic moves of competitors.
3. *Interfunctional coordination.* Companies integrate all functions, share information, and organize themselves to provide added value for the customers.
4. *Profit orientation.* Companies focus on making profits in both the short term and the long term.

To achieve business success, companies must search constantly for future markets, in addition to actively serving the markets of today. The primary responsibility of marketing is to scan the relevant business environment for future opportunities (such as what bundle of products and services to offer to whom, at what price, at what time, and in which market segments) and to provide insight into the needs of current and future customers and the intentions of competitors.

Presented in Fig. 8.4 is the marketing process, which is iterative in nature. This process defines opportunities (unsatisfied needs) in the marketplace; the products or services with features to satisfy these needs; and the product (service) pricing, distribution, and communications strategies to reach the target market segments. Market segments refer to specific groups of customers who share similar purchasing preferences.

For companies to succeed in the marketplace, marketing must be a core value, central to the company's strategy formulation and execution. Through marketing, companies identify and satisfy the needs of customers and achieve long-term profitability by

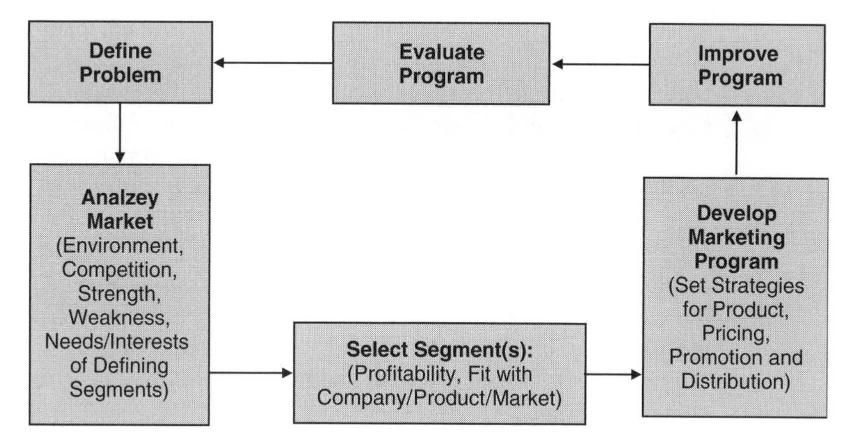

Figure 8.4 Marketing process.

attracting and retaining customers. Specific tasks undertaken to attain these objectives include (1) interacting with and understanding the market and its customers, (2) planning long-term marketing strategies, and (3) implementing short-term tactical marketing programs.

The effectiveness of a marketing program is often measured by two metrics: (1) how attractive the company's services are to the target customers and (2) how successfully the company can satisfy and retain these target customers. Figure 8.5 illustrates the marketing effectiveness diagram.

The marketing program of a company is regarded as a *total success* if both the customer retention and service attractiveness to customers are high. This is when high profitability can be achieved at a maximum growth rate. The marketing program scores a *partial success* if the service attractiveness to the customer is high, but the customer retention is low. Although lost customers are typically replaced by new customers, the total customer base will experience little growth. *Partial failure* of a marketing program is when service attractiveness to the customer is low but the customer retention is high. Under this scenario, business remains stagnant because it relies on loyal customers to

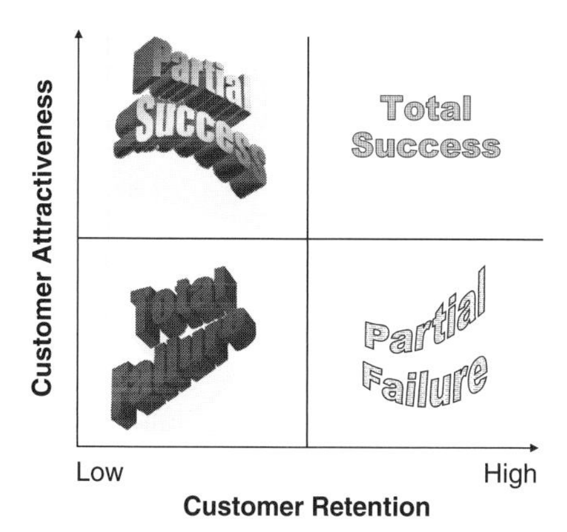

Figure 8.5 Marketing effectiveness diagram.

repeatedly buy mature, noncompetitive services. The company's sales may slow down or fall as few new customers are added. The marketing program is a *total failure* if both service attractiveness to the customer and the customer retention level are low. Customers are leaving, and the company's sales are falling.

Marketing strategies are implemented at the corporate, business, and operational levels. Top management provides inputs to identify future opportunities, and addresses such questions as what business the company is in and what business the company should be in. The business managers then specify their ideas, bring out products or services, and strive to create and maintain a sustainable competitive advantage in the marketplace. At the operational level, managers and support personnel conduct the planning for specific marketing programs, and implement and control marketing efforts related to segmentation, product (service), pricing, distribution channels, and communications.

8.2.3 Products and Services Marketing

Companies offer value propositions to the marketplace in the forms of products, services and/or a combination of both. Products represent tangible physical entities (jet engines, iPad, Droid smartphone, computers, etc.) that embody specific features and functionalities to deliver value to their end users. Usually, the design, production and consumption of products take place at separate times and locations, even though the consumer's feedback is aggressively solicited as inputs to the design process. Services, by contrast, are intangible knowledge-intensive bundles of value aimed at causing a transformation of the end user's state in knowledge (business consulting, education, information search, operations), wealth (financial advisement, tax services), health (physicians and hospital services, insurance), enjoyment (video rentals, entertainment services), physical movement (car rentals, travel, relocation, transportation services), and others. In general, the design, production and consumption of services occur at the same time and location and consumers participate actively in the whole process. Other companies deliver product-enhancing services to support the utilization of their products (maintenance, problem-solving, operations). In general, services differ from products in the following four ways.

Lack of Ownership. Consumers who purchase a product gain ownership over the product and use it for as long as they want during the product's life cycle. Consumers who purchase a specific service will obtain only the rights to use it during a specific period of time (e.g., services related to financial investing, banking, legal advisement, education and transportation). Because the life cycle of services is usually shorter than that of products, consumers tend to be more sensitive to the quality of services than to the quality of products.

Intangibility. Products are tangible in that they possess measurable physical features, such as size, weight, and color, whereas services are considered intangible, even though both products and services deliver long-lasting values that are useful to consumers. Customer experience derived from a product is typically affected by its primary value proposition (e.g., price, product features, product quality, frequency of use, support services). Customer experience derived from a service is affected by a different set of value propositions (e.g., price, importance of the change of state made possible by the use of service, attitude of vendor's service staff, service quality, physical environment in which service is delivered and consumed, etc.). Customer satisfaction is harder to accomplish with services, as customer's expectations may be subjectively divergent due to an increased level of person-to-person interactions in the delivery and consumption of services.

Concurrency in Production and Consumption. Products are usually manufactured at a factory location, transported to another location, and displayed at a retail store location, to be purchased and used by customers at an entirely different time and location. Services, by contrast, are typically produced and consumed at the same time and at the same place (e.g., financial advisement, seeing a performance). Thus, services cannot be separated from the service providers. Products may be stored, whereas services may not. Services last a specific time and cannot be stored for later use. Again because of this time constraint, consumers demand more.

Variability in Service Quality. The quality of service depends, to a large extent, on the interpersonal skills, attitude, and service knowledge and the support process in place to resolve customer's problems. It is very difficult to make each service experience identical.

Because of these differences, the marketing of services needs to be more attentive to factors such as physical evidence, process, and people, as compared to the marketing of products.

8.2.4 Key Elements in Marketing

Those who plan and implement marketing programs are called *marketers*. Marketers consider various influence factors and make diverse decisions to penetrate a specific marketplace (Ferrell and Hartline 2007). Marketers pay attention to several key elements of marketing, such as the market itself, the environment, the customer, the product (service), pricing, promotion, and distribution.

The *market* is made of buyers who are expected to purchase certain services and also buy substitutes that offer similar values. Of great importance to the marketers is the size of the market, measured in millions of dollars per year, its growth rate, its location characteristics, and its requirements. The market must be large, stable enough with a reasonable growth rate, and relatively easy to reach and serve in order for it to be a worthwhile target for the marketers.

The *environment* refers to competitors, barriers to entry, rules and regulations, resources, and other such factors affecting the marketers' success in a given market segment. Marketers must also understand the opportunities and threats present in the environment.

The *customers* consist of all potential buyers of a given service. Companies need to understand why they buy, how they buy, who makes the decision to buy, who will use the service, in what specific way the use of the service will contribute value to the user, what might be the buyer's preference related to support service and warranty, what other service features the customer may want, as well as other factors. A company that uses marketing strategies to understand its customers' needs can then implement service strategies that build and maintain competitive advantages in the marketplace. (Appendix 8.10.1 contains additional sample customer-survey questions.)

To become customer-oriented, companies need to (1) define the generic needs of customers through research (such as the buyer's perception of the cost, status and value of an financial advisement service), (2) identify the target customers by segmentation (including which selected groups of customers have shared needs), (3) differentiate services and communications (for example, offering special reasons for customers to buy through unique service attributes or unique communications), and (4) bring about differentiated values for customers.

Consumers and business customers have different service needs. Services are not of the same value to all their intended end users, nor do they require the same depth of knowledge and technology to deliver. Because rational end users are willing to pay proportionally more for high-value services, marketing efforts must be devoted to services in relation to their value to the end users. This measure affects corporate profitability derivable from the marketplace. Figure 8.6 presents some service examples based on value and the extent of knowledge and technology needed to deliver them.

There are seven key marketing elements for services which are described below. Companies will need to adopt suitable marketing strategies to optimally promote awareness, build brand image to attract their targeted consumers and business customers, affect their buying decisions, and retain them.

The *product (service)* symbolizes the actual "bundle of benefits" that is offered to customers by the marketers. Factors considered include functional attributes, appropriateness to customer needs, distinguished features over competition, service-line strategy, and service-to-market fit.

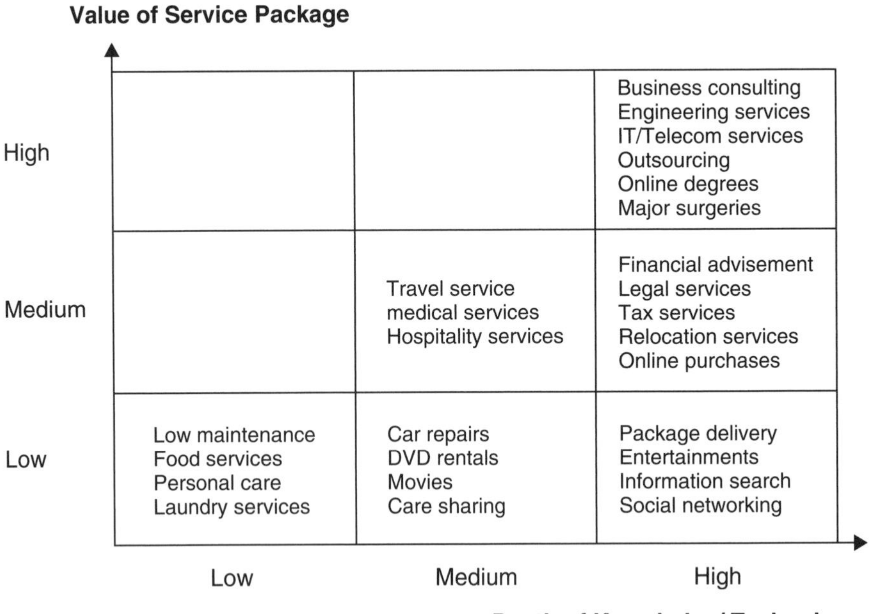

Figure 8.6 Service value versus knowledge/technology.

Promotion and *communication* consider strategies of service and brand promotion, options to use a push–pull strategy, selection of advertising media, and the choice of promotional intensity. These ensure that the selected means for communication are compatible with the target market segments.

The *pricing strategy* concerns itself with (1) the choice of either a skimming or a penetrating strategy to set the price, (2) the use of value-added pricing, and (3) the fit of a chosen pricing strategy to the target segment.

The *placement (distribution) strategy* defines such areas as (1) the service delivery options of either an intensive, exclusive, or selective distribution system; (2) the company's relationship with dealers; and (3) the changes in distribution systems.

The *physical evidence* refers to the physical setting (e.g., store appearance, layout and color, dress of service staff, service equipment, service brochures, etc.) that affects customer experience.

The *process design* specifies the applicable operations policies and procedures to effectively serve the customers as related to order processing, logistics, inventory planning, franchising policy, sales training, and flow of activities in delivering services.

The *people* represent the key factor influencing customer experience, which is affected by the attitude and helpful and considerate behavior of the customer-facing service staff.

These elements characterize the multidimensional nature of marketing and are centered on customers, who are the focus of any successful marketing program.

While there are seven controllable variables in the marketing of services, the marketing mix for products consists of only the first four Ps. The last three Ps, namely physical evidence, process design, and people, are added to account for the significant impact these factors exert on the marketing of services (Collier 1991).

To help the reader understand the market and the customer, market forecast and market segmentation are introduced in sections 8.3 and 8.4. The seven marketing mix elements depicted in Fig. 8.7 are addressed in detail in section 8.5.

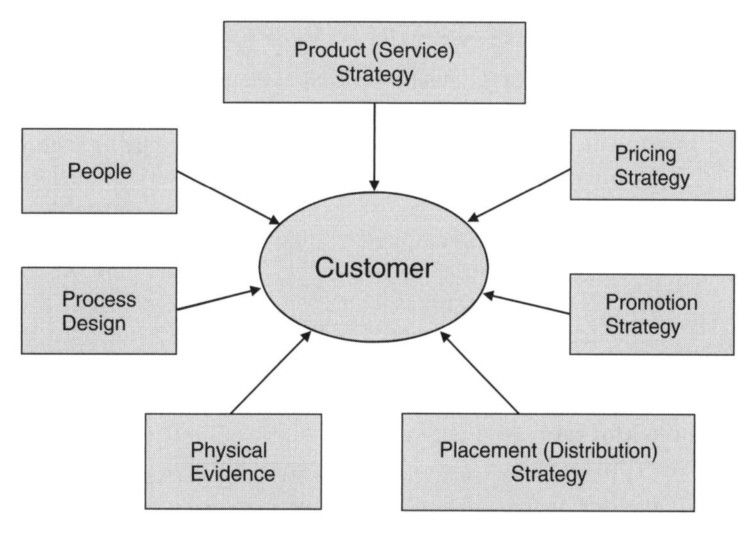

Figure 8.7 Marketing mix.

Example 8.1 To us engineers and technologists, products and services are the key "bundles of value" we create as the primary basis for any enterprise to become profitable over time. In the absence of such "bundles of values" the enterprise has nothing meaningful to offer to the marketplace. Things we do and say are usually reproducible and verifiable. We the engineers are clearly the backbones of any enterprise.

Please explain why any enterprise needs a marketing team. Why should we engineers/technologists bother wasting our time to collaborate with these marketing guys who talk a funny language, use imprecise concepts, invoke data that they can hardly prove, wave their hands, and project such an air of optimism that can seem rather arrogant at times? Why should we engineers/ technologists respect this "fast-talking bunch" and treat them as our equals in the first place?

ANSWER 8.1 For any service enterprise to succeed in the marketplace, it needs innovative services that customers are willing to buy. There are indeed two aspects in this business model. Engineering and technologists are important contributors to the development of innovative services, which are low cost, reliable, and easy to deliver and support. This core service element is the key part of any business venture.

However, without marketing, engineering would not know which services features to emphasize that would satisfy the needs of customers while offering meaningful differentiations to the competition. Furthermore, marketing creates customer awareness through the right advertising channels; such awareness is critical to achieving commercial success by any service. Marketing also divides the customer base into segments so that they can be better served, communicates with customers constantly to understand their future needs, manages the customer relationships to discover unique usage patterns that might be important to the development of new services, and tracks the competition to mitigate potential threats.

Successful service enterprises need to demand that marketing and engineering are collaborating effectively in order to assure sustainable profitability.

8.3 MARKET FORECAST—FOUR-STEP PROCESS

The purpose of conducting a market forecast is to define the characteristics of the target market as to, for example, size, stability, growth rate, and serviceability. Market size and growth rate must be large enough to warrant further consideration by marketers. Any future market is always uncertain due to potential changes in end-user behavior, global economics, new technologies, competition, and economic and political conditions.

Barnett (2009) emphasizes that the key to successfully forecasting market size is to understand the underlying forces behind the demand. Barnett proposes a four-step process, as follows: (1) define the market, (2) segment the market, (3) determine the segment drivers and model its changes, and (4) conduct a sensitivity analysis. These steps are explained next.

8.3.1 Define the Market

On the basis of customer interviews, the market should be defined broadly to include the principal service to be marketed, its "bundle of benefits" to customers, and service substitution.

8.3.2 Segment the Market

In segmentation, the potential customers for the principal service are divided into homogeneous subgroups (segments) whose members have similar service preferences and buying behavior. (Market segmentation is elucidated in detail in section 8.4.)

8.3.3 Determine the Segment Drivers and Model its Changes

Segment drivers may be composed of macroeconomic factors, such as the increase in white-collar workers and in population, as well as industry-specific factors, including the industrial growth rate and business climate. Possible sources of information related to segment drivers are industrial associations, governments, industrial experts, marketing data and service providers, and specialized market studies.

8.3.4 Conduct a Sensitivity Analysis

Sensitivity analyses are conducted to test assumptions. Monte Carlo simulations may be performed to generate the maximum—most likely—and minimum total market demand values, as well as an assessment of the risks involved. (See section 6.6.)

The following are illustrative examples in which the total market demand for a service is estimated by (1) defining the industrial segments that purchased the service in the past, (2) determining the future growth rates of these industrial segments, and (3) calculating the total market demand for the service with these industrial segment growth rates as the segment drivers. The assumption here is that the service demand of a given industrial segment is in direct proportion to its segment growth rate.

For example, to predict the demand of electricity in future years, a utility company has subdivided its consumers into three segments: industrial, commercial, and residential. The need for electricity by the industrial segment depends on its future production level and business climate. The electricity demand by the commercial segment is related to retail sales that in turn are negatively affected by retail stores consolidating and by growing Internet sales. (Web-based sales increased by 28.5 percent to $14.8 billion in 2002.) The residential electricity demand is affected positively by new home sales and home sizes and negatively by the increased energy efficiency of home appliances.

A second example is a paper-producing company that has determined the total market demand for uncoated white paper by deconstructing the market into end-use segments such as business forms, commercial printing, reprographics, envelopes, stationery, tablets, and books. The drivers in each segment are then modeled in terms of macroeconomic and industrial factors, using regression analysis and statistics. Examples of applicable drivers include growth in the use of electronic technology, white-collar workers, the present level of economic activity, the growing use of personal printers, population growth, demand growth induced by price reduction, and the practice of paying bills online and not by checks stuffed into envelopes.

Market forecast is a difficult but critical first step to take when defining a marketing program. Companies routinely engage both internal and external resources to assess the principal characteristics of the target market for their services.

8.4 MARKET SEGMENTATION

Once it is determined that the target market is worth pursuing (i.e., the market is large and stable enough with a high growth rate), then it is useful to understand the potential customers so that they can be served well. Market segmentation is a process whereby companies recognize the differences between various customer groups and define the representative group behaviors. Members in each of these customer groups respond to product or service offerings in similar manners and have comparable preferences with respect to the price–quality ratio, reliability, and service requirements.

8.4.1 Purpose of Market Segmentation

By dividing consumers into groups that have similar preferences and behaviors with respect to the services being marketed, companies achieve four specific purposes: (1) matching services to appropriate customer groups, (2) creating suitable channels of distribution to reach the customer groups, (3) uncovering new customer groups that may not have been served sufficiently, and (4) focusing on niches that have been neglected by competitors.

Overall, segmentation allows companies to realize the following benefits: formulating applicable marketing strategies and objectives, formulating and implementing marketing programs that address the needs of the different customer groups, tracking changes in buying behavior over time, understanding the enterprise's competitive position in the marketplace, recognizing opportunities and threats, and utilizing marketing resources effectively.

8.4.2 Steps in Marketing Segmentation

Displayed in Fig. 8.8 is a segmentation flow diagram that illustrates the key steps in segmenting a market.

Companies need to classify consumers into segments by understanding their individual, institutional, and service-related characteristics. *Individual* characteristics include culture, demographics, location, socioeconomic factors, lifestyles, family life cycles, and personalities. *Institutional* characteristics include the type of business, its size, and extent of global reach. *Service-related* characteristics include type of use (original equipment manufacturer (OEM) versus end user), usage level, service knowledge, brand preference, and brand loyalty. Also to be studied are benefits sought by consumers such as psychological and emotional benefits, functional performance, and price. Instead of focusing on the traditional demographis traits (e.g., age, sex, education levels and income), companies need to pay attention to nondemographic traits (e.g., values, tastes and preferences) that are more likely to influence purchase decisions (Yankelovich and Meer 2006).

Millions of consumers purchase cars every year. To some, cars are a status symbol; to others, cars are simply a means of transportation. A large number of car buyers emphasize safety and reliability, while others focus on fuel economy. Socioeconomic factors, demographics, personalities, and family life are all known to influence the behavior of car buyers. These consumers are extensively segmented by all major carmakers.

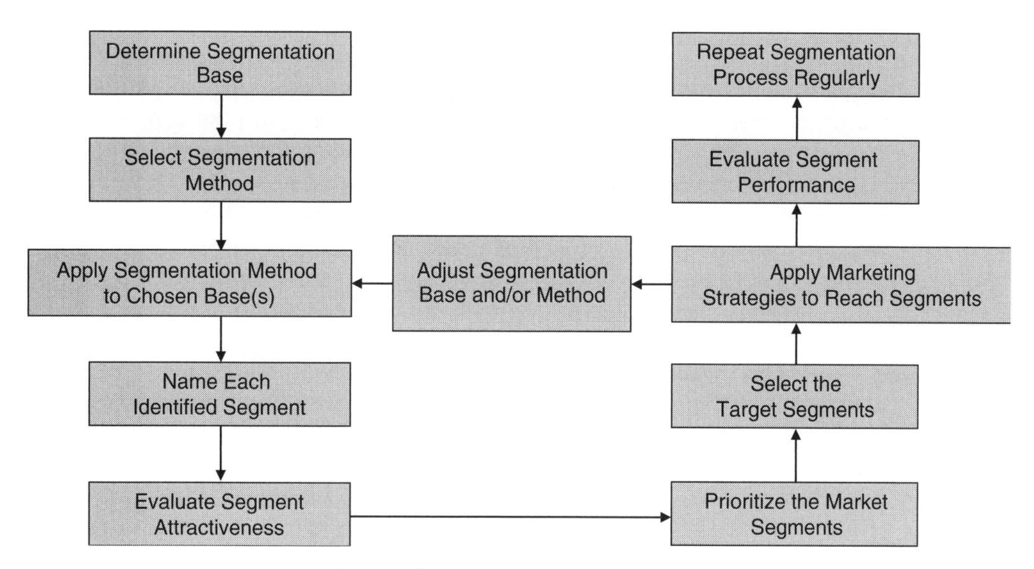

Figure 8.8 Segmentation flow diagram.

8.4.3 Criteria for Market Segmentation

To be effective, the segmentation of a market needs to satisfy several requirements.

The segmentation should be measurable. It should result in readily identifiable customer groups. The identified customer groups should also be homogeneous. Each group's members should possess more or less unified value perceptions and display compatible behavioral patterns. These customer groups are reachable by promotion and distribution means. Above all, the segments should be large enough in size to justify marketing efforts, and they should have a high-growth rate to allow the company to achieve long-term profitability.

8.4.4 Pitfalls of Market Segmentation

There are pitfalls to market segmentation. Certain *old economy* companies adopt the asset-rich business strategy of pursuing the scale of economy advantages. For these companies, a potential pitfall is oversegmentation, because the selected segments may be too small or fragmented to serve effectually. Such an oversegmentation is not a pitfall for other *knowledge economy* companies that form partnerships to establish supply chains for "build-to-order" services. To foster differentiation, knowledge economy companies pursue service customization as the basis for their business strategies (Grenci and Watts 2007). Examples of these products and services are minibrewers, computer systems, custom cosmetics, financial advisement, and architectural design services.

For other companies, overconcentration (lack of balance between segments) could have a negative impact on their overall marketing effectiveness (Hartley 2006).

Market segmentation is a prerequisite to developing a workable marketing program. Knowledge derived from customer groups serves as valuable inputs to service design, pricing, advertising, and customer services, all of which are yet to be finalized.

8.4.5 Rediscovery of Market Segmentation

Segmentation is designed to identify groups of customers who are particularly receptive to a particular brand or service, as well as sufficiently numerous and lucrative to justify pursuit. Customer's traits such as values, tastes, and preferences are more likely to influence their purchase decisions than age, sex, educational background, and income level (Yankelovich and Meer 2006). Segmentation must be done regularly, as customers' buying patterns (based on needs, attitudes, and behavior) may change over

Table 8.1 Modified "Gravity of Decision Spectrum" for Services

	Decision	Issues to address	Consumers' concerns	To find out by Segmentation	Examples of Services
1	Shallowest level	Whether to make small improvements to existing services How to select targets of a media campaign Whether to change prices	How relevant and believable the new service claims are How to evaluate a given service Whether to switch services	Buying and usage behavior Willingness to pay a small premium for higher quality Degree of brand loyalty	Information search (Google), car repairs, car rentals, video rentals, personal care (hair cuts), entertainments, food services (Starbucks), package delivery,
2	Middle level	How to position the brand Which segments to pursue Whether to change the service fundamentally Whether to develop an entirely new service	Whether to visit a clinic about a medical condition Whether to switch one's brand of service Whether to sign on an online degree program	Whether the consumers being studied are do-it-yourself or do-it-for me types Consumer's need (better service, convenience, functionality) their social status, self-image and lifestyle	Online degree programs, travel services, eBay purchases, physician services
3	Deepest level	Whether to revise the business model in response to powerful social and economical forces changing how people live their lives and how corporate business strategies are revised	Choosing a course of medical treatment Deciding where to live	Core value and beliefs related to the buying decision	Consulting (business, outsourcing, financial advisement, legal), relocation, major medical services

Adopted and modified from Yankelovich and Meer (2006)

time and can be reshaped by market conditions (economics, new consumer niches, and new technologies). Effective segmentation focuses on one or two issues and must be reconsidered as soon as they have lost their relevance. Yankelovich and Meer (2006) suggested a "Gravity of Decision Spectrum" to focus on the relationship of consumers to a product or service category:

1. *The shallow end.* Consumers seek products and services they think will save them time, effort and money. Segmentation should focus on price sensitivity, habits, and impulsiveness of target consumers.
2. *In the middle.* Customers buy big-ticket items (cars and electronics). Segmentation focuses on concern about quality, design, complexity, and the status the service might confer.
3. *The deepest end.* Customers' emotional investment is great and their core values are engaged. These core values are often in conflict with market values. Segmentation needs to expose these tensions. Health care is an example.

We have modified the original "What Is in Stake Diagram" by Yankelovich and Meer (2006) for services as presented in Table 8.1.

Example 8.2 A company has divided the market for its existing services into three segments: (1) mass-market applications, (2) applications requiring a quality service, even though consumers continue buying on price, and (3) critical applications to which both quality and reliability are important. Advise the company on the marketing mix that should be applied to these three segments.

ANSWER 8.2 For the company to be successful, a different set of marketing strategies needs to be applied to each of these segments, as suggested in Table 8.2.

Table 8.2 Marketing Strategies for Specific Segments

	Segment 1	Segment 2	Segment 3
Price	Low	Low	High
Product/service	Standard	Quality	Quality and reliability
Promotion	Broad	Limited	Focused
Place	Multiple distributions	Multiple distributions	Direct
Physical evidence	Not important	Low emphasis	High emphasis
Process	Highly efficient	Efficient	Standard
People	Standard	Standard	Dedicated

Multiple distributions are recommended, including mass-merchandise department stores for wide distribution. Direct distribution should include catalogs, specialty stores, and upscale department stores.

8.5 MARKETING MIX (SEVEN PS)

8.5.1 Product (Service) Strategy

The product and service strategy takes the center stage in any marketing program (Edvardsson, Gustafsson, Kristensson, and Magnusson 2006). If marketed properly, products and services that offer unique and valuable functional features to consumers are expected to enjoy a strong marketplace acceptance.

Products (services) may be generally classified as either industrial or consumer oriented. Their characteristics are different, as displayed in Table 8.3. Marketing programs for consumer products/services are quite different from those for industrial products/services, even though the same basic marketing approach applies to both (Miller and Palmer 2000; Kirk 2003).

A good marketing program must take into account the consumer's perception of services. Indeed, consumers perceive services in different ways than the producers and marketers do. When buying services, consumers look for "bundles of benefits" that satisfy their immediate wants. Services that producers regard to be different because of physical embodiment (e.g., input materials), production process, or functional characteristics may, in fact, be equivalent in how consumers perceive them, provided that the same or similar benefits are advantageous (substitute services) to the customers. Table 8.4 contains illustrative examples of these different perceptions.

Table 8.3 Industrial Versus Consumer Products/Services

	Industrial Products/Services	Consumer Products/Services
1. Number of buyers	Few	Many
2. Target end users	Employers	Individual
3. Nature of products/ services	Tailor-made, technical	Commodity, non technical
4. Buyer sophistication	High	Low
5. Buying factors	Technical, quality, price, delivery, service	Price, convenience, packaging, brand
6. Consumption	OEM parts for reselling, own consumption	Direct consumption
7. Producer end-user contact	Low	High
8. Time lag between demand and supply	Large	Small
9. Segmentation techniques	SIC (standard industrial classification), size, geography, end user, decision level	Demographics, lifestyle, geography, ethnic, religious, neighborhood, behavior
10. Classification of goods	Raw materials, fabricated parts, capital goods, accessory equipment, MRO supplies	Convenience (household supplies, foods), shopping (cameras, refrigerators), specialty (foods, brand-name clothing)

MRO is the abbreviation for maintenance, repair, and operations

Table 8.4 Service Perceptions

Services	Vendors' Perceptions	Consumers' Perceptions
Major surgeries	Sequences of diagnostic tests Surgical procedure Medications, emergency steps	Hope of recovering while enduring pains and suffering
Financial Advisement	Models, diversification strategy Economic projections Projected risks in global economy	Chances of preserving capital and making money

Companies must define competition based on the way customers perceive their services. Note that services that appear to be physically different to marketers may appear to be the same to users.

Product (service) strategy must also be established with respect to competition. A company may decide to market premium services, characterized by having features that are outstanding or superior to those offered by the competition. Such outstanding service features may be possible because of the company's innovative capabilities, technological superiority, and other core competencies. Companies with such "high-road" brands (see Fig. 8.14) tend to enjoy and sustain high profitability. Other companies may elect to make commodity-type (value) services with commonly available features so that they compete head-on against their competitors on the basis of price, service, distribution, and customer relations management. They pursue the option of "low-road" services. Service positioning is the step that addresses such competitive issues related to services.

Product (Service) Composition. When customers buy and use a product (service), they realize specific benefits from the core product (service) as well as a number of supplemental services that support the core. Lovelock and Wirtz (2006) introduce the *Flower of Service* model to illustrate the composition of such a service, illustrated in Fig. 8.9. Although this model was created for hotels, it applies to numerous other services as well, even though not all supplemental services listed are of equal importance to others. In the clockwise direction, the first four supplemental services are normally expected from any core, whereas the last four may vary, depending on the specific core services at hand.

1. *Information*. Information must be timely and accurate in specifying service hours, locations, price, usage instructions, conditions of sales, warnings, directions to service site, receipts, tickets, documentation, tracking of package delivery by FedEx (front-line employees, technological enablers—Web site, video kiosks, brochures).
2. *Order-taking*. Make the process easy and timely for customers to receive confirmation, assure the rights to be accepted, and get receipts.
3. *Billing*. Must be timely, accurate, itemized, and clear in order to promote fast payment. Set up a system to encourage self-billing and promote fast checkout.

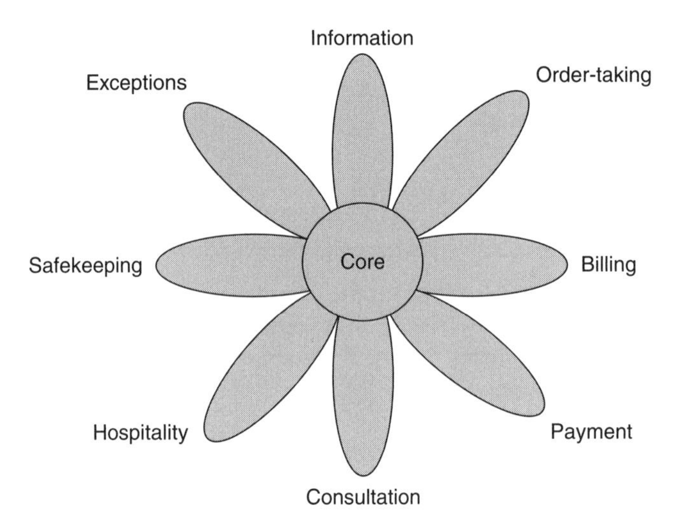

Figure 8.9 Flower of Service model. (Adapted from Lovelock & Wirtz 2006)

4. *Payment*. Make it easy and convenient for customers, such as using allowing use of credit cards.
5. *Consultation.* Must be customized to the situation at hand for dispensing personal advice, counseling, tutoring/training or business consulting, and free-of-charge, as way to promote sales.
6. *Hospitality.* Must be friendly in interacting with customers. Airlines offer departure and arrival lounges. Physical appearance of such facilities is important
7. *Safekeeping.* This refers to safes in hotel rooms, parking spaces for cars on site, coatrooms, child care and pet care facilities, and a concern for customer safety.
8. *Exceptions.* These include special requests, problem solving, complaints handling, and restitution. Oftentimes, supplemental services are used to define differentiable grades of service, such as premium, standard, or economy. For the service to be successful in the marketplace, the attitudes and approach must be fresh and the strategies well formed.

Product (Service) Positioning. An important question that companies should answer is, which service attributes should be included? A *perceptual map* is a useful tool to position the company's services in relation to existing competitive services in the marketplace. It enables companies to select the correct set of service attributes to maximize its marketing advantages. It also articulates customer preferences and identifies gaps; these are useful steps in positioning new entries or repositioning existing services.

Figure 8.10 is an example of a perceptual map for four-year B. S. degree programs in engineering considering the service attributes of value and cost. Value is defined as knowledge and skills gained, plus the intangible benefits (e.g., social networking, alumni organization activities, school reputations) as related to the institutions involved. Costs include tuition (out of state), living expense, and others required to complete the four-year programs. Only the relative magnitudes of the attributes are emphasized in

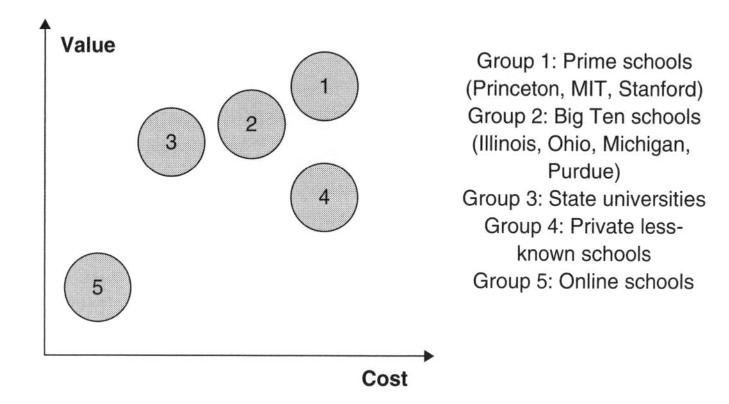

Figure 8.10 Perceptual map.

such a map. However, the map helps to identify which degree programs are in direct competition and which ones are not. It is also possible to link customer segments to these pairs of service attributes, thus enabling companies to refine their advertising strategies for these customer segments.

Services with more than two important attributes are readily mapped into an n-dimensional perceptual map. A service (e.g., S1) is designated by a single point having the coordinates F_1, F_2, F_3 through F_n, with each representing an independent service attribute. This representation is complete if the elements of the attributes set $(F_1, F_2, \ldots, F_n)$ are mutually exclusive and collectively exhaustive. For example, for automobiles, these attributes include price, styling, fuel economy, driving comfort, safety, brand prestige, power, and longer-term dependability (number of problems per 100 three-year-old vehicles). The spacing between two neighboring points (each identifying a service) as depicted in this n-dimensional map is equal to the square root of the sum of the individual attribute differences, squared. Table 8.5 is an example of the description of services with six distinguishable attributes.

Product (Service) Life Cycles. Every service goes through four important stages throughout its useful life (Saaksvuori and Immonen 2008; Moon 2005):

1. The *initiation stage*—service testing, market development, and advertising.
2. The *growth stage*—service promotion, market acceptance, and profit growth.

Table 8.5 Inputs to Six-Dimensional Perceptual Map

Service	F_1	F_2	F_3	F_4	F_5	F_6
S1						
S2						
S3						
S4						
Your Service						
S5						
S6						
S7						

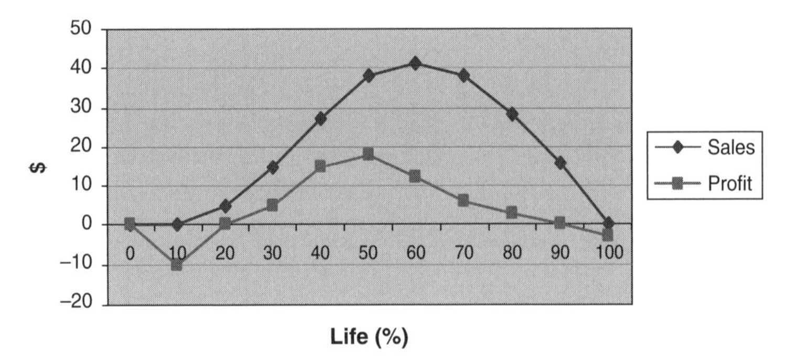

Figure 8.11 Service life cycle.

3. The *stagnation stage*—price competition, substitution, and new technologies.
4. The *decline stage*—cash-cow strategy with no more investment.

Companies need to understand which phase a given service is in. (See the example of service life cycle in Fig. 8.11.) From the standpoint of the service life cycle, an important service development strategy is to sequence the introduction of new services so that a high average level of profitability can be maintained for the company over time.

Engineering managers are particularly qualified to constantly come up with innovative services so that their employers may introduce these services at the right intervals.

Product (Service) Portfolio. Another service strategy issue is related to the types of services concurrently being marketed. With the exception of a few, most companies market a group of services at the same time, referred to as a *service portfolio* (Neufeld 2005).

Services in a portfolio are usually not "created equal." From the company's standpoints of profitability and market-share position, some are more valuable than others. Boston Consulting Group (BCG) of Boston, Massachusetts, developed a portfolio matrix based on the two measures of growth rate and market share. (See Fig. 8.12.) According to this classification scheme, services are regarded as *stars* if they enjoy high growth rate and high market share and *question marks* if they have high growth rate, but low market share. *Cash cows* are those services with low growth rate and high market share. Services are designated as *dogs* if both growth rate and market share are low.

Figure 8.12 indicates that companies need to differentiate the services they market by strategically emphasizing some and deemphasizing others, according to the responses from the marketplace. For example, a useful strategy to manage a service portfolio is to milk the *cows* to provide capital for building *question marks* into *stars* that will eventually become *cash cows*. Divest the *dogs*.

Services and Brands. Numerous high-tech companies operate in a "service-centric" business model, in that they market services on price and performance. Recent market studies suggest that the success of technology-based services in the marketplace is not purely dependent on the price–performance ratio, but also on the trust, reliability, and promised values the customers perceive in a given brand.

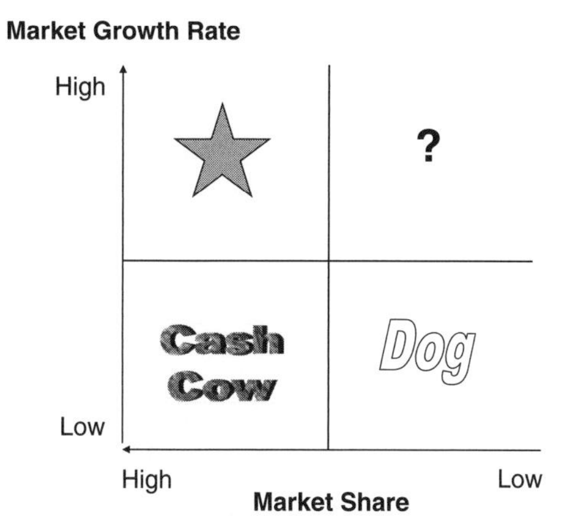

Figure 8.12 Product (service) portfolio.

According to Dev (2008), brand is "a distinct identity that differentiates a relevant, enduring and credible promise of value associated with a product, service, or organization that indicates the source of that promise."

The brand of a company is more than a name. It stands for all of the images and experience (e.g., product, service, interactions, and relations) that customers associate with the organization. It is a link forged between the company and the customers. It is a bridge for the company to strengthen relationships with customers, according to Keller (2000).

A promise of value is an expectation of the customer that the company is committed to deliver. Examples of such promises of value from several companies are listed in Table 8.6. They must be relevant to the enduring needs of the targeted customers and made credible by the persistent commitments of the company. To be competitive, the promise of value must be distinguishable from those offered by other brands.

Research by Ward, Light, and Goldstine (1999) indicates that customers consider questions at five levels when purchasing both high-tech and consumer types of services. These questions may be grouped into a brand pyramid, as illustrated in Fig. 8.13.

Technology-oriented buyers are typically focused on questions at levels 1 and 2. However, higher-level business managers who make purchase decisions are also known to address questions at levels 3 to 5. These decision makers are interested in what the product or service will do for them, not just how it works. Consequently, to project a trustworthy and reliable image, to build strong relationships with customers, and to

Table 8.6 Promises of Value

Corporate Brands	Promises of Value
IBM	Superior service and support
Apple	Simple and easy to use
Lucent	Newest technologies
Gateway	Friendly service

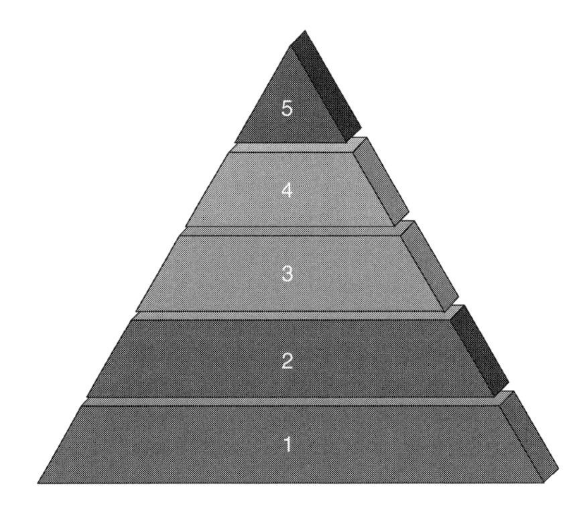

- Level-5: What is the personality of the brand (aggressiveness, conservatism, etc.)?
- Level-4: Does the value offered by the service reflect that favored by the customers (family values, achievement)?
- Level-3: What are the psychological or emotional benefits of using the service in question? How will the customers feel when they experience its technical benefits?
- Level-2: What are the technical benefits to customers (solutions to problems, cost saving benefits, and speed of production)?
- Level-1: How does the service work (service characteristics, technological feature, and functional performance)?

Figure 8.13 Brand pyramid.

invigorate brand loyalty and customer retention, companies are well advised to pay attention to questions at all five levels. This is the emphasis of brand management.

Brand is a major asset that must be properly managed and constantly strengthened. Useful inputs for brand management are typically secured from customer feedback. Once the market is properly segmented, the promise of values is specifically designed to address the needs of the target segments involved. Actions are then taken to deliver the stated promise of values, and results are constantly measured to monitor progress.

Brand is evaluated on how well it is doing with respect to a number of metrics: delivering according to the customer's desire, relevance to the customers, value to the customers, positioning, service portfolio management, and integration of marketing efforts, management, support, and monitoring.

In the past, brand management has been focused primarily on points of difference, such as how a given brand differs from the other competing brands within the same category. Maytag is known to emphasize "dependability." Tide® focuses on "whitening power." BMW stresses "superior handling." Keller, Sternthal, and Tybout (2002) suggested that attention should be paid to points of parity and the applicable frame of reference, in addition to the points of difference, when marketing a given brand. Emphasizing the frame of reference is intended to help customers recognize the brand category comprising all of the competing brands. Focusing on the points of parity will ensure that customers recognize a given brand as a member of the identified brand category.

Brand may be classified with respect to the two dimensions of category and relative market share. The brand category is defined as *premium* if the category is dominated by premium brands—those with high values to customers. Examples of premium brands include BMW, Mercedes Benz, Jaguar, and other luxury and specialty cars that each has special, high-value attributes. The brand category is defined as *value* if it is dominated by value brands—those with basic, minimum, low-end attributes. Examples of value brands include Chevy, Saturn, and other compact and four-door family cars. Gillette markets its Mach-3 Turbo shaving system as a premium brand, whereas the cheap disposable razors from its own company as well as its competition are the value brands. The relative market share refers to the percentage of market share a given brand is able to attain.

In Fig. 8.14, brands are grouped into four classes: high-road, low-road, hitchhiker, and dead-end brands (Dev 2008). ROS is return on sales, which is defined as net income divided by the sales revenue. (See Section 7.4.1.)

High-road brands are those with services that offer premium features, options, qualities, and functions to command high selling prices while attaining a leadership position in the market share. Examples of such high-road brands are Coca-Cola, Frito-Lay, and Nabisco. These brands enjoy excellent profitability that may be sustained for long periods. The key success factors for high-road brands are technological innovation (constantly adding new service features and values), time to market, flexible manufacturing, and advertising.

Low-road brands are those that offer value brands while enjoying a high market-share position. Because of marketplace competition and a lack of distinguishable service

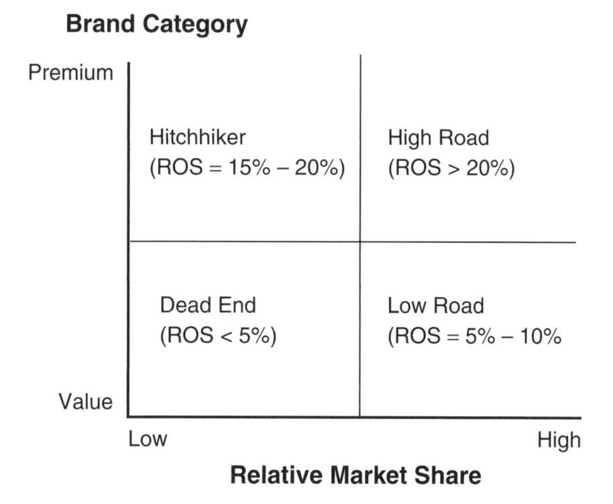

Figure 8.14 Brand classes. (Adapted from Berry & Settman, 2007)

features, these brands can be successfully managed by emphasizing cost reduction, production efficiency, service simplification, and distribution effectiveness.

Hitchhiker brands are those with premium service values and low market share. For these brands to become high-road brands, management must emphasize cost reduction, flexible manufacturing, and service innovation.

Dead-end brands are value brands with low market share. These brands attain only marginal profitability. There are several strategies to grow the profitability of these brands: (1) reduce the price to penetrate the market and thus move these brands to the low-road category, (2) increase the scale of economies by applying the "string-of-pearls" strategy: producing and marketing several services together to cut costs, and (3) introduce a superior, premium service to "trump" this brand into the hitchhiker category. Failing all of these attempts, dead-end brands should be discontinued. Table 8.7 summarizes the strategies that deal with these four classes of brands.

Table 8.7 Strategic Options for Brands*

Brands	Strategic Options
High road	Apply R&D to constantly innovate to make services premium—adding new service features and changing forms and functions Expand service lines (service proliferation) Initiate media campaign Invest in capital Decrease time to market Flexible manufacturing Direct store delivery to preoccupy shelf space
Low road	Pursue cost reduction aggressively Lessen service proliferation (SKUs) by simplifying service types and designs Consolidate production facilities to improve efficiency and cut wastes Use realized cost savings to slash price Consider ways to add premium services (advancing to high road)
Hitchhiker	Apply R&D to constantly innovate to make services premium—adding new service features and changing forms and functions Reduce costs Reduce time to market Use flexible manufacturing Initiate media campaign Consider capital investment
Dead end	Cut price (advancing to low road) Outsource in areas with economies of scale Apply the "string-of-pearls" strategy to enhance scale "Trump" the category by introducing a superior, premium service that resets consumer's expectation (advancing to hitchhiker) Do not spend on marketing Make no capital investment

**Source*: Vijay Vishwanath and Jonathan Mark. "Your Brand's Best Strategy." *Harvard Business Review* (May–June 1997).

The preceding discussion on service brands should assist engineering managers in understanding the significant value added by brands to the success of the company's marketing program (Bedbury with Fenichell 2003; Wheeler 2009).

Such an understanding should make it easier for them to channel their support efforts to actively strengthen the company's brand strategy.

Berry and Seltman (2007) presented a model to build brand equity. This model is depicted in Fig. 8.15 "Organization's presented brand" is the desired brand image advocated by the service company. The brand image is promoted through advertising, brand name, logo, Web sites, employee uniforms and facilities design. This strategy creates "brand awareness" among target customers. Brand awareness is the customer's ability to recognize and recall a brand. "Brand meaning" is the customer's dominant perception of the brand. Perception is built on the personal experience customers have in dealing with the service organization. Customer experience will also influence what other people say and write (word of mouth, blogs, message boards, etc.) about the brand—these *external brand communications* have a secondary influence on both brand awareness and brand meaning, as depicted in Fig. 8.15. Together, brand awareness and brand meaning form the brand equity, which is the degree of marketing advantage or disadvantage the organization has over its competition in the marketplace.

Based on this model, building a strong brand equity requires (1) dedicated effort in brand presentation (advertising, logo, Web sites, messages, physical appearance of facilities), (2) commitment to effect a satisfactory customer experience (employee training, employee compensation, employee attitudes, efficient support processes, pleasant working environment, etc.), and (3) a customer feedback system to allow efficient mitigation of any negative external communications (unfavorable publicity, ethics, corporate citizenship). In services, the customer's experience is typically generated by interacting with service people. The brand name is associated with the entire service organization, including all service people and products.

The Mayo Clinic is a world-renowned hospital headquartered in Rochester, Minnesota. Its brand name is built on three specific strategies: (1) unique skills and clinical outcome, (2) a team medicine model to solve patient's problems collaboratively, and (3) actual experience that exceeds expectations. The Mayo Clinic, by emphasizing the value of services it offers and by creating a system capable of meeting and surpassing

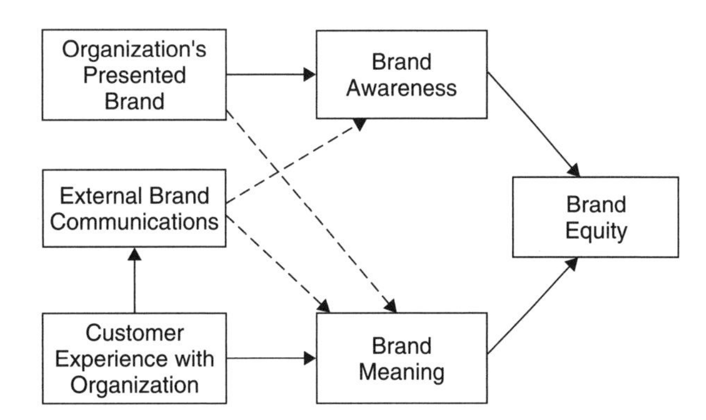

Figure 8.15 Model of building brand equity. (Adapted from Berry & Settman, 2007)

expectations, was able to turn its patients into marketers for the clinic. Because the Mayo Clinic treats its patients well; patients will talk to forty others about their good experience at the clinic. The marketing promotion effected by their own patients has been possible only because the outcomes the Mayo Clinic projected exceeded patient's expectations. Just meeting expectation is not enough to generate a word of mouth campaign (Berry and Seltman, 2007).

Brand equity is the sum of a customer's assessment of the brand's intangible qualities, positive or negative. According to Rust et al. (2004), emphasizing only the building of brand equity misses valuable opportunities, often to the detriment of the company. Instead of focusing on brand equity, companies should focus on customer equity.

The logic is simple. Customers may change their perception about a specific brand. However, if these customers can be encouraged to attach their loyalty to another brand within the company, their relationship with the company will continue. Customer lifetime value is the net profit derived from a customer during the time when the customer has a relationship with the company.

Brand perception can be influenced by geography. For example, South Americans may have a vastly different perception about an American brand than people in the United States.

The seven strategies to pursue customer-centric branding strategies, as recommended by Rust et al. (2004) are as follows:

1. Focus on customer relationships ahead of branding. Manage key accounts of profitable and important customers and apply appropriate branding strategies to these customer groups or segments.
2. Build brand around customer segments, not the other way around.
3. Make the brand as narrow as possible. In a customer-centered approach, the brand should be able to satisfy as small a customer segment and be economically feasible.
4. If the customers are similar, then different brands may be advertised to them. An example is Disney, which offers movies, hotels, and amusement parks all to the same customers (the young and young at heart, all who want to be entertained).
5. Be ready to hand off customers to other brands in the same company. For example, encourage customers of Fairfield Inn to trade up to the Marriot brand (Marriot owns Fairfield inn brand).
6. Do not take heroic measures. If a brand is no longer viable, do not spend time and resources attempting to save it. Let it go.
7. Change how brand equity is measured.

The long-term goal is to create and cultivate profitable, long-term relationships with customers. Brand management is only one tool of many that can be used toward accomplishing this goal.

Example 8.3 Forecasting future market conditions and technologies is a difficult, but necessary, skill for companies striving to sustain business success. Looking out for emerging technologies should be the primary role of engineering and technology managers.

What might be a good strategy for engineering and technology managers to become sensitized to forecasting technology and scanning emerging technologies so that they fulfill their important role of serving as technology "gatekeepers"?

ANSWER 8.3 Different engineering and technology managers will have different preferences in fulfilling this important role of forecasting. One possibility is to adopt the following logical sequence of steps:

1. Compose a "wish list" of technologies that would make the company's current products cheaper, faster, and better. Define desirable new product features based on customer inputs and the technologies required for their development. Define new product concepts and the requisite technologies that might be compatible with the current product lines marketed by the company.
2. Understand some of the emerging technologies noted in the literature.
3. Determine the useful technologies that might be available during the next five to ten years to support the current products, product enrichments, or new product concepts.
4. Assess the development activities associated with these useful technologies in universities, start-ups, technology incubator firms, contract research companies, or other organizations, both domestic and global, to gauge their quality and readiness for commercialization.
5. Make specific recommendations in a timely manner to secure the supply of such new technologies for enhancing the commercial success of the company's products.

Example 8.4 There is a strategic method called *second brand strategy*. Explain what is unique about this strategy. Under what conditions would this strategy be best applied, and what is the purpose it is intended to achieve?

ANSWER 8.4 The *second brand strategy* refers to the creation of a new brand, which is to be distinguished from the primary brands of the company, in order to market similar products at much lower prices. Companies pursue this strategy to avoid or minimize the cannibalization effect on the primary brands and to counter the aggressive selling efforts of new competitors who enter the marketplace with low-price products.

This strategy would work well for companies that have financial staying power, are market leaders interested in protecting their market share positions, and want to pursue this strategy as a short-term solution to counter the market penetration efforts of new competitors and to drive away new competition, while preserving the loyalty of their current customers. As soon as the competitors become disenchanted and disappear from the scene, the company would stop the second brand strategy immediately.

Engineering Contributions to Product/Service and Brand Strategy. The product/service is a key element in the marketing mix. Engineers and engineering managers have major opportunities to add value by (1) understanding the customers' perceptions of services, (2) designing services with features that are wanted by customers,

(3) helping to position the company's services strategically to derive marketing advantage, (4) practicing innovations in the design, development, production, reliability, serviceability, and maintenance of services to differentiate them from others, (5) sustaining the company's long-term profitability by creating a constant flow of new services for introduction on a timely basis, (6) assisting in managing companies' service portfolios by adding premium features to some and reducing costs to others, and (7) ensuring commercial success of the high-road and hitchhiker brands in the marketplace.

In the knowledge economy of the twenty-first century, time to market is an increasingly important competitive factor. Once the desirable set of service attributes is known from market research, those companies that bring the suitable services to the market first will enjoy preemptive selling advantages and will recover the service development costs faster than others.

Engineering managers should also be well prepared to contribute in shortening the services' time to market by utilizing advanced technologies to create modular design, eliminate prototyping, whittle away design changes, foster parts interchangeability, ensure quality control, and produce other innovations.

8.5.2 Pricing Strategy

Price is a very important service attribute (Dolan and Gourville 2005). Companies pay a great deal of attention to the setting of service prices. Setting the price too high will discourage consumers from buying, whereas setting it too low will not assure profitability for the company. Generally speaking, the two major strategies for setting the service prices are the skimming strategy and the penetration strategy.

Skimming and Penetration Strategies. Companies applying the *skimming strategy* set the service price at the premium levels initially and then cut the service price in time to reach additional customers. In other words, they "skim the cream" first. An example is the marketing of a new book with hardcover copies selling at a high price (e.g., $29.95), followed by the paperback version at a low price ($4.95). New technology services are also typically sold at high prices initially in the absence of competition. As competitors enter the market with services of similar features, service prices are lowered accordingly.

In contrast, companies pursuing the *penetration strategy* set service prices low to penetrate a new market and rapidly acquire a large market share. A high market-share position sets forth a barrier of entry for other potential competitors. In general, companies use a penetration pricing strategy to enter an existing, but highly competitive market. An example is the marketing of Japanese motorcycles in the United States.

Factors Affecting Price. In setting service prices, besides using the skimming and penetration strategies, companies broadly consider a number of other factors: financial aspects, service characteristics, marketplace characteristics, distribution and production capabilities, price–quality relationship, and the relative position of power. These factors are described in more detail as follows:

Financial aspects—The more solid the company's financial position, the more capable it is at initially setting the service price low. Companies strong in finance stay afloat for a long period of time even with low profitability. Companies that desire high, short-term profitability tend to set the service prices high.

Service characteristics—The service price may be set in direct proportion to the value and importance of the service to users, as well as the income levels of its target customers. Usually, a new service in its early life cycle sells at a high price; the company benefits from the service's novelty.

Marketplace characteristics—Companies set service prices in reverse proportion to the level of competition in the marketplace. The level of competition refers to the number of direct competitors, the number of indirect competitors marketing substitution services that offer similar value to customers, and the competitive counterstrategies (speed and intensity) that these competitors may exercise. Companies tend to set the service price high if the barriers to market entry are high. The barriers to market entry depend on lead time and resources—technical and financial, patents, cost structure, and production experience. In addition, services with inelastic price-demand characteristics tend to carry a high price. A service has inelastic price-demand characteristics if a large price increase induces a small change in the quantity of the service demanded in the marketplace.

Distribution and production capabilities—Service availability to consumers depends on the company's service distribution capabilities. With strong distribution channels in place, companies may set the service price high, as quickly making services available to consumers represents a competitive strength.

Sales volume impacts the company's production experience. Companies with extensive production experience are known to produce services at a low unit cost. A lower service unit cost enables these companies to set a lower service price to gain market share.

The Boston Consultant Group studied manufacturing operations and discovered that there is a correlation between production volume and service unit cost. For every doubling of the production volume, the unit cost is whittled down by about 15 percent—or the 85 percent experience curve. (See Fig. 8.16. Note that the horizontal axis in the figure is nonlinear.)

Companies with a faster time-to-market strategy are able to accumulate production experience more quickly, attain a lower service unit cost sooner, and sustain company profitability for longer periods of time.

Price–quality relationship—One important consideration in setting the service price is the perceived cost–quality relationship by customers. There is substantial evidence in business literature to indicate that customers tend to believe that "low-priced items cannot be good." Price is perceived to be an indicator of quality.

Service prices should not be set too low. There is a price threshold below which customers may raise questions regarding the quality, as indicated in Fig. 8.17. The demand curve quantity T illustrates a normal price-demand relationship in the absence of a price threshold, whereas the demand curve quantity A contains a price threshold at about $30 per unit, below which the demand for the services starts to drop off as the perception of poor quality related to low price sets in.

Relative position of power—Consumer services are typically marketed by a few major companies to a very large number of customers. For innumerable industrial services with high technological contents, the number of both producers and

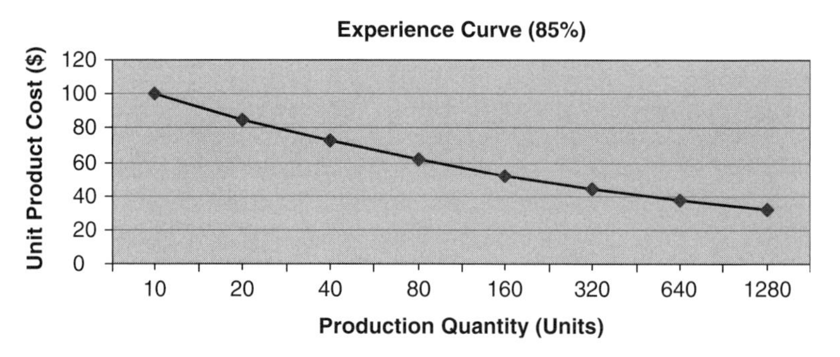

Figure 8.16 Experience curve.

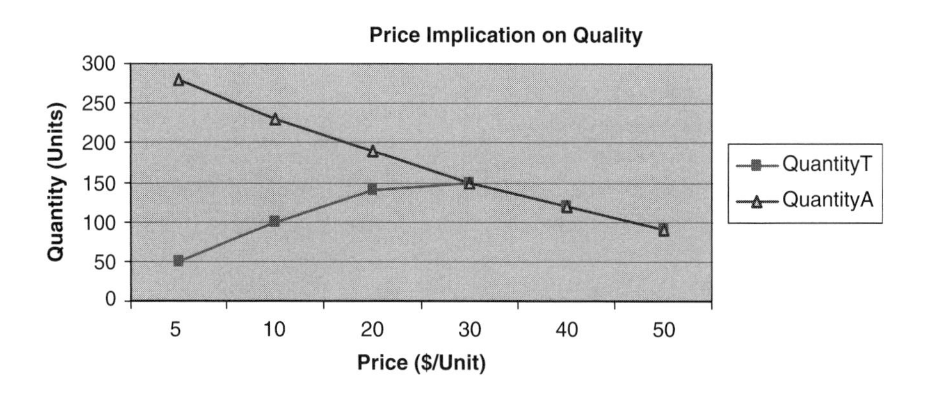

Figure 8.17 Price–quality relationships.

customers may be limited. The greater the number of sellers there are available for a given service, the weaker each seller's position in the marketplace will be. Similarly, the more buyers there are for a specific service, the weaker the buyers' relative market position will be.

Less competition makes either sellers or buyers more powerful. The relative position power between buyers (customers) and sellers (producers) has an impact on service pricing, as illustrated in Fig. 8.18. The final price offered by the sellers and accepted by the buyers is usually arrived at by a suitable negotiation or auction process.

If both buyers and sellers are strong—for example, when the U.S. government (customer) procures fighter airplanes from the defense industry (producer)—a final price is typically reached by a *negotiation* made up of a series of offers and counteroffers. A typical pricing arrangement may be cost plus a fixed percentage of gross margins.

When the sellers are strong (e.g., selling an original master painting, a porcelain vase from the Ming Dynasty, or some other type of unique physical asset) and the

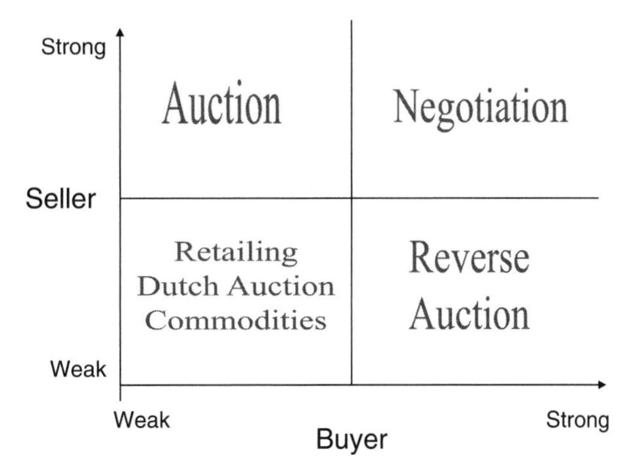

Figure 8.18 Processes of setting service price.

buyers are weak, sellers tend to take advantage of their dominant supplier position by employing an auction. An *auction* is a bidding process by which buyers are forced to compete against each other by committing themselves to consecutively higher prices, with the final price being set by the highest winning bid.

If buyers are in a strong position (e.g., due to large transaction volumes), they force weak sellers to compete against each other in a reverse auction. A *reverse auction* requires the prequalified sellers to submit increasingly lower bid prices within a fixed period of time. The lowest bid defines the final price and the ultimate winner of the sales contract (Smock 2003). Some large companies employ such pricing tactics to purchase high-volume supply items such as computers, paper and pencils, tires, batteries, and maintenance, repair, and operations (MRO) goods.

Finally, when both sellers and buyers are weak, services are usually not differentiable, and the service prices are highly depressed and fixed. Examples include various commodity products sold in retail stores. Some sellers (e.g., Land's End) may activate a Dutch auction to compete. In a *Dutch auction*, sellers slash the product prices by certain percentage at a regular time interval (e.g., every week) until the products are sold or taken off the market. In this case, buyers compete against other "sight-unseen" buyers to seize the lowest possible selling prices.

The Internet has made many of these pricing processes much more practical and efficient to implement (Chaffey, Ellis-Chadwick, Johnston and Mayer 2009). Because of its ability to allow sellers and buyers to rapidly reach other buyers and sellers, respectively, the Internet tends to weaken the relative power positions of both the sellers and the buyers, causing products to become increasingly commoditized, thus depressing product prices and intensifying competition.

Table 8.8 enumerates a number of other factors that have an impact on setting the service price.

Pricing Methods. In setting service prices, companies broadly consider a number of factors. Several of these methods are briefly discussed next.

Table 8.8 Factors Affecting Service Price

Factor	Skimming	Penetration
Demand	Inelastic	Elastic
	Users know little about service	Familiar service
	Market segments with different price elasticity	Absence of high-price segment
Competition	Few competitors attracts competition	Keep out competition
	Market entry difficult	Market entry easy
Objective	Risk aversion	Risk taking
	Go for profits	Go for market share
Service	Establish high-volume image	Image less important
	Service needs to be tested	Few technical service problems
	Short service life cycle	Long service life cycle
Price	Easy to go down later	Tough to increase later
	More room to maneuver	Little room to maneuver
Distribution and promotion	No previous experience	Existing distribution system and promotion program
	Need gross margin to finance its development	
Financing	Low investment	High investment
	Faster profits	Slower profits
Production	Little economy of scale	High economy of scale
	Little knowledge of costs	Good knowledge of costs

Cost-oriented—Some companies set prices by adding a well-defined markup percentage to the service cost. This is to ensure that all services sold generate an equal amount of contribution margin to the company's profitability:

$$\text{Price} = \text{Cost} + \text{Markup (e.g., 35 percent of cost)} \tag{8.1}$$

Cost-plus contracts are often used for industrial services related to R&D, military procurements, unique machine tools, and other uses. Small sellers use cost-plus pricing to ensure a fair return while minimizing cost factor risks. Larger buyers favor this type of pricing so that they can push for vendor cost reduction via experience. Larger buyers may optionally offer to help absorb the cost risks related to inflation.

On the one hand, often, sellers and buyers enter a target-incentive contract, which prescribes that, if actual costs are lower than the target costs, sellers and buyers split the savings at a specific ratio. On the other hand, if the target costs are exceeded, then both parties pay a fixed percentage of the excess; the buyers pay no more than a predetermined ceiling price.

Profit-oriented—Other companies prefer to require that all services contribute a fixed amount of profit. This pricing method ensures that sellers realize a predetermined return on investment goal:

$$\text{Price} = \text{Cost} + \text{Profits (e.g., in term of ROI)} \tag{8.2}$$

Market-oriented—Some companies set prices of certain industrial services, such as those requiring customization, to what the buyers are willing to pay. For example, the companies strive to negotiate for the highest price possible and take advantage of the fact that service and pricing information may not be easily accessible. The continued advancement of Web-based communication tools tends to make information just one click away, rendering this type of pricing method no longer practical in today's marketplace.

Companies may also price the services slightly below *the next-best alternative* services available to the customer. The companies that have exhaustively studied the next-best alternative services available to their customers garner advantages in price negotiations.

Competitive bidding is used often by governments and large buyers. Usually three bids are needed for procurements exceeding a specific dollar value. Sealed bids are opened at a predetermined date, and the lowest bidder is typically the winner. Some industrial companies may engage in *negotiated bidding*, wherein they continue negotiating with the lowest one or two bidders for additional price concessions after the bidding process (e.g., a reverse auction) has been completed.

Valued-added pricing—Companies with extensive application know-how related to their industrial services may set service prices in proportion to the services' expected value to the customer. The service's value to the customer depends on the realizable improvement in quality, productivity enhancement, cost reduction, profitability increase, and other such benefits attributable to the use of the service. Producers set the service prices high if there is a large value added by these services to the customer.

Competition-oriented—A common pricing method is to set prices at the same level as those of the competition. Doing so induces a head-on competition in the marketplace. In oligopolistic markets (typically dominated by one or two major producers or sellers and participated in by several other smaller followers), the market leader sets the price.

One well-known example of a competition-oriented pricing practice is target pricing. *Target pricing* was initiated and applied by a number of Japanese companies. Some American companies have now started to successfully apply this method. Target pricing is briefly addressed as one of several external benchmarking strategies in section 5.3. The process of target pricing (see Fig. 8.19) is as follows:

1. Determine the market prices of services that are similar or equivalent to the new service planned by the company. Find all service attributes customers may desire. This is usually accomplished by a multifunctional team composed of representatives of such disciplines as design, engineering, production, service, reliability, and marketing. Select a service price (e.g., at 80 percent of the market price) that makes the company's new service competitive in the marketplace. This is then the target service price.
2. Define a gross margin that the company must have in order to remain in business.
3. Calculate the maximum cost of goods sold (CGS) by subtracting the gross margin from the target service price. This is the target service cost, which must not be exceeded.
4. Conduct a detailed cost analysis to determine the costs of all materials, parts, subassemblies, engineering, and other activities required to make the new service. Usually, the sum of these individual costs will exceed the target service

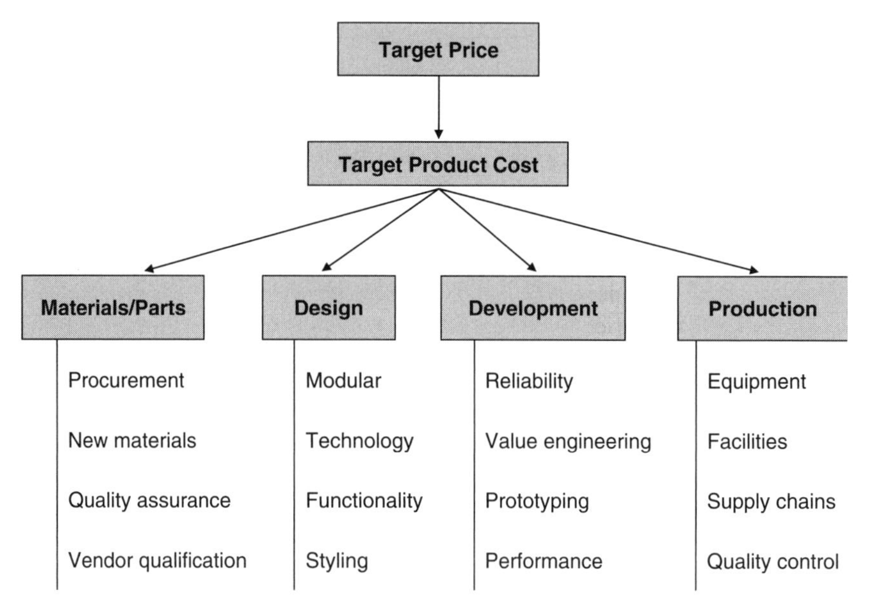

Figure 8.19 Target pricing.

cost previously defined. Apply innovations in service design, manufacturing, procurement, outsourcing, and other cost reduction techniques to bring the CGS down to or below the target service cost level.

5. Initiate the development process for the new service only if the target service cost goal can be met.

The target pricing method ensures that the company's new service can be sold in the marketplace at the predetermined competitive price, with features desired by consumers, to generate a well-defined profitability for the company. This method systematically evaluates low-risk, high-return investment opportunities because it forces the company to invest only when the commercial success of the service is more or less assured. Furthermore, it also focuses the company's service innovations on finding ways to meet specific and well-defined target service cost goals. It avoids the potential of wasting its precious intellectual talents in chasing ideas with no practical value.

Numerous companies use the pricing methods just discussed. Service prices are usually set by the marketing department in consultation with business managers. Engineering managers are advised to become aware of these methods, but to defer pricing decisions and related discussions to the marketing department.

Pricing and Psychology of Consumption. Recent studies indicate that buyers are more likely to consume a service when they are aware of its cost. The more they consume, the more they will buy again and thus become repeat customers. One useful way to induce them to repeatedly consume the services is to remind them of the costs committed through the choice of payment methods. This is based on the assumption that the more often the customers are reminded of the payments, the more they feel guilty if they do not fully utilize the services they have paid for.

Gourville and Soman (2002) point out that time payment better induces regular consumption of a service than lump-sum prepayment (e.g., prepaid season tickets) at the same total value. This is because the time payments remind the buyers of the costs periodically and thus invoke the *sunk-cost effect* on a regular basis. The psychology of the sunk-cost effect is that consumers feel compelled to use services they have paid for to avoid the embarrassing feeling that they have wasted their money.

Credit card payments are less effective in inducing consumption than cash payments because cash payments require the buyers to take out currency notes and count them one by one; thus, they experience the pain of making payments.

In price-bundling situations, the more clearly the individual prices of services are itemized, the better the perceived sunk-cost effect will be. Breaking down large payments into a number of smaller ones, thus clearly highlighting the costs of individual services sold in the bundle, can magnify this effect.

Studies of membership rates at commercial wellness and fitness centers support this logic. It has been well documented that those members who pay the membership fees once a year use the facilities only occasionally. These members are the least likely to renew, in comparison with those who pay on a monthly basis. Similar observations are made in sports events in which holders of season tickets show up less frequently than those who buy tickets for specific sets of events.

These examples point out that companies can induce customers to become repeat customers by focusing on ways to encourage consumption. Only consumption lets customers experience the benefits of the services they have purchased. Without such favorable experience, they may not feel that they have good reasons to buy the services again in the future. Hence, besides providing a good bundle of value made up of price, service features, convenient and efficient order processing and delivery, and quick-response after-sales services, companies should devise ways to stimulate consumption as a strategy to cultivate and retain repeat customers.

Example 8.5 The company has been selling a number of services to customers. It is about to launch a new service with features far superior to any services currently on the market. One option is to price this new service at a large premium above the current price range so that the heavy development expenses can be readily recouped. The other option is to set the price comparable with the existing range in order to retain customer loyalty. What is your pricing advice to the company?

ANSWER 8.5 Hold a focus group to find out the potential response of customers to the new service's features. Are these features of real value to them? How much more are they willing to pay for these features? Exciting new features from the manufacturer's viewpoints may not be as exciting to customers. Should customers appreciate the new features, then it is advisable to apply the skimming strategy and set a high price for the new service. This is also the principle of value pricing. Furthermore, doing so will avoid cannibalizing the existing services of the company.

An efficacious promotional campaign is essential to heighten service awareness. Keep monitoring the response of the market. If the market response is poor, cut down the service price slowly to induce more demand.

8.5.3 Marketing Communications (Promotion)

Marketing communication is intended to make the target customers aware of the features and benefits of the company's services. Service promotion follows a well-planned process (see Fig. 8.20) for who says what to whom, in what way, through what channel, and with what resulting effect.

Communication Process. Companies select communicators who are publicly recognized and who have trustworthy images, as these speakers tend to induce public acceptance of their messages. Examples include Bob Dole for Viagra®, Tiger Woods for golf products, and Yao Ming for basketball-related items.

Messages may be in various forms, including slogans. A slogan is a brief phrase used to get the consumer's attention and acts as a mnemonic aid. Successful slogans typically represent a symbolization of service features in terms of the customer's wants and needs (such as information, persuasion, and education). Examples include "The Power of Dreams" (Honda), "Reach Out and Touch Someone" (AT&T), "Empowering Customers to Achieve More" (SAP), "Talents Knows No Boundaries, Nor should Your Education" (Kaplan University), "All Good News, All The Time," (GNN), "Impossible is Nothing" (Adidas), "Leave the Rest to Us" (Lunesta Medications), "So Easy a Caveman Could Do It" (GEICO), and "The Internet in Your Pocket" (Apple Inc.).

Channels of communication are specific means to foster market communications. In general, there are two channels of communications: the marketer controlled and the consumer controlled. The *marketer-controlled* channels include advertisements placed in trade journals, national television programs, and distribution of specific service brochures, promotion by technical salespeople, industrial exhibitions, and direct-mail marketing. The *consumer-controlled* channels include interpersonal communications by word of mouth, news reports, and others sources of information perceived to contain no conflicts of interest.

The audience is the target for marketing communications. When selecting communications channels to reach specific consumers, the consumers' characteristics, media habits, and service knowledge must be taken into account. Consumers' characteristics include socioeconomic status, demographics, lifestyle, and psychology. For industrial customers, characteristics include big versus small firms, large versus small market shares, and stable versus unstable financial position. Media habits point to sources of

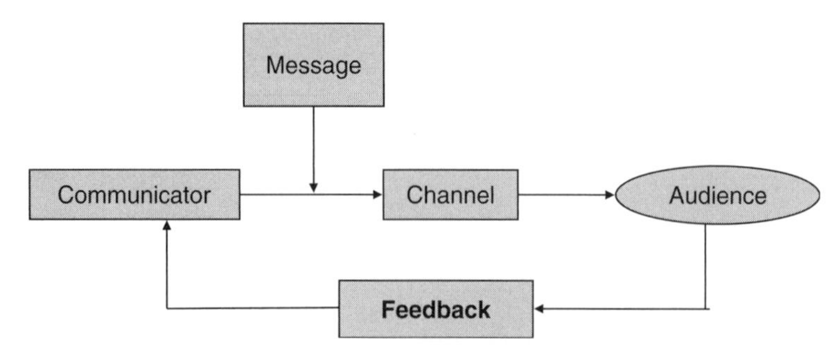

Figure 8.20 Marketing communications.

information preferred by the customers (e.g., types of magazines and TV programs). Service knowledge is the consumers' understanding and appreciation of the values offered by the service packages.

Some companies invest a considerable amount of effort into educating their consumers. A case in point is the known practice of some drug companies of sponsoring large-scale clinical studies conducted by universities and other independent organizations. The purpose of such funded studies is to generate results for publication in technical journals from which consumers gain service knowledge in ways preferred by the sponsoring drug companies.

The impact of marketing communications on services may be short term or long term. The short-term impact is related to recall, recognition, awareness, and purchase intention with respect to the services in question. The long-term impact is reflected in the purchase by customers and brand loyalty with repeat purchase. Several factors influence the effectiveness of marketing communication, such as timing, price, service availability, responses by competition, service warranty conditions, and service.

Marketing communication brings about heightened service awareness. An improved familiarity with the service induces more people to buy the service at the current price, thus causing an up-shift of the service demand curve (as illustrated in Fig. 8.21).

Promotion Strategy. Service promotion may be pursued by either causing the consumers to want to pull the services from the supply chain or pushing the services to the consumers through the supply chains. A large number of companies practice both strategies.

In a *pull* strategy, the consumers go to retail stores to query about the services because they have been informed of the service values by advertisements and other promotional efforts of the sellers. In this case, the product or service is presold to the consumers, who practically pull the products/services through its distribution channels (see Fig. 8.22).

In exercising a *push* strategy, sellers introduce incentive programs (e.g., factory rebate, sales bonus, telemarketing, rebate selling, door-to-door sales, or discount

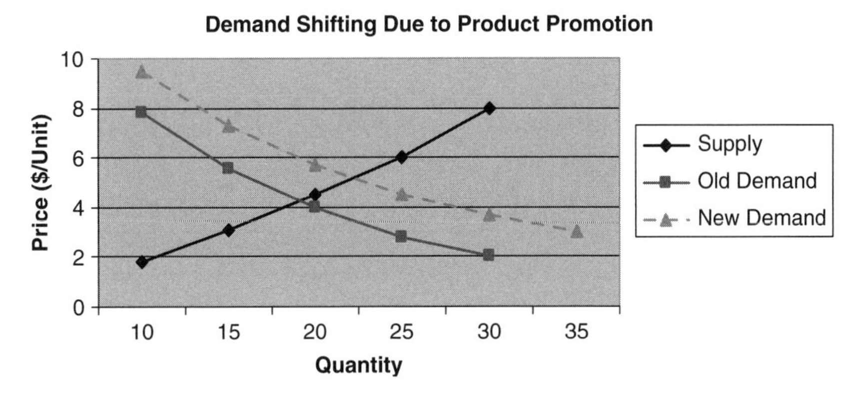

Figure 8.21 Up-shift of the service demand curve.

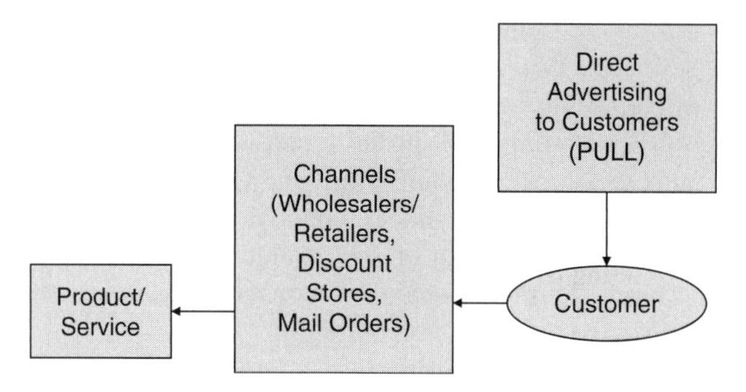

Figure 8.22 Pull strategy.

coupons) to push services onto the consumers. Figure 8.23 illustrates the push strategy. Table 8.9 compares these two promotional strategies.

Promotion of High-Tech and Consumer Services. High-tech and consumer services are promoted differently. To bring the most convincing marketing messages to the intended users, marketers for high-tech and consumer services use different channels. Table 8.10 summarizes the major differences in tactics applied in marketing these services.

Internet-Enabled Communications Options. Communications among sellers, intermediaries (e.g., distribution partners), and buyers have been significantly intensified by the Internet (Zimmerman 2000). Figure 8.24 presents four specific modes of communication.

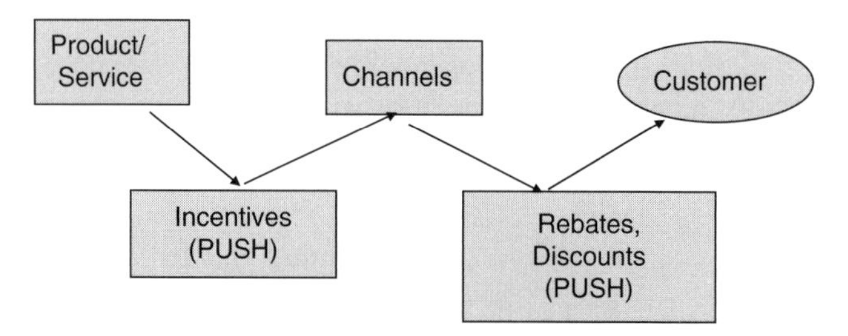

Figure 8.23 The Push strategy.

Table 8.9 Comparison of Pull and Push Promotional Strategies

	Push	Pull
Communication	Personal selling	Mass advertising
Price	High	Low
Service's need of special support	High	Low
Distribution	Selective	Broad

Table 8.10 Promotion of Services

	High-Tech Services	Consumer Services
Marketing costs	Low	High
Consumer segments	More	Less
Focus	More	Less
Advertising	Less important	More important
Marketing channels	Trade shows	TV
	Users groups	Print media
	Trade journals	Internet
	Internet	Radio
Brand	Important	Critically important

Source: Scott Ward, Larry Light, and Jonathan Goldstine, "What High-Tech Managers Need to Know about Brands." *Harvard Business Review* (July–August 1999).

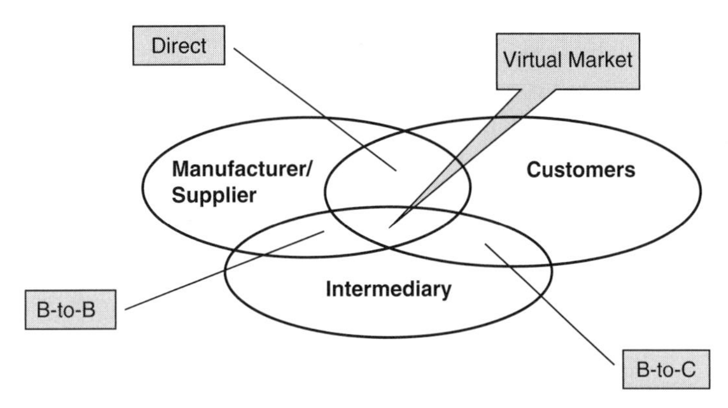

Figure 8.24 Modes of communications enhanced by the Internet. (Adapted from Zimmerman, 2000)

Manufacturers and suppliers usually set up the intranet to communicate with intermediaries (business to business, or B-to-B). Intermediaries may create their own Web sites and other tools to communicate with customers in a business-to-customers (B-to-C) mode. A direct communication between manufacturers, suppliers, and customers can be readily fostered by the companies Web sites, call centers, and other means for order processing, inquiry coordination, problem solving, and additional mission-oriented activities such as customer surveys, focus groups, and service testing.

The virtual market is a segment of the Internet domain wherein third-party portals (e.g., Google™, Yahoo®, and other search engines), auction sites (e.g., eBay®), and eMarketplace (e.g., ChemConnect®) actively provide channels to access information useful to all parties involved (manufacturers, suppliers, intermediaries, and customers).

In the B-to-B markets, businesses buy essentially two kinds of goods from other businesses: manufacturing inputs (raw materials, equipment, and components) and operational inputs (MRO goods, office supplies, spare parts, travel services, computer systems, cleaning, and other services). They buy these goods by systematic sourcing or spot sourcing. Electronic hubs provide the useful functions of aggregation and matching.

Contextual Marketing. Companies invest effort into creating Web sites that offer product information and facilitate sales transactions. However, studies show that these efforts have not yet returned the high profitability generally anticipated from such a marketing approach. The basic reason is that it remains unpredictable how frequently new and repeat customers visit these Web sites and then actually place orders.

A new way of thinking is offered by Kenny and Marshall (2000), who suggest that the focus of e-commerce should be shifted from contents to context. They believe that contextual marketing (bringing the marketing message directly to the customer at the point of need) is the key. A number of contextual marketing examples are described next.

Johnson and Johnson's banner ads for Tylenol® (an over-the-counter pain-killing drug) show up on e-brokers' Web sites whenever the Dow Jones Industrial Average falls by more than 100 points on a given business day. They anticipate that investors will have headaches and thus will need Tylenol when they see their stocks lose money. The marketing message is brought out in the correct context as a way to reinforce its relevance and to offer transactional convenience at the right time.

CNET and ZDNET Web sites attract diverse visitors interested in computers. Instead of placing banner ads in these CNET and ZDNET sites to redirect visitors to its own Web site, Dell offers product information directly within the CNET and ZDNET Web sites. Dell piggybacks on CNET's and ZDNET's relationship with its computer-savvy customers in order to promote Dell's own customer-acquisition economics. Doing so holds the customer's attention on computer design and offers competitive design choices and speedy order processing at the optimal moment. Here, the tactic used by Dell is to insert itself into a preexisting customer relationship at the right time and place.

Several search engines (e.g., Google, Yahoo, and AOL®) practice contextual marketing. When a user conducts a keyword search, the output is typically placed in a left-aligned column under the heading "Matching Sites" and rank-ordered according to hit frequency. Often, several items under the heading "Sponsored Links" are placed on top of the "Matching Sites" column. These are paid advertisements related to the keywords entered by the user. They are there to offer contextual marketing messages relevant to the expressed interest of the user.

As Web-based technologies continue to advance, the Internet will become more accessible by many more users from almost anywhere and, at any time, causing them to become overwhelmed by information and choices. Bringing the right marketing information to the customer at the point of need is likely to become a critical success factor for various companies in marketing communications.

The Proper Approach to Communicate. In general, there are three ways to promote products or services, mass advertising, marketing to segment of one, and middle of the road marketing (Nunes and Merrihue 2007).

1. *Mass advertising.* People resent being exposed to mass advertising, which often is unconnected to their lives or interests. In a recent survey, two thirds of respondents said they felt constantly bombarded by ads and half said the ads they saw

had little or no relevance to their lives. More than 60 percent of respondents look forward to new technologies that would block advertisements. At the present, people use a number of tactics to avoid mass advertising. Examples include:

- On-demand technology or digital video recorders to fast-forward through advertisements or to skip them completely.
- Mobile devices to bypass advertisements—free versions of popular shows.
- Software to block spam mails and pop-up ads.
- Answering machines, caller ID, and the "Do Not Call" registry to avoid unsolicited interruptions at home.

The future does not look good for mass advertising. What is the way around this problem?

2. *Marketing to segment of one.* The other extreme is one-to-one marketing by offering customized products and services through targeted outreach to microsegmented customers. This "marketing to segment of one" approach requires a significant upfront investment, including (1) implementing customer relationship management software applications, (2) filtering, enhancing, and cleaning customer data, and (3) personalizing interactions—e-mail, billing, offers, etc.). These activities take time and coordination from multiple units of the organization (marketing, customer services, sales, IT), which can be daunting. What happens if the individual does not open the envelopes, pick up the phone, or click on a box? The customers concerns around privacy issues must also be taken into account.
3. *Middle-of-the-road approach.* This approach targets broad groups of customers with messages that cannot be turned off. For example:

   - *Catch people in the bottlenecks in public space*. Target customers in the "bottlenecks," the places where they cannot help but pay attention, such as riding in an elevator (wireless screens), a taxi (backseat video screen), an escalator (revolving stairs), flying from on city to another (in-flight movies channels, in-flight magazines, ads wrapped on airline tray tables), or using a public restroom (stall-mount messages). People see taxi rooftop screen advertising nearby stores and restaurants, when the roving taxi is GPS-linked. The focus is on attracting the attention of an on-the-go but temporarily captive audience.
   - *Use a Trojan Horse.* Advertise on frequently encountered materials such as paper coffee cups, paycheck stubs, pizza boxes, take-out food containers, and garbage trucks. This advertising strategy is to infiltrate private spaces with mobile ads.
   - *Target people at play*. An example is to program the GPS of golf carts to alternatively advertise the golf course and selected advertising materials. The strategy is to reach people while they are pursuing an active leisure activity.
   - *Get people to play games.* Display interactive posters in public spaces (mall, bus stations) so that customers can try out the products (e.g., MP3 players). The focus here is to engage people in ways that require them to interact physically with an ad or product.

Service innovations do require a high degree of innovativeness to reach customers while purposefully challenging them.

8.5.4 Placement (Distribution) Strategy

Numerous organizations are involved in moving products and services from the points of production to the points of consumption. As indicated in Fig. 8.25, some companies may engage intermediaries (e.g., wholesalers and retailers) to distribute their services, while others may elect to interact directly with their customers.

Distribution channels serve very useful functions, including

- *Transportation*—Overcoming the spatial gap between the producer and the user.
- *Inventory*—Bridging the time gap between production and usage.
- *Allocation*—Assigning quantity and lot size.
- *Assortment*—Grouping compatible services for the convenience of users, to allow technical representatives to sell several service lines at the same time.
- *Financing*—Facilitating timely possession of services.
- *Communication*—Providing service information and feedback to and from consumers.

In recent years, distribution channels for some services have experienced significant changes due to upgraded logistics, transportation technologies, and advancements in communications technologies. For example, the Internet has enabled countless producers to deliver digitized services—books, newspapers, magazines, music services, video services, and travel services—directly to consumers, thus bypassing the traditional intermediaries. Because of the use of sophisticated Web sites from which extensive service catalogs may be accessed, retail stores have also lost some of their traditional importance for selling physical goods—clothing, cars, appliances, and others.

Furthermore, logistics companies such as United Parcel Service (UPS) are constantly improving their transportation capabilities and satellite-based communications system technologies in order to deliver physical goods anywhere in the world and to constantly track the status of consumer orders.

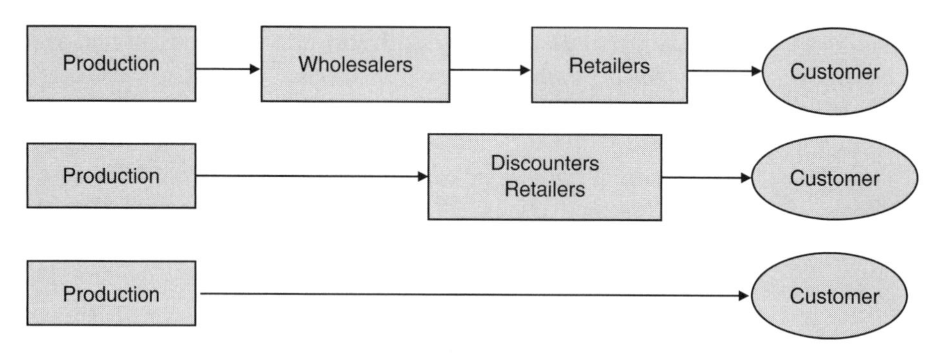

Figure 8.25 Distribution channels.

Warehouse design is expected to increasingly involve gantry robots and complex process optimization for constantly improving the efficiency of automated high volume operations.

Types of Distribution. Traditionally, distribution is classified as either intensive, exclusive, or selective.

In *intensive distribution*, products are stocked in diverse outlets, such as hardware stores, department stores, and catalog rooms, for wide distribution. This type of distribution is particularly suitable for consumer products of low technology and differentiation features. Examples include films, calculators, electric fans, books, and CDs.

With *exclusive distribution*, certain services are distributed only through exclusively designated outlets. This allows producers to retain more control over price policy, promotion, credit, and service, as well as to augment the image of the services. Examples include dealerships for specific cars and qualified service centers for brand-name investment services (Fidelity, Vanguard, etc.).

The *selective distribution* is suitable for certain other products, the sales and service of which require special technical know-how and training. Examples include electronic instruments, high-tech equipment, and insurance brokerage, and custom software.

Organizational Structures. In order to enhance distribution effectiveness, some companies elect to exercise more control over the supply chains by integrating forward. Others have elected to integrate backward.

A *forwardly integrated* organization strives to control the distribution channels leading to the customers. For the purpose of securing a larger market share and exercising more direct control, a producer may attempt to build its own retail outlets. Doing so allows the producer to gain a direct access to customers and thus to benefit from their feedback.

A *backwardly integrated* organization seeks to control the value chain leading backward to production. For example, some retailers or wholesalers may attempt to own specific production facilities or to outsource production for creating private-label products in order to market products with their own brand names, reduce costs, ensure supply, and control quality.

Impact of E-Commerce on Distribution. The Internet has significantly modified the traditional classifications of distribution. A variety of consumer services, as well as certain high-technology services, are now marketed directly through the company's Web sites, including order processing and after-sales services. As a consequence, quite a number of intermediate companies currently involved in distributions—wholesalers, discounters, and some retail stores—are gradually being forced out by the Internet-enabled e-commerce companies and by the increased involvement of efficient and fast-responding logistics providers.

One immediate impact of e-commerce is the reduction of the final service price and service delivery schedule, both of which are beneficial to end users.

Example 8.6 Customers' wants and needs are regionally different for products/services intended for global markets. How can a centralized, concurrent engineering team design a product/service that will serve as the common "platform" for global markets?

ANSWER 8.6 Insofar as products are concerned, one option is to segregate the mechanical aspects (functionality) of the products from their aesthetic aspects (look and feel). General Motors is accomplishing this challenging objective by

- Building identical assembly plants for Buick® cars at four global locations.
- Outsourcing major subassemblies to local industries to lessen import duties and to satisfy local content laws.
- Standardizing the technical specifications so that parts supplied by one region can be readily rerouted for use by another, in order to balance loads due to market demand, labor disputes, governmental regulations, and other unpredictable events.
- Modifying design to account for local market conditions relative to cultural preference (e.g., car names in local languages, styling preferences, purchase habits, colors, etc.).
- Retaining centralized concurrent engineering approach to facilitate global business strategy and scale of economy, while being flexible enough to adjust to local needs.

With respect to services, the option could be to centrally develop and specify the core service element that contains novel and strategically differentiable features. This core service element is then supplemented by support elements (e.g., payment methods, delivery, information access, order processing, billing, exception, customer support, etc.), which are made locally adjustable.

Example 8.7 The company plans to enter a new global market. It has three services currently selling well in its home country. The company's brand name is strong and internationally well recognized.

Current market research indicates that the segments for these three services in the targeted global market are of different size, growth rate, and profitability for the foreseeable future. Other service characteristics are included in Table 8.11.

Which one service should be selected to penetrate the targeted global market? Why? If the company has the required resources to market all three services in the targeted global market, in what priority order should the company proceed?

Table 8.11 Product Characteristics

	Service A	Service B	Service C
Segment size (dollars)	Small	Medium	High
Segment growth rate	Medium	Medium	Low
Profitability	High	High	Medium
Service value to customers	High	Medium	Medium
Brand strength	High	High	High
Delivery/distribution efficiency	Low	High	Medium
After-sales support activities	Medium	Medium	High

ANSWER 8.7 To enter a global market, the company must examine two key questions: (1) How attractive is the target market segment to the company? and (2) How acceptable is the service offered to the customers in the target segment?

The attractiveness of a market segment to a company is generally defined by three factors: segment size, segment growth rate, and profitability. By using the information presented in the table, it becomes clear that the ranking based on "attractiveness" should be Service B first, with Service A and C sharing the second spot.

How acceptable the company's service is to the customers depends on the service value as perceived by the customers, the brand strength of the product, the delivery or distribution efficiency that affects the service's availability to the customers, and the ease with which customers obtain the needed after-sales support activities. Based on the "acceptability" criterion, the ranking of these products is service B, C, and A.

Since both the "attractiveness" and "acceptability" criteria are equally important, we need to come up with a combination ranking, which says that the company should select Service B as its first choice to enter the global market, followed by Service C and then Service A.

8.5.5 Physical Evidence

Physical evidence refers to the physical setting in which a service is offered, delivered, or consumed. Customers form a brand image of the service vendor by observing the physical layout, decor and color, design details, equipment capabilities, the status of facilities that support the service delivery and enhance the communication of messages (e.g., service availability, customer satisfaction) to other potential customers, and by judging the extent to which the physical setting meets or exceeds their expectations.

Singapore Airlines have maintained the same uniform for its stewardesses for twenty-five years. McDonald's and Citibank design its retail branches to look and operate the same way in any location.

8.5.6 Process Design

By process we refer to the chosen work procedure, interunit collaboration, and the flow of activities by which services are delivered, service consumption is facilitated, and customers' problems are solved, as measured by such metrics as speed, convenience, efficiency, and the extent of empowerment enjoyed by the customer-facing staff. Customers gain an overall impression regarding the extent of the process being customer-focused.

8.5.7 People

Numerous people are directly or indirectly involved in the production and consumption of services (knowledge workers, employees, management and other customers), who may add to the value of the service offering. Following the value profit chain model (Heskett et al. 2002), recruiting the right customer-facing staff, proper training in interpersonal skills, aptitude, and service knowledge, empowering them to take care of customers, and compensating them well are all essential steps toward achieving customer satisfaction and corporate profitability.

Example 8.8 In this Chapter, we talked about the marketing mix (7Ps), which include: (1) product/service, (2) price, (3) promotion (advertising), (4) placement (distribution), (5) physical evidence, (6) process, and (7) people.

When marketing products, it is usually sufficient to focus on the first four of these seven marketing elements. However, all seven marketing elements are deemed important when marketing services. What are the underlying reasons for the last three (e.g., 5, 6, and 7) marketing elements to be particularly important for marketing services?

ANSWER 8.8 Companies address all seven marketing elements (7Ps) when marketing services. The principal reason for this is that service (e.g., healthcare, business consultation, financial services, leisure and travel, insurance, and others) requires a much higher degree of customization in the process of specifying, producing, delivering, and offering after-sales services than products (e.g., automobiles, computers, appliances). Because of these service-specific characteristics, customers are exposed to the vendor's performance in (1) physical evidence, (2) process, and (3) people to a much greater extent, making these elements more important in affecting the customer's satisfaction, than in marketing products.

Thus, to be customer focused, service companies pay more attention to (1) physical evidence related to facility design, office layout, and employee uniforms, (2) processes in problem solving and conflict resolutions affecting customers, and (3) people by choosing, training, and monitoring customer-facing staff to assure friendliness and customer-centered attention.

8.6 CUSTOMERS

Customers are important to any service or product company. Company's marketing program needs to focus on the targeted customers, understand them, practice the right techniques to acquire customers, create emotional bond with them, improve interactions and intensify customer loyalty, and continue to expand the number of satisfied customers by asking the right customer survey question.

8.6.1 Customer Focus

Customer focus is aimed at knowing the real needs of customers—past, present, and future. It requires the collaboration of diverse employees as well as functioning support organizations to make it happen. Based on a study of the Royal Bank of Canada case, Gulati and Oldroyd (2005) suggest a four-stage coordination process for service companies to become customer focused: (1) communal, (2) serial, (3) symbiotic, and (4) integral coordination (see Table 8.12).

Getting close to customer is a journey the entire company must take, not just the marketers and customer-facing staff. It requires corporate leadership and commitment to get useful results.

Table 8.12 Process Leading to Customer Focus

	Stage 1 Communal Coordination	Stage 2 Serial Coordination	Stage 3 Symbiotic Coordination	Stage 4 Integral Coordination
1 The Primary organizational objective	Collation of information	Gaining insight into customers from past behavior	Developing an understanding of likely future behavior	Real time response to customer's needs
2 The coordination requirement	Communal coordination between a neutral information owner and the sources of customer information	Serial coordination between the neutral collator of information analytics experts, and line organizations	Symbiotic coordination between the neutral collator of information, analytics experts, and line organization	Integral coordination among all of the company's employees across divisions, geographies, and other boundaries
3 The locus of leadership	Corporate strategy leaders and information technology	Corporate strategy leaders, the neutral entity that collates information (such as IT) analytics experts and marketing	Corporate leaders customer segment managers, and/or pivots that move information vertically and horizontally within the organization	Corporate leaders and cross-business integrators

Source: Gulati and Oldroyd (2005)

8.6.2 Customer Acquisition in Business Markets

The benefits derived by customers belong to four categories:

1. Tangible financial benefits.
2. Nontangible financial benefits—Conducting pilot projects, money back guarantees for nonperformance, and pay-for-performance contracts.
3. Tangible nonfinancial benefits—Corporate reputations, global scale, and innovation capabilities.
4. Nontangible nonfinancial benefits—something the vendor does extra for consumers to improve convenience, customer relationship, services offered above and beyond contracts.

According to Narayandas (2005), to acquire customers, companies must be on par with rivals on tangible financial benefits. They need to use tangible nonfinancial benefits to differentiate, shift customer's focus from tangible benefits to non-tangible nonfinancial ones (e.g., free services that reduce customers operating expenses; suggest ways to make customer's process more efficient, thus earning the trust of the customers).

Consumer markets and business markets are quite different. Table 8.13 illustrates the difference.

Table 8.13 Differences between Consumer and Business Markets

	Characteristics	Consumer Markets	Business Markets
1	Segmentation of Market	Critical	Not important—"Segment of one"
	Branding	Important	Not important
	Focus of communication	Novelty of features	Solving customer's specific problems
2	Number of buyers	Large	Small
3	Transaction value	Small	Large
4	Production	Mass-production	Customized to individual needs
5	Value	Customer's perception	Defined by customer's usage
6	Sales process	Brief	Elaborate (long and complex)
7	Retailing strategy	Important	Not important
8	Focus of Sales Efforts	End users	Group of decision makers

Source: Narayandas, 2005

When marketing to business customers, companies need to know that business decisions are typically made by a team of people. Each team member could have one or more specific needs. Knowing who these people are and what each of them is looking for is of critical importance to the seller. The vendor must become sensitive to these needs and be prepared to address all of them adequately.

8.6.3 Moments of Truth in Customer Service

The *moments of truth* are times when a customer invests a significant amount of emotional energy in the outcome of a service transaction with vendors. Examples of such moments include (1) customer has a problem—encountering an unexpected difficulty in applications, receiving a hold on a check, for example, (2) customer has a need to get a quick answer, (3) customer receives financial advice (good or bad), and other advice. If these moments are handled positively, customers are likely to create an emotional bond with the vendor and subsequently increase their future commitment to the vendor's services. The study of Beaujean et al. (2006) shows that the customer experience related to other humdrum service interactions that involve unelevated emotional energy is of little consequence.

Management must therefore support and develop front-line staff to enable them to handle such "moments of truth" by way of empowerment, nurturing the right service mindsets, and acquiring the needed service knowledge, while de-emphasizing the efficiency improvement of other humdrum transactions. Developing deeper and long-lasting relationships with customers is key to sustaining long-term profitability (Beaujean et al. 2006).

8.6.4 Customer Interactions and Loyalty

The interactions between customers and companies play a very important role in securing marketing success. Creating a pleasant experience for customers (in order processing, service information dissimilation, inquiry coordination, problem solving, after-sales service, and market surveys) is crucial for customer retention. Winning customer cooperation in offering much-needed feedback is vital to the company's service development. Management of customer relations is thus an important corporate responsibility.

Some lessons from the past are noteworthy. Customer interactions are not limited to marketing. Various other functions of the company are involved, including service, accounts receivable, legal, engineering, manufacturing, and shipping. Empowered employees can act on behalf of the company to satisfy customer requirements. Adequate support infrastructure must be established to enable employees to perform the tasks in an innovative and customer-responsive manner.

The major payoff of a successful customer interaction program is customer loyalty. Customer loyalty contributes to company profitability. Studies indicate that increasing customer retention rate by 5 percent raises profits by 25 to 95 percent (Reichhold and Shefter 2000). Loyal customers are valuable because they buy more, refer their pleasant experience to new customers, and offer consultations to these new customers at no cost to companies.

Dell created a *Customer Experience Council* to monitor the effectiveness of the programs that were geared to build customer loyalty. The council determined that customer loyalty is attained by (1) how the company fulfills the customer's orders, (2) how well the products perform, and the effectiveness and promptness of past sales service or support (or both). Other studies have indicated that there are five primary determinants of loyalty: (1) quality of customer support, (2) on-time delivery, (3) compelling service presentations, (4) convenient and reasonably priced shipping and handling, and (5) clear and trustworthy privacy policies.

To build customer loyalty, the customer interaction strategy must be focused on creating trust. Amazon.com is viewed by many as one of the most reliable and trustworthy Web sites on the Internet. It registers user preference, becomes smarter over time at offering services tailored to each user, provides one-click convenience for purchasing items, and delivers the ordered services free of errors. It is reported that 59 percent of Amazon.com sales are derived from repeat customers, roughly twice the rate of typical "bricks-and-mortar" bookstores.

Vanguard Group, a company that markets index-based mutual funds, offers timely and high-quality financial advice on its Web site and does not attempt to hard sell any specific service. Its customer interaction strategy is focused on building trust. "You cannot buy trust with advertising or salesmanship; you have to earn it by always acting in the best interest of customers," says Jack Brennan, Vanguard CEO.

eBay is known to have over 50 percent of its new customers referred by loyal customers who also serve as helpers to them. One major concern in the business of auctioning used merchandise is reliability and fraud prevention. eBay asks each buyer and seller to rate each other after every transaction. The ratings are posted on the Web site. Every member's reputation becomes public record. Furthermore, eBay insures the first $200 for each transaction and holds the money in escrow until the buyer is satisfied with the received product.

How is trust related to profitability? Studies show that, in some businesses, customers must typically stay on board for at least two to three years just for the companies to recoup their initial customer-acquisition investment. In other words, for companies to achieve profitability, customers must be loyal enough to stay beyond this break-even period. A large percentage of customer defects before many new companies reach this break-even point. Table 8.14 lists statistics related to customer acquisition cost, years to break even, and percentage of customers who defect before the break-even point.

A large number of companies are successful in planning and implementing strategies to interact effectively with customers. According to Dorf, Peppers, and Rogers (2002), these companies identify and prioritize customers, define their needs, and customize services to fit these needs. They reap the benefits of increased cross selling, reduced customer attrition, enhanced customer satisfaction, minimized transaction costs, and sped-up cycle times. Companies known for their success in relationship marketing include Pitney Bowes, Wells Fargo, 3M, Owens Corning, British Airways, Hewlett-Packard, and American Express.

Customer loyalty is won, not by technology, but through the delivery of a consistently superior customer experience. It requires a well-designed customer interaction strategy that is supported by companies with a firm corporate commitment.

8.6.5 Customer Feedback—The Ultimate Question

A great number of customer-satisfaction surveys contain too many questions. According to Reichheld (2006), the ultimate question to ask is: "How likely are you to recommend this company to a friend or colleague?" Score the results in a 0 to 10 scale and classify the responses as follows:

- Loyal promoters (9s and 10s)
- Customers who do business passively with the company (7s and 8s)
- Detractors (6s and below)

The *net promoter score* (NPS) is defined as the percentage of promoters minus the percentage of detractors. For example, if the promoters are 35 percent, passive customers are 50 percent and detractors are 15 percent, then the NPS is 20 percent. Based on a survey conducted by Bain & Company, companies with NPS in the range of 50 percent to 80 percent, are superior in achieving good profits (see Table 8.15).

The key to growing a business is to have more promoters and less detractors. Since the company already has a relationship with these "promoter" customers, they should solicit information from these customers in order to understand exactly where the company is succeeding and how to apply this information toward ensuring the companies ongoing success. They should contact detractors as well, to find out what

Table 8.14 Customer-Loyalty Related Statistics

Products	Acquisition Cost per Customer (dollar)	Years to Break Even	Percentage of Customers Defecting before Breakeven Point (percent)
Consumer electronics and appliances	56	4+	60+
Groceries	84	1.7	40
Apparel	53	1.1	15

Source: Frederik F. Reichhold and Phil Shelfter, "E-Loyalty: Your Secret Weapon on the Web," *Harvard Business Review* (July–August 2000).

Table 8.15 NPS Scores of Selected U.S. Companies

NPS Score of Selected Companies	
USAA	82%
HomeBanc	81%
Harley-Davidson	81%
Costco	79%
Amazon.com	73%
Chick-fil-A	72%
eBay	71%
Vanguard	70%
SAS	66%
Apple	66%
Intuit (TurboTax)	58%
Cisco	57%
FedEx	56%
Southwest Airlines	51%
American Express	50%
Commerce Bank	50%
Dell	50%
Adobe	48%
Electronic Arts	48%

Source: Reichheld (2006)

the company can do to improve their service offerings. Reihheld (2006) believes that business growth can only be sustained over a long period based on the loyalty of satisfied customers and customer-initiated promotions. To achieve success, companies must always practice the Golden Rule: Treat others as you would want to be treated in return.

This concept appears to be consistent with the *value profit chain* model (section 8.5.7) in that satisfied customers will not only increase spending for themselves, but also recommend that their friends follow suit. The way to ensure customer satisfaction is to adopt company policies that make for happy and loyal employees, who will in turn provide excellent customer service leading to long-term corporate profitability.

In this Internet age, one additional way of getting to know customers better is via word-of-mouth marketing (Servovitz 2006). According to PEW Internet reports, 32 million people are regularly posting content to message boards. Since people are psychologically more inclined to talk about an "extraordinary" experience that happened in their daily life than "average" ones, these message boards could serve as useful screening filters, passing their self-selected comfort thresholds, for best and worst practices in service marketing. Assuming companies routinely strive for product/service differentiation, to have something unique to stand out from the competition, then they could gain useful feedback and valuable insights in what customers like and dislike from systematically mining these public message boards. Examples of differentiable offerings include: (a) Offer a drink to barbershop customers, (b) Practice a generous return policy at Nordstrom, (c) Pick-up at the homes of Enterprise Rent-a-car customers, (d) Answer a call on the first ring, (e) Solve a problem on the first phone call, (f) Introduce extra service efforts to help customers, among others.

Example 8.9 Over the years, Company XYZ spent a considerable amount of effort in developing and testing a new drug intended for reducing the LDL (bad cholesterol) and raising HDL (good cholesterol) of patients with cardiovascular disease. After having passed the phases 1, 2, and 3 trials, the drug received FDA approval for marketing to the public. There are some known drugs already in the marketplace for this type of disorder, which affects millions of people in the U.S. alone. The size of the overall market for cardiovascular drugs is estimated to be about $25 billion annually. Devise a marketing plan to bring this new drug into the U.S. marketplace.

ANSWER 8.9 The marketing plan should consists of four parts, corresponding to the 4Ps of marketing products, namely, product, price, placement (distribution), and promotion

As a new product, the drug in question offers the useful features of lowering LDL and raising HDL, a very powerful combination to combat heart disease, based on the current state of clinical knowledge. A direct competitor is Lipitor, which has the same combination effect as the new drug. Furthermore, Crestor, together with Niacin, is also known to achieve this combination effect. It is important for the company to delineate any differences this new drug might have regarding (1) side effects, (2) potential of long-term health hazards, (3) lower frequency of taking the drug, (4) longer effectiveness. These differences must be clinically verifiable by ways of large-scale clinical studies.

Pricing is an important issue to some patients. The new drug should be retail-priced at a level slightly lower than its current competitors. Aggressive contracts should be entered into with major insurance carriers, mail-order drug companies, AARP prescription drugs program, and others to allow volume-based discounts.

Promotion is rather critical for the new drug entering an existing market. Key targets are patients, physicians, and insurance carriers. Patients need to become aware of the unique benefits of this new drug via TV, magazine articles, and Web-based advertisement, so that they could "pull" this drug from the supply chains. The message should emphasize its distinguishing features in view of the existing competition. Physicians must be convinced via trade shows, clinical studies, and publications of its merits, so that they would be willing to suggest/prescribe this new drug and allow brochures to be distributed through their offices. Frequent publication of supporting articles in highly reputable journals such as *New England Journal of Medicine, Journal of American Medical Association, Journal of Cardiology, Circulation,* and others will gain the attention of physicians. Insurance carriers need to be convinced of the benefits of this new drug in order for them to be willing to place it in their formularies.

Distribution of this new drug would follow the usual wholesale channels, as it is a prescription drug authorized by physicians and available only from pharmacies.

Company XYZ should monitor the market constantly and adjust its marketing program accordingly.

8.7 OTHER FACTORS AFFECTING MARKETING SUCCESS

There are several other factors that may affect the marketing success of any company.

8.7.1 Alliances and Partnerships

Nowadays, companies realize increasingly that they do not always have or cannot cultivate internally, with the resource and time constraints under which they have to work, all competencies needed to compete in the world markets. Either because market access may be unattainable, the technology unaffordable, resources unavailable, time to market too long, or because other barriers may exist, an individual company may find it increasingly difficult to compete alone. Partnerships and alliances have become more and more important for numerous companies in order to compete effectually and to constantly deliver value to customers.

For companies to succeed in the marketplace under these circumstances, the marketing concepts must penetrate to all members of the partnerships and alliances. (Steinhilker 2008). Two examples illustrate the working of such partnerships and alliances:

1. Calyx and Corolla formed a network of partnerships to provide the seamless delivery of fresh flowers from grower to final customer in one-fourth of the time required by the traditional channels.
2. Dell teamed up with parts manufacturers and assembled computer systems in their own plants. Then Dell linked with logistics partners to deliver custom-designed personal computers (PCs) within three days to customers anywhere in the world. The success of the Dell model is clearly reflected in its 18.6 percent worldwide market share of PCs in the first quarter of 2004, ahead of Hewlett-Packard, IBM, Fujitsu Siemens, Acer, and others.

Partners must appreciate that mutual gain results only when all members of the alliance embrace the marketing concept and come to recognize the importance of creating superior customer value by joining hands.

8.7.2 Organizational Effectiveness

Marketing success is influenced by how effectively the company operates. In general, organizations with less rigid structure have a higher likelihood of becoming more customers focused, technologically innovative, and market responsive. Certainly, any conflicts between internal functions—manufacturing, design, engineering, and marketing—must be minimized. Technology for mass customization requires an integration of R&D, procurement, customer relations management, and supply chain management to achieve a high degree of customer satisfaction.

Above all, company management must apply discretionary resources (e.g., R&D, production capacity, human resources, organizational expertise, and information services) to the right combination of strategies (e.g., marketing, service, distribution, promotion, and price) so that maximum strategic marketing leverage can be achieved to capture opportunities offered in the marketplace. Figure 8.26 illustrates this core concept of organizational effectiveness.

Example 8.10 Engineering refined the design specifications of a service as originally recommended by marketing. Manufacturing made further changes to the service design in order to fabricate the service automatically. Unfortunately, the service did not sell in sufficient quantities to make it a success. Explain the possible reasons.

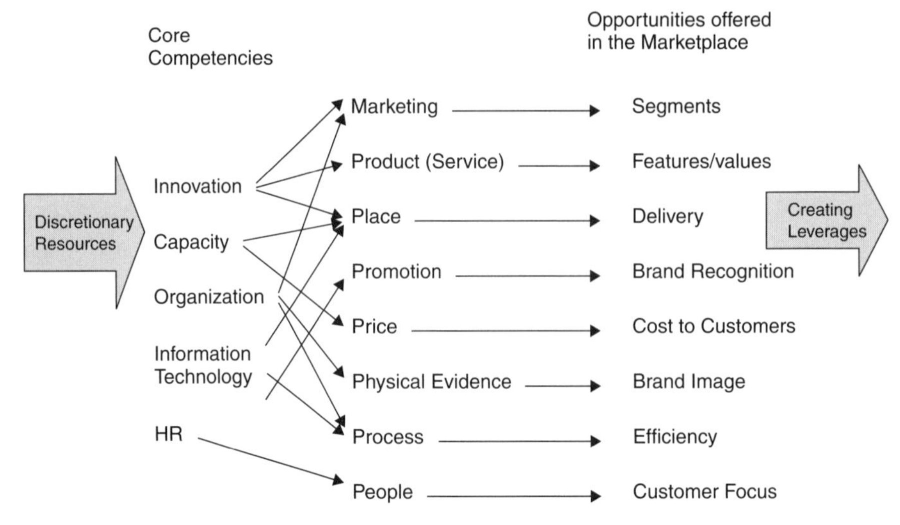

Figure 8.26 Organizational effectiveness.

ANSWER 8.10 The service features defined initially by the marketing department may not be exactly what the majority of customers want. Service testing is a critical step to fine-tune the service design. The selected method of production does not readily accommodate an adjustment of service features, even if they are identified by feedback from the marketplace. The manufacturing of the product should be based on demand to assure market acceptance, not based on production technology, which is aimed at cost reduction.

Skipping the service testing step and applying the automatic production method too soon are two likely reasons for the noted failure.

8.8 CONCLUSIONS

This chapter covers a large number of important issues related to the marketing of the company's products and services. Engineering managers should understand the overall objectives of the firm's marketing efforts and become sensitive to various marketing issues affecting engineering. They should become well versed with marketing terminology and elements of the marketing mix—namely, product (service), price, promotion, distribution, physical evidence, process, and people. It is important for engineering managers to accept the fact that marketing of services involves a lot of uncertainties associated with consumers' perceptions, competitive analyses, market forecasting, and people-related issues. They need to wholeheartedly adopt customer orientation in planning and implementing all engineering programs. They must strive to work closely with marketing personnel and remain supportive of the overall marketing efforts by providing high-quality engineering/technology inputs to the firm's marketing program.

Obviously, the engineering/technology inputs most useful to marketing are related to services and associated production and delivery activities. These include specifying and designing service features to be of value to customers; generating innovative ideas to offer premium service features that demand high margins; utilizing technologies to confer competitive advantages in time to market, quality, reliability, and convenience; and delivering after-sales support activities needed to ensure customer satisfaction.

Engineers are also expected to control costs by improving and managing the service production process, resources (labor and materials), and quality control, and by estimating service cost accurately with the activity-based costing (ABC) method. Engineers may also get involved in training salespeople, making presentations before customers, conducting industrial exhibits, and evaluating customer feedback related to new service features.

Having learned the marketing concepts and been exposed to the complex marketing issues reviewed in this chapter, engineering managers will be able to appreciate the difficult but critically important functions of marketing and can become more effectual in interacting with marketing management.

Marketing and innovations are two principal activities of any service-based and profit-seeking organization. Engineers already know how to innovate. If they also learn how to market, this combination of capabilities will surely enable them to become major contributors in any service organization.

8.9 REFERENCES

Barnett, F. W. 2009. "Four Steps to Forecast Total Market Demand." Download PDF. *Harvard Business School Press* (March 3).

Baron, Steve, Kim Cassidy, Kim Harris, and Toni Hilton 2009. *Service Marketing: Text and Cases*, 3rd ed. New York: Palgrave MacMillan.

Beaujean, Marc, Jonathan Davidson, and Stacey Madge 2006. "The 'Moment of Truth' in Customer Service." *McKinsey Quarterly* (1).

Bedbury, S., with S. Fenichell. 2003. *A New Brand World: 8 Principles for Achieving Brand Leadership in the 21st Century*. New York: Penguin.

Berry, Leonard L and Kent D. Seltman. 2007. "Building a Strong Services Brand: Lessons from Mayo Clinic." *Business Horizon* 50: 199–209.

Chaffey, D., F. Ellis-Chadwick, K. Johnston, and Richard Mayer. 2009. Internet Marketing: Strategy, Implementation and Practice, 4th ed. Upper Saddle River, NJ: Prentice Hall.

Dev, C. S. 2008. "The Corporate Brand, Help or Hindrance." *Harvard Business Review* 86, (2), pp. 49–58 (February 1).

Dolan, R. J., and John T. Gourville. 2005. "Principles of Pricing." Harvard Business School Note #506021. (September 22).

Dorf, B., D. Peppers, and Martha Rogers. 2002. "Is Your Company Ready for One-to-One Marketing?" Harvard Business School OnPoint Article # 8954. Harvard Business School Press.

Edvardsson, B., Anders Gustafsson, Per Kristensson, and Peter Magnusson 2006. *Involving Customers in the New Service Development*. London: Imperial College Press.

Ennew, Christine, and Nigel Waite. 2006. *Financial Services Marketing: An International Guide to Principles and Practices*. Woburn, MA: Butterworth-Heinemann.

Ferrell, O. C., and Michael Hartline. 2007. *Marketing Strategy*, 4th ed. Cincinnati, OH: South-Western College Publishing.

French, Thomas D. and Trond Riiber Knudsen. 2007. "Marketing in Transition." *McKinsey Quarterly* (3).

Grenci, R. T., and C. A. Watts. 2007. "Maximizing Customer Value via Mass-Customized E-Consumer Services." Business Horizon Article # BH226, Indiana University.

Griffin, Jill. 2002. "Customer Loyalty: How to earn It, How to Keep It," Rev. Sub Edition. San Francisco: Jossey-Bass.

Gourville, J., and D. Soman. 2002. "Pricing and the Psychology of Consumption." *Harvard Business Review* (September).

Gulati, Ranjay, and James B. Oldroyd. 2005. "The Quest for Customer Focus." *Harvard Business Review* 83 (4) (April).

Harrison, Tina. 2006. "Financial Services Marketing." *Financial Times Management*.

Hartley, R. 2006. *Marketing Mistakes and Successes*. 10th ed. Hoboken, NJ: John Wiley & Sons, Inc.

Harwood, Tracy, Anne J. Broderick, and Tony Garry 2008. *Relations Marketing*. New York: McGraw-Hill Higher Education.

Heskett, James L. W., Earl Sasser, and Leonard A. Schlessinger. 2002. *The Value Profit Chain: Treating Employees like Customers and Customers like Employees*. New York: Free Press.

Jones, Rene F. 2006. *Power Marketing of Architectural Services: A Critical Look at the Services Provided by Architects and Designers*. Bloomington, IN: Trafford Publishing.

Keller, K. L. 2000. "The Brand Report Card." *Harvard Business Review* (January–February).

Keller, K. L., B. Sternthal, and A. Tybout. 2002. "Three Questions You Need to Ask about Your Brand." *Harvard Business Review* (September).

Kenny, D., and J. E. Marshall. 2000. "Contextual Marketing: The Real Business of the Internet." *Harvard Business Review* (November–December).

Kirk, B. C. 2003. *Lessons from a Chief Marketing Officer: What It Takes to Win in Consumer Marketing*. New York: McGraw-Hill.

Kotabe, M., and K. Helsen. 2007. *Global Marketing Management*, 4th ed. Hoboken, NJ: John Wiley & Sons, Inc.

Kotler, P., D. C. Jain, and S. Maesincee. 2002. *Marketing Moves: A New Approach to Profits, Growth and Renewal*. Cambridge, MA: Harvard Business School Press.

Kotler, Philip, Thomas Hayes, and Paul N. Bloom. 2002. *Marketing Professional Services*, 2nd ed. Upper Saddle River, NJ: Prentice Hall.

Kotler, P., and K. Keller. 2008. *Marketing Management*, 13th ed. Upper Saddle River, NJ: Prentice Hall.

Lovelock, Christopher, and Jochen Wirtz. 2006. *Service Marketing*, 6th ed., Upper Saddle River, NJ: Prentice Hall.

Lovelock, Christopher, Jochen Wirtz, and Patricia Chew. 2008. *Essentials of Services Marketing*. Upper Saddle River, NJ: Pearson Education.

Miller P., and R. Palmer. 2000. *Nuts, Bolts and Magnetrons: A Practical Guide for Industrial Marketers*. New York: John Wiley & Sons, Inc.

Monk, David, and Alastair Moyers. 2008. "Marketing Legal Services: Succeeding in the New Legal Marketplace." The Law Society.

Moon, Y. 2005. "Break Free From the Product Life Cycle." Harvard Business School Note #Ro5o5E (May 1).

Narayandas, Das. 2005. "Building Loyalty in Business Markets." *Harvard Business Review* (September).

Neufeld, D. 2005. "RBC Investments: Portfolio Planning Initiative." Revised. Ivey School of Business Case #905E05. University of Western Ontario, Ontario, Canada.

Nunes, Paul F., and Jeffrey Merrihue. 2007. "The Continuing Power of Mass Advertising." *MIT Sloan Management Review* 48 (2): (Winter).

Paley, N. 2000. *Marketing for the Nonmarketing Executive: An Integrated Resource Management Guide for the 21st Century*. Boca Raton, FL: CRC Press.

Palmer, Adrian. 2007. *Principles of Services Marketing*, 5th ed. New York: McGraw-Hill Higher Education.

Reichhold F. F., and P. Shefter. 2000. "E-Loyalty: Your Secret Weapon on the Web." *Harvard Business Review* (July–August).

Reichheld, Fred. 2006. "The Ultimate Question." Boston: Harvard Business School Press.

Reid, Robert D., and David C. Bojanic. 2005. "Hospitality Marketing Management," 4th ed., New York: John Wiley & Sons, Inc.

Rust, Roland T., Valarie A. Zeithaml, and Katherine N. Lemon. 2004. "Customer-Centered Brand Management." *Harvard Business Review* (September).

Saaksvuori, A., and A. Immonen. 2008. *Product Life Cycle Management*, 3rd ed. New York: Springer.

Servovitz, Andy (2006), *Word of Mouth Marketing: How Smart Companies Get People Talking*, Kaplan Publishing.

Smock, D. A. 2003. "Ten Commandments of Reverse Auctions." *Purchasing* 132 (2): (February).

Steinhilker, S. 2008. *Strategic Alliances: Three Ways to Make Them Work*. Cambridge, MA: Harvard Business School Press.

Strouse, Karen. 2004. *Customer-Centered: Telecommunications Services Marketing*. Boston: Artech House.

Thomas, Richard K. 2008. *Health Sciences Marketing*. New York: Springer.

Vargo, S. L. and R. E. Lusch. 2010. "A Service Perspective of Marketing, Operations and Value Creation," Chapter 15 in Salvendy, Gavriel and Waldemar Karwowski (Eds), *Introduction to Service Engineering*, John Wiley (January).

Waugh, Troy. 2001. *101 Marketing Strategies for Accounting, Law, Consulting, and Professional Services Firms*. New York: John Wiley & Sons, Inc.

Wheeler, A. 2009. *Designing Brand Identity: An Essential Guide for the Whole Branding Team*, 3rd ed. New York: John Wiley & Sons, Inc.

Yankelovich, Daniel, and David Meer. 2006. "Rediscovering Market Segmentation." *Harvard Business Review* (February):122–131

Zeithaml, Valarie A., Mary Jo Bitner, and Dwayne D. Greimer. 2005. *Service Marketing*, 4th ed. New York: McGraw-Hill/Irwin (E-marketing).

8.10 APPENDICES

8.10.1 Consumer Survey and Market Research

To market consumer products, companies need to have a very detailed understanding of their customers, just as companies marketing industrial products also need to understand their customers, although to a much lesser extent.

When dealing with customers, the key questions typically concern what, how, where, when, why, and who.

For example, what functions does the product serve? What are the criteria to buy the product (price, color, size)? What is the value to the customer? What do they really want from the product (psychological, functional, and other benefits)?

How do customers compare products? How do customers decide what to buy? How is the product used? How much are customers willing to pay for it? How much do they buy? How would the distribution mode and service center location affect the customers' buying decisions?

Where is the purchase decision made (e.g., what is the customer's position in the company or household)? Where do they receive information? Where do they buy their products (retail store, mail order, department store, Internet, etc.)?

When do they buy it (weekly, monthly, special occasions)?

Why do they prefer one brand over the other (performance, price, convenience, packaging, colors, service, etc.)?

Who are the customers (age, background, sex, geographic location, members of social groups, etc.)? Who buys the competitor's products? Who does the buying (wife, husband, children, purchasing agent, engineers, others)? Who makes what decision for whom (decision-making units)?

To understand the behavior of consumers in making purchase decisions, companies focus on customers' habits in purchasing, consuming, and information gathering. Who buys, how often, where, how much, when, and at what price? Who consumes, on what occasions, how do they consume, in what quantities, where, when, and with what other products? What media do they use (industrial exhibits, trade journals, TV, newspaper, radio, Internet, etc.), and when?

It is also useful for companies to understand the process by which consumers make their purchase decisions. This process typically encompasses the steps of need arousal (problems to solve; discovery from neutral sources; Jones the bragging next-door neighbor; etc.), information search (online resources are now one click away), and evaluation and decision making (comparative shopping, making tradeoffs, brand versus product, and price versus quality and features).

8.11 QUESTIONS

8.1 Study the Ivey Business Case #98A006 "Fastlane Technologies, Incorporated" (Rev. November 1999), and answer the following questions. The case materials may be purchased by contacting its publisher at www.custserve@hbsp.harvard.edu or at 1-800-545-7685.

A. What is this case about?
B. What are the major strengths and weakness of Fastlane?
C. How does the buying process for scripting language like FINAL differ from that of an application tool like FLYTE?
D. What are the advantages and disadvantages of focusing on FINAL versus focusing on application tools?
E. What channel strategy would you recommend for Fastlane?
F. What other recommendations would you make to the board?
G. What are the lessons you have learned from having studied this case?

8.2 Study the Harvard Business School Case "Starbucks: Delivering Customer Service," Case # 9-504-016 (Rev. July 10, 2006), and answer the following questions. The case materials may be purchased by contacting its publisher at www.custserve@hbsp.harvard.edu or at 1-800-545-7685.

A. What is this case about? What type of company is Starbucks in the first place (selling coffee, selling service, selling . . .)?

B. What factors accounted for the extraordinary success of Starbucks in the early 1990s? What are so compelling about the Starbucks' value proposition? What brand image did Starbucks develop during this period?

C. Why had its customer satisfaction scores declined when measured in 2002? Had the company's service declined, or is it simply measuring satisfaction the wrong way? How does the Starbucks of 2002 differ from the Starbucks of 1992?

D. Describe the ideal Starbucks customer from a profitability standpoint. What would it take to ensure that this customer is highly satisfied? How valuable is a highly satisfied customer to Starbucks?

E. Should Starbucks make the $40 million investment in labor in the stores? What's the goal of this investment? Is it possible for a mega-brand to deliver customer intimacy?

F. When using the $40 million to add more store labor, what kind of labor is most useful to achieve the specific objective at hand? Should the $40 million be invested only for adding more labor? What might be some of these non-labor upgrades Starbucks should pursue?

G. What are the lessons you have learned from having studied this case?

8.3 How can service-development costs be minimized by entering the market late?

8.4 Is it better to market a new service quickly and then upgrade the design later or to incorporate all possible design modifications or improvements before launching the service?

8.5 ABC Company wishes to enter a new market arena on the basis of its strength in core technologies and financial staying power. However, the market arena in question is currently dominated by a major competitor with 80 percent of the market share, and a number of smaller competitors are each focused on small niche segments. How should ABC Company enter this market?

8.6 A company makes a range of services and sells to several large, loyal customers to achieve a healthy market share. A new competitor has emerged to offer equivalent services at much lower prices. What should the company do?

8.7 The company wants to produce a new service for a high-end consumer market. It is known that customers in this market are difficult to identify and are geographically dispersed. How should the company plan for service distribution and promotion?

8.8 The company wishes to sell its current service in a new market segment. At the same time, it wants to launch a new service in the existing market segment. How should the company handle the service promotion?

8.9 Organizational effectiveness is necessary for any corporate leader to attain company leverage and hence strategic differentiation in the marketplace. Explain in your own words how corporate leaders should go about realizing this much-needed organizational effectiveness?

8.10 What are the bases for tradeoffs between conflicting wants and needs of different customers with respect to the same product? How important is it to emphasize service quality when a new and unique service is launched?

8.11 Over the years, Company XYZ spent a considerable amount of efforts in developing and testing a new drug intended for reducing LDL (bad cholesterol) and raising HDL (good cholesterol) of patients with cardiovascular disease. After having passed the Phases 1, 2, and 3 trials, Company XYZ received FDA approval for marketing the drug to the public. There are some known drugs already in the marketplace for treating this type of disease, which affects millions of people in the U.S. alone. The size of the overall market for cardiovascular drugs is estimated to be around $25 billion annually. Devise a marketing plan to bring this new drug into the U.S. marketplace.

PART III

SSME Leadership in the New Millennium

Part III of this book addresses six major topics: service engineers as managers and leaders (Chapter 9), ethics (Chapter 10), knowledge management (Chapter 11), innovation in services (Chapter 12), operational excellence (Chapter 13), and globalization (Chapter 14). These discussions provide additional building blocks to prepare service systems professionals to assume technology and business leadership positions and to meet the challenges of the new millennium.

Engineers are known to possess a strong set of skills that enable them to do extraordinarily well in certain types of managerial work. However, some may also exhibit weaknesses that prevent them from becoming effective leaders in service organizations, or even from being able to survive as professionals in industry. The expected norms of effective leaders are described. Steps are reviewed that enable service systems professionals to enhance their leadership qualities and attune themselves to the value-centered business acumen. Certain outlined steps should be of great value to those services systems professionals who want to become better prepared to devise new services based on technology, integrate technologies into their organizations, and lead technology-based enterprises.

Many tried and true rules are included that serve as suitable guidelines for service systems professionals to become excellent leaders. Above all, service systems professionals are expected to lead with a vision of how to apply (1) company core competencies to bring about strategic differentiation and operational excellence, (2) insights into how to capture opportunities offered by emerging technologies and global markets, and (3) innovations in making services better, faster, and cheaper, so that they constantly improve customer satisfaction. The concepts of value addition, customer focus, open innovations, knowledge management, time to market, mass customization, supply chains, and enterprise resources integration are also discussed.

Although engineers are ranked high in trustworthiness and integrity (ahead of business executive, bankers, accountants, and lawyers), [Gallup Poll 2008, (see Table

10.3)], it is important for all service professionals to remain vigilant in observing a code of ethics along with taking seriously other topics related to ethics.

To many service enterprises, corporate knowledge and know-how are the backbones of the competitive base. Service systems professionals need to know how to create, preserve, maintain, update, and disseminate corporate knowledge, especially the tacit type, so that it can be optimally utilized to add value to their customers. Industrial best practices are explained. To be effective in knowledge management, companies need to set priorities, commit resources (time and facilities), and implement programs that encourage frequent interactions between domains experts (such as in communities of practices or centers of excellence).

Service enterprises strive to create strategic differentiations in the forms of innovative service features, better delivery modes, faster processes of providing customer supports, and higher quality of customer experience. What it takes for service systems professionals to foster creative thinking methods, promote open innovations paradigms, lead teams with creative professionals, and manage innovation projects to commercial success are elucidated.

Operational excellence is important to any service enterprise. Service systems professionals need to be versed in applying known tools such as Lean Six Sigma, FMEA, and value stream mapping to promote productivity. The changes wrought by the Internet are transforming most aspects of company business, including service design, information dissemination, service delivery, and customer support. As processor design, software development, and transmission hardware technologies continue to advance, their roles in business will surely grow in size and scope. Progressive service systems professionals need to be aware of Web-based tools, Web services and SOA-based developments that could be applied effectively to promote service customization, expedite new service to market, manage supply chains, foster creativity and innovations, and improve customer support.

Globalization is further expanding the perspective of service systems engineers and leaders with respect to divergence in culture, business practices, and value. Globalization is a major business trend that will affect many service enterprises in the next decades. Service systems professionals must be sensitized to the issues involved and prepare themselves to be able to seize the new technological and business opportunities offered in the global emerging markets.

Service systems engineers and leaders will face challenges in the new millennium. What these specific challenges are, how service systems professional need to prepare to meet these challenges, and how to optimally make use of the location-specific opportunities to derive competitive advantages will be examined. The United Nations has predicted that by the year 2020, four of the five biggest national economies will be located in Asia. It is important for service systems professionals to explore prudent corporate strategies for service enterprises in the pursuit of globalization, while minimizing any detrimental impact on the environment and maintaining acceptable human rights and labor conditions.

How should service systems professionals prepare themselves to add value in the new millennium? What are the success factors for service professionals in the decades ahead? These questions are addressed in the final chapter of the book. Globalization will open up ample opportunities for those who know how to properly prepare and equip themselves with the required global mindset, knowledge, and savvy.

The overall design base of this text is illustrated in the "Three-Decker Leadership-Building Architecture," as all chapters come together in support of building an SSME leadership. To further foster the leadership roles of service systems professionals, a six-dimensional model is proposed to emphasize the inside, outside, present, future, local and global dimensions. The management challenges for service systems professionals in these dimensions are presented.

Chapter 9

Service Systems Management and Engineering Leaders

9.1 INTRODUCTION

Engineering managers lead teams, groups, units, or enterprises to generate products and services that are highly technical in nature. The importance of developing engineering managers is well recognized. Edmundson (2005) reports that 85 percent of engineering managers surveyed believe that the development of new engineering management talent is crucial to the survival and growth of their companies' businesses.

Technical talents are important prerequisites for becoming leaders, as most engineering professionals do not readily accept superiors whose engineering credentials they do not respect. For this reason, countless organizations choose only top-flight engineers as engineering managers.

Typically, engineers are well trained for certain managerial functions. They are known to have the following skills and attributes:

- Thinking logically, methodically, and objectively while making unemotional decisions based on facts.
- Analyzing problems and defining technologically feasible solutions.
- Understanding what motivates other engineers.
- Evaluating work with highly technological contents.
- Planning for the future, taking into account technology, productivity, and cost effectiveness.
- Discussing technical information with customers.
- Possessing technical expertise that enhances high quality leadership.

Statistics indicate that, whereas engineers are highly trained professionals, as a group, they play only a limited leadership role in U.S. industry and economy. A survey conducted by *BusinessWeek* in 1990 indicated that only 26 percent of CEOs in the top 1000 companies had their first degrees in engineering; about 46.7 percent were business majors.

Does the engineering mindset represent a disadvantage for top-level managerial jobs? Some people say there may be certain personal attributes, traits, and habits that tend to cause engineers to be poor managers. The appendix of this chapter addresses some of these potential drawbacks.

The chapter covers: (1) a new competency model for service leaders, (2) total leadership, (3) leading change, (4) leadership styles, qualities, and attributes, (5) leaders and managers, (6) the factors affecting the promotion of engineers to managers, (7) the differences in work done by engineers and managers, (8) unique contributions expected of engineering managers. (9) leadership skills for the twenty-first century, and (10) a career strategy for the new millennium. The chapter concludes with "Take Charge" and "Get Success" formulae to inspire engineers to be proactive in managing their own professional lives.

9.2 NEW COMPETENCY MODEL FOR SERVICE LEADERS

Morrison et al. (2007) suggested a two-part competency model for engineering leaders, which includes leadership fundamentals and leadership brand. I have modified their model to fit the needs of service leaders. This new competency model for service leaders consists of a broad set of competencies, as follows:

- *Strategic differentiation*. Exhibit vision for the future and utilize competencies (in-house and external) to bring about new service opportunities as well as meet and exceed customer expectations. It may be essential to pursue open innovations by actively involving customers, supply chains members, and/or external experts in order to speed up the offering of new services in the marketplace.
- *Operational excellence*. Apply resources well to achieve efficiency and productivity by communicating effectively, introducing the best industrial engineering techniques (e.g., Six Sigma, lean thinking, value stream mapping, FMEA, and others), setting challenging standards, resolving potential conflicts between business units to optimize corporate performance, monitoring projects, and rewarding employees for positive attitudes and accomplishments. Pursue the "value profit chain model" by establishing a friendly working environment and treating employees well, who will, in turn, service customers better, in order to sustain long-term enterprise profitability.
- *Leadership fundamentals*. Earn and maintain creditability, demonstrate high ethical standards, take good care of followers, select and develop employees, ask good questions, and keep focusing on self-improvement techniques.

This competency model prescribes a large set of leadership skills and activities. Several comments in support of this model are noteworthy, as related to innovation, operations, networking, communications and followership.

9.2.1 Innovations

To create a differentiation in the marketplace, leaders envision new services that please customers. They lead the divergent and then convergent process of generating ideas, selecting novel ones, assessing their marketplace reactions, proposing and then implementing the service innovation project to develop the new service, considering the aspects related to design, financing, delivery, support, marketing/positioning, and sales. Leaders need to possess both technical and business management training to achieve innovative success. The topics covered in this text are included to facilitate such a broad-based multidisciplinary training.

9.2.2 Attention to Operations

Strategic differentiation is critical to the long-term health of a service enterprise, whereas operational excellence is vital to its short-term performance. Good leaders pay attention to both. Stadler (2007) conducted an extensive survey to determine how the best companies in Europe differ from good ones, as measured by their corporate financial performance over a period of time. They advocate three corporate specific strategies for growth:

1. *Exploit before explore*. Great companies grow by efficiently exploiting the fullest potential of existing assets and resources, rather than innovate their ways to growth. Developing in-house innovations has shown to be less attractive than acquiring external innovations to grow business.
2. *Diversify business portfolio*. Firms with multiple but related businesses fare much better over time. Diversification works well only if economies of scale advantage are gained. Conglomeration is a poor corporate strategy. Single businesses will not last, as all services have a finite life span.
3. *Be conservative about change*. Make changes incrementally based on careful planning and implementation.

Indeed, taking care of the enterprise's short-term performance is necessary before expending efforts to plan for the long-term future. For service organizations, the back-office operations must be systematically streamlined, standardized, and monitored for constantly improving speed, reliability, and cost-effectiveness. Service systems engineers should regularly update their knowledge of industrial best practices. Customer-facing employees must also be selected, trained, and monitored to make sure that they have the right mindsets, possess the required skills, and have access to relevant technologies in order to properly service customers

It is important to note that over time, companies will need to innovate—be it through in-house or open innovation channels—in order to avoid their services becoming increasingly commoditized. Such drives on the renewal of service offerings should be attempted at a frequency commensurate with changing market demand, competitive landscapes, and available implementation capabilities.

9.2.3 Networking

Networking is defined as steps taken in creating a fabric of personal contacts that will provide support, feedback, insights, resources, and information (Ibarra and Hunter 2007). To be effective, leaders engage in all three forms of networking—namely, operational, personal, and strategic, as shown in Table 9.1.

People are usually busy with day-to-day and short-term activities, so the need for an operational network is obvious. Personal networks are also relatively self-evident, as feedback and referrals are constantly needed. Strategic networks, while not being commonly perceived to be critically important, are essential in shaping the leader's strategic plans for the future.

9.2.4 Communications

Leaders need to communicate effectively, using simple, direct, and precise terms, according to Hamm (2006). A leader's real job is to inspire the organization to take

Table 9.1 Forms of Networking

	Operational	Personal	Strategic
1 Purpose	Efficiency and productivity of current operations	Development (personal and professional)Referrals to information and contacts	Future priorities and challenges, gain support from stakeholders
2 Contacts and Focus	Internal contacts focusing on current demands	External contacts on current and future interests	Both internal and external contacts toward the future
3 Target membership	Usual internal contacts as related to functions and organizational structure	Must be selectively developed	Must be developed in accordance with strategic context
4 Network attributes and key behaviors	Depth: Building strong working relationships	Breath: Reaching out to those who can make referrals	Leverage: Creating links that will matter

Adopted and Modified from Herminia (2007)

responsibility for creating a better future. They need to avoid the use of generalized terms and clichés, as these may mean different things to different people. Improved communications in five specific areas have the most profound impact on company's performance: structural changes, business priorities (time management), financial results, employee feedback, and corporate culture (mission, vision, and value statements).

9.2.5 Know Your Followers

To be a good leader, one needs numerous competent, loyal, and reliable followers. Kellerman (2008) believes that leaders need to understand their followers' needs and behaviors. There is a new topology for determining and appreciating differences among subordinates. These distinctions have critical implications for how leaders and managers fulfill their roles.

Traditionally, followers occupy lower positions in a business and have less power, authority and influence than their superiors. However, nowadays, knowledge workers at low corporate ranks have superior technological skills, and may be able to network with others to get things done, making leaders more and more dependent on them for success.

Types of Followers. The nature of the superior—subordinate relationship is determined by the degree of involvement the follower is engaged in their work. Followers "level of engagement" is categorized based on a continuum, ranging from "feeling and doing absolutely nothing" to "being passionately committed and deeply involved." Followers belong to one of these six types:

1. *Isolates*. Completely detached, usually invisible to top managers, isolates do their regular jobs with zero enthusiasm. They can drag down the organization.

Solutions can include training and development for lack of job satisfaction, or alternative scheduling for those suffering from job stress. If there is no viable solution, isolates should be removed from the company.

2. *Bystanders.* They observe but do not participate, and are implicitly in support of the status quo. They are aware of what is going on, but for their own self preservation, they remain quiet. They are silent but productive workers, who do not take initiatives. Leaders should offer incentive to cause them to become more engaged.
3. *Participants*. They are engaged in some way. They have an independent mind and they are either for or against the leader's position. Leaders need to understand this worker in order to determine whether this person is a supporter or an opponent.
4. *Activists*. They feel strongly about their leaders and organizations. They are eager, energetic, and engaged. They invest heavily in people and processes and they work hard either for or against the leader. These people are in the inner circle and appear as strong supporters of the leader.
5. *Diehards*. They are ready to go down for their cause. They show all-consuming dedication to a person or an issue they deem worthy.

Attitudes and options do not matter for isolates and bystanders. The remaining three types have strong opinions.

Classification of Followers. The followers may be classified as follows:

1. Followers who do something are always preferred over those who do nothing. Those who do nothing are not recommended.
2. Good followers support leaders who are good (ethical, effective) and oppose bad leaders.
3. Bad followers will do nothing to support good leaders and they will only support unethical and ineffective leaders.

Leaders have authority, power, and influence. Followers do not have authority, but they do have power and influence.

Concerns of Followers. Followers have concerns, and therefore it is important for leaders to make an effort to understand these concerns. Followers are generally concerned with: (1) the quality of their interpersonal relationships with their superiors, (2) their passion for the organization's mission, (3) their salary, (4) their titles, and (5) their other benefits. Knowing these will help leaders to address these concerns.

Degree of Engagement. The degree in which the followers become engaged is readily assessed. Questions might include (1) Do followers participate in meetings and proceedings? (2) Do they demonstrate engagement by pursuing dialogues, asking good questions, and generating new ideas? (3) Have they checked out—peeking away at their Blackberries or keeping a close eye on the clock?

Companies need leaders and followers who are collectively attuned to advancing the company's goals. Leaders and followers are not there to undermine each other based on personal issues or to gain advantages that are not in the best interest of the company (Bennis 2009). There are two good insights for us to review:

1. Spending too much effort to determine who is an ally and who is an enemy makes little sense within a company environment.
2. Companies are not an open community, thus followers are not encouraged to pursue the ideologies and programs of special interests groups. Unlike a free society, where special interests are looked after by countless independent people who derive satisfaction by being able to pursue ideologies regardless of cost or consequence, corporate culture directs resources (leaders and followers) toward achieving specific goals.

Strategies in Dealing with Followers. Based on Kellerman's classification methodology of followers, we have devised a strategic model, shown in Fig. 9.1, to suggest options of continuously improving support from followers:

- *Move* the negative followers out to stop them from working against the leader. If needed, hire new people to replace them.
- *Try* to influence the neutral fence-sitters to become supporters.
- *Invite* the positive followers to become members of one's "inner circle."
- *Ignore* the isolates and bystanders, as they do not impact the leader's cause. Move them out if they bring down the morale of an organization.

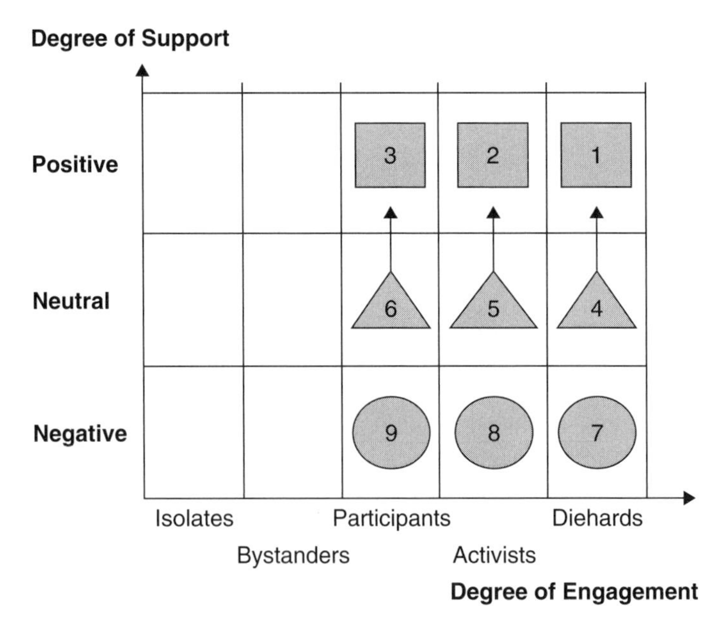

Figure 9.1 Options to gain supports from followers.

9.3 TOTAL LEADERSHIP

The New Competency Model just described emphasizes the work of service leaders. Friedman (2008) adds one more important aspect to the discussion. He believes that all leaders should strive for *four-way wins* by achieving a proper balance between the domains of work, home, community, and self. Such a balance is defined as *total leadership*. Three steps must be taken to pursue the total leadership concept.

1. Take a clear view of what one wants from and what one can contribute to each domain involved: Work, home, community and self, now and in the future. Assess the situation by thoughtfully considering the people who are most important to you and the expectations that you have for each other.
2. Design and implement carefully crafted experiments systematically—try out something new to see the impact on all four domains. If the experiment does not work beyond one or two domains, stop and readjust the design. If the experiment works, continue adding more domains.
3. Consider self-interests including physical and emotional health and intellectual and spiritual growth.

The following specific processes are recommended:

1. Define core values, leadership vision, and the current alignment of your actions and values—clarifying what is important.
2. Identify the most important people—Key stakeholders in all domains (work, home, community and self) and the performance expectations you have of one another and verify what and how much your key stakeholders actually need from you (most likely they need much less).
3. Focus own attention intelligently to spur innovative actions. By knowing what is most important to you, and having a more complete picture of your inner circle, you begin to see new ways of improving life for yourself, and for the people around you.
4. Conduct experiments and try them out during a controlled period of time. The best experiments are changes that your stakeholders wish for as much as, if not more than, you do.

Freidman (2008) suggests nine general types of experiments in order to test the boundaries:

1. *Having self-reflection regularly*. Keep a record of your thoughts, activities, and feelings for a month to see how various actions influence your performance and quality of life. Assess personal and professional goals, thereby increasing self-awareness and maintaining priorities. Track the times of day when you feel most engaged or most lethargic.
2. *Planning and organizing your time.* Use a PDA for all activities, share your schedule with someone else, and prepare for the week on Sunday evening.
3. *Rejuvenating and restoring.* Attend to body, mind, and spirit so that daily activities are undertaken with renewed power, focus, and commitment. Make time

for reading a novel. Engage in activities that improve emotional and spiritual health (yoga, meditation, etc.).

4. *Appreciating and caring.* It is important to have fun with people—do things, care for others, and appreciate relationships. These activities increase trust between people and connects them at the most basic human level. Join a book club or health club with co-workers. Help your child complete their homework. Devote one day a month to community service.
5. *Focusing and concentrating.* A total leader is physically present, psychologically present, or both, when attention to important stakeholders is required. This can mean declining other opportunities or obligations in order to prioritize needs. A leader shows respect and remains accessible to people of importance they encounter in different domains. Turn off digital communication devices at a set time. Set aside a specific time to focus on one thing or person. Review e-mails only at preset times during the day.
6. *Revealing and engaging.* Share more of yourself and spend time listening so others can better support your values and the steps you want to take toward your leadership vision. By enhancing communication between different aspects of life, you demonstrate respect for the whole person. For example, have weekly conversations about religion with your spouse. Describe your vision to others. Mentor a new employee.
7. *Time shifting and re-placing*. Work remotely or during different hours to increase flexibility and thus better fit in community, family, and personal activities while increasing efficiency, questioning traditional assumptions, and trying new ways to get things done. Work from home. Take music lesions during your lunch hour. Do work during your commute.
8. *Delegating and developing*. Reallocate tasks in ways that increase trust, free up time and develop skills in yourself and others, work smarter by reducing or eliminating low-priority activities. Hire a personal assistant. Have a subordinate take on some of your responsibilities.
9. *Exploring and venturing.* Take steps toward a new job, career, or other activity that better aligns your work, home, community, and self with your core values and aspirations. Take on new roles at work, such as a cross-functional assignments. Try a new coaching style. Join the board of your child's day care center.

Many of the aforementioned options suggested by Friedman (2008) are rather obvious. They are listed here to encourage innovative thinking, when exploring new ways of experimentation to better satisfy the needs of all stakeholders. The goal is that through ongoing experiments, one can find the healthy balance between all four domains necessary for achieving success as defined by the total leadership model.

9.4 LEADING CHANGE

Service businesses must reinvent themselves from time to time to help cope with newer, and more challenging market environments. Effecting change can be difficult because of the natural human resistance put up by workers in the trenches of the business.

(Kotter 2007). Leaders who successfully transform businesses need to correctly implement the following eight steps in the right order (see Section 4.8.2):

1. *Establish a sense of urgency for change*. Examine market and competitive realities and identify threats and opportunities.
2. *Form a powerful guiding coalition.* Effect changes by coordinating the group's team effort.
3. *Create a vision (a future state).* Develop strategies for achieving the vision. Do not try to "boil the ocean," in order to be meaninglessly ambitious. Do not let minutiae to obscure the bigger picture.
4. *Communicate the new vision and strategies widely*. Secure awareness and acceptance.
5. *Empower others to act on the vision*. Remove obstacles (structures, people) and encouraging risk taking.
6. *Plan for and achieving short-term wins.* Recognize employees contributing to such wins.
7. *Consolidate improvements and producing still more change*. Introduce modifications to structures, systems, and policies, hiring and promoting right people to implement the vision, and enhance the process with new projects and change agents.
8. *Institutionalize new approaches*. Advertise the corporate success due to these new gains and ensure leadership development and succession.

Some of these steps have already been included in the new competency model. Others are of a more tactical nature, to ensure that momentum and leadership creditability are practically maintained in the changing process. It is important to have patience, as "the Chinese Great Wall was not build in one day," things do take some efforts.

9.5 LEADERSHIP STYLES, QUALITIES, AND ATTRIBUTES

9.5.1 Leadership Styles

There are five major styles of leadership that are classified according to the attributes of either a *concern for people* or an *emphasis on tasks*. These are defined as follows:

1. *The nice guy*—Places too much value on social acceptance while neglecting technical tasks.
2. *The loser*—Does not obtain acceptance from others, nor get the job done.
3. *The compromiser*—Balances both the needs of people and task factors.
4. *The task master*—Is interested in getting the job done right without concern for human feelings.
5. *The ideal manager*—Gets the job done and at the same time makes everyone happy.

Figure 9.2 illustrates these styles. The principal style of leadership exhibited by a leader is largely determined by his or her personal characteristics derived from traits such as childhood experiences, parental impact, work habits, value systems, and others.

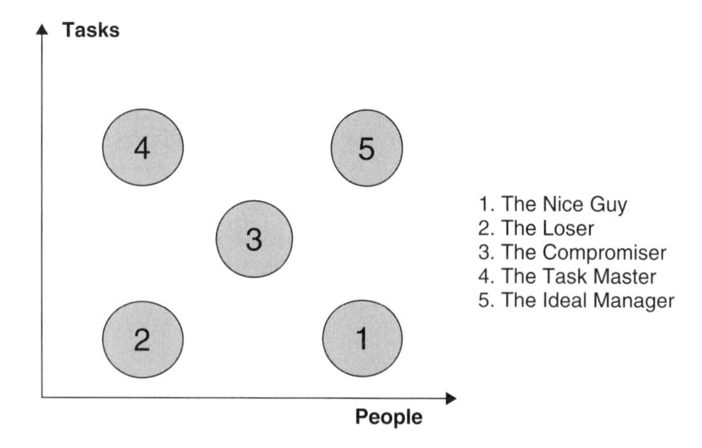

Figure 9.2 Leadership styles.

Leadership style is effective or ineffective, flexible or inflexible. Most leaders practice more than one style in their leadership. Different styles are applicable with different people at different times (Cohen 2002). Engineering managers must vary their styles according to the needs of the employees (depending on whether the employees are experienced or novices) and the situation at hand. One cannot allow the perfect to be the enemy of the effective and of the necessary. The main thing is to keep the main thing being the main thing.

The consulting firm Hay Group conducted a survey on leadership styles with a sample of 3,871 executives, from a worldwide database of 20,000 executives. According to Goleman (2000), the following styles were identified by the survey as particularly useful for generating results: authoritative ("come with me"), affiliate ("people come first"), democratic ("what do you think"), and coaching ("try this"). Goleman also advises that leaders should minimize the use of two additional styles, namely, coercive ("do what I tell you") and pacesetting ("do as I do now"). Practicing the paradigm of "my way or the highway" is not recommended.

Effective leaders are said to switch constantly between the authoritative, affiliative, democratic, and coaching styles, depending on the situation at hand. The key concept here is *flexibility*. Those engineers who are weak in any of these four styles are advised to seek self-improvement.

Example 9.1 The NASA space shuttle *Columbia* broke up during its descent on February 1, 2003, killing all crew members on board. Immediately after its launch, it was made known to NASA Mission Control Center that a piece of solid foam broke off from the shuttle's external fuel tank during its ascent and smashed into its left wing at high speed. During *Columbia*'s sixteen-day space journey, NASA management dismissed the significance of this mishap and did nothing. Instead, NASA insisted that the incident was a routine affair, as foam pieces had broken off in a number of previous space flights. Furthermore, NASA overturned the requests of its lower-level engineers to obtain U.S. Department of Defense pictures of the wing that were photographed by military satellites. Low-level engineers e-mailed a query to their superiors about the possibility of getting the astronauts to take a space walk, which was safe and easy to do, to inspect the wing. That e-mail was never answered.

NASA management lost two precious opportunities to avert the disaster. Furthermore *Columbia* could have had its mission extended by another three to four weeks, allowing the Shuttle *Atlantis,* whose launch schedule could have been accelerated to February 10 (from its original launch date of March 1) to rendezvous and rescue the crew. The other option would have been for two *Columbia* spacewalkers to repair the damaged wing with heavy tools and metal scraps scavenged from the crew compartment.

The *Columbia* Accident Investigation Board (CAIB) conducted interviews, shot foam pieces experimentally at test wing targets, and analyzed a large volume of data. Its final report pointed clearly to a NASA management failure, attributable to several specifically named managers.

What were the principal modes of management failure deemed responsible for the shuttle *Columbia* disaster?

ANSWER 9.1 The *Columbia* shuttle disaster was the result of several factors (Langewiesche 2003):

1. *Culture*. NASA claims that it is a "badgeless society," meaning that it does not matter what title of position and responsibility is on one's name badge; everyone is equal when it comes to concern about shuttle safety. NASA has an open-door and free communications policy.

   The opposite was true, according to Langewiesche. He found that a large number of NASA employees were afraid to speak up. In fact, Langewiesche claims in his article that many employees feared that their position in the organization predetermined how they themselves were viewed, as well as how their opinion was welcomed or accepted by the higher-ups. Many employees felt NASA's leadership style to be coercive, not democratic.

2. *Overconfidence and personal arrogance.* The NASA managers had shown arrogance and insularity, while exhibiting a tough and domineering management style. Langewiesche says that the CAIB report singled out a specific NASA decision maker as "intellectually arrogant and an abysmally failed manager." Moreover, Langewiesche claims that the NASA managers demonstrated imperious and self-convinced attitudes, suffered from a lack of curiosity, and believed in themselves blindly. It is not difficult to see and understand that, as previous shuttles had survived frequent foam strikes at lift-off, managers could blithely accept the notion that there would be no fatal risk this time.

   Ultimately, it was conclusively substantiated by CAIB tests that the falling foam punched a hole about 10 inches wide into the wing's leading edge. This hole allowed the hot gases of reentry to enter the shuttle compartment and burn it from the inside. Unfortunately, there was no other past experience upon which the higher-ups could base their decisions, although they perhaps should have erred on the cautious side, rather than the less costly side.

3. *Pressure to meet deadline.* The NASA administrator set stringent performance goals related to the International Space Station Project. The strict deadline

for completing the "core," of which *Columbia* was a part, was February 19, 2004. As a consequence, organizational and bureaucratic concerns weighed heavily on the managers' minds.

A combination of the foregoing three factors led the key managers to stubbornly believe that the foam strike was insignificant and also led them to forgo the opportunity to collect more data and hence initiate any necessary emergency steps to save the crew. It was a colossal engineering management failure.

9.5.2 Self-Awareness—A Part of Emotional Intelligence

According to Goleman (1998), all effective leaders have a high degree of emotional intelligence. There are five components of emotional intelligence: self-awareness, self-regulation, motivation, empathy, and social skills. Each of these components can be learned and enhanced by coaching, observation, training, and practice.

Sun Tzu (2008) says: "Know oneself and your opponents, win all battles ever waged." Knowing yourself is the first step toward "self-awareness." In addition, understanding others, spending time analyzing their behaviors, watching how they interact, gives you a winning advantage over your opponents and ensures success in all battles. The key lies in operating from your strengths to achieve true excellence. Drucker (2005) suggests seven useful pointers:

1. *Strengths/weaknesses*. Within a two- to three-year time period, analyze your own strength and weaknesses through feedback analysis from trials and experiments Focus on applying strengths (Roberts et al. 2005), improve on them, and fill in gaps in skills. Make sure not to show intellectual arrogance (being bright is not a substitute for knowledge), as it is self-defeating. Recognize your own weaknesses and make corrections (e.g., manners, planning without follow-up-execution). Do not waste time on improving areas of incompetence; it is more effective to improve on your strengths.
2. *Method of learning*. Everyone has his or her own method of learning. Some prefer reading, others take copious notes, others prefer direct action, and others like to think things through quietly or talk them out. Identify your most effective and preferred way of learning and practice these methods regularly.
3. *Styles of working*. How do you work with others, in what relationships (team members or alone, as coaches, as a decision maker, or as an advisor)? Do you work well under stress, or would you need a highly structured and predictable environment? Do not try to change yourself. Work hard to cultivate the way you perform, be selective in doing work which makes optimal use of your strengths.
4. *Definition of own values*. Values of individuals must be in line with that of the organization. Use the mirror test to check on the ethics-related issues: What kind of person do I want to see in the mirror in the morning? Here are some examples of nonethics values differences:
   - Company favors to promote from within (encourage loyalty and offer rewards to contributors) or to hire from the outside (getting fresh blood.).

- Company favors incremental improvements in drugs (helping physicians to do better jobs steadily) or focuses on an occasional "breakthroughs" (scientific discovery).
- Business should run on a short-term operational emphasis (daily operations to focus on performance) or a long-term horizon (health).

5. *The right place.* Where to make the greatest contributions? The decision should be based on three questions: (a) What are my strengths? (b) What is my style of working (team players or solo player, large or small organization, decision maker or adviser)? and (c) What are my values?

   Sometimes there may be a conflict between a person's value and his strengths. In such cases, just quit the work and pursue other opportunities with aligned values.

6. *Which contributions.* To identify the best contributions to make, ask the following questions:

   - What does the situation require?
   - Given my strengths, my way of performing and my values, how can I make the greatest contribution to what needs to be done?
   - What results have to be achieved to make a difference? The recommended selection criteria are as follows:

     - Results should be hard to achieve—requiring stretching, but within reach. Do not aim at something very difficult to accomplish—this is futile behavior.
     - Results should be meaningful, they should make a difference.
     - Results should be visible and measurable.

7. *Responsibility for relationships*. Managing oneself requires taking responsibility for relationships:

   - *Accept others as they are, as they have their own strengths, values, and ways of doing things.* You need to understand these factors about your coworkers. The secret of managing a superior is to study their work habits so that you can align yourself with their methods to increase the productivity of the company. Learn how to make the best use of your coworkers' strengths, values, and ways of doing things, so that working relationships are well maintained.
   - *Take responsibility for communications.* Organizations are built on trust, not on force. Trust is built on an understanding of each other's roles.

Service systems engineers and leaders are advised to take note as successful leadership also depends on following the previously noted criteria.

9.5.3 Inspirational Leadership Qualities

Leaders typically have vision, energy, authority, and strategic direction. However, leaders will not succeed if there are no followers.

Followers are hard to find in these "empowered" times. Goffee and Jones (2001) point out that there are four specific qualities needed by leaders who want to be truly inspirational:

1. *Approachability and humanity*. It is a human quality to show personal humility and vulnerability. Doing so will demonstrate authenticity, promote trust building, enhance solidarity, and foster collaboration. Acknowledging one's own shortcomings opens up opportunities for improvement. (See the lesson contained in "The Wisdom of the Mountain," Appendix 9.14.2.) However, Goffee and Jones advise that the weaknesses shown should be tangential flaws (such as hardworking habits) only and not fatal ones or character-related issues.
2. *Tact*. It is the ability to know when and how to act based on intuition. It is a situation sensor that is capable of reading underlying currents, detecting subtle cues, and gauging unexpressed feelings. (See the lesson contained in "The Sound of the Forest," Appendix 9.14.3.) This quality is widespread among excellent business leaders. However, one needs to make sure that reality testing is done frequently with trustworthy friends or confidants, to avoid disasters due to one's own inability to evaluate faulty situations.
3. *Tough empathy*. It is the ability to establish a balance between respect for individuals and the task at hand, in order to impel leaders to take risks and to care about the people and the work they do. Leaders do well when they close the distance between themselves and their employees. Leaders must give employees what they need by helping, coaching, and participating in what they do. (See the lesson contained in "The Wheel and the Light," Appendix 9.14.4.)
4. *Uniqueness*. It is the ability to maximize the benefits derivable from the leader's own uniqueness (e.g., dress style, physical appearance, imagination, loyalty, expertise, handshake, humor, etc.) and to use this separateness for motivating others to perform better.

Leaders need to be themselves, acquire more skills, and apply these four qualities to fit their own personality styles in order to successfully inspire others.

9.5.4 Leadership Attributes

Leaders are those who have special knowledge, are accessible, exhibit charisma (the natural ability to attract followers), and possess the authority to delegate. Table 9.2 contains a set of basic guidelines for engineering managers to lead.

Various research articles on leadership indicate that efficacious managers possess a set of common attributes. They have unquestionable character. Their creditability is high because of their technical skills, ethics, fairness, and moral standards. They master the functions of management, such as planning, organizing, leading, and controlling. They constantly perfect their skills in dealing with people, managing time, and controlling their own stress. They communicate effectually both in oral and written forms (Mai and Akerson 2003). They have learned to listen well. (See the lesson contained in "The Sound of the Forest," Appendix 9.14.3.) They are full of energy and in good health. They are enthusiastic and positive about things and people. They are self-motivated,

Table 9.2 Guidelines for Leading as Engineering Managers

1. Prepare oneself (e.g., study the rules, policies, and objectives of your organization)
2. Understand all jobs under one's direction
3. Be observant of things going on around oneself (e.g., managing by walking around)
4. Pay attention to details
5. Pose questions
6. Keep things in perspective (e.g., avoid being too close to the trees and unable to see the forest)
7. Be anticipatory of the future conditions

flexible, and independent. They take the initiative to originate actions to influence events. They take prudent risks (Nelson 1999). They have superior conceptual skills in reviewing data, solving problems, taking action, and planning strategically. They are both persistent and persuasive in achieving the goals they set out to achieve. Because of these attributes, they leave behind good impressions and build confidence in others' minds. Effective leaders attract those who are willing to follow them.

In fact, these attributes are not dissimilar to the profile of top executives commonly noted in business literature. (See Table 9.3.)

Managers and leaders have different mental orientations. These mental orientations make leaders as important as managers in adding value to their organizations. Leadership talents are defined as having a natural predisposition for recurring patterns of thoughts, feelings, and behaviors that can be applied productively. The Gallup Organization (www.gallup.com) has identified twenty key leadership talents by interviewing

Table 9.3 Profile of Top Executives

1. Are able to work with people
2. Possess social poise (self-assured and confident)
3. Are considerate of others
4. Are tactful and diplomatic
5. Practice self-control
6. Are able to analyze facts, to understand and solve problems
7. Are able to make decisions
8. Are able to maintain high standards
9. Are tolerant and patient
10. Are honest and objective
11. Are able to organize time and priorities
12. Are able to delegate
13. Are able to generate enthusiasm
14. Are able to be persuasive
15. Possess a great concern for communication

more than 40,000 leaders and top-tier managers over a period of thirty years (Hughes, Ginnett and Curphy 2008). These twenty leadership talents are outlined in the following four groups:

A. Ability to provide direction
   1. Vision: is able to build and project beneficial images.
   2. Concept: is able to give the best explanation for most events.
   3. Focus: is goal oriented.
B. Drive to execute—Related to motivation
   4. Ego driven: defines oneself as significant.
   5. Competitive: has the desire to win.
   6. Achiever: is energetic.
   7. Courageous: relishes challenges.
   8. Activator: is proactive.
C. Capacity to develop relationship with others
   9. Relater: is able to build trust and be caring.
   10. Developer: desires to help people grow.
   11. Multi-relater: has a wide circle of relationships.
   12. Individuality perceiver: recognizes people's individuality.
   13. Stimulator: is able to create good feelings in others.
   14. Team leader: is able to get people to help each other.
D. Management system—Related to management abilities
   15. Performance orientated: is results oriented.
   16. Disciplined: is able to structure time and work environment.
   17. Responsible and ethical: is able to take psychological ownership of one's own behavior.
   18. Arranger: is able to coordinate people and their activities.
   19. Operational: is able to administer systems that help people to be more productive.
   20. Strategic thinker: is able to do what-if? Thinking and create paths to future goals.

It is obvious that a good leader must be a good manager, but a good manager may not be a good leader if he or she lacks some of the leadership talents indicated.

Bass and Bass (2008) state that a successful leader is characterized by the attributes outlined in Table 9.4.

Harvard Business School Press (2005) suggests eleven attributes that an effective leader must possess. These are delineated in Table 9.5.

The twenty leadership talents described by Gallup, the ten characteristics of successful leaders by Bass and Bass (2008) (Table 9.4), and the eleven attributes of effective leaders by HBSP (2005) (Table 9.5) are consistent with one another. In fact, all of them asymptotically converge to a finite set of common leadership attributes that all future engineering leaders should feel comfortable assimilating. Engineers who

Table 9.4 Leadership Attributes

1.	A strong drive for responsibility and task completion
2.	Vigor and persistence in pursuit of goals
3.	Adventuresome and original in problem solving
4.	Drive to exercise initiative in social situations
5.	Self-confidence and sense of personal identity
6.	Willingness to accept consequences of decisions and actions
7.	Readiness to absorb interpersonal stress
8.	Willingness to tolerate frustration and delay
9.	Ability to influence another person's behavior
10.	Capacity to structure social interaction systems to suit the purpose at hand

Table 9.5 Leadership Attributes (HBSP)

1	Communicate well
2	Are good listeners
3	Are approachable
4	Delegate
5	Lead by example
6	Read situations and people well
7	Are good teachers
8	Care about people and display it
9	Are fair, honest, and consistent
10	Know how to criticize
11	Know how to accept criticism

aspire to become leaders are encouraged to understand and display these attributes so that, over time, the leadership talents and attributes will become second nature to them.

Example 9.2 Highly talented technical professionals may have academic training (advanced degrees), experience (company tenure), professional credentials (societal committee activities, awards, business connections), and accomplishments (patents, publications, completion of major projects) superior to the engineering manager. They could be difficult to manage. What are some of their characteristics and working habits? What strategies are effective in managing them?

ANSWER 9.2 Highly talented technical professionals tend to have the following work-related preferences:

- They favor individual assignments with clearly recognizable responsibility. They do not prefer teamwork, wherein individual contributions may be crowded out.
- They tend to strive for perfection, as they view the technological output as a reflection of themselves.

- Technical professionals typically become easily frustrated by unexpected changes in program priority or resource allocation strategies for approved action steps deemed essential to achieve the program objectives.
- They hate management jargon.
- They find happiness in technical work, without being constrained by other nontechnical concerns or the involvement of low-skill people.
- They are readily turned off by administrative details, restrictive policies and guidelines, poor quality decisions based on questionable data and assumptions, excessive reporting requirements, and overly tight management control.
- These workers assign high value to independence, self-motivation, self-direction, and fairness.
- They demonstrate a reserved attitude in social interactions.

Highly talented technical professionals may be managed by adopting the following strategies:

1. Decide on objectives of technical programs or assignments, define the funding priority, understand the reasons for decisions, and secure the company's commitment to the chosen programs.
2. Assign technology programs and activities to specific individuals by clearly communicating the objectives, budget constraints, expected results, and other details. Suggest specific ways to measure outcome.
3. Solicit comments from the individual technical professionals involved regarding project value, interest, readiness to perform tasks, and other issues that may have been of concern to them.
4. Invite the individual technical professionals to:
   - Outline specific technical methods to accomplish the program or activity at hand.
   - Produce an action plan and define budget requirements (accounting for man-hours, equipment, supplies, computation resources, outside resources, etc.).
   - Specify preliminary milestones of when interim results are to be reported out (monthly, biweekly etc.).
   - Define deliverables.
5. Review and accept the plans. Authorize the individuals to commence programs and assignments.
6. Be available for any unexpected problems encountered by the individuals pursuing these assignments (e.g., practicing the concept of management by exception). Must be helpful and leave sufficient room for independent work by the individuals.
7. Acknowledge receipt of the final report after it is submitted. Read the report, review results with individuals, and invite comments on any work extension needed or desired and on how to enhance the management aspects of the program.

8. Evaluate the work performed in terms of its expected value to the organization and offer feedback, including any responses from top management and other parties affected by the accomplished programs or assignments.
9. Praise the individuals appropriately whenever good work is done, by, for example, practicing the concept of motivation by positive reinforcement and recognition. Offer improvement suggestions if performance is to be upgraded.
10. Seek and arrange opportunities for the individuals to make prepared presentations on technical programs and assignments during review meetings with upper management groups.
11. Document tasks and evaluations (including specific contributions made and their significance to the company) to be in a position to provide an instant report to upper management, if required, and to form a basis for the annual appraisals of the individuals involved.

9.5.5 Self-Motivation

Jones (2008) points out that there are six common characteristics shared among star performers in business and sports:

1. Love the pressure—the ability to remain cool in high-stress situations.
2. Being self-directed.
3. Have a secondary passion in life, so that they can switch their involvement on and off.
4. Having an eye on the long-term goals and remaining focused on them. They accomplish a series of short-term goals to maintain momentum.
5. Use competition to self-improve and reinvent oneself.
6. The will to win, in spite of a lot of unavoidable failures along the way.

Although the athletic talents, self-motivation, dedication, perseverance, goal-orientation, and strong will to win are responsible for a variety of great successes in individual sports, they are by no means sufficient in achieving business successes, which require the willing collaboration of thousands of others, who have the unique skills and insights to provide invaluable support. The challenges to business leaders are thus to lead on building teams and securing broad-based collaborations.

9.6 LEADERS AND MANAGERS

Companies need both leaders and managers. According to Kotter (2007), management focuses on the steps of (1) choosing goals and targets typically set by existing corporate strategies, (2) planning action steps to achieve the goals, (3) allocating required resources, (4) creating organizational structures, and (5) exercising control, measuring results, and correcting performance to ensure the attainment of goals. The primary duty of management is to keep the organization functioning properly. Management planning is deductive in nature, since it aims at creating orderly results.

Leadership, by contrast, focuses on the steps of (1) setting a vision and direction for the future in response to changes imposed on the organization externally, (2) creating strategies to produce the changes needed to achieve the vision, (3) designing action steps to accomplish the strategic goals, (4) aligning people and forming coalitions, and (5) motivating and inspiring people to move in the right direction. The function of leadership is to produce change. Setting direction by leadership is inductive in nature, as it attempts to produce a new vision and new strategies. Vision is the picture of what the company will be in the long run with respect to its business portfolio, market position, technological prowess, and company culture.

The emphasis placed and approach taken by managers are different from those of leaders. Table 9.6 summarizes such differences, based on published work by Zaleznik (2004).

Table 9.6 Differences between Managers and Leaders*

Characteristics	Managers	Leaders
Focus	Do things the right way	Do the right thing
	Administration, problem solving	Direction setting
	Reconcile differences	Creativity and innovation
	Seek compromises	
	Maintain balance of power	
Emphasis	Rationality and control	Innovative approach
	Accept and maintain status quo	Challenge status quo
	Put out fires	Blaze new trails
Targets	Goals, resources, structures, people	Ideas
Orientation	Tasks, affairs	Risk taking
	Persistence	Imagination
	Short-term view	Long-term perspective
Success factors	Tough-mindedness	Perceptual capability
	Hard work	
	Tolerance	
	Goodwill	
	Analytical capability	
Points of inquiry	How and when	What and why
Preference	Order, harmony	Chaos, lack of structure
Aspiration	Classic good soldier	Own person
Favor	Routine	Unstructured
	Follow established procedure	
Approach with people	Using established rules	Intuitive and empathetic
Personality	Team player	Individualist
Relevance	Necessary	Essential
Thrust	Blend in	Stand out
	Bring about compromise	Lead changes
	Achieve win–win	
Mentality	"If it isn't broken, don't fix it."	"When it isn't broken, this may be the only time you can fix it."

*Adapted from Abraham Zaleznik, "Managers and Leaders: Are They Different?" HBR Classic Article # R0401G. *Harvard Business Review* (January 1, 2004).

Kotter (1990) further believes that strong leadership with weak management is no better—and is sometimes actually worse—than the reverse. A competitive organization needs both strong leadership and strong management and should use each to balance the other. Management is needed to ensure that complex organizations operate properly, including attaining incremental improvements. Leadership is needed to cope with changes that will be thrust on the organization due to advancements in technology, changes in market environments, and competition at the international level.

The transition from managers to leaders has been the subject of a few studies (Bennis 2003; Zenger and Folkman 2009). Engineers interested in becoming managers and leaders should also review the seventy-eight questions raised by Clark-Epstein (2006), study the twelve principles delineated by Gronfeldt and Strother (2005), and master the six competencies described by Pelus and Horth (2002). They may benefit from the lessons learned by other leaders (Krames 2005). Those who regard General Electric's Jack Welch as a professional to emulate have several resources to consult (Welch and Bryne 2003; Krames 2001; Slater 2001).

Example 9.3 Every engineering manager has strengths and weaknesses. The key to continuous improvement is, of course, to identify one's own weaknesses and do something about them. How should an engineering manager study oneself and systematically discover opportunities for improvement?

ANSWER 9.3 Henry Mintzberg (2003), in his article "The Manager's Job: Folklore and Fact," proposed a large number of insightful questions for managerial self-study. These questions are worth studying.

9.7 FACTORS AFFECTING THE PROMOTION TO MANAGER

Generally speaking, there are three basic prerequisites for engineers to receive promotions to the managerial ranks. These prerequisites will be discussed next.

9.7.1 Competence in Current Technical Assignments

The engineer must be able to master the duties and responsibilities of his or her current position, have the respect of his or her coworkers, and receive favorable recommendations from his or her superiors.

9.7.2 Readiness and Desire to Be Manager

The candidate must have demonstrated the readiness to handle larger and more challenging responsibilities, as well as have gained the required skills and knowledge via courses, seminars, on-the-job training, professional activities, teamwork, volunteer work, tasks related to proposal development, feasibility studies, technology assessment, and other avenues. The candidate must possess the skills to manage time and have the desire to seek leadership positions and opportunities to exercise power and manage people, spearhead change, and resolve conflicts. The candidate must also be already perceived by others as a competent, responsible, reliable, fair, and visionary person to take on critical and broad-based leadership role.

9.7.3 Good Match with Organizational Needs

Being competent and ready for promotion to managerial positions is a necessary, but not sufficient, condition for being promoted. The candidate's ambition, desires, and capabilities must also be a good match with the current and long-term needs of the company. Well-organized enterprises constantly need new managers and leaders because of the dynamics in the business environment, advancements of technology, competition in the marketplace, and the mobility of people.

Managerial competency may be classified according to the categories illustrated in Fig. 9.3. Specifically, it is helpful for engineers who aspire to become managers to focus on the following general capabilities:

- *Engineering management skills*. These skills include the engineering management functions of planning, organizing, leading, and controlling (see previous chapters), the ability to work with people, and excellent communication skills.
- *Power base formation*. Candidates must nurture the ability to build personal power by technical know-how, experience, and networking. Promoting and marketing one's own achievements are important, following the well-known saying, "Early to bed, early to rise, work like hell, and advertise!"
- *Assertiveness*. Candidates should demonstrate the ability to become and remain assertive in exercising judgment and making decisions. They should be proficient in resolving conflicts and problems of a technical, political, conceptual, and people-centered nature.

As an example, Ciampa (2005) suggests the key factors affecting the promotion to higher level positions being alliances and political savvy, whereas superior performance is normally expected. Table 9.7 summarizes the related comments regarding these two factors.

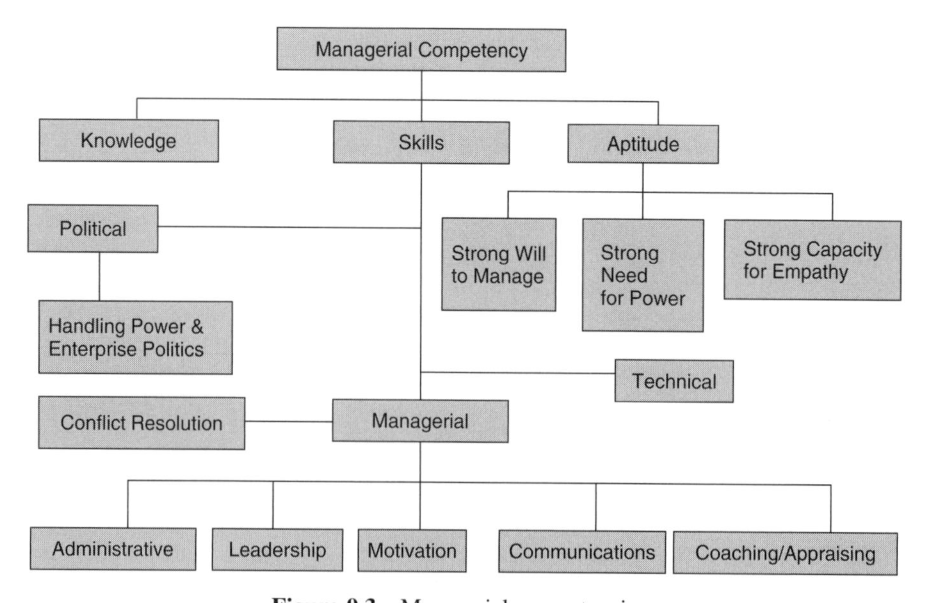

Figure 9.3 Managerial competencies.

Table 9.7 Critical Performance Factors

	The good candidate	The better candidate
Management Savvy	Know what is required operationally for short-term results Motivates others to do it Uses time well Prioritizes among issues that are all important Frequently delegates tasks Has a history of developing subordinates and exporting talent Organizes and mobilizes talent toward most significant problems Pushes people to achieve more than they think they can	Avoids jumping in personally to solve problems others can handle Makes the right judgments about what to expend energy on Maintains control of the key decision and a full pipeline of talented people Makes people feel appreciated and stay loyal
Political Intelligence	Accurately reads political currents Understands patterns of relationships quickly in an unfamiliar environment Build relationships with peers and subordinates Makes sure the CEO and the board know what he or she is capable of doing	Isn't labeled "political" Recognizes how relationships are likely to affect early success Gets peers and subordinates to go out of their way to help Doesn't seem self-serving
Personal Style	Is a star performer Is intense and driven to excel Is hardworking, usually putting in more time and effort than peers do Enthusiastically backs initiatives that will help the business succeed Is a leader among peers Understands new ways of doing things and makes important connections	Makes success look effortless Allows other's performance to be recognized, too Manages energy to stay on the "rested edge" and to avoid the "ragged edge" Knows when to hold back and when to let go Enables peers to improve their performance Stays grounded and makes sure basic needs are met while mastering new concepts

Source: Ciampa, 2005

Example 9.4 According to H. C. Aucoin (2002), engineering managers must possess a set of skills not taught in a typical college engineering curriculum. These skills include the ability to do seven things:

1. Deal with ambiguity and uncertainty
2. Lead (people related and technological)

3. Take risks
4. Delegate
5. Be a team builder
6. Communicate
7. Initiate

How do you propose that an engineer acquire these skills?

ANSWER 9.4 To acquire the previously described skills, engineers could follow these steps:

1. Understand why each skill set is important, and verify its importance by talking with trusted partners (e.g., parents, close friends, relatives, professional acquaintance, and mentors). Other useful steps to enhance understanding and building leadership skills are to:

   A. Browse technical, business, and managerial publications (e.g., technical journals, *BusinessWeek*, *Fortune*, *Wall Street Journal*, and *Harvard Business Review*).
   B. Keep informed of new advancements in the field, such as business strategies, market expansion, technologies, innovations, customer relations management, enterprise integration systems, supply chain management, business models, lean manufacturing, and e-business.
   C. Absorb new concepts and practices, and become proficient in identifying *best practices, success factors*, and other benchmarks.
   D. Recognize new opportunities in technologies, business, and products potentially valuable to the organization.

2. Understand the metrics (standards) for measuring and monitoring progress made in acquiring these skill sets.
3. Conceive a plan for action, including specific action steps and milestones.
4. Make a commitment by setting aside time and effort to implement the plan.
5. Take courses and training seminars, observe the experienced managers and leaders in action, and ask questions of qualified people to acquire the specific techniques needed to facilitate technical and managerial growth and build and maintain skills.

   Training options include programs at professional societies, companies offering training services, American Management Association courses, and university-based training programs.
6. Seek proactively opportunities to practice the learned techniques. Volunteer for team assignments, become an officer in a student organization. Do volunteer work in church, scouting organizations, benefits, the United Way, the Rotary Club, or political groups. Spend time in professional societies or industrial committees. Join Toastmasters International to practice public speaking skills.

9.8 LEADERSHIP SKILLS FOR THE TWENTY-FIRST CENTURY

John C. Maxwell said: "A leader is one who knows the way, goes the way and shows the way."* To become superior leaders, engineering managers are advised to focus on the following eight attributes, according to Cohen (2002):

1. *Maintain absolute integrity.* Any doubt about the leader's integrity will be reflected in the trust others place in the leader.
2. *Be knowledgeable.* The leader should be good at what it takes to get the job done.
3. *Declare expectations.* The leader should let people know which direction to go in and what results are expected.
4. *Display unwavering commitment.* The leader must demonstrate his or her clear commitment.
5. *Get out in front.* The leader needs to build, establish, and maintain a strong positive image. The leader should get out of the office and see what is going on (e.g., to the plant floor, marketplace, customer service center, and technology labs). The leader should also get out in front to be seen, so that others know their manager is committed.

   General MacArthur gave this advice to a young battalion commander during World War II: "Major, when the signal comes to go over the top, if you go first, before your men, your battalion will follow you. Moreover, they will never doubt your leadership or courage in the future."*
6. *Expect positive results.* Show self-confidence and work to get favorable results.
7. *Take care of people.* This is the basic reciprocity doctrine of Confucius: "If you take care of people, people will take care of you." Starbucks is said to practice this doctrine by taking care of their employees first, then customers, and finally shareholders.
8. *Put duty before self-interests.* The mission and the employees must be more important than one's own self-interests.

The preceding list of attributes neglects to include the all-important quality of strategic thinking and the leader's capability to create vision. Without vision, a jperson with these attributes is merely a hard-working taskmaster who is responsible, goal oriented, and socially assertive.

The twenty-first-century arena is characterized by a high pace and level of education, low patience, and compliance with authority, close relationships with customers, and a brisk speed to market. Continued changes due to global competition, environment, technological advances, and population diversity are expected to be very rapid. Jack Welch, a former CEO of General Electric, said, "When the rate of change on the outside exceeds the rate of change on the inside, the end is in sight" (Krames 2002).

**Source*: http://quotations.about.com/cs/inspirationquotes/a/Leadership10.htm.

*Source unknown, but is attributed to MacArthur.

Business success in the twenty-first century requires global connectivity, obsession with customer satisfaction, enhanced performance of people and technologies, alternative organizational frameworks, real-time responses, and enduring self-examination.

In the new century, leaders need a new set of skills to help exercise their leaderships, centered on results. The leadership set includes these ten skills:

1. *Leading* with a strategic focus and vision. Advance and articulate a value proposition that represents a proposed model to add value to companies' stakeholders.
2. *Managing* multiple points of view simultaneously, such as those from customers, suppliers, shareholders, and employees. Remain flexible and adaptable in dealing with technology, working with people, and forming business networks, and being capable of negotiating for solutions that are acceptable to parties involved.
3. *Keeping* high-level goals in sight, while managing and tracking day-to-day success. Keep the spirit of the enterprise alive.
4. *Fostering* productive changes. The "boiled frog syndrome" is explained as follows: If you put a frog in a pan full of cold water and slowly turn up the heat, the frog will boil to death rather than jump out. If you drop a frog in a pan full of boiling water, however, it will jump out immediately. The moral of the story is that, if people do not sense a significant need to change, they may not get out of their comfort zones and change what they are doing. Usually, effective leaders are needed to convince people of the need to change.
5. *Being* inspirational, technologically savvy, entrepreneurial, and devoted to service. By focusing on the needs of customers.
6. *Investing* in a business model that guides employees' decision making at all levels. So that all of them are empowered to contribute.
7. *Devising* and maintaining transformational knowledge systems. So that corporate knowledge is constantly developed and applied.
8. *Accessing* relevant information rapidly in light of the explosion of available information. In order to remain competitive.
9. *Understanding* how global business practices have evolved.In order to take advantages of new opportunities offered by global economies of scale and scope.
10. *Learning* quickly while not relying on what is already known or understood. In order to be able to think out-of-box.

In his May 1999 speech at the University of Richmond, Virginia, Gen. H. Norman Schwarzkopf, U.S. Army (Ret.), pointed out that leadership is a whole combination of different ingredients; but, by far, the single most important ingredient of leadership is character. During the last 100 years, about 99 percent of leadership failures have been due to faulty character, not incompetence. His remarks are particularly pertinent today in view of the unethical practices committed by leaders such as financier Bernie Madoff, former governor of Illinois Rod Blagojevich, Jeffrey Skilling at Enron, and Bernie Ebbers at WorldCom.

Engineering managers need to have business savvy in order to lead in the twenty-first century. The combination of technological know-how and business savvy is powerful. These five guidelines may help engineering managers to acquire the needed business acumen:

1. *Become* well versed in the business issues faced by the company. This includes a thorough understanding of the corporate vision; the company's priorities, strengths, and weaknesses; the current market position; business processes; and engineering and technology factors driving shareholder value. In other words, constantly sharpen one's own business sense.
2. *Know* how to define proper metrics to measure the financial and cost performance of the company. These include income statements, balance sheets, funds flow statements, and various ratio analyses.
3. *Recognize* that technology is to be viewed as a tool to achieve business success. That is, technology can make a business become more profitable and productive. Delivering value to customers remains the key to achieving business success.
4. *Be* able to recommend suitable emerging technologies to enhance shareholder value and to mobilize resources (including the engagement of networked partners) to turn these visionary goals into reality.
5. *Be* persistent in pursuing vision, which is based on legacy and not on activity. Winston Churchill said, "The further backward you look, the further forward you can see" (Jackson 2004).

Example 9.5 Negotiating for agreements between employees, departments, suppliers, production partners, and networked distributors is part of a manager's job. Explain the guidelines for effective negotiations.

ANSWER 9.5 There are numerous excellent books on negotiations. Walker (2003); Shell 2006; Mahotra and Bazerman 2008) offer a summary of key guidelines for effective negotiation:

1. Focus on the merits by attacking the underlying issues involved, not the opponent or his or her position.
2. Look for creative solutions with which you both can win.
3. Prepare yourself well beforehand. The prenegotiation preparation centers on standards that suggest the best deal and available alternatives.

   Sample questions related to preparation include the following: (1) How much are the other vendors selling that brass dish for? (2) What do your competitors charge for the service you are offering? (3) How much does a person with your experience get paid? (4) What is your best alternative to a negotiated agreement (BATNA)?
4. Raise questions to find out what your opponent really wants, and prepare clever arguments to support what you need.

   The *parable of the orange* says that two parties each want an orange and agree finally to split it in half. But it turns out that one side simply wanted the juice and the other side wanted the rind. If only they had worked together to solve the problem, each side could have gotten what it wanted. According to Walker, situations similar to that in the parable of the orange pop up a lot.

Be prepared to ask questions pertaining to who, where, what, why, and how, as they tend to drive your opponent to disclose more information than the yes–no questions (e.g., How did you arrive at that figure?). Posing questions to your opponent is also useful for fending off your opponent's questions to you that you may not be prepared to answer. Table 9.8 presents some examples of such exchanges.

5. Listen intently, as the power rests with the listeners. Silence is one of the best weapons available to negotiators. Keeping silent will force the opponent to talk more and, as a consequence, to revise his or her position and reveal useful information in the process.
6. Make use of the principle of consistency. The *principle of consistency* is that people have the need to appear reasonable. This can be used skillfully to make your opponent feel that, to appear reasonable, he or she needs to use your standards that have been determined during your prenegotiation preparations. The more authoritative your standards seem, the better. An example of such a standard is the price charged for similar goods by the competitors of your opponent.
7. Let your opponent make the opening offer. Studies indicate that people often underestimate their own strengths and exaggerate those of the rivals.
8. Take a psychological test (e.g., the Thomas-Kilmann Conflict Mode Instrument) to understand your own style, be it "competitor" or "collaborator." Taking such a test will help define any aspects of your negotiation style that should be fine-tuned.
9. Be aware of some tactics employed by your opponents. Some may flinch at your proposals on purpose. Others, with the intent to mislead, may exaggerate things they do not really care about. At the close of negotiation, some opponents may say something like, "Wow, you did a fantastic job negotiating that. You were brilliant." Yet others may take advantage of *the Columbo effect* by lulling you into underestimating them and becoming overconfident.
10. Practice makes perfect. Effectual negotiation is 10 percent technique and 90 percent attitude. Attitude is affected by realism, intelligence, and self-respect.

Table 9.8 Sample Questions and Answers

Question	Answer
What is the most you would pay if you had to?	If you think that no agreement between us is possible, perhaps we should get someone trustworthy to arbitrate.
If your company agrees to be merged into ours, how many of your employees can be laid off to achieve economics of scale?	Which of your branch offices would you be keeping and which would you close?

9.9 UNIQUE CONTRIBUTIONS EXPECTED OF SSME LEADERS

In what way is the technical background of a manager important to today's executives? If an engineering manager does only what a typical nontechnical manager does, then the engineering manager does not earn his or her keep. Specifically, what can a technically trained manager do that a nontechnical manager cannot? Technological intuition and innovation are the areas engineering managers can and should excel at (Tucker 2008; Bohm 2004, Millson, and Wileman 2007).

How are innovation and creativity measured? A commonly accepted performance yardstick is counting the number of patents a company receives. More recently, a new measure was proposed to assess the relative value of patents. If a new patent application cites certain prior art patents as background on which the new patent is based, then the prior art patents are regarded as *forward citations*. The value of a patent is said to be directly proportional to its number of forward citations, as more forward citations indicate a broader significance.

For the moment, let us stay with the patent number as a measure of innovation and creativity (Durham 2009). According to statistics published by the U.S. Patent and Trademark Office, five of the top ten global companies receiving the most U.S. patents in 2008 are Japanese. (Figure 9.4 shows these comparisons.) One of several likely reasons for this outcome is that, compared with their American counterparts, Japanese companies may have appreciated to a greater extent the strategic importance of patents to their long-term competitiveness and hence devoted more management attention and critical resources to innovation and creativity. U.S. managers do have their work cut out for them in the years ahead.

Besides the aspect of relative competitiveness, the new era is full of challenges due to rapid advancement in technologies, Internet-based communications techniques, and globalization. The need for technological leadership is becoming increasingly pronounced (Hamel 2002). The areas in which engineering managers are expected to make significant contributions are discussed in the next sections.

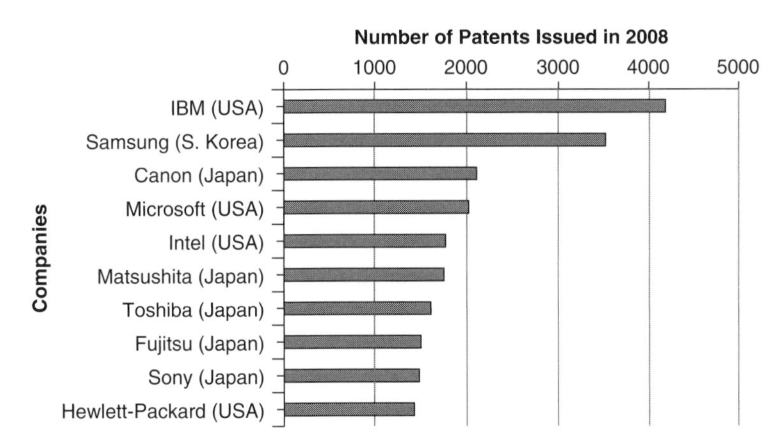

Figure 9.4 Ranking of patents awards, 2008.

9.9.1 Technologists as Gatekeepers

A gatekeeper's primary job is to inspect and authorize the entry of people or materials into a gated organization. Technically capable people are usually entrusted with this important corporate activity to systematically scout, evaluate, and introduce new technologies for use by the enterprise.

Engineering managers are in the best position to mobilize capable technologists who understand the new or emerging technologies available in the marketplace and their relevant value to company's products, operations and services. Capable technologists can also bring in and selectively apply new technologies.

Example 9.6 What are some new characteristics of knowledge workers in the twenty-first century?

ANSWER 9.6 Knowledge workers in the new century may have these characteristics:

- Free agents can now sell their skills around the world via the Internet; this was impossible to do not too long ago.
- Professional groups are likely to offer the senses of identity and community, health insurance, and other benefits needed by free agents who float from one company to another.
- Each employee may have as many as twenty different jobs throughout a career of forty-five years (an average of 2.5 years per job). They tend to constantly bargain for better deals within their organizations (e.g., stock options, a sign-up bonus, new projects, Thursdays off, an August sabbatical, etc.).
- Workers will seek to acquire a broad set of marketable skills, as companies will continue to outsource white-collar jobs and spread centers of excellence around the world to seek advantages in cost, speed, and expertise.
- Managers and professional workers need to be flexible and adaptable to organizational changes and become cosmopolitan, equally at ease both at home and abroad.

9.9.2 Technological Intuition

According to Thurow (1987), nontechnical managers do not have enough background to develop intuitions about which of the possible technologies now on the horizon are apt to advance further and which are apt to be discarded.

Nontechnical managers cannot judge the merits of revolutionary changes in technology; this is known as a factor of *ignorance*. They have no choice but to procrastinate and wait for it to become clear which technology is the best; this is known as a factor of *risk aversion*. By the time the answer is clear, foreign firms may already have a two- to three-year lead in understanding and employing the new technologies. Thus, the nontechnical managers cannot exert technological leadership.

This is also evident in diverse technical startup companies, which, by the way, are typically headed by technically talented people. These talented entrepreneurs are able to invent new technology to serve as the basis for a new business. Eventually, some of these startup companies are bought up by big companies that cannot help flourish the technologies on their own.

Technological intuition is most needed in the strategic planning process conducted by a company. Hence, engineering managers have an important role to play here.

9.9.3 Technological Innovations—Lead in Strategic Differentiation

Another area of technological leadership expected of engineering managers is in the management of technological innovation (White and Wright 2003; Von Stamm 2008; Gaynor 2002). Technological innovation is the process by which technological ideas are generated, strengthened, and transformed into new products, processes, and services that are used to make a profit and establish a marketplace advantage.

The following statistics were published by Pearce and Robinson (2008):

1. Out of fifty-eight initial product ideas, only twelve survive the business-analysis screening for compatibility with the company's mission and long-term objectives. This step uses 8 percent of the total development time.
2. Only seven of the twelve ideas remain after an evaluation of their potential. This step uses 9 percent of the total development time.
3. Three of the seven remaining ideas survive after development work is completed. This step uses 41 percent of the total development time.
4. Only two of the original ideas remain past the pilot and field testing involved in the commercialization step, which uses 19 percent of the total development time.
5. Eventually, only one idea results in a commercially successful product. This step uses 23 percent of the total development time.

Within the product development process (see Fig. 9.5), the most time-consuming and resource-intensive steps involve development and testing. Engineering managers can make significant contributions to shorten development time and reduce costs, while assuring technical quality.

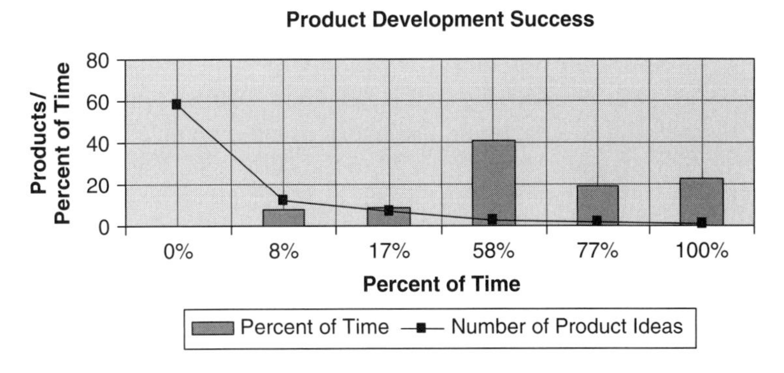

Figure 9.5 Product development process.

In heading up concurrent engineering teams in product development, engineering managers can excel by

- Asking pertinent technological questions
- Applying their interdisciplinary background to set technological priorities
- Incorporating new technologies to achieve competitive advantages and to satisfy customers' needs

Exerting strong technological leadership is where technically trained managers must shine. This is the uniquely attractive niche for engineering managers. Engineers do not have serious competition from nontechnical majors here, as it is relatively easy for engineers to learn how to manage, but not so easy for nontechnical managers to learn engineering. However, those engineering managers who cannot exercise technological leadership will be no better than nontechnical managers as far as the value added to their companies is concerned.

Innovation requires knowledge, ingenuity, and predisposition. Innovation cannot succeed without hard work. Purposeful work demands diligence, persistence, and commitment. "Nothing great was ever achieved without enthusiasm," said Ralph Waldo Emerson (Andrews, Biggs and Seidel 1996). Innovations need to be built on the company's strengths and core technologies. They should focus on opportunities that are "temperamentally fit"—that is, exciting and attractive to the innovators—inspiring them to do the required hard work. In addition, innovation must be market driven and focused on customers.

Engineering managers can benefit from taking a systematic approach to enhance individual innovation. Such an approach could include the following four elements:

1. *Analyzing* innovative opportunities systematically. Focus on (a) the unexpected successes or failures of the company and its competitors, (b) incongruities in processes (production, distribution, and customer behavior), (c) process needs, (d) changes in industry and market structures, (e) changes in demographics, (f) changes in meaning and perception, and (g) new knowledge.
2. *Being* observant (asking, looking, and listening). The types of questions to pose may include these: (a) What engineering processes or technologies from the past should be kept because they have future value? (b) What past engineering and technological practices should be modified to be more relevant? (c) What activities should be eliminated because they have no future value? (d) What needs to be performed to ensure future success?
3. *Recognizing* that innovations must be simple and focused, application specific to the present marketplace, and pinpointed on satisfying a need and producing an end result useful to the customers. It is not wise to innovate for the distant future markets, which may or may not materialize.
4. *Starting* small scale and aiming at producing a series of small, but useful incremental values. Focus on areas for which knowledge and expertise are available.

Equally important for engineering managers is to exercise leadership in fostering corporate innovations. Managing group innovation is closely linked to managing group creativity. Implementing some well-established techniques, such as those enumerated in the following list, may enhance the creativity of groups:

1. *Brainstorming (for groups of six to twelve people).* Many important corporate business or engineering issues may be cast in the form of problems. Examples of such issues include product design simplification, product or component cost reduction, and improvement of operations.

   By using brainstorming techniques, the leader defines a specific problem and requests each participant in the group to take a turn proposing possible solutions. No criticisms are allowed during these exchanges in order not to impede the free flow of ideas. After all of the ideas are generated and recorded, the group then carefully evaluates each solution and jointly defines the best solution to the problem at hand.
2. *Nominal group technique (small groups).* The leader defines a specific problem. Each member is encouraged to generate as many written solutions as possible during the group meeting. Each member is then invited to present his or her solution and to elucidate the relevant rationale behind the proposed solution. No criticism is allowed.

   After all of the proposed solutions are presented and recorded, each solution is thoroughly discussed, evaluated, and criticized. The participants are then requested to anonymously rank all of the solutions in writing. The final results are presented to management for action.
3. *Delphi technique (for identifying future trends).* The leader defines a specific problem and a set of questions and sends them to a panel of geographically dispersed domain experts who do not have contact with one another. Each expert then answers the questions individually and anonymously. A summary of all of the answers is documented by the leader and sent back to the experts. By reviewing the comments and the possible criticisms, the experts, again anonymously, modify their original answers. No one knows who proposed or criticized what specific solutions. The focus is on the merits of the ideas, not on the personality of those who advanced the ideas. The leader again prepares a summary and returns it to the experts, offering additional explanations and justifications. In the end, every solution is justified. Each time the experts respond to a summary, it is called a *wave*. After the third wave, a summary is prepared and the leader makes a forecast. This method is time consuming, but it is particularly useful for predicting the future course that a company may take in technology and business.

Engineering managers ought to be well versed in many of these techniques to manage and promote creativity and innovation both from individuals and groups.

9.9.4 Specific Contributions—Lead in Operational Excellence

Engineering managers are capable of adding value to their employers in diverse ways. The following are seven additional broad-based contributions expected of engineering managers:

1. Use of specific new technologies in product design—novel use of materials, parts, subassemblies, production technologies, and other components.

2. Application of Web-based technologies to e-transform the enterprise in order to achieve refinements in process efficiency, quality, speed, or customer satisfaction.
3. Selection of enterprise integration tools for expediting business information collection, transmission, and processing in order to realize speed, cost, and quality advantages.
4. Alignment of networking partners to secure competitive advantages in supply chains, production systems, and customer service.
5. Looking out for new technology-based tools that could facilitate serving customers better, cheaper, and faster, with products that have a larger degree of customization.
6. Employment of new technologies and innovations to add value to stakeholders other than customers (e.g., investors, employees, suppliers, and the communities in which the companies operate).
7. Scanning literature to constantly learn the best practices of technology management in the industry.

9.10 CAREER STRATEGIES FOR THE TWENTY-FIRST CENTURY

Today's leaders adopt specific guidelines in managing their own careers. Among many such guidelines published in the literature on career management, the eight specific guidelines offered by Kacena (2002) shown in Table 9.9 are particularly comprehensive and insightful. Fernandez (1999) offers a slightly different set of career strategies, shown in Table 9.10.

Engineering managers are advised to refine their own career strategies constantly by using the aforementioned inputs as general guidelines.

9.11 "TAKE CHARGE" TO "GET SUCCESS" FORMULAE

According to Roach (1998), everyone must strive to become exceptional and to "take charge" of underutilized potential:

T Time should be taken to reflect on your strengths and weaknesses, as well as to do something about the weaknesses.

A Attitude must be fostered and modified as needed.

K Knowledge must be updated to keep yourself marketable.

E Empathy and consideration in caring for others' feelings should be strived for.

C Communication must be constantly improved and perfected.

H Health and humor must be diligently nurtured.

A Appearance must be properly maintained.

R Respect yourself and others and live one day at a time—enjoy life; why simply wait?

G Goals should be set for yourself and family by, for example, creating a five-year plan.

E Empower the possibilities by finding ways to delegate, assist, entrust, and praise others and by being generous and giving to others.

Table 9.9 Guidelines for Managing Your Own Career

1. Think, speak, act, and walk like an entrepreneur. Adopt an entrepreneurial mindset, as if your own investment is involved. Accept the notion that jobs exist so that problems can be solved.
2. Make chaos a friend. Embrace change as an opportunity for growth. The mantra for today's career advancement is "eager to stay, yet ready to leave."
3. Don't be afraid to break the rules. Attempt to be a visionary, as innovation is possible only in cultures that tolerate mistakes. Be detail oriented.
4. Know your own strengths and weaknesses. Set high standards for yourself and affiliates. Be very competitive. Market yourself. Express commitment, passion, and excitement about your own work.
5. Be nonlinear. Radical career shifts will become commonplace. Companies are ignoring specialization in favor of adaptability, cross-functionality, people skills, and a rock-solid customer focus. Follow the new paradigm that anyone who does not know how to do something must learn or partner with someone who does know.
6. Maintain balance. Set your own priorities with respect to health, family, and business—in that order—and have fun. "Earn a living, make a life."
7. Stay connected. Building alliances by networking is essential. Establish reciprocal relationships with colleagues, clients, customers, and competitors. Seek to be helpful and supportive so that they become resources for ideas, skills, and knowledge.
8. Always keep your options open. Keep abreast of the market, nurture skills that are marketable, stay professionally active, and avoid becoming complacent.

Table 9.10 Career Strategies

1. Balance your priorities between your job, personal interests, family, and the community. This is the key to having a full and meaningful life.
2. Cultivate a broad business background through education, diversified work experience, and success skills. Ranked most important among success skills are integrity and persistence. Persistence means unwillingness to accept defeat.
3. Learn leadership from proven leaders. Observe successful people and learn from their behavior.
4. Learn what the company and industry are really about. Understand the company values, aspirations, brand character, market position, organizational structure and culture, and qualities of people.
5. Make an impact. Strive to contribute from the basis of knowledge and attitude. Ask how the world would be different and just a bit better because of your efforts.

These ten rules are generic in nature. I have come up with another set of ten rules that are more relevant to the various issues discussed in this text, the "Get Success" formula:

G **Get** connected to empower personal and business networks.
E **Embrace** own mistakes and learn from them.
T **Take** the lead in teams to motivate and empower people.
S **Secure** diversified experience and knowledge.
U **Understand** personal strengths and weaknesses by conducting SWOT analyses.
C **Create** a personal strategic and operational plan.
C **Cultivate** an open mind toward things different and foreign.
E **Exemplify** own truthful and authentic self.
S **Strengthen** own practices and adhere to ethics and integrity rules.
S **Sustain** best efforts all the time.

Service systems engineers and leaders are advised to keep the "Take Charge" and "Get Success" formulae in mind, when they move ahead in their careers.

9.12 CONCLUSIONS

In this chapter we discussed a variety of aspects related to service systems management and engineering leadership, including styles, skills, attributes, and managerial competencies. Many rules of thumb are derived from experience. They are common sense heuristics that are all reasonable and intuitively correct.

Knowing what is needed to become (1) an effective engineer, (2) a good engineering manager, or (3) an excellent engineering leader is a very good start. The next step is to learn the skills and capabilities to shape one's own attitudes, and to acquire the attributes needed to become an effectual systems engineer, good engineering manager, or excellent engineering leader. The third step is to lead and contribute in creating competitive advantages in strategic differentiation and operational excellence for the service enterprises.

To be successful, one must practice, practice, and practice until the preferred behavior becomes second nature.

9.13 REFERENCES

Andrews, R., Mary Biggs, and Michael Seidel 1996. *The Columbia World of Quotations on CD-ROM*. New York: Columbia University Press.

Bass, B.M., and R. Bass. 2008. *The Bass Handbook of Leadership: Theory, Research and Managerial Applications*, 4th ed. New York: Free Press.

Bennis, W. 2003. *Learning to Lead: A Workbook on Becoming a Leader*, 3rd ed. New York: Basic Books.

Bennis, Warren. 2009. *On Becoming a Leader*, 4th ed. New York: Basics Books.

Bohm, D. 2004. *On Creativity*, 2nd ed. London: Routledge.

Certo, S. 2007. *Supervision: Concepts and Skill Building*, 6th ed. New York: McGraw-Hill/Irwin.

Cohen, W. A. 2002. "The Art of the Successful Leader." *Financial Service Advisor* (July–August).

Ciampa, Dan. 2005. "Almost Ready: How Leaders Move Up." *Harvard Business Review* 83 (1).

Clark-Epstein, C. 2006. *78 Important Questions Every Leader Should Ask and Answer*. New York: AMACOM.

Drucker, Peter. 2005. "Managing Oneself," *Harvard Business Review* 83 (1). Excerpt from Peter Drucker (1999) *Management Challenges for the 21st Century*. New York: HarperCollins.

Durham, A. L. 2009. *Patent Law Essentials*. 3rd ed. Santa Barbara, CA: Praeger.

Edmundson, Louis. 2005. Technology and the Human Touch. (Download HTML), Academic Exchange Quarterly— article was originally published on June 22, 2001.

Fernandez, V. 1999. "Career Strategies for the 21st Century." *Executive Speeches* (June–July).

Friedman, Stewart D. 2008. *Total Leadership: Be a Better Leader, Have a Richer Life*. Cambridge, MA: Harvard Business School Press, p 272 .

Gaynor, G. H. 2002. *Innovation by Design: What It Takes to Keep Your Company on the Cutting Edge*. New York: AMACOM.

Goffee, R., and G. Jones. 2001. "*Why Should Anyone Be Led by You?*" HBR onPoint Article # 5890. Harvard Business School Press.

Goleman, D. 2000. "Leadership That Gets Results." *Harvard Business Review* (March–April).

Goleman, D. 2004. "What Makes a Leader." HBR Classic Article #R0401H. Harvard Business School Press.

Gronfeldt, S., and J. B. Strother. 2005. *Service Leadership: The Quest for Competitive Advantage*. Newbury Park, CA: Sage Publications.

Hamel, G. 2002. *Leading the Revolution: How to Thrive in Turbulent Times by Making Innovation a Way of Life*. Cambridge, MA: Harvard Business School Press.

Hamm, John. 2006. "The Five Message Leaders Must Manage." *Harvard Business Review* 84 (5).

Harvard Business School Press. 2005. *Becoming an Effective Leader. Results Driven Manager Series*. Cambridge, MA: Harvard Business School Press.

Hughes, R., R. Ginnett, and Gordon Curphy 2008. *Leadership: Enhancing the Lessons of Experience*, 6th ed. New York: McGraw-Hill/Irwin.

Ibarra, Herminia, and Mark Hunter. 2007. "How Leaders Create and Use Networks." *Harvard Business Review* 85 (1).

Jackson, Michael A. 2004. *Look Back to Get Ahead: Life Lessons from History's Heroes*. New York: Seaver Books.

Jones, Graham. 2008. "How the Best of the Best Get Better and Better." *Harvard Business Review* 86 (6).

Kacena, J. F. 2002. "New Leadership Directions." *The Journal of Business Strategy* (March–April).

Kellerman, Barbara. 2008. "Followership: How Followers are Creating Change and Changing Leaders." Cambridge, MA: Harvard Business School Press.

Kotter, John P. 2007. "Leading Change." *Harvard Business Review* 85 (1).

Krames, J. A. 2001. *The Jack Welch Lexicon of Leadership*. New York: McGraw-Hill.

Krames, J. A. 2005. *What the Best CEOs Know: Seven Exceptional Leaders and Their Lessons for Transforming Any Business*, Paperback. New York: McGraw-Hill.

Langewiesche, W. 2003. "Columbia's Last Flight." *The Atlantic Monthly* (November).

Lowendahl, Bente R., 2005. "Strategic Management of Professional Firms", 3rd edition, Copenhagen Business School Press, Denmark.

Mahotra, D., and M. Bazerman. 2008. *Negotiation Genius: How to Overcome Obstacles and Achieve Brilliant Results at the Bargaining Table and Beyond*. New York: Bantam.

Mai R., and A. Akerson. 2003. *The Leader as Communicator: Strategies and Tactics to Build Loyalty, Focus Effort and Spark Creativity*. New York: AMACOM.

Millson, M., and David Wileman. 2007. *The Strategy of Managing Innovation and Technology*. Upper Saddle River, NJ: Prentice Hall.

Mintzberg, H. 2003. "The Manager's Job: Folklore and Fact." HBR OnPoint Enhanced Edition Article # 5429. Harvard Business School Press (November 1).

Morrison, Mike, Rueben Mark, Rebecca Ray, George Manderlink, and Dave Ulrich. 2007. "The Very Model of Modern Senior Manager." *Harvard Business Review* 85 (1).

Nelson, B. 1999. *1001 Ways to Take Initiative at Work*. New York: Workman Publishing.

Pearce, A. J., and R. B. Robinson, Jr. 2008. *Strategic Management*. 11th ed. New York: McGraw-Hill/Irwin.

Pelus, C. J., and D. M. Horth. 2002. *The Leader's Edge: Six Creative Competencies for Navigating Complex Challenges*. San Francisco: Jossey-Bass.

Roach, M. B. 1998. "Take Charge." Speech before the 17th Turbomachinery Symposium at Dallas, Texas, November 9.

Roberts, Laura Morgan, Gretchen Spreitzer, Jane Dutton, Robert Quinn, Emily Heaphy, and Brainna Barker. 2005. "How to Play to Your Strengths." *Harvard Business Review* 83 (1).

Schwarzkopf, N. 1999. "Leaders for the 21st Century." *Vital Speeches of the Day* (June 15).

Shell, G. R. 2006. *Bargaining for Advantage: Negotiation Strategies for Reasonable People*. 2nd ed. New York: Penguin.

Slater, R. 2001. *Get Better or Get Beaten*, 2nd ed. New York: McGraw Hill.

Stadler, Christian. 2007. "The Four Principles of Enduring Success." *Harvard Business Review* 85 (7/8).

Sun Tzu. 2009. *The Art of War*. Paperback. CreateSpace.

Tucker, R. B. 2008. *Driving Growth through Innovation: How Leading Firms Are Transforming Their Futures*. 2nd rev. and updated ed. San Francisco: Berrett-Koehler Publishers.

Thurow, L. 1987. "The Tasks at Hand." *Wall Street Journal*, June 12, p. 46D.

Von Stamm, B. 2008. *Managing Innovation, Design and Creativity*. 2nd ed. Hoboken, NJ: John Wiley & Sons, Inc.

Walker, B. 2003. "Take It or Leave It: The Only Guide to Negotiating You Will Ever Need." *Inc*. (August).

Welch, J., and J. A. Byrne. 2003. *Jack: Straight from the Gut*. Paperback. Business Plus.

White, S. P., with G. P. Wright. 2003. *New Ideas about New Ideas: Insight on Creativity from the World's Leading Innovators*. New York: Basic Books.

Zaleznik, A. 2004. "Managers and Leaders: Are They Different?" HBR Classic Article # R0401G, *Harvard Business Review* (January 1).

Zenger J. H., and J. Folkman. 2009. *The Extraordinary Leader: Turning Good Managers into Great Leaders*, 2nd ed. New York: McGraw-Hill.

9.14 APPENDICES

9.14.1 Tips on Coping for First-Time Supervisors and Managers

It typically takes two to three years for a first-time supervisor or manager to become fully effectual. Here are eight tips that can help the novice to cope during this initially challenging period:

1. Organize the office so that important files and project folders can be located readily.
2. Get a good perspective from one's superior in term of priorities, strategic plans, previous problems, and vision to operate the unit or department. Do the homework to learn the new languages: finance, marketing, manufacturing, and customer service. Acquire the business perspectives of markets, cost-price position, product distribution, supply chain management, enterprise integration application, and customer relation management.
3. Obtain training in evaluating staff performance, managing time, and developing multidisciplinary teams.
4. Ready yourself mentally to delegate responsibilities while maintaining control in order to achieve results through people.
5. Communicate your expectations to staff, both individually and in groups, and solicit feedback.
6. Foster relationships with peer managers in other departments.
7. Build the relationship with your own superior.
8. Start practicing and polishing your own management styles in order to become increasingly effective.

Certo (2007) offers nine identifiable people, technical, and administrative skills for frontline supervisors in the new century:

- **People skills**
  1. Communication—ability to adjust own style to correspond with individuals' needs and an ability to listen to other people.
  2. Teamwork—ability to take into account the diversity of other people's backgrounds.
  3. Coaching skills—ability to assist other people.
- **Technical skills**
  4. Business skills—ability to assess business performance of others.
  5. Continuous improvement—ability to constantly update and refine one's own technical skills.
  6. Technologically savvy—ability to use modern office equipment.
- **Administrative skills**
  7. Project management skills—ability to plan and implement new ideas.
  8. Writing and documentation skills—ability to write reports and keep management informed.
  9. Resource management skills—ability to network with those who control resources.

Over time, first-time supervisors and managers need to demonstrate the ability to build team spirit, shape a work environment that fosters self-motivation, solve people and technical problems, and make decisions by integrating technical issues with business issues affecting the company.

9.14.2 The Wisdom of the Mountain

For more than twenty years, Lao-li studied under the great master Hwan at a mountaintop temple. He struggled and struggled, but could not reach enlightenment. One day, the sight of a falling cherry blossom spoke to his heart. "I can no longer fight my destiny," he reflected. "Like the cherry blossom, I must gracefully resign myself to my lot." Lao-li determined to retreat down the mountain, giving up his hope of enlightenment.

Lao-li looked for Hwan to tell him of his decision. Hwan surmised his intention to resign and said to Lao-li, "Tomorrow, I will join you on your journey down the mountain." The next morning, the two started walking downhill. Looking out into the vastness surrounding the mountain peak, Hwan asked, "Tell me, Lao-li, what do you see?" "Master, I see the sun beginning to wake just below the horizon, meandering hills and mountains that go on for miles, and couched in the valley below, a lake and an old town." Hwan smiled, and they continued their long descent.

Hour after hour, as the sun crossed the sky, they pursued their journey, stopping only as they reached the foot of the mountain. Again, Hwan asked Lao-li to tell him what he saw. "Master, in the distance I see roosters as they run around barns, cows asleep in sprouting meadows, old ones basking in the late afternoon sun, and children romping by a brook."

The master, remaining silent, continued to walk until they reached the gate to the town. There the master gestured to Lao-li, and together they sat under an old tree. "What did you learn today, Lao-li?" asked the master. "Perhaps this is the last wisdom I will impart to you." Silence was Lao-li's response.

At last, after a long silence, the master continued. "The road to enlightenment is like the journey down the mountain. It comes only to those who realize that what one sees at the top of the mountain is not what one sees at the bottom. Without this wisdom, we close our minds to all that we cannot view from our position and so limit our capability to grow and improve. But with this wisdom, Lao-li, there comes an awakening. We recognize that alone one sees only so much—which is not much at all. This is the wisdom that opens our minds to betterment, knocks down prejudices, and teaches us to respect what at first we cannot view. Never forget this last lesson, Lao-li: What you cannot see can be seen from a different part of the mountain."

When the master stopped speaking, Lao-li looked out to the horizon. As the sun set before him, it seemed to rise in his heart. Lao-li turned to his master, but Hwan was gone.

So the old Chinese tale ends. But it has been said that Lao-li returned to the top of the mountain to live out his life. He became a great enlightened one himself.*

9.14.3 The Sound of the Forest

Back in third-century China, the King of Tsao sent his son, Prince Tai, to the temple to learn under the great master Pan Ku how to become a good ruler. When the prince arrived at the mountaintop temple, the master sent him to the Ming-Li Forest. After one year, the prince was to return to the temple to describe the sound of the forest.

**Source*: W. Chan Kim and R. A. Mauborgne, "Parables of Leadership," *Harvard Business Review* (July–August 1992).

When the prince returned after one year, he told the master, "I could hear the cuckoos sing, the leaves rustle, the hummingbirds hum, the crickets chirp, the grass blow, the bees buzz, and the wind whisper and holler." When the prince had finished, the master told him to go back to the forest to listen more. The prince was puzzled by the master's request. Had he not discerned every sound already?

For days and nights on end, the young prince sat alone in the forest listening. But he heard no sounds other than those he had already described. Then, one morning as the prince sat silently beneath the trees, he started to discern faint sounds unlike those he had ever heard before. The more acutely he listened, the clearer the sounds became. The feeling of enlightenment enveloped the boy. He reflected, "These must be the sounds the master wished me to discern."

When the prince returned to the temple, the master asked him what more he had heard. "Master," responded the prince reverently, "when I listened most closely, I could hear the unheard—the sound of the flowers opening, the sound of sun warming the earth, and the sound of the grass drinking the morning dew."

The master nodded approvingly. "To hear the unheard," remarked Pan Ku, "is a necessary discipline to be a good ruler. For only when a ruler has learned to listen closely to the people's hearts, hearing their feelings not communicated, pains unexpressed, and complaints unspoken, can he hope to inspire confidence in his people, understand when something is wrong, and meet the true needs of his citizens. The demise of states comes when leaders listen only to superficial words and do not penetrate deeply into the souls of the people to hear their true opinions, feelings, and desires."*

9.14.4 The Wheel and the Light

At the beginning of the Han dynasty, Liu Bang, the Chinese emperor, had just unified China into a strong country. To commemorate this event, Liu Bang invited all of his high-ranking political advisors, military officials, and scholars to a grand celebration.

At the center table sat Liu Bang with his three staff chiefs, Xiao He, the logistics master; Han Xin, the military chief; and Chang Yang, the political and diplomatic strategist.

At another table sat Cheng Cen, the master scholar, whom Liu Bang consulted often during the war years. With him sat three of Cheng's disciples.

Everyone enjoyed the event. Food and entertainment were plentiful. Midway through the festivals, one of Cheng's disciples discreetly raised a question. "Master, look at the central table. Xiao He had the best supply logistics. Han Xin's military tactics were beyond reproach. Chang Yang had the winning diplomatic tactics. Their contributions were clearly visible and readily understood. How is it, then, that Liu Bang is the emperor?"

The master smiled and asked his disciples to imagine the wheel of a chariot. "What determines the strength of a wheel in carrying a chariot forward?" he inquired. "Is it the sturdiness of the spokes, Master?" one disciple responded. "But then, why is it that two wheels made of identical spokes differ in strength?" After a moment, the master

Source: W. Chan Kim and R. A. Mauborgne, "Parables of Leadership," *Harvard Business Review* (July–August, 1992).

continued, "See beyond what is seen. Never forget that a wheel is made not only of spokes, but also of the space between the spokes. Sturdy spokes poorly placed make a weak wheel. Whether their full potential is realized depends on the harmony between them. The essence of wheel making lies in the craftsman's ability to conceive and build the space that holds and balances the spokes within the wheel. Think now, who is the craftsman here?"

A glimmer of moonlight was visible behind the door. Silence reigned until one disciple said, "But Master, how does a craftsman secure the harmony between the spokes?" "Think of sunlight," replied the master. "The sun nurtures and vitalizes the trees and flowers. It does so by giving away its light. But in the end, in which direction do they all grow? So it is with a master craftsman like Liu Bang. After placing individuals in positions that fully realize their potential, he secures harmony among them by giving them all credit for their distinctive achievements. And in the end, as the trees and flowers grow toward the giver, the sun, individuals grow toward Liu Bang with devotion."*

9.15 QUESTIONS

9.1 Nissan Motor's CEO Carlo Ghosn became a world-renowned automotive executive because he was credited for having successfully turned around Nissan a few years back. That turnaround efforts were was described in the Harvard Business School case "Nissan Motor Co., Ltd, 2002" Case # 9-303-042 (Rev. December 9, 2002). Review this case and answer the following questions. The case materials may be purchased by contacting its publisher at www.custserve@hbsp.harvard.edu or at 1-800-545-7685.

A. When the Nissan Revival Plan was developed, which activities initiated by Carlo Ghosn and Nissan employees were related to the management functions of planning, organizing, leading and controlling?

B. In what ways was the Nissan Revival Plan unique?

C. What were the major factors responsible for the success of the Nissan Revival Plan?

D. When Carlo Ghson arrived in Japan to head up Nissan, he was all by himself. In your view, which of Ghosn's leadership characteristics were the primary reasons for his success?

E. If you were to head up the Nissan product development programs, what additional efforts would you want to initiate to make Nissan even more competitive with respect to its Japanese and U.S. automotive rivals?

F. For meeting the new challenges ahead, Nissan must now develop new plans. In your opinion, what should be included in these new plans? Be as specific as you can.

G. If you were the CEO of Ford Motor Company in 2006, would you entertain the idea of talking with Nissan's Carlo Ghson at this time to look for ways to collaborate? Why and why not?

H. What are the lessons you have learned from having studied this case?

**Source*: W. Chan Kim and R. A. Mauborgne, "Parables of Leadership," *Harvard Business Review* (July–August, 1992).

9.2 Engineering managers may be called on to resolve conflicts between employees, departments, vendors, and business partners, as well as to handle customer complaints. What are the recommended guidelines for handling complaints? Please elaborate.

9.3 Hoffman (1989) believes that a management education program should have three elements:

1. *Behavioral*—people skills, motivation, team building, communications, and delegation
2. *Cognitive*—production, marketing, finance, and control
3. *Environmental*—markets, competition, customers, political, social, and economical environment in which the organization operates

The importance of the first two elements should be self-evident. Explain why the third element, the environmental, is important.

9.4 How is engineering management different from management in general?

9.5 How does a manager or leader become a good superior? What should the superior do and not do?

9.6 Does the job of managing a high-technology function (e.g., an engineering design department) differ from that of managing a low-technology function (e.g., a hotel)? Explain the specific details of the jobs.

9.7 What rules and principles can guide managers to have successful people management skills?

9.8 There are "unwritten laws of engineering" that recommend acceptable conduct and behavior for engineers in industry. How important are these unwritten laws to individual engineers, and where can these laws be located?

9.9 Some engineers and managers are known to have more difficulties in interpersonal relations than others. These difficulties may arise due to personality, chemistry, value system, priority, tolerance, competition, and other such factors. How can they improve their interpersonal skills?

9.10 In your opinion, what are the characteristics common to many future engineering leaders? Please explain.

9.11 Recently, you have been promoted to the position of engineering manager in a company known to be progressive in developing new services. You are under some pressure to quickly come up with service concepts that could be developed as new services to enter the marketplace, as a periodical introduction of such new services is deemed to be critically important for sustaining its long-term competitiveness and corporate profitability.

Two of your competitors have just introduced their brand-new services, X1 and X2, which offer unique features not available in other existing services. Their sales prices are around $100 per unit. Based on the marketing intelligence reports available to you, services X1 and X2 appear to have gained very good traction in the marketplace. Your engineers believe that a new service similar to X1 and X2 could be rather readily developed, based on company's in-house capabilities and those of existing business supply chains, without the risks of infringing on the patent-protected designs of both X1 and X2. Intuitively, you believe that developing new services Y1 or Y2 by your company to compete against X1 or X2 might indeed be a smart strategy to pursue.

How do you plan to evaluate the option of developing the services Y1 or Y2? What is the low-risk strategy to develop services Y1 or Y2, in order to attain the highest probability of achieving profitability for your company?

9.12 In your own words please describe the principal lessons you have learned from this chapter.

9.13 It was suggested by a number of researchers that a SSME educational program should have three elements: (1) behavioral—people skills, motivation, team building, communications, and delegation, (2) cognitive—production, marketing, finance, and control, (3) environmental—markets, competition, customers, political, social, and economical environment in which the organization operates.

The importance of the first two elements should be self-evident. Explain why the third element, the environmental, is important.

9.14 How can engineering managers make the best use of tools such as the Myers-Briggs Type Indicator (MBTI) to assist in selecting project leaders or assigning employees to teams to ensure the likelihood of avoiding personality conflicts that could otherwise hinder team success?

Chapter 10

Ethics in Service Systems Management and Engineering

10.1 INTRODUCTION

Recent U.S. corporate scandals have raised serious questions about ethics in the workplace. The past cases of Global Crossing and Adelphia involved falsifying financial data or misusing corporate funds. Enron was charged for having used off-the-book deals to hide billions of dollars in debts. It was the seventh-largest company in the United States before it filed for bankruptcy in December 2001. Jeffrey Skilling, the former president and CEO of Enron, was subsequently convicted of conspiracy and fraud in May 2006 and sentenced to serve a twenty-four-year prison term. In 2009, Bernard Madoff was convicted after using a Ponzi scheme to cheat a total of $50 billion from his investors. A Ponzi scheme is one in which money from new investors is used to pay off early investors to create the appearance of legitimate returns. Madoff was subsequently sentenced to serve 150 years in prison. Steps have been taken by FASB (Financial Accounting Standards Board) to tighten the financial audit guidelines for the future. In 2003, two top editors of *The New York Times* resigned because they failed to reign in one of their staffers who fabricated or plagiarized three dozen stories over a six-month period. Exposures of these cases in the news media bring about public anxieties and apprehensions. But the focus of the public's attention on ethics in the workplace should help motivate managers to avoid willful wrongdoing (Elliot and Schroth 2002).

As the markets in the new millennium become more dynamic and business relationships increasingly intertwine, opportunities for conflicts of interest and ethical dilemmas are likely to become more prevalent. This chapter addresses some of the issues and solution strategies related to ethics in business and engineering management (Seebauer and Barry 2001; Johnson 1991; Boylan 2000; Boatright 2000).

Before delving into a discussion of ethics, two pieces of background information might be useful. Patterson and Kim (1992) report that on the basis of interviews with 5,000 people across the United States in 1992, 90 percent of the people surveyed decide alone—without the influence of church, governments, or family—what is moral and what is not in their lives. Seventy-four percent of those surveyed said they would steal. Sixty-four percent said that they would lie when it suited them. If any of these numbers are believable, then the moral standards of the general public in the United States are on shaky ground. Let us hope that the sample of 5,000 people picked in the aforementioned survey might have included a few too many bad apples.

Recently, the Pew Research Center for the People and the Press in Washington, D.C., conducted a survey by asking the following question: "Would you say most business executives try to obey the laws governing their professions, or do they try to find a way around the laws?" Thirty-five percent of respondents said they obey the laws, 58 percent said they find a way around the laws, and 7 percent offered no answer. In terms of whose interests the companies put first, Table 10.1 summarizes the drastic difference between the public's perceptions of what current practices are and what the perceived right emphasis should be.

In a recent *Gallup poll*, the public was asked to assess the honesty and ethical standards among professions. The sum of the percentages of responses for "very high" and "high" are added together to provide a composite percentage for purposes of ranking. Table 10.2 presents some ranked professions, plus business executives and a few others for purposes of comparison. Over the period of 2000 to 2008, nurses,

Table 10.1 Whose Interests Do Companies Put First?

	Current Practice (percent)	Right Emphasis (percent)
Top executives	43	3
Stockholders	37	14
Employees	3	31
Customers	5	27
Communities	5	19
None	2	N/A
Don't know	5	N/A

Source: PEW Research Center for the People and the Press (2002), see questions 19 and 20 in the questionnaire section.

Table 10.2 Gallup Poll on Honesty and Ethics (2008 versus 2000)

Profession	Very High and High (Percent of Responses) 2008	Very High and High (Percent of Response) 2000
Nurses	84	79
Druggists/pharmacists	70	67
Medical doctors	64	63
Elementary/high school teachers	65	62
Clergy	56	60
Policemen	56	55
Bankers	23	37
Businessmen	12	23
Stockbrokers	12	19
Lawyers	18	17
Car Salesmen	7	7

Source: Gallup Poll. Subscribers may access its detailed breakdown scores, www.gallup.com/poll/1654/Honesty-Ethics-Professions.aspx.

druggists, and medical doctors have seen a small gain in the public perception of their professional honesty and ethics, whereas bankers, businessmen, and stockbrokers suffered quite a significant loss in perceived degree in honesty and ethics.

In a 2006 Gallup poll, engineers and college teachers have scored quite high in honesty and ethics in comparison with business executives, bankers and brokers; see Table 10.3.

There are numerous situations in which engineers and engineering managers might encounter problems with ethics. These situations include public safety and welfare, risks, health and environment, conflicts of interest, truthfulness, integrity, choice of a job, loyalty, gift giving and taking, confidentiality, industrial espionage, trade secrets, discrimination, and professional responsibility.

Generally speaking, there are microethical and macroethical issues. In microethics, the focus is on the relationships among service engineers and their coworkers, clients, and employers. In macroethics, service engineers are concerned with the collective social responsibility of the profession in relation to, for example, product liability, sustainable development, globalization, and the impact of technology. Sustainable development refers to industrial practices that minimize harmful impacts on the environment,

Table 10.3 Gallup Poll on Honesty and Ethics (2006)

Profession	Very High and High (Percent of Responses) 2006
Nurses	84
Druggists/pharmacists	73
Veterinarians	71
Medical doctors	69
Dentists	62
Engineers	61
Clergy	58
College teachers	58
Policemen	54
Psychiatrists	38
Bankers	37
Chiropractors	36
Journalists	26
State governors	22
Lawyers	18
Business Executives	18
Stockbrokers	17
Senators	15
Congress men	14
Insurance Salesmen	13
HMO managers	12
Advertising practitioners	11
Car Salesmen	7

Source: Gallup Poll (2006), www.gallup.com/poll/1654/Honesty-Ethics-Professions.aspx

while maximizing the efficiency of energy and material utilization. This chapter will address both types of issues.

In the research literature, two basic approaches are taken to handle ethical issues: addressing the general philosophy underlying a particular outlook and examining specific cases in order to draw out lessons related to ethics. There are weaknesses in both approaches. The philosophical approach lacks a connectedness to the real-world environment, thus producing no guidelines to deal with actual situations. The case-based approach produces only "school solutions" after the fact. These might or might not generate any useful lessons to be learned (Brown 2005; Vesilind and Gunn 2008). Since there are many well-known cases related to ethics, this second approach is selected for use in this chapter.

Not all problems in ethics have practical solutions, just as not all product/service design problems have feasible solutions. Furthermore, as more and more companies pursue globalization, an additional concern is that what is "normal" in one culture might not translate acceptably to another (Kline 2005).

In this chapter, the deliberations about ethics proceed with basic definitions, followed by discussions about engineering ethics that are then extended to business ethics. The chapter concludes with general guidelines for dealing with ethical issues.

10.2 ETHICS IN THE WORKPLACE

Ethics is important to corporations because companies with strong, positive overall reputations attract and keep the best customers, employees, suppliers, and investors. These companies also avoid the trouble of litigation, fines, recalls, bankruptcy proceedings, and antitrust suits. Corporations are made of people; what people say and do create the corporate reputation.

A U.S. Roper Poll (by GFK Roper Consulting) conducted in 2005 revealed that 72 percent of respondents believed wrongdoing was widespread in industry. Corporations take steps to solve ethics-related issues. According to Schwartz (2002), more than 90 percent of large U.S. corporations have formulated company-specific codes of ethics. This trend is expanding to corporations in other countries as well. (See Table 10.4.)

A code of ethics is a formal, written document that contains normative guidelines for behavior. Publishing and implementing such a code allows a corporation to provide a set of consistent standards for employees to follow, both to avoid any adverse legal consequences due to possible wrongdoing and to promote a wholesome public image.

Table 10.4 Percentage of Companies with Codes of Ethics

Countries	Percentage of Large Corporations with Codes of Ethics
United States	90
Canada	85
United Kingdom	57
Germany	51
France	30

Source: Mark S. Schwartz, "A Code of Ethics for Corporate Code of Ethics," *Journal of Business Ethics* (November–December 2002), pp. 27–43.

10.2.1 Universal Moral Standards

Whether a code of ethics contains right or wrong guidelines depends on the benchmark standards used to make the assessment. Schwartz (2002) has assembled a set of universal moral standards that appear to be intuitively correct. (See Table 10.5.)

By using such a set of universal moral standards as a yardstick, different codes of ethics can be compared and evaluated.

Ethics in the workplace can be discussed within three different scopes—ethics in engineering, management, and business—depending on the expected complexity of the situation and the potential impact on stakeholders.

10.2.2 Ethical Minds

Ordinary people are said to have five different minds (Gardner 2009):

1. *Disciplined mind*—building up knowledge in disciplined ways.
2. *Synthesizing mind*—screening and collecting knowledge.
3. *Creating mind*—to generate new ideas.
4. *Respectful mind*—respecting the relationships with others.
5. *Ethical mind*—value ethics above others—whistle-blowers.

Ethical mindset is critical to the successful practice of ethics standards. These people have a very strong inner desire to conduct themselves ethically. They firmly believe that all actions taken need to be right for the organization from the ethics standpoint. They practice rigorous self-honesty. They also consult trusted advisers inside and outside the organization and get feedback from genuinely independent sources to make sure that the actions in questions are completely ethical from multiple perspectives.

Vitell et al. (2003) defined *Perception of the Relative Importance of Ethics and Social Responsibility* (PRESOR) as the perceptions of the relative importance of ethics and social responsibility. They suggested that:

- Individuals who are more democratic and who believe that it is acceptable to disagree with their superiors tend to be more likely to believe that PRESOR is critical.
- Those who prefer to avoid risks will place greater importance on the long-term benefits of ethics and social responsibility.

Table 10.5 Universal Moral Standards

No.	Standards	Contents
1	Trustworthiness	Honesty, integrity, reliability, loyalty
2	Respect	Respect for human rights
3	Responsibility	Accountability
4	Fairness	Process, impartiality, and equity
5	Caring	Avoiding unnecessary harm
6	Citizenship	Obeying laws and protecting the environment

Source: Mark S. Schwartz, "A Code of Ethics for Corporate Codes of Ethics," *Journal of Business Ethics* (November–December 2002), pp. 27–43.

- Those who exhibit more loyalty and are more likely to follow social norms are in favor of PRESOR.
- Those who believe in universal moral principles are in favor of PRESOR. Those who reject the notion of universal moral absolutes are less likely to believe in PRESOR
- Those in high power positions tend to rely on established rules and policies for guidance and see any behaviors as ethical as long as there is no specific policy against them (they would not have done anything unlawful, even if questionable from an ethics standpoint).
- Those in lower power positions place more importance on personal ethics and behave ethically even in the absence of specific ethical policies.
- Strong corporate ethical values tend to lead to greater perceived importance of PRESOR.
- Enforcing a code of ethics tends to bring about a greater perceived importance for PRESOR.
- Those with high individualism and masculinity are not reported to be associated with low PRESOR (the negative correlations were not verified).

Ethical mindsets need to be actively nurtured by way of best practice examples, policies, corporate and personal values, social interactions, and recognition.

10.2.3 Engineering Ethics

Engineers play key roles in the advancement, production, and operation of technology. They should therefore assume a degree of responsibility for the consequences of applied technologies. Martin and Schinzinger (2004), Harris, Pritchard, Rabins (2008), and Davis (1998) address different aspects related to engineering ethics.

Engineering companies publish codes of ethics to guide engineers in performing their work in ethically and socially responsible manners. Many engineering societies, such as the National Society of Professional Engineers (NSPE), American Society of Civil Engineers (ASCE), American Society of Mechanical Engineers (ASME), and the Institute of Electrical and Electronics Engineers (IEEE), have also issued codes of ethics to provide discipline-specific guidelines. In addition, the files of NSPE's Board of Ethical Review contain various types of issues in engineering ethics and their respective resolutions.

Pinkus, Shuman, Humon, and Wolfe (1997) have noted that there are three principles of engineering ethics: competency, responsibility, and public stewardship. *Competency* means that engineers are obliged to know as much as is reasonably possible about the technology with which they work. They should be honest and candid enough to acknowledge their own deficiencies and seek assistance from others to fill in any gaps. *Responsibility* requires that engineers voice their concerns when an ethical dilemma is identified. Responsible organizations must then evaluate these concerns promptly. With respect to *public stewardship,* engineers must understand the risks associated with the technology that they apply and deploy.

A key concept in engineering ethics is the notion of *professional responsibility,* a type of moral obligation arising from the special responsibility possessed by an individual engineer. According to Baura (2006), this moral responsibility requires that engineers exercise judgment and care to achieve and maintain a desirable state of affairs, as well as to protect the public health and safety.

Many U.S. engineering schools are now actively involved in teaching ethics in undergraduate curricula. ABET, the Accreditation Board of Engineering and Technology, reviews and accredits countless engineering school programs in the United States on a regular basis. ABET has defined a specific program outcome related to ethics in its *Engineering Criteria 2000*. This program prescribes that graduates are to have an understanding of professional and ethical responsibility and the broad education necessary to understand the impact of engineering solutions in a global and societal context. The ABET *Engineering Criteria 2000* has been implemented since the fall of 2001.

Teaching ethics to engineering students is aimed at achieving four outcomes: (1) raising ethical sensitivity, (2) increasing knowledge about relevant standards of conduct, (3) improving ethical judgment (e.g., by way of discussions of real-world cases), and (4) strengthening the ethical willpower to enable a greater capability to act ethically. Teaching ethics illustrates real-world complex relationships between technology development and the welfare of individuals, society, and the environment. It also promotes an understanding of the professional nature of engineering and of the responsibilities associated with a professional career. It enhances the engineers' abilities to analyze situations that raise questions about ethics and to articulate reasonable ways to respond to ethical dilemmas.

Table 10.6 lists major engineering ethics cases reported in engineering and business literature. Other more recent cases may be found at the Online Ethics Center of National Academy of Engineeing (www.onlineethics.com). The case method of teaching ethics

Table 10.6 Sample Engineering Ethics Cases

No.	Case	References
A	Hooker Chemical—Love Canal, 1978	Gary Whitney. "Case Study: Hooker Chemical and Plastics," in T. Donaldson (editor), *Case Studies in Business Ethics.* Englewood Cliffs, NJ: Prentice-Hall (1984).
B	The Collapse of Walkways at Hyatt Regency Hotel in Kansas City, 1981	*http://ethics.tamu.edu/ethics/hyatt/hyatt1.htm,* E. Phrang and R. Marshall. "Collapse of the Kansas City Hyatt Regency Walkways," *Civil Engineering ASCE*, July 1982.
C	DC-10 Cargo-Door Accident Case near Paris, 1974	P. French. "What Is Hamlet to McDonnell-Douglas or McDonnell-Douglas to Hamlet: DC-10," *Business and Professional Ethics Journal* 1 (2) (1982). J. Fielder and D. Birsch. *The DC-10 Case: A Study in Applied Ethics, Technology and Society*, Albany, NY: State University of New York Press, 1992.
D	Spiro Agnew and Construction Kickback in Maryland, 1973	Richard M. Cohen and Jules Witcover. "A Heartbeat Away: the Investigation and Resignation of Vice President Spiro Agnew," New York: Viking, 1974.
E	Space Shuttle *Challenger* Explosion, 1986	R. L. B. Pinkus, L. Shuman, N. P. Hummon, and H. Wolfe, *Engineering Ethics: Balancing Cost, Schedule and Risk, Lessons Learned from the Space Shuttle*, New York: Cambridge University Press, 1997.

encourages students to express ethical opinions, prompts them to identify ethical issues, and helps them to formulate and justify decisions in an effectual manner. The case method also seeks to strengthen students' sense of the practical context of ethics.

A. The Hooker Chemical Case. Located in Niagara Falls, New York, Love Canal was 10 feet deep, 60 feet wide, and 3,000 feet long and surrounded by a virtually impervious clay soil. Hooker Chemical obtained permission to use the canal for dumping waste chemicals in 1942 and subsequently acquired a strip of land 200 feet wide with the canal in the center. Hooker sealed chemical wastes into steel drums, dropped them into the canal, and covered them with a layer of clay. Approximately 22,000 tons were deposited from 1942 to 1953. At the time, there were no federal or state regulations governing the dumping of chemical wastes.

In 1953, Hooker closed the dump and sold the land to the city's board of education for one dollar after the board threatened to condemn the land. Hooker included a clause in the deed to the board describing the past use of the land and required that the board assume the risk of liability for any future claims that might result from the buried chemicals.

Subsequently, the board built a school on part of the land and sold the remainder to a developer who built homes for families. None of the homes was built directly over the canal. In 1978, traces of chemicals were noted on the surface of the land. Residents complained of increased rates of miscarriage, risks of birth defects, increased urinary tract infections, and other health-related problems. In August 1978, President Carter declared Love Canal a limited disaster area. Federal investigators took 5,000 soil samples but failed to establish a direct link with the buried chemicals.

B. The Hyatt Regency Walkways Case. During a dance party in July 1981, two walkways suspended over the atrium of the hotel lobby collapsed, killing 114 people and injuring 185 others. Detailed investigations indicated that the mechanical supports for the walkways were insufficiently designed for the anticipated loads. The license of the engineering firm responsible for the design was revoked.

C. The DC-10 Case. In March 1974, a Turkish Airlines DC-10 crashed near Paris due to a failed cargo door and the collapse of the passenger compartment floor. Investigators noted that all of the hydraulic and electric control systems were severed. A total of 346 passengers and crew members died.

The problems of the DC-10's cargo door and the buckling of the passenger compartment floor were noticed when McDonnell-Douglas pressure-tested the airplane in 1970. In 1972, an American Airlines DC-10 experienced similar problems while flying over Windsor, Ontario. (That plane was brought down safely.)

McDonnell-Douglas used an electric latching system and installed three parallel redundant control systems, instead of the industrywide standard of four. The design engineers placed all three systems in the leading edge of the wing, rather than channeling them through different sections of the airplane to provide redundancy.

Records also verified that the Turkish Airlines DC-10 went in for cargo-door modifications in July 1972. Although three McDonnell-Douglas inspectors stamped the maintenance papers, the actual work to modify the cargo door was never done.

D. Spiro Agnew and the Construction Kickback in Maryland. In 1961, an engineering firm in Baltimore noticed that county government contracts were being awarded to firms that maintained close connections with governmental officials. The firm started to donate to the campaign funds of Spiro Agnew, who was running for county executive at the time. Agnew won, and his contractual relationship with the firm continued. From that point on, this engineering firm would provide 5 percent of its contract value as a cash kickback to Agnew. This practice continued over a long period, even when Agnew was later the governor of Maryland and then vice president of the United States. The total payment made to Agnew was alleged to be more than $100,000. This engineering firm accumulated the needed cash by giving key employees "bonuses" and requiring them to pay the money back partly in cash. The firm also contrived "loans" to individual employees, who then channeled cash back to the company treasury.

In October 1973, Spiro Agnew resigned as vice president after a no-contest plea. The court fined him $10,000 and sentenced him to unsupervised probation for three years.

E. The Space Shuttle Challenger Case. In the well-known space shuttle *Challenger* explosion case, the night before the launch, Morton Thiokol engineers had identified the potential danger of launching the shuttle in temperatures of less than 53 degrees Fahrenheit. NASA management challenged the recommendation. During an off-line discussion among Morton Thiokol participants in Utah, a vice president of engineering was the only one among four to hold out for a launch delay. A senior vice president told him bluntly, "It's time to take off your engineering hat and put on your management hat." The vice president capitulated, and the launch went forward, resulting in the disaster on record. This episode and the anecdote about engineering and management "hats" are now widespread in the literature on engineering ethics.

10.2.4 Management Ethics

Should there be any difference between engineering ethics and management ethics? Not a lot, except that managers must consider broader issues and deal with ethical situations more complicated than those typically encountered by engineers. Pontiff (2008) and Johnson (2006) include some discussion on management ethics.

In the space shuttle *Challenger* case, the Morton Thiokol management standpoint was as follows: The company was a contractor to NASA and had clearly expressed its technical concerns related to the launch. Even though the recommendation was not fully supported by available data, the company did fully discharge its moral and ethical responsibility in the O-ring issue. The proper role for Morton Thiokol was indeed to respect the view of its client. If NASA, as the paying client, decided to launch the shuttle anyway, then the responsibility for any negative consequences rested entirely with NASA.

From the NASA management standpoint, the O-ring issue was technically not supportable by data. NASA had its mission goals to fulfill. Under such circumstances, someone made a decision under uncertainty based on gut feeling and personal experience in risk assessment. Unfortunately, the decision turned out to be wrong. However, management is paid to make such hard decisions under uncertainty.

Generally speaking, engineering managers consider factors related to the well-being of the organization, such as cost, schedule, employee morale, customers, supply chains, investors, public image, local communities, health and safety, social and environmental impacts, market development, profitability, and globalization. Engineers, by contrast, focus on technical matters that fall within their professional engineering practices, such as product design, production, technology, public health and safety, and environmental impact.

Example 10.1 The ocean liner SS *United States* was a luxurious ship in the 1950s. It had approximately 500,000 square feet of harmful asbestos insulation. Initial estimates indicated that it would cost $100 million to have it refurbished in the United States. In 1992, it was towed to Turkey, where the cost of removing the asbestos was cited at only $2 million. Turkish officials, however, refused to allow the removal because of the danger of exposure to the cancer-causing asbestos. In October 1993, the ship was towed to the Black Sea port of Sebastopol, where laws were lax and the removal of asbestos would cost less than $2 million.

Do you approve of the program that removed asbestos from the SS *United States* at Sebastopol? Why or why not?

ANSWER 10.1 The ethically correct answer is to disapprove. The removal of asbestos at Sebastopol was economically attractive for the ship owner, because the local laws did not require specific safeguards and cautious processes needed for effectively protecting the health of workers. Thus, the cost saving was derived by taking advantage of the ignorance of the local people and the inadequacy of the local laws. The program has the potential for damaging the health of workers at Sebastopol, a blatant violation of core human rights. It must be rejected without reservation.

10.2.5 Ethics in Business

As engineering managers move up the corporate ladder, they get more and more involved in influential decision making that extends beyond the traditional domains of engineering and technology. Guidelines related to ethics in business kick in at that time to help shape the decision making with business implications.

Such situations arise from the fact that business management must take care of the broad interests of five stakeholder groups: shareholders, customers, employees, suppliers, and people in local communities in which the company operates. Business managers must also remain consistent in their professional responsibilities and ethical standards. Note that all of these stakeholders are members of the society at large, but they represent only a small part of it. (See Fig. 10.1.) Thus, inherently, there will be situations in which choices made to pursue the interests of these five stakeholder groups might clash with the interests of the remaining members of society. Although an ethical company must attempt to eliminate or minimize such clashes, conflicts of these types are likely to occur.

In fact, actual situations in real environments are far worse than this. As indicated previously in Table 10.1, questions have been raised concerning the conflicts of interest reflected in the management of numerous U.S. corporations. The U.S. public perceives that countless companies unethically put the top managers' own interests ahead of the five major stakeholder groups. Congress passed the Sarbanes–Oxley Act of 2002,

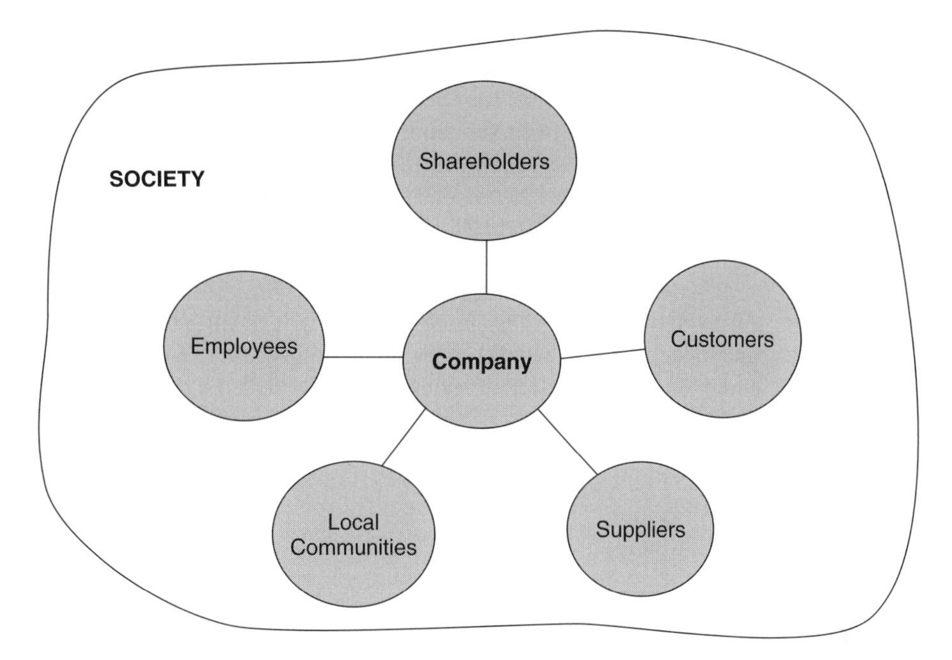

Figure 10.1 Stakeholders of a company.

the Public Company Accounting Reform and Investor Protection Act, to more tightly regulate some of the questionable accounting practices.

Michael (2006) pointed out that rules, in general, are not good enough, because they are by nature deficient in covering all encompassing situations. We need to shape the ethical mindsets to enable the individuals to make ethical decisions on their own, not by simply following the law of rules.

Confucius said: "If the people be led by laws, and uniformity sought to be given them by punishments, they will try to avoid the punishment, but have no sense of shame. If they be led by virtue, and uniformity sought to be given them by the rules of property, they will have the sense of shame, and moreover will become good" (Confucius 1971).

Mober (2007) suggested that practical wisdom enables people to make tough moral choices from among an almost limitless series of possibilities in business. Practical wisdom is the disposition toward cleverness in crafting morally excellent responses to, or in anticipation of, challenging particularizes. It has four components: knowledge, emotion, thinking, and motivation.

Shaw (2007), DeGeorge (2005), and Howard and Korver (2008) offer additional viewpoints related to business ethics.

Arthur Andersen. A case in point is Arthur Andersen, which was at one time the most highly regarded accounting firm in the United States for its standards of integrity and staff quality (Boyd 2004). The firm had created a strong normative culture, introduced standardized rules of operations, and trained staff to be obedient and efficient. Leaders exhibited impeccable integrity.

Subsequent leaders chose to diversify the accounting firm into consulting and reset the corporate vision to pursue revenues. Heavy infighting erupted between the auditing group and the fast-growing, more-profitable consulting group, causing the independence of the audit to be completely corrupted by the quest to get ever-more consulting income from audit clients. Auditors become less and less independent because of consulting. There was also no communications channel set up for employees to report questionable practices and to meet the needs of whistle-blowers.

Many questionable practices were introduced at the firm. For example, audit partners were asked to arrange for consultants to make pitches to audit clients. Audit partners were compensated according to the level of consulting services sold to an audit client and according to the degree of satisfaction expressed by the audit client. Audit partners were financially penalized if involved in a restatement of a client's financial position, such as charging entertainment expenses to audit fees, golfing with audit clients, and offering summer employment to the children of senior officers of audit clients. One audit partner who had objected to Enron's accounting policies had been removed from the audit team when Enron's management complained to the firm about his stance. In fact, the firm's creed was to "keep the client happy above all other considerations" (Boyd 2004).

Eventually, the company became so corrupted by greed that it completely lost its moral compass. Greed and lack of leadership derailed and eventually destroyed the firm. Thereafter, the demise of Arthur Andersen was just a matter of time. It sought bankruptcy shortly after Enron. The sad story of Arthur Andersen illustrates the demise of a once respected company that destroyed itself because of unethical business practices, valuing greed over truth.

General Electric. General Electric has had an exemplary corporate ethics program. Heineman (2007) reports that GE's high performance with high integrity has been sustained by the company's culture. GE's norms and values must be shared across the board so that everyone wants to win the right way. A number of GE's core principles and key practices are noteworthy:

- Demonstrating consistent and committed leadership—communicating the guiding principles, implementing a set of practices that have real consequences, demonstrating consistent leadership behaviors, issuing a clear warning of "one strike you're out," training employees on ethics issues and practices, and offering frequent consultation on specific issues.
- Building standards into business processes—specifying integrity metrics that are beyond formal financial and legal rules and setting up global standards that are tighter than those allowable by local practices.
- Encouraging finance, legal and human resources to be both partners and guardians.
- Giving employees a voice—allowing them to report integrity-related issues anonymously.
- Monitoring transactions and auditing activities constantly.
- Holding business leaders accountable for integrity-related misconduct.

Over the years, a few GE executives and business unit leaders have been discharged due to their proven negligence in monitoring and enforcing this corporate ethics program. This ethics program has served the company well.

10.3 GUIDELINES FOR MAKING TOUGH ETHICAL DECISIONS

Everyone could use guidelines when making tough ethical decisions. Peter Drucker, a well-known business management professor, recommends a *mirror test*. Ask yourself, "What kind of person do I want to see when I shave ... or put on my lipstick in the morning?" (Seglin 2005). This method might not deter those violators who are self-serving and who apply double standards to justify what they do.

Norman Augustine, former CEO of Lockheed Martin, has proposed four questions to gauge how ethical a course of action is:

1. Is it legal?
2. If someone else did it to you, would you think it was fair?
3. Would you be content if it appeared on the front page of your hometown newspaper?
4. Would you like your mother to see you do it?

If you answer yes to all four questions, then, according to Augustine, whatever you are about to do is probably ethical*. Following this method of screening, one is then advised to always have the cell phone numbers of a lawyer, editor of the hometown newspaper, and one's own mother on hand, plus a well-calibrated, unbiased "barometer" of fairness.

Badaracco (2004) believes that character is forged in situations where responsibilities come into conflict with values. These situations are called *defining moments.* At defining moments, managers must choose between right and right. Badaracco suggests a set of questions for individuals, managers of working groups, and executives of companies to answer when evaluating such defining moments. These questions are described in the next sections.

(Badaracco (2004) includes various examples in his article to demonstrate the application of these questions to real-world situations.)

10.3.1 Questions for Individuals Facing Defining Moments

What feelings and intuitions come into conflict in this situation? Which of the values in conflict are most deeply rooted in my life? What combination of expediency and shrewdness coupled with imagination and boldness will help me implement my personal understanding of what is right?

10.3.2 Questions between Right and Wrong That Managers of Working Groups Must Answer

What are the other persuasive interpretations of the ethics of this situation? What point of view is the most likely to win a contest of interpretations inside of my organization and influence the thinking of other people? Have I orchestrated a process that can manifest the values I care about in my organization?

**Source*: "Business & Ethics," by Norman Augustine, accessible at www.focusonethics.com/excerptinsearchofethics.html.

10.3.3 Questions That Confront Company Executives

Have I done all that I can to secure my position and the strength of my organization? Have I thought creatively and boldly about my organization's role in society and its relationship with stockholders? What combination of shrewdness, creativity, and tenacity will help me transform my vision into reality?

Some managers are constantly required to resolve the conflict between discharging their responsibility to maximize shareholder value, on the one hand, and behave in an ethical manner, on the other. Bagley (2003) suggests a simple decision tree to guide managers in making ethical decisions in any corporate projects. (See Fig. 10.2.)

Bagley advises managers to raise three questions: (1) Is it legal? (2) Does it maximize shareholder value?, and (3) Is it ethical? Managers should refuse to pursue any project or take any action if the answer to any of these questions is no. However, managers should decide to carry out projects that are legal and ethical, even if the projects do not maximize shareholder value, as long as pursuing them could benefit the other stakeholders of the company. (See Fig. 10.1.) The argument proceeds as follows: Managers work for the best interests of the company, and these best interests might not always require them to maximize shareholder value.

The logic of this decision-tree model is rather compelling. But a major weakness of this model is that, for some courses of action, the answers to the questions related to shareholder values and ethics will likely be "maybe" instead of a clear-cut yes or no.

In the *Challenger* case, Morton Thiokol engineers and managers argued for a delay of the launch due to a suspected O-ring failure under low-temperature conditions. In

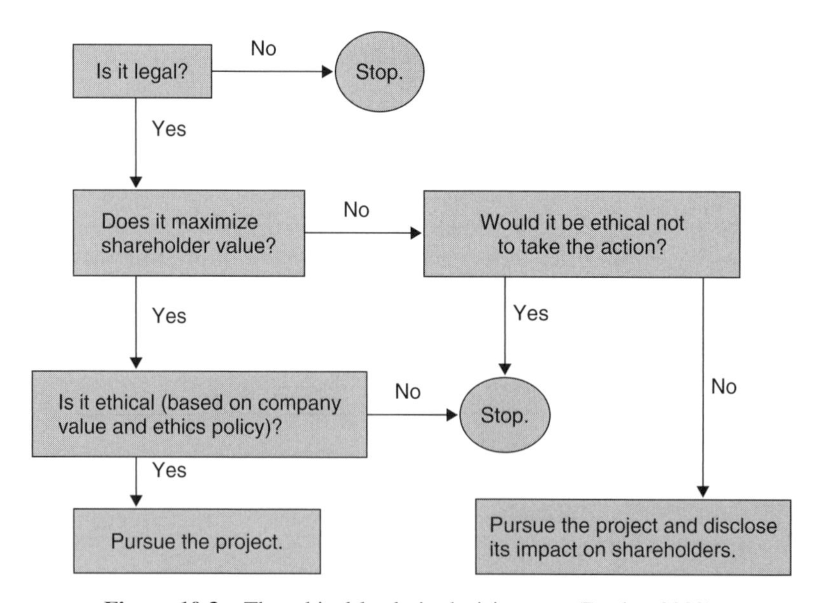

Figure 10.2 The ethical leader's decision tree (Bagley 2003).

contrast, NASA managers were under pressure to proceed with the launch as scheduled. Numerous questions were raised. Did Morton Thiokol have to prove that the flight was unsafe in order for the launch to be delayed? There were no data that could conclusively substantiate the recommendation of a launch delay. Should NASA managers have allowed factors other than engineering judgment to influence the flight-schedule decision? From the following two options, which stand should NASA have taken? (1) Don't fly if it cannot be shown to be safe, or (2) Fly, unless it can be shown to be unsafe. Certainly, the NASA program objective to build the shuttle for about half of the originally proposed cost might also have contributed to the decisions that resulted in disaster.

According to the decision-tree model presented in Fig. 10.2, it is clear that NASA had the legal authority to decide. When it comes to the second question, shareholder value, the answer was not clear-cut. Two options were under consideration at the time:

1. *Launch the shuttle as scheduled.* The benefits of a successful launch were expected to be the continued enhancement of the value of NASA programs to the American people and to the scientific communities, and the preservation of public trust and confidence in the capabilities of NASA management. However, if the launch were unsuccessful, then there would be a significant sacrifice of human life, the destruction of physical assets, and the loss of the public's goodwill. Perhaps some personnel changes at NASA would have become necessary under that scenario.

   The key problem was that neither Morton Thiokol nor NASA had data to confirm that the launch would definitely be unsafe. It was only the best judgment of Morton Thiokol engineers that the launch might be unsafe. The expected cost of a launch failure could not be quantitatively estimated because there was no reliable number for the probability of the O-ring to fail at low temperatures.
2. *Delay the launch until temperatures were right.* The cost of such a delay would have been the loss of public confidence in NASA managerial competency and certain operational and equipment maintenance costs associated with a delay.

Depending on the assumed value of the probability of occurrence for the O-ring failure, the expected cost for Option 1 could change. Thus, the question related to shareholder value in Fig. 10.2 should be answered "maybe." To move forward from this point on, it would require the introduction of an assumption (usually unproven and likely relying on gut feeling or an extrapolation from past experience) related to the probability for the O-rings to fail. Determining an accurate probability becomes a judgment call. Apparently, NASA management believed that the probability for the O-rings to fail was extremely low.

NASA's assumption of an extremely low probability for the O-rings to fail also led to a yes answer to the third question related to ethics in Fig. 10.2. If this probability of O-ring failure were high, then it would have been unethical for NASA to decide on a course of action that would have been likely to lead to the demise of the shuttle crew. An academic ethicist (Werhane 1991) evaluated the *Challenger* disaster and concluded that it was the result of four kinds of difficulties: (1) different perceptions and priorities of engineers and management at Thiokol and at NASA, (2) a preoccupation with roles and responsibilities on the part of engineers and managers, (3) contrasting corporate cultures at Thiokol and its parent, Morton, and (4) a failure both by engineers and managers to exercise individual moral responsibility. For a detailed discussion of the space shuttle *Challenger* case, see Vaughan (1997).

Although the decision-tree model is generally useful, its application can become complicated when conflicts of interest arise among the stakeholders. What happens if value is added to one group of stakeholders at the expense of others? The following examples illustrate this type of situation:

- Some U.S. company boards routinely approve bonus and stock options to top managers to offer them extra incentives at the expense of shareholders. This practice contributes to the distorted corporate emphasis indicated in Table 10.1. Such practices are not common in other industrialized countries.
- When involved in mergers and acquisitions, some top managers negotiate for special separation contracts to benefit themselves at the expense of the surviving company.
- Companies outsource manufacturing operations to developing countries where environmental and safety regulations are typically less stringent than in the United States; this achieves cost advantages for the companies at the expense of the communities in which they operate and of the employees in the developing countries.
- Companies sell unsafe automobiles and drug products to consumers, causing deaths due to accidents or side effects, while realizing sales revenue and profitability for themselves.
- The U.S. Securities and Exchange Commission (SEC) announced its settlement with ten Wall Street security firms to restore public trust. These firms agreed, without admitting any wrongdoing, to pay a fine of $1.4 billion for giving biased stock investment advice that caused investors to lose money during the recent past and delivered huge financial benefits to the firms involved.

The troubling point here is that all decision makers involved in the preceding cases were fully aware of the ethical implications related to their actions. Thus, knowing what is ethical and unethical is clearly only the first step. The decision tree in Fig. 10.2 offers a logical road map to force a needed critical review of all management decisions. However, additional steps beyond making knowledge available are required to ensure that all decisions carried out are ethical at all times and under all circumstances.

10.4 CORPORATE ETHICS PROGRAMS

Building an effective corporate ethics program requires leadership, commitment, planning, and execution. According to Nirvana (1997), companies need to have clear statements of their vision and values. They should have an organizational code of ethics. Creating an ethics officer position and forming an ethics committee helps communicate the company's ethics strategy, coordinate employee training, maintain a help line to offer confidential advice as needed under specific circumstances, monitor and track activities with ethics implications, and take action to reward good and punish unethical practices.

More than 90 percent of all Fortune 500 companies have codes of ethics, and 70 percent have statements of vision. Companies such as Procter & Gamble, IBM, Johnson & Johnson, Texas Instruments, John Deere, Cummins Engine, Eaton, and Dow Corning publish codes of ethics online. The Center for the Study of Ethics in the Professions at the Illinois Institute of Technology received a grant from the National Science Foundation to design and maintain a Web site, which stores 2000 codes of ethics online, including those issued by engineering associations (Section 10.2.3) and corporations. Its URL is http://ethics.iit.edu/codes/coe.html.

A corporation's statement of ethics serves as a behavioral compass for the employees. Kinni (2003) recommends a set of guidelines to formulate a code of ethics. Companies must first establish their values and direction. To be relevant, a statement of ethics must be connected to the core direction in which the organization is going. The code of ethics should be centered on such values as – integrity and trustworthiness, instead of being merely a compliance document. It should include specifics unique to the business under consideration or details about the company philosophy. The company should then go public with the code of ethics so that all employees, customers, suppliers, and any other stakeholders are fully informed of it. The code of ethics must be updated regularly to help guide the day-to-day behavior and decision making of the employees. The overall effectiveness of a statement of ethics depends on the company leaders' commitment to its disciplined enforcement. (Examples of ethics statements can be found in Murphy 1998.)

It is useful to issue guidelines to handle potentially unethical situations and to set clear standards of conduct applicable to daily responsibilities. Employees are advised to (1) analyze carefully the situation at hand, (2) list all possible failings and downsides of the potentially unethical practice in question, (3) compile all possible benefits that could accrue if the practice in question ceases and is admitted now, rather than discovered by someone else later, (4) issue a memo, and (5) attempt to make a full disclosure to coworkers, the superior, the superior's superior, the company president, the customers, the public, and the media.

Besides defining what the employees are expected to do, it is equally important to spell out what they should not do. Of specific value is the description of actions that are deemed unethical. Setting such lower bounds ensures that there is no ambiguity in interpreting what is not allowed. As sample unethical cases from both internal and external sources are continuously added to the code of ethics, the resulting casebook will become an increasingly important benchmark reference.

Honeywell has put teeth in its ethical principles by making it mandatory for all employees to adhere to the company's code of conduct. Starbucks Corporation introduced its Framework for a Code of Conduct for coffee-producing countries in order to standardize ethics practices among coffee retailers, exporters, and growers.

Some people believe that it is ineffective for companies to self-police their own codes. Instead, there should be an independent monitoring organization of ethical practices. To deter wrongdoing, the penalties must be high and the enforcement disciplined. Among examples of such penalties are dismissal with charges and forfeiture of all pension rights, a jail term with no opportunities for parole, severe financial penalties, and denial of reentry to industries involved.

There should probably also be more recognition from society for good actions taken by companies or individuals. Making ethically sound decisions might have been taken for granted by innumerable people for a long time. In 1998, *Deloitte/Management Magazine* created a Business Ethics Award program in New Zealand and has recognized the following companies with best practices in ethics since that time: New Zealand Post in 2008, New Zealand Aluminum in 2007, Snowy Peak in 2006, and Honda New Zealand in 2005. More awards of this type should be established to honor such people and to reflect society's appreciation for these exemplary ethical behaviors!

In managing corporate ethics programs, the key is to make sure that everyone is honest and will disclose fully all of the details of every situation in question. A well-known publisher suggests a single question as the basis for assessing an ethical situation: *"If what I just said or neglected to say, did or neglected to do, saw and failed to report, or heard and failed to mention, were disclosed openly by someone else in reputable communications channels, would it embarrass me, my organization, or my family?"* If the answer is yes, then the action or inaction in question is unethical. This question is likely to be useful for honest people with self-respect. However, embarrassment is a personal perception based on value. It may not be as useful to white-collar crooks who are driven by greed and who are willing and able to circumvent the laws to act unethically. Generally speaking, individuals who offer a full disclosure to all concerned are usually ethical (Johnson and Phillips 2003).

The commitment of top management to the corporate ethics program is of critical importance to its success. A case in point is Johnson & Johnson. The company is known for its Credo Challenges sessions in which employees and managers talk about ethics related to current business problems and offer criticisms of existing policies and ideas for improvement. The company achieved excellent business results by using this industrial best practice in the field of ethics.

An opposite case in point is Enron, which had a sixty-five-page code of ethics. Yet some top Enron managers entered into special business deals with off-balance-sheet financing that resulted in a falsely inflated corporate profitability that permitted selected management personnel to cash out stock options while siphoning out special bonus payments to individuals, all eventually at the expense of the company shareholders. Reports of internal whistle-blowers were simply ignored by Enron top managers who elected to take no corrective actions. It was clear to everyone involved that these were unethical and illegal management actions. Allegedly acting as its partner in crime, the auditing firm Arthur Andersen was also accused of committing the criminal offense of willfully destroying Enron papers relevant to the case.

In order to promote the global adaptation of uniform standards regarding ethics and integrity, the United Nations initiated in June 2000 the *Global Compact*, which contains the following ten principles (see www.unglobalcompact.org):

1. Support and respect the protection of international human rights with their sphere of influence.
2. Make sure that each corporation is not complicit in human rights abuses.
3. Uphold freedom of association and the recognition of the right to collective bargaining.
4. Eliminate all forms of forced and compulsory labor.
5. Support the effective abolition of child labor.
6. Eliminate discrimination with respect to employment and occupation.

7. Support a precautionary approach to environmental challenges.
8. Undertake initiatives to promote greater environmental responsibility.
9. Encourage the development and diffusion of environmentally friendly technologies
10. Work against corruption in all its form, including extortion and bribery.

As of January 2009, this Global Compact has gained a total membership of 4,700 businesses in 120 countries. However, no ethics program will be effective unless company top management supports its implementation.

Example 10.2 A global job migration trend is starting to emerge. Significant numbers of jobs are being exported from the United States and other developed countries to such developing countries as Mexico, China, India, the Philippines, Ireland, and Bangladesh. Typical jobs are related to software programming, call-center service, financial accounting, tax preparation, selected R&D, and claims processing. Other types of jobs may be involved in the future. The principal driving force behind the job migration trend is cost: A comparable quality of workmanship can be accomplished in Third World countries at about one-third of the cost of paying workers in developed countries.

Does Corporate America display an apparent indifference to its workforce at home? Is it unethical for American companies to outsource work in search of cost-effectiveness at the expense of American workers?

ANSWER 10.2 Corporate America is legally empowered to seek cost-effectiveness in creating and marketing products and services, as long as it is doing so in compliance with laws and commonly accepted ethics standards. Because of the free market system that the United States practices, Corporate America does not have the obligation to guarantee jobs for any sector of American workers, be they engineers, software programmers, accountants, claims-processing clerks, call-center service personnel, or any others. Individual workers, though, need to keep their own skills marketable at all times. If certain sectors of employment appear to be declining because of global competition, the workers in these sectors must be able to learn new skills quickly in order to keep themselves competitive in the job market. Past cases with textile, steel, and agriculture workers are typical examples of decline due to globalization.

The job-protection concept might be appropriate in a socialist system, wherein the government exercises control over the economy, but it is not appropriate in a free-market economy. By outsourcing, Corporate America does not display an apparent indifference to its workforce at home. It is not unethical for American companies to outsource work.

10.5 AFFIRMATIVE ACTION AND WORKFORCE DIVERSITY

Related to corporate ethics is the issue of affirmative action. The law is very clear on this issue. In 1964, the Civil Rights Act was enacted to ban discrimination on the basis of a person's color, race, national origin, religion, or sex. The Equal Employment Opportunity Act was passed in 1972, empowering the Equal Employment Opportunity

Commission (EEOC) to enforce Title VII of the Civil Rights Act. The EEOC directives do not allow for discrimination in hiring, placement, training, advancement, compensation, or other activities against any person on the basis of race, color, religion, national origin, gender, age, disability, marital status, or any other such characteristic protected by law. These acts have been enforced, and progress in affirmative action has been made over the last several decades.

For the past few decades, a number of private colleges and universities have introduced race-sensitive admissions policies to increase the number of African-American, Hispanic, and American Indian students they enroll. Park (2008) defends such policies on the ground that these institutions have the right to define the preferred composition of their student bodies. In recent years, however, there have been more state or institutional actions taken against such policies. In 1996, California abolished affirmative-action programs and decided to stop using race and gender as college admission factors. Washington State followed the California example in 1998. In 2000, Florida joined this anti-affirmative action movement.

Massachusetts Institute of Technology, Cambridge, Massachusetts, once excluded white students from its annual summer mathematics and science programs for incoming freshmen and high school students. In 2002, a complaint was filed with the Department of Education's Office of Civil Rights. Subsequently, MIT opened the programs to all students.

The University of Michigan had a well-known undergraduate admissions policy that allowed twelve points for a perfect SAT score and added twenty points if the applicant was black, Hispanic, or American Indian, but assigned no points for being white or Asian. Figure 10.3 illustrates the changing race and ethnicity composition of undergraduate students admitted to the University of Michigan. The U.S. Supreme Court announced its famous split decision in 2003: The vote was against the point system, but in favor of a race-based admissions policy for the University of Michigan Law School to use in the absence of a point system. Proponents on both sides of this

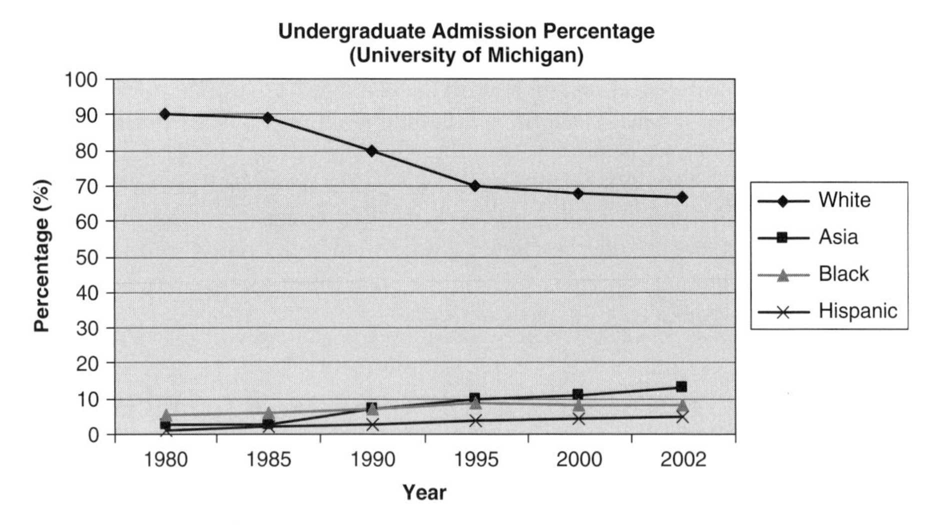

Figure 10.3 Undergraduate admission at University of Michigan.

issue were able to claim victory. Is it fair to those white students whose admissions were denied just because of their race and ethnicity? Questions like this remain unresolved.

Even after companies allow more minorities and women to gain entry, there is still an atmosphere of tension, instability, and distrust between white and nonwhite managers, according to Carver, Thomas, Ely, et al. (2005). Minority managers have a high turnover rate and they encounter deep-rooted, complex, and highly personal attitudes and assumptions in their coworkers.

Thomas (2000) argues that companies need to move beyond affirmative action and to strive for equal opportunities for everyone in the workforce as a way of creating competitive advantage in a global economy. Globalization might indeed precipitate workforce diversification and force everyone to work with everyone else, regardless of values, culture, race, or gender. Thomas and Ely (1996) indicate that companies should use the integration paradigm to promote equal opportunities for all people, promote open discussion of diverse cultural issues, eliminate all forms of domination (by hierarchies, race, gender, etc.) that inhibit full contribution, and secure organizational trust.

This debate will rage on. In the meantime, engineering managers are well-advised to constantly treat everyone and every situation honestly and fairly and to value each person's contributions properly.

10.6 GLOBAL ISSUES OF ETHICS

The problems of ethics are broader in complexity and scope for companies with a global reach than those for companies that operate domestically. This is due to the fact that the values, business practices, laws, environmental and safety standards, and other related references for making ethical decisions differ depending on the countries involved.

Engineering managers encounter problems with global ethics implications in a number of ways. Some problems related to product safety, plant operations, environmental discharges, work rules, and child labor are to be expected. Managers might need to interact with local governments with respect to permits, customs, transportation services, and procurement of parts and materials. They might also need to set up local supply chains and require all local participants to comply with specific codes of conduct.

Take environmental standards as an example. Countries that are economically active consume a disproportionate amount of energy and natural resources per capita compared with emerging countries. According to the World Bank's *Little Green Data Book* 2009, the United States uses sixteen times as much energy per person as India. About 15 percent of the world's population living in rich countries is responsible for 50 percent of the overall carbon dioxide emissions, the principal gas behind global warming. Figure 10.4 exhibits a detailed account of CO_2 remission per person in 2002. The question may be asked whether it is ethical and socially responsible for the economically active countries to pollute more than others.

Another example is related to values and business practices. Figure 10.5 displays the study results of Transparency International, a Berlin-based organization. According to this worldwide study, the least ethically corrupted country is Finland, followed by Singapore and then Great Britain, Hong Kong, Germany, and the United States.

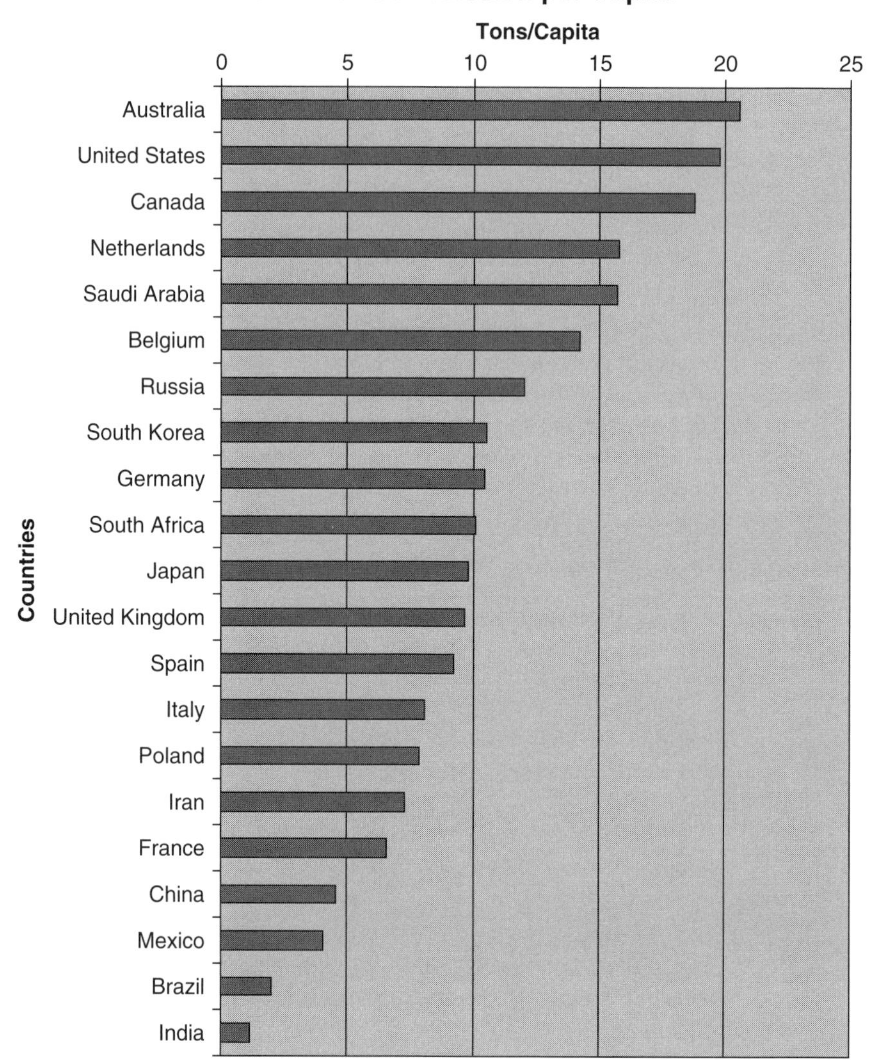

Figure 10.4 Carbon dioxide emission by countries. (Source: U.S. Department of Energy, 2006)

Yet another problem that has global ethics implications has to do with the law. In 1977, the U.S. Congress enacted the Foreign Corrupt Practices Act, which requires U.S. parent companies and their foreign subsidiaries to abide by certain business standards (Hinsey, Subramanian, and Kalka 2001). This act has three provisions:

1. *Recordkeeping and disclosure provision.* This provision requires that companies shall "... make and keep books, records and accounts, which, in reasonable details, accurately and fairly reflect the transactions and disposition of the assets

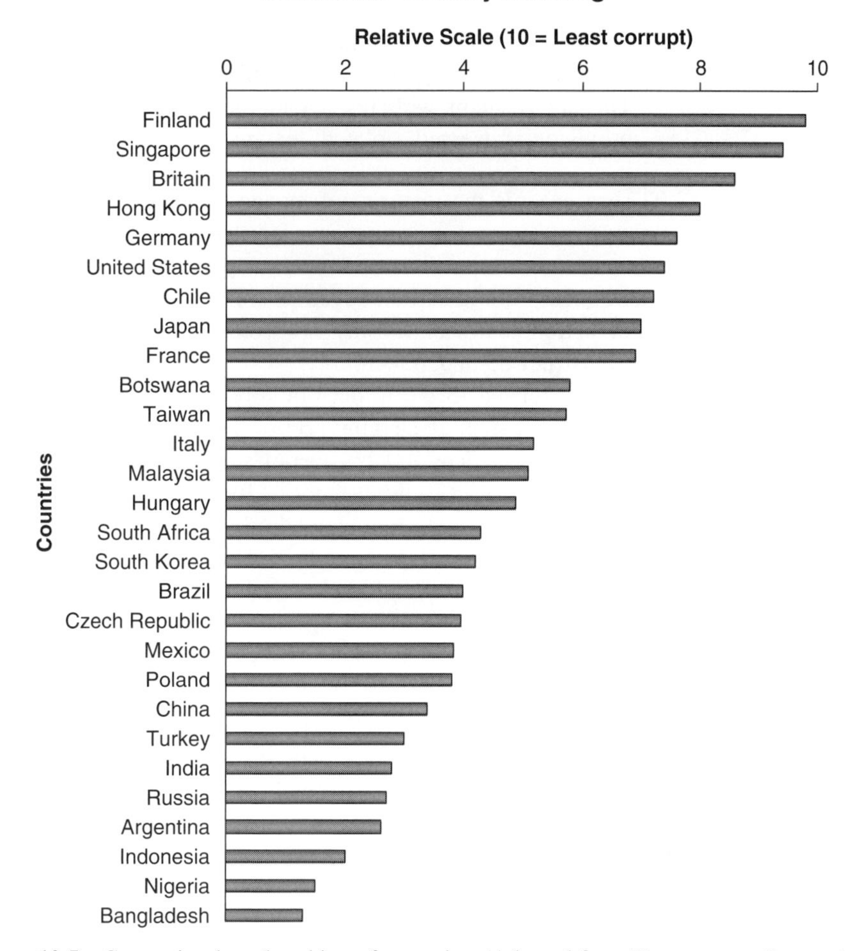

Figure 10.5 Corruption-based ranking of countries. (Adapted from Transparency International, Berlin Germany)

of the issuers.*" Not recording and disclosing payments (such as bribes) made to foreign recipients is an offense clearly prosecutable under this provision.

2. *Internal controls provision.* This provision prescribes that companies have audit committees composed of independent, outside members of their boards of directors to provide independent financial audits. Not exercising proper managerial control in the use of funds (such as accepting bribes or other illegal payments) is a violation of this provision.
3. *Antibribery provision.* This provision prohibits payments to foreign officials, foreign political candidates, or foreign political parties intended to corrupt those

**Source*: Anti-Bribery and Books & Records Provisions of the Foreign Corrupt Practices Act (November 10, 1998) accessible at www.usdoj.gov/criminal/fraud/docs/statute.html.

recipients who act in favor of the companies. Doing so is against the act even if it does not violate the laws of the respective countries involved. However, "grease payments" in the form of entertainment expenses and small gift items to minor officials are generally allowed for the purpose of facilitating transactions. The act is silent with respect to the actual dollar amount above which this provision is deemed violated, leaving such interpretations to the courts.

The act regards the integrity of management as a material factor. Engineers and engineering managers need to become familiar with and sensitized to these provisions when engaged in interactions with foreign personnel. When in foreign countries, the applicable local laws must be obeyed as well. Thus, it should be self-evident that any course of action that violates any laws, either of the home country or of the host country, must not be undertaken.

The issue related to the bribery of foreign public officials is a troublesome one. Some researchers, such as Deming (2006), have questioned the positive impact of the Foreign Corrupt Practices Act on overall standards of international business conduct, particularly with respect to the bribery of foreign public officials.

Problems involving environmental discharge or child labor might cause engineering managers to be fearful of another law: The Alien Tort Claims Act has been used to prosecute U.S. companies for alleged human rights violations and environmental degradation abroad. Olsen (2002) reports three such industrial cases.

Even if all of the actions taken by engineering mangers are lawful, it is possible that not all lawful business decisions they make are ethical. It is usually quite complex to determine whether a given course of action taken in a host country is ethical or not, because the contrasts between the cultures, values, and customs of the host and home countries come into play. Examples cited by Donaldson (1996) illustrate the differences:

- Indonesians tolerate the bribery of public officers. Bribery is considered unethical and unlawful in the United States.
- Belgians do not find insider trading morally repugnant. Insider trading is a criminal offense in the United States. For example, Martha Stewart was indicted and subsequently convicted for having committed conspiracy, making false statements, and obstruction related to her suspicious sales of ImClone Systems stock. She resigned as CEO of her company and was sentenced to serve time in prison.
- In some countries, loyalty to a community, a family, an organization, or a society is the foundation of all ethical behavior. The Japanese people define business ethics in terms of loyalty to their companies, their business networks, and their nation. The Japanese are group-oriented. In contrast, Americans place a higher value on liberty than loyalty and emphasize equality, fairness, and individual freedom over group achievements. Americans are focused on individualism.
- The notion of a right or *entitlement* evolved with the rise of democracy in post-Renaissance Europe and the United States. This term does not exist in either Confucian or Buddhist traditions.

- Low wages might be considered by workers in wealthy countries (e.g., the United States and Europe) as an exploitation of the workforce, but developing nations might be acting ethically when they accept low wages to induce foreign investment to improve their living standards.
- Some developing-country governments might want to use more fertilizer to enhance crop yields to combat food supply shortage problems, even though doing so would mean accepting a relatively high level of thermal water pollution.
- A manager at a large U.S. specialty products company in China caught an employee stealing. She followed the company's practice and turned the employee over to the provincial authorities. The provincial authorities executed him.
- In Japan, people who do business together exchange gifts, sometimes very expensive ones, as part of a long-standing tradition. Any foreign managers wanting to do business there will need to accept such practices as given.
- Managers in Hong Kong have a higher tolerance for some forms of bribery than their Western counterparts, but they have a much lower tolerance for the failure to acknowledge the work of a subordinate. In some parts of the Far East, stealing credit from a subordinate for work or ideas is the most unethical activity.

  In the case of *The New York Times*, a staff writer stole credit by plagiarizing the writings of freelance writers and other sources. He did so for an extended period. In spite of multiple warnings, the responsible editors did nothing to stop this staff writer from publishing. Finally, the staff writer resigned, and both editors were forced out[*].
- Some Indian companies offer employees the opportunity for one of their children to get a job with the company once the child has completed a certain grade level in school. The company honors this commitment even when other applicants are more qualified. This perk would be regarded as nepotism in the United States, as it is against the principle of equal opportunity that jobs should go to the best-qualified applicants. However, some U.S. universities do reserve certain admission quotas for children of alumni, major donors, and members of specific minorities.
- The Swiss are known for their time sensitivity, whereas South Americans are known for their time laxity.
- Forty percent of managers in the United States believe that the primary goal of a company is to make a profit, while only 33 percent in England, 35 percent in Australia, 11 percent in Singapore, and 8 percent in Japan share this belief.

  There is a significant divergence in the perceived goals of a company from the manager's perspective. Once a company achieves its financial objectives, how should the money be distributed? Table 10.1 illustrates the emphasis of U.S. managers versus those of the general public in the United States. The self-centered approach of some U.S. managers could cause tensions and conflicts in values when dealing with foreign managers.

There is no international consensus on standards of business conduct (Williams 2000). Donaldson (1996) offers three basic principles as guidelines: (1) observing core

[*]*Source*: Low Point in the 152-year History of the Newspaper at www.pbs.org/newshour/bb/media/jan-june03/nytimes_06-05.html.

human values, (2) showing respect for local traditions, and (3) focusing on the context when deciding what is right and wrong. Core human values determine the absolute moral threshold for all business activities. These values include the right to good health, economic advancement, and an improved standard of living. One must respect human dignity and recognize a person's value as a human being, and not treat others simply as tools. A good yardstick to use is a form of the Golden Rule, which says, "Do not do unto others what you do not want done to yourself".

This principle requires that customers be treated well through the production of safe products, and employees through the maintenance of a safe workplace. Also, the local environment should be protected. Companies must avoid employing children and thereby preventing them from receiving a basic education. Local economic and education systems ought to be supported. Companies should forgo those business relationships that violate rights to health, education, and safety, and which prevent the development of an adequate standard of living.

Donaldson (1996) classifies ethics conflicts into two types: conflicts due to relative economic development and conflicts due to cultural tradition. Ethical situations of the first type are related to wages, safety, and the environment; they arise from a low level of economic development in the host country in comparison with the developed country. To determine whether a given course of action is ethical, Donaldson suggests that the following question be raised: "Would the course of action under consideration be acceptable in my home country if our economic development were at the same stage as that of the host country?" If the answer is yes, then the course of action is ethically acceptable. For example, if a specific developing country is currently at the stage comparable to the United States in the 1970s, then U.S. rules and regulations related to wages, safety, and the environment in practice during the 1970s, not those in the 2000s, should be used to assess any situation involving current ethical conflicts in that developing country.

From time to time, courses of action must be taken even if they bring about conflicts due to cultural tradition if companies want to do business in a given host country. Generally, cultural tradition is to be respected and accepted. Saudi Arabia is known not to allow women to serve as corporate managers; most women there work in education and health care. Most foreigners who do business in Japan have now generally accepted the Japanese gift-giving tradition. Of course, compromises made in tradition-based conflicts must not violate core human values.

Some companies use a specific gift-giving strategy in order to be lawful, ethical, and compatible with local cultural tradition while promoting goodwill and fostering close working relationships. They elect to present two sets of gifts: a big and very expensive company-to-company gift and several small personal gifts, each being, for example, under $25. Past practices have shown that such an approach seems to work out well for the parties involved.

Again, when handling global problems of ethics, the following are the basic questions: "Is it legal with respect to laws in both the host and home countries?" "Does the planned course of action violate basic human values?" "Is it consistent with the local cultural norms?" "Is there a creative way to reconcile the ethics issues at hand?" Fig. 10.6 presents the decision-tree diagram for global problems of ethics.

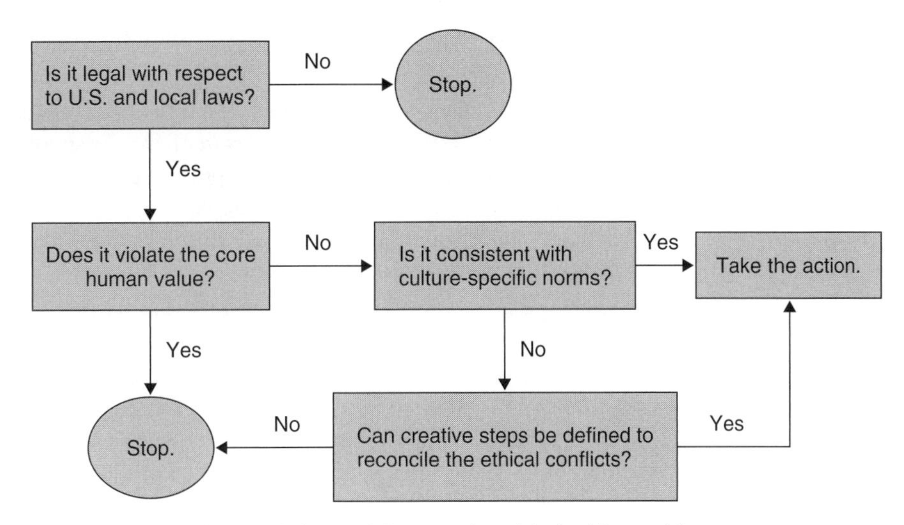

Figure 10.6 Decision tree for global ethics problems.

Cavanagh (2004) studied three strategies regarding global business ethics, namely, regulations, codes of ethics and self-restraints. Regulations are multilateral treaties and international agreements. Global codes are global codes of business conduct (e.g., UN Global Compact). Self-restraint is the voluntary self-restraint of individual executives and firms. Table 10.7 outlines the advantages and disadvantages of each.

Lau and Wong (2006) described a global ethics case involving a German subsidiary plant operating in Hong Kong. Ethics issues were noted in conjunction with "referral payments" and other such business practices due to a divergence in the legacy cultural and business practices involved. Eventually, the authors came up with a set

Table 10.7 Alternative Strategies of Providing Global Ethical Guidelines for Business

#	Strategies	Advantages	Disadvantages
1.	Regulation	Provides clear guidelines, which are applicable to all. Lessens advantages of "free riders." Provides enforcement methods.	Difficult to negotiate, ineffective without monitors and sanctions. Manager's dislike for global controls. A nation can remain outside agreement.
2.	Global codes	Focuses managers on ethical issues. Easier to support because not coercive. Initiates cooperation and communication.	Many countries and firms do not participate. U.S. executives fear costly lawsuits. No mandatory monitors or sanctions.
3.	Self-restraint	Enables stakeholders to identify ethical firms. Managers able to anticipate regulations. Imposes no "red tape" or limiting Restrictions. Useful if no regulation or code in place.	Inadequate to deal with some issues. Provides spotty long-term results. Gives financial advantages to unethical firms.

Source: Cavanagh (2004)

of recommended steps to handle global ethics issues by emphasizing transparency, management commitment and rigorous enforcement:

- *Specify a set of code of ethics.* Include the statement of company values, the form and limitations of "referral payments" from clients or suppliers, the proper ways of using "quanxi" and other metrics for preferred ethical practices, disciplinary procedures and penalties for staff violations, and written pledge from all staff to abide by the codes. Improve the code of ethics regularly based on industrial best practices and advisement of external experts.
- *Communicate the code of ethics.* Inform business partners, suppliers, clients and others doing business with the company of the contents of the code.
- *Implement the code with determination.* Increase the transparency of interactions with clients and suppliers, as well as staff recruitment and reward processes, conducting frequent internal audits, creating open communications channels to facilitate anonymous feedback regarding ethics issues, making senior management available for consultations concerning ethics issues, and training staff to self-review ethics-related issues by conducting five tests:
  1. *Smell test*—Does it smell OK?
  2. *Sleep test*—Will it keep me awake at night?
  3. *Newspaper test*—How would it look on the front page of the newspaper?
  4. *Mirror test*—How do I feel when I look in the mirror in the morning?
  5. *Mother test*—How would Mom react if I tell her what I am doing? (People are said to have special relationships with their mothers, to whom they are least likely to lie, in comparison with fathers, mentors, teachers and other such respectful individuals.)

In handling challenging situations involving ethics in global settings, engineering managers need to uphold core human values, account for the relevant local traditions, and be creative in problem solving to come up with a suitable and ethically acceptable course of action without violating laws (Tichy and McGill 2003).

Example 10.3 In the past, Levi Strauss & Co. engaged numerous contract manufacturers in Third World countries to produce athletic shoes. The company discovered in 1992 that two sewing subcontractors in Bangladesh employed children under the age of fourteen, in violation of company rules. This presented a dilemma: On the one hand, the local economic condition was such that the children were contributing significantly to their family income. If they were discharged instantly, the lack of these income streams would cause economic hardship to the families involved. On the other hand, allowing the children to continue working clearly violated the stated company ethics policy prohibiting the employment of child labor. What should Levi Strauss have done?

ANSWER 10.3 In fact, Levi Strauss came up with a creative solution to this ethics dilemma. The children were sent back to school with tuition, books, and uniforms fully paid by the company. In addition, they continued to receive their full wages. They were promised jobs after they completed their education at the age of fourteen. This creative solution satisfied both Levi Strauss and the children's families.

This is a practical example of the creative steps indicated in Fig. 10.6.

Example 10.4 Your company is among several U.S. suppliers actively competing for a major equipment project of a state-owned enterprise in Shanghai worth about $20 million. You are introduced to a Chinese consultant who offers to help. This consultant claims that he has contacts inside the enterprise's midlevel project evaluation teams and its top-level decision-making groups and that he can find out sensitive information for you, such as the equipment design features in the competitor's proposals. He can also relate back to you the enterprise's hints for improvements to your proposal and help move your project along in a multiple-round bidding process. If you do not win, you owe him nothing. However, if your project succeeds, you must pay him 5 percent of the final project value as a consulting fee. Your peers tell you that such an arrangement is quite normal in China and that a large part of the consulting fee goes directly to staff people in that enterprise and to the enterprise itself. Those who have rejected such help in the past have seen major equipment contracts awarded to less fussy competitors.

What should you do?

ANSWER 10.4 Some Chinese state-owned enterprises use their own high-level employees to serve as such "consultants," with the ultimate purpose of driving down the final project cost. With the tacit support of the enterprise, each consultant forms a team consisting of employees situated in various departments and groups within the enterprise. In offering advice, they pass along information prepared to exert increased competitive pressure among the foreign suppliers. They also collect design intelligence so that the enterprise knows what critical questions to raise in project negotiations. It is one of those tools to help an unsophisticated technological buyer get the bid.

If your company needs the project to penetrate the Chinese market and views the winning of this project to be critically important for your company's future growth in the global market, then you should proceed to engage the consultant, as not doing so will render your company less competitive.

If your company regards this project to be useful but not critically important to your future success, then stay on the high ground of morality and reject the services of the consultant.

Example 10.5 You are a manager of a joint venture in Russia. One day you discover that your most senior officer in Russia has been "borrowing" equipment from the company and using it in his other businesses (Puffer and McCarthy 1995). When you confront him, the Russian partner defends his actions. After all, as a part owner of both companies, isn't he entitled to share in the equipment?

How do you propose to resolve this conflict?

ANSWER 10.5 You should tell the Russian partner that the joint venture would be very happy to assist his other companies in any way. Then you should proceed to send him a bill for lease expenses of the equipment at a commonly accepted market rate. Tell him that, in the future, you would like to preapprove any such leases to make sure that the affected equipment is not needed in your own shop at the time.

10.7 PHILOSOPHICAL APPROACH OF ADDRESSING ETHICS ISSUES

In the introduction section of this chapter, we mentioned that one of the two approaches to handling ethical issues is to address the general philosophy underlying a particular ethical outlook. Some academicians are very much in favor of this approach. In order to offer a broad-based exposure to some of the philosophical treatments of ethics, we have collected a sample of ten references:

1. *Leadership Ethics* (Ciulla 2005)
2. *International Business Norms* (Windsor 2004)
3. *Integrity, Responsibility and Affinity in Banking Ethics* (Cowton 2002)
4. *Business Ethics and Stakeholder Theory* (Cragg 2002)
5. *Global Ethics* (Madsen 2004)
6. *Deception and Information Withholding* (Carson 2001)
7. *Business Ethics: The Law of Rules* (Michael 2006)
8. *Practical Wisdom and Business Ethics* (Moberg 2007)
9. *Mutual Advantages of Economic and Business Ethics* (Luetge 2005)
10. *Ethical Undercurrent of Pension Fund Management* (Ryan and Dennis 2003)

The concept of moral rationality is discussed by Paine (2003). Moral rationality has four components: (1) risk reduction—ethical practices reduce risks, (2) organizational integration—better ethical actions facilitate a better collaboration among members within a company, (3) market reputation—ethical practices enhance brand image, and (4) communication reputation—ethical practices reduce friction with communities and raise a company's reputation. Paine (2003) thinks that moral rationality provides the driving force for high-performing companies to want to increase practice of ethical competencies, in addition to the economic and technical competencies they have historically sought to cultivate. This is a futuristic view of business ethics.

Interested readers should also consult additional articles related to ethics published in the current issues of *Business Ethics Quarterly* and *Business Ethics: a European Review*.

10.8 CONCLUSIONS

Ethics in the workplace is a fundamental requirement for all members of a corporation—individual employees, managers of working groups, and executives. Everyone must diligently observe universal moral standards and basic human values, act legally, respect the local cultural traditions and practices, and disclose the courses of action taken extensively.

Most situations in the workplace involving ethics originate from a conflict between responsibilities and values. These situations must be carefully analyzed by the involved decision maker to arrive at a proper course of action. A number of models are included in this chapter to assist engineers, managers, and executives in carrying out such analyses.

Creating codes of ethics is helpful. Employees become more sensitive if they are exposed to more real-world cases with ethical implications. Knowing what is right and what is wrong is usually not sufficient in preventing unethical outcomes. Of critical importance to the success of a corporate ethics program is the disciplined enforcement of the code of ethics supported by top management. Good behavior should be recognized and rewarded promptly. Unacceptable behavior must be punished fully and in a timely manner.

The key to ethical behavior is honesty, fairness, and full disclosure. When selecting employees, engineering managers need to focus on their characters, as skills can more readily be imparted. Managers themselves need to be technically outstanding, managerially competent, and ethically sensitive.

10.9 REFERENCES

Badaracco, J. L. Jr. 2004. "Personal Values and Professional Responsibilities." Harvard Business School Note # 304070, Harvard Business School Press (January 12).

Bagley, C. E. 2003. "The Ethical Leader's Decision Tree." *Harvard Business Review*, Reprint F0302C (February).

Baura, G. 2006. *Engineering Ethics: An Industrial Perspective*. New York: Academic Press.

Boatright, J. R. 2008. *Ethics and the Conduct of Business*, 6th ed. Upper Saddle River, NJ: Prentice Hall.

Boyd, Colin. 2004. "The Last Straw." *Business Ethics Quarterly* 14 (3).

Brown, M. T. 2005. *Corporate Integrity: Rethinking Organizational Ethics ad Leadership*. London: Cambridge University Press.

Carson, Thomas. 2001. "Deception and Withholding Information in Sales." *Business Ethics Quarterly* 11 (2).

Carver, K., David A. Thomas, Robin R. Ely, Sylvia A. Hewlett, Carolyn Buck Luce, Cornel West and Ancella B. Livers. 2005. *Required Reading for White Executives*, 2nd ed. Boston: Harvard Business OnPoint Collection# 2205 (November 1).

Cavanagh. Gerald F. 2004. "Global Business Ethics—Regulation, Code or Self-Restraint." *Business Ethics Quarterly* 14 (4).

Ciulla, Joanne B. 2005. "The State of Leadership Ethics and the Work that Lies Before Us." *Business Ethics: A European Review* 14 (4).

Confucius. 1971. "Confucian Analects, the Great Learning and the Doctrine of Man 146." James Legge Translations. Mineola, NY: Dover Publications.

Cowton, C. J. 2002. "Integrity, Responsibility and Affinity: Three Aspects of Ethics in Banking." *Business Ethics: A European Review* 11 (4).

Cragg, Wesley. 2002. "Business Ethics and Stakeholder Theory." *Business Ethics Quarterly* 12 (2).

Davis, M. 1998. *Thinking Like an Engineer: Studies in the Ethics of a Profession*. New York: Oxford University Press.

DeGeorge, R. 2005. *Business Ethics with CD-Rom*. 6th ed. Upper Saddle River, NJ: Prentice Hall.

Deming, S. H. 2006. *The Foreign Corrupt Practices Act and the New International Norms*. American Bar Association.

Donaldson, T. 1984. *Case Studies in Business Ethics*. Englewood Cliffs, NJ: Prentice Hall.

Donaldson, T. 1996. "Values in Tension: Ethics Away from Home." *Harvard Business Review* (September–October).

Elliot A. L., and R. J. Schroth. 2002. *How Companies Lie: Why Enron Is Just the Tip of the Iceberg*. New York: Crown Business.

Ferrell, O. C., John Fraedrich, and Linda Ferrell. 2007. "Business Ethics: Ethical Decision Making and Cases." 7th ed. Boston: Houghton Mifflin Company.

Gardner, Howard, 2009. "Five Minds for the Future." Boston: Harvard Business School Press.

Global Compact 2003. "The Global Compact—Corporate Citizenship in the World Economy." United Nations (see www.unglobalcompact.org).

Harris, C. E. Jr., M. S. Pritchard, and M. J. Rabins. 2008. *Engineering Ethics: Concepts and Cases with CD-Rom*, 4th ed. Belmont, CA: Wadsworth.

Heineman, Ben W. Jr. 2007. "Avoiding Integrity Land Mines." *Harvard Business Review* 85 (4).

Hinsey, J. IV., G. Subramanian and M. Kalka. 2001. "Global Approaches to Anti-corruption." Harvard Business School Note # 902962, Harvard Business Press (November 14).

Howard, R. A. and Clinton D. Korver 2008. *Ethics for the Real World: Creating a Personal Code to Guide Decision in Work and Life*. Boston: Harvard Business School Press.

Johnson, C. E. 2006. "Ethics in Workplace: Tools and Tactics for Organizational Transformation." Newbury Park, CA: Sage Publications, Inc.

Johnson, L., and B. Phillips. 2003. *Absolute Honesty: Building a Corporate Culture That Values Straight Talk and Rewards Integrity*. New York: AMACOM.

Kinni, T. 2003. "Words to Work By: Crafting Meaningful Corporate Ethics Statements." *Harvard Business Management Update*, No. C0301E.

Kline, J. M. 2005. *Ethics for International Business: Decision-Making in a Global Political Economy*. London: Routledge.

Kline, J. M. 2005. *Ethics for International Business: Decision-Making in a Global Political Economy*. London: Routledge.

Lau, Amy, and Raymond Wong. 2006. Ethical Leadership and Its Challenges in the Era of Globalization—Teaching Note. Boston: Harvard Business School Case HKU623.

Luetga, Christoph. 2005. "Economic Ethics, Business Ethics and the Idea of Mutual Advantages." *Business Ethics: A European Review* 14 (2).

Martin M. W., and R. Schinzinger. 2004. *Ethics in Engineering*, 4th ed. New York: McGraw-Hill/ Science/ Engineering/Math.

Michael, Michael L. 2006. "Business Ethics: the Law of Rules." *Business Ethics Quarterly* 16 (4): 475–504.

Moberg, Dennis, J. 2007. "Practical Wisdom and Business Ethics." *Business Ethics Quarterly* 17 (3).

Murphy, P. E. (ed.). 1998. *Eighty Exemplary Ethics Statements*. Notre Dame, IN: University of Notre Dame Press.

National Academy of Engineering. 2004. *Emerging Technologies and Ethical Issues in Engineering: Papers from a Workshop*," illustrated edition. Washington, DC: National Academies Press.

Nirvana, F. 1997. "Twelve Steps to Building a Best-Practice Ethics Program." *Workforce* (September).

Olsen, J. E. 2002. "Global Ethics and the Alien Tort Claims Act: A Summary of Three Cases within the Oil and Gas Industry." *Management Decision* 40 (7): 720–724.

Paine, Lynn Sharp. 2003. "Value Shift: Why Companies Must Merge Social and Financial Imperatives Achieve Superior Performance." New York: McGraw-Hill.

Park, A. 2008. "Making Diversity a Business Advantage." Harvard Management Update Article # U0804A (April 1).

Patterson J., and P. Kim. 1992. *The Day America Told the Truth*. New York: Dutton/Plume.

Pinkus, R. L. B., L. J. Shuman, N. P. Hummon, and H. Wolfe. 1997. *Engineering Ethics: Balancing Cost, Schedule and Risk, Lessons Learned from the Space Shuttle*. New York: Cambridge University Press.

Pontiff, S. 2008. *Ethical Choices: How Managers Perceive Their Decision-Making Experience in the Face of Ethical Dilemmas*. Saarbrücken, Germany: VDM Verlag.

Puffer, S., and McCarthy, D. J. 1995. "Pinning the Common Ground in Russian and American Business Ethics." *California Management Review* 57 (2) (Winter): 29–46.

Ryan, Lori Verstegen, and Bryan Dennis. 2003. "The Ethical Undercurrent of Pension Fund Management: Establish a Research Agenda." *Business Ethics Quarterly* 13 (3).

Schwartz, M. S. 2002. "A Code of Ethics for Corporate Code of Ethics." *Journal of Business Ethics* (November–December): 27–43.

Seebauer, E. G., and R. L. Barry. 2001. *Fundamentals of Ethics for Scientists and Engineers*. New York: Oxford University Press.

Seglin, J. L. 2005. "How to Make Tough Ethical Calls." Harvard Management Update article # U0504C, Harvard Business School Publishing.

Shaw, William H. 2007. *Business Ethics*, 6th ed. Belmont, CA: Wadsworth.

Singer, Peter. 2002. *One World: The Ethics of Globalization*. New Haven and London: Yale University Press.

Thomas, R. R., Jr. 2000. "From Affirmative Actions to Affirming Diversity." Harvard Business OnPoint Article # 391X, Harvard Business School Press.

Thomas, D. A., and R. J. Ely. 1996. "Making Differences Matter: A New Paradigm for Managing Diversity." *Harvard Business Review* (September–October).

Tichy, N. M., and A. R. McGill (eds.). 2003. *The Ethical Challenge: How to Lead with Unyielding Integrity*. San Francisco: Jossey-Bass.

Toffler, Barbara Ley, with Jennifer Reingold. 2003. "Financial Accounting: Ambition, Greed and the Fall of Arthur Andersen." New York: Broadway Books.

Vaughan, D. 1997. *The Challenger Launch Decision: Risky Technology Culture and Deviance at NASA*.(paperback) Chicago: University of Chicago Press.

Vesilind, P. Aarne, and A. S. Gunn. 2008. *Engineering, Ethics and the Environment*. London: Cambridge University Press.

Vitell, Scott, J., Joseph G. P. Paolillo, and James L. Thomas. 2003. "The Perceived Role of Ethics and Social Responsibility: A Study of Marketing Professionals." *Business Ethics Quarterly* 13 (1).

Werhane, P. H. 1991. "Engineers and Management: The Challenge of the Challenger Incident." *Journal of Business Ethics* 10: 605–616.

Williams, O. F. (ed.). 2000. *Global Codes of Conduct: An Idea Whose Time Has Come*. Notre Dame, IN: University of Notre Dame Press.

Windsor, Duane. 2004. "The Development of International Business Norms." *Business Ethics Quarterly* 14 (4).

World Bank. 2009. *The Little Green Data Book 2009*. Washington, DC: The World Bank.

10.10 QUESTIONS

10.1 Smith, an unemployed engineer who recently received certification as an engineer intern from the State Board of Registration for Engineers and Land Surveyors, was seeking employment with a consulting firm. Engineer A, a principal with a large consulting firm, contacted Smith. After a long deliberation concerning such matters as working conditions, salary, and benefits, Engineer A offered, and Smith accepted, a position with the firm. Thereafter, Smith canceled several other job interviews.

Two days later, in a meeting with other principals of the firm, it was agreed by the firm's management (including Engineer A) that the vacancy should be filled by an engineering technician, not a graduate engineer. A week and a half later, Engineer A contacted Smith and rescinded the firm's offer. Did the actions of Engineer A in his relations with Smith constitute unethical conduct? Why or why not?*

10.2 As the business manager of your company, you are visiting several companies in Africa to promote new business. At the tail end of several successful rounds of negotiation, you are invited to attend a family banquet hosted by one of your potential business partners. This invitation represents a genuine sign of friendship and a commitment to good-faith business dealings in the future. Would you be offended if the host wants you to pay for the food and drinks you enjoyed at the banquet when you depart? Explain your answer.

10.3 Cindy Jones, a chemical engineer with considerable experience in offset printing processes, was hired recently as an engineering supervisor by Company A. Before that, she had been working as a research chemist for a competing firm, Company B, where she invented a new formula and manufacturing process for press blankets. Jones's technique makes the blanket less prone to failure and produces better print quality. These press blankets are being marketed by Company B with great success.

When Jones was hired, there was no discussion about the new offset blanket during the interview. Jones was interested in moving into management; Company A had no openings available, whereas Company B was seeking to add managerial personnel with superior technical background.

One day soon after she had started her new job, Jones received an unexpected invitation to a staff meeting from the director of engineering. The meeting agenda focused on the formulas and manufacturing processes for offset blankets. What should Cindy Jones do?**

10.4 Sara King is a member of the International Union of Operating Engineers. Through the union, she has secured a new job to operate a truck with an end loader at the XYZ Construction Company.

About two hours into her new job, the truck began to boil because of a leaky radiator. She stopped the truck and went to look for water.

About 100 feet ahead, Sara spotted a 5-gallon pail. On the way to get the pail, she happened to pass Joe Dow, an old union man, who was tending an air compressor. Joe Dow shouted, "Where are you going?" When Sara told him, Joe Dow replied, "I've got news for you. You are not going to get that pail. Understand? If you want to work on this job, you'd better start acting like a union worker, or I'll report you to the master mechanic. You'd better get back on the truck and wait for the foreman to get a couple of laborers to help you. Remember, if you stop your truck because of a boiling radiator and there's no pail within 40 feet of where you happen to stop, it's not your job to get a container."

Sara did not want any trouble. So she went back to the truck and waited for the foreman. It was two hours before the foreman came. In the meantime, seven other dump trucks and their drivers were idle.

When the foreman finally did come, Sara explained the situation to him. The foreman said, "I'll get you a couple of laborers to draw some water." Sara explained further that she could easily have gotten the water herself earlier, but the operator at the air compressor

---

Source: Adopted from the files of National Society of Professional Engineers Board of Ethics Review.

***Source*: Condensed and adopted from T. M. Garrett, R. C. Baumhart, T. V. Purcell, and P. Roets, *Cases in Business Ethics*, New York: Appleton-Century-Crofts, 1968

had told her to lay off. The foreman answered, "That's the way things are on this job. I don't want any trouble, so I do what the union people want."

Sara encountered other similar incidents as she continued on the job. The basic idea was always the same. Various craft unions decided on a lot of unreasonable restrictions that made a full day's work unproductive. The XYZ Construction Company had entered a cost-plus contract with the client, a steel company. So the more the employees loafed on the job and raised the cost, the more money XYZ Construction Company made. The steel company client was the one bearing the costs. In the long run, the consuming public ended up paying for this labor waste, which contributed to the increasing cost of steel. Are there any ethics involved in this problem?*

10.5 Quick Meal is an international fast-food chain that operates in a number of countries. Company management wants to apply a uniform standard of business ethics, modeled after U.S. practices, to all of its stores worldwide.

When Quick Meal opened a new store in Country X, initially the local government cooperated fully.

Then the government changed hands and a corrupt group took over. Shortly thereafter, Quick Meal noticed that the general manager of the new store in Country X was providing free food and other concessions to governmental officials "under the table." The general manager was an American married to a local national. He was trying to get an "in" with the new government.

Store profits were still high, but Quick Meal decided to fire the general manager. The officials of the new government intervened and told Quick Meal to keep him or they would confiscate the local store.

Quick Meal stuck to its decision and let the general manager go. The new government followed through with its threats and took away the local store.

A few years later, the government of Country X changed hands again. Although Quick Meal was promised some indemnity, there was still a considerable financial loss to the company. In spite of the fact that these losses were written off, some of the Quick Meal stockholders were unhappy with the company's decision regarding the general manager. What should Quick Meal have done?**

10.6 Jane is a member of the board of directors of Power Company Z, which is considering the construction of a new power plant.

Coal-fired power plants emit sulfur dioxide into the atmosphere. Ambient air containing a high-concentration of sulfur dioxide is known to create acid rain, which damages crops and erodes some metals (e.g., nickel and cooper). If No. 2 oil is used as fuel instead, the sulfur dioxide emission of the power plant could be significantly reduced. However, replacing coal with oil will raise the fuel cost by about 20 percent.

Some directors believe that any increased costs would have to be reflected in higher prices. An increase in electricity prices would create problems for the company. For example, the Public Utilities Commission might delay approving a rate increase. Consumers might react negatively to the price increase, which could hurt the company's public image.

Other directors are convinced that the company should not use methods that would increase expenses. They point out that diverse industries and motor vehicles are far more guilty of causing air pollution than the power industry. As one director put it, "Why should

---

*Condensed and adopted from T. M. Garrett, R. C. Baumhart, T. V. Purcell, and P. Roets, *Cases in Business Ethics*, New York: Appleton-Century-Crofts, 1968.

**Condensed and adopted from T. M. Garrett, R. C. Baumhart, T. V. Purcell, and P. Roets, *Cases in Business Ethics*, New York: Appleton-Century-Crofts, 1968.

we be leaders in this area when it is going to cost either stockholders or consumers a great deal of money?"

Jane knows that fuel represents only one-seventh of the total cost of generating and distributing electricity. She thinks that the company has an obligation to protect public health as long as it can stay reasonably profitable. Further, she thinks that the company should not allow purely business considerations to dominate its decisions in an area of such critical importance.

What do you think Jane should do?*

10.7 Company A recently bought a rock-crushing unit from Company B. The unit was expected to produce 750 tons of crushed rock per hour but has in practice been producing only 500 tons per hour.

Paul, president of Company A, complains to Gordon, sales manager of Company B, about the fact that he is now unable to fill contracts, which he secured based on the expected capacity of the machine. In some instances, he has been required to buy crushed rock—at retail prices—to satisfy his contract obligations. Furthermore, he is not able to repay the loan with the expected higher income from the increased production. Paul threatens to sue Company B unless it returns half of the purchase price of the equipment.

Frank, the foreman of Company A's new rock-crushing installation, and Elmer, the company's chief engineer, are not happy with the new equipment. However, they are not sure that Company B is at fault. The contract for the new equipment specified that the unit should be able to crush 750 tons of properly graded limestone per hour. Company B had samples supplied by Company A and based its promise of performance on those tests. Paul had been using stone taken from several different company quarries. Both Frank and Elmer had objected to this since much of this stone was harder than that in the sample given to Company B. The equipment had not broken down, but it was not able to deliver the specified capacity.

Frank and Elmer reviewed this matter and decided to present the problem to Paul. If Company B fought Paul's suit, Frank and Elmer would certainly be called on to testify. Moreover, they both felt that Company B has a right to know that Company A has been using a harder rock than that used in the tests.

Paul listened to Frank and Elmer, but was not convinced that he ought to inform Company B of the difference in rock hardness. Paul thought that the performance guarantee covered the crushing of rock for any and all of the company's quarries.

What course of action do you suggest for Company A?**

---

*Condensed and adopted from T. M. Garrett, R. C. Baumhart, T. V. Purcell, and P. Roets, *Cases in Business Ethics*." New York: Appleton-Century-Crofts, 1968.

**Condensed and adopted from T. M. Garrett, R. C. Baumhart, T. V. Purcell, and P. Roets, *Cases in Business Ethics*," New York: Appleton-Century-Crofts, 1968.

Chapter 11

Knowledge Management

11.1 INTRODUCTION

In today's business environment wherein the future is changing and uncertain, the one sure source of lasting competitive advantage is knowledge. When markets shift, technologies proliferate, competitors multiply and products/services become obsolete almost overnight, successful companies are those that constantly build new knowledge, disseminate it widely throughout the organization, and quickly embody it in new technologies and products (Nonaka 2007).

Statistics point out that in the next ten years, 43 percent of the U.S. workforce will be eligible for retirement (du Plessis 2005). Hundred of the largest companies can expect to lose 50 percent of their senior management in the next five years. How to preserve some of the critical knowledge accumulated through experience over time is a matter of profound importance to numerous companies. The half-life of new knowledge is becoming ever shorter. Knowledge management is a corporate function on the identification, creation, organization and distribution of knowledge for reuse, awareness and learning across the organization. Knowledge workers and managers bear the principal responsibility of preserving and fostering the companywide reuse of corporate knowledge.

Knowledge management is important but very difficult to accomplish, because a considerable portion of corporate knowledge is tacit. Tacit knowledge includes all knowledge that is known at a nonverbal level and does not lend itself to being described or translated into formal, codified categories. Tacit knowledge is like "deep smarts," which Leonard and Swap (2005) describe as "the ability to comprehend complex, interactive relationships and make swift, expert decisions based on both the system as well as component-level understanding." As pointed out before, "We can know more than we can tell" (Polanyi 1983). It is the efficient application of corporate tacit knowledge that will yield competitive advantages.

In this chapter, we will review the basics of knowledge management, difficulties and challenges in knowledge management, a large number of industrial practices of knowledge management, and the future of knowledge management. Conclusions are offered.

11.2 BASICS OF KNOWLEDGE MANAGEMENT (KM)

11.2.1 Data, Information, Knowledge, and Wisdom

Data embody the recordings of observed phenomena, be they from experiments, surveys or analyses. Users process (e.g., manipulate, interpret, or categorize) data and ask targeted questions in order to extract information from data. Information could be tested for validity, then codified and applied to a specific scenario as knowledge. Certain portions of this knowledge are of the explicit type, the rest are of tacit nature. Although the explicit knowledge are usually easy to document, preserve, transmit and share; the tacit knowledge resides in expert's head and is thus very difficult to preserve and transfer. Users further process available knowledge by ways of combination, extrapolation, pattern recognition, analogies, and contrasting to extend its applications to other situations in order to gain wisdom. *Deep smarts* are as close as we get to wisdom.

Knowledge might contain a number of elements, such as work procedures, problem-solving methodologies, decision-making strategies, lessons learned, skills, experience, insights, heuristics, rules of thumb, perceptions, know-how, intuition, and others. The concept of knowledge element is well defined in the literature and used by the author in a U.S. patent (Chang 1996).

The ultimate goal of knowledge management is to apply wisdom to generate competitive advantages.

11.2.2 Tacit and Explicit Knowledge

A portion of what we know can be readily codified as explicit knowledge. This type of knowledge is contained in patent literature, solution manuals, operations procedures, and other forms of written documents, as well as voice and graphics recordings. Knowledge of this type can be systematically preserved, transmitted, and retrieved by companies and individuals. The other portion of our knowledge is tacit in nature. It consists of that part of knowledge related to our know-how, images, patterns stored in our heads, hidden relationships we envision, rules of engagement we surmise, hunches, and intuitions that are not readily expressed in words or recordable in texts. Most of such tacit knowledge has been generated by keen observations, lateral thinking, scenario analysis, experimentation, trial and error, and hands-on experience.

Some literature proposes the iceberg analogy to describe the explicit part (the visible portion above the water line) and the implicit part (the invisible portion staying below the water line) of the knowledge (Eucker 2007). Iceberg has a fixed mass ratio of 1 to 6 between the two parts. In reality, the ratio of the explicit to implicit parts of our knowledge varies greatly from one application domain to another. For plant operational processes, there usually exists a large amount of documented information, and this ratio could be 1 to 1. For strategic planning and service innovation, very little can be preserved as explicit knowledge, so this ratio might be 1 to 5.

Explicit knowledge is helpful in efficiently handling routine projects and non-creative tasks. Tacit knowledge is needed to solve new problems and deal with new ambiguity. In fact, according to some, having explicit knowledge alone will cause nothing to get accomplished. Tacit knowledge makes explicit knowledge useful.

11.2.3 Preserving and Disseminating Explicit Knowledge

Explicit knowledge, as documented in manuals, work procedures, best practices, and lessons learned, is relatively easy to preserve and to share. It requires a firm commitment of the company's senior management (resources, management attention), a proper alignment of knowledge management efforts with specific business objectives (values and outcomes), and an establishment of knowledge-oriented company culture (expectations of all participants).

Sharp (2003) said many industrial firms have been quite successful in preserving and distributing explicit knowledge. Shell International Exploration & Oil (SIEO) is known to have spent $6 million in 1997 to establish three technology-focused communities (i.e., surface, subsurface, and wells) and ten other support communities (e.g., finance, procurement, human resources, health, safety, environment) for documenting and distributing internal best practices and learned lessons to 10,000 geographically dispersed employees The focus was on using what was already known (best practices and empirical know-how to solve specific problems in the field) more effectively. The company reported a $300 million benefit realized from this knowledge management program in 2000. Chrysler Europe brought about the "Engineering Book of Knowledge" as a Web-based central depository of leading practices and lessons learned in engineering. Dow Chemical initiated a systematic effort of reviewing its 30,000 patents to determine their market value, leading to a drastic increase in licensing royalty income and a concurrent reduction of patent maintenance expenditures. A water utility company decided to scan customer contracts and make them available to all of its field agents, to facilitate service. CIGNA produced a database to assist its agents in more accurately evaluating new clients (risk profiles, earthquake faults, sprinkler systems, etc.). Hewlett-Packard introduced Web-based "Knowledge Links," which contain information about product generational processes from the perspectives of marketing, R&D, engineering and manufacturing, in order to advise managers in various HP divisions. Xerox documented its internal best practices to promote employee productivity. Ontario Power Generation streamlined and standardized document management and introduced simplified e-mail communications standards. Workers' time was saved when looking for applicable knowledge to solve problems in the field.

Saving explicit knowledge and distributing it widely to encourage employee reuse will produce benefits in the form of (1) shortened time and efforts to solve repetitive problems—typically, employees would spend 50 percent to 60 percent of their time looking for applicable knowledge needed in their jobs, (2) avoiding the need to "reinvent the wheel," thus not wasting company resources and introducing no possible new problems, and (3) establishment of the current state of knowledge, upon which a continuous improvement can be initiated. Preserving and transmitting explicit knowledge is indeed useful, but it is not sufficient in creating and maintaining the long-term competitiveness of companies in the marketplace.

11.2.4 Creating and Sharing Tacit Knowledge

Tacit knowledge is the source of a firm's competitive advantage. Von Krogh et al. (2000) suggested five specific steps to enable tacit knowledge creation:

1. Instill a knowledge vision.
2. Manage conversations.
3. Mobilize knowledge activities.
4. Specify the right context.
5. Globalize local knowledge.

It is believed that knowledge creation is a social and small-team activity, based on relationships and mutual trusts among the members. Tacit knowledge is produced by interaction between trustworthy people in small groups—the micro-communities.

Two factors influence the willingness of employees to share their tacit knowledge: organizational commitment and trust in coworkers (Lin 2007). Organizational commitment refers to the strength of an employee's identification and involvement with a particular organization and his positive response toward coworkers. *Trust* is an expression of faith and confidence that the organization or individuals will be fair, reliable, ethical, competent and nonthreatening. Employees with trust in coworkers believe that their coworkers will follow the Golden Rule, which would cause the sharing of knowledge to be reciprocated by others. Thus, management is advised to implement policies that are open, fair and conducive to encouraging tacit knowledge sharing, to offer rewards and compensation to those whose knowledge-sharing efforts are exemplar, and to introduce communications channels that encourage employee feedback about the knowledge-sharing program. Certainly, it is also useful to sponsor occasions for enhancing the social interactions among employees as a way to promote mutual trust.

It should also be useful to establish a setting for sharing knowledge and eliminating communication "filters" by allowing open communications to people at all levels. Politics, turf, and implementation responsibilities can squelch ideas in traditional communications channels. Alagasco, a North Birmingham, Alabama-based natural gas company, provides a direct link to the CEO via cards posted throughout the organization. USAA, an insurance and financial services company, has an online system enabling employees to share opinions and ideas spontaneously and anonymously.

In an insurance company, actuaries, underwriters, and claims adjusters often have different views of the insurance business, primarily because there is a lack of efficient knowledge sharing among them (Bondura 2007). Breaking down the functional silos is essential in facilitating knowledge sharing.

Because tacit knowledge resides in people's heads and is not readily preserved in explicit form, its successful transfer requires a unique approach. Domain experts need to be encouraged to share their implicit knowledge. Novices must have the strong desire to want to learn from them. Grudzewski and Hejduck (2010) discuss the difficulties involved in transferring technology in service industry. Intel is known to have applied a number of techniques to bring about such a transfer (Eucker, 2007):

- Conduct face-to-face meetings to exhaustively review the process and design documents in order to pull implicit knowledge into the explicit-information package.
- Use video to capture intricate procedures.
- Conduct forums, meetings, internal conferences and supplier days to facilitate learning from domain experts.
- Coach the experts and novices involved to become effective.
- Rotate assignments of experts as a way for novices to gain specific experience.
- Form teams to engage domain experts in applying knowledge (Seeley 2000).
- Simulate specific processes to help expose hidden knowledge nuggets of importance.

11.2.5 Enhancement of Explicit Knowledge by Tacit Knowledge

Tacit knowledge, once properly preserved, might contribute to the creation of new explicit knowledge, in the form of new processes, procedures, heuristics of problem solving, and rules of assessing markets, competition and customers. Figure 11.1 describes the generation of wisdom by capable people who creatively and innovatively apply information with true understanding.

A company preserves its existing explicit knowledge in its knowledge base. Employees access this knowledge and utilize it in new projects. Upgraded understanding then can lead to improved work procedures and new explicit knowledge, which can be fed back into the company knowledge base. This process is recycled to allow the knowledge base to be updated continuously.

Tacit knowledge, by contrast, might be externalized by performing deep thinking and reflection and by engaging in communities of professionals, wherein workers interact with other trustworthy participants. The new tacit knowledge so produced will reside primarily in the head of participants and could be transmitted to others through additional interactions (e.g., job rotations, expert networks, question-and-answer forums, and teamwork).

According to Nonaka (2007), some of the tacit knowledge could be converted to explicit knowledge, in the form of operational procedures, design, or work processes—the process of articulation. Tacit knowledge can also be generated by using explicit knowledge intuitively or creatively—the internalization process.

For such knowledge creation processes to work successfully, company management must provide adequate resources to set up and update the knowledge base, to encourage employees to communicate with one another and to use knowledge widely and seek its improvement, to promote the creation of new tacit knowledge through an efficacious incentive policy, and to reward the conversion of some of this new tacit knowledge to new explicit knowledge (see Fig. 11.2).

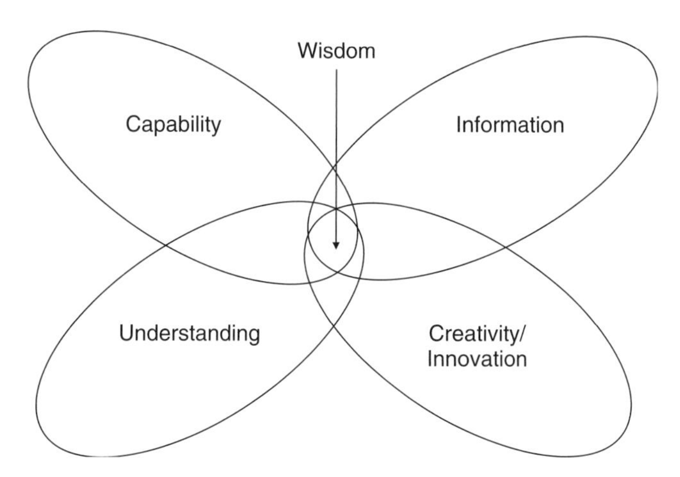

Figure 11.1 Source of wisdom.

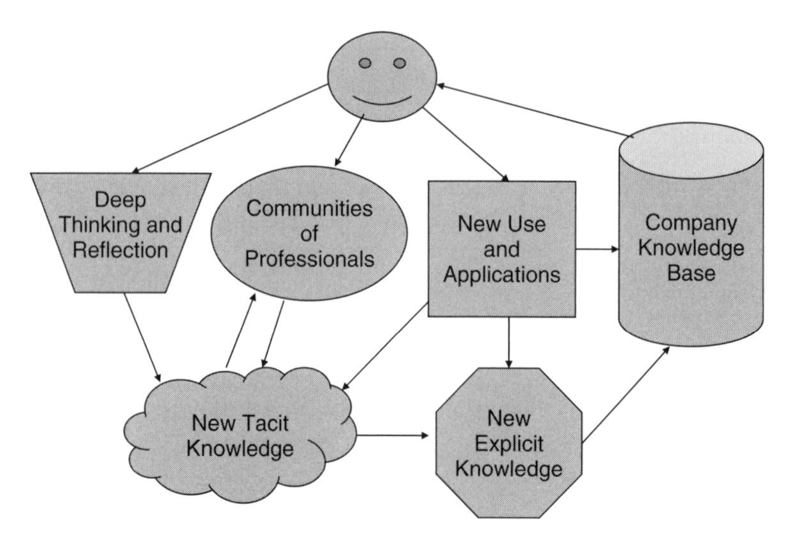

Figure 11.2 The creation of tacit and explicit knowledge.

11.2.6 Communities of Practice (CoPs)

Communities of practice (CoPs) are teamlike structures established for the purpose of facilitating communications between knowledge workers who exhibit a common interest in sharing and creating knowledge within a specific subject domain.

CoPs can be business-specific by area with an international link among various units. They focus on managing documents, offering project support tools, and building worker competency. CoPs can also be cross-divisional, to avoid reinventing the wheel. The emphasis is on e-learning, promoting learning by the workers themselves, and linking with domain experts and best-practice forums. CoPs can also be built to foster collaboration with clients, to follow up on client projects, or to learn the future needs of clients and partners.

Typically, CoPs follow a certain life cycle. Before the startup phase, a team of knowledge workers will prepare a business plan. The plan details the subject matter, aims, activities, and potential benefits to the company and its potential membership, financial and organizational details, technological infrastructure, and metrics to assess its effectiveness. Specific activities might include sharing tacit knowledge, processing quick questions, and processing knowledge assets (Sieber and Andreu, 2004).

The benefits of CoPs are in general difficult to estimate quantitatively. CoPs might cultivate members' capability of developing and combining new knowledge, speed up the learning curve for some, expedite certain business processes, or locate experts in specific domains.

John Deere, an equipment manufacturer in Moline, Illinois, relies on CoPs to drive innovation, efficiency, and lifelong learning. Since 2002, John Deere has built 300 CoPs on various topics (from Six Sigma to mergers and acquisitions) (Sauve 2007). Communications are being facilitated via e-mails, face-to-face meetings, or online conferences. The CoPs are centered on practice areas; they cover best practices, training, mentoring, and peer resources. There are a few success factors for creating useful CoPs in companies (Sauve 2007):

- Focus on core values of the organization, concentrate on topics important to the business and community members, and precipitate real dialog about cutting-edge issues.
- Commit management support (resources, attention and priority).
- Empower well-respected leaders to coordinate.
- Encourage passionate employees and thought leaders to participate (allow time and provide incentive to knowledge sharers).
- Enhance trust and good working relationships among members.
- Foster thinking and information sharing.
- Make tacit knowledge easy to access and use (job rotations, an experts network, internal discussion forums, etc.).

One of the key issues related to CoPs is how to measure success and monitor progress. CoPs can be assessed by using some crude numbers. The efficiency ratio is defined as the ratio between the knowledge contributed to the community and the knowledge consumed in that community. Members in a COP undertake activities such as uploading documents, reading documents online, downloading copies of documents, subscribing to discussions, and so on. Each document uploading is assigned three points. Each deletion of a document is two points. Every member accumulates points depending on his/her activities in the community. The sum of all points accumulated by members in a CoP is called a global score. The global score divided by the total number of actions taken produces the efficiency ratio. The higher the ratio, the more "productive" the community is. The effectiveness ratio is the average activity by user, which is equal to global score divided by total number of members in the community. Figure 11.3 plots the relative positions of a set of CoPs, wherein the radii of the bubbles are representing the effectiveness ratios (Sieber and Andreu 2004).

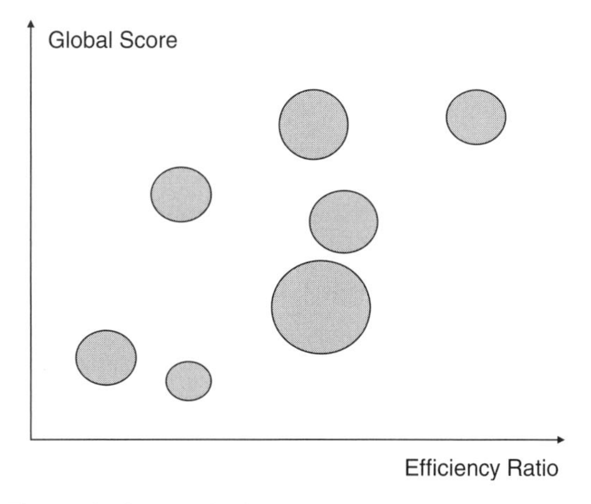

Figure 11.3 Monitoring the communities of practice.

Table 11.1 Corporate Functions versus Implementation Activities

	Planning	Cost Reduction	Value Engineering	Quality	IT/Software Application
Innovations					
Marketing					
Finance					
Technology					
Communications					
Customer services					
Media					

Although the global score is important for measuring the basic knowledge transfer activities, it does not say anything about the value of such transfers. Transferring a lot of knowledge that is of low value to the company is not necessarily a good thing.

Assume that the company's principal activities are organized by functions, such as those included as examples in the left-hand column of Table 11.1. We further assume that any of those functions can follow implementation activities (in the first row). Then the company management could assign a weight (e.g., 10 being most valuable, 0 being least valuable) to each function/activity to illustrate the importance of the associated knowledge transfer. The weights could be changed periodically to reflect the changed priorities set by the company in view of its business strategies and marketplace conditions.

By multiplying the global score with the individual weight assigned to each of these function/activity pairs, to which the CoPs belong, the company would gain a better understanding of the relative value contributed by the knowledge transfer in those CoPs. Publishing such weighted scores on a regular basis would encourage employees to focus on CoPs related to high-value function/activity pairs. Doing so would make CoPs more responsive to the company's priorities at that point in time.

Another issue related to CoPs is how to capture and document the tacit knowledge generated therein so that it can be made widely available to others. Xerox, a copier company based in Rochester, New York, encouraged its equipment troubleshooters to brainstorm and communicate problems through storytelling in CoPs. The company initiated the Eureka Project to preserve resourceful ideas, passed them through a number of quality-control steps (similar to peer reviews for technical and scientific journal articles), and published the winning tips with the originator's name. Doing so allowed the individual idea contributors to gain social and intellectual capital. As of 2000, Eureka had 30,000 records and saved Xerox $100 million in costs (Bobrow, and Whalen 2002).

When properly organized, communities of professionals appear to be able to cultivate and produce tacit knowledge that is of critical importance to companies interested in developing and maintaining long-term competitiveness.

Example 11.1 A successful knowledge management program requires that both explicit and tacit knowledge be properly produced, preserved and disseminated throughout the organization. To transfer tacit knowledge, companies need to foster the formation of networks of professionals, based on the premise that the more often experts and knowledge users interact, the more likely tacit knowledge will be transferred. If you were responsible for promoting the transfer of tacit knowledge via networks, what types of guiding principles would you offer in building such networks?

ANSWER 11.1 For tacit knowledge to be effectively transferred, experts and knowledge seekers need to interact with mutual trust. Knowledge seekers must possess an in-depth understanding of a given subject domain, not merely a passing familiarity, to enable asking insightful questions that can draw out the tacit knowledge.

Networks designed for tacit knowledge are to be composed of people with shared knowledge and interests in the subject matters involved, the communities of practices (CoPs). People in such groups migrated to one another according to the *self-similarity principle*. The self-similarity principle states that people tend to make friends with those who have similar backgrounds, work experiences, and outlooks. Networks formed based on this principle might not be particularly useful to promote innovation. However, the common technical languages they use tend to make everyone "at home" and thus readily bond them together to build mutual trust and understanding. People with similar training, backgrounds, and work experiences tend to be able to exchange ideas at deep levels than would be possible otherwise. Only those people who have a thorough understanding of a subject domain would be able to identify and assess the real value of any tacit knowledge that would be transferred.

The tacit-knowledge exchanges involved insights, perceptions, and hunches and recognized patterns that would become a natural outcome of such in-depth exchanges. Companies need to foster the formation of such networks based on the self-similarity principle.

11.2.7 Key Requirements of Knowledge Management

Knowledge management must be aligned with the strategic vision of the company and be focused on serving a specific useful business purpose. For professional consulting firms, whose only output is knowledge, the purpose of knowledge management is self-evident. For other companies, it might not be so clear-cut. Companies are advised to define the specific business goals (e.g., promoting service innovation, adopting breakthrough technologies earlier, improving processes, or serving customers better than the competition) for which knowledge management is expected to make a difference.

From the operational standpoint, the key requirements are accumulating knowledge and mobilizing it (Gupta and Govindarajan 2000). The tasks of accumulating knowledge can be accomplished by (1) creation (learning by doing), (2) acquisition (internalizing external knowledge), and (3) retention (minimizing the loss). Moreover, the tasks of mobilizing knowledge can be accomplished by (4) identification (uncovering opportunities for knowledge sharing), (5) outflow (motivating experts to share with others), (6) transmission (channels for knowledge transfer), and (7) inflow (motivating recipients to accept and use incoming knowledge).

To succeed in knowledge management, companies need to overcome various pathologies and pitfalls known to be associated with each of these seven elements. Table 11.2 enumerates some of such pitfalls and remedies.

Table 11.2 Knowledge Management Pitfalls and Remedies

#	KM Elements	Pitfalls	Solutions
1	Creation	Staff complacency. Lack of management support. Lack of employee empowerment.	Select staff well. Apply group-based incentives. Empower staff.
2	Acquisition	Lack of initiative. Not using external knowledge effectively.	Encourage risk taking. Empower staff.
3	Retention	High turnover. Loss of proprietary knowledge to competition.	No layoff, but reduce workweeks.
4	Identification	Misguided notion that high performers have nothing to learn and low performers have no good ideas to share. No effect to distinguish superior ideas from good and not-so-good ones.	Publish group performance data to encourage sharing of best practices. Encourage and screen new ideas. Apply group-based incentives.
5	Outflow	Some people resist sharing due to self-interest and power hoarding. Incentives are tied to internal relative performance.	Recognize "knowledge givers" as heroes.
6	Transmission	Communications channels are not compatible with knowledge types.	Facilitate face-to-face meetings, transfer of people. Keep group small to encourage interaction. Pay group visits to share experience, coaching.
7	Inflow	NIH (not invented here) syndrome—egos prevent some from recognizing peers' superiority	Publish group performance data. Recognize stronger performers and induce weak performers to change. Use training.

Sources: Gupta and Govindarajan (2000)

In addition, there is a need to maintain the knowledge base. Knowledge has a half-life because the context in which it is applied might change over time. Typically, the firm's knowledge base includes its technological competences, its understanding of customers and suppliers, and its way of doing business. The rapid change in the marketplace (e.g., competition, technology, customer preferences, governmental regulations, supply chain partners) requires that the firm keep its knowledge base current. An outdated knowledge base is worthless.

Of particular interest is the approach that Nucor Steel (Gupta and Govindarajan 2000) took to promote knowledge sharing and knowledge inflow—by publishing group performance data to induce intra-group competition, by assigning group-based incentives to discourage the control of knowledge by individuals, and by recognizing "knowledge givers" and weeding out weak performers to further enhance the social dimension.

Example 11.2 For many companies, finding and leveraging expertise is of paramount importance. However, the right expertise needs to be defined in the context of specific problems that are to be solved. Companies should strive to characterize expertise in ways that are sensitive to different contexts. If you were to come up with a practical way of finding and leveraging expertise, how would you do it?

ANSWER 11.2 A useful start is to assemble a cross-functional team to define objectives and to brainstorm options based on resources and data available in-house and best practices in the industry. Note that it is important for expertise seekers to quickly identify people with the expertise that can assist in solving a problem. In order to assess the expertise in context, knowledge seekers need to know about the experts involved, including training, past publications, and experience.

To accomplish such a difficult task, Idinopoulos and Kempler (2003) suggest a three-part system: user interface, search engine, and databases. The user interface needs to be customized to the company's special needs. The search engine can typically be licensed from a known vendor. The databases must be set up based on information available in the company. Figure 11.4 illustrates the components of such a system.

Specifically, the "accounting" database includes project hours, project assignments and descriptions, and colleagues involved. The "document" database stores internal documents, memorandums, and project plans. The "patent" database saves patents filed. The "expertise directory" database describes expertise, language capabilities and computer skills. The "professional development" database delineates continuing education, training and workshops. The "recruiting" database lists academic experience, previous work experience, previous publications and professional certifications. The "human resources" database depicts titles, roles, geographic locations and contact information.

Search results should include name, contact information, training, and experience related to the keywords specifying the sought-after expertise. Expertise seekers can then interview the individuals involved to make a final choice.

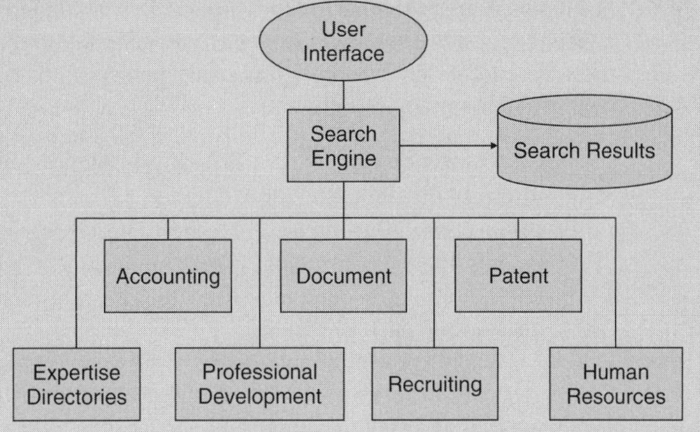

Figure 11.4 Expertise search and retrieval system. (Adapted from Idinopoulos & Kempler, 2003)

11.3 MANAGEMENT OF KM PROGRAMS

11.3.1 Planning for Knowledge Management

There are many steps to take to align knowledge management with corporate strategies. It is important to identify which knowledge area has an impact on the business to assure that there is a strong link between KM and business performance. The following six steps are recommended (Sieber and Andreu 2004).

1. Determine the business context, the strategies, and goals of the company.
2. Identify the knowledge areas that are relevant to the business priorities.
3. Define the critical performance indicators (such as customer success, profitability index, innovative outcomes, time to market, return on investment, etc.).
4. Analyze the impact of knowledge areas on the key performance indicators already identified.
5. Analyze the status of selected knowledge areas and identify needs for improvement.
6. Formulate the knowledge management action plan.

Usually, companies establish balanced scorecard measures to guide the definition of key performance indicators. Step 3 assures that all knowledge management projects are fully aligned with the company's overall strategies.

11.3.2 Comprehensive Approach to Knowledge Management

Knowledge management needs to examine issues related to generation, preservation, and dissemination of both explicit and tacit knowledge so that workers can make intelligent decisions in real time. During these processes, communication plays a critical role. Asking penetrating questions, listening intensively for insights, telling clearly the critical knowledge elements involved, and writing down the lessons learned and know-how gained are essential for the success in knowledge management. This is the core concept of *knowledge management by communicating*. The goal is to make the best use of corporate knowledge by as many people as possible when making decisions to benefit the company rapidly:

- *Generation of knowledge*. Testing, simulation, observation, and discussion are useful in creating new knowledge.
- *Preservation of knowledge*. Explicit knowledge can be readily preserved in documents (patents, manuals, procedures, project reports, and others) and continuously refined. Companies need to have a systematic way of categorizing and distributing such knowledge.

  Tacit knowledge needs to be preserved by a number of interactive strategies: (1) conduct meetings, conferences, forums, supplier days, and other such gatherings, wherein domain experts can provide tacit knowledge to solve specific problems or issues, (2) form teams to involve domain experts so that they can offer their tacit knowledge and add value, and (3) rotate assignments to encourage the sharing of domain expertise by others.

- *Dissemination of knowledge.* The use of modern IT tools is quite practical to allow an easy access to the stored knowledge by anyone so authorized. A push system is deemed useful if the system pushes relevant knowledge to users, who need specific information, instead of waiting for the users to pull the applicable knowledge from the system.

  A reward system should be included in the knowledge management program to induce participation and encourage those who actively contribute to the success of the program.

11.3.3 Critical Success Factors

Many factors affect the success or failure of a corporate knowledge management initiative (du Plessis 2007):

- Senior management buy-in
- Commitment of sufficient resources
- Corporate culture aligned with knowledge management objectives
- Motivation of employees to share knowledge
- Collaboration of teamwork
- Communication within the organization (cross-fertilization)
- Use of appropriate technologies
- Sense of urgency

Company leadership must assign a high priority to knowledge management by announcing its strategic importance, define specific knowledge management objectives that are related to company profitability, appoint a champion to lead the way, build multichannel technology systems to allow knowledge sharing, monitor progress and make improvements, and incorporate a reward system to induce widespread participation.

Besides supporting knowledge management activities, leaders also need to align the organizational culture to implementation of the KM plan. Organizational culture is defined as a set of implicit assumptions held by employees that determine how they behave and respond to the organization's environment (Shein 1985). Specifically, in the context of knowledge management, organizational culture determines the social context (norms and practices) that defines who is expected to control what knowledge, as well as who must share it and who can hoard it (Delong and Fahey 2000). In a knowledge-friendly culture, employees would view knowledge management positively and would not be reluctant to share knowledge, for fear of losing their jobs.

Another critical success factor is the willingness of employees who possess tacit knowledge to participate. Companies need to offer incentives and stimulus to encourage such participation. Other critical factors involving communications, collaboration and technology and urgency are self-evident.

Because tacit knowledge represents a significant portion of the corporate wisdom that is worth preserving, and because tacit knowledge is difficult to codify, a successful knowledge management system needs to promote the creation of linkages between knowledge experts and knowledge users to foster the just-in-time transfer of technical, process, business or customer know-how, and insights.

11.3.4 Knowledge Management Obstacles

A large number of corporate knowledge management initiatives failed in the past. Surveys conducted by American Productivity & Quality Center (APQC) concluded that there are generally five obstacles for developing a successful knowledge management program:

1. *Lack of a business purpose.* Knowledge management programs should be focused on solving specific business problems.
2. *Poor planning and inadequate resources.* Companies tend to focus on developing a pilot program, but neglect the planning and resources requirements of the rollout.
3. *Lack of accountability.* If there are no champions identified to spearhead the knowledge management effort, it will likely suffer a decline due to a lack of accountability.
4. *Lack of customization.* To be useful, the knowledge management system should not be one-size-fits-all. Instead, it should be closely tailored to its diversified needs of its diverse users.
5. *Wrong focus.* Traditional knowledge management strategies pursue the *publishing model,* in which knowledge is acquired, preserved and distributed. In this model, knowledge is treated as a stock (Fahey and Prusak 1998). This model is built on a questionable assumption that important knowledge can be effectively extracted from experts. The reality is quite the opposite. The contents of what people communicate depend on who is on the receiving end of the exchange. It is part of human nature to hoard knowledge for a number of reasons, among them job security and the desire to appear valuable to the organization.

   Alternatively, the *brokering model* has shown some success in promoting knowledge transfer. According to this model, the knowledge management strategies should focus on creating convenient linkages between those who possess knowledge and those who need knowledge. By allowing the individuals to control their own knowledge and to decide in what ways their knowledge is to be transferred, this model would encourage experts to become more willing to share their expertise (Gilmour 2003).

11.3.5 Known Mistakes in Knowledge Management

Research has shown that a variety of companies failed to achieve success in knowledge management because of certain major mistakes they made (Fahey and Prusak 1998; Rogers 2007):

- No unified definition of knowledge, which is linked to data and information
- Focusing on knowledge stock rather than on knowledge flow
- Neglecting tacit knowledge
- Separating knowledge from its proper application context
- Deemphasizing thinking and reasoning
- Ignoring the importance of human interface

- Not being a part of the business process
- Incompatible company culture
- Misguided notion of focusing on databases and IT, software and centralized storage
- Expecting the CoPs to function autonomously

11.4 KNOWLEDGE MANAGEMENT PRACTICES IN SERVICE AND MANUFACTURING SECTORS

Some companies have practiced knowledge management with great success. Shown below are few examples. Professional consulting firms (e.g., McKinsey, Accenture, Bain, Ernst & Young, IBM) are particularly keen on knowledge management, because their only output is knowledge.

11.4.1 McKinsey & Company

McKinsey & Company is a global management-consulting firm. Its revenue was $5.33 billion in 2007 with ninety office locations in fifty countries. Its consulting services have expanded to seventeen industrial practices (e.g., auto, chemical, power, financial, media, high-tech, medical, telecommunications) and six functional practices (corporate finance, marketing and sales, operations, organization, strategy, and business technology office). The company has a very pragmatic view regarding knowledge: "Knowledge is the lifeblood of McKinsey."

McKinsey spearheaded the notion that successful consultants need to have broad generalist perspective with an in-depth industry or functional specialty. It believes that "knowledge is only valuable when it is between the ears of consultants and applied to clients' problems." The company shifted from developing knowledge to building individual and team capabilities by adopting the "engage-explore-apply-share" approach.

McKinsey's knowledge management process relies heavily on personal networks, staff transfer and the norms of asking everyone to help others when called on (Barlett 2000).

Over the years, the company has initiated a number of knowledge management practices to accomplish the complex objectives of developing knowledge, sharing knowledge to serve clients, advertising the company's expertise to future clients, and making its experts visible both inside and outside McKinsey. The following ten practices are particularly noteworthy:

1. *Centers of competence*. Fifteen of these centers were established (e.g., in strategy, organization, marketing, change management, systems and others) to help prepare and invigorate consultants and to ensure the continued renewal of the firm's intellectual resources.
2. *Staff publications*. Employees were encouraged to publish their key findings to gain visibility for the company. Examples: *In Search of Excellence* by Peters and Waterman and *The Mind of the Strategist* by Kenichi Ohmae.

3. *Practice bulletins*. These are two-page summaries of important new ideas that identified the experts who could provide additional details.
4. *Practice Development Network (PDNet)*. A Web-based system that contains the core knowledge of McKinsey practices in the form of thousands of PD documents written by consultants.
5. *Firm Practice Information Systems (FPIS)*. It is a computerized database of client engagements.
6. *Knowledge Resources Directory (KRD)*. It is a small booklet containing the firm's experts and key document titles by practice area. Consultants can reach any experts, when needed, to receive advisement within a preset time frame on urgent issues of importance to a client.
7. *White papers*. These papers focus on major issues of long-term importance such as the corporation of the future, creating and managing of strategic growth, and capturing global opportunities.
8. *Practice olympics*. Teams of two to six people of any office are encouraged to develop ideas that grew out of recent client engagements for regional and firmwide competition. In its second year (1995) this event attracted more than fifteen teams and involved about 15 percent of McKinsey employees.
9. *McKinsey Global Institute*. A research arm is focused on studying implications of changes in global economy or business.
10. *McKinsey Quarterly*. This journal contains management and business articles written by both McKinsey experts as well as outside researchers. It has been in circulation since June 1965.

McKinsey & Company is regarded as one of the most successful consulting firms in the world, whose knowledge management practices are leading-edge examples in the industry.

11.4.2 Accenture

Accenture is a $23.4 billion consulting firm in 2008 with 100,000 people, based in Bermuda. It specializes in business consulting, alliances, technology, venture capital, and outsourcing (Meister and Davenport 2005). In July 2001, the company builds the Knowledge Xchange, which offers a document repository, expertise directory, and topic area and search functionality. Collaboration, while not widely used, is a required capability for those who can't find what they need or are working in an area that requires expertise; see Fig. 11.5.

The Accenture Portal is the single point entry to the system. Such a simple interface offers multiple ways to access needed content. The search capability allows users to query the systems. The browse function allows the user to browse via links to content, managed-content pages and other sites. The collaboration function facilitates the user's direct access to peers and domain experts for discussions. The managed-topic pages are managed by experts and governed through subject ownership. Intention pages are designed to align the user's intention with proper links and Web pages. The knowledge content store is a central depository of Accenture knowledge. The other enablement contents bring content from other sources to the user.

A total of 7000 databases are built which catalogue knowledge capital and client experience (Rao, 2005). Examples of items contained in each database include: (a) Proposals, (b) Client deliverables (sanitized, if required), (c) Methodologies, (d) Thought

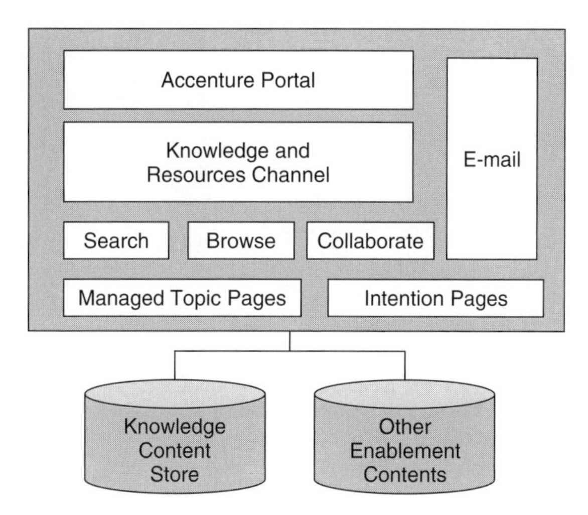

Figure 11.5 Accenture's knowledge management system. (Adapted from Meister & Davenport, 2005)

leadership and white papers, (e) Links to external information, (f) Project plan, and (g) Links to experts. A group of 300 people globally coordinate activities to ensure that the right knowledge is brought to the right users at the right time.

11.4.3 Ernst & Young

Ernst & Young, one of the Big Four accounting and consulting firms in the United States, is based in New York City, with $21.4 billion in sales and 114,000 employees, offering management consulting services at 700 locations in 140 countries in 2009. The management consulting industry is characterized by growth, services integration, and globalization; it relies heavily on managing expertise and knowledge to compete.

The company's unique knowledge management system consists of the Knowledge Web, which is linked to five separate supporting units (Chard and Sarvary 1997). The Knowledge Web is the basic interface between the firm's knowledge base and its consultants. It contains (1) InfoLink (linkages to external databases), (2) discussion database, (3) practice area database, (4) PowerPacks (highly filtered knowledge about a specific topic), and (5) repositories of large unfiltered documents. Consultants access the Knowledge Web to add or retrieve knowledge (see Fig. 11.6). The five supporting units are:

1. *Center for Business Knowledge* (Cleveland, Ohio) collects, stores and synthesizes the firm's internally generated knowledge as well as external knowledge and information. The center employs 200 people.
2. *Center for Business Transformation* (Dallas, Texas) translates knowledge into methods and automated tools. It employs 120 people.
3. *Center for Business Innovation* (Boston, Massachusetts) is responsible for generating new knowledge. It has twenty workers.
4. *Knowledge Networks* represent a unit that provides a searchable directory of experts. Twelve workers take care of this unit.
5. *Organizational Infrastructure* offers information related to culture, performance, and training.

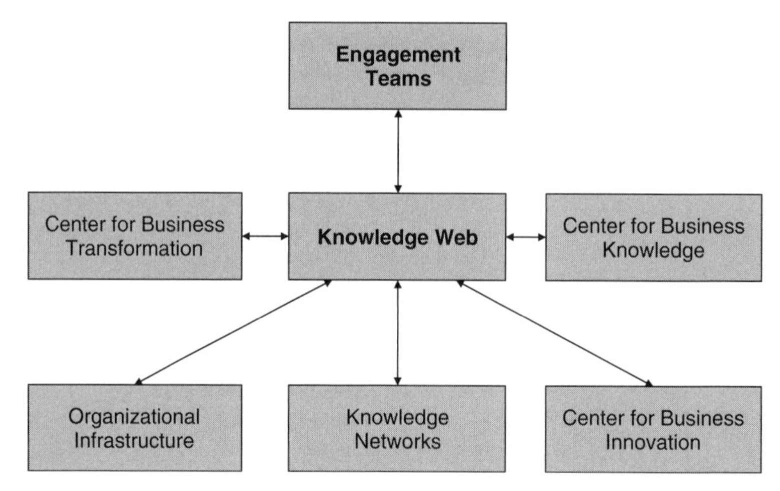

Figure 11.6 Ernst & Young's knowledge management system. (Adapted from Chad and Gavin, 1997)

This knowledge management system is regarded as one of the best in the management consulting industry.

11.4.4 Booz Allen Hamilton

Booz Allen Hamilton, Inc., has two divisions. Its Worldwide Technology Business (WTB) advises the U.S. government on management and technology. Its Worldwide Commercial Business (WCB) provides management-consulting services to major companies globally. In 2009, it employed 20,000 people and has an annual sales revenue of $4 billion.

Christensen and Baird (1997) studied the knowledge management system of its WCB division. The WCB division had $600 million revenue in 1996, employed 2,000 consultants, and maintained thirty offices worldwide. It offered integrated solutions to clients.

Its Knowledge Engine has three thrusts. The organization is designed as a matrix type, in which the columns were functional practices and the rows were industry practices. Table 11.3 illustrates this matrix-organizational design. Most partners and staff of the firm will focus on building knowledge through research, training and assignment rotations in both an industry practice and a functional practice.

Table 11.3 Matrix Organization at Booz Allen Hamilton

	Strategic Leadership	Operations	Information Technology
Financial services			
Communications			
Media			
Technology			

Furthermore, special-interest groups were formed to identify and consolidate what was known within the firm on particular subjects, to suggest new services based on such insights, and to assemble experts for selling these new services, when needed. There were twelve such special-interest groups, covering topics such as new business models, innovation, sourcing, conglomerates in emerging markets, and strategic war gaming. In addition, the company introduced the Web-based Knowledge-On-Line (KOL) that enabled keyword searches by all consultants to access information on industries, technologies, markets, past clients, and client reports without confidential information.

Consulting companies know well that their clients monitor which firms do the best work on what types of problems. As a way to showcase company expertise, Booz Allen Hamilton launched the *Journal of Strategy & Business* in 1997, which achieved a circulation of 60,000 and received the Folio Editorial Excellence Award from the business publishing industry.

11.4.5 Arthur D. Little

Like many other management-consulting firms, Arthur D. Little builds its business on employees' knowledge and experience. To be effective, all client-facing employees must have instant access to knowledge, skills, and experience relevant to their assignments at hand (Chait 1999). For this purpose, the company initiated a Lotus Notes application, named "ADL Link," to connect employees globally. It includes the databases for skills depository of staff, client database, methodologies and tools, and practices and business groups (strategies, functions, operations and organizations).

Three factors are deemed important for the company to achieve success in knowledge management:

1. *Ensuring vision and alignment*—Link knowledge management to specific organizational objectives and strategies and to align staff with the vision and make corporate commitments to the resulting vision, which is to be communicated widely to the staff.
2. *Managing four domains*—These are contents, culture, process and infrastructure.

   - *Contents*—Identify knowledge elements, their relative importance to individuals, groups and corporate objectives, and the context in which to utilize them. Constantly revising contents to make them relevant and useful. Study the usage patterns, advertise the availability of these contents to encourage additional use, update the contents regularly.
   - *Culture*—Link knowledge management to cultures and values.
   - *Process*—(1) process to capture, evaluate, cleanse, store, provide and use knowledge; (2) specific KM roles are assigned to customer-facing groups (knowledge advocate and knowledge steward set up communities of interests); and (c) process guiding the KM operations and implementation.
   - *Infrastructure*—hardware systems and training programs.

3. *Creating an effective plan*—The plan prioritizes knowledge management tasks to make sure the KM system grows and progresses in accordance with the corporate needs.

11.4.6 Katzenbach Partners

Katzenbach Partners, a small management consulting firm, was formed in 1998. At its largest, the firm employed 150 consultants in four major US cities and realized an annual sales revenue of $50 million in 2007. The firm became a part of the Booz company in 2009.

Blumenstein and Burgelman (2007) studied the knowledge management system of Katzenbach Partners. The company thinks that the needs of knowledge seekers are best addressed through a conversation, which can then be supplemented with documents.

For promoting the access of documents, the company built a novel hybrid knowledge management system composed of two parts:

1. *Formal*—Documents are classified with predefined subjects (e.g., change management, strategic planning) and types (e.g., case study, client presentation).
2. *Informal*—Users can also classify a given document using informal "tags" (e.g., client name, project code, topic, final, initial) in order to facilitate their personal retrieval. All tags are made public. Since any one document could be associated with tags of multiple users, this informal document classification system produces a more complete picture of a document's contents than by the tags of the document's original author.

   The novelty of this system lies in the added convenience and flexibility accorded by the informal tags that users can add.

11.4.7 IBM

At IBM, storytelling is recognized as a powerful means of discovering and transferring knowledge (Dalkir 2005). Its story-telling approach has four stages: (1) elicitation through interviews, and observation, (2) decomposition of the story to analyze the inherent values, rules and beliefs, (3) intervention to design and enhance the story, (4) deployment of the story. Values are moral principles and standards. Rules are the code of disciplines that drives or conforms behavior. Beliefs are the collection of ideas that a community regards as true.

IBM created several tools to link knowledge seekers to domain experts. Such linkages reduce time, save costs and improve performance in the field (Powers 2006):

- *Knowledge view*. This contains repositories of intellectual capital, key resources and discussion forums available to those who sell and deliver consulting work in the IBM Business Consulting Services unit. In 2004, it had more than 1 million assets read. Consultants find content quicker, and the capital sharing increased by 59 percent.
- *Worldwide asset reuse*. In 2004, it captured 384 anecdotal success stories that demonstrated significant business impact.
- *Xtreme leverage*. This is a knowledge sharing and collaboration tool aimed at IBM software sellers. The portal contains intellectual capitals, expertise location and community facilities for each of IBM's five global brands. Experts include helping sellers as a part of their jobs. In this system, there are 400 communities composed of salespeople, business partners, and clients.

Reducing the time to find an expert from one week to less than eight hours upgraded IBM productivity by $50 million a year.

- *BluePages*. This is a searchable directory that is enabled with instant messaging and e-mail linkage. About 84 percent of IBM's 300,000 employees are registered in this directory.
- *Knowledge Point*. This provides IBM consultants with research and expertise location.
- *Collaboration Central*. This is a companywide portal for collaboration guidelines, tools, best practices and team rooms. The team rooms are online collaboration spaces for remote teams to share information and work collaboratively.
- *Thinkspace*. This is a site for employees to submit innovative ideas that are evaluated monthly.
- *Learning@IBM.* This provides streams of profile-driven learning directly to the learners' desktops to ensure that employees focus on learning that is relevant to their specific jobs.

Clearly, IBM's knowledge management strategy is to work concurrently in documenting intellectual assets (patents, manuals, procedures, and success stories), encouraging all employees to learn and to innovate, and inducing domain experts to share, all focused on deriving productivity and business results.

Example 11.3 A healthcare organization is planning to build a new knowledge management system with its initial emphasis on enhancing the reuse of existing corporate knowledge regarding work procedures, methodologies of problem solving, and other established practices. The company wants to make sure that knowledge, skills and experience must be available to the right people in the right places, at the right times, so that these knowledge elements are leveraged and multiplied in value. As the focus is on the instant access to the up-to-date employee knowledge, such a knowledge management system must be dynamic in nature. How would you recommend that such a system be designed?

ANSWER 11.3 Since this new knowledge management system focuses dynamically on up-to-date knowledge of the explicit type, the key design figure must be making communications easy and frequent. A case in point is the Web-based Knowledge Management Exchange system introduced by GlaxoWellcome Canada (Duffy 2000).

GlaxoWellcome Canada was a part of GlaxoWellcome, which merged with Smithkline Beecham in 2000 to form GlaxoSmithkleine (GSK). The Knowledge Management Exchange of the GlaxoWellcome Canada contains the following five elements:

1. *I Want to Share*. This knowledge base allows employees to share what they have already learned. Company should offer recognition of and provide incentives to these knowledge sharers.
2. *What We Already Know*. This depository contains the existing answers to various questions already on file. Knowledge seekers are encouraged to search for possible solutions to problems at hand, in order to save time. Company should take notes of which answers delivered positive outcome.

3. *Does Anyone Know?* Knowledge seekers are encouraged to post questions on this blackboard so that experts within the company can respond. Again, companies should monitor and reward those whose responses are proven to be useful.
4. *Tip-of-the-Week Archive.* This area provides a collection of past e-mail tips on specific topics and assistance to employees in the use of the Knowledge Exchange.
5. *Feedback.* This area solicits general questions, comments and other input from employees for the purpose of improving the Exchange continuously.

The company should reward active participants and monitor the performance of all of these elements in order to augment the overall effectiveness of the system.

11.4.8 British Petroleum (BP)

British Petroleum (BP) advocates the idea that domain experts need to allocate 15 percent to 20 percent of their time to cross-unit knowledge-sharing activities (horizontal) and the remainder to business unit performance (vertical) (Hanson and von Oetinger 2001). The horizontal knowledge-sharing activities are composed of the following four types:

1. Collaboration (assist in organizing knowledge meetings to solve problems)
2. Connection (help in locating domain experts)
3. Give (offer knowledge-based advice)
4. Take (obtain specific assistance from others)

This horizontal model allows knowledge users to communicate across units directly to knowledge experts, to reach people with in-depth useful knowledge, and to achieve speed in solving problems by having avoided the potential silo effect in bureaucratic companies.

The value added by the horizontal knowledge-transfer activities are summarized as follows:

- Increasing efficiency through the transfer of best practices across units
- Improving the quality of decisions through peer advice
- Growing revenue through shared expertise
- Formulating new business opportunities through the cross-pollination of ideas
- Making bold strategic moves through the promise of well-coordinated implementation

This horizontal knowledge sharing facilitates the decentralization of decision making to people who possess the right working knowledge. Learning is best achieved through decentralized and horizontal networking.

11.4.9 Chrysler Group

The Chrysler Group went through a number of major changes in recent years. It merged with Daimler Benz in 1998 to form DaimlerChrysler and then became independent from that joint venture in 2007. It went bankrupt in April 2009 and received a federal bailout of $6.6 Billion to restructure itself. More recently, Fiat committed investment funds and acquired some ownership of the post-bankruptcy Chrysler.

The knowledge management practices at DaimlerChrysler was studied by Rukstad and Coughlan (2001). They report that its system consists of six key elements (Fig. 11.7 illustrates the relationship between these key elements).

The "KM processes" element includes knowledge domains, communities of practice, knowledge sharing and development processes, and interconnections. The "business strategic framework" element contains strategies/objectives, core competencies, leadership in the new economy, and strategy formulation. The "KM strategic framework" element encompasses KM strategy, KM concepts and principles, leadership in KM, and culture. The "organizational context" element specifies funding, people development, rewards and recognition, and technology—communication and repository. The "KM support structures" element delineates the KM board and chief knowledge officer, support team, and KM toolkit. Finally, the "KM assessment/renewal" element provides activity assessment, knowledge and learning metrics, relation to business outcomes, and renewal processes.

Compared to other automotive companies, such as Ford, General Motors and Toyota, the DaimlerChrysler's knowledge management appears to be the most advanced one.

11.4.10 Hill & Knowlton

Hill & Knowlton offers public relations, product launching, advertising and other related services. It established a Web site, based on the Collaborative Workspace of Vignette, Austin, Texas (Mark and Meister 2004). It contains several *cabinets,* including (1) news you can use, (2) HK directory, (3) channels related to e-mail archives, schedules, and work-in-progress documents, and (4) "drop zone," where information was first reviewed by an editor before being made available on the site.

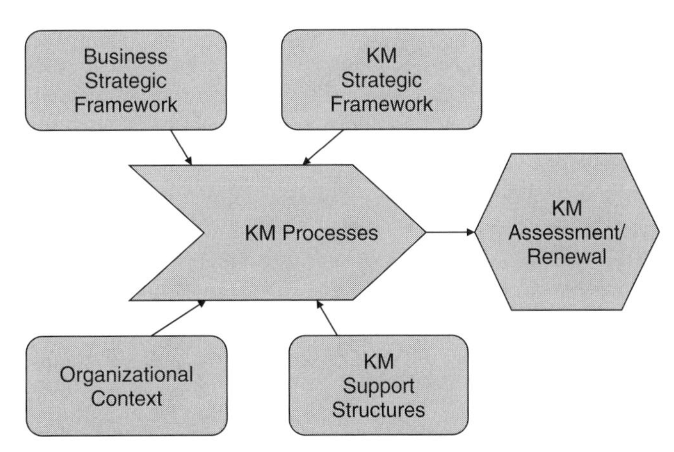

Figure 11.7 Elements of knowledge management system at DaimlerChrysler. (Adapted from Rutstad & Coughlan, 2001)

The uniqueness of the company's knowledge management strategy lies in its systematic way of extracting unstructured knowledge from e-mails, presentation notes, spreadsheets, bios, pictures, video clips, conference notes, and other less formal documents. It also preserves and makes available structured knowledge contained in text documents, project files, case studies, research reports and news clippings, as well as fosters contacts between domain experts and knowledge users. It is to be noted that the unstructured knowledge is still a part of the explicit knowledge, not the tacit knowledge defined earlier in this chapter.

To encourage participation, Hill & Knowlton award beenz—a digital S&H Green Stamp, which could be redeemed for magazine subscriptions and CDs—to knowledge contributors. Some employees tend to hold onto them, suggesting that the beenx possess intrinsic values as a rough proxy for an employees' perception of worth to the company.

11.4.11 Xerox

Xerox used KANA IQ of Kana Software Inc., Menlo Park, California, to streamline its online customer service for its customers of printers, copiers and a few other products and services in more than 100 countries (Anonymous 2007). A Pareto analysis showed that 19 percent of existing product contents was required to support 90 percent of the call volume. KANA IQ ported this 19 percent key content and translated it into eight languages. The knowledge base supports virtually every market worldwide in the most common global languages. It is said to have reduced cost-per-solution and made Xerox's customers happier.

Xerox institutionalized several tools to facilitate knowledge creation and transfer, such as (1) Online Knowledge Universe—a catalog of best practices, and chat rooms for CoPs, (2) Company Yellow Pages, and (3) Knowledge Street for promoting knowledge sharing (Dalkir 2005). Storytelling at informal get-togethers and offsite meetings was also strongly encouraged. Its Eureka system captures the tacit knowledge in these stories and makes them more widely available. The innovators are rewarded by having their names displayed prominently next to their respective contributions.

11.4.12 Healthcare Partners—Massachusetts

The order-entry systems at Healthcare Partners is a useful knowledge-based decision support system that is constantly updated to offer advice and comments to physicians who decide on prescriptions for patients, based on patient records, a drug-interactions database, latest prescription guidelines, and other relevant resources. The system is said to have reduced errors and saved money for hospital systems (Davenport and Glaser 2002).

Example 11.4 Knowledge is power, certainly a very useful one for companies to derive competitive advantages in the marketplace. We know for a fact that 80 percent of what people know is tacit in nature. As a consequence, the great majority of all company knowledge management programs are focused on the preservation, creation, dissemination, and application of tacit knowledge.

What are the known best industrial practices in creating, preserving, distributing, and applying tacit knowledge?

ANSWER 11.4 Managing explicit knowledge is relatively easy. Some companies are doing well in this respect by requiring the documentation of staff performance, work procedures, test reports, project summaries, and other disclosures on a regular basis. These documents can be digitized, searched, and accessed by others in the organization in relatively efficient manners.

Tacit knowledge, by contrast, is rather difficult to produce, preserve, distribute, and apply. The best practice today is to create opportunities to allow knowledge seekers, who have a problem to solve and know how to ask the right questions in the proper context, to interact with knowledge givers, who are willing to share the applicable knowledge.

One such opportunity is the formation of communities of practices, which are comprised of domain experts and others who have profound interests in the domain subjects involved. Companies invest time and resources to encourage the formation of such CoPs, so that knowledge in selected domains of value is constantly produced, transferred, and updated by the interactions between their members, as well as between knowledge givers and knowledge seekers. Another opportunity is for the company to rotate the domain experts to different sites, so that more knowledge seekers can benefit from the resulting interactions. In general, companies are advised to promote a culture that favors those domain experts who are willing to externalize their thoughts, understanding, and insights and those open-minded knowledge seekers who are able to accept them.

11.5 NEW FRONTIERS OF KNOWLEDGE MANAGEMENT

Desouza (2005) edited a contributed volume titled "New Frontiers of Knowledge Management," which contains a collection of twelve chapters, covering the following topics:

1. Science and technology knowledge management
2. Knowledge visualization
3. Personalizing knowledge-delivery services
4. Knowledge security in organizations
5. Knowledge markets
6. Software artifacts for knowledge management
7. Ubiquitous computing in networked organizations
8. Collaborative enterprise
9. Knowledge flow dynamics
10. Knowledge integration in teams
11. The roles of incentives in knowledge transfer
12. Innocuous knowledge management

As evident from this list, the field of knowledge management is being researched on some theoretical fronts. From a practical standpoint, we could use more studies regarding the creation, valuation, utilization, preservation, and transmission of tacit knowledge, the key tenet in knowledge management, by ways of communities of professionals and through articulation of explicit knowledge. Furthermore, while experts utilize the articulation process to convert some explicit knowledge to tacit knowledge in their minds, it ought to be worthwhile for academic researchers to devise systematical ways of *knowledge mining* the existing explicit knowledge bases so that useful tacit knowledge can be extracted from them. Let us hope that future advancements in knowledge management will allow industrial practitioners to generate and utilize tacit knowledge more fruitfully.

11.6 CONCLUSIONS

In this chapter, we reviewed the basics of knowledge management. In addition, we discussed various industrial practices in knowledge management. Each industrial practice example offers specific ways of devising knowledge creation, preservation, and transfer. These practices serve as a useful description of the background in knowledge management.

As of today, most companies are still struggling to capture, update, and reuse the wealth of expertise, ideas, and latent insights that lie scattered across or deeply embedded in their organization. The industrial examples discussed in this chapter have shown that with corporate leadership, adequate planning, proper alignment with company strategy, innovative reward programs, conducive culture, and firm commitment to implementation, knowledge management can be successful in producing the expected benefits. Figure 11.8 illustrates the design of a comprehensive knowledge management system that allows the instant access of both explicit and tacit knowledge that is constantly collected and improved.

Knowledge management (KM) is important to various firms, more critical to some than to others. In summarizing the key observations derived from the industrial examples reviewed, the following are noteworthy:

- It is important to align KM with company strategy, so that the outcomes of KM are contributing to the specific goals defined beforehand.

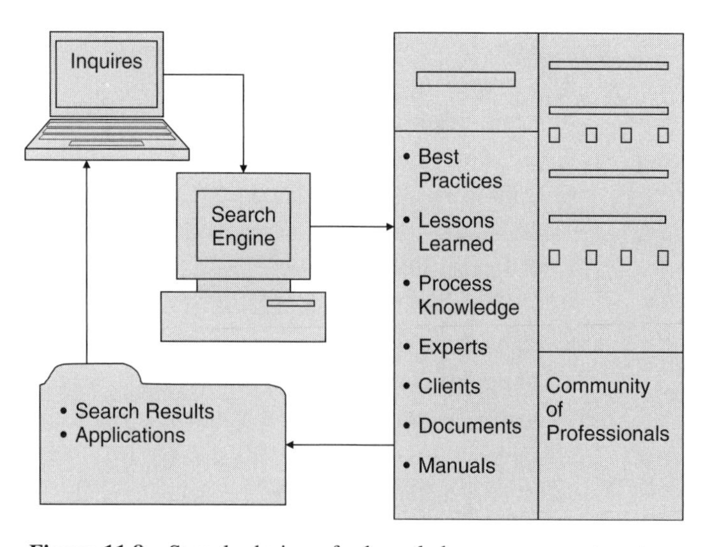

Figure 11.8 Sample design of a knowledge management system.

- KM requires top management support and commitment.
- KM needs to be promoted by group-based incentive programs, which tends to discourage individualism in favor of team performance.
- The preservation of explicit knowledge is generally easy to accomplish. The generation of new knowledge needs to be fostered by the company culture and compensation programs.
- Having good monitoring systems helps to track KM activities.
- Various IT and Web-based tools are available to enable the implementation of KM programs. However, focusing on the enabling tools alone will lead to no success, as KM has a large social dimension to it.
- Workers seeking knowledge should "pull" the required knowledge, as they know the important issues to them. "Pushing" knowledge by domain experts onto workers is not workable because of a potential divergence in the perceived needs.
- Allow multiple communications channels (face-to-face meetings, conferences, seminars, e-mail, video-conferences, etc.) to facilitate the interactions between domain experts and knowledge seekers.
- KM must be mission-oriented in order to be useful. Companies could focus on one or more specific objectives, such as innovation, transfer of process knowledge, customer relationship management, and so on.
- KM is more suitable to large enterprises whose employees are geographically dispersed and whose diverse products serve many different customers. It is not that critical for small companies, wherein everyone knows everyone else and they talk to one another daily.
- Publicize company expertise as a way to attract clients (This is what *Journal of Strategy and Business* and *McKinsey Quarterly* are doing).

It is believed that if some of these observations are taken into consideration, particularly with respect to the generating and sharing of tacit knowledge in a timely manner, knowledge management programs can be properly designed and implemented to achieve the expected benefits to the enterprises.

11.7 REFERENCES

Anomalous. 2000. *Knowledge Management: Four Obstacles to Overcome*. Boston: Harvard Management Update.

Anonymous. 2007. Leveraging Knowledge at Xerox, *Knowledge Management World* 16 (3): P. S3.

Barlett, Christopher A. 2000. "McKinsey & Company: Managing Knowledge and Learning." Harvard Business School Case # 9-396-357, Revision: January 4.

Blumenstein, Brooks, and Robert A. Burgelman. 2007. Knowledge Management at Katzenbach Partners LLC. Palo Alto, CA: Stanford Graduate School of Business Case SM-162.

Bobrow, Daniel G. and Jack Whalen. 2002. "Community Knowledge Sharing: The Eureka Story," *Journal of Society for Organizational Learning*, 4, (2).

Bondura, Mike. 2007. Nine Ways to Share Knowledge for Gain. *Bests' Review* 107 (11): 78.

Chait, Laurence P. 1999. "Arthur D. Little Knowledge Management Systems." *Journal of Business Strategy* (March–April).

Chang, C. M. 1996. *Knowledge Based Diagnostic Advisory System and Method for an Air Separation Plant*, U.S. Patent 5,557,549, September 17.

Chard, Ann Marie, and Miklos Sarvary. 1997. "Knowledge Management at Ernst & Young." Stanford University Case M-291, September.

Christensen, Clayton M., and Bret Baird. 1997. "Cultivating Capabilities to Innovate: Booz Allen Hamilton." Harvard Business School Case # 9-698-027.

Dalkir, Kimiz. 2005. "Knowledge Management in Theory and Practice," Burlington, MA: Elsevier Butterworth-Heinenmann.

Davenport, Thomas H., and John Glaser. 2002. "Just-in-time Delivery Comes to Knowledge Management." *Harvard Business Review* (July).

Delong, D. W., and L. Fahey. 2000. "Diagnosing Cultural Barriers to Knowledge Management." *Academy of Management Executive* 14 (4): 113–127.

Desouza, Kevin C. (ed.), 2005. *New Frontiers of Knowledge Management*. New York: Palgrave MacMillian.

Du Plessis, Marina. 2005. "Drivers of Knowledge Management in Corporate Environment," *International Journal of Informaiton Management*, 25 (3), 193–202.

Du Plessis, Marina. 2007. "Knowledge Management: What Makes Complex Implementation Successful?" *Journal of Knowledge Management* 11 (2): 91.

Duffy, Jan 2000. "Knowledge Exchange at GlaxoWellcome." *Information Management Journal*, 34 (3).

Eucker, Tom R. 2007. "Maintaining Levels of Experience at Intel." *Knowledge Management Review* 10 (3) (July/August).

Fahey, Liam, and Laurence Prusak. 1998. "The Eleven Deadliest Sins of Knowledge Management." *California Management Review* 1.40 (3) (Spring).

Gilmour, David 2003. How to Fix Knowledge Management, *Harvard Business Review* (October).

Grudzewski, W. M. and I. K. Hejduk. 2010. "Technology Transfer Streams in Service Industry," Chapter 29 in Salvendy, Gavriel and Waldemar Karwowski (Eds), *Introduction to Service Engineering*, John Wiley (January).

Gupta, Anil K., and Vijay Govindarajan. 2000. "Knowledge Management's Social Dimension: Lessons from Nucor Steel." *MIT Sloan Management Review*, 42 (1): 71.

Hanson, Morten T., and Bolko von Oetinger. 2001. "Introducing T-shaped Managers: Knowledge Management's Next Generation." *Harvard Business Review* (March–April).

Idinopoulos, Michael, and Lee Kempler. 2003. "Do You Know Who Your Experts Are?" *McKinsey Quarterly* 4.

Leonard, Dorothy, and Walter Swap. 2005. *Deep Smarts: How to Cultivate and Transfer Enduring Business Wisdom*, Boston: Harvard Business School Press.

Lin, Chieh-Peng. 2007. "To Share or Not to Share—Modeling Tacit Knowledge Sharing." *Journal of Business Ethics*, 70: 411–428.

Mark, Ken, and Darren Meister. 2004 "Hill & Knowlton: Knowledge Management," Case # 904E03, Richard Ivey School of Business, the University of Western Ontario, London, Ontario, Canada.

Meister, Darren, and Tom Davenport, 2005. "Knowledge Management at Accenture," Richard Ivey School of Business case # 905E18, the University of Western Ontario, London, Ontario, Canada.

Nonaka, Ikujiro 2007. "The Knowledge-Creating Company." *Harvard Business Review* 85 (7/8).

Polanyi, Michael. 1983. *The Tacit Dimension*. Glouchester, MA: Peter Smith Publisher, Inc.

Powers, Vicki. 2006. "IBM's Knowledge Management Strategy." *Knowledge Management World* 15 (7): 16.

Rao, Madanmohan. 2005. *Knowledge Management Tools and Techniques: Practitioners and Experts Evaluate KM Solutions*, Elsevier Butterworth-Heinemann, MA: Burlington.

Rogers, Ed. 2007. "The Top Ten Management Myths." *Knowledge Management Review* 10 (2) (May/June).

Rukstad, Michael, and Peter Coughlan. 2001. "DaimlerChrysler Knowledge Management Strategy," Harvard Business School Case # 9-702-412, Boston.

Sauve, Eric. 2007. "Informal Knowledge Transfer." *T + D* 61 (3) (March).

Seeley, Charles P. 2001. "Subject Matter Expert: Setting up Effective Knowledge–Sharing Teams." *Management Review* 4 (3) (July/August).

Sharp, D. 2003. "Knowledge Management Today: Challenges and Opportunities." *Information Systems Management* 20 (2): 32–37.

Shein, E. H. 1983. *Organizational Culture and Leadership*. San Francisco: Jossey-Bass.

Sieber, Sandra, and Rafael Andreu. 2004. "Knowledge Management at Siemens Spain, University of Nararra," Case # IES 135, available at Harvard Business School Publishing, Boston.

Von Krogh, George, Kazuo Ichijo, and Ikujiro Nonaka. 2000. *Enabling Knowledge Creation; How to Unlock the Mystery of Tacit Knowledge and Release the Power of Innovation*. Oxford, England: Oxford University Press.

11.8 QUESTIONS

11.1 Organizational culture defines the ways people behave in different situations and make decisions following their deeply held values, beliefs and assumptions. Organizational culture is said to be a key factor that affects the success of any knowledge management program. Why is it so and how can culture be modified to be in support of the knowledge management objectives?

11.2 Communities of practice (CoPs) are regarded to be a key organizational design by which willing and able experts could interact to produce, exchange, and improve tacit knowledge. Which are the three most critical success factors for achieving success in CoPs? How should companies use them?

11.3 Connections between people are essential for the implementation of any knowledge management program. What are the best practices in creating connections, so that knowledge exchange can take place effectively?

11.4 People need to collaborate with others in collecting, preserving, and disseminating knowledge, both explicit and tacit, in order to leverage knowledge and realize its value. Name a few proven techniques that enhance collaboration among people.

11.5 Explicit knowledge differs from tacit knowledge in that it is readily externalized by the experts in the form of written documents, patents, procedure manuals, test reports, and project files, whereas tacit knowledge consists of insights, intuition, heuristics, rules of thumb, and perceptions, and is not easily explained or expressed. Tacit knowledge is regarded as more valuable, as it forms the foundation for creating competitiveness in the marketplace. Under what conditions are the aggressive collection, improvement, and dissemination of explicit knowledge still useful to companies?

11.6 Sharing knowledge willingly and ably by experts is the central tenet in any knowledge management system. How can companies motivate able experts to become more willing to share their knowledge?

11.7 There is a saying: "You get what you measure." To promote knowledge management, companies must devise criteria and techniques to monitor its progress. Explain the best practices in measuring progress in knowledge management.

11.8 Study the Harvard Business School Case "Siemens ShareNet: Building a Knowledge Network," Case #9-603-036 (November 2, 2002) and answer the following questions. The case materials may be purchased by contacting its publisher at 1-800-545-7685, or at http://www.custserve@hbsp.harvard.edu.

A. What is the case about?

B. How did Siemens build its ShareNet Systems? How novel is its design?

C. How are data, information, knowledge and wisdom defined?

D. What are the challenges ShareNet faces going forward?

E. What are the key lessons learned from this case?

Chapter 12

Innovations in Services

12.1 INTRODUCTION

Innovation is defined as the act of coming up with something new that adds value to customers and service providers alike. George Bernard Shaw said: "Imagination is the beginning of creation. You image what you desire, you will what you imagine and at last you create what you will."* Innovations are of tremendous importance to all service-sector companies that are constantly in competition for customers and market shares. In the absence of innovations, service providers will soon lose their relative corporate strengths and profitability in today's highly competitive business environment.

Innovations require creative ideas, which, in turn, must be generated beforehand. Some but not all creative ideas will lead to innovations. Because creativity precedes innovations, it is important to review various ways in which creative ideas may be efficaciously generated. Certain thinking strategies may be especially useful in engaging the creative minds of all service systems engineers to enable them "thinking out of the box."

In this chapter, we will elucidate first creativity and creative thinking strategies, and then address the fundamentals of innovations, including value chain, processes, and the keys to high impact innovations. Established practices of managing innovations are then elaborated, including organizational settings, business dimensions, best practice, and some additional guidelines. These discussions will then be followed by selected innovation examples in various service industries. Finally, conclusions are offered.

12.2 CREATIVITY AND CREATIVE THINKING STRATEGIES

Creativity and innovation confer business advantages important to any enterprises facing a plethora of competition. Creativity focuses on the generation of new ideas; regardless of how useful these new ideas might be in the short- or long-term. Innovation, by contrast, centers on adding value to enterprises by implementing and marketing selected new ideas. The extent of creativity is measured by the number of new ideas generated, whereas the extent of innovation is judged by the practical value it brings about. Not all novel ideas lead to adding value. Great ideas with poor implementation will bring forth outcomes of no or only minor significance. Innovation

---

Source: /www.quotationspage.com/quote/1656.html.

is recognized to be the foundation of sustainable competitiveness and profitability. Its importance is well recognized in the business world.

Creativity comes from workers such as scientists, engineers, architects, designers, educators, artists, musicians, and entertainers who devise new ideas, new technologies, new contents and/or new business models. Some individuals are more creative than others in generating new ideas. Nikola Tesla said: "I do not think that there's no thrill that can go through the human heart like that felt by the inventor who sees some creation of the brain unfolding to success."*

There are certain prerequisite for individuals to become creative. They need to have curious and inquisitive minds, handle ambiguity and uncertainties well, and be highly motivated to act. Training can cultivate the methodologies with which new ideas are generated. Training is also helpful to minimize the constraining effects of past experience and procedure known to inhibit "out-of-the-box" thinking, thus improving their propensity in coming up with new ideas.

Literature research indicates that in general an individual needs about ten years of incubation time to amass sufficient knowledge in order to become creative. Even with sufficient knowledge at their fingertips, many individuals do not become creative overnight. Creativity has always been regarded as desirable but difficult to attain.

12.2.1 The Creative Process

Albert Einstein said: "Discovery consists of seeing what everybody has seen and thinking what nobody had thought."** For pursuing creativity and innovations, workers go through a creative process. This process is usually initiated by some stimuli, which prompt the individuals to look for new ideas beyond the conventional. Individuals are then applying these new ideas to data they have available in order to gain information, which, in turn, produces knowledge. The newly gained knowledge leads to wisdom, which, when applied properly, delivers innovative outcomes. The innovative outcomes benefit the society at large. Figure 12.1 illustrates this process schematically. Specifically, this creative process consists of six steps, reviewed in the following sections.

Excite the Inquisitive and Curious Minds with Stimuli. A creative worker needs to have an inquisitive and curious mind, which responses to internal and external stimuli. Some individuals are readily excited by external stimuli, such as rewards, recognition, challenges, and desire to contribute to an important cause, a workplace friendly to new ideas, teams of other innovative people, free access to knowledge sources, opportunities for cross-pollination of ideas, and organizational support for innovative work (Florida et al. 2005; HBSP 2003A) . Others are predominantly driven by their internal stimuli, such as interest and passion.

Workers must have the mettle to be creative. They need to lay building blocks to creativity; these include resourcefulness, ability to think outside of the rules, playfulness, focus on exploring possibilities, being accepting, ability to accept failure and

**Source*: www.brainyquote.com/quotes/quotes/n/nikolatesl127569.htm
***Source*: www.useful-information.info/quotations/famous_quotes_one.html

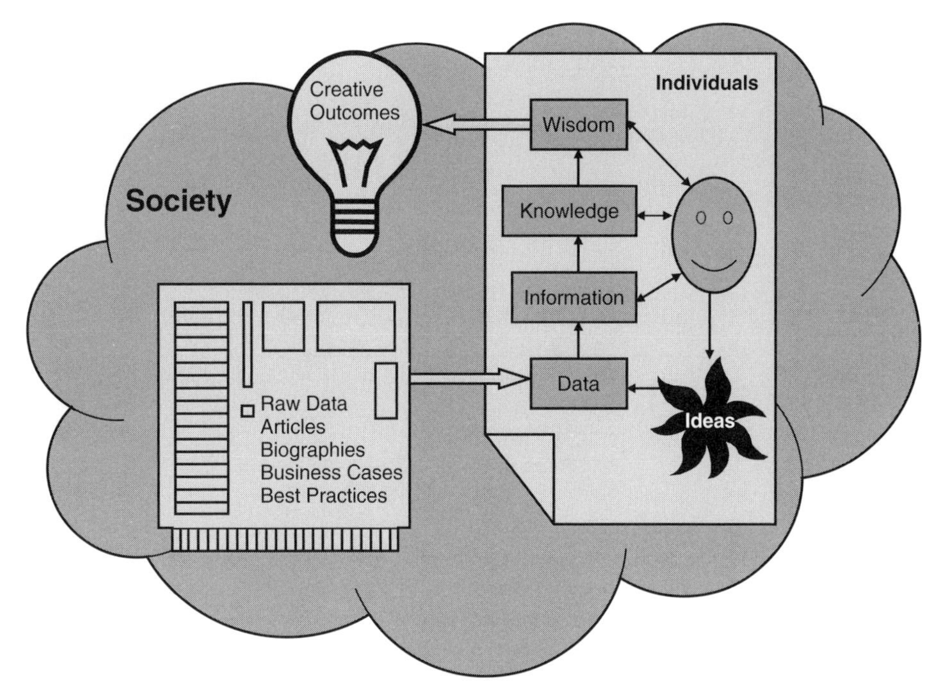

Figure 12.1 The creative process. (Adapted from Amablile, 1998)

learn from it, intelligent risk taking, active listening, receptivity to ideas, collaboration, tolerance, flexibility, persistence, having an inner focus, and recognizing creative potential in self (HBSP 2003A).

The components of individual creativity are identified to be motivation, creative thinking skills and expertise (Amabile 1998), as illustrated in Figure 12.2. In addition, individuals must devote significant personal efforts in order to be productive, as a lot of new ideas may have to be abandoned before some good ones will eventually shine through. They need to be mentally flexible to tolerate ambiguities and uncertainties. Accordingly, individuals may also be excited to become more creative through workplace practices and conditions, for examples, by (1) matching the right people with right assignments, (2) giving freedom to decide on the means, and (3) providing sufficient time and resources to pursue creative ideas (Amabile 1998).

Nurture Different Thinking Strategies to Produce New Ideas Beyond the Conventional. The experience an individual accumulates over time is known to have an inhibiting impact on his/her ability of generating new ideas. This is because human minds have the tendency to start from things known to unknown and from familiar to unfamiliar. As most individuals strive from effectiveness, past experience has a natural tendency to railroad their thinking patterns so that the outcomes become conventional and traditional, being fully compatible with their past experience.

The constraints placed by past experience on workers are well recognized. It is indicated that to enable workers to broaden knowledge, see old problems in new ways, generate fresh perspectives, and become free from their past experience, companies

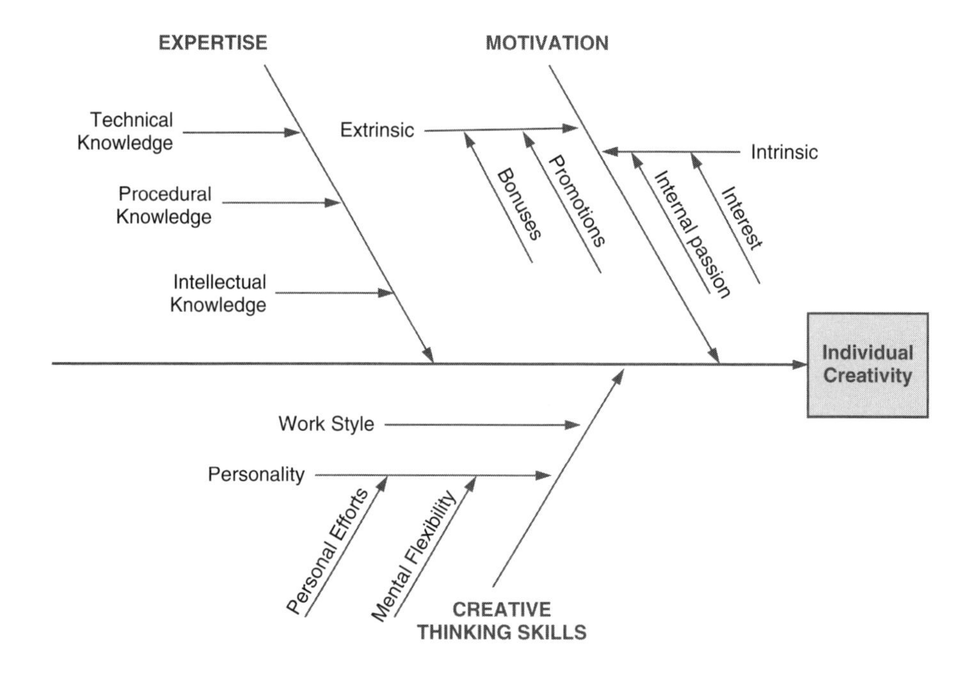

Figure 12.2 Components of individual creativity. (Amabile 1998.)

should take some weird steps, such as (1) recruiting people who are not blinded by preconceptions, (2) encouraging workers to defy superiors and peers, and (3) supporting risky projects with people whose ideas are not biased by past successes (Sutton 2001).

Knowledge brokering is defined as using olds ideas in new places, new ways, and new combinations. Thomas Edison said, "To invent, you need a good imagination and a pile of junk."* By imaging new uses for old ideas and testing promising one, new innovations can be generated (Hargadon et al. 2000).

Another way for individuals to think freely is to get trainings in utilizing divergent thinking strategies. Some thinking strategies have been recommended in Michalko (1998). A modified set of thinking strategies will be explained in the following section.

Gain New Information and Perspectives by Evaluating Data with New Ideas. Ideas are spectacles through which individuals look at data in order to see information. Here, information is defined as the outcome from having evaluated data, in the forms of relationships between data, parameters of importance, applicable assumptions, relevant constraints, range of applicability, and other such knowledge constructs. This outcome constitutes a "state of being informed." An individual will ferret out different information from a given set of data if different ideas are utilized in the process. Similarly, the same set of data looked at by different people with different ideas will yield different information. Thus, individuals with more new ideas are capable of producing more different information from the same set of data everyone else has.

**Source*: www.brainyquote.com/quotes/quotes/t/thomasaed125362.html.

Guided by a given new idea, a series of penetrating questions may be used to discover new information from a given set of data. Figure 12.3 illustrate examples of such *deep-learning* questions (Chang 2006).

The first-level questions are those involving the use of what, why, which, when, who, and how. These questions are intended to carve out the knowledge territory to mine. The second-level questions would involve: (1) Why so? (2) What happen if we . . . ? (3) What else is possible? (4) How about trying this..? (5) Why not try to combine . . . ? (6) Would it make sense, if . . . ? (7) Which would be better, if we . . . ? (8) What else do we need to know, if we like to pursue . . . .? These questions are digging for knowledge constructs, analyzing the findings and understanding their meanings to the issue at hand. The third-level questions would involve: (1) Given the situation as we are in today, are there any ways we could do this and that together..? (2) In view of that experience, should we not explore this combination . . . ? (3) In order to build advantages in cost and performance, would it not be better for us to focus on . . . ? These questions are focused on forming combinations, and on synthesizing and generating new ideas. At this stage, individuals will draw on one's own libraries of pertinent "best practices," "patterns," "rules," "heuristics," and "insights," to envisage specific new ideas, based on a selective combinations of others, to address the identified needs at the time.

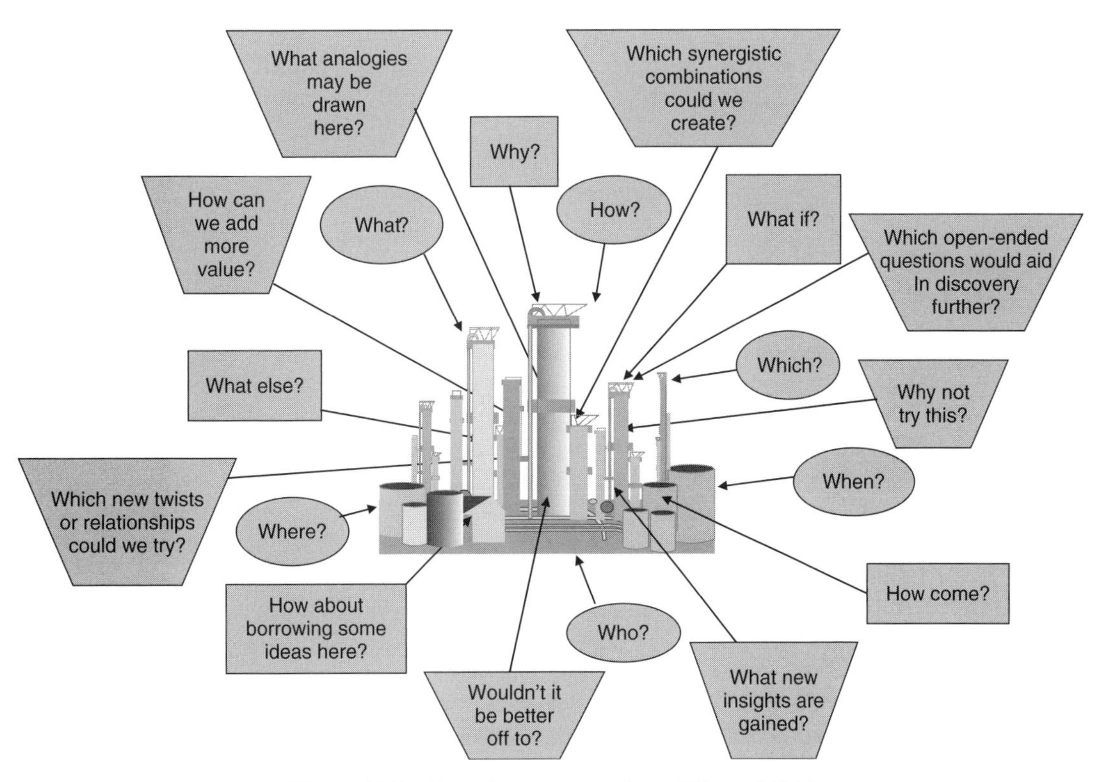

Figure 12.3 Deep-learning questions. (Chang 2006.)

Voltaire said: "Judge a man by his questions rather his answers." Good insights come from asking good questions. The creative thinking mantra are "what if?" "what else?" and "why?" Great creative thinkers raise more questions, as one needs to dig deeper until magic insights reveal themselves.

Apply Insights and Interpretations to Glean New Knowledge from Information. Knowledge is "the sum or range of what has been perceived, discovered, or learned," (Ferguson et. al. 2005). Information needs to be further processed in context with respect to the problem, issue, or circumstance at hand so that useful knowledge can be derived from it. It requires analysis, testing, judgment, and proper interpretations to glean new knowledge.

Bertrand Russell said: "The greatest challenges to any thinker is stating the problem in a way that will allow a solution."* A practically doable strategy to uncover knowledge from information is to first post the right questions and then to extract the respective answers from the information available via analysis. Examples of such questions could include the following: (1) Given the information we have today, couldn't we find out .....? or (2) With what we know today, would it be possible to ...? (Ferguson et al. 2005).

Grow New Wisdom from Processing and Distilling New Knowledge. Wisdom is the understanding of "what is true, right and lasting." Wisdom is like *deep smarts* defined in (Leonard et al. 2004). It may include rules, patterns, cause-effect relationships, concepts, paradigms, scenarios, and other such internalized deep-level knowledge nuggets with potential transferability from one domain to another. To reach this level of intellectual understanding, individuals need to further process the knowledge so gained, so that wisdom can be obtained.

Reflective learning is a process by which the individuals compare the newly gained knowledge with their past understanding in order to derive inferences and gain a deeper appreciation of what has been learned (Brockbank et al. 2002). Examples of reflective questions include: (1) In what ways is the new knowledge contradictory to or supportive of one's prior beliefs? (2) What is really unique about the new knowledge that is worth preserving in the context of known customer preferences, marketplace competition, current state of emerging technologies, and others? (3) What novel perspectives are implied by the new knowledge? (4) How else could the new knowledge be viewed and interpreted? (5) Which new opportunities are suggested by the new knowledge? (6) What actions are called for by this new knowledge? (7) What does the new knowledge say about one's creative thinking strategies?

Empower New Wisdom to Procreate Creative and Innovative Outcomes. When wisdom is diligently accumulated and properly applied to conceptualize the future, creative outcomes are likely to be delivered in the form of unique solutions to the problem/issue at hand.

Although creativity is often an individual act, a significant number of innovations are products of creative groups. Group creativity is enhanced by the processes of divergent and convergent thinking. Divergent thinking brings about new possibilities through

**Source*: www.quotationspage.com/quote/32858.html

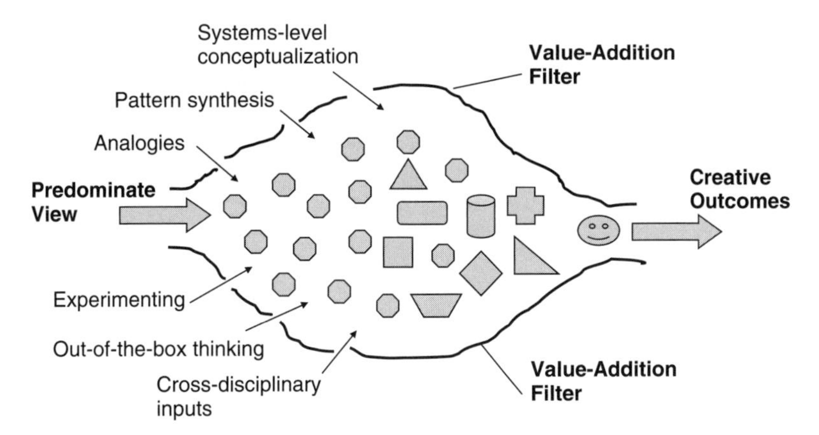

Figure 12.4 Divergent and convergent thinking processes.

the synthesis of new ideas generated by the group members, whereas convergent thinking channels to useful innovations. Figure 12.4 illustrates the divergent and convergent thinking processes.

During the convergent thinking phase, individuals may call for questions such as: (1) Which functions are essential to the customer, and which ones are nice to have? (2) What factor will affect the customer's purchase decision? (3) What are the constraints with respect to price, size, weight, and other such factors in view of the competition in the marketplace? (4) What must be satisfied in order to be compatible with the product lines of the company? (HBSP 2003B)

Members of these groups need to have varied skills, interests, and background and are curious and open-minded and collaborative, in order to maximize the beneficial impact of cross-pollination of different ideas. Enterprises need to make such groups innovative by (1) conferring confidence, (2) enabling communications, (3) delegating responsibility, (4) committing support, (5) defining challenging assignments, and (6) monitoring progress by milestones (Biolos 1996). Group interactions are essential in combining the individual wisdoms to yield innovative outcomes.

The key steps of the creative process leading to innovations are summarized in Figure 12.5.

It is believed that enterprises will do well if this creative process is aggressively promoted to catch lightning in a bottle in coming up with "blow-the-roof-off" inventions, by enhancing both the individual creativity and group innovations.

12.2.2 Thinking Strategies

One of the key steps in the previously described model for creative process involves the generation of new ideas. There needs to be new creative ideas, before new innovations may be realized.

A law of innovation was proposed that says that for every 1,000 ideas formulated, only one or two will prove to be both feasible and marketable (Hamel 2002). Individual creativity is necessary, but not sufficient for generating value. Creativity is not enough

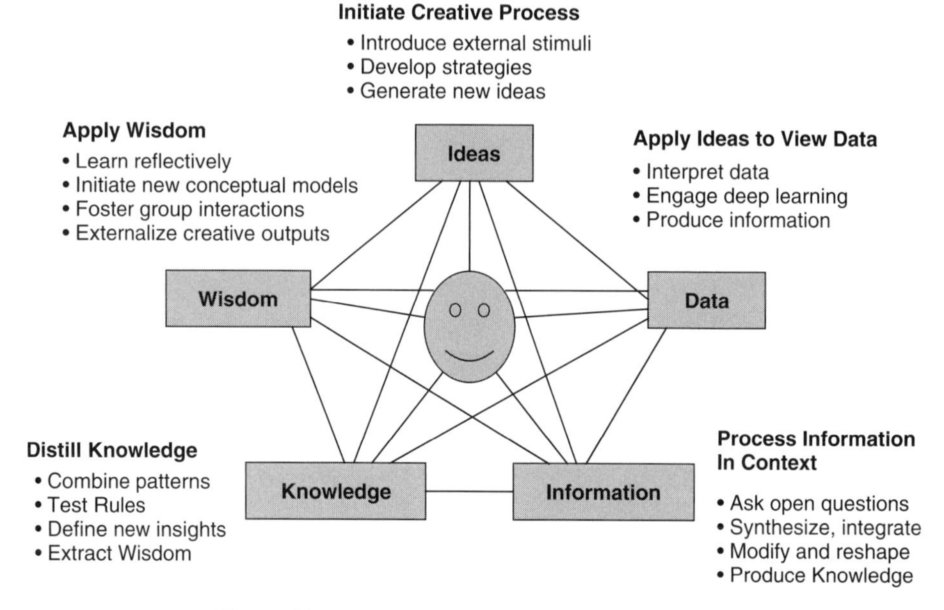

Figure 12.5 Creative process leading to innovation.

and ideas are useless until they are put to work (Levitt 2002). We need a very large number of new ideas before useful innovations can be expected. Creativity is essential.

Individuals formulate ideas based on their existing patterns of thinking; they think in ways they were taught to think. Their experience tends to conform the ideas they generate, inhibiting new types of ideas from surfacing up. Creativity implies a divergence from past experience and procedure. Creative people could benefit from some training in methodologies in order to make individuals becoming less constrained by their respective past experience and working procedure, both of which tend to "channel" their thought process toward something familiar and readily acceptable. Figure 12.6 illustrates this conventional process of generating stagnate ideas.

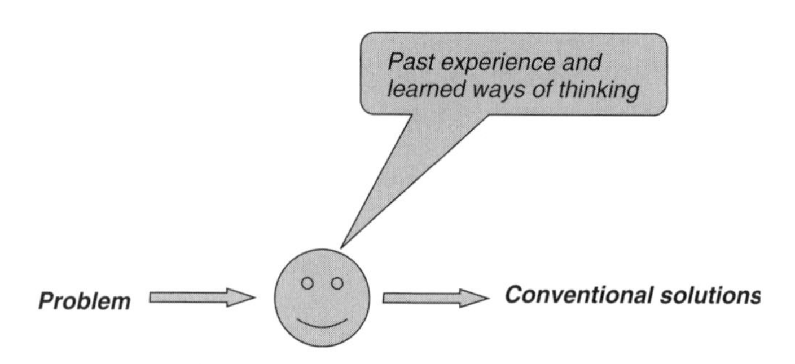

Figure 12.6 Conventional thinking patterns.

Excessive experience might, in fact, have a negative impact on the individual's capability of thinking out of the box. Most people search for solutions in the nearer context of the problem as they are led by already fixed thinking structures; functional fixedness based on experiences of former projects can block the way to innovative solutions (Herstatt et al. 2005). Thus, one of the key steps to enhance the individual's thinking capability is indeed to nurture different ways of thinking, beyond the conventional.

To generate ideas, workers are advised to (1) search the literature, (2) look at all sides of the problem, (3) talk with people who are familiar with the problem, (4) play with the problem, and (5) ignore the accepted wisdom (Shapiro 2003).

Several thinking strategies were proposed by Michael Michalko (1998) based on multiple-year research on past inventors and thinkers, as well as the field practices of selected creative people. Since not all strategies contained in Michalko (1998) are equally effective and important, I simplified and reorganized them into the following six, shown in Figure 12.7.

Explore Metaphors and Analogies. Thomas Edison said: "The inventor has a logical mind that sees analogies" (Schwartz, 2004). The ability to perceive similarities and analogies is one of the most fundamental aspects of human cognition. Analogies are known to be useful in promoting breakthrough ideas. Substantial innovations often result from transferring problem solutions from one industry or domain to another. Although an innovation can be based on a new scientific or technical discovery, the recombining nature of innovations is more dominant (Hargadon 2003). A systematic method of applying analogies is advocated in (Herstatt et al. 2005), as depicted in Figure 12.8.

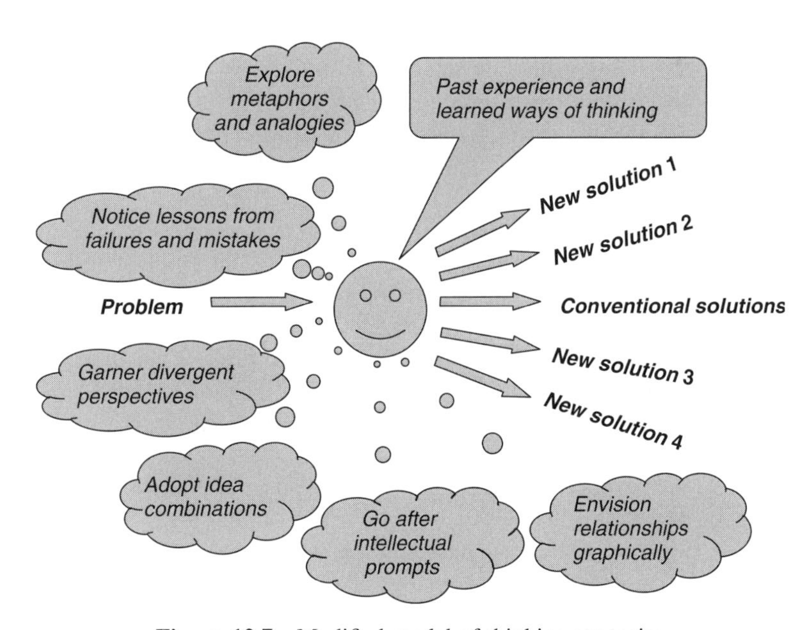

Figure 12.7 Modified model of thinking strategies.

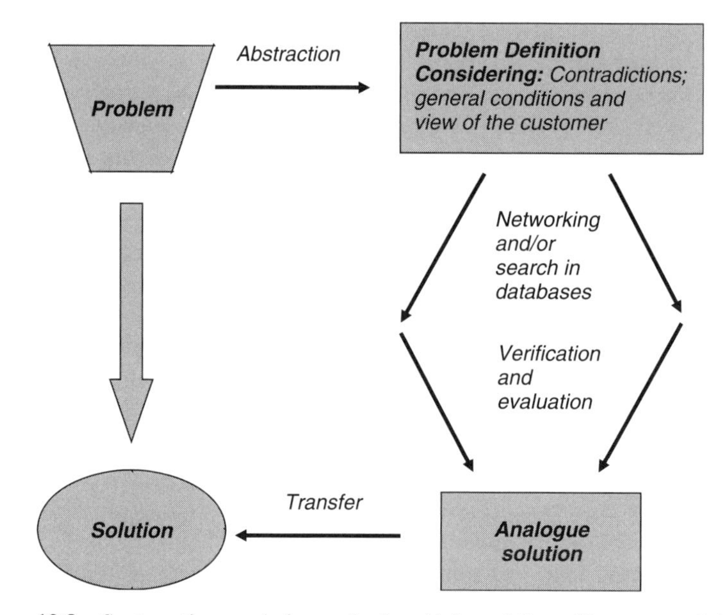

Figure 12.8 Systematic search for analogies. (Adapted from Herstatt, et al. 2005.)

Nikola Tesla invented the AC motor by pursuing the analogy of the Sun rotating around the Earth to motor's magnetic field rotating inside the motor. Leonard da Vinci discovered the fundamental property of sound traveling in waves by relating the sound of a bell to the water waves generating by a falling stone. Samuel Morse periodically boosted the weak electromagnetic signals for them to become strong enough to travel coast to coast by invoking the analogy of exchanging horses at relay stations.

Recognizing metaphors is also a useful way to discover new ideas. Alexander Graham Bell conceived the telephone by understanding the inner working of the ear and the movement of a stout piece of membrane to move steel. Thomas Edison invented phonograph by combining a toy funnel and the motions of a paper man and sound vibrations.

A number of years ago, parallel computers were invented to process computationally intensive tasks concurrently in different processors, in order to arrive at the final solutions in the shortest period of time possible. In analogy, modern-day supply chain systems parcel out the tasks of producing specific components service packages to different vendors and requiring them to design, manufacture, and quality control their respective parts concurrently at global locations, so that the final service package may be assembled, tested, and delivered in very short period of time.

It is useful to apply known analogies to unrelated domains by raising questions such as: (1) What simple analogies can be applied to create new ones so as to solve the problems at hand? (2) What does this remind me of? (3) How can a known technology in one domain be repurposed for a new problem space?

Notice Lessons from Failures and Mistakes. Winston Churchill said: "Success consists of going from failure to failure without loss of enthusiasm." Charles Goodyear said: "What is hidden and unknown and cannot be discovered by scientific research will most likely be discovered by accident, if at all, by the one who is most observing of everything related thereto."*

Various unexpected discoveries were made in the past by inventors, who searched for reasons for failures in their original attempts. Question may be asked: How can we find new discoveries by finding out why something (e.g., chance events, accidental outcome) behaves strangely? Finding out what is good about the failures and mistakes will allow new learning opportunities to be preserved. When finding something interesting, individuals should pursue them immediately—even if it would cause them to detour from the original path of discover—in order to benefit from the opportunities created by accidental discovery.

Garner Divergent Perspectives. Albert Einstein said: "To raise new questions, new possibilities, to regard old questions from a new angle, requires creative imagination and marks real advance."** Individuals need to set aside their traditional mindsets and look at the broader picture. It takes a tectonic paradigm shift in order to "see around the corners." This is the reason why oftentimes outsiders can offer a fresh point of view that the closed community cannot. Individuals need to cajole themselves into seeing things from a variety of different perspectives.

Cosmetic products offer different value prepositions to different groups of stakeholders. They are chemicals, colors, and powders to the producers, hope for a younger and more attractive look to its consumers, tax revenue sources to governments, and marketing challenges to vendors competing for the same customers. Companies pay more heed to customers in order to benefit from their perspectives.

China's one-child policy created a lot of "little emperors" in the families, as both parents and grandparents spoiled them, allowing them to do whatever they wanted, including eating junk foods. They generated a widespread public health concern in China. Looking at this concern from another perspective, these little emperors contrived golden opportunities to some innovative food companies, which made big fortunes by marketing nutrition-enriched drinks to Chinese kids while pleasing Chinese parents and grandparents.

Restructuring or restating a problem in different ways will shepherd to different understanding. Formulating a problem in diverse ways helps avoid seeing things in routine manners. Post the question: "In what way might I ...?" to start a problem statement. Applying thought experiments to change the level of abstraction, from local to global and from components to systems. Separating components from their systems will promote the generation of alternative perspectives.

Lateral thinking was advocated in (Debono 1970) as a way to break loose for "vertical" or "analytical" thinking paradigm. Individuals could garner different perspectives of looking at a problem by (1) exploring the interface in cross-disciplinary domains for opportunities, (2) envisioning problems from different angles to think outside the box, (3) nudging the conceptualization level up from elements to systems, or (4) going after new forward-looking ways to do things familiar. Table 12.1 lists examples of open-ended questions that could be suggested from different perspectives.

**Source*: www.brainyquote.com/quotes/quotes/w/winstonchu108950.html.

***Source*: www.brainyquote.com/quotes/quotes/a/alberteins130625.html

Table 12.1 Examples of Open-ended Questions from Different Perspectives

#	Open-ended Exploratory Question from Different Perspectives
1	May the problem at hand be reframed in a new way to lead to unexpected possibilities?
2	What new possibilities can be envisioned from the problem at hand?
3	Which industries, domains, or fields could be bridged to find new opportunities?
4	What types of prototype could be built to try out an idea?
5	What type of missing need there is?
6	Is the newly defined problem worth solving (e.g., market size, profitability)?
7	What problem may be lie inside of the problem at hand?
8	What is wrong in something, and how can an enhanced replacement be devised?
9	How can a technique in one area be applied to a new domain?
10	How can something be made to work better, once its reason of working becomes understood?
11	How can insights be layered on others to create new ones?
12	What may be the outcome, if a systematic thinking process is adopted?
13	What "future states" may be imagined that differs significantly from the "current state"?
14	What is next?
15	What can be learned if a higher-level broad-based perspective is adopted—instead of looking at individuals trees, glancing at the forest?
16	Which nuggets of wisdom (e.g., business trend, scenarios, rules, application practices, ways of looking at the available knowledge, reasons for something that works, and how analogies work in different contexts, etc.) could or should be replaced, modified, or reapplied?

Adopt Idea Combinations. Combining ideas from different domains is known to render new possibilities, by ways of integrating and synthesizing known patterns, concepts and relationships. Combining things in novel ways is a technique to conceive new ideas. Questions may be raised: (a) What changes may be made to the mental model through a combination of new information, combined with own experience? (b) Which conceptual models or images could be combined so that cross-disciplinary combinatorial invention may be possible at their intersections?

Go after Intellectual Prompts. Intellectual prompts may come from reading literature, making observations, doing experiments, traveling, and/or interacting with creative people in groups. By employing the open innovation paradigm (Chesbrough 2003; Rigby and Zook 2002; Huston and Sakkab 2006) involving global inputs (Santos et al. 2004), one could realize the benefits of global scope of economies in cultivating new ideas.

"Read thousands of pages and walk thousands of miles" is a known Chinese strategy to promote creativity by constantly seeking intellectual prompts.

Envision Relationship Graphically. "One picture is worth a thousand words," as the Chinese would say. Making thought visible helps display relationships in different ways other than in words and mathematics. An active deployment of graphic tools would

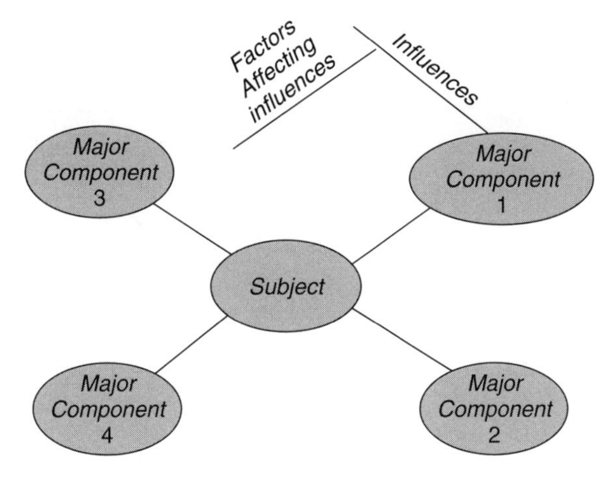

Figure 12.9 System-mapping diagram.

derive a better conceptualization of ideas, patterns, relationships, and insights. Questions may be asked: (1) How can it be visualized in pictures, props, and in graphical forms? (2) How thought experiments and mind games may be activated to probe *what if*?

Fishbone diagrams are good tools to uncover relationship, as exemplified in Figure 12.2. System mapping is another tool that is widely known to promote idea generation. Figure 12.9 reveals an example in which components, influences and major factors are linked together. The influences and factors can then be readily prioritized for analysis and solutions for each factor can then be found by brainstorming.

The combinations of these six thinking strategies should lead to a bubbling of new creative ideas, as illustrated in Figure 12.10.

Creativity precedes innovation, which, in turn, builds competitiveness and sustains profitability. The ENGAGE(Explore, Notice, Garner, Adopt, Go After and Envison) model (Chang 2008A) delineates the creative process, leading to innovations and the specific thinking strategies deemed useful in promoting the generation of new creative ideas. The central tenet of this model is that individuals need to leave no stone unturned in the quest of generating new ideas and then to converge on a selected few to shape the innovative outcomes. In order to apply these thinking strategies to produce innovative service ideas, an iterative routine, named *combinatorial, heuristic and normatively guided* technique, was proposed and successfully tested by Chang (2008B) and Chang (2010), shown in Figure 12.11.

David Starr Jordan said: "Wisdom is knowing what to do next, virtue is doing it."* It is believed that by aggressively following the previously described models, both in university education and professional development environments, individuals would be able to become creative and innovative sooner, allowing them to make better contributions to their professions, to their enterprises, and to the society at large.

* *Source*: www.brainyquote.com/quotes/quotes/d/davidstarr106080.html.

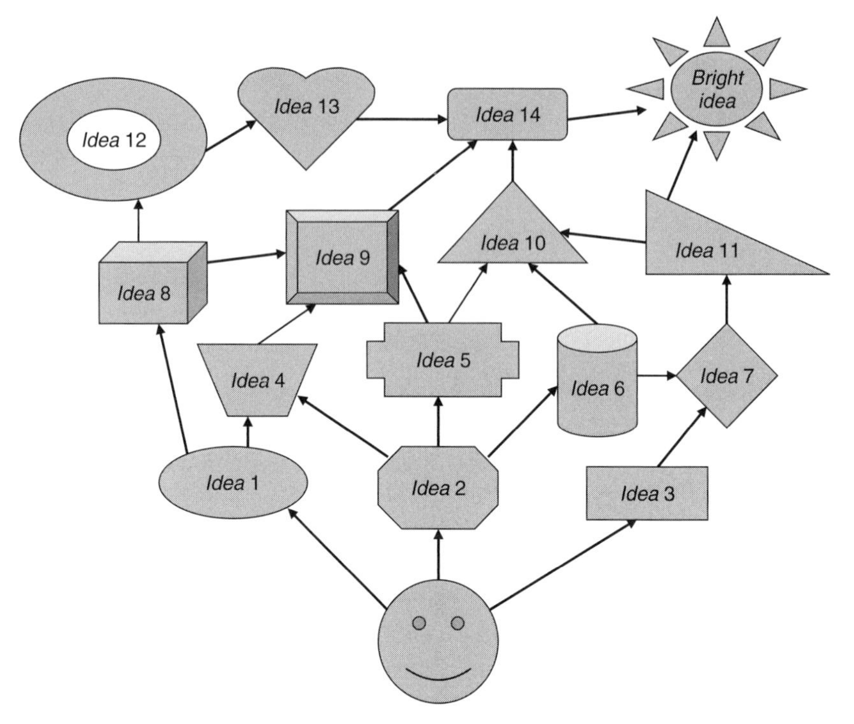

Figure 12.10 Generation of new ideas.

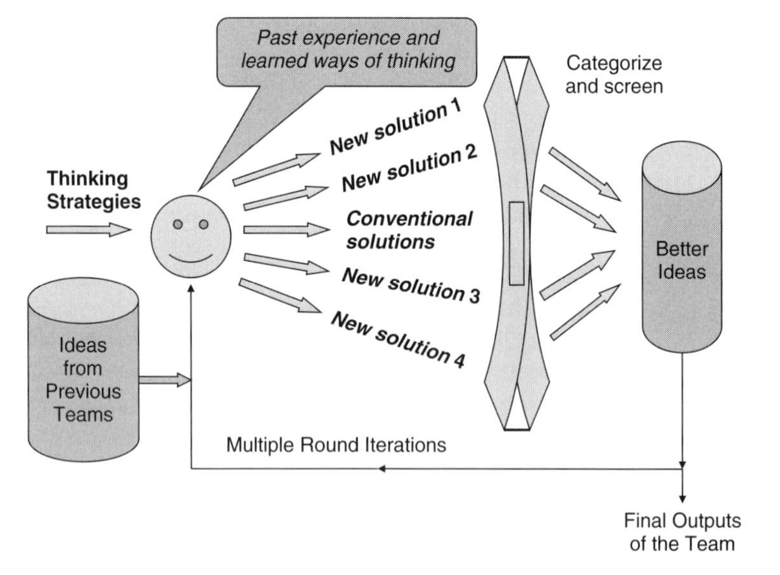

Figure 12.11 Iterative routine to generate innovative service ideas.

12.3 FUNDAMENTALS OF INNOVATION

In this section, we address the fundamentals of innovation, made up of the innovation value chain, the innovation process, open and closed innovations, and the categories of innovation in practice.

12.3.1 The Innovation Value Chain

The innovation value chain (Hanson et. al 2007) consists of three sequential processes: idea generation, idea development, and the diffusion of the invented concepts. Some companies may be strong in one and weak in another. The ultimate success of a company's innovation pursuits depends not on its strengths in any of these processes but on the weaknesses in any of them. Like in a mechanical chain, the overall strength is determined by the weakest link in the chain. Companies must understand its own strengths and weaknesses in all these processes and take actions to strengthen its weak processes, in order to achieve acceptable outcome in its innovation value chain.

Idea Generation. Thomas Edison said: "To have a great idea, have a lot of them."* Innovative ideas may come from creative people of different sources. Table 12.2 presents a group of known sources of *big ideas,* based survey conducted by IBM Business Consulting Services (Violino 2006).

The concepts of open innovation and open-market innovation are well-known in the business literature [Chesbrough 2003; Rigby et. al 2002]. Benefits of treating customers and suppliers as strategic partners are obvious. For example, physicians may know about opportunities for new capabilities to make their care delivery more effective. Ulwick (2002) and Leonard (2002) suggest that service providers ask customers for the outcomes they want, not for particularly service features they think they need. Christensen (1997) explains the rationale of not asking existing customers solely for ideas, thus likely missing some simpler needs of lower-end customers in rapidly growing segments.

Table 12.2 Sources of Big Ideas

Source	Percent of Respondents
Employees	41%
Business partners	38%
Customers	37%
Consultants	22%
Competitors	20%
Associations, trade groups	18%
Sales or service units	18%
Internal R&D	17%
Academia	13%

Multiple choices allowed

Data: IBM Business Consulting Services

**Source*: www.brainyquote.com/quotes/quotes/t/thomasaed149040.html.

Idea generation may be improved by building an external solution network. As practiced by Proctor & Gamble (Huston et. al 2006), customer needs are first translated into technology briefs that include the problems to be solved. These briefs are then sent through the company's solution network, which include technology scouts, suppliers, research labs, and retailers to solicit possible solutions.

Idea generation may also be enhanced by creating a solution-seeking Web site (e.g., Eli Lilly's InnoCentive—www.innocentive.com). Over 80,000 research participants (called *solvers*) in more than seventy countries are currently registered at the site to help its clients (called *seekers*) to find solutions to difficult R&D challenges. The clients include Dow Chemical, P&G, Eli Lilly, and twenty-plus others. When seekers confront a particularly difficult research challenges, they post their requirements to InnoCentive's solver network and offer a bounty to anyone who finds a solution. The ultimate price for the solution winner is $20,000.

Alternatively, companies may set up a discovery network. Siemens is known to have set up a fifteen-person scouting unit in Berkeley, California, to cultivate personal relationships with scientists, engineering doctoral students, venture capitalists, and entrepreneurs, as well as government labs and corporate research centers. The unit keeps track of emerging technologies and business ideas and feed back their learning constantly to allow Siemens to selectively take advantage of such insights in order to achieve time-to-market, technological leadership advantages. Similarly, Intuit sent out a ten-member team to learn and observe how small business owners do their books. The result is the *Simple Start* edition, which became a best seller for Intuit.

When promoting the generation of idea, having a diversity of unique contacts is the key, not the number of similar sources. Others have also promoted the concept of open innovations (e.g., Wolpert 2000; Rigby and Zook 2002).

Certainly, companies may also build internal cross-unit networks, in order to tap into the special talents from within. Proctor & Gamble formed thirty communities of practices (CoPs) (see Chapter 11), each focused on an area of expertise, to encourage interactions between people of diversified background. These communities solve specific problems that were brought to them. Members of these communities participate in monthly technology summits with business representatives.

Example 12.1 Transposing an old known idea to a new unrelated field is a well-known technique to innovate. How can company management foster this specific technique of innovation?

ANSWER 12.1 For such a technique to be workable, companies need to motivate people with diversified experience and knowledge to collaborate extensively and have a risk-tolerant organization in place to timely recognize and reward superior performance.

Hargadon and Sutton (2000) suggest that an effective innovation system can be built by the right organization and with the right attitude. Their research points out that the best innovators systematically use old, otherwise disconnected ideas as the raw materials for new innovations. Robert Fulton applied the old steam engine idea that was used in mines for seventy-five years to power boats.

Successful innovators of companies systematically go through the knowledge-brokering cycle of identifying good ideas, keeping ideas alive, imaging new uses for old ideas, and testing promising ones. Thomas Edison said: "To invent, you need a good imagination and a pile of junk."*

Idea Conversion. There are two strategies that could promote the success of the idea generation process: multichannel funding and safe haven.

Shell Oil created a twenty-five-person unit, GameChanger, with an annual budget of $40 million. This unit evaluates and funds business ideas submitted by Shell employees, typically to the tune of $300,000 to $500,000 per project. Successful projects are funded by Shell divisions subsequently.

Safe havens are business units designed to shield new business ideas and projects from the short-term thinking and budget constraints that pervade in line organizations, without isolating them. Line managers participate in the board that governs these safe havens, which are allowed to operate autonomously in separate geographical locations.

Idea Diffusion. A champion is very much needed to preach and broadcast the value of new ideas, and to spearhead the diffusion process throughout a large organization.

Manage and Monitor the Value Chain. One way to constantly cultivate the innovation value chain is to set up performance indicators for each of its three component processes and then monitor the progress made steadily. For idea generation, these indicators may simply include the number of high-quality ideas produced within, across units and from outside the firm. For idea conversion, the idea selection process may be evaluated by the percentage of all ideas generated that end up being selected and funded, whereas the development process could be gauged by the percentage of funded ideas that lead to revenues, and the number of months to first sale. For idea diffusion, the percentage of penetration in desired markets, channels, customer groups, and number of months to full diffusion (Hansen et al. 2007).

There are a lot of "best practices" in innovations available in the literature. But one size does not fit all. The value chain concept and its management allow the proper choices to be made as to which best practices are to be introduced when and where in the organization, so that the overall innovation process is efficaciously advanced.

12.3.2 Innovation Processes

According to Kaplan and Norton (2004), the basic innovation processes comprise the following four processes: (1) identify opportunities for new services, (2) evaluate and prioritize ideas for development, (3) design and develop the new services, and (4) bring the new services to market. These are reviewed next.

Identify Opportunities. The objectives are to anticipate future customer needs and then to discover and design new, more customizable or safer services. The key metrics to measure progress are (1) time spent with key customers learning about their future opportunities and needs, (2) number or percent of new service projects launched, based on client input, (3) number of new concepts presented for development, and (4) number of new value-added services identified.

**Source*: www.brainyquote.com/quotes/quotes/t/thomasaed125362.html.

Innovation opportunities of great value are generally one of the following: (1) ideas that enhance functionality aspects and performance attributes, (2) ideas that shorten the time-to-market requirements, (3) ideas that extend existing service to applications beyond the one initially targeted, and (4) ideas that slash service prices for customers.

Evaluate and Prioritize Ideas for Development. New ideas need to be carefully evaluated and prioritized in view of the respective potential market values, timelines, risk profiles, and resources required for development. The potential market value may be estimated using probabilistic models. The possibilities of joint development should also be considered

Design and Develop the New Services. Countless companies utilize the stage-gate process to guide the development process of products, so that specific objectives are met at predetermined stages. Each gate represents a go/no-go decision in view of the evolution of technologies, customer preferences, competitive actions and governmental regulations. The stage-gate process can also be used to provide a discipline for the development process of services, although it is less formal than in developing products. Figure 12.12 illustrates the stage-gate process that is modified for services.

Evenson and Dubberly (2010) discuss ways to design services in order to create an experience advantage. Ko and Nof (2010) suggest ways to design collaborative e-service systems. Kim and Meiren (2010) review the development process of new services. Partarakis et. al. (2010) suggests specific ways to design and develop web-based services.

In developing services, customer input is sought at all stages. The design/development efforts is successful if the service so developed has the functional features desired by the customer in the marketplace, possesses the acceptable service quality, and is deliverable at a cost which enables satisfactory profit margins to be earned. The more a company understands its customers' specific behavior and needs, the less likely its services innovations can be readily copied by the competition.

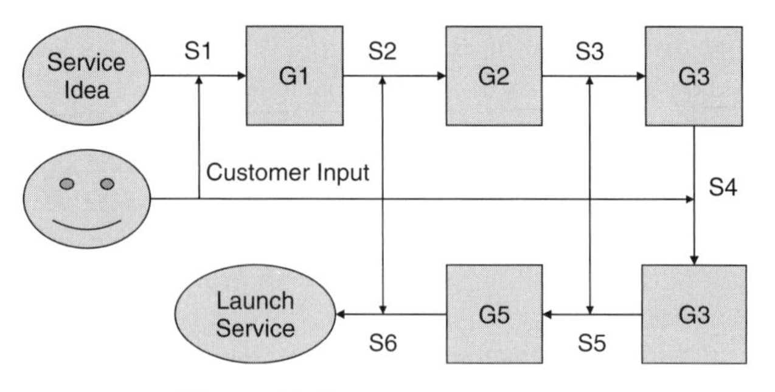

Figure 12.12 Stage-gate process.

Bring the New Service to Market. After the launch of the newer service, companies actively provide after-sale support to assure that the delivered service produces the anticipated value to the targeted customers. This involves communications, problem diagnosis, troubleshooting, and remedial actions, if any.

According the Berry et al. (2006), services are characterized by three factors:

1. Service-providing staff affects the customer experience directly.
2. Service-providing production is decentralized to foster convenience to customers.
3. Service innovations are intangible and, hence, do not usually carry a brand name.

When marketing services, two key dimensions may be used to categorize them: consumption of service, concurrent or separate from its production, and key service value—contents; delivery mode, and cost.

For a service provider to be successful in the marketplace, it must stand out, whether through a unique service benefit, a novel delivery mode, or a low cost. These must be the foci of company innovation drives. Figure 12.13 illustrates the service consumption patterns and benefits.

Examples of superior service contents include the following: Starbucks is known for its unique "store atmosphere" that is enhanced by the physical layout, pleasant employee interactions, and the special coffee drinks and foods it serves. Cirque de Soleil offers a unique combination of highly skilled acrobats and dances, not seen before. Barnes & Noble designs a seating area for customers to browse its books while enjoying a drink.

Examples of achieving differentiation by offering a superior delivery mode are the following: Ball Memorial Hospital redesigned its emergency room reception area to facilitate the rapid processing of patients and to make them more comfortable.

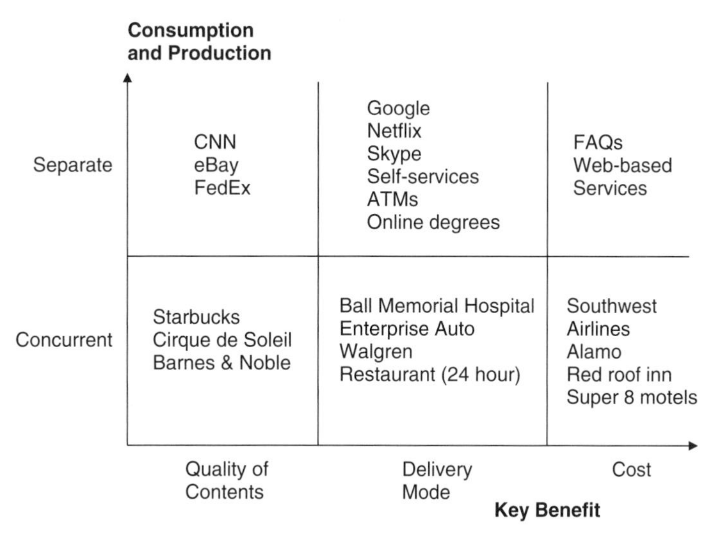

Figure 12.13 Service consumption patterns and benefits. (Adapted and modified from Berry et al. 2006)

Enterprise Auto locates its car rental stores to where local customers are in order to better serve those who have short-term transportation needs due to car repairs. A lot of Walgreens stores are open twenty-four hours a day and are conveniently located at major cross-sections of main streets in a city. Similarly, Denny's and Perkins are known to be open twenty-four hours for the convenience of some night folks. Online degree programs represent the innovation disruption in delivering university education, spearheaded by the University of Phoenix, which became the world's largest university with over 60,000 online students. Online degree programs are known to please customers who would otherwise not able to gain the educational experience.

Other services distinguish themselves by lower prices. Southwest Airlines offers a significant discount in airfare by using a common aircraft type to cut down maintenance cost, flying a limited number of service routes to optimize the utilization of equipment, eliminating seat assignment to speed up its loading operation at the gates, and reducing on-board food services to cut costs. Alamo is a car renting company that is known to have placed its rental offices far away from the airport to save operational costs. Hilton Hotels is known to have offered groups of hotels at lower price ranges in order to broaden the customer niches it can serve.

Christensen and Anthony (2004) point out specific examples in healthcare, education, and law of achieving marketplace acceptance by lowering prices. Specifically, adjusting downward the service features to align with the affordability and needs of lower-end customers segments represents *innovative disruptions.* Such disruptions are possible once the patterns and rules are discovered in various routine services work procedures. Using paralegals to take care of less demanding tasks of a law firm will allow "good enough" legal service to be provided to more people at a lower cost. Producing software to promote self-help in legal matters is another viable option. Nurses can follow the diagnostic procedures specified by physicians. In Minnesota, QuickMedx kiosks are located in a local grocery store and run by a nurse practitioner, who can dispense prescription medication for certain common illnesses within 15 minutes in the absence of a physician.

CNN is known for its programming strategy of offering news highlights on a 24-hour basis. eBay is the primary online market for buying and selling almost anything. FedEx prides itself on being the on-time delivery package service that is extremely dependable.

Google provides a superb delivery of search results in speed and relevance. Netflix initiated a unique way of communicating with and delivering rental DVDs to its customers. The location of bank ATMs enhances customer loyalty by enabling convenient 24-hour access to deposit and withdrawal transactions. Other services offer Web-based frequently asked questions and self-service at no cost to customers, in order to enhance customer experience.

For both concurrent and separate services, companies (1) focus on creating a culture that supports employee performance and continuous innovation, (2) offer a superior service benefit to customers, (3) simplify service designs to raise customer's affordability for services and (4) empower a champion to lead the innovation process in securing a strategic differentiation. Furthermore, for concurrent services, companies (5) emphasize the development of scalable business model in order to enhance productivity, (6) train employees and motivate them to service customers well, and (7) invest in the service "factory" by offering a pleasant customer experience (e.g., physical layout, unique benefit, pleasant employee interactions). For separate services, companies (8) innovate to raise the level of operational excellence, and (9) build brand to overcome uncertainties perceived by customers in selecting "sight-unseen" new innovative services (Berry et. al 2006).

12.3.3 Open and Closed Innovations

How could innovations be promoted by networking? Personal networks are extremely important for innovative pursuits. Companies should foster innovation by creating networks that facilitate the interactions between employees and others with diversified background.

A successful network delivers three advantages: (1) private information, which is not readily available from public sources, (2) diverse skills sets, which greatly compliment one's own knowledge and experience, and (3) information brokers, who are linked to useful networks to access information if needed (Uzzi and Dunlop 2005). For a network to offer these advantages, its members should preferably have diversified knowledge and experience, be able to serve as linkages to other networks of diversified people, and can offer private information when trust has been established.

In general, people tend to form their network by predominantly following the *self-similarity principle*. The self-similarity principles states that people tend to make friends with those who have similar background, work experience, and outlooks. They also should not form networks by the *proximity principles,* thereby becoming closely connected with people who work in nearby offices or neighborhoods. The drawbacks of such self-similarity or proximity networks lie in their lack of diversity in opinions, values, judgments, and world views. Too much similarity restricts one's access to discrepant information.

Instead, people should pursue the *shared activities principle* to form networks by making friends through high-stakes group activities of common interest. Examples of such group activities include community service ventures, interdepartmental initiatives, voluntary associations, cross-functional teams, charitable foundations, for-profit boards, and others. The mutual trusts cultivated among participants in such activities will form a very solid foundation for personal interactions and knowledge interchanges.

Important for innovation is the feedback secured from networks that are composed of trustworthy people with diversified perspectives.

MacCormak and Forbach (2008) point out that companies which are successful in pursing innovation via collaboration pay attention to four Ps:

1. *People*. Select people on soft skills (communications) and train partners.
2. *Processes*. Teams from different cultures have different strengths and working methods, which need to be matched with their assigned tasks. Frequent communications is a must to resolve problems.
3. *Platform*. Autilize the same infrastructure (tools and standards).
4. *Program*. Implement a coherent program to continuously monitor and cultivate collaborations efforts (not as individual projects).

The *creation nets* suggested by Brown and Hagel (2006) foster the notion that when companies look outside their own boundaries, they can gain better access to ideas, knowledge and technology than they would have if they relied solely on their own resources. As knowledge and technology change rapidly, companies should allow a lot of participants from diverse institutional settings to collaborate and build on one another's work to pursue distributed, collaborate and cumulative innovation – all under the guidance of a network organizer –in order to remain competitive. Two specific examples illustrate the effectiveness of the open innovation paradigm.

The Boeing 787 Dreamliner project involves 50 partners at 130 locations. The key for competitive advantages lies in the collaboration of such a large number of partners, so that their individual expertise can be effectively merged to conceive innovative outputs. The Chinese conglomerate Li & Fung Group is known to have engaged multiple and alternating contractors to work together for certain projects and then disband them after the tasks are completed. By tapping into its various contractors, Li & Fund has been able to access and take advantage of the current and diversified technologies and business ideas, to a much greater extent than possible when listening only to direct suppliers.

Example 12.2 In the new millennium, innovative ideas, rather than physical assets, will enable companies to compete efficaciously in a global-knowledge economy. Usually, innovative ideas come from knowledge workers who are inventive, independent, and mobile. No single company is capable of chaining down these workers, as they are happy to be there, but ready to move on at any time. It is likely to be a major challenge for engineering or technology managers to foster innovation on a continued basis in such an environment.

What might be a good strategy for engineering and technology managers to adopt in order to secure a constant flow of innovations potent enough to sustain the relative competitiveness of their employers?

ANSWER 12.2 During the last century, a number of well-known companies (IBM, AT&T, Dupont, GE, Merck, and others) achieved remarkable business success by emphasizing R&D in-house and innovation on the inside. They proudly advertised the number of U.S. patents they received per year as an indication of their inventive power. They kept a large number of experts on their payroll to foster innovations. Many of these giants have since left their historical mission of inventive discovery. Some have also abandoned the past practice of not sharing with others those inventions that did not fit their respective corporate strategies at the time.

Companies in the knowledge economy have implemented a flexible technology strategy with great success. Known examples include Microsoft, Cisco, Dell, and Pfizer. Because skilled workers are mobile, companies can no longer count on in-depth development of innovations on the inside. In order to secure a constant inflow of creative ideas, they pursue outside innovations (e.g., by acquisition, joint venture, or contract research) deemed useful to foster their corporate objectives. The emphasis has been shifted from in-depth innovation within a discipline to innovation with breadth and integration across disciplines.

Proctor & Gamble (P&G) is known to be an aggressive acquirer of creative ideas from the outside. In 2001, 10 percent of P&G products came from outside sources, and this percentage was expected to rise to 50 percent by 2006. P&G also decided to make a patented technology available to outsiders, including competitors, if it is not used by at least one internal business unit within three years (Chesbrough 2003).

To meet the new challenge of creating a constant flow of creative ideas, engineering and technology managers must scan promising innovations on the outside (universities, startups, competitors, and others) and integrate them for profitable internal applications. Any inside innovations that do not conform to the corporate objectives are to be aggressively marketed to outside companies to generate licensing revenues.

A. G. Lafley, CEO of Proctor & Gamble, noted in the company's 2009 Annual Report: "More than half of all product innovation coming from P&G includes at least one major component from an outside partner." The implementation of this unique innovation strategy had been admirably successful over the years.

12.3.4 Categories of Innovation in Practice

There are basically two categories of innovations: breakthrough innovations and incremental innovations. Breakthrough innovations are those which can, out of the gate, produce significant sales revenue and transform entire industries. Examples include search (Google), electronic commerce (eBay and e-markets), on-line education (University of Phoenix), and social networking (Facebook, Twitter, LinkedIn). Incremental innovations are those which represent small but distinct improvements to the existing service offerings. Examples include online services added by Barnes and Nobles, electronic reading of and wireless access to books using Kindle offered by Amazon.com, and the useful package tracking capabilities introduced by FedEx. While breakthrough innovations are highly desirable in producing positive business results, the small incremental innovations are quite valuable as well, if they can be introduced rapidly and frequently.

Some companies are known to be more innovative than others. Business Week and Boston Consulting Group teamed up in defining the most innovative companies in the world (Anonymous 2009). In 2009, the top ten innovative companies were (1) Apple, (2) Goggle, (3) Toyota Motor, (4) Microsoft, (5) Nintendo, (6) IBM, (7) Hewlett-Packard, (8) Research in Motion, (9) Nokia, and (10) Wal-Mart Stores. A number of useful innovation practices have been defined in their combined study.

In 1990s, innovation was about technology and control of quality, cost, and efficiency. Today, it is about creativity and growth. In general there are three types of innovations, as outlined here:

A. *Technology innovations.* Blackberry made by Research in Motion, Ltd. and iPad by Apple.
B. *Business model innovations.* Virginia Group applies hip lifestyle brand to airlines, financial services, and health insurance; Bharti Tele-Ventures, Ltd. outsources everything except marketing and customer management. FedEx is known to have implemented four specific business model innovations:
   1. Branching out into new businesses in order to burrow deeper into the supply chains of customer.
   2. Expanding into fast-growth markets (e.g., China).
   3. Improving package-tracking capabilities and other customer oriented features.
   4. Adopting "here, there, and everywhere" strategy (e.g., 1,450 Kinko's outlets) (Boyle 2007).

Kroger, a traditional grocery chain with annual sales of $76 billion as of the end of 2008 and 6400 stores in thirty-one states expanded its businesses to include credit cards, Kroger-branded ATM machines, mortgage, home equity lines of credit and insurance coverage (identify theft, home, life and pet) in order to "drive more people to the store and bring them back."

Wal-Mart, another grocery chain, added a prepaid Visa debit card and other money services and introduced "MoneyCenter" alcoves in store.

C. *Process innovations.* Southwest Airlines is known for operational improvement, for example.

According to Anonymous (2009), there are four innovation obstacles:

1. Slow development time.
2. Lack of coordination (flexible organizational setting, new leadership positions, diversity of team membership).
3. Lack of culture (rewards, metrics to monitor progress, risk management—Culture is defined by the CEO.)
4. Getting good consumer insights (observing what customers do and understanding the local culture, behaviors).

For companies to achieve success in innovations, these obstacles must be overcome.

12.4 INNOVATION MANAGEMENT

In this section, we will discuss the organizational settings that are conducive to innovations, business dimensions to focus, best practice for managing creativity, and some additional guidelines for managing innovations. Tidd and Hull (2010) discuss the fundamentals of managing service innovations.

12.4.1 Organizational Settings Conducive to Innovations

Brown (2005) argues for an open "pull-type" of organizational form to foster innovation. When demand is well known, the traditional push system (top-down, centrally managed, routine operation) is best in efficiency. However, when future demand is not readily forecast, then the pull system is much more agile. It consists of diverse partners, each having special skills and technology, that can be pulled together for a given new demand. Table 12.3 illustrates the difference between the organizational forms of *push* and *pull* types.

Companies interested in fostering innovation are strongly advised to adapt the pull type of form so that they can increase opportunities for collaboration, build closer relationship with customers, receive more rapid feedback, attain greater scalability and network with deep sources of competitive advantages. Examples of pull platforms in the service industries include Amazon.com, Netflix, Expedia, and the hospital's emergency ward.

Since 2002, John Deere has established 300 communities of practices (CoPs) to drive innovations by facilitating connections among knowledge workers (Suave 2007). In each of these practice-based communities, which cover best practices, training, mentoring, and peer resources, subject matter experts are available via e-mail, face-to-face meeting, or online conferencing.

Table 12.3 Push versus Pull Systems (Brown 2005)

#	Items	Push	Pull
1	Basic assumption	Demand is foreseeable	Demand is not readily forecast
2	Key characteristics	Program defined by manuals and procedures	Platform being open ended Loosely coupled modules
3	Operational style	Rigid	Agile
4	Key requirements	Discipline	Open-mindedness
5	Worker orientation	Closely directed employees following orders	Self-directed networked creators
6	Advantages	Tight control	Emphasis delegation
		Efficiency	Flexibility
7	Disadvantages	Inhibit innovation in closed environments	Foster open innovation with partners

IDEO, in Palo Alto, California, is a product design company known internationally for its innovative skills. The company follows the strategy of observation, brainstorming, prototyping, and implementation to pursue innovations. As pointed out by Kelley (2005), its success in innovation is primarily due to ten capabilities that it systematically nurtures and strengthens:

1. Observe and study customer behavior.
2. Conduct experiments to continuously learn.
3. Bring findings in other industries to the project at hand.
4. Have perseverance to overcome difficulties.
5. Facilitate collaboration among team members.
6. Organize the right set of team members.
7. Design to meet customer's needs.
8. Promote creativity by changing the physical environment.
9. Anticipate customers' needs.
10. Motivate the team.

A clear focus on customers, together with experimentation and internal collaboration represents IDEO's model for innovative success. The company's motto is: "Fail often to succeed early."

SRI International, Palo Alto, California, is also known to have come up with innovative products, such as a compute mouse, HDTV, robotic surgery, and others. The company follows the following five disciplines to innovate (Carlson and Wilmot 2006):

1. *Define important customer needs and value.* Qualify needs by using the *value factor analysis* (see Appendix 12.8.1).

2. *Formulate innovation plan to create value*. Make a specific value proposition, formulate approach, get ideas from brainstorming, and continue to refine by iteration (see Appendix 12.8.2).
3. *Lead by champions.* Involve leaders to pursue various elements concurrently.
4. *Assemble innovations teams.* Select the right people.
5. *Align organization*. Offer support and empowerment.

This model is similar to that practiced at IDEO in that it also emphasizes understanding the important needs of customers, brainstorming ideas, creating experiments or prototypes, iterating continuously to achieve superior value for customers, and sharing rewards and recognition between all team members, while securing the full commitment of their organizations.

Example 12.3 It is known that managers have a direct influence on the outcome of the company's innovation program. How can company management effectively foster innovation?

ANSWER 12.3 There are quite a few actions that company management could take to promote the cross-disciplinary collaboration of people with diversified skills, knowledge, and experience and thus enhance the probability of success in the company's innovation programs.

As indicated in Mendonca et al. (2007), Bill Campbell offers eleven tips to managers in order to foster innovations:

1. Empower engineers.
2. Establish innovation culture.
3. Lead by people who care about building durability and lasting value.
4. Focus on things that are really differentiable.
5. Support innovations (fellowship, awards, bonuses, extra vacations, etc.).
6. Be technology centered.
7. Involve marketing to understand the problems.
8. Set high expectations.
9. Take risks and anticipate some failures.
10. Hire creative people.
11. Screen ideas and projects from the viewpoint of a venture capitalist.

12.4.2 Business Dimensions to Focus

Sawhney et al. (2006) point out that there are a number of business dimensions along which innovation could bring about corporate competitiveness. Neglecting some of these dimensions may cause companies to miss valuable growth opportunities:

- *Service offering*—the functional features
- *Platform*—modular components of the service, which can be used to bring forth related new services
- *Solutions*—the real value customers get

- *Customers*—future needs
- *Customer experience*—the pleasure conveyed to customers by the interactions
- *Value capture*—profitability to service provider
- *Processes*—streaming the production process to be cost-efficient
- *Organization*—changing organization to augment customer supports
- *Supply chains*—engaging suppliers to upgrade offering
- *Presence*—changing the touch points to enrich customer experience
- *Networking*—intelligent offerings via networks
- *Brand*—creating brand to reduce uncertainties perceived by customers

From the customers' perspectives, there are nine value dimensions that are important. These are features, price, delivery, customer experience, convenience, reliability, risk, cycle time and quality (Fig. 12.14). Customer value is enhanced by having achieved satisfaction along one or more of these dimensions. Innovations along these nine dimensions tend to bring about competitiveness in the marketplace efficaciously. Figure 12.14 illustrates these nine dimensions and the foci of innovations that are needed to achieve each.

It is not necessary that companies must pursue innovations in all these dimensions in order to succeed in the marketplace. Oftentimes, it might be sufficient to focus on one or two dimensions that have been neglected by the competition. The following examples demonstrate the effectiveness of this innovation approach:

- *JetBlue Airways* offers a better customer experience by having live satellite television, leather seats and fashionably clad flight attendants.

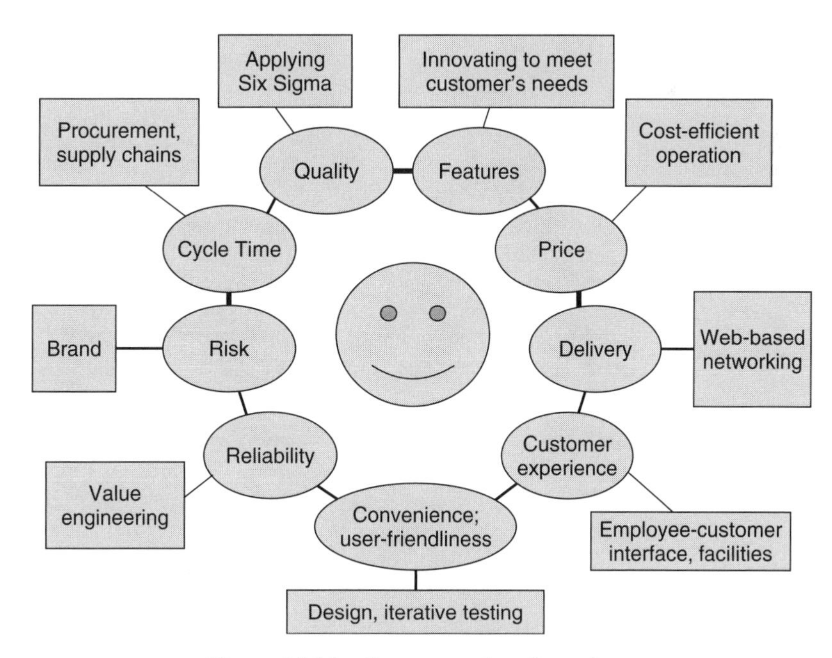

Figure 12.14 Customer value dimensions.

- *Kaiser Permanente* offers comfortable waiting rooms, lobbies with clear directions, and large exam rooms with three or more people and curtains for privacy to cultivate customer experience (Nussbaum 2004).
- *Cemex* shortened the time window from three hours to twenty minutes for delivering ready-to-pour concrete by utilizing GPS systems and computer in its fleet of trucks.
- *MinuteClinic* locates its kiosks in stores such as Target and CVS to offer a menu of service to diagnose about two dozen straightforward ailments, including strep throat and pinkeye. The nurse practitioners who staff the kiosks can reliably diagnose the conditions in less than fifteen minutes and write a prescription that customers can fill at the in-store pharmacy. The value dimensions emphasized here are customer convenience and speed.
- *Starbucks* offers a pleasant environment for conversion, reading or relaxation, even though the coffee is selling at $4.00 a cup.
- *The Home Depot Inc.* focuses on the needs of "do-it-yourself" customers.

When innovating along the value dimension of service features, it is important to understand the real needs of customers. Services incorporating excessive performance features overshoot the customers' needs, while demanding high price, could lead to poor marketplace performance. Scaling down the service features oftentimes reduce price and meet a greater majority of lower-strata customers. Intuit was successful in simplifying its Quicken product to better fit the needs of small business customers and scored a great success in the marketplace. Figure 12.15 displays the possibilities of overshooting the market and undercutting the main players in the marketplace by being "disruptive."

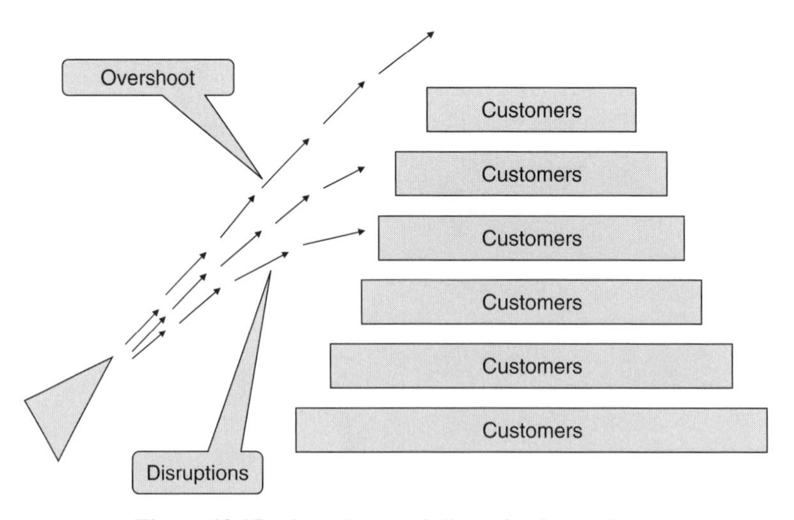

Figure 12.15 Overshoot and disruptive innovations.

12.4.3 Best Practices in Managing Creative People

Companies have numerous types of resources, such as raw materials, logistic systems, and political influence. One of the most valuable resources is creative capital. Creative capital is embodied in creative thinkers, who pioneer new technologies, give birth to new industries, and power economic growth, and foster relationships between these thinkers and customers, suppliers, and sales and support staff. This creative class is said to make up about one-third of the U.S. workforce (Florida et al. 2005).

Over the years, SAS Institute, a software development company located in Cary, North Carolina, was able to advance three principles for managing such a creative class.

A. Help employees do their best work by keeping them intellectually engaged and by removing distractions. Employees' minds are stimulated by increasing challenging work, attendance at professional conferences, opportunities to author and publish journal articles, the use of updated tools, and the financial support for additional training. Hassles are minimized based on employee surveys. Various on-site benefits are made available (day care center, basketball court, exercise room, college advisement services, home care aids, dry cleaning, haircut and auto detailing, and others). Management should allow a flexible time schedule for employees.

B. Make managers responsible for sparking creativity and eliminate arbitrary distinctions between "managers" and "technologists.". To spark creativity, managers post a lot of questions, bring groups of people together to facilitate the exchange of ideas and spur innovations, and procure the materials and tools needed by employees. Since managers also do hands-on work, they earn the respect of technologists. There is no penalty for making honest mistakes, as experimentation is crucial for breakthroughs.

Example 12.4 Managers need to encourage creativity. However, a variety of companies take steps that unintentionally kill creativity. Name a few of such practices that kill employee creativity.

ANSWER 12.4 Creativity is known to be affected by three components: (1) expertise, (2) creative thinking skills, and (3) motivation. Expertise and creative thinking skills are, by and large, a part of individual capabilities, which are not readily influenced by management practices. There are two kinds of motivations: extrinsic and intrinsic. Managerial practices have an impact on both. There are at least seven ways management could inadvertently de-motivate individuals to become creative (Amabile 1998):

1. Do not match people with the right assignments that utilize their expertise and skills, and stretch their abilities.
2. Fail to define project goals clearly and change them often. Leave no freedom for the individual to select methods of accomplishing specific project objectives. Autonomy around work process fosters creativity and enhances the intrinsic motivation.
3. Impose fake deadline or impossibly tight ones and constrain resources needed to pursue creativity.

4. Set up teams with members who do not have diversified experience and expertise, do not collaborate well, and do not build trust among themselves.
5. Lack positive managerial encouragement (e.g., remaining silent when outcome is good, taking long time to evaluate outcomes, creating an atmosphere of fear by criticizing failures, etc.).
6. Demonstrate no role models to promote all three creativity components.
7. Do not fostering an organizationwide policy of information sharing and employee collaboration.

C. Engaging Customers as Creative Partners to the company can deliver superior products/services. The company tracks and organizes customer inputs and suggestions that are obtained though its Web site and phone lines, its annual Web-based SASware Ballot, and its annual user's conference. SAS software developers, consultants, and technical support staff interact with customers, learn their future needs and invent new solutions.

Separately, Selden and MacMillan (2006) emphasize the customer-centric research to promote innovations based on knowing who the customers are and what they want, by establishing a deep relationship with core customers, and then extending the number of customers beyond the core to new ones. In so doing, companies gain knowledge that is often opaque to competitors, promote employees loyalty and diminish turnover, and attain innovations that add true value to the companies.

Example 12.5 In today's highly competitive marketplace, companies are striving to be customer-focused. In addition, many of them pursue open innovation strategies in order to offer services that are useful to future customers. If you were in charge of turning customer input into innovation, how would you proceed?

ANSWER 12.5 Soliciting input from customers to foster innovation is indeed important for those companies that want to benefit from the customers' understanding of their future needs as well as their insights regarding new ways to satisfy such needs.

Simply asking the customers what they would like to see in new services may not work. This is because most customers have a limited frame of reference and they only know what they have experienced. They are not in a position to externalize what they do not know about emergent technologies, new materials, and the like. Furthermore, some customers may merely recite the missing features that other service providers already offer. Following such advice will lead to incremental development of "me-too" services. The service ideas of "lead users," those who have an advanced understanding of a service and are experts in its use, may be of limited general appeal, because they are not the average users. Customers may, in fact, not always welcome "new and improved" service features and functionalities. It is well-known that vendors continue to upgrade the versions of their software with new features that most customers do not need but are forced to buy. Industrial survey indicates that customers typically use less than 10 percent of the software's overall capability. Innovation along this strategy creates resentment among customers.

However, there is a methodology for capturing customer input that could be conducive to corporate innovation (Ulwick 2002):

1. *Plan outcome-based customer interviews.* The company should lay out the underlying process or activities associated with a service (e.g., a specific medical procedure) and hold interviews with a diversified set of customers who can judge the value of the service from the cost standpoint.
2. *Capture desired outcome.* Appoint a moderator who could focus the customers on the underlying processes, not on solutions of a given service. Encourage customers to specify the improvements needed for each step in the process, the reasons for wanting such improvements, and the respective quantitative measures (time, number, frequency), so that the outcome statements can be used later for benchmarking, competitive analysis, and concept evaluation. In general, it may take three 2-hour sessions to complete such outcome-based interviews.
3. *Process the outcome.* The company organizes the outcomes and urges customers to rate them for relative importance and satisfaction.
4. *Utillize the outcome to initiate planning and innovation activities.* The outcome may be used to jump-start innovation by identify new areas for service development, updating market segmentation, and defining desirable competitive position for the company.

The key novelty of this approach is to query customers for the new value or benefit they like to seek (not to name specific solutions). The company then develops solutions in specific forms that offer the desirable value to these customers.

12.4.4 Additional Guidelines for Managing Innovations

Three additional guidelines are offered below to facilitate the management of innovations in established companies, develop new services by focusing on service complements, and target innovation for the emerging markets.

Strategic Innovations for Established Companies. Established companies are known to encounter problems in pursuing breakthrough innovations. These problems are typically related to company culture, business practices, and management mindsets.

Built on the business case of Analog Devices, Inc., Boston, Massachusetts, Govindarajan and Trimble (2005) suggest three specific guidelines for established companies to succeed in bringing new innovations to market: (1) forget, (2) borrow, and (3) learn. Companies need to forget the old, albeit successful, ways of doing business, to borrow from within the needed capabilities and resources to support the new innovative ventures, and to learn new ways to overcome uncertainties through careful planning.

Service Complements. When managing the development of a new service innovation, it is also important to assess the status of the critical complements that are required to bring value to the marketplace (Adner 2006). Offering a Ferrari in a world without gasoline or highways will produce no marketplace success. Besides assessing the uncertainties of developing the target service itself, companies must coordinate with complementary innovators and align all value chain partners to deliver the expected service performance in time.

Emerging Markets. When pursuing the emerging markets, service companies need to promote the effectiveness of their strategic innovation by closely tailoring services to local needs. To accomplish this challenging objective, Anderson and Markides (2007) advocate the use of the four-A framework: Affordability, Acceptability, Availability and Awareness.

- Affordability – Make sure that the service offerings can be afforded by the poorest consumers in the regions.
- Acceptabiltiy – Align with the needs of local consumers and distributors so that the services are readily accepted locally.
- Avaialbiltiy – Conceive ways to distribute services to the most isolated communities.
- Awareness – Use alternative communication means and modes to build service awareness in hard-to-reach consumers.

12.4.5 Protection of Inventions and Innovations

Inventions and innovations need to be properly protected. They form the valuable intellectual properties that strengthen the relative competitiveness of service companies. Intellectual properties, including innovative ideas, may be protected by (1) trade secret, (2) copyrights, (3) trademarks, and (4) patents.

Trade secrets are valuable insights, work procedures, methodologies, know-how, or design that the inventors keep secret in order to realize competitive advantages. *Copyrights* protect the works of authorships (e.g., text, drawing, music, video, etc.) fixed on a tangible medium (e.g., paper, CD, etc.) and are in effect for the duration of author's life plus seventy years. *Trademarks* are graphic symbols that indicate the source of goods or services. They remain in effect as long as they are being used by trademarks owners.

Patents are contracts between the inventor and the government. Inventors are required to disclose their novel ideas in details (e.g., composition of matter, process, products, or improvements) so that the society at large will benefit from such novel and nonobvious ideas in the long run, in exchange for the right to exclude others from making, using, selling, offering for sale, and importing the invention into the United States for twenty years, measured from the date of patent filing.

12.4.6 Patents

Filing for a patent application is usually proceeded by an evaluation of (1) IP (intellectual property) strategy, (2) novelty—patentability, and (3) commercial value. Inventors (and their employers) may pursue an offensive, defensive, or spoiler IP strategy. An offensive strategy is one in which patents are sought to carve out a new service domain for generating revenues from a not-yet-served market segment. A defensive strategy, by contrast, focuses on protecting the inventors' own core competencies from being encroached by competitors. The spoiler strategy is one in which patents are filed to prevent the competitors from seeking patent protection for certain novel matter, process, or design, even though these areas are of no direct commercial interest to the inventors.

The issue of novelty is typically addressed via a prior art search of the patentable ideas involved. Internet sources such as USPTO and EPO are quite convenient for inventors to access. To be patentable, ideas must be novel, nonobvious, and useful. To be nonobvious, the ideas should not have been obvious to persons having ordinary skills in the art. Visualizing the circumstances in which implementing the ideas at hand could generate unexpected failures is a good way to demonstrate the idea's nature of being nonobvious. Commercial value is generally a very important decision factor for filing a patent, as related to the alignment with corporate strategy, compatibility with current service offerings, and other such considerations.

A patent is a written document that consists of the following sections: (1) background of the invention, (2) summary of the invention, (3) brief description of the drawings, (4) description of the invention, (5) claims, and (6) abstract. Inventors supply materials, data, and information, which are then absorbed into all sections. Claims are usually prepared by patent lawyers.

Claims represent the most important part of the patent. Using legal language, claims describe the specifics that inventors could enforce to exclude others from practicing the inventive contents comprising of composition, process, product design or improvements. Inventors try to prepare claims in as broad a scope as possible, whereas U.S. patent examiners strive to whittle it down as narrow as possible.

A formal patent application represents a costly and time-consuming endeavor. As an alternative, inventors may first elect to file a provisionary patent with the U.S. Patent Office in order to retain the right to file a formal and detailed patent within one year of the original filing date, the earliest priority date. Provisional patents are useful to some inventors who wish to preserve the priority date so that they may negotiate with potential users and explore its commercial value of the invention during this one-year period of time, and then decide if a formal patent application should be filed.

To facilitate the filing of patents in foreign countries, inventors may elect to file a PCT registration within one year of the original priority date. They can then follow up with the filing of actual patents in any of the 139 signatory countries that participate in this system. The PCT system does not offer patents, as patents must be granted by the individual countries or regions (such as EU).

Public Disclosures of Potentially Novel Ideas. Inventors must pay special attention to the rules regarding public disclosure. Public disclosure is the practice of one of the followings: filing a provisionary patent application, publishing papers in conferences, defending thesis in open sessions, discussing ideas in departmental meetings, submitting abstracts of grant proposals to funding agencies, describing ideas in Web pages, and claiming inventorship of the patentable ideas by other means. If there has been no public disclosure of the patentable ideas before filing a U.S. patent, then the inventor has one year of time to file patent applications for protecting the same ideas in foreign countries. However, if a public disclosure has been made prior to filing a U.S. patent, then the inventor can only file foreign applications to protect the same ideas within one year of the public disclosure. It is thus very important to not disclose potentially patentable ideas to anyone carelessly.

Documentation and Inventorship. There are strict rules for establishing the concept of the patentable ideas. It is generally recommended that bound notebooks with continuously numbered pages are used to document the novel ideas conceived, the date of idea conception, the names of inventors involved. This page must be signed by all inventors and co-signed by a witness in order to be legally acceptable.

Whether all inventors that sign such pages are a part of the eventual inventorship of a patent depends ultimately on the claims the patent contains. Inventors whose ideas form the basis of at least one of the claims in the patent are co-inventors. All co-inventors have the same rights to make use of a granted patent (e.g. selling, enforcing, offering license, creating joint ventures, signing co-development contracts, etc.), without the consent of the remainder co-inventors, unless the rights issue is clarified and agreed to by all co-inventors before the patent applications.

12.5 SELECTED INNOVATION PRACTICES IN THE SERVICE SECTORS

Service sectors could use many more innovations than what is available today. The management of service innovations is somewhat difficult, because the service sectors are so divergent in business contents and types. In the following, a number of specific practices are described in which a selected group of service companies have been successful in fostering the innovation of their service offerings. A close study of such practices could inspire new innovation approaches which may be useful to other service companies.

12.5.1 Innovation in Communications, Financial, and Technical Business Services

Based on a study of Canadian service industries in 1998, several innovation strategies are common among three services industries, communications, financial services, and technical business services: (1) focusing on existing customers and service quality to maintain or increase market share, (2) treating customers as the key sources of innovation ideas, (3) emphasizing the use of copyrights and trademarks, instead of patents, to protect intellectual properties, and (4) deemphasizing the drive to raise productivity or employee skill level (Gellatly and Peters 1999).

Companies in the service industries face different pressures from the marketplace. As a consequence, they tend to drive their innovation processes differently. Innovation is pursued for different reasons, in different ways, and to meet different industry-specific objectives:

- The communication industry perceives as its major threats the advancements in production technologies and regulatory constraints. They orientate their innovation strategy toward the use of high-quality suppliers and the acquisition of new technologies. They focus on improving service quality and reliability. They get innovation ideas from suppliers and the availability of new technologies (e.g., conversion to fiber-optics and digital-based technologies). They regard legislative restrictions (e.g., restricting the nature and contents of services) as an innovation obstacle to overcome. In this industry, competition is viewed as less intense than in others.
- The financial services industry regards as a major uncertainty the service substitutability by consumers and price competition. They understand the need of

customizing service and remaining price-competitive. Thus, they focus on treating their employees well (compensation plans, skills development) and hiring qualified staff. As a result, their innovation strategies are focused on reducing labor costs, speeding up delivery, protecting know-how by trademarks, and getting innovation ideas from the competitors.

- The technical business services industry faces uncertainties in predicting the future actions by customers and competitors and dealing with product obsolescence. They need to emphasize service quality and customer experience. They are typically oriented toward technology development and refinement through in-house R&D. Their innovation drives tend to concentrate on service flexibility to diversified customers, as well as service quality, adaptability, user friendliness, speed of delivery, and accessibility. In-house R&D is a major source of ideas for innovation, which is typically protected by patents. Companies in this industry may face some major hurtles related to financial and manpower resources.

  When planning for innovation strategies, it is thus important to consider the main market pressure that the industry is under in order to maximize the benefits derivable from the innovative efforts.

12.5.2 Innovations in the Insurance Industry

The insurance sector is generally known to be a slow adopter of new innovations, such as the Internet and wireless data communication devices. This is partly due to its business nature of being low volume and low frequency. However, development of IT technologies, globalization, customer's demand for transparency leading to insurance products commoditization, and industrial consolidation will induce more changes in the future. Besides cost control, the insurance sector is expected to embrace new business designs and to promote offerings, marketing ability, product distributions, and risk assessment.

Examples of some of these new innovations in the insurance sector include the following:

- *Progressive Insurance*, Mayfield Village, Ohio, introduced the *Immediate Response Claims Service*, which was later revised to become Immediate Response Vehicles, to enable an adjuster to write a check on the spot of an auto accident, thus lower the cost and enrich a positive customer experience. Oftentimes, the agent was able to impress the other party in the auto collision so much that he or she became a Progressive customer.
- *Progressive* launched the first insurance Web site to offer online policy purchasing in real time and to pilot "usage based" insurance via GPS and cellular data modems. Customers pay insurance premiums that are based on the actual behavior of the driver.
- *IBM* developed Components of a new Insurance Application Architecture (IAA) to promote the creation of the next generation of insurance applications. These components make use of the service-oriented architecture to connect hardware and software elements into an underlying *Enterprise Service Bus*, which fosters the development of new insurance business design, strategies, and opportunities.

- *Jefferson Pilot Financial*, Greensboro, North Carolina, and Grange Mutual, Columbus, Ohio, installed new Web portals for its agents and broker customers in order to support their distribution channels and to decrease barrier of using the systems.
- *Nationwide Insurance, Columbus, Ohio,* connected several data sources to allow a single view of its customers and thereby enriched customer experience and minimized processing errors.

12.5.3 Innovations in the Food Industry

The food industry is increasingly adopting new technologies to slash cost, streamline operations, and raise customer satisfaction (Steward et al. 2002). Wal-Mart pioneered the practices of linking across the supply chain and using information technology to respond more promptly to the marketplace. The company became the nation's largest food retailer in 2001.

Several examples of innovations in the food industry are outlined as follows:

A. *Efficient Consumer Response* (ECR). ECR was launched by the grocery retails and industry trade associations in 1992 to do the following:

   1. Manage the mix of products on retail store shelves and increase sales and product turnover.

   2. Decrease out-of-stocks percentages.

   3. Eliminate inefficiencies between supply chain partners and minimize problems leading to inefficient trade promotions.

   4. Increase the success rate of new food products.

B. *Efficient Foodservice Response (EFR).* EFR was establishes by the food service industry in 1996 to promote the use of standard product identification codes. Bar coding produces 1 error in 3 million, whereas manual key entries register 1 in 300. Tyson Foods bar codes nearly 100 percent of its 4,000 products.

C. *eFS Network.* The eFS Network was created in July 2000 by the food industry leaders such as McDonald's, Sysco, Cargill, and Tyson Foods to increase an Internet-based, industrywide marketplace for food service companies.

   To induce sales, stores devise special tactics to cause shoppers to stay longer and walk more slowly through the aisles. They use smells and sights to trigger shoppers' appetite. Here are eight of the most well-known tactics the stores use:

   1. Place candy and magazines near the checkout counters so that bored, cranky, and hungry shoppers are tempted to grab them.

   2. Breakfast cereals are located directly across from the candy section so that kids can check out the candy while Mom is filling up the cart with cereal.

   3. Sugary cereals for kids are always on the lower shelves, at a child's eye level and within reach of little arms.

   4. Most folks assume that the items located at the end of the aisles where shoppers do their U-turns into the next aisle are full of sales items. This is usually not the case. That very valuable real estate is full of special items, as suppliers pay the stores more to have their merchandise put there.

5. Move items regularly throughout the stores so shoppers walk around longer and buy more stuff.
6. New products are placed near the top selling items in order to encourage shoppers to try them.
7. Keep the shoppers staying longer in stores by offering services such as a coffee shop, cafeteria-style buffet, pharmacy, photo processing, and others, while playing slow music.
8. Attract all of shoppers' senses by placing cookies with mouthwatering smell and rotisserie chicks at the store's entrance section. Use food samples to trigger shoppers' appetite.

Studies have shown that as a consequence of implementing such insightful tactics, shoppers usually buy 40 percent more than what they set out to buy.

12.5.4 Major Hurdles to Innovations in Healthcare

According to US Government (Anonymous 2008), the costs of U.S. healthcare are around $2.4 trillion (17 percent of GDP) per year in 2008, among the highest in the world. Hospital activities accounted for $400 billion of the excessive costs. About 300,000 people were killed by hospital "medical errors" in the past few years (2004 – 2008). Third-party pay providers control reimbursement and favor physicians who follow "recipes" that are established by the innovation-killing peer-review processes.

Duke University Medical Center initiated an innovative program for people with congestive heart failure. In only one year, the program cut costs by 40 percent and resulted in substantially improved patients' health that hospital visits and usage plummeted. As a consequence, the hospital's income dropped accordingly, as the providers (insurance companies) pay only for treating sick people. The third-party pay providers in our current system do not reward innovations in health care (Herzinger 2007).

Entrepreneurs avoid healthcare delivery because providers (governments, insurance companies) set prices, specify procedures, and define the types of patients to be covered.

To change the system in favor of innovation, we need a consumer-driven health care system in which consumers can reward new and innovative developments that improve health.

Example 12.6 Innovations are of fundamental importance to any company. However, innovations entail risks. Management must be prepared to anticipate risks and devise ways to mitigate them. What kind of strategies should company consider to manage risks related to their innovative pursuits?

ANSWER 12.6 In general, management needs to follow the following general principles to pursue service innovations:

1. Define a clear challenge in terms of a customer need (not a business need) that is worthy satisfying.
2. Identify a process to pursue innovations that include multidisciplinary participation and sources of cutting-edge ideas.

3. **3.** Favor concepts that combine multiple elements of innovations (e.g., business model, IT platform, and channel) to increase impact and distinctiveness.
4. **4.** Utilize techniques and structures that counterbalance the forces of risk aversion. Uncertainties are inherent in service innovations. Rae (2005) suggests a few best practices to mitigate them:
   - Remove the naysayers in leadership positions.
   - Make go/no go decision by determining if the company can survive the worst-case outcome.
   - Make exploratory funds available for trying new concepts.
   - Motivate innovation team by inducing competition.
   - Set up special environments for allowing exploration of new opportunities far afield from the main business line (e.g., IBM's Emerging Business Opportunities).

12.5.5 Innovation in Car Sharing Services

A new car sharing service, Zipcar, was praised by Keegan (2009) as the best new idea in business. Analogues to the time-sharing concept practiced in rental properties business, the car sharing model is enabled by three technologies: high-speed Internet, mobile broadband and global positioning systems (GPS).

Customers sign up by paying a basic annual membership fee of $50. They make reservations (at Zipcar website or via a wireless iPhone application), go to unlock the car parked in a GPS-directed neighborhood, and drive away. If needed, they fill up free using a special charge card available in the car. They return the car and sign off. Billing is done automatically. A daily or hourly rate is charged (including insurance, gas, maintenance and 24/7 roadside assistance) for using the car up to 180 miles per day and customers may modify the reservation details, if there are delays or extensions. The intended customers include upwardly mobile professionals, who are technically savvy in the use of smartphones, computers, social networks and other digital devices and services.

The American Automobile Association (AAA) estimates that owning a car costs about $8000 a year. Customers of Zipcar claim that they save an average of $600 per month. This is partially due to a reduction of miles driven by about 44%, as paying by hours provides a strong incentive for customers to cut back on driving. Furthermore, each shared care takes up to 20 cars off the road. As a consequence, the company estimated that there will be a reduction of about 50% of Carbon Dioxide emission.

As of September 2009, Zipcar claimed to have 325,000 individual members and 8,500 corporate members. It's car sharing program is active in a number of major cities in the US, Canada and England

12.5.6 Innovations in the Airline Industry

Jetblue Airways offered passengers in 2009 an "all-you-can-jet" pass for $599 in which passengers can book an unlimited amount of flights between any of its 56 destinations within a one-month span. Flights must be booked within three days of the departure date. Taxes and fees are included. This service innovation is analogous to the business models of (1) all you can eat (cafeterias), (2) all you can exercise (fitness centers), (3) all you can drive (rental cars), (4) all you can learn (online education, such as www.eduFire.com), (5) all you can read anywhere (global wireless reading service, such as Amazon.com's Kindle), and (6) all you can view (DVD rentals). It is much better than the one offered by Air Canada in 2007, in which passengers can buy an unlimited flight pass starting at $1667 per month.

12.6 CONCLUSIONS

In this chapter, we first introduced the creative process and several creative thinking strategies to promote the generation of new ideas that precede innovations. We then elaborated on the fundamentals in innovations (value chains, processes, keys), and the strategies and practices of managing innovations (organization, business dimensions, best practices, examples, guidelines), followed by innovation examples in service sectors.

The examples of service innovations reviewed in this chapter represent the initial results of a long innovation drive of the service industry. The specific cases discussed illustrate the various new opportunities that remain open for additional innovations in technology, business model and/or process. Much more is yet to come.

The objective of this chapter is to lay the basic foundation that should enable students to start cultivating innovations. The sooner they acquire the essential skills and understanding relation to innovations, the sooner they will become productive in making creative contributions to the service sectors. There is a significant opportunity for those who are willing and technologically able to innovate in the service industry.

12.7 REFERENCES

Adner, Ron. 2006. "Match Your Innovation Strategy to Your Innovation Ecosystem." *Harvard Business Review* (April).

Amabile, Teresa M. 1998. "How to Kill Creativity." *Harvard Business Review* (September–October).

Anderson, Jamie and Costas Markides. 2007. "Strategic Innovation at the Base of the Pyramid," *MIT Sloan Management Review*, 49(1).

Anonymous. 2008 "National Healthcare Expenditure Projections 2008-2018," Center of Medicare and Medicaid Services, US Department of Human and Health Services, Washington, D. C. www.cms.hhs.gov/NationalHealthExpendData/downloads/proj2008.pdf.

Anonymous. 2009, "The 25 Most Innovative Companies," Business Week (April 20), p. 46.

Berry, Leonard L., Venkatesh Shankar, Janet Turner Parish, Susan Cadwallader, and Thomas Dotzel 2006. "Creating New Markets through Service Innovations," *MIT Sloan Management Review* 47(2): 56–63.

Biolos, Jim. 1996. "Six Steps Toward Making a Team Innovative." *Harvard Management Update* (August).

Boyle, Matthew. 2007 "Kinko's Chief: Willing to Take Short Term Pain," Fortune 500 (January 18).

Brockbank, Anne, Ian McGill, and Nic Beech (eds.) 2002. *Reflective Learning in Practice*. Burlington, VT: Gower Publishing Company.

Brown, John Seely 2005. "The Next Frontier of Innovation." *McKinsey Quarterly* (3).

Brown, John Seely and John Hagel III. 2006. "Creation Nets: Getting the Most from Open Innovation," *McKinsey Quarterly*, Number 2.

Carlson, Curtis R., and Williams W. Wilmot. 2006. *Innovation: The Five Disciplines for Creating What Customers Want*. New York: Crown Business.

Chang, C. M. 2006. "Deep Learning to Acquire Personal Tacit Knowledge: The Wellspring of Innovation." Proceedings of IAMOT 2006 Conference, Beijing, China (May 3–7, 2006).

Chang, C. M. 2008A. "Engaging the Creative Minds: The ENGAGE Models." *International Journal of Innovation and Technology Management* 5 (1): 1–17.

Chang, C. M. 2008B. "Collaborative, Heuristic and Normatively Guided Techniques to Creativity." Paper presented at and published in the Proceedings of the PICMET (Portland International Conference on Management of Engineering and Technology) Conference, Cape Town, South Africa (July 27–31, 2008).

Chang, C. M. 2010. "Developing Novel and Marketable Service Ideas." *International Journal of Innovation and Technology Management*, (Accepted for publication).

Chesbrough, Henry 2003. *Open Innovation: The New Imperative for Creating and Profiting from Technology*. Boston: Harvard University School Press.

Christensen, Clayton 1997. *The Innovator's Dilemma: When New Technologies Cause Great Firms to Fail*. Boston: Harvard Business School Press.

Christensen, Clayton M., and Scott D. Anthony. 2004. "Cheaper, Faster, Easier: Disruption in the Service Sector." *Harvard Business Review, Strategy & Innovation Newsletter* (January–February).

DeBono, Edward. 1971. *Lateral Thinking for Management: A Handbook of Creativity*. New York: American Management Association.

Evenson, S. and H. Dubberly. 2010. "Designing for Service – Creating an Experience Advantage," Chapter 19 in Salvendy, Gavriel and Waldemar Karwowski (Eds), *Introduction to Service Engineering*, John Wiley (January).

Ferguson, Glover, Sanjay Mathur, and Baiju Shah 2005. "Evolving from Information to Insight." *MIT Sloan Management Review* 46 (2).

Florida, Richard, and Jim Goodnight. 2005. "Managing for Creativity." *Harvard Business Review* (July–August).

Frei, Frances X. (2008), "New Service Design Exercise," Harvard Business School case #9-605-053, Boston.

Gellatly, Guy, and Valerie Peters. 1999. "Understanding the Innovation Process: Innovation in Dynamic Service Industries." Research Paper No. 127. Analytical Studies Branch, Ottawa: Statistics Canada.

Govindarajan, Vijay, and Chris Trimble. 2005. *Ten Rules for Strategic Innovations: From Idea to Execution*. Boston: Harvard Business School Press.

Hamel, G. 2002. *Leading the Revolution*. New York: Plume Books.

Hansen, Morten T., and Julian Birkinshaw 2007. "The Innovation Value Chain." *Harvard Business Review* (June): 121–130

Hargadon Andrew, and Robert I. Sutton. 2000. "Building an Innovation Factory." *Harvard Business Review* (May–June).

Hargadon, A. 2003. *How Breakthroughs Happen: The Surprising Truth about How Companies Innovate*. Boston: Harvard Business School Press.

Harvard Business School Press 2003A. Idea Generation: Opening the Genie's Bottle." *Chapter 3 in "Managing Creativity and Innovation*. Boston: Harvard Business School Publishing.

Harvard Business School Press 2003B. "Creativity and Creative Groups; Two Keys to Innovation." Chapter 3 in *Managing Creativity and Innovation*. Boston: Harvard Business School Publishing.

Herstatt, Cornelius, and Katharina Kalogerakis. 2005. "How to Use Analogies for Breakthrough Innovations." *International Journal of Innovation and Technology Management* 2 (3).

Herzlinger, Regina F. 2007. "Where Are the Innovators in Health Care?" *Wall Street Journal* (eastern ed.) (July19), p. A15.

Huston, Larry, and Nabil Sakkab. 2006. "Connect and Develop: Inside Proctor & Gamble's New Model for Innovation." *Harvard Business Review* (March): 1–8.

Kaplan, Robert S., and David P. Norton. 2004. "Innovation Processes." Chapter 6 in *Strategy Maps: Converting Intangible Assets into Tangible Outcomes*. Boston: Harvard Business School Press.

Keegan, Paul. 2009. "The Best New Idea in Business," *Fortune*, 160 (5), p. 42.

Kelley, Tom. 2005. *The Ten Faces of Innovations: IDEO's Strategies for Beating the Devils" Advocates and Driving Creativity Throughout your Organization*. New York: Currency-Doubleday.

Kepner, C. H., and B. B. Tregoe. 1981. The New Rational Manager. Princeton, NJ: Princeton Research Press.

Kim, K. J. and T. Meiren. 2010. "New Service Development Process," Chapter 12 in Salvendy, Gavriel and Waldemar Karwowski (Eds), *Introduction to Service Engineering*, John Wiley (January).

Kim, W. C., and R. Mauborgne. 1997. "Value Innovation: The Strategic Logic of High Growth." *Harvard Business Review* (January–February): 91–101.

Ko, H. S. and S. Y, Nof. 2010. "Design of Collaborative e-Service Systems," Chapter 11 in Salvendy, Gavriel and Waldemar Karwowski (Eds), *Introduction to Service Engineering*, John Wiley (January).

Leonard, Dorothy. 2002. "The Limitations of Listening." *Harvard Business Review* (January): 93.

Leonard, Dorothy, and Walter Swap. 2004. *Deep Smarts: How to Cultivate and Transfer Enduring Business Wisdom*. Boston: Harvard Business School Press.

Levitt, Theodore. 2002. "Creativity is Not Enough." *Harvard Business Review*. (August).

MacCormak, Alan, and Theodore Forbach 2008. "Learning the Fine Art of Global Collaboration." *Harvard Business Review* (January).

Maslow A. H. 1954. *Motivation and Personality*. New York: Harper and Row.

Mendonca, Lenny T., and Kevin D. Sneader. 2007. "Coaching Innovation: An Interview with Intuit's Bill Campbell." *McKinsey Quarterly* (1).

Michalko, Michael. 1998. *Cracking Creativity: The Secrets of Creative Genius*. Berkeley, CA: Ten Speed Press.

Nussbaum, B. 2004. "The Power of Design." *BusinessWeek* (May 14): 86.

Partarakis, N., C. Doulgeraki, M. Antona and C. Stephanidis. 2010. "Designing Web-based Services," Chapter 22, and "The Development of Web-based Services," Chapter 24, in Salvendy, Gavriel and Waldemar Karwowski (Eds), *Introduction to Service Engineering*, John Wiley (January).

Rao, Jeneanne. 2005. "The Keys to High-Impact Innovation." *Business Week OnLine* (September 27).

Rigby Darrell, and Chris Zook. 2002. "Open-Market Innovation." Harvard Business Review (October).

Santos Jose, Yves Doz, and Peter Williamson 2004. "Is Your Innovation Process Global?" *MIT Sloan Management Review* (Summer).

Sauve, Eric. 2007. "Informal Knowledge Transfer." *T + D* (Training and Development Magazine) 61 (3): 22.

Sawhney, Mohanbir, Robert C. Wolcott, and Inigo Arroniz. 2006. "The 12 Different Ways for Companies to Innovate." *MIT Sloan Management Review* 47 (3): 75–81.

Schwartz, Evan. 2004. "*Juice: the Creative Fuel that Drives World-Class Inventors*," Boston: Harvard Business Press, p. 111.

Selden, Larry, and Ian C. MacMillan. 2006. "Manage Customer-Centric Innovation Systematically." *Harvard Business Review* (April).

Shapiro, Albert. 1985. "Creativity and the Management of Creative Professionals." *Research-Technology Management* (March–April).

Stewart, Hayden, and Steve Martinez. 2002. "Innovations by Food Companies— Key to Growth and Profitability." *Food Review* 25 (1): 28–33.

Sutton, Robert I. 2001. "The Weird Rules of Creativity." *Harvard Business Review* (September).

Tidd J. and F. Hull. 2010. "Managing Service Innovation," Chapter 27 in Salvendy, Gavriel and Waldemar Karwowski (Eds), *Introduction to Service Engineering*, John Wiley (January).

Ulwick, Anthony. 2002. "Turning Customer Input into Innovation." *Harvard Business Review*, (January): 91–97.

Uzzi, Brian, and Shannon Dunlop. 2005. "How to Build Your Network." *Harvard Business Review* (December).

Violino, Bob. 2006. "Top-down Innovation." *Optimize* (May) www.optimizemag.com.

Wolpert, John D. 2002. "Breaking Out of the Innovation Box." *Harvard Business Review* (August).

12.8 APPENDICES

Included below are three specific innovation-related appendices of importance. The value factor analysis is potentially very useful to those who need to rank-order competing service offerings. The innovation plan of Stanford Research Institute International (SRI) represents an excellent example of how innovations should be planned. The last appendix offers a set of valuable tools to help assessing the organization's strengths and weakness in pursuing innovations.

12.8.1 Value Factor Analysis

Carlson and Williams (2006) introduce the *value factor analysis,* which utilizes the Kepner–Tregoe rational decision analysis method (Kepner and Tregoe 1981) to evaluate competing products/services. For a given product/service, the value factor is defined as the product of quality and convenience, divided by costs:

$$\text{Value factor} = \text{Quality} \times \text{Convenience}/\text{Cost}$$

The quality of a product/service to customers depends on the quality attributes, the relative importance of these attributes to customers, and to the extent these quality attributes are deemed satisfactory to customers. The quality measure is then determined by the total score of the relative importance times the satisfaction score.

Similarly, the convenience of a product/service to customers depends on the convenience attributes, the relative importance of these convenience attributes to customers, and the extent to which these convenience attributes are deemed satisfactory to customers. The convenience measure is then determined by the total score of the relative importance times the satisfaction score.

In general, companies strive to produce products/services at the lowest cost possible in order to maximize profitability, as price is mostly defined by the marketplace. The cost of a product/service depends on the cost attributes, their relative importance to customers, and the actual expenses committed by the producer. Again, the cost measure is then determined by the total score of the relative importance times the actual expenses.

Table 12.A1 illustrates the computation of value factors of a package delivery service offered by three competitors. In this illustrative example, service B appears to offer the best value package in view of its quality, convenience, and cost attributes.

This method is generically useful to provide a relative ranking among the competitive offerings, provided that the quality, convenience and cost attributes pertinent to the service under considerations are selected properly.

Table 12.A1 Value Factor Analysis

#	Quality Attributes	Customer Importance	A Satisfaction	B Satisfaction	C Satisfaction
1	On-time and correct delivery	5	5	4	3
2	Fast delivery	4	4	5	4
3	Package damage/insurance processing	5	3	4	5
4	Friendly drivers and service agents	5	4	4	5
5	Real-time online package tracking	4	5	4	3
6	Fast response to customer inquiries	4	5	4	3
7	Green delivery trucks (diesel, hybrid)	3	4	4	5
	Total Quality		128	124	120
	Convenience Attributes	**Customer Importance**	**A Satisfaction**	**B Satisfaction**	**C Satisfaction**
1	Fast check-in of packages	5	5	3	4
2	Easy forms to fill out	4	5	5	4
3	24/7 sites to accept packages	5	4	5	3
4	Bundled service offerings	3	5	4	3
5	Global services to all major cities	3	3	5	4
6	Services to rural areas by third parties	3	4	5	3
7	Recovery of nondeliverable packages	4	5	3	4
	Total Convenience		121	114	97
	Cost Attributes	**Customer Importance**	**A Expense**	**B Expense**	**C Expense**
1	Fleet operations (planes and trucks)	5	$$	$$$	$$$
2	Labor	3	$$$	$$	$
3	Logistics and distribution	3	$$$	$$	$$$$
4	Packaging materials	3	$$	$$	$
5	Emergency responses	4	$$$	$$	$$$
	Total Cost		46	41	45
	Value Factor		337	344	259

$ = 1 base unit of cost
$$ = 2 base units of cost

Adopted and modified from Carlson and Wilmot 2006, p. 294

12.8.2 SRI International Innovation Plan

SRI International pioneers an innovation process that consists of the following planning elements:

1. *Vision*. Share the mission statement and business objectives.
2. *Need*. Overall market space (size, players, business models, disruptions); Market segments (customer, size, growth); the important need (where is the pain?) and customer value.
3. *Approach*. Product/service description, (the pain killer); development plan; keys to success that allow differentiations (technology human business, partnerships); business model (how money is to be made), product/service positioning (how service is to be sold and to whom); financial plan (investments, revenues, profits, timetables), staffing plan; risk mitigation plan.
4. *Benefits*. Customer benefits per costs, including value factor analysis, investor benefits (profits, POI), and employee benefits (profit sharing, equity).
5. *Competition or alternatives*. Competitors (now and in the future); competitive advantages (IP); and barrier to entry (partnerships). The keys to success in creating competitive differentiations are to be found in technology (patents), relationship, business model, and process novelty and distribution methodology.
6. *Recycle*. Continuously improve.

Customer value plays a central role in bringing about successful innovations. It is thus useful to post three questions:

1. Who are your customers?
2. What is the customer value you provide, and how do you measure it? The customer value must be derived from satisfying an important need of customers. The innovation process will fail if this need is only interesting, but not important, to customers.
3. What innovation best practices do you use to rapidly, effectively, and systematically build new customer value?

In general, services add value by enabling customers to do the following:

- *Obtain* new and useful information and knowledge (education, consulting, newspaper, TV programs, journals, search engines, e-portal, e-markets).
- *Improve* personal health and wellness (healthcare, fitness centers, drugs, vitamin).
- *Strengthen* financial wealth and independence (investment, portfolio management, retirements planning, tax advisory).
- *Remove* a deficiency (fix a roof leak, solve a car problem, overcome a trouble, eliminate a bottleneck, resolve an equipment malfunction).
- *Acquire* a pleasant experience (music, travel, entertainment show).
- *Gain* a convenience (supermarket, 24/hour convenience stores, sending packages long-distances, transportation services)

The relative importance of these value-creating functions depends on the segments of the target customers and the next best alternatives available to them.

Emotions related value may include ego being associated with a brand-name service and prestige related to the early use of a high-tech service.

Values are refined systematically by iterating and compounding ideas. Ideas are produced in teams, tested, and improved on, as illustrated in Figure 12.A1.

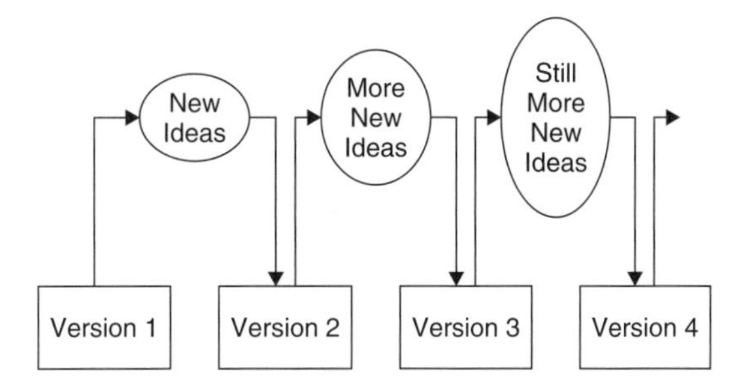

Figure 12.A1 Process of improving customer value

New ideas may be derived by holding brainstorming session of cross-functional teams. For such teams to be productive in coming up with new ideas, they must meet the following basic requirements:

1. Shared strategic vision.
2. Unique, complementary skills (clarified roles for each), divergent perspectives.
3. Build the smallest team possible (less than ten to twenty people).
4. Shared rewards.
5. Mutual trust.
6. Constant communications, because team members are motivated by achievement, empowerment, and involvement.

Figure 12.A2 illustrates this model, which is linked together by constant and respectful communications. The use of project planning tools and the Kanban communications system are encouraged. All team members need to share the same vision, which must align with organizational mission and goals, is clear, compelling, and forceful, can be stated easily, and encompasses all team members. Trust is built on respect for others, with absolute integrity and a generosity of spirit. Team members need to have complementary skills so that all roles and responsibilities are clearly laid out and understood. The group must share all rewards and recognition, so that group performance, instead of individual superstar performance, is encouraged.

The value of a service to customers forms a hierarchy, similar to Maslow's hierarchy of needs (Maslow 1954); see section 4.5. Carlson and Wilmot (2006) suggest such a model, as indicated in Figure 12.A3, using Kodak film products as an example.

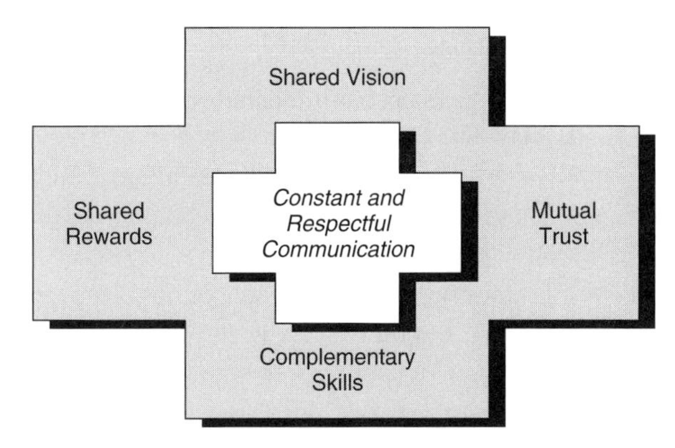

Figure 12.A2 Collaborating teams.

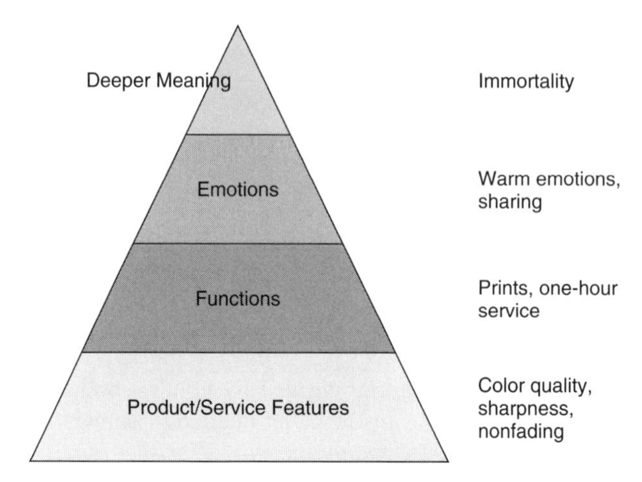

Figure 12.A3 Hierarchy of customer value.

Amazon's wireless book reading device Kindle offers an alternative service with low cost, convenience and minimum storage space requirements for books. Best services touch customers at multiple levels.

12.8.3 Self-Assessment Tool in Innovation Management

To manage innovation effectively, service systems engineers and leaders need to pay attention to a large number of managerial practices. Tidd and Bessant (2009) offer a useful self-assessment tool to help determine the work habit, behavior, and readiness of the organization in managing innovation in products and processes. This tool contains a large number of questions, which we have adopted and modified to better fit to service innovations.

Answer each question in the five groups presented below with a score of 1 (not true at all) to 7 (very true) and then determine the average score of each group by summing the eight individual scores and dividing the sum by 8. The five group average scores represent the self-assessment results of the organization in corporate strategy, process capabilities, organizational structure, sources of innovative ideas, and corporate renewal and learning.

Group 1—Corporate Strategy.

1.1 Our top team has a shared vision of how the company will grow through innovation.

1.2 There is top management commitment and support for innovation.

1.3 There is a clear link between innovation projects we carry out and the overall strategy of the business.

1.4 Employees have a clear idea of how innovation can help the company compete.

1.5 Our innovation strategy is clearly communicated so everyone knows the targets for improvement.

1.6 People know what our distinctive competence is—what gives the company a competitive edge.

1.7 We look ahead in a structured way (using forecasting tools and techniques) to try and imagine future threats and opportunities.

1.8 We have processes in place to review new technologies or market developments and what they mean for our company's strategy.

Group 2—Process Capabilities.

2.1 We have processes in place to help us manage new service development effectively from idea to launch.

2.2 Our innovation projects are usually completed on time and within budget.

2.3 We have proven mechanisms to make sure everyone (not just marketing) understands customer needs.

2.4 We have workable mechanisms for managing process change from idea through to successful implementation.

2.5 We systematically search for new service ideas.

2.6 We have mechanisms in place to ensure early involvement of all departments in developing new services/processes.

2.7 We have a clear system for choosing innovative projects.

2.8 There is sufficient flexibility in our system for service development to allow small "fast-track" projects to happen.

Group 3—Organizational Structure.

3.1 Our organizational structure does not stifle innovation but helps it to happen.

3.2 People work well together across departmental boundaries.

3.3 People are involved in suggesting ideas for improvements to services and processes.

3.4 Our organizational structure helps us to make decision rapidly.

3.5 Communication is effective and works top-down, bottom-up, and across the organization.

3.6 Our reward and recognition system supports innovation.

3.7 We have a supportive climate for new ideas—people do not have to leave the organization to make them happen.

3.8 We work well in teams.

Group 4—Sources of Innovative Ideas.

4.1 We work closely with our customers in exploring and developing new service concepts.

4.2 We work closely with "lead users" to come up with new services.

4.3 We have good win–relationships with our suppliers.

4.4 We try to form external networks of people who can help us with specialist knowledge.

4.5 We are good at understanding the needs of our customers/end users.

4.6 We work well with universities and other research centers to produce new knowledge of value to us.

4.7 We collaborate with other firms to invent new services or processes.

4.8 We work closely with local and national systems to communicate our needs for skills.

Group 5—Corporate Renewal and Learning.

5.1 We systematically benchmark our services with other firms in the industry.

5.2 There is a strong commitment to training and developing people.

5.3 We take time to extract lessons from our projects in order to upgrade our performance next time.

5.4 We learn from our mistakes.

5.5 We meet and share our experience with other firms to help us learn.

5.6 We are good at capturing what we have learned so that others in the organization can make use of it.

5.7 We are good at learning from other organizations.

5.8 We use measurement to help identify where and when we can augment our innovation management.

Since the five group average scores should all be ideally at 7, the resultant self-assessment scores will illustrate the areas in need of further improvement.

12.9 QUESTIONS

12.1 The 3M Company is a diversified technology company that aggressively pursued innovation. What is unique about the innovation strategy of this company?

12.2 What are the relationships between creativity and innovation?

12.3 Discuss the steps of the inventing procedure commonly followed by inventors.

12.4 Problem-solving is a fundamental skill that all systems service engineers and managers must possess. Its use in industry is rather frequent. What steps should be included in a generic problem-solving process?

12.5 Service systems engineers needs to access new ideas from time to time. Which sources should they tap into in order to obtain new ideas?

12.6 Service enterprises need creative employees to attain strategic differentiation. These creative employees need to be properly managed. What are the useful guidelines for managing creative people?

12.7 Creativity is a highly desirable personality trait most y service enterprises are looking for in their new employees. Service systems engineers and managers are special people who possess the basic training in engineering/scientific disciplines with a bend in entrepreneurship. What in your opinion are the important characteristics of such a creative person?

12.8 There are numerous techniques used in engineering to promote creative thinking. Service systems engineers and managers should employ them whenever appropriate. Name some of these creativity methods in engineering.

12.9 The performance of any company can be promoted by specifying a set of metrics that can be readily measured and then monitoring them rigorously. To promote the creativity and innovation of a service enterprise, one needs to define a set of metrics for determining its innovative performance. Name a number of such metrics.

12.10 Creativity and innovation will usually result in some changes to be introduced into the organization, such as new production process, new marketing strategies, new procedure of offering customer supports, and others. Not everyone likes changes, and some will, in fact, resist them. Service systems engineers and managers need to be prepared to manage changes. Explain the principal reasons for resistance to change in organizations.

12.11 When managing innovation, service systems engineers and leaders might be able to take advantage of specific practices that were beneficial to other managers in the past. Name a few useful lessons about managing innovation that can be gleaned by scanning the applicable business literature.

12.12 The management of innovation demands a broad and balanced approach. What are the factors deemed to be critical for achieving success in innovation management?

12.13 Service systems engineers and leaders face a variety of challenges when managing innovation. What are some of these top challenges in managing innovation?

12.14 Marketing and innovation are said to be two significant competencies of any enterprise.

A. Explain, why these two competencies have been singled out as significant.

B. Why would those engineering managers, who possess demonstrable skills and insights in both of these competencies, be perceived to be particularly valuable to corporate enterprises?

C. What specific activities could an engineering manager who possesses demonstrable skills and insights in both of these competencies actually pursue in order to maximize the value he/she produces for the employer?

12.15 The topic of cross-function teams was discussed in this text. It was postulated that nowadays teams are the most appropriate organizational structure for managers to deploy in promoting and realizing innovations, be it in conjunction with new service/product development, new process improvement, service design, problem solving, or other such efforts needed on a continuous basis to meet the challenges of the future. In fact, without innovation, companies will soon cease to have long-term profitability in the absence of relative competitiveness in the marketplace. Thus, it is imperative that future managers know how to lead teams effectively. Explain what is required for teams to become innovative and to be able to build synergies that are otherwise not readily attainable.

12.16 Study the Harvard Business School Case "Innovation at Progressive (A) Pay-as-you-go Insurance," Case #9-602-175 (Rev. April 29, 2004) and answer the following questions. The case materials may be purchased by contacting its publisher at 1-800-545-7685, or at http://www.custserve@hbsp.harvard.edu.

A. What is the case about?

B. Progressive Insurance has been rather successful in competing against its peers, such as State Farm, Geico, Allstate and others. Explain how unique the "Immediate Response" concept is.

C. Progressive Insurance practiced another novel service design concept, the "Comparison Quote." Explain why this concept benefitted the company.

D. Progressive Insurance piloted the new concept of "Pay-as-you-go" service in Texas successfully. Explain how this service concept works and in what way its implementation would be beneficial to both customers and the company.

E. More recently, Progressive Insurance decided to temporarily withdraw the "Pay-as-you-go" program from the market, although it reserves the right to reactivate it at a later time. Explain the reasons for this temporary withdrawal.

F. When Progressive Insurance decides to reactive the "pay-as-you-go" program, how should it be done differently than before?

G. What are the key lessons learned from this case?

Chapter 13

Operational Excellence—Lean Six Sigma, Web-Based Applications, and SOA

13.1 INTRODUCTION

The goal of achieving operational excellence for any service enterprise is to reduce cost, improve productivity, and enhance efficiency. Operational excellence contributes to the bottom line.

Services are composed of a variety of components. The core components are typically developed/designed with customer input and involve a high degree of customization. The processes of producing these components are not standardized. The majority of the support components, on the other hand, usually consist of procedures and activities that could be standardized to a great extent. They present the best opportunities for cost reduction, quality improvement and process optimization to derive operational excellence. McLauglin (2010) discusses the overall issues related to service operation and management.

Service systems engineers and leaders are expected to apply diligently the engineering management functions of planning, organizing, leading and controlling their projects, teams and programs to continuously improve productivity. They also need to become aware of specific distinctive methodologies that could aid in their attempt to achieve operational excellence. This chapter is focused on three sets of such tools, Lean Six Sigma in services, web-based applications packages, and well as web services and service-oriented architecture (SOA) approach.

Lean Six Sigma is a well-known waste-reduction methodology that has been applied with great success in the manufacturing industry. The Lean principle focuses on process speed. Six Sigma applies the DMAIC (define, measure, analyze, improve, and control) procedure to enhance the quality of standardized work processes. Together, Lean and Six Sigma strive to simplify work complexity. This methodology requires that a service enterprise follow a well-defined procedure to speed up the processes and eliminate waste in the process of delivering services. Thus, Lean Six Sigma for services will be covered first.

We then move to the use of external skills and capabilities in achieving operational excellence. The software industry has offered numerous self-contained applications, which can be bought and installed in-house to promote efficiency or leased from application service providers on a pay-as-you-go basis. These self-contained software tools deliver useful assistance in project management, communications, collaborative design, and other such business/engineering functions. These are the Web-based applications, which will be discussed next. Table 13.1 contains examples of the specific benefits derived by diverse companies from utilizing such Web-based applications.

More recently, the IT community has been promoting the *service-oriented architecture* (SOA) approach in performing business computing. Proponents of this approach think that service enterprises will start abandoning their in-house data centers in favor of buying computing services from the outside. As the software products become readily available on different platforms (virtualization); computer hardware becomes more interactive to form an operating unit (grid computing); and both legacy and new software products become more interactive (Web services), the end of corporate computing is in sight. There will be IT utilities, similar to the electric utilities, that supply business computing to service enterprises. They will allow businesses to quickly assemble the right set of software modules and perform the required computing to get results, then disassemble the modules thereafter, readying them for use by someone else. Experts predict that this approach will lead to great cost reduction in business computing, faster time to market, and more flexibility in computing capabilities to meet the changing needs of the marketplace. The SOA approach will be discussed as well.

Conclusions are offered to remind systems engineers and leaders of the need to make the best use of these divergent tools in pursuing operational excellence.

13.2 THE NEW REVOLUTION IN PRODUCTIVITY

Operational excellence is built largely on the productivity of the processes being practiced by service enterprises. Process delays, waste, inefficiencies, and redundancies are the major barriers to reaching high levels of operational excellence. On the horizon is a new revolution, which has the propensity of changing productivity in countless organizations (Merrified et al. 2008).

To pursue operational excellence requires that companies understand their current processes and the activities that support these processes. Value stream mapping (Nash and Poling 2008) and Lean Six Sigma (George 2003) are proven tools that can be efficaciously applied. This approach is similar to the business capacity analysis suggested in Merrified et al. (2008). There are three steps to take:

1. **Describe** business operations in terms of desired outcome (output value), not the work people do and the ways work is to be performed.
2. **Identify** the activities supporting these desired outcomes. This activity-centered approach is analogous to that used in activity-based costing (ABC), in which cost is assigned to a specific outcome, for which activities are required. The activities can be classified into several groups:

   - *Primary*—support critical outcomes of importance to the company, such as those producing differential advantages, enhancing customer loyalty, or driving the key performance indicators (e.g., costs, quality, and time to market).
   - *Standard*—shared by other parts of the company (e.g., monthly progress reports, invention disclosure, project management, financial accounting, procurement, human resource management, legal affairs). These activities can be

Table 13.1 Benefits Derived from Using Web Technologies

Value	Specific Benefits	Company	Web-Based Applications
Cost reduction	Cut $375 million annually from training budget and another $20 million in travel expenses	IBM	Collaborated and enhanced skills online and used Web conferencing
	Saved $25 million over 10 years to build and test new Stealth fighter plane	Lockheed Martin	Used Web to link 80 major suppliers
	Saved $180 million via auctions in 2002 alone	GM	Used electronic auctions to unload vehicles to auto dealers at the end of their leases
	Cut costs by $20 million annually over two years	GE	Automatically sifted through data in commercial loans to weed out poor loan applicants
	Trimmed $100 million a year	Yellow	Analyzed 60,000 orders daily to figure out the exact number of workers needed at 325 facilities
	Saved $10 million annually in training costs and rolled out new services more efficiently	Kinko's	Replaced 51 employee training sites with e-learning in 2002
Customer satisfaction	Promoted customer satisfaction and increased sales by 15 percent	Gilbane	Managed large and complex construction projects over the Web
Productivity gain	Increases sales productivity by 25 percent	Taylor-Made	Update sales and inventory figures via Web by 100 salespeople using handheld devices
Time reduction	Decreased time to solve problems to months instead of two to three years	Eli Lilly	Posted problem on Web site and rewarded best ideas with cash prices
	Reduced gasoline station renovation time by 50 percent	BP America	Used Web to link builders, architects, and suppliers to revamp 10,000 BP America stations
Inventory reduction	Cut inventory from 15 percent of sales to 12 percent	Whirlpool	Used Web to link factory with sales operations, suppliers, and key retail partners
	Scaled down shoes ordered on speculation from 30 percent to 3 percent	Nike	Used Web to link Nike with manufacturing partners
	Cut inventory by 30 percent	Shiseido	Used Web to link sales outlets to sales staff and factory floor
Others	Provided more options, slashed delivery time by one-third, and cut overstock.	BMW	Used Web to link dealership, suppliers, and factories and to build to order most cars in Europe
	Whittled down steel cost by $17 a ton and the inventory and delivery times each by 50 percent	Posco	Used Web to plan steel production and track orders
	Increased output by 40 percent	Dell	Processed orders from the Web and automated product assembly operations with robots

Source: Heather Green, "The Web Smart 50," Business Week, November 24. 2003.

better performed by engaging Internet-based technology application packages (Web-based enablers).

- *Shifted*—can be transferred to customers (self-service, access to Web sites, user consortiums, etc.), suppliers (inventory control), and external experts.

3. **Define** the capabilities supporting each of these activities, which are needed to produce the desired outcomes. Doing so will allow an easy optimization of which in-house capabilities are to be nurtured and utilized and which ones are to be outsourced.

Such a business capacity analysis or a value stream mapping analysis will also discover the major sources of inefficiency, such as (1) duplicate processes, (2) excessive waste in time and resources in conducting specific activities, and (3) activities of no or low value. So far, the outlined approach would produce the generally expected improvement in companies' understanding of the business processes and their supporting activities.

The business capacity analysis outcome of Harvard Pilgrim, a health insurance company, is noteworthy. Taking the previous approach, the company decided to do the following:

- Transfer about 40 percent of the company's noncore activities to outside vendors. Examples include pharmacy-benefit management, disease management, behavior health management, and claims processing.
- Engage outside experts to perform data-mining analysis in order to discover patients who might be in early stages of developing heart disease and diabetes, so that these patients could be enrolled in preventive care or disease management programs before their conditions become serious.
- Focus on distinctive activities that produce competitive advantages to the company. Those include customer service, development of new products, pricing health insurance, attracting doctors into the network, selling to large groups and marketing to individual policyholders.

The next step for service enterprises to drastically raise productivity is to prepare software modules that perform as many of these activities as possible, with the assistance of SOA vendors. Vendors would employ Web services to modularize both legacy software components and new software components using the SOA approach so that these modules can interact with other such modules and be made operable in Internet-based computer networks by specific multilayered designs. The new gain in productivity is predicated on the promise that by making software components more modular, companies can swap, buy, or sell specific in-house processes to create a radically more efficient "plug-and-play" business enabled by Internet-based computer networks that distribute them to work instantly on different platforms and be assembled into any user-defined packages. By having transformed the enterprise from a collection of proprietary operations into a collection of plug-and-play activities, the enterprise's productivity is expected to improve drastically.

Gartner, Inc. reported that about 50 percent of mission-critical systems had installed SOA systems in 2007 and that this percentage was expected to rise to 80 percent by 2010. Forrester Research reported that 60 percent of enterprises in North America,

Europe and the Asian-Pacific regions it surveyed had either built SOA or planned to start SOA in 2007, and 40 percent of current SOA users were satisfied with the SOA outcomes.

More recently, Gartner Inc. discovered in its 2008 survey that this upward pointing trend of building SOA had been reversed (Anonymous 2008). While SOA adoption is nearly universal in Europe, it is only moderate in North America and lagging in Asia.

In the following sections, Lean Six Sigma for services will be discussed, followed by the use of Internet-based technological applications (Web-based enablers), then Web services and service-oriented architecture (SOA).

13.3 LEAN SIX SIGMA FOR SERVICES—THE SERVICE MODEL

Six Sigma was originally introduced at Motorola in the 1980s for large-volume production processes, which are highly standardized. The purpose is to eliminate waste by achieving near perfect results (Biolos 2002). Six Sigma quality means that there will be no more than 3.4 defects per million (see section 5.11.3). Billions of dollars of savings were achieved through Six Sigma by companies such as General Electric, AlliedSignal, and others.

In recent years, the Lean principle, which is focused on process speed, was added, creating Lean Six Sigma that is capable of enhancing the speed and quality while reducing the complexity of processes and activities. By emphasizing flow of information and the interaction between people, Lean Six Sigma can be applied to various nonmanufacturing processes such as procurement, accounts receivable, sales, call centers, surgical suites, government offices, R&D, and others. Voehl and Elshennawy (2010) review the characteristics of lean services. A combined progress in all three metrics leads to a faster pace in delivering high-quality services at low costs to positively affect customers' satisfaction. Lean Six Sigma has been shown to:

- Slash service costs by 30 to 60 percent.
- Decrease service delivery time by 50 percent.
- Expand capacity by 20 percent without adding staff.

Several cases of applying Lean Six Sigma are published in (George 2003)—for example, Lockheed Martin, Bank One, City of Fort Wayne and Stanford Hospitals and Clinics. Lean Six Sigma may be usefully applied to services by following the seven-step procedure that constitutes the mnemonic of our SERVICE model:

1. *Study* customer value. Select those tasks/processes and activities that are likely to lead to a better satisfaction of customers. Lovelock and Writz (2006) illustrate the service value package as consisting of a core benefit supported by eight supplemental service elements; see Fig. 13.1. For a service to be appreciated by customers, both the core and its supplemental elements must be fresh and well formed, because all of them will affect the overall perception gained by the customers. This model fits well with many service value packages, although not all supplemental service elements are equally important to different core services. Some supplemental service elements engage the customers directly, thus having a more profound impact on influencing the customer's perception than others. A useful way to prioritize the target activities for applying Lean

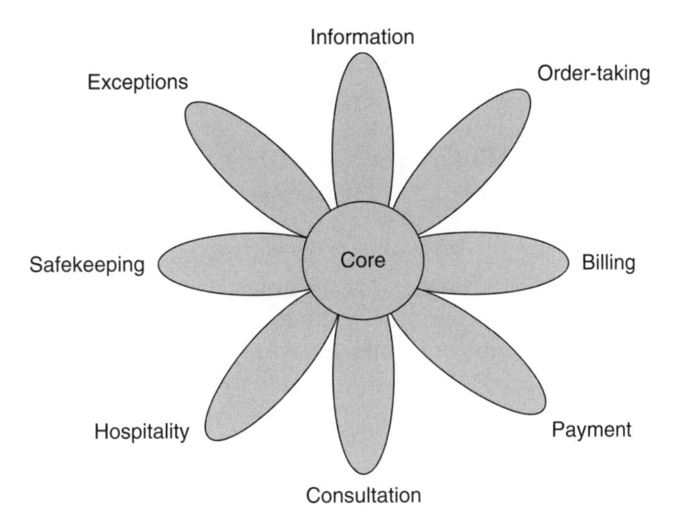

Figure 13.1 Service elements.

Six Sigma is to then look the core and the supplemental elements of the service value packages offered by the company.

Processes that go into a service can also be classified into the types of highly customized, mass customized, and standardized. Among these three types, the best candidates for applying Lean Six Sigma are the standardized processes, such as payroll and benefit processing, credit card account services, and fast-food services, according to Biolos (2002).

2. *Evaluate* these service elements and map their relative values to customers. Take a cross-functional view to screen out activities that do not add value and to identify activities that cause processing delays (Nash and Poling 2008).
3. *Refine* work flow diagrams to preserve process data and materials for analysis.
4. *Validate* and improve on cycle time, and use the pull strategy to expand Lean Six Sigma activities. By ways of promotion and management endorsements, company employees are encouraged to voluntarily participate, so that the Lean Six Sigma projects are pulled through all units of the company. In order to encourage employee buy-in and acceptance, try to start with a meaningful but self-contained project that does not require changes of other processes, and then advertise the resulting successes aggressively.
5. *Institutionalize* the drive for continuous improvement—applying the DMAIC process of Six Sigma to the target process/activities to upgrade quality. (Recall that DMAIC stands for the steps of define measure, analyze, improve, and control.)

   Define service defect and the ways to measure it. From the customer's standpoint, a service defect is a flaw in a process that results in a lower level of customer satisfaction, high service turn around time or a lost customer.

Consumers' Checkbook, an independent rating organization, uses the following ratings to rate banking services: (1) overall, (2) pleasantness of staff, (3) Knowledge of staff, (4) Speed of service, (5) Reasonableness of fee policies, (6) convenience of hours, and (7) clarity of written communications (Checkbook 2009).

Failure mode and effect analysis (FMEA) (Stamatis 2003; McDermot et al. 2008) (see section 5.11.2) and root cause analysis (Latino et al. 2006; Anderson et al. 2006) may be engaged to find out the root causes of these service defects. There could be as many as half-dozen contributors to a specific service defect. For example: The commission checks of an insurance company were often inaccurate. It turned out that a few root causes were responsible: (1) complex commission rules, (2) lack of knowledge on part of a check processing staff, and (3) confusing procedures for processing checks.

Besides Six Sigma, the use of Internet–based package applications, Web services, and service-oriented architecture (SOA), which are the subjects discussed in the following sections, may also be beneficially considered.

6. *Call* out the complexities hidden in the process and activities, using both value stream mapping and Six Sigma. Complexities are brought about by service differentiations. Standardizing the supplemental service elements may be useful to eliminate complexities. Since the core service elements are usually customized and thus highly differentiated, optimizing the service offerings by eliminating those services that do not contribute to profitability will help. The outcome of eliminating complexities will be the reduction of costs and wastes.
7. *Excel* in delivering Six Sigma level of service quality at lean speed and low cost. Lean Six Sigma is an important productivity tool for various service operations. Some lessons learned in applying Lean Six Sigma may be found in (George 2003). Service systems engineers and leaders need to understand its strategic importance to service enterprises and implement the SERVICE model properly.

Example 13.1 Delivering services at the Six Sigma quality level, at fast pace and at low cost is desirable to any service enterprise. However planning and executing Lean Six Sigma projects have not always been shown to be easy. What are the industrial best practices in Lean Six Sigma projects in the service sectors?

ANSWER 13.1 Based on lessons learned over the years, there are a number of best practices that appear to be commonly responsible for the significant successes gained in various successful Lean Six Sigma projects:

- *Secure* a visible and influential champion for the project.
- *Gain* corporate commitment regarding the alignment with business objectives, resource authorization and management attention.
- *Define* project accountabilities and allocate adequate resources (e.g., certified Black Belts).
- *Select* participants with potential to be "future leaders of the company."
- *Plan* the projects carefully (e.g., selecting target service processes based on the Pareto principle, training participants to use the right methodologies, and defining metrics and benchmarks to monitor progresses, etc.).

- *Organize* the "activities in progress" to minimize cycle time.
- *Track* results rigorously and advertise aggressively to encourage "pulling" of new Lean Six Sigma projects through the organization.
- *Institutionalize* the process to enable future improvement and innovation on a continuous basis.

13.4 INTERNET-BASED APPLICATIONS FOR SERVICE MANAGEMENT

Customers, back-office workers, and supply chain partners are the three major groups of stakeholders that form the backbone of a progressive enterprise. Their participation in the business activities of the enterprise makes it possible to generate and deliver customer value through integrated business applications. Enablers for Web-based management are software tools for managing these stakeholders. These tools are used for customer relationship management, enterprise integration and procurement, and supply chain management.

Engineering managers need to become well versed in the use of such tools, which are becoming increasingly available in the marketplace.

13.4.1 Customer Relationship Management

Customer relationship management (CRM) focuses on the process of understanding and satisfying customers' present and future requirements. Engineers are usually engaged in some part of customer relationship management activities (Dyche 2001; Nykamp 2001; Burnett 2000). They are involved when the products of the company are of high technological content and require product customization, operations assistance, problem solving, application support, maintenance and repair assistance, spare parts, and other such services.

All customers are important, but some are more important than others, depending on the magnitude of their lifetime value (LTV). LTV is the total revenue realized by the company from the purchases made by a customer over the period in which this customer continues to buy from the company. High-LTV customers should be managed differently from low-LTV customers.

It is known that a number of U.S. corporations have set up special corporate desks, each dedicated to specific major customers and manned by full-time employees who pay personal attention to all of the inquiries initiated by these high-LTV customers. At the other end of the spectrum, companies have set up extranets so that minor, low-LTV customers are encouraged to help themselves by accessing Web-based resources. Communications links with low-LTV customers may be via e-mail, text chat, Web-based callback request, and voice over the Internet. These extranets may contain product catalogs, installation and repair instructions, an FAQ (frequently asked questions) section, a news bulletin, discussion groups, chat rooms, users' groups, and other features. Surveys may be given to solicit customer feedback on product quality, desirable new features, service improvements, and other pertinent issues. Customer service centers are useful in coordinating the inquiries of minor- or medium-LTV customers who prefer human voices. An efficacious customer service center is expected to respond to customer inquiries within a very short period (say, roughly twenty minutes).

Customer relationship management has gained importance in recent years as a result of competition, globalization, the high cost of customer acquisition, and high customer turnover (Schmitt and Schmitt 2003). Surveys point out that U.S. corporations lose one-half of their customers every five years. The following statistics support the importance of this corporate function:

- It costs five times more to sell to a new customer than to sell to an existing one.
- A typical dissatisfied customer tells eight to ten people about the bad experience.
- A company can boost its profit by 85 percent by increasing its annual customer retention by only 5 percent.
- The odds of selling a product to a new customer are 15 percent; to an existing customer, 50 percent.
- Seventy percent of complaining customers will do business with the company again if past service deficiencies are quickly resolved.

By managing customer relationships properly, an Internet-enhanced enterprise hopes to use existing relationships to increase revenue and utilize integrated information to offer excellent customer service. The enterprise can also introduce more repeatable sales processes and procedures, conceive new value and instill loyalty, and implement a more proactive solution strategy. The key success factor for CRM is the efficient integration of the following front-end activities:

- *Sales.* Capture customer information, link sales and service functions, and cross-sell and up-sell other company products.
- *Direct marketing.* Market new products and fulfill new orders.
- *Customer service.* Access customer profiles in real time, add new value, instill loyalty, and deliver support functions.
- *Call centers.* Respond with a human voice.
- *Field service.* Deliver the needed maintenance or repair services and other special activities.
- *Retention management.* Induce customer loyalty.

Charles Schwab has used a customer relationship management system marketed by Siebel. Table 13.2 delineates the various dimensions of customer relationship management. Its specific functions are as follows:

- Obtain detailed information about customers' behaviors, preferences, needs, and buying patterns.
- Use that information to customize the relationship with the respective customer.
- Apply that information to set prices, determine needs and desires of customers, and negotiate terms of purchase.

Fidelity Investments, Principal Financial Group, and several other mutual-fund companies have started using Web-based "e-401(k)" programs. The logic for doing so is compelling. When a customer places a phone call to a service center, it costs the company about $9 to respond. This cost is only $2 if the customer gets the required services from an automated voice-mail system. The cost is further lessened to 10 cents if the customer accesses a Web-based self-service system.

Table 13.2 Dimensions of Customer Relationship Management

Dimension	Customer Relationship Management
Advertising	Provide information in response to specific customer inquiries
Targeting	Identify and respond to specific customer behaviors and preferences
Promotions and offering discounts	Tailored individually to the customer
Distribution channels	Direct or through intermediaries; customer's choice
Pricing of products and services	Negotiated with each customer
New product features	Created in response to customer demands
Measurements used to manage the customer relationship	Customer intention; total value of the individual customer relationship

Autodesk Inc. is known to have applied an innovative Web-enabled corporate help desk in providing high-tech help to customers. The company was able to whittle away the cost associated with internal phone-based support by 70 percent. Uniglobe.com uses a Web-enabled call center for travel services. A useful strategy is to build a self-service knowledge base that allows customers to perform self-help tasks. Another strategy is to embed e-support capabilities into products. Hewlett Packard has done that.

iSky (www.isky.com) in Laurel, Maryland, provides an Internet-based real-time customer-care system that involves customer acquisition, retention, and optimization. It allows end customers to communicate with companies through any available media—phone, e-mail, and others. It offers self-help services by accessing the client's e-commerce site and using database and FAQ lookups, automated e-mail, and interactive voice response. It can move to human assistance from an agent. The human-assisted customer care can be accessed through e-mail, text chat, voice-over Internet protocol, telephone callback, and the traditional teleservices support of phone or fax.

Waiting for customers to call for service is not what customer relationship management is all about. The company must also understand customers, anticipate their future needs, and take action to satisfy those needs. *Data mining* is a useful tool for profiling customers (by segmentation) and predicting their purchasing behaviors (Rud 2001; Kamber and Han 2005; Pendharker 2003). Questions explored in data-mining analyses may include the following:

- Who are my best customers, and can I acquire more of them?
- How can I increase business with my best customers?
- Who are my worst customers, and should I salvage those relationships? If so, how?
- Why are some customers leaving?
- Are there other products or services that I can provide?
- How can I avoid acquiring unprofitable customers?
- What are the characteristics of prospective loyal customers?

- Which products and services are the prospective customers looking at or inquiring about?
- Which products has a single customer purchased most often?
- Are there customer complaints about product features, service quality, prices, or other issues?

Customer relationship management must determine which customers have the tendency to purchase what products and take proactive steps to promote these products to them (e.g., cross-selling and up-selling in order to derive increased business benefits).

A case in point is Caterpillar (Berry and Linoff 1999; Chopoorian, Witherell, Khalil, and Ahmed 2001). In 1997, this company set up a database of 100,000 client companies with known purchase records for fleets of truck engines from Caterpillar and its competitors. It then defined which client companies would have the best potential to buy from Caterpillar in the future by studying a predictive model based on the use of data-mining techniques. Caterpillar initiated a direct-mail campaign to the resulting 10,000 to 20,000 client companies regarded as having high potential to purchase truck engines from Caterpillar. The resulting response rate to the direct mail campaign was 37 percent in 1999, compared with 16 percent in 1996 and the industrial average of only 3 to 5 percent for similar mailings. The conversion rate of non-Caterpillar clients was around 40 percent. As a consequence, Caterpillar's truck engine market share went up to 30.4 percent in 1999, from 23 percent in 1997.

In addition, the company built an intranet to post detailed information about each client company that included all past correspondences. This allowed its national and regional account managers to custom tailor communications packages on-line for individual client companies. All sales proposals for truck engine packages were Web-based. Within forty-eight hours, a personalized package, complete with the salesperson's signature, could be sent. The Caterpillar case is a good example of how data mining can be applied to serve the business needs of the company.

Example 13.2 A 2001 issue of *Technology Review* reviewed ten emerging technologies that will affect the world profoundly (Zacks 2001). Data mining was one of them. What are the ten, and why is data mining so important?

ANSWER 13.2 The ten technologies cited to have a profound impact on the world economy and on how we live and work are (1) brain–machine interface, (2) the flexible transistor, (3) data mining, (4) digital rights management, (5) biometrics, (6) natural-language processing, (7) microphotonics, (8) untangling code, (9) robot design, and (10) microfluidics.

Data mining refers to the analysis and nontrivial extraction of information from databases for the purpose of discovering new and valuable knowledge. The knowledge is in the form of patterns and rules, derived from relationships between data elements. Data mining emphasizes the use of computationally intensive machine learning tools in the analysis of large and complex databases. As processor speed is being increased constantly, difficult data-mining solutions are becoming attainable in a time frame that has become more and more practical to business decision makers (Berry and Linoff 2004).

Implementing data mining usually involves the following stages: (1) setting specific goals for data-mining efforts, (2) data collection (data type), (3) data preparation (data segmentation, formatting, and quality control), (4) analysis and prediction (employing specific tools such as neural networks, decision trees, logistic expression, and visualization to build predictive models), and (5) measurement and feedback (implementing models to bring forth results and taking corresponding business actions).

An important application of data mining is customer relationship management. Customer data are systematically analyzed to define customer preferences and characteristics so that products and services can be customized to grow better business results.

13.4.2 Enterprise Integration and Resource Planning

To achieve sustainable profitability, companies must plan, align, execute, and control all basic business processes. By streamlining all transactions and effecting data exchange between operations, companies can minimize inventories, shorten the cycle time to market, cut costs, upgrade overall operational efficiency, and support customer service.

Enterprise resource planning (ERP) software is designed to integrate into one information system all back-office operations, such as manufacturing, engineering, finance, distribution, procurement, decision support, knowledge management, marketing and sales, and other internal business functions (Meyerson 2001; Wagner and Monk 2008; O'Leary 2000). This building block is at the foundation of a progressive enterprise. The capability of ERP and integration becomes particularly important for companies having numerous sites and multinational operations. Reviewed next are three specific components in enterprise integration systems:

1. Procurement resources management
2. Marketing and sales management
3. Decision support and knowledge management

Procurement Resources Management. Procurement deals with the acquisition of typical operating resources, such as office supplies, services, travel, computer equipment and software, MRO (maintenance, repair, and operations) supplies, fuels, and training. Procurement activities involve identifying and evaluating vendors, selecting specific products, placing orders, and resolving problems and issues after purchase.

Traditional procurement follows a number of manually implemented and time-consuming steps that are subject to frequent human errors. These steps include the following:

1. Identify vendors who make engineered parts needed for the company's equipment and who satisfy a specific set of requirements defined by the company.
2. Exchange and review technical documents, such as drawings and product specifications.
3. Solicit bids and evaluate available offers.
4. Process the purchase orders within the company and obtain approvals.

5. Register orders and subsequent payment transactions into the company's back-office systems, involving accounts payable and the factory-receiving department.
6. Resolve human errors and transaction mistakes, such as wrong parts ordered, shipment dates entered incorrectly, and receiving locations improperly specified.

In contrast, Web-based procurement makes this process significantly more efficient, decreases administrative costs per item purchased, and is adaptable to the evolution of business models because it can do the following:

- Use online catalogs provided by suppliers.
- Make use of an electronic data interchange (EDI) network to process documents, a network that currently has 400,000 members in active participation.
- Utilize CAD software to communicate through drawings.
- Monitor inventory levels and automate the procurement process.

Online procurement is clearly the current trend. The objectives are to diminish order processing costs and cycle time, strengthen the corporate procurement capabilities, facilitate self-service, integrate with back-office systems, and raise the strategic importance of procurement.

In newer e-procurement models, reverse auction has been advanced as a tactic to procure well-defined items. Reverse auction works in favor of companies that enjoy a dominant buyer's position because of their large transaction volume. (See section 8.5.2.) Companies engaged in reverse auction first prequalify selected vendors on the basis of product quality, financial strength, company reputation, and other factors. Then they publish the specifications of the goods needed and invite prequalified vendors to submit multiple rounds of bids before a predetermined date. The vendor with the lowest bid is awarded the supply contract. It is called *reverse auction* because the bid price is lowered in each consecutive round. The Internet greatly facilitates the process of reverse auction.

Microsoft is said to have spent millions of dollars designing MS Market, an online ordering system. It eliminates all paperwork; handles $1 billion in supplies; works in a distributed procurement environment; is particularly suitable for high-volume, low-dollar value transactions; processes orders linked to supply sources; and permits order status tracking. One important advantage of using such a procurement system is the elimination of rules and hidden procedures that greatly slow down the procurement process. MS Market scales down the purchase cycle from eight to three days and slices overhead costs by about 90 percent. It averages 100 orders and 6,000 transactions per day.

Canadian Imperial Bank of Commerce (CIBC) has 1,400 branch offices, 40,000 employees, and 14,000 suppliers. Using a software system marketed by Ariba, Inc. CIBC employees are able to buy online from preferred suppliers. CIBC decreased its procurement costs by 50 percent.

The following are additional examples of e-procurement software products available in the marketplace:

- eProcurement (www.purchasingnet.com) is a Web-based procurement system produced by PurchasingNet, Inc. (Red Bank, New Jersey). It accomplishes the following: (1) uses XML to support supplier catalogs, purchase orders, and invoices; (2) automates stock requisition by scanning inventory records and issuing requisition orders if preset inventory levels are met; (3) upgrades browser interface; and (4) integrates front and back office functionality. eProcurement operates on a Windows environment and is compatible with SQL and Oracle databases. Purchasing Net claims that the use of this system brought about savings of 5 to 15 percent in purchasing costs. A potentially useful application of the software is for the procurement of MRO (maintenance, repair and operations) parts, which are typically low-volume, high-variety, and noncore items.

- Purchase Manager (www.Verian.com). Verian Technologies (Charlotte, North Carolina) markets this Web-based procurement system, which has a total of fourteen modules. It automates purchasing with attachments (e.g., drawings and documents), processes orders, manages inventory, issues reports, performs audits, and tracks the status of customer orders. The company claims that use of this software whittled down its clients' purchase order costs by 80 percent (from $150 to $30 per order).

Marketing and Sales Management. To service customers well, companies need to devise an easy ordering process, add value for the customers, increase sales force effectiveness, and coordinate team building. The following processes are readily automated by the use of self-service centers:

- Sales configuration
- Pricing
- Quote and proposal generation
- Commission and contract management
- Order entry management
- Product promotion

Whirlpool makes home appliances such as washers, dryers, dishwashers, dehumidifiers, microwave ovens, ranges, refrigerators, and air conditioners. It uses the Trilogy system (at www.trilogy.com) with success to integrate pricing management, product management, sales, commissions, promotions, contract management, and channel management. Other industrial users include Ford Motor Company, Xerox, Cooper Industries, Fisher Control International, Goodyear Tire & Rubber, Lucent Technologies and NCR.

Decision Support and Knowledge Management. The functions of decision support and knowledge management focus on (1) business analysis, such as filtering, reporting, what-if analyses, forecasting, and risk analysis, (2) data capture and storage, including data warehousing, data mining, and query processing, (3) decision support, such as expert systems, case-based reasoning modules, and intelligent knowledge modules, and (4) data dissemination via proper means, including wireless or mobile front ends.

About 70 percent of Fortune 1000 companies have or will soon possess ERP integration capabilities; among them are Coca-Cola, Cisco, Hershey Foods, Eli Lilly, Alcoa, and Compaq. ERP vendors include Sap, Oracle, PeopleSoft, J. D. Edwards, and Baan.

Example 13.3 What is a value chain? In what ways does the Internet affect the traditional value chain?

ANSWER 13.3 A value chain designates the ways a firm organizes its business activities to sell products and services. The primary activities of a value chain include the following:

- Identifying customers through market research and customer satisfaction surveys.
- Designing by concept research, product design, engineering, and test marketing.
- Purchasing materials and supplies by selecting vendors, ensuring quality, and keeping up timeliness of delivery.
- Manufacturing by fabrication, assembly, testing, and packaging.
- Marketing and selling through advertising, promoting, pricing, and monitoring sales and distribution channels.
- Delivering through warehousing, materials handling, and monitoring timeliness of delivery.
- Providing after-sales services and support through installation, testing, maintenance, repair, warranty replacement, replacement parts, and problem solving.

The support activities of a value chain include the following:

- Financing and administration—accounting, bill payment, borrowing, adherence to regulations, and compliance with laws.
- Human resource activities of recruiting, training, compensation, and benefits administration.
- Use of technology to streamline processes, facilitate communications, and eliminate wastes.

The Internet has exerted a strong influence on value chains. It caused disintermediation (e.g., the elimination of intermediate steps) and re-intermediation (e.g., adding intermediate steps) (see Fig. 13.2).

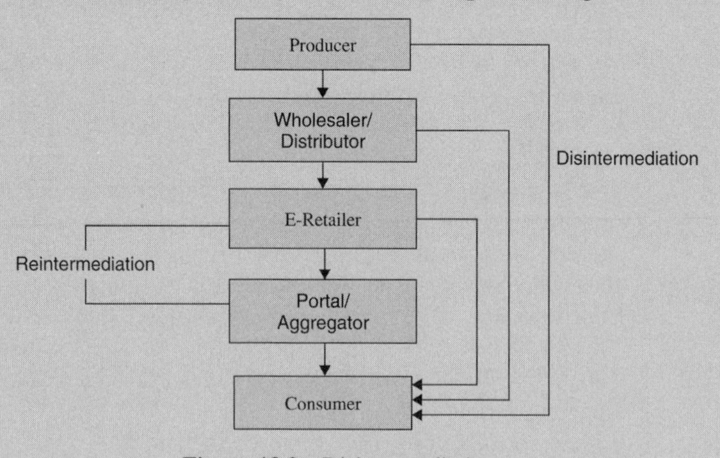

Figure 13.2 Disintermediation.

13.4.3 Supply Chain Management

Supply chain management involves the formation and maintenance of a complex network of relationships with business partners to the sources of, for example, raw materials; the manufacture and storage of intermediate products and finished goods; and the delivery of products. It focuses on managing the partners of the flows of materials (tangible goods), information (demand forecast, order processing, and order status reporting), and finance (credit-card information, credit terms, payment schedule, and title ownership arrangements).

The goals of supply chain management are to lessen the time to market, reduce the cost to distribute, and supply the right products at the right time (Franzelle 2002; Simchi-Levi, Kaminsky, and Simchi-Levi 2007; Ayers 2006). Achieving these challenging goals requires an integration of various operations, including market-demands forecasting; resources and capacity utilization subject to imposed constraints and real-time scheduling; and the optimization of multiple-objective functions involving cost, time, service, and quality. Web-based tools are available to assist in the management of such a complex supply chain. Examples of software products for supply chain management include Advanced Planning and Optimization by *SAP* and Rhythm by I2 Technologies.

The Visteon division of Ford makes chassis components and drive assemblies. It uses Rhythm by I2 Technologies to integrate intercompany processes. The benefits realized include a streamlined production schedule, the reduction of inventory by 15 percent, the removal of bottlenecks, and a shortened response time to orders.

There are a number of additional Web-based supply chain management enablers, as illustrated in the next list.

EDI Linkage. For automotive companies, the supply of engineered parts in a multiple-tiered vendor system requires a major managerial effort. Companies in other industries also recognize the efficiency gain and cost reduction that can be realized by streamlining the supply process.

Surveys indicate that over 70 percent of best-in-class companies have established electronic data interchange (EDI) systems with their suppliers to expedite communications, minimize cycle time and error rates, standardize information transactions, and decrease purchasing expenditures. Users of EDI systems to manage their supply chains include General Motors, Ford, DaimlerChrysler, Sun Microsystems, Haggar Clothing, Wrangler, Xerox, and Honeywell.

SAP ERP (www.sap.com). SAP ERP is a commercial, off-the-shelf enterprise resource planning software package marketed by SAP, Newtown Square, Pennsylvania. It supports twenty-five specific functions related to accounting and finance, production planning and material management, human resources, and sales and distribution (Hernandez, Montinez and Koegh 2005; Williams 2000; Stengl and Ematinger 2001). It also supports order entry, facilitates supply-planning processes, and assists in customer service. Users of this system claim that the order entry process can be shortened from eighteen days to one day and its financial close cycle from eight days to four days.

Microsoft is said to have spent ten months and $25 million to install mySAP ERP to manage its 25,000 employees and fifty subsidiaries. It claims to have realized an annual cost reduction of $18 million. Nestlé (maker of drinks, sweets, pharmaceuticals, and foods) uses the mySAP ERP application suite to manage its 498 factories, 210,000 employees, and operations in sixty-nine countries.

Rhom & Haas company spent $300 million in 1999 to convert the company's sixty-four different information systems to SAS over a four-year period. By 2003, the company was able to trim 4,000 workers and save $500 million in operating expenses. It realized an increase in the after-tax profit margin from 2 to 8 percent by 2005. Other companies that use mySAP ERP include Owens Corning, Colgate-Palmolive, and Warner-Lambert.

Supply Management (www.i2.com). The Supply Management software is marketed by i2 Technologies, Dallas Texas. Toyota Motors buys over $1 billion a year in service parts, and its top twenty-five to thirty suppliers deliver about 80 percent of the total. It also plans to save $100 million in three years by increasing the level of ordering accuracy and inventory management (Anonymous 2000). Toyota Motors uses the supply chain system of i2, which consists of the optimization of fulfillment, logistics, production, revenue, profit, and spending, to build links with its dealers and suppliers. The specific goals are to (1) enable the suppliers to meet the company's annual demand forecast, which is updated quarterly for the needs of its 1500 dealers in the United States; (2) track delivery dates and lead time on-line and identify production bottlenecks; and (3) facilitate communications between the parties.

SAS SRM (www.sas.com). This Suppler Relationship Management software is marketed by SAS, Cary, North Carolina. The minimum-cost solution to a packaging, inventory, and distribution problem of United Sugars was found using the SAS SRM solution. It involved 80 plants comprising packaging, storage, distribution, and transportation points; 250 sugar products; and 200 customers. The optimization model consists of 220,000 nodes, 1 million arcs, 3000 non-arc variables, and 26,000 linear side constraints. To generate solutions, the primal–dual predictor–corrector interior-point algorithm of the NETFLOW procedure is used, which performs the task in about 2.5 hours operating in a Windows environment.

The resulting optimal solution specifies the minimum cost schedule and defines the assets of packaging, distribution, and inventory that are needed to satisfy customer demand. The solution also allows United Sugars to reoptimize storage capacities, perform what-if analyses, maximize inventory turns, handle customer sourcing requirements, and issue suitable reports for users to view with Web browsers.

Transportation (www.kleinschmidt.com). Kleinschmidt, Inc. (Deerfield, Illinois) offers this program as a real-time Web-based supply chain solution for North American Railroad customers. It allows rail shippers and consignees to view rail shipment information and predict equipment arrivals. Specifically, its capabilities include the following:

- Offering an in-transit visibility of inbound rail shipments via the Internet.
- Providing secure access.
- Conducting searches by purchase order or STCC (Standard Transportation Commodity Code).

- Monitoring outbound shipments by plant and other criteria.
- Registering shipments from origin to destination.
- Estimating the time of arrival (ETA) based on historical data.
- Notifying customers by e-mail.

Many companies employ enablers for Web-based management to derive efficiency and cost benefits. Other tools are also being constantly utilized to augment the company's productivity and competitiveness. Engineering managers need to follow the literature closely so that they can offer the best advice in the selection and implementation of these advanced tools to their employers at any given time.

Example 13.4 What is so special about Dell that caused this company to score increased sales in a worldwide-depressed PC market in 2000–2004, while other major players like IBM, Hewlett-Packard, and Gateway suffered significant declines in revenue?

ANSWER 13.4 Dell should be examined from the standpoints of business models, operational excellence, and leadership qualities. At one point in time, Dell's stock had risen 29,600 percent to peak at $55 in 2000, although its price was around $35–$40 in 2004 and about $15 in 2010 due to the general weakness of the U.S. economy in recent years.

Dell's success is due to several innovative business strategies (Magretta 1998; Sheridan 1999):

- *Direct model.* Dell invented the *direct model*, in which it sells *build-to-order* (BTO) computers directly to business or individual consumers. Bypassing several intermediaries in the distribution channel significantly reduces costs. It also nurtures a close relationship with consumers, resulting in a better capture of the shifts in consumer needs, which are constantly being influenced by the rapid change in technologies.
- *Virtual integration.* Dell's success is also made possible by an excellent execution of its *virtual integration* strategy, whereby it coordinates all of its supply chain partners to maximize value to its customers. Dell focuses on areas where value can be added, such as supply chain management, product assembly, supplier selection for component changes and quality assurance, information flow, customer relationship management, segment gross margin, and cash-flow control.

  In the rapidly changing marketplace of computers, where technologies of processors, monitors, hard drives, other hardware components, and software all advance very swiftly, the old-fashioned business paradigm of "vertical integration" (i.e., we-have-to-develop-everything in house) is no longer applicable. Only buying, not building, and switching suppliers often will permit a company to introduce new products quickly enough to meet the customer's demands.

  The competitive advantage of Dell lies in its capability of meeting customers' needs faster and more efficiently than any other company can.
- *Operational excellence.* Dell operates three assembly plants (Austin, Texas; Limerick, Ireland; and Penning, Malaysia) located near parts suppliers. It forms partnerships with Sony for monitors, Airborne and UPS for logistics, unnamed

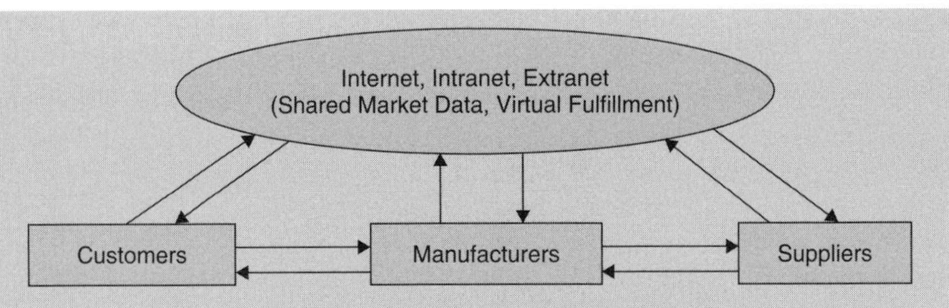

Figure 13.3 Dell model.

third-party maintainers (TPMs) for after-sale service, plus others for additional computer components. Dell uses the in-depth expertise of TPMs to its advantage. Dell focuses on the management of information flow between a few selected partners to increase the velocity of inventory movement by sharing its design database and methodologies and by maintaining an intranet (www.premier.dell.com). Suppliers are constantly informed of the changing marketplace demand. (See Fig. 13.3.)

The company has established a Platinum Council to meet and talk with customers constantly. From the Dell Web page, customers can design computers by selecting various components. Once the order is submitted, Dell will ship out the completed product in three days. Ninety-seven percent of orders are shipped directly.

Dell realizes $10 million per day in sales from the Web page, although its major revenue comes from large corporations. Customers can track order status and receive technical support on-line by keying in a tag code for the purchased computer. Dell's R&D is focused, not on creating new technologies, but on improving user experience, selecting relevant technologies, fostering ease of use, and facilitating cost reduction. Dell achieves eight-day cash-flow conversion cycles, as customers pay by credit card and Dell pays suppliers via accounts payable.

In response to customer needs, Dell builds stability into its computers to make them "intergenerationally consistent" over numerous years. In its plants, Dell uses one-person build cells to assemble computers and makes its own central-processing units. It maintains a six-day inventory for parts, but no finished-goods inventory. Employees participate in profit sharing. Bonuses typically make up about 20 percent of a worker's total pay and are tied to certain key metrics, one of which is 15 percent year-to-year sales improvement.

- *Leadership qualities.* Dell believes that the most important leadership qualities in information economy are the following two:

  1. An ability to process information and make decisions in real time to sense customer value shifts and to respond accordingly
  2. Proficiency in managing a tightly controlled value chain formed by business partners contributing expertise. Integration of direct sales channels with the back-end supply chain is a key step.

13.4.4 Project Management

Project management deals with the management of human and physical resources required to attain a well-defined project objective on time and within budget (Lewis 2006; Kerzner 2009). Companies may focus on civil construction, capital equipment, research and development, new product or process design, cost reduction, equipment retrofit, quality improvement, customer and market research, and other projects with high engineering content. Project team members may consist of company employees, contractors, suppliers, or customers.

Project management has become more difficult as business development scopes enlarge, the contents of projects expand beyond organizational boundaries, and project team members become geographically dispersed. Globalization and new business practices make project variables such as contractor interfaces and increased uncertainties for on-time and on-budget deliverables more complicated. A new collaborative work style demands a different project management paradigm—from *command-and-control* to *quick collaboration*—as well as an increased effectiveness that demands just-in-time development and Internet-speed deployment.

The best practices in project management focus on five key management issues: tasks and resources, cost, risk, communications, and knowledge.

Task and Resource Management. The first and foremost step in managing a project is for the project manager to define the overall project objective and the associated budgetary, time, and other constraints, in consultation with company management. In the project-planning phase, tasks are formulated for implementation in parallel or in sequence, and their starting and completion dates are properly defined. Based on skills, experience, and availability, suitable people are assigned to these tasks, thereby establishing individual responsibilities and accountabilities. A Gantt chart or PERT (project evaluation and review technique) diagram is then used to graphically model the overall schedule and activities of the project. Tasks on the critical path are well defined, as these must be managed carefully. At that point, the anticipated completion date and budget for the project are iteratively finalized. The project manager initiates the project after having secured the approval of the project plan by company management and the commitment of all team members involved. The project manager also specifies the communications and data and document-transfer guidelines, as well as policies related to conflict resolution, problem solving, progress reporting, change management, and other issues. Training sessions are held to ensure that team members can utilize all of the project management tools efficiently.

Cost Management. Tools commonly used to manage costs include spreadsheet programs integrated with databases and automatic time entry systems. When incurred, actual expenses related to purchased items and capital assets are reported instantly. The total, cumulative cost incurred for the project is then determined and plotted graphically to compare with the planned budget. Progress reports are issued to keep company management as well as client organizations informed of the project status. When necessary, remedial actions are initiated in a timely manner to minimize the impact of possible budgetary and other deviations.

Risk Management. The successful completion of diverse projects is inherently subject to uncertainties beyond the control of the project manager or team members. Examples of such uncertainties include weather conditions, labor strikes, priority shifts, delays in the delivery of materials, and the unavailability of key team members due to emergencies. Forward-thinking project managers should be aware of the relative impact of such uncertainties and be prepared to respond effectively.

Most project management software tools permit what-if analyses that determine the overall project outcome (e.g., schedule and budget) when one or more input variables (e.g., task duration, task expenditure, and task initiation date) deviate from the plan, while keeping all remaining input variables unchanged. The results of such analyses allow the project manager to understand the relative sensitivity of the overall project outcome to these deviations. By rank ordering potential plan deviations, the project manager is able to proactively devise suitable contingency plans that minimize adverse impacts on the overall project outcome. Past experience in handling similar projects is helpful in managing this type of project risk.

However, it should be mentioned that a what-if analysis has a fundamental deficiency: It varies only a small number of input variables while it keeps a very large number of remaining input variables constant. This deficiency is not consistent with how projects behave in the real world. Real-world projects have uncertainties associated with each and every project task, although these uncertainties are larger in some cases than in others. The overall project outcome model would be a truly realistic one if all input variables were allowed to vary within their respective ranges. The Monte Carlo simulation technique is capable of permitting all input variables of a risky project to vary simultaneously. (See section 6.6.)

In recent years, companies in a large number of industries have started to apply the Monte Carlo simulation technique to manage risky projects on personal computers. All input variables may be modeled by a set of three values: the minimum, the most likely, and the maximum. The most likely value of an input variable corresponds to its value as originally planned in the conventional deterministic model, whereas the minimum and maximum values are the variable's possible lower and upper bounds, respectively. For each input variable, its lower and upper bound must be defined carefully to account for its range of potential variation. Each input variable is then modeled by a probability distribution function—triangular, Gaussian, beta, Poisson, or another—that portrays how the value of this input variable changes with probability and the highest probability being assigned to its most likely value. Upon initiation of the Monte Carlo simulation technique, all input variables are concurrently varied between their minimum and maximum values. The resulting project outcome as to schedule and budget is represented by a probability distribution function that has a set of minimum, most likely, and maximum values of its own.

By varying both the minimum and maximum values of specific input variables and the shapes of the distribution functions representing them, the project manager is able to ascertain their relative impacts on the overall project outcome. What the Monte Carlo simulation technique can deliver that the traditional what–if analysis cannot is two pieces of new information: (1) the absolute minimum and maximum values for the overall project outcome as to schedule and budget and (2) the probability (e.g., 80 percent) that the project will be completed within a specific budget and on or before a specific completion date. Both pieces of information are valuable to the project manager.

In fact, special cases are readily recovered by using the Monte Carlo simulation analysis. If all input variables are single valued (e.g., setting the minimum and maximum values of all input variables equal to their respective most likely values), then the overall project outcome will also be single valued. This result is equivalent to that produced by an ordinary deterministic project management tool. If one or more input variables are modeled by their three-valued sets while the majority of other input variables remain single valued, then the project outcome will correspond to that of a traditional what-if analysis.

Engineers and engineering managers need to familiarize themselves with advanced analysis techniques, such as the Monte Carlo simulation technique, which is also readily applicable to other engineering subjects involving uncertainties, such as system reliability, technology forecasting, and decision analysis.

Communication. Communication between team members is a key factor for success in any project. As project numbers increase and project scope expands, team members may become diversified in composition and dispersed in geographical locations. Team members communicate with one another through voice, text, video, data, and documents. Specific technologies must be actively applied to make communication between team members faster, easier, cheaper, of higher quality, and more productive. An open communication policy tends to encourage collaboration, which is essential for securing the success of any project.

Knowledge Management. Managing knowledge within a project is a job of critical importance. At the end of a project, all team members usually gather to celebrate, give recognition to those who performed well, congratulate each other for the success achieved, and then immediately dash out to take on the next assignment. There may be debriefing meetings that summarize what went well and what did not go so well. In general, preserving the learning from each project is not always done systematically (Stokes 2001).

Oftentimes, there is a significant amount of knowledge worth preserving. Project management practices that yield positive results need to be documented. Mission-critical data are worth saving. Contingency plans that were successfully activated to eliminate unexpected risk reflect corporate expertise that needs to be recorded. Specific lessons learned from each project should be preserved.

Also important is the need to store useful knowledge in forms that make it readily reusable, such as searchable databases. Knowledge that is saved to add value, but is not frequently applied, represents a waste of resources.

Software tools are available to facilitate various activities related to project management. What capabilities should one look for in a Web-based project management tool? Obviously, the project management tool must assist the project team, whose members may reside in distant locations and work in different time zones, in accomplishing the project objectives in time and within budget, while facilitating instant communications and data sharing. England and Finney (2007) offer one of the most comprehensive set of selection guidelines for Web-based project management tools, among those available in the literature. According to England and Finney (2007), a good Web-based project management tool should be able to do the following tasks:

- Offer easy access to the project status of assignments, costs, and the timeline in real time, via a browser-based client.
- Have a portal for project-related information.
- Import from and export to other project management tools.
- Run programs on a Web server.
- Manage contact information about every team member.
- Map projects and assign tasks to team members (Gantt charts, histograms, and PERT charts).
- Assign different views to allow each team member to see only the information and assignments pertinent to him or her.
- Enable members to update tasks, add notes to tasks, attach documents to the list, and send e-mails to others.
- Permit easy installation and use.
- Offer sample templates for specific project topics that are readily modified to fit individual needs.
- Possess the capability of tracking any number of tasks and milestones with associated links to Web sites, documents, and every other project.
- Hold on-line, text-based chats for virtual meetings or threaded discussions for collaboration between team members.
- Track resources allocation—load-leveling capabilities and membership rosters integrated with external directories.
- Organize user interfaces.
- Perform scenario analyses (what-if).
- Facilitate risk-adjusted modeling.
- Document project outcome and learning.

Project management software tools have migrated from the infrastructure of centralized minicomputers, to PC desktops and local area networks (LAN), and, finally, to Web-based tools (Sommerhoff 2000). In the Web environment, project management tools involve the following:

- E-mail
- Web pages to increase speed and availability of information
- GroupWare for project tracking and scheduling
- File transfer protocol (FTP) for transmission of documents and data
- Shared databases of multimedia formats
- Remote accessibility
- Integration with wireless and palm-top technologies

There are still two major constraints currently present in Web-based project management tools: bandwidth and infrastructure for multimedia transmission and data security.

Project management tools are advancing steadily. Enumerated here are the major project management tools available to date:

1. *Primavera P6 Professional* (www.oracle.com). This project management software is marketed by Oracle, Redwood Shores, CA. It has several modules useful for project management, including Project Planner to manage activities and resources, PrimeContract to manage costs, Evolve to manage the portfolio, and TeamPlay to promote the collaboration of participants.

   Toronto Transportation Commission (Toronto, Canada) successfully applied Project Planner by Primavera to manage fifty projects, each worth $2 million to $1 billion and having 100 to 600 activities, involving a total of 9,000 people. A single master schedule is updated across a series of PCs linked to four servers on a LAN. Data are then converted to PERT charts and distributed to all involved. The software displays the links between each project and its associated activities. Communications are facilitated by e-mail, Web pages, and other forms of Internet-based tools to keep everyone focused on the project goals. Resources to be deployed are categorized by discipline, workload, work hours available, and planned work to optimize resource loading for meeting project deadlines. Report generation may be in the form of paper, e-mail, and posting on TV or the Web. Data can be used for risk management purposes. The software allows project managers to keep in mind the big picture of all of their projects without getting lost in the details, but it also permits them to drill down for details if needed.

   Dick Corporation (Pittsburgh, Pennsylvania) utilized Primavera's Project Planner to manage the construction of a 115,000-square-foot addition to St. Francis Health System Hospital while the hospital was still operating. The software assisted in scheduling and coordinating all pertinent resources, including contractors, hospital personnel, and special equipment.

2. *Collaboration Workspace CW* (www.sword-ctspace.com). Sword-Ctspace in San Francisco markets this software, which provides an online collaboration workspace particularly suitable for managing construction projects in the building industry. Built on the Oracle database using Application Server 3.0 operating in a Windows environment, this software plans actions; tracks activities; coordinates the work of engineers, architects, and contractors; and automates various project components. Team members are connected to a password-protected Web site that contains project documentation, schedules, drawings, charts, and other project data files. Project managers have access to a messaging system and can monitor communications between team members.

   This software has been used by the San Francisco Public Works Department, Sun Microsystems, Stanford University, Fidelity Investment, and Charles Schwab.

3. *ProjectGrid* (www.ProjectGrid.com). This software of ProjectGrid.com, Inc., Columbus, Ohio, provides Web sites on a subscription basis to host project-specific information. It functions as an online communications management tool for project management activities. It is capable of managing (1) financial

information, (2) bid calendars, (3) project schedules, (4) work in progress, (5) site visit reports, (6) design drawings, (7) change orders, (8) contract invoices, (9) file downloads, (10) contract information, (11) administrative functions, (12) digital photo logs, and (13) other data. Team members access a password-secured Web site to receive updated project information. This program may be particularly attractive to small and midsize companies.

4. *Prolog* (www.meridiansystems.com). This software by Meridian Systems (Folsom, California) has five modules to aid in project management: (1) document management, (2) cost control, (3) field management, (4) purchasing control, and (5) reports and queries. Through browsers, team members can access design drawings, job-site photos, project schedules, and over 400 reports. Prolog is linkable to "e-commerce" and "collaboration" modules of Projecttalk.com to facilitate Web-based communication and collaboration among team members. It is regarded as a standard in the AEC (architects, engineering, and construction) industry and is widely used by architects, engineers, general contractors, and public institutions.
5. *Elite 3E* (www.elite.com). This financial and practice management tool by elite.com (Los Angles, California) connects team members with subcontractors and clients. It can be accessed via standard Web browsers and wireless Palm connections. It does full project budgeting, tracking, and reporting, and claims to be easy to learn and convenient to use. It offers convenient access to subcontractors and clients and can be activated through subscription.
6. *Microsoft Project® 2007 and Project Central* (www.microsoft.com). Project plans written on the desktop Microsoft Project 2007 program can be uploaded into Project Central, which resides on a Web server for storage in a linkable database (such as Microsoft SQL server, Oracle database, or others). Its capabilities include (1) creating Gantt charts for project and individual team members, (2) handling multiple project tasks and timelines, (3) producing summary reports, (4) updating project status and information by "time sheets," and (5) offering online help and tutorials (Chatfield and Johnson 2007). In January 2010, Microsoft released its Project 2010 (Beta) software.

   Team members can access project information and view Gantt charts, multiple project tasks, timelines, and summary reports through a standard Web browser. Project status information can be updated through time sheets that permit authorized team members to add tasks for themselves or others. Rules can be entered that, when triggered, will alert the project manager of unexpected deviations from the plan.
7. *Others*. Additional project management tools include (1) TeamCenter Project by Siemens (www.plm.automation.siemens.com/en_us) and (2) Basecamp Project Management by 37 Signals, Inc. (www.37signals.com).

As evidenced by the preceding discussion, numerous Web-based project management tools are strong in some aspects and weak in others. There appears to be no single tool capable of satisfying all project management requirements. A growing trend in the field of project management is to lease advanced tools to manage specific projects. Some application service providers (ASPs) market Web-based project management tools. Leasing project management tools is more cost-effective, meets the specific project needs better, and consumes less time for the company, compared with the option of either buying expensive software tools that have to be customized or building them in-house.

Application service providers are also capable of offering various additional services useful to the management of projects. These services include the following:

- *Collaborative capabilities*—Web site hosting, e-mail, fax, Internet chat session, news group, message boards, scheduling online meetings, audio- and videoconferencing, electronic calendars, online paging, and access control (checking in and out).
- *Work process management*—estimating and budgeting, task assignment and scheduling, job progress reports, purchase and procurement, accounting, and facilities management.
- *Work-flow capabilities*—document exchange (e.g., using PDF format), document management (review and editing of CAD models, photos, and drawings; control of file versions), data up- and download, approvals, change notification, transmittals, submittals, meeting minutes, and correspondence management.

Example 13.5 In the Information Age, every knowledge worker must know how to obtain useful information quickly from the Internet with a minimum amount of effort. As the amount of information continues to increase, information overload is a distinct possibility. Explain what one must know and do to surf the Internet effectively, find what one needs, and keep oneself current in selected topic-specific areas.

ANSWER 13.5 The following are commonsense knowledge chunks related to extracting topic-specific knowledge from Internet-based information resources:

1. Hardware and software

To access the Internet for an information search, one needs to have the use of a computer with the following capabilities:

- Modem (built-in or external) capable of dialing a local phone number.
- Windows or other operating systems (Unix, etc.).
- Web browser software program (Microsoft Internet Explorer, Mozilla Firefox, Netscape Navigator) that enables the viewing of and searching for information
- Word-processing software.
- Screen capture software that captures a photo, drawing, table, or graph contained in a document.
- Internet connectivity.

The user's computer must have a gateway to the Internet. A subscription-based Internet service provider (ISP) usually provides this gateway. An ISP-specific software program is then installed on the computer to check the user's identification, to automatically dial one of several local phone numbers, and to establish a gateway to the Web site maintained by the ISP. From this Web site, users may access other services (e.g., news reports, advertisements, etc.) offered by the ISP or may surf the Internet.

The speed of Internet access depends on the modem and the transmission line between the client (user's computer) and server (ISP's server computer). Examples of bandwidths include the following:

- 14.4-kbps modem 14.4 kbps
- 28.8-kbps modem 28.8 kbps
- Digital (ISDN) phone line 128 kbps
- T1 (dedicated connection) 1,500 kbps
- T3 (fiber optic backbone) 45,000 kbps

For students enrolled in a college-based degree program, computers with the aforementioned capabilities and built-in access to the Internet (including high-speed connectivity) are widely available.

2. Access to the Internet

Upon turning on an Internet-linked computer, the browser software icon is typically visible on the desktop. Double-click the browser software icon to start the browser program, type the URL of a database Web site or that of a search engine into the URL address bar, and hit "Enter" to activate. Different engines have different prompts.

Digitized engineering, management, and scientific information (e.g., publications in journals, periodicals, newspapers, conference proceedings, and books) is typically searchable from databases. The outcome of a search from a database is a number of engineering, management, or scientific articles (either abstract or full text) related to the search topic. Examples of full-text databases include the following:

- ABI/Inform (business periodicals and academic journals of management)
- Business/Industry (900 business periodicals)
- Business Index ASAP (Infotrac—business, management, and trade publications, as well as local newspapers)
- Disclosure Global Access (U.S. corporate financial reports)
- Dow Jones Interactive (*The Wall Street Journal* and other business news)
- Gale Business Resources (448,000 U.S. and international companies)
- IEEE Xplore (100 IEEE journals and transactions, including those on engineering management)

One additional database worth mentioning is Compendex Plus (an engineering index), which offers abstracts but no full-text documents.

Information of a commercial nature (e.g., companies or people offering expertise in specific domains, including products and services) is readily accessed through search engines. Search engines deliver a number of hyperlinked Web sites that offer additional information on products, services, and expertise related to the search topic.

3. Conduct search

Upon reaching the home page of either a database or a search engine, the user enters a set of keywords to specify the search topic in the search

Table 13.3 Internet-based Search (July 1, 2009)

Keywords	Google.com Hits	Business Index/ASAP Hits
Service Design and Development	127,000,000	34
"Service" "Design" and "Development"	137,000,000	34
Service AND Design AND Development	130,000,000	852
Service Design AND Service Development	180,000,000	5
Service "Design and Development"	7,080,000	8
"Service Design" and "Service Development"	23,600	1
"Service Design and Development"	444,800	2

box. Results produced are usually ranked on the basis of a combination of keyword matches, how often keywords appear in a document, and how often other sites link to a page. Some Web sites allows additional searches within the just-obtained search results to further refine the results, or "hits."

Experience has shown that it is advisable for the user to conduct topic-specific searches with several databases or search engines, as each might locate different resources.

The outcome of a search depends largely on how the keywords are entered. The use of Boolean operators (e.g., AND, OR, AND NOT, NEAR) and quotation marks around the key words will help narrow the search, as demonstrated in the example contained in Table 13.3.

To attain optimal search results, users are advised to consult the search help files available in each database or search engine to understand how to perform advanced searches constrained by domain, time horizon, and other criteria.

Newer search techniques are being advanced that make use of natural language and navigation engines. The purpose is to produce faster and more relevant search results.

4. Evaluation and file saving

Usually, most of the user's search results are at best only marginally related to the search topic. This initial list may contain thousands of documents. Additional manual editing is often required to narrow down the list further.

One way to do so is to scan the title of the document to gauge its relevance to the search topic and then open only those sites that appear to be relevant. Save the relevant information by selecting, copying, and pasting it into a word-processing file.

Some documents contain photos, tables graphs, or drawings that cannot be copied by the standard select copy and paste steps. However, these can easily be transferred into a word-processing file by using a screen capture software program.

5. Organization of knowledge

The saved word-processing files should be organized according to search topic, database or search engine, and search dates. This should preserve source information and minimize the additional search efforts needed to remain current on the topics.

Read the saved documents to extract important knowledge chunks, and produce summaries to review and update regularly as a step toward personal knowledge management.

13.4.5 Corporate Innovation

Continual innovation is the key to success for all enterprises. As a result of new products marketed and new patents issued, Microsoft, Cisco, IBM, DaimlerChrysler, General Electric, 3M, DuPont, Pfizer, DEKA corp., Rubbermaid, and other companies have been recognized as innovative.

Among the few best practices that foster corporate innovation are (a) recognizing innovation as a key to corporate survival in the long run, (b) committing sufficient resources to pursue innovative activities, (c) encouraging innovation from all organizational units in addition to R&D, and (d) selectively implementing innovative ideas to ensure business viability (Zairi 1999; Gulati, Sawhney, and Paoni 2002; Robert 1997).

Web-based tools are available to support the generation of innovative ideas. Support is available in the form of access to published information by using advanced search engines on the Internet. Support is achieved by exposing problems via Internet-based communications in order to engage more employees in finding solutions. Support is also obtained by using Web-based software tools to guide the generation of innovative ideas.

Invention Machine, Inc. (Boston, www.invention-machine.com), offers solutions to problems related to intellectual assets management. To facilitate innovations, its software applies semantic processing technology that has been perfected using the *theory of inventive problem solving* (TRIZ). The central principle of this theory is value analysis: To upgrade a product, place a value on each component, and then organize research according to those values. Invention Machine software searches deep into the Internet, analyzes large volumes of text, breaks down sentences, and reorganizes the contents into problems and solutions. Its Knowledgist program offers access to over 700 Web sites grouped by industry, reviews search results, sorts information according to user-defined parameters, and generates a structured index to allow an efficient review. Its Cobra program can "read" documents, pinpoint desired information, and extract only the needed sections for review.

Engineers using Invention Machine software are said to have significantly increased their innovative outputs, as measured by the number of original ideas advanced over a period of time. At least 500 companies are currently utilizing this software. Innovative products known to have been brought into being by the software include stronger parts for FormulaOne racing cars, better adhesives for Dow Corning, and a new type of filter for oil drilling equipment.

Assistance in generating innovation might be obtained from Dynamic Thinking (www.dynamicthinking.com), which specializes in providing a framework for understanding how ideas are formulated, promoted, and implemented.

The website yet2.com (www.yet2.com) offers a marketplace for licensable technologies. Performing patent searches online (www.uspto.gov) is already a common practice. Writing a provisional patent application can be assisted by using the software Patentwizard 2.0 (www.patentwizard.com). Filing a provisional patent application for a new innovation is usually advantageous, as it allows the inventor to test the market for the yet-to-be-patented idea or product for a period of one year before a formal patent application is submitted, thus preserving the inventor's original filing date.

These examples illustrates that some Web-based tools may augment the innovation process at any company. Indeed, the speed of access to information on the Internet will continue to increase. Search engines will become more powerful and intelligent. Knowledge extraction techniques will be further perfected. For the innovation process to be productive, however, these technological enablers are not enough. Employees have to be trained and proper methods must be incorporated into the workplace as a part of daily work life. Successful inventors must be properly recognized and adequately rewarded for innovation to flourish in a company.

Example 13.6 Describe some well-known e-business models that have been proposed and pursued in recent years by the knowledge companies.

ANSWER 13.6 Business models are methods of doing business by which an enterprise becomes and remains profitable. Typically, firms combine several models to pursue a Web strategy. E-business models are generally of the following specific types:

1. *Brokerage model.* A Web site is set up to (1) facilitate transactions (e.g., B2B, B2C, C2C), (2) charge a fee per transaction, (3) perform buy–sell fulfillment functions, (4) report results, and (5) provide access to information needed by the parties involved. (See Fig. 13.4.)

   Examples of such sites are the following:

   - E-Trade (financial securities fulfillment)
   - Expedia (travel services)
   - CarsDirect (autos)
   - MySimon (search assistance for the best price)
   - Priceline (demand collection, reverse auction, name your price)
   - ChemConnects (chemicals market exchange)
   - Mercata (aggregate buyers to benefit from volume-based purchasing)
   - DigitalMarket (connecting preferred distributors and buyers)
     - Yahoo's Store (set up merchants in a virtual mall for buyers to reach)
     - Amazon ZShops (transaction services, track orders, billing, and collection services)

2. *Informediary model.* The Web site offers information of value to individuals and businesses, provides free access to the site and free e-mails for personal communications, and collects users' profiles for sales. (See Fig. 13.5.) The principal income is derived from the sales of consumer profiles to businesses.

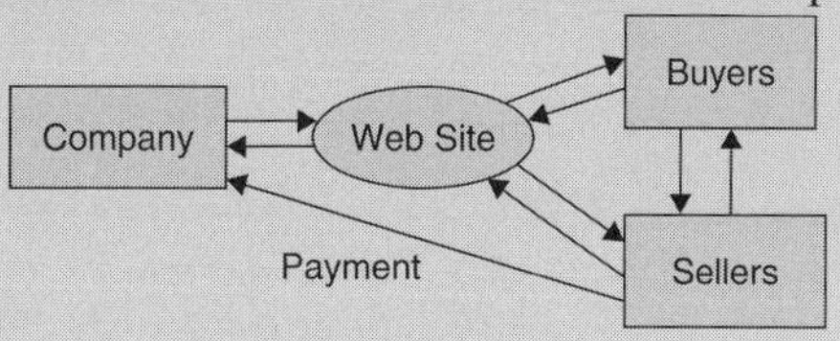

Figure 13.4 Brokerage model.

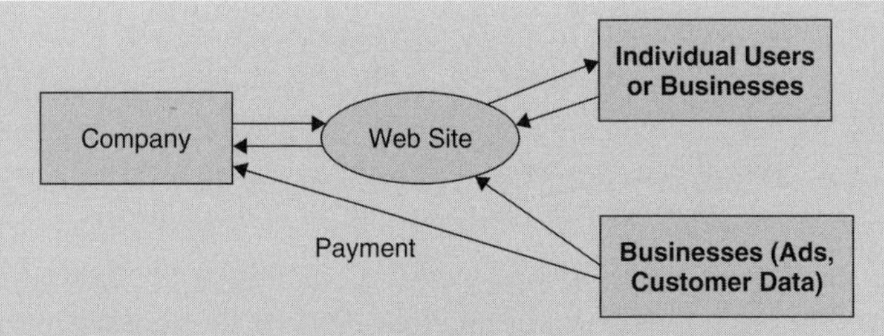

Figure 13.5 Informediary model.

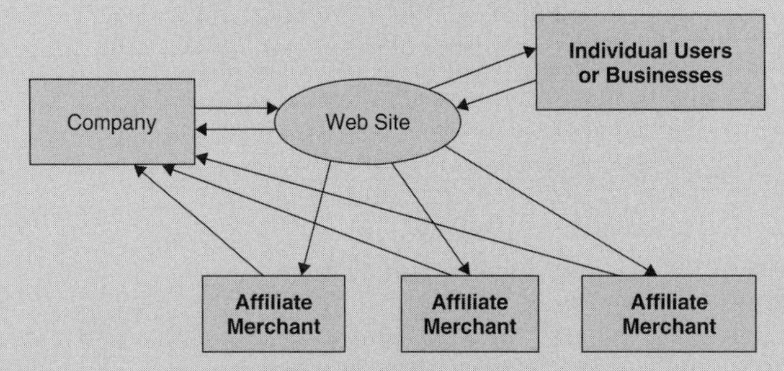

Figure 13.6 Affiliate model.

Examples of such sites are the following:

- NYTimes (uses registration to collect user information, free view of contents)
- Emachines (free hardware in exchange for detailed information about Web surfing and purchasing habits)
- Gomez (provides consumers with useful information about Web sites)
- Netzero (provides free Internet access)

3. *Affiliate model.* The company sets up a network of affiliate merchants, each having specific products to sell. The Web site offers free information to consumers, who can point and click to reach a specific affiliate merchant. The site provides fulfillment services. The affiliate merchant pays a fee per concluded purchase. This is a pay-for-performance model. (See Fig. 13.6.) Examples of such sites include (1) BeFree, (2) Affiliateworld, and (3) I-Revenue.net.
4. *Merchant/manufacturer model.* The company promotes goods (self-produced or vendor-supplied) made by partners for sale through its Web site, which offers information access to customers and facilitates all online fulfillment functions. The company may also maintain bricks-and-mortar storefronts to serve local customers ("surf and turf"). Examples of such sites include the following:

   - Gap
   - Lands' End
   - Virtual merchants (Amazon)

- Intel
- Apple
- Gateway

Under certain circumstances, channel conflicts may show up for businesses that maintain both online and physical storefronts.

5. *Advertising model.* The company establishes a Web site that offers contents free to consumers (e.g., news, search capabilities, chat rooms, stock quotes, forums, articles on specific topics, entertainment, etc.). The company's principal income is advertisements placed and paid for by merchants or businesses. In this model, the company works more or less like a radio or TV station; hence, it is also referred to as a broadcaster model. Examples of such sites include Google, AltaVista, and other search engines.
6. *Subscription model.* Users pay a subscription fee to access the Web site for premium news or specific reports, although some news is free. Examples of companies maintaining such sites include *The Wall Street Journal, Consumer Reports*, and TimeWarner.
7. *Others.* Engineering managers are advised to watch out for new models being constantly advanced. Diverse businesses may use more than one model in combinations to achieve Web success.

13.5 WEB SERVICES

In recent years, Web services have become increasingly popular as a computing concept. Pautasso (2010) introduces the Web service technologies. In general, Web services have several common components that make them capable of interacting with human users or with other applications:

- Simple Object Access Protocol (SOAP) provides the envelope for sending the Web services messages.
- The Web Services Definition Language (WSDL) forms the basis for Web services. A service provider describes its Web services using WSDL (e.g., XML, Java, and others).
- Universal Description, Discovery, and Integration (UDDI). Registries can be searched to quickly, easily and dynamically find and use Web services.

There are two specific types of Web services. The first type is associated with application services that are delivered to human users over the Web. The second type is related to applications modules that are made accessible to other applications over the Internet through XML-based protocols.

Several IT organizations have started offering the first type of Web services. Amazon.com, Sun Microsystems, Microsoft, Salesforce.com, and others are known to be the vendors for such services (Huckman et al. 2008).

- *Storage*. Simple Storage Service (S3) by Amazon.com provides a basic interface for storing and retrieving data from anywhere on the Web. To ensure secure storage, each object uploaded to S3 will be duplicated and multi copies are stored in multiple locations.

- *Computing*. Elastic Cloud Computing (EC2) by Amazon.com represents a Web service that provided resizable computing capacity in the cloud. Clients could run programs on Amazon's computers. The Sun Grid utility services offered by Sun Microsystems provides clients with the computing power, also on a pay-as-you-go basis.
- *Database*. SimpleDB by Amazon.com enables real-time lookup and query of structured data.
- *Simple queue service*. SQS of Amazon.com is a scalable, hosted queue, which can be used to store data messages sent between applications, thus allowing clients to move data between application components that perform different tasks. In other words, it allows components of an application to send messages to other components and create a place to store these SQS requests and have the consuming components pick them up.
- *Flexible payment services* (FPS). This service is offered by Amazon.com for clients to collect funds from their customers using credit cards, while allowing the clients to define payment conditions.
- *Online marketplace*. Salesforce.com offers Exchange as an online marketplace, allowing clients to swap and sell applications. It's the modules of sales, service, marketing and partners.
- *Build and host Web applications*. Google offers App Engine, which allows clients to build and host Web applications on Google's scalable infrastructure. Windows Live by Microsoft offers the service of an e-mail program, news headlines, blog and audio feeds, and a Web page builder.
- *Premium support*. Assistance is offered by Amazon.com to clients regarding on S3, EC2 and SQS.

In addition, companies such as IBM, Sun Microsystems, and Hewlett-Packard offer Web-based infrastructure and software solutions to business clients. The proper use of these types of Internet-based technology services will likely to raise the operational productivity of many service enterprises.

The second type of Web services focuses on creating modularized applications that interact with other modularized applications, following specific standards regarding language, data description, data exchange, and connection protocols. Because everyone shares the same standards, these modularized applications are able to talk freely with other applications (McAfee 2005). Thus, this second type of Web services allows the construction of modular and interchangeable building blocks of software, which are most useful as the basic layer components of an SOA (see Fig. 13.7).

13.6 SERVICE-ORIENTED ARCHITECTURE (SOA)

Since 2003, service-oriented architecture (SOA) and Web services have become a major new development in the distributed computing world. Innumerable publications in the literature promote the business value that may be created by employing this new and emergent computing approach. Zhang and Bernardini (2010) address the design issues related to SOA. Ratakonda et. al. (2010) discuss the complex problems related to the global delivery of SOA solutions.

It is important for SSE engineers and leaders to understand how SOA is defined, what elements SOA is composed of, what benefits and risks might be involved in the

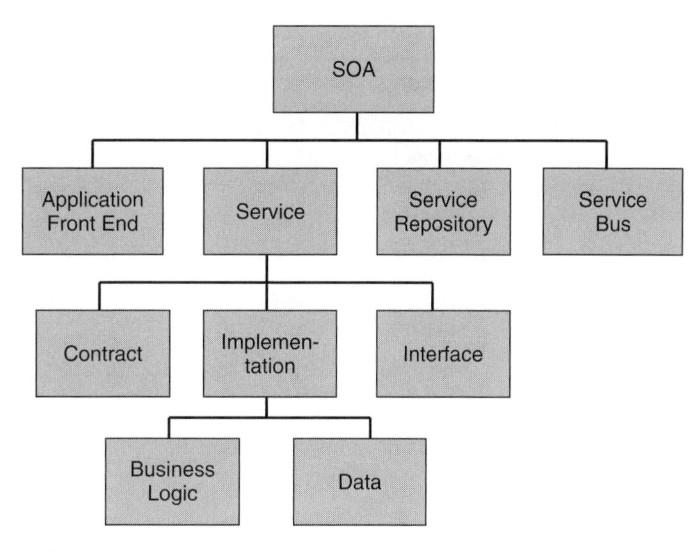

Figure 13.7 Structural elements of an SOA. (Adapted from Krafig et al. 2005)

implementation of SOA, which SOA applications cases have become available, and, above all, what might be the potential impact SOA on operational excellence to services enterprises.

13.6.1 Definition

In the literature, service-oriented architecture (SOA) has been defined by different people in different ways (Erickson and Siau 2008). Table 13.4 presents some of these definitions. Currently, there is no consensus among various groups regarding a unified definition for SOA.

SOA is defined from a technological standpoint in (Krafig et al. 2005) as a service architecture that is based on the key concepts of an application front end, service, service repository, and service bus. A service consists of a contract, one or more interfaces, and an implementation. Accordingly, SOA may have up to four layers, as shown in Fig. 13.7

1. *Enterprise layer*—application front ends (for communicating with end users) and public enterprise services (enabling cross enterprise integration).
2. *Process layer*—containing process-centric services.
3. *Intermediary layer*—providing technological gateways, facades, adapters.
4. *Basic layers*—providing the basic services of the SOA (business logic, data, proxies for other company's public enterprise services).

For airlines, Fig. 13.8 illustrates its SOA structures (a fully developed SOA with four layers).

XML-based Web services (based on SOAP and WSDL) are not the only viable technology platform for an SOA, which is generally not dependent on any particular technology platform.

Table 13.4 SOA Definitions

#	Definition of SOA	Proposed by
1	SOA is an architectural style that supports service orientation	Open Group
2	SOA is a paradigm for organizing and utilizing distributed capabilities that may be under the control of different ownership domains	Organization for the Advancement of Structured Information Standards (OASIS)
3	SOA is an architectural style for a community of providers and consumers of services to achieve mutual value	The Object Management Group (OMG)
4	SOA is a form of distributed systems architecture that is typically characterized by a logical view, a message orientation, a description orientation, granularity, and platform neutrality	Worldwide Web Consortium (W3C)
5	SOA is an architectural style whose goal is to achieve loose coupling among interacting software agents	XML.com
6	SOA is an evolution of distributed computing based on the request/reply paradigm for synchronous and asynchronous applications	Javaworld.com
7	SOA describes a style of architecture that treats software components as a set of services	IBM

Source: Erickson and Siau (2008).

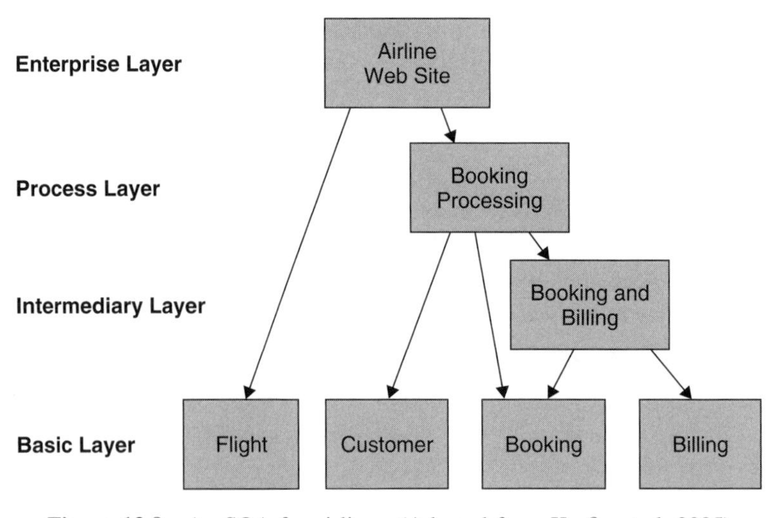

Figure 13.8 An SOA for airlines. (Adapted from Krafig et al. 2005)

Erl (2007) offers the following definition for SOA: "SOA establishes an architectural model that aims to enhance the efficiency, agility, and productivity of an enterprise by positioning services as the primary means through which solution logic is represented in support of the realization of strategic goals associated with service-oriented computing."

This definition seems to encompass the various common elements contained in the previous seven divergent definitions. It tends to emphasize the commonality among them. It appears to be the best broad-based description of what SOA is all about.

13.6.2 Implementation of SOA

SOA contains three principle elements, according to (Carr 2005):

1. *Virtualization*. It eliminates the platform-related restrictions to allow applications to run on different platforms.
2. *Grid computing*. It allows a number of diverse hardware components (servers, disk drivers, etc.) to interact readily and to form an operating unit.
3. *Web services*. It makes both legacy and new software to interact effectively by standardizing the interface between applications (input format, language of transmitting messages, and other interfacing protocols) thus turning them into modules that can be assembled and disassembled easily.

As a result, SOA is expected to deliver the following useful functionalities:

- *Composite applications.* SOA supports the creation of composite applications that bring together numerous business processes from multiple systems into a simple user interface. With SOA companies will be empowered to change business processes with greater confidence, predictability and frequency, leading to better agility and results.

  The services provided by vendors are shared among applications. Web services create a common platform that enables SOA.
- *Agility*. Rapid change with SOA, change inherently becomes faster and easier, because the architecture is centered on the concept of services that are designed to be interoperable and used across different business processes and different business contexts.
- *Reusability*. The service-oriented approach delivers IT systems as a set of reusable services than can be assembled easily to create a composite application that automates a business process. This "assembly approach" accelerates the time to market for new applications and reduces IT costs.

SOA allows services to be reused across the organization and assembled on the fly. In so doing, SOA produces rapid business and technology responses to changes in the marketplace.

To minimize the operating risk associated with the implementation of SOA computing, there are five phases of SOA implementation that must be managed properly.

1. Defines the desirable future state and the roadmap to achieve it (e.g., set the vision and plan the project).

2. Design an initial set of enterprise software services with obvious business values and impact to gain corporate acceptance.
3. Set up a library of enterprise services that are useful to a vast majority of corporate users.
4. Institutionalize the SOA program to further align with the business organization.
5. Continue to innovate to enhance the value of SOA to the organization.

SSE engineers and managers who are interested in becoming versed in the computational/technological issues related to SOA design and implementation should consult the "Prentice Hall Service-Oriented Computing Series From Thomas Erl," which include (1) concepts, technology and design (Erl 2005), (2) service design (Erl 2007), (3) service pattern (Erl 2009) and (4) design and versioning (Erl et al. 2008), as well as other suitable technology resources.

13.6.3 Benefits of SOA

Because of the aforementioned set of functionality, SOA is expected to produce a number of benefits to its corporate users:

- Foster time-to-market for new applications and help organizations achieve competitive advantage by rapidly seizing new opportunities and responding to unexpected threats. A reduction in time needed to deliver new services is made possible by the highly reconfigurable implementation of business processes.
- Flexible applications lead to flexible business processes, making the organization more agile and adaptable to the pace of changing business environment. The integration of applications is streamlined.
- Service reuse brings about greater efficiency and lowers maintenance costs, making IT more efficient in supporting business initiatives. A better utilization of development resources is assured by the reuse of existing software assets.
- The ability to respond quickly to new regulatory requirements or specific compliance issues helps companies avoid governmental penalties.

These potential benefits are directly useful to service enterprises which strive to achieve operational excellence.

13.6.4 Potential Causes for SOA Failures

SOA is a mainstream IT initiative. However it may fail due to a number of deficiencies listed here, some of them are management related:

- Not communicated properly to enable the wide use of SOA by everyone.
- Lack of trust in the quality offered by the SOA software involved.
- Insufficient and difficult testing of SOA components. An insufficiently tested but reusable service can amplify risks, as a failure in a crucial service can ripple through all the applications and other service that access it. Thus, proper testing of services is a critical requirement with respect to functionality, performance, security policies, interoperability, and boundary conditions.
- Service-level agreement between providers and consumers is inadequate.

- Change of service infrastructure has impact on availability and performance of the service.
- Requirements for services are not clearly visible.
- Services designed are not properly governed and designed for reuse.
- Support personnel is lacking.

Quite a few of these identified deficiencies are related to project management. SSE engineers and leaders should pay attention to them when implementing SOA.

13.6.5 Critical Success Factors

Rogers (2008) conducted a survey among IT managers and enterprise architects to define the critical success factors for SOA program:

- *Business alignment*. Focus on enterprise business goals and strategies, business support and involvement, value measurement, and align funding model).
- *Organizational change management.* Align resources, implement recognition/incentive program, skills and education.
- *Communications*. Communications are necessary for advocacy, awareness, visibility, and discovery, and progress reporting.
- *Trust*. Trust is key for performance, availability, and reliability of services managed to SLAs (service level agreements) and contracts, security, quality.
- *Architecture*. Institutionalize enterprise SOA reference architecture and standards. Implement cross-enterprise services management discipline.
- *Scale and sustainability*. This includes SOA program scalability, technical scalability, service granularity.
- *Governance*. Introduce strong SOA governance program at early stages of initiative, integrate SOA and overall IT governance processes, specify runtime practices and design policies, and define and assign key roles and responsibilities.

  It is rather clear from this list that it takes a determined corporate commitment to pursue SOA initiatives and to achieve useful outcome.

13.6.6 SOA Applications Cases

A European-based telecommunications and Internet provider doubled its corporate size in 2005 by merging with two other companies. The organization needed to become more flexible in managing its business lines, reducing staff costs, and ensuring high levels of subscriber support. Hewlett-Packard, one of several major SOA service providers, implemented HP SOA Management and HP SOA Systinet programs to better align the company's IT project management and the organizations' business strategies. The SOA initiative was focused on managing its assets and monitoring service performance. The following programs were introduced:

- HP SOA Systinet—governance
- HP SOA Policy Enforcer and HP Diagnostics—service monitoring
- HP Business Availability Center for SOA—service availability

Benefits realized include (1) 33 percent increase in services uptime, (2) 50 percent reduction in time to market for new services, and (3) 30 percent savings in total service management costs. The average annual benefits had been in excess of $5,900,000 (Anonymous 2008). Krafzig et al. (2005) contains additional SAO applications cases of interest in the service industry:

- Deutsche Post AG (German post office)
- Winterthur Group (insurance)
- Credit Suisse (financial services)
- Halifax Bank of Scotland (financial services)

Interested readers should consult these references.

Example 13.7 Peachtree Healthcare, Marietta, Georgia, manages a group of community hospitals that are affiliated with a total of 4,000 physicians who care for 1 million patients per year. Over the years it had grown rapidly. The CEO, Max, is under pressure to introduce increased level of standardization of the work Peachtree does, to correct the mismatched equipment it currently has, and to fix the IT systems that break down often. The CEO was presented two options to modernize the IT system (Glaser 2007):

1. Install a complete new monolithic system, which will cost about $1 billion. Such a system is well proven in performance.
2. Design an SOA system, assisted by SOA vendor, that could lead to increased productivity in speed, patient care, and quality over the long run. The SOA system will be the very first in healthcare industry, and there is risk of unknown magnitude involved.

A panel of four experts was asked to offer recommendations as to what the CEO should do.

- The CEO of another healthcare system recommends that Peachtree should stick with the installation of a traditional IT system, without imposing strict operating standards on physicians.
- A vice president of an airline company suggests that it may be prudent for Peachtree to consider installing SOA piecewise, by replacing the broken parts of the old IT system first and then updating the whole system in stages. This is to lessen the uncertainties associated with employing a brand new SOA technology.
- The vice president of an SOA consultancy believes that Peachtree should go ahead with SOA in full force, as SOA is a widely acceptable technology by a large majority of enterprises in the developed world, although not yet in healthcare industry.
- A business professor argues that standardization is a bad thing for physicians and that Peachtree should stay away from using SOA.

What is your opinion on this issue?

ANSWER 13.7 All four suggestions have specific merits. On the one hand, risk management is a major concern in healthcare organization, as it may involve life and death. On the other hand, management must consider the use of new technology prudently in order to capture its potential long-term value to an organization, while gaining a widespread acceptance by its key stakeholders, including physicians.

Since the development of SOA goes through several expansion stages—Fundamental SOA (two layers), networked SOA (three layers), and Process-enabled SOA (four layers) (Krafzig et al. 2005)—the best option moving forward should be one with a stage-wise installation of SOA. Doing so will allow the organization to experience the new benefits in a timely manner, to gain the buy-in from physicians based on new results, to minimize the financial burden to the organization, and to preserve the option of changing course in midstream if SOA prototypes do not work out.

13.6.7 The Future of SOA and the IT Utilities

Currently, countless service companies have invested in building and maintaining in-house computer facilities to provide the needed business analysis and other computing activities in support of their business needs. In the future, many of these computing needs may be satisfied by running software hosted by major vendors. This is similar to today's industrial and residential consumers who buy electricity from electric utilities, rather than producing electricity with their own small-scale generators.

Proponents of SOA envision that the creation of an IT utility, just like the electric utility, is just around the corner and that this SOA approach will lead to the end of corporate computing (Carr 2005) because of its obvious impact on costs, speed and process flexibility of business computing.

Example 13.8 There has been some talk in the business literature about "The End of Corporate Computing" (Carr 2005) because of the anticipated expansion of SOA and the widespread use of Web services in the future. An IT utility is to be set up, just like the evolution of the electric utility in the last century, from which individual businesses and consumers will "buy" IT services, instead of creating their own local computing sites.

The success of an IT utility, built on SOA and Web services, is predicated on the premise of a superior scale of economies and the unprecedented computing flexibility/agility it offers to its clients, thus significantly improving their productivity in business computing. Please explain the basis for the claimed scale of economies and the flexibility/agility characteristics of such a future IT utility.

ANSWER 13.8 The key drivers for the expanded scale of economies are due to the fact that in the recent past, thousands of independent data centers were built by enterprises, all using virtually the same hardware. Many of these facilities run at just 10 percent to 35 percent of their processing powers. Studies also show that desktop computers use only about 5 percent of its capacities and that 50 percent to 60 percent of a typical company's data center storage capacities is wasted. According to a 2003 survey, 60 percent of the IT staffing budget of an average U.S. company goes to routine support and maintenance functions. These data centers run, for the most part, similar software programs as well, thus adding more to the diseconomies they represent. An IT utility allows clients to assemble and disassemble modularized and reusable software components, which may reside in a variety of different platforms at global locations, will be potentially very cost-attractive to its clients on a "pay-as-you-go" basis.

Furthermore, self-contained software programs are typically designed to be mission-specific. They do not allow flexible changes dictated by the changing business or market circumstances. Such inflexibility will lead to longer time-to-market disadvantages and higher costs related to their modifications. SOA will change the inflexible IT silos into flexible and aligned business-service building blocks giving IT the foundation to become more responsive and agile. The traditional applications and business processes are being transformed into re-usable shared services. SOA makes it easy for companies to develop, reuse, and modify business-critical services cost-effectively.

Once the IT utility becomes available, clients can readily buy and sell modularized software components, access the proper one on the fly, and be able to perform business computing customized to the instantaneous needs of the marketplace. Such flexibility/agility would be of tremendous competitive value to service enterprises.

13.7 CONCLUSIONS

As the service sectors become increasingly competitive on a global scale, service systems engineers and leaders will need to continue to drive for operational excellence. Besides applying the standard engineering management functions of planning, organizing, leading and controlling to projects, teams, processes and activities, they need to become versed in utilizing tools discussed in this chapter, such as *Lean Six Sigma*, Web-based applications software, Web services and the low cost and flexible computing capabilities offered by an emergent SOA-based IT utility.

Each of these tools discussed in this chapter has its rightful place in service enterprises, to be applied under the right circumstances. Reducing process speed is the primary drive of the Lean principle. Service quality is emphasized by the standardization approach pursued in Six Sigma. The combination of Lean and Six Sigma will simplify work complexity. The application of *value stream mapping* helps to identify wastes, which represent work that does not add value to the customers. Cost is reduced by having screened out the wastes and standardized the work that does add value. Operational excellence is accomplished by having increased service speed, raised service quality, and minimized service costs. Operational excellence can be further enhanced by an increase in productivity, if Web-based applications software are selectively employed on a pay-as-you-go basis.

Service contains a number of support elements, most of which are amendable to standardization, the premise of Six Sigma. The core service elements are exceptions; they cannot be readily standardized because of the diversity in customers' needs. The use of Web services and the emergent SOA based computing model appears to offer an unprecedentedly high degree of operational flexibility, in addition to business computing agility and software reusability, that would facilitate the optimization of these core service elements to achieve an improved level of service customization.

These tools need to be considered by service systems engineers and leaders when they strive to find optimized approaches to effectively deal with problems or opportunities in the future. It is advisable that they keep themselves current with the future developments in the emergent business computing domains.

13.8 REFERENCES

Anderson Djorn and Tom Fagerhaug. 2006. *Root Cause Analysis: Simplified Tools and Techniques*, 2nd ed. ASQ Productivity Press (June 15).

Anonymous. 2000. "i2 Chosen by Leading Automakers for e-Business, Initiatives: DaimlerChrysler, Toyota and Volkswagen each select i2 solutions to power private marketplaces" *Business Wire*, Dallas (August 23). www.thefreelibrary.com/i2+Chosen+by+Leading+Automakers+for+e-Business+Initiatives%3B . . . -a064493089.

Anonymous. 2008. "Reducing Operations Costs and Improving Customer Experience with HP SOA Management," IDC Business Value Spotlight # 08C5613 (August).

Anonymous. 2008 "2008 SIA User Survey: Adoption Trends and Characteristics," Gartner Inc. www.gartner.com/DisplayDocument?ref=g_search&id=765720&subref=simplesearch.

Ayers, J. B. (ed). 2006. *Handbook of Supply Chain Management*, 2nd ed. Boca Raton, FL: St. Lucie Press.

Berry, M. J. A., and Gordon Linoff. 2004. *Data Mining Techniques for Marketing, Sales and Customer Relationship Management*. 2nd ed. New York: Wiley Computer Publishing.

Berry M. J. A, and G. Linoff. 1999. *Mastering Data Mining: The Art and Science of Customer Relationship Management*. New York: John Wiley & Sons, Inc.

Biolos, Jim. 2002. "Six Sigma Meets the Service Economy," *Harvard Management Update* (November).

Brunet, P., C. Hoffmann, and D. Roller (eds.). 2000. *CAD Tools and Algorithms for Product Design*. New York: Springer.

Burnett, K. 2000. *The Handbook of Key Customer Relationship Management: The Definitive Guide to Winning, Managing and Developing Key Account Business*. Upper Saddle River, NJ: Financial Times Prentice Hall.

Carr, Nicholas G. 2005. "The End of Corporate Computing." *MIT Sloan Management Review*, Vol 46 (3) (Spring).

Chatfield, D. S., and T. Johnson. 2007. *Microsoft Project 2007 Step by Step*. Richmond, WA: Microsoft Press.

Checkbook. 2009 "How Banks Rate for Constomer Service," www.checkbook.org/sitemap/Washington_DC/Ratings_And_Articles/Banks/detail.cfm?uKey=2092

Chopoorian, J. A., R. Witherell, O. E. M. Khalil, and M. Ahmed. 2001. "Mind Your Business by Mining Your Data." *SAM Advanced Management Journal* 66 (2) (Spring): 45–51.

Dyche, J. 2001. *The CRM Handbook: A Business Guide to Customer Relationship Management*. Boston: Addison-Wesley.

Erickson, John, and Ken Siau. 2008. "Web Services, Service-Oriented computing, and Service-Oriented Architecture: Separating Hype from Reality." *Journal of Database Management* 19 (3).

England, E., and Andy Finney. 2007. *Managing Interactive Media: Project Management for Web and Digital Media.* Boston: Addison-Wesley (August 30).

Erl, Thomas. 2005. *Service Oriented Architecture (SOA) Concepts, Technology and Design.* Upper Saddle River, NJ: Prentice Hall.

Erl, Thomas. 2007. *SOA: Principles of Service Design.* Upper Saddle River, NJ: Prentice Hall, page 38.

Erl, Thomas. 2009. *SOA: Design Pattern.* Upper Saddle River, NJ: Prentice Hall.

Erl, Thomas, Amish Karmarkar, Priscilla Walmsley, and Hugo Hass. 2008. Web Services Contract Design and Versioning for SOA. Upper Saddle River, NJ: Prentice Hall.

Franzelle, E. 2002. *Supply Chain Strategy: The Logistics of Supply Chain Management*. New York: McGraw-Hill.

George, Michael. 2003. Lean Six Sigma for Services: How to Use Lean Speed and Six Sigma Quality to Improve Services and Transactions. New York: McGraw-Hill.

Glaser, John P. 2007. "Too Far Ahead of the IT Curve." *Harvard Business Review* 85 (7/8).

Green, H. 2003. "The Web Smart 50." *BusinessWeek* (November 24).

Gulati, R., M. Sawhney, and A. Paoni (eds.). 2002. *Kellogg on Technology and Innovation*. Hoboken, NJ: John Wiley & Sons, Inc.

Hernandez, J. A., Franklin Montinez, and James Koegh. 2005. *The SAP R/3 Handbook*, 3rd ed. New York: McGraw-Hill.

Huckman, Robert, S., Cary P. Pisano, and Liz Kind. 2008. "Amazon Web Services," Harvard Business School Case #9-609-048.

Kamber, Micheline, and Jiawei Han. 2005. *Data Mining: Concepts and Technologies,* 2nd ed. San Francisco: Morgan Kaufman.

Kerzner, H. 2009. *Project Management: A Systems Approach to Planning, Scheduling and Controlling*, 10th ed. Hoboken, NJ: John Wiley & Sons, Inc.

Krafzig, Dirk, Karl Banke, and Dirk Slama. 2005. *Enterprise SOA—Service Oriented Architecture Best Practices.* Upper Saddle River, NJ: Prentice Hall Professional Technical Reference.

Latino, Robert J., and Kenneth C. Latino. 2006. "Root Cause Analysis: Improving Performance for Bottom-line Results," 3rd ed. Boca Raton, FL: CRC Press.

Lewis, James P. 2006. *Fundamentals of Project Management (WorkSmart)* 3rd ed. New York: AMACOM.

Lovelock, Christopher, and Jochen Wirtz. 2006. "Service Marketing—People, Technology and Strategy," 6th ed. Upper Saddle River, NJ: Pearson Prentice Hall.

Magretta, J. 1988. "The Power of Virtual Integration: An Interview with Dell Computer's Michael Dell." *Harvard Business Review* (March–April): 72.

Martin, James W. 2006. *Lean Six Sigma for Supply Chain Management—The Ten Step Solution Process.* New York: McGraw-Hill Professional (January 1).

McAfee, Andrew. 2005. "Will Web Service Really Transform Collaboration?" *MIT Sloan Management Review* 46 (2).

McDermot, Robin E., Raymond J. Mikulak, and Michael R. Beauregard. 2008. *The Basics to FMEA,* 2nd ed. New York: Productivity Press.

McLauglin, S. 2010. "Service Operations and Management," Chapter 14 in Salvendy, Gavriel and Waldemar Karwowski (Eds), *Introduction to Service Engineering*, John Wiley (January).

Merrifield, Ric, Jack Calhoun, and Dennis Stevens. 2008. "The Next Revolution in Productivity," *Harvard Business Review* 86 (6) (June).

Meyerson, J. M. 2001. *Enterprise Systems Integration*, 2nd ed. Boca Raton, FL: Auerbach.

Nash, Mark, and Sheila Poling. 2008. *Value Stream Mapping: The Complete Guide to Production and Transactional Mapping.* New York: Productivity Press.

Nykamp, M. 2001. *The Customer Differential: The Complete Guide to Implementing Customer Relationship Management*. New York: AMACOM.

O'Leary, D. E. 2000. *Enterprise Resource Planning Systems: Systems, Life Cycles, Electronic Commerce, and Risk*. New York: Cambridge University Press.

Pautasso, C. 2010. "Web Service Technology," Chapter 23 in Salvendy, Gavriel and Waldemar Karwowski (Eds), *Introduction to Service Engineering*, John Wiley (January).

Pendharkar, P. C. 2003. *Managing Data Mining Technologies in Organization: Techniques and Applications*. Hershey, PA: Idea Group Publishing.

Ratakonda, K., Y. M. Chee, D. Openheim and F. Bernardini. 2010. "Streamlining the Delivery of Complex SOA Solutions with Global Resources," Chapter 28 in Salvendy, Gavriel and Waldemar Karwowski (Eds), *Introduction to Service Engineering*, John Wiley (January).

Robert, M. 1997. *Strategy Pure and Simple II: How Winning Companies Dominate Their Competitors*. New York: McGraw-Hill.

Rogers, Sandra. 2008. "A Study in Critical Success Factor for SOA," *IDC— Analyze the Future* (September).

Rud, O. P. 2001. *Data Mining Cookbook: Modeling Data for Marketing, Risk and Customer Relationship Management*. New York: John Wiley & Sons, Inc.

Schmitt, B. H., and Bernd Schmitt. 2003. *Customer Experience Management: A Revolutionary Approach to Connecting with Your Customers*. Hoboken, NJ: John Wiley & Sons, Inc.

Sheridan, J. H. 1999. "Focus on Flow." *Industry Week* (October 18), p. 46.

Simchi-Levi, D., P. Kaminsky, and E. Simchi-Levi. 2007. *Designing and Managing the Supply Chain.* 3rd ed. Boston: McGraw-Hill/Irwin.

Sommerhoff, E. W. 2000. "E-Commerce: Managing Design and Construction Online." *Facilities Design and Management*, October.

Stamatis, D. H. 2003. "Failure Mode and Effect Analysis: FMEA From Theory to Execution." 2nd ed., Milwaukee, WI: ASQ Quality Press.

Stengl, B., and R. Ematinger. 2001. *SAP R/3 Plant Maintenance: Making It Work for Your Business*. New York: Addison-Wesley.

Stokes, M. (ed.). 2001. *Managing Engineering Knowledge: MOKA—Methodology for Knowledge Based Engineering Applications*. New York: ASME Press.

Voehl, F. and A. Elshennawy. 2010. "Lean Service," Chapter 18 in Salvendy, Gavriel and Waldemar Karwowski (Eds), *Introduction to Service Engineering*, John Wiley (January).

Wagner, Bretand, and Ellen Monk. 2008. *Enterprise Resource Planning,* 3rd ed., Course Technology. Boston:

Williams, G. C. 2000. *Implementing SAP R/3 Sales and Distribution*. New York: McGraw-Hill.

Zacks, R. 2001. "Ten Emerging Technologies that Will Change the World." *Technology Review* 104 (January): 97.

Zairi, M. 1999. *Best Practice: Process Innovation Management*. Woburn, MA: Butterworth-Heinenman.

Zhang, L. J. and F. Bernardini. 2010. "Design of Service-Oriented Architecture (SOA)," Chapter 10 in Salvendy, Gavriel and Waldemar Karwowski (Eds), *Introduction to Service Engineering*, John Wiley (January).

13.9 QUESTIONS

13.1 How are URL, domain name, and search engines defined? Use examples to explain the relationship between them. How can one make use of Web pages to promote business?

13.2 What are Internet, intranet, and extranet? How are they being used by numerous large and small companies today?

13.3 What are the standard markup languages used in the design of Web pages?

13.4 What are some of the legal issues related to the Internet and Web-based business transactions that remain unresolved at this time?

13.5 In implementing a computerized maintenance management system to reduce maintenance costs, what steps are taken?

13.6 What is Web mining, and how significant is it in generating useful results to support management decision making?

13.7 For the development of software products, the SCM (software configuration management) process is closely followed as a way to ensure performance and reliability while controlling costs. Explain what SCM can do and in what ways it is important that both developers and intended customers insist on SCM.

13.8 Although marketing and sales are not functions of engineering, they have a direct impact on product development and customer relationship management. Which Web-based applications are currently available to facilitate marketing and sales?

13.9 The business environment in the new millennium will continue to be fast paced, Internet enhanced and globally oriented. Name a few factors that will affect the business successes of any companies in such a challenging environment.

13.10 The Design for Lean Six Sigma (DFLSS) is a methodology known to be particularly useful to designing new services that are in close alignment with customer and business needs. Explain the key phases the DFISS methodology goes through.

13.11 Services are known to have a variety of wastes, which, if not removed, will increase costs and erode service quality, leading to customer dissatisfaction. Name a few of the typical wastes encountered in service offerings.

13.12 The Lean principle focuses on the improvement of process speed. It is thus particularly useful to service enterprises, which are in need of shortening its customer response time. Explain the basic concepts involved in Lean to improve process speed.

13.13 There are two types of Web services. The first type offered software applications that are accessible to human users. The second type provides software applications that can be accessed by other applications. Explain the basic requirements of building Web services to create applications, which can be accessed by human users.

13.14 As the SOA service vendors are likely to be consolidated over time, an IT utility would emerge. Under that scenario, most businesses will buy computing services instead of maintaining their own in-house computing data centers, much like how business, commercial and residential customers are buy electricity today. What are the potential concerns to service enterprises, which become dependent on the IT utility, insofar as operation and financial risks are concerned?

13.15 Study the Harvard Business School Case "Cleveland Clinic" Case #9-607-143 (Rev. September 26, 2007) and answer the following questions. The case materials may be purchased by contacting its publisher at 1-800-545-7685, or at www.custserve @hbsp.harvard.edu.

A. What is the case about?

B. What specific practices have contributed to creating its culture of excellence and innovations?

C. Which aspects of the Cleveland Clinic practices are unique? Would these practices be applicable to companies in other industries?

D. Are there any obstacles for Cleveland Clinic to move forward?

E. What are the key lessons learned from this case?

Chapter 14

Globalization

14.1 INTRODUCTION

Globalization is defined by the International Monetary Fund as the growing economic interdependence of countries worldwide through the increasing volume and variety of cross-border exchanges in goods, services, capital, and technologies.

Globalization is not a new phenomenon. International trade and commerce have a very long history. In recent years, the growth of the world economy and the migration of goods, services, capital, people, and technologies across borders has dramatically increased. The rapid expansion of the digital economy has also helped to accelerate the pace of globalization (Collier and Dollar 2002; Langhorne 2001). Some American companies are expanding to reach new global markets and foreign resources.

The world economy has become increasingly global, as current markets are more interconnected than ever before. Instead of only a few countries handling the trade of most currencies and goods, now many more countries play a part. American companies are actively pursuing markets in Asia, Europe, Latin America, and other regions. Some foreign-owned companies have achieved more sales outside their home countries than in their respective domestic markets (Govindarajan and Gupta 2001; Steger 2002).

Global mindset is defined as the ability to establish, interpret, and implement criteria for business performance that are not dependent on the assumptions of a single country, culture or context (Begley and Boyd 2003). The truly globalized corporations think globally (recognizing when it is beneficial to establish a consistent global standard for products and services), think locally (deepening the company's understanding of the local and cultural differences), and think both globally and locally (recognizing situations in which demands from both global and local elements are compelling). Both global consistency and local responsiveness are important.

John Zeglis, then president of AT&T, said in 1999, "There are two kinds of companies in the future: those that go global and those that go bankrupt."

In the past, global corporations too often demanded special treatment for their export businesses, pushed back on environmental regulations, sought to avoid taxes, and resisted costly labor market rules. The foreign direct investment (FDI) they contributed boosted economic performance by endowing developing countries with new skills, new technologies, and new jobs, all of which increased the standard of living of the host countries. According to Oppenheim (2004), the right way forward is for the global corporations to promote policies that strengthen the stability of the host countries and for the developing countries to deconstruct the barriers and restrictions that hinder FDI. Multinational corporations should become allies of social progress, so that the economic growth of host countries is accelerated, while allowing the multinational corporations to build enduring competitive advantages based on established trust and legitimacy.

In this chapter, we will explore various management issues related to globalization (Rao 2001; Sullivan 2002). We shall focus on steps that service systems engineers and managers can pursue to take advantage of the value-adding opportunities offered by globalization.

14.2 GLOBAL TRADE AND COMMERCE

Global trade and commerce are complex in today's environment. It can be described from several different perspectives. First of all, United Nations, as an international body promoting the peace and wellbeing of all nations, has set specific goals of reducing poverty via globalization, the extent of which is clearly reflected in the statistics it produces. Multi-national enterprises have become increasingly important to global trade and commerce. The ownership of a variety of global companies are becoming more global in its self, reflecting the investors readiness in taking on larger uncertainties associated with global businesses. Globalization is driven by a number of specific forces, which are worth recognizing. From time to time, globalization will also be affected by catastrophic events, albeit only for a limited period of time. This section offers discussion on all these factors.

14.2.1 United Nations Statistics and Goals

Statistics published by the World Trade Organization (WTO) indicate that world trade, defined as the total value of exports, has increased about 500 percent in the past 30 years at a compound growth rate of 6.15 percent a year. Specifically, world trade grew sixfold in real terms from 1980 to 2007, and its share of global gross domestic product (GDP) has risen from 36 percent to 55 percent. Between 1990 and 2004, cross-border financial assets have increased from 58 percent to 131 percent of global GDP.

It is well-known that those countries that were open to global trade grew twice as fast as those that remain relatively closed to trade. At least two African nations, Nigeria and Tanzania, have chosen to rely on protectionism, foreign aid, and inefficient public policy. Today, they remain at the 1960s economic development levels of Malaysia, Thailand, and Indonesia. In recent years, Latin America has started to embrace market liberalization. It has abandoned its old policies of a dominant state presence in the economy, import substitution, and domestic industry protection. The results of these changes are encouraging, and more countries are expected to open their markets in response to increased globalization.

According to the World Bank (2001), the world output is projected to increase 33 percent from \$30 trillion in 2001 to \$40 trillion in 2010. A more recent Word Bank press release (World Bank 2008) stated that the world output had increased to \$59 trillion in 2006. Furthermore, based on purchasing power parity (PPP), China was ranked as the second largest economy in the world and five of the twelve largest economies were developing economies in 2006. Since the onset of the global recession of 2007 – 2009, the world output has now been forecast to decline for the first time in 2009 and to experience a slow recovery in 2010 (World Bank 2009).

The disposable income in regions such as China, India, Southeast Asia, and Latin America had doubled during the growth period of 2001 – 2006. About 300 million people (roughly the population of the United States) joined the thriving worldwide middle class at the same time. World Bank statistics also showed that from 1990 to 2000, only about 800 million people moved out of absolute poverty, which is defined as having income of less than a U.S. dollar a day. As of 2001, 50 percent of the world population lived on less than $2 a day. Eighty percent of the global population lives on less than 20 percent of all global income. The recession of 2007 – 2009 brought about a major hardship to countless people in the developing countries because of the induced unemployment there and this situation has attained crisis proportion.

The United Nations has declared that one of its goals is to decrease the number of people in absolute poverty by 50 percent by 2015. Globalization is regarded as a key process in achieving this meaningful goal (Lamberton 2002).

The world economy has become interdependent in recent decades. The worldwide integration of national economies—through the trade of goods, services, capital, and technologies—has become broad and deep. Another indicator of global trade activities is the steadily increasing number of strategic alliances formed across the twenty-nine industrialized countries that are members of the Organization of Economic Cooperation and Development (OECD). This trend is expected to continue.

14.2.2 Multinational Enterprises

Multinational enterprises (MNEs) operate in more than one country. These enterprises play important roles in the global economy:

- Holding 90 percent of all technology and product patents worldwide.
- Conducting 70 percent of world trade, 30 percent of which is intracompany.
- Pursuing diversified businesses, such as (1) mining, (2) refining and distributing oil, gasoline, diesel and jet fuel, (3) building energy plants, (4) extracting minerals, (5) making and selling autos, airplanes, communication satellites, computers, home electronics, chemicals, medicines, and biotechnology products, (6) harvesting wood and making paper, and (7) growing crops and processing and distributing food products.
- Inducing governments to form treaties and trading blocs among the European Union, the North American Free Trade Agreement (NAFTA), the WTO, the Multilateral Agreement on Investment, and the Uruguay round of the General Agreement on Tariffs and Trade (GATT). These treaties tend to provide great power and authority for multinational enterprises to pursue globalization, thus increasingly undercutting the authority and power of national governments and local communities.

The 500 largest MNEs are responsible for 80 percent of all foreign direct investments. Of these MNEs, 443 are located in only three regions: the United States, the European Union, and Japan. (A detailed distribution of these major MNEs is presented in Fig. 14.1.) About 80 percent of the global trade has been between NAFTA (United States, Canada, and Mexico), the European Union, and Asia (including Japan).

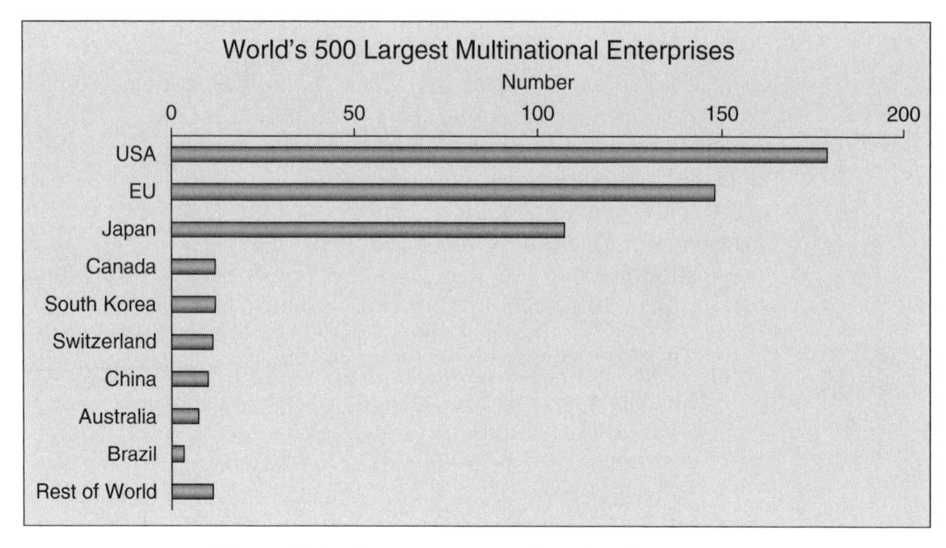

Figure 14.1 Current concentration of major MNES.

If we take a longer-term view, we cannot afford to ignore the forecast made by the World Bank (1992). That forecast said that by 2020, the largest economies in the world are projected to be China, the United States, Japan, India, and Indonesia.

In the future, economic growth rates in emerging markets are predicted to be three to ten times that of the United States. About 50 percent of worldwide GDP (gross domestic product) would be generated in emerging markets. Consequently, we should expect that the situation just described (i.e., the MNEs having only regional operations for now and the home bases of the 500 largest MNEs being concentrated in the triad regions) will surely change in the years to come. The roles played by the emerging countries in Asia, such as China, India, and Indonesia, could become substantial indeed. The extent of trade and commerce globalization is expected to further increase.

Goldman Sachs, an investment firm based in New York, studied the gross domestic product (GDP) of both the G6 and BRIC nations. The G6 is composed of the United States, Japan, Germany, France, Britain, and Italy. Canada is normally part of the G7 but is excluded in this study because its GDP is only about 3 percent of the G7 total. BRIC is an abbreviation for Brazil, Russia, India, and China. Goldman Sachs predicts that by 2037, the total GDP of BRIC will match that of the G6. (See Fig. 14.2.)

Should that prediction hold true, it would mean that the center of the next phase of globalization would move from the G6 nations to BRIC nations.

This type of forecast is, of course, valid only in the absence of any disruptive events, such as wars, global economic recessions, or natural disasters. Nevertheless, it does foretell the emergence of some developing economies and the increased degree of globalization in the years ahead.

14.2.3 Ownership of Global Companies

The five major stakeholders of any company are customers, employees, suppliers, investors, and the communities in which the company operates.

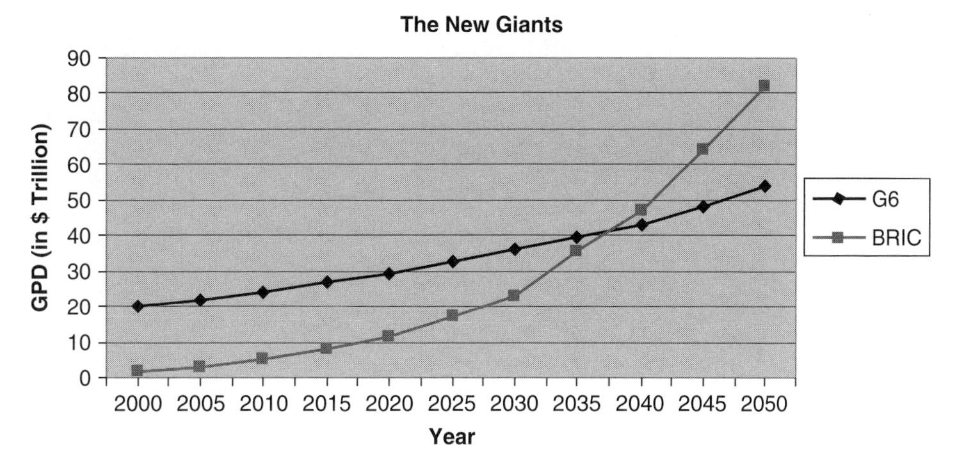

Figure 14.2 GDP forecasts for G6 and BRIC nations. (Adapted from Wilson and Parashothaman, 2003.)

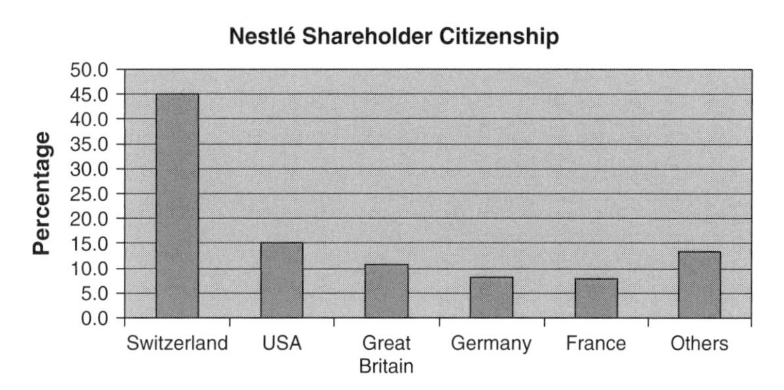

Figure 14.3 Ownership of Nestlé.

A large number of global companies manufacture products designed to reach global customers, employ workers from different countries, source materials and components from suppliers in global markets, interact with local communities at global locations, and have global shareholders. In recent years, countless countries have been setting up stock exchanges and security markets to attract foreign or domestic investments. Nowadays, it is easy for an investor to become a shareholder of any global company that is traded on one of several public stock exchanges.

As illustrated in Fig. 14.3, the ownership of Nestlé, a well-known multinational enterprise, is quite global indeed. No single Nestlé shareholder owns more than 3 percent of the company stock. This trend of global ownership is expected to continue as the capital markets become more accessible to investors residing in various countries. Over time, companies will diligently apply innovative global marketing strategies to sell products and services to global customers, in order to generate value for a global ownership.

Example 14.1 What are some practical reasons for a company to expand into international business?

ANSWER 14.1 The reasons for a company to expand into global markets are plentiful. The following are the primary four:

1. *Desire to expand markets (finding new customers).* For companies whose products have been selling in a saturated domestic market at home, expanding into global markets represents an attractive opportunity to find new customers.

   Theodore Levitt (the late *Harvard Business Review* editor) proposed the idea that the characteristics of some products are converging, making them more and more universal, thus allowing companies to market efficiently to the global marketplace. Companies might be able to confer advantages based on the global economies of scale. However, in order to do well, companies need to understand the local customers, business practices, and cultures of global customers.
2. *Search for natural resources.* Companies pursue foreign investments to avail themselves of resources that might otherwise not be readily available. U.S. investments in Saudi Arabia and various U.S. offshore natural gas exploration projects are the prime examples.
3. *Proximity to customers.* Companies expand into the global marketplace to be closer to their customers, for the sole purpose of understanding and serving them better, faster, and cheaper. Some products require customization in order to enrich the value offered to the customer. Customer satisfaction will increasingly become a key competitive focus. Companies that understand their customers more thoroughly and customize their products will have a significant advantage in the marketplace.
4. *Labor savings.* Today, certain developing countries offer skilled labor at a fraction of the cost needed to hire similar workers in the home countries of numerous major companies. Mexico is a prime example. Score of U.S. companies have set up manufacturing shops in Mexico. The products made there are shipped back to the United States for distribution and marketing. Several other countries, such as China, India, the Philippines, and Thailand, are also candidates for companies to realize labor savings in certain types of products or services.

Example 14.2 When companies attempt to pursue global markets, what common entry strategies are deployed?

ANSWER 14.2 There are several common entry strategies into global markets. In general, companies are well-advised to first study the relative attractiveness of the target market (e.g., specific segments in different countries) by considering factors such as profitability, market size, and market growth. In addition, companies need to assess the degree of acceptance of their products in the targeted marketplaces (e.g., brand name, competitive position, market access). Once the most favorable product-market-segment pairs are selected, companies can pursue global markets by:

- Exporting
- Licensing or contracting manufacturing
- Forming joint ventures with local partners
- Creating a foreign branch of the company
- Establishing a foreign subsidiary of the company

14.2.4 Globalization Drivers

Numerous driving forces are present in the global economy. Companies might consider different factors, such as market reach, cost, competition, and government, before choosing to go global.

Market drivers include worldwide increases in per-capita income that result in greater purchasing power and an increasing demand for goods worldwide.

Another market driver is the convergence in lifestyles, tastes, aspirations, and expectations of consumers. An additional market driver is increased global travel, which brings about a new class of global consumers. A further market driver is the creation of larger consumer markets in emerging countries. More than 90 percent of the world's population is outside the United States. Companies need to pursue globalization to reach those extended markets (Johansson 2000; Cundiff and Hilger 1988).

To be located close to customers is an important corporate marketing strategy. Since 1980s, Japanese automotive firms (Toyota, Honda and Nissan) have been reliably expanding production in Mexico and US (Rogers and Gereffi 2007). Concurrently, US automotive firms (GM, Ford and Chrysler) have been cutting back productions in the US and Canada while expanding productions in Mexico and China. Within North America, it was noted that the US share of automobile production has been decreasing constantly, whereas that in Canada and Mexico had increased in the recent years. For the year 2009, analysts forecast that 12.6 million cars will be sold in China and only 10.5 million cars in the US.

Cost drivers include lower manufacturing and production costs (primarily lower labor costs), economies of scale, accelerating technological innovations, and upgraded transportation and logistics. Some companies seek a cost advantage as the primary motive to go global.

Competitive drivers include (1) global competitors with speed and flexibility, (2) the increased formation of global strategic alliances, resulting in a proliferation of partnerships with suppliers, customers, and competitors (Calpan 2002), and (3) more countries becoming attractive battlegrounds. Creating competitive advantages is the principal goal for some companies to pursue globalization.

Government drivers of globalization include the emergence of trading blocks (EU, NAFTA), large-scale privatization (Brazil, China, etc.), and a reduction of trade barriers (WTO). Companies go global to take advantage of the benefits made possible by these official or semiofficial government bodies.

Ernst & Young conducted a survey of more than 300 CEOs in 1993. The top ten drivers in the global race were recognized as follows:

1. Increased speed of delivery to customers
2. Enriched ties with strategic partners abroad
3. Enhanced support of domestic customers' international operations

4. Meeting of cultural needs of foreign customers
5. Access to new technologies
6. Avoidance of overseas protectionism
7. Reach for lower taxes and government benefits
8. Access to foreign technical and management talent
9. Utilization of low-cost labor
10. Avoidance of domestic regulatory constraints

These results are consistent with the survey outcome obtained by Pricewaterhouse-Cooper in 2006 (Nelson 2006) in that the primary drivers for OECD (Organization for Economic Co-operation and Development) multinationals to pursue globalization are to access new customers, to better service their existing customers, and to do businesses in the BRIC (Brazil, Russia, India and China) economies. Indeed, numerous forces of significant magnitude are driving companies toward globalization.

14.2.5 Impact of Catastrophic Events on Globalization

Globalization may be negatively affected by catastrophic events to varying degrees. This section discusses a few of such examples before and after 2007.

Notable Events Before 2007. In recent years, several major events have had a profound impact on the world economy, political stability, and peace. Included are the terrorist attacks on the Twin Towers of the World Trade Center in New York City on September 11, 2001; the Iraqi war in 2003; and the Severe Acute Respiratory Syndrome (SARS) outbreak in South China in 2003 that spread rapidly to Hong Kong, Singapore, Toronto, Taiwan, and other locations. The immediate consequences of each of these events have been an increase in the cost of doing international business and changes in business relationships between the United States and other countries. A number of projected factors could exert a cooling effect on globalization:

- Since the war against terrorism might be protracted, insurance premiums might be raised because of heightened security concerns.
- Increased security risks reduce the willingness of businesspeople to travel internationally and might precipitate a reduction in team performance, collaboration, information sharing, and knowledge management.
- A higher return might be demanded to compensate for increased investment uncertainties.
- Heightened border inspections could slow cargo movements and force companies to stock more inventories (such as spare parts).
- Tighter U.S. immigration policies could curtail the inflow of skilled and blue-collar workers (e.g., from Mexico and Canada to the United States).
- Time horizons for international projects might be shortened when companies make new foreign direct investment.
- The availability of global equities might drop because fewer investors are willing to take the added risks involved. Foreign direct investment to specific countries regarded as posing a relatively high business risk (e.g., India, Pakistan, the Philippines, parts of South America and Southeast Asia, most of the Arab world, and Russia) might be cut.

- Because of disagreements with U.S. foreign policy, some businesspeople from developing countries might become reluctant to make deals with American businesses.

There are countless specific examples of the rising costs and uncertainties of doing international business. U.S. expatriates are leaving Indonesia because of the radical Islamic unrest against U.S. and British interests there. Cargo-laden trucks are taking seven hours to cross the Laredo, Texas, border crossing, compared with only two hours before the September 11, 2001, terrorist attacks. Delphi Automotive Systems, which operates fifty-six plants in Mexico, scheduled 200 trucks a day to bring products into the United States before the terrorist attacks. Now, the company ships parts in smaller lots more frequently so that it can redirect shipments to planes, boats, or helicopters if the transportation situation so requires. It might take several years for these effects to dissipate and for the world economy to resume a normal growth pattern. Even though Delphi Automotive systems since declared bankruptcy in 2005 due to financial reasons and reemerged from it on 2009, the company's handling of the border crossing problems remains an exemplar response.

The Global Recession of 2007 to 2009. The 2007 to 2009 global recession deserves a special mention. From 2004 to 2006, U.S. housing prices increased rapidly as a result of several related events: (1) banks encouraged would-be home buyers with poor or questionable credit scores to take variable-rate loans at low or no down payment, (2) builders added excessively large numbers of new homes (3) credit rating agencies assigned more favorable ratings than justifiable to companies marketing mortgage-based securities, and (4) greedy hedge-fund specialists repackaged such securities and aggressively marketed them to drive up their market valuation. When some of the mortgage loans were reset at their regular higher interest rates, a great number of homeowners with poor creditworthiness could not pay, leading to delinquencies and accelerated foreclosure. Eventually, the U.S. housing market bubble burst in 2007, causing nationwide housing prices to tank significantly. In many cases, the market price of the home dropped well below the balance of the mortgage, causing these homeowners to become "under water." Big losses were incurred by the affected banks, hedge funds, and insurance and financial institutions that held these toxic mortgage loans, or that insured or bought mortgage-secured securities. These losses made them less creditworthy, further limiting the banks' abilities to borrow money from other banks and making them less capable of lending money to businesses and consumers.

In the United States, consumer spending accounts for about 70 percent of the GDP. As consumers spent less, businesses suffered declines in sales, which necessitated cuts to production and eventually lay off workers. Indeed, most sectors—materials, energy, industrials, materials, technology, and telecommunications—were continuing to post strong declines. Loss of jobs and tight consumer credit further compounded the problem of lower sales and reduced profitability for businesses. This vicious cycle caused a variety of businesses to run into liquidity problems and to declare bankruptcy, which in turn further increased the unemployment level in the United States. During this period, consumer confidence continued to slide. The United States was officially declared in recession in November 2007. The net worth of U.S. households decreased from $64.2 trillion in the second quarter of 2007 to $51.48 trillion in the first quarter of 2009 (see Fig. 14.4).

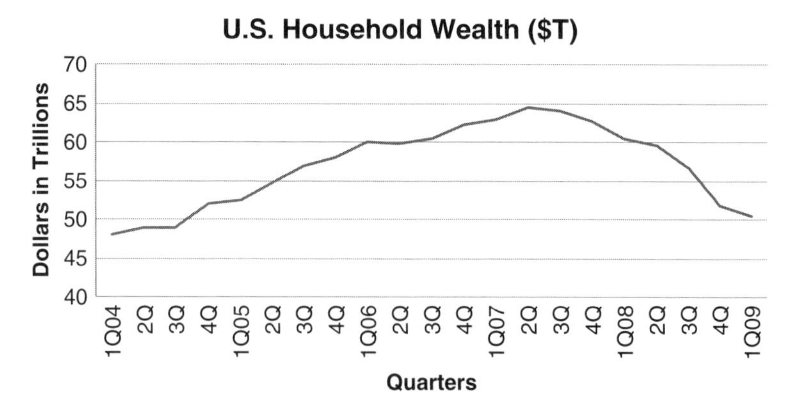

Figure 14.4 Major Change in U.S. household wealth.

In September 2008, Lehman Brothers collapsed and sent the global credit crisis into a tail spin. JPMorgan Chase acquired the failing Bear Stearns and Washington Mutual. Bank of America swallowed the near-bankrupt Merrill Lynch. Wells Fargo took over Wachovia Bank. In the same month, the federal government placed the lenders Fannie Mae and Freddie Mac into conservatorship of the Federal Housing Finance Agency. In October 2008, the U.S. Congress approved the Troubled Asset Relief Program (TARP) Act, authorizing $700 billion to buy up the toxic assets held by the financial institutions. Since then, some big-name players, such as Citigroup, Goldman Sachs, Bank of America, Morgan Stanley, JPMorgan Chase, Wells Fargo, General Motors, and Chrysler, have received bailout money in exchange for stock warrants.

While receiving governmental bailout money, Bank of America and Merrill Lynch distributed bonuses worth a total of $6.5 billion to their top managers. The case is being investigated by New York State Attorney General, as of this writing. At the same time, Bank of America allowed its vice chairman to decorate his office with an $80,000 rug. American International Group (AIG), which lost $61.7 billion in the fourth quarter of 2008, distributed $165 million in bonuses to its top managers in March 2009 after having received bailout money in the amount of $170 billion from the federal government. Also in March 2009, financier Bernard Madoff confessed to operating a Ponzi scheme, which paid earlier investors with money received from new investors. He stole $50 billion from his investment clients, as the scheme went undetected by the Securities and Exchange Commission for more than ten years. In June 2009, Madoff was sentenced to 150 years in prison for his wicked deeds. During 2008 to 2009, over 160 regional banks in the U.S. were shut down by the federal government.

In February 2009, the U.S. Congress approved the American Recovery and Reinvestment Act (ARRA 2009), providing $787 billion to expand consumer loans, help homeowners to pay mortgages, invest in infrastructure to create new jobs, digitize healthcare records, reduce taxes, and support the development of green energy, among other programs.

The U.S. unemployment rate, which stood at 6 percent in January 2007, 8.1 percent in February 2009, and 9.8 percent in September 2009, was expected to reach 10 percent plus by the end of 2009. As of March 2009, 5.3 million workers were receiving unemployment benefits. The U.S. recession quickly spilled over to Asia and Europe and precipitated a severe global economic crisis, due to international trade and interdependency of finance and investments across international borders. Although the United States focused on the use of stimulus as the principal means of spurring a recovery, France and Germany believed that the excess of banks and financial markets were responsible for the crisis. China, India, Brazil, and Russia—the BRIC countries—called for stricter control of the highly unregulated U.S. hedge funds and credit derivatives. Tough controls over U.S. credit-rating agencies, which gave strong ratings to weak securities before, also was urged. The United States was demanding that other governments commit 2 percent of their respective GDP to financial stimulus, but only China, Saudi Arabia, and Spain responded. The other countries strongly urged the United States to reform its financial regulatory system and to exert stricter oversight going forward.

The recovery from this global recession is already somewhat in sight, as of this writing; some prominent economists think that it might take up to two years for the global economy to return to normal again. Looking forward, it is reasonable to assume that several new conditions will be established in the post-crisis global economy (Davis 2009):

- *Less leverage*. Money is less easy to come by than before. The risk premium will be higher for new capital in the future. Companies should focus on productivity to gain attractive returns on equity. Instead of relying on innovations in financial engineering, they should go back to those in real engineering.
- *More government*. New regulations will demand better transparency and financial disclosure. There will be increased global coordination between governments. Government ownership of specific businesses will be the norm rather than the exception, including shaping the executive compensation policies. Some industries will become more regulated than before—capitalism will be on sabbatical.
- *Reduced consumption*. The future economy will be less dependent on consumer spending, which will make up a percentage of GDP less than 70 percent. Businesses need to adjust to the potential change of this key driving force accordingly.
- *Eastward shift to Asia*. The globalization center will shift to Asia, where some big national economies will continue to gain a respectable growth rate in the post-crisis period.

The terrorist attacks on 9/11, the Iraqi war in 2003, and the SARS epidemic in 2003 were enemies from the outside. They are relatively easy to target, conquer and neutralize. Responsible for the global recession of 2007 to 2009 is greed, corruption, collusion, dishonesty, and incompetence—enemies from within, found hiding in unsuspected places, making them much more difficult to detect, regulate, control, and eradicate.

Globalization is manifested in international trade, business alliances and cross-border flow of investment capital and technology, and other types of collaboration between nations. The 2007 to 2009 global recession is expected to slow down the globalization pace, at least for a while, until the financial systems are stabilized and consumers and investors regain their confidence. In fact, some countries might become increasingly inclined to enact protectionist measures. However, global integration is likely to continue in the long run.

14.3 THE GREAT PHILOSOPHICAL DEBATE ABOUT GLOBALIZATION

In 2007, the International Monetary Fund published a study titled "Globalization and Inequality," which pointed out that because of globalization, income growth has been evident in all regions and income groups. However, income inequality has increased mainly in middle- and high-income countries, and less so in low-income countries. The rich have gotten a lot richer. The poor have become less poor but have stayed more or less where they were in terms of their share of total income. The middle class has been squeezed (Gumbel 2007).

Not everyone favors globalization. It produces winners and losers. Generally speaking, government leaders of both large and small countries are in favor of globalization because of its heightened opportunities for foreign direct investment, transfer of technology and best practices, and trade benefits. Business leaders are strong advocates of globalization for various reasons, including the following:

- The greater flow of trade and investments stimulates economic growth.
- Rising output brings about employment and income, which means higher living standards for consumers.
- Higher living standards facilitate a greater social willingness to devote resources to the environment, education, healthcare, and other social goals.
- Global competition keeps domestic business competitive and innovative, which in turn brings about higher quality output and productivity.
- Rapidly developing economies tend to generate a new middle class that is the bulwark of support for personal liberty and economic freedom.

Opponents of all stripes and creeds blame globalization for many of the world's ills (Sassen 1998). They are primarily from three major groups: labor union members, human rights activists, and environmentalists.

Labor unions want to protect local jobs in industrialized countries, as globalization will likely induce multinational enterprises to transfer manufacturing and other high-paying jobs to developing regions in search of cost competitiveness (Phillips 1998). They also raise the issues of child labor and forced labor in poor countries, citing past incidents of exploitation by multinational enterprises.

The human rights groups say local economic growth induced by globalization in some emerging countries might allow their respective dictatorships to stay in power longer, thus indirectly supporting continued suppression of the people. Globalization would become an inadvertent coalition partner in crime against humanity (Brysk 2002).

The environmentalists think that by relocating the manufacturing operations to developing countries with lower environmental control standards, the multinational enterprises are essentially exporting pollution and other environmentally unacceptable practices to the poor countries, thus causing irreversible damage (Asheghian and Ebrahimi 1990).

These three groups are united in their opposition to globalization. Their reasoning is further summarized as follows:

- Globalization is a conspiracy of big companies exploiting small countries. It concentrates market power in the hands of a few large corporations, allowing them to trample over smaller commercial rivals and flourish at the expense of small companies and consumers. Globalization is akin to companies without rules.

- Globalization promotes the suppression of human rights in developing countries.
- Globalization destroys the environment.
- Globalization spreads terrorism, narcotics, disease, and money laundering (Condon 2002; Kugler and Frost 2001; Horowith 2001).
- Globalization lowers labor standards and turns developing nations' workers into slaves.
- Information technology (IT) is a tool for evil in globalization.
- Globalization takes away jobs from the United States.
- Globalization undermines cultural diversity.
- Globalization widens the gap between the rich and the poor.

Some of the arguments in opposition to the antiglobalization views are enumerated next:

- *Representation.* Numerous antiglobalization demands reflect the values of young, middle-class U.S. and EU consumers. They might not be the true representative voice for the Third World countries they claim to speak for.
- *Dominance.* Globalization does not mean the triumph of giant companies over small ones. A case in point is Nokia versus Motorola. Nokia is small in size and dynamic in marketing strategy; this attests to the fact that corporate size is not a requirement for global success. Globalization does shift the balance of advantages from local incumbents (big or small) to foreign challengers. Protection barriers such as the high cost of capital and the difficulty of acquiring new technology are gradually removed over time.
- *Environment.* It is partly correct that globalization might indeed affect the environmental conditions in some developing countries. Competing to attract foreign investment could accelerate the establishment of production plants that generate carbon dioxide, toxic wastes, and other environmentally unacceptable discharges.

  It is true that any pollution discharged into the air by a production plant is bad for the environment. In the United States, various environmental regulations for reducing harmful emissions were enacted only after many tough struggles between big business and government. The key issue is how to balance the value created by greenness with that produced by wealth and economic progress. While the United States can afford to go green now, its current environmental standards might not be appropriate to impose on other countries that are in situations comparable to the United States in the 1970s.

  Thus, the acceptable degree of greenness for a given country is not to be decided by rich countries' environmentalists. For India and other countries, wealth generation might be more pressing in the short run than environmental greenness. This is why the governments of various developing countries welcome globalization and do not share the views of environmentalists.

  While such a hair-raising debate rages on, some progressive companies are known to have taken commendable steps on their own. Levi Strauss established ethical manufacturing standards for its overseas operations. Home Depot adopted an ecofriendly lumber supply program. Starbucks buys coffee from farmers who preserve forests.

- *Labor standards.* The claim that globalization diminishes labor standards is a questionable one. The key issues involved are wages, work conditions, and child labor.

  Most foreign direct investment (FDI) is value-driven and does not primarily chase after low wages. For example, the United States has a positive FDI, meaning that the total amount of foreign investment in the United States is larger than the U.S. investment abroad. Clearly, this surplus FDI is not driving down U.S. labor standards.

  The governments of developing countries oppose the imposition of the current U.S. and EU labor standards onto their regions, as doing so will cause them to lose the wage advantages they currently enjoy. Imposing an external wage standard that is not locally sustainable can be harmful, as evidenced in Germany. After reunification, West Germany imposed its high wage standards on East Germany. The result was an economic disaster: There was zero growth and high unemployment in the East. The governments of developing countries argue that the *Asian Tigers* (e.g., Taiwan, South Korea, Hong Kong, and Singapore) have convincingly shown the road to prosperity for developing economies. Each of these countries started out with low wages and cheap exports and then allowed wages and income to rise gradually in concert with the increased value of products and the betterment of their workers' skills. Local wages must be sustainable in local economies. As expected, all developing countries insist on speaking for themselves and want more investment, freer trade, and better enforcement of local laws, not the imposition of foreign wage standards.

  In general, global companies do provide higher wages to their workers than their local rivals. Some global companies have started paying attention to workplace conditions as well. Gap and Nike are said to have adopted codes of conduct for their overseas plants.

  Child labor is commonly accepted on American farms today and was legal during the long period when the United States was a developing country. Imposing twenty-first-century labor standards on today's developing countries thus runs the risk of appearing hypocritical. For countless families in developing countries, child labor may be a major source of income, just as it is on American farms today and was for others many years ago.

- *Human rights.* The argument that globalization supports human rights suppression is a questionable one. According to Maslow's hierarchy of needs, once a person's physiological needs (clothing, shelter, transportation, and other subsistence needs) are met, the next higher level of needs (social acceptance, peer recognition, and self-actualization) become activated in search of continued personal satisfaction. Accordingly, the local population will most likely seek more freedom of speech, rights of assembly, and respect for human rights over time, but only after their basic subsistence needs are met. An example is Taiwan, which transformed itself peacefully from a dictatorship to a democracy through its rapid advancement of global trade and economy.

- *Side effects.* Globalization promotes free trade and the exchange of goods, services, information, money, and technologies across international borders. Indeed, there are no workable solutions for minimizing the detrimental side effects of increased flows of terrorism, narcotics, disease, and money laundering, unless the governments involved are committed to forcefully combating them. In addition, an aggressive implementation of some of the following four programs might help:
  1. Enhance educational training for the poor (postsecondary, vocational).
  2. Make social services more widely available.
  3. Adopt policies to strengthen the productive capabilities of all, including the low-income groups.
  4. Set up safety nets (e.g., social security, unemployment insurance) for those who are in need.

It is true that what one believes depends on where one stands. Winners and losers have different views on globalization. In the United States, the steel industry was in deep trouble, due mainly to cheap imports. The U.S. textile and farm industries needed government subsidies to survive. On the other hand, U.S. high-technology industries (e.g., electronics, computers, airplanes, appliances, consumer goods, telecommunications equipment, and banking services) are benefiting tremendously from an open global market.

Globalization is an inevitable and unstoppable trend that unfortunately causes dislocations. A prudent approach should be to go forward with globalization while initiating steps to minimize its detrimental effects. Examples of such steps include urging MNEs to support education and job-skills retraining in emerging countries, encouraging MNEs to adopt responsible environmental and labor practices when producing products in developing countries, and promoting democracy and respect for human rights.

14.4 NEW OPPORTUNITIES OFFERED BY GLOBALIZATION

Globalization offers unique opportunities for the creation of value. These globalization-specific opportunities are not open to those businesses that do not globalize. It would appear obvious that businesses in pursuit of globalization should explore these opportunities to the fullest.

Ghenmawat (2007) suggests that global companies pursue one or more of the "AAA Triangle" strategies, which consist of the following:

- *Adaptation*—seeking to boost revenues and market share by maximizing a firm's local relevance (e.g., empower local units in each country).
- *Aggregation*—attempting to deliver economies of scale by creating regional or global operations (standardizing products and services and grouping production processes).
- *Arbitrage*—exploiting the differences between national and regional markets. locating different parts of the supply chain in different places (e.g., call centers in India, factories in China and retail shops in EU).

According to Gupta and Govindarajan (2001), there are five globalization-specific opportunities: (1) adapt to local market differences, (2) exploit economies of global scale, (3) exploit economies of global scope, (4) tap optimal locations for activities and resources, and (5) maximize knowledge transfer across locations. Each of these opportunities will be discussed next.

14.4.1 Adapting to Local Market Differences

Among local markets, there are major differences in language, culture, income levels, customer preferences, distribution systems, business practices, and marketing environment. Companies need to adapt their products, services, and processes accordingly.

BusinessWeek has North American, Asian, and EU editions. Baskin Robbins introduced ice cream flavored with green tea in Japan. Coke markets Asian tea (sokenbicha), English tea (kochakaden), coffee drinks not offered by local competitors, and fermented milk drinks in Japan.

The fast-food chain McDonald's serves veggie burgers in India, both tea and coffee in Britain, and beer in Germany and France.

Anheuser-Busch Inc. successfully marketed its premium and regular beer products in aluminum cans in diverse countries except China. After a careful market study, the company adapted to local conditions by switching to glass bottles and marketing its premium-grade beer in large bottles to restaurants and regular-grade beer in small bottles to average consumers who buy from the local supermarkets. Budweiser scored a huge sales success in China, because this marketing strategy added value to all three parties involved. Chinese customers like to show off by ordering premium beer in large bottles when inviting friends to eat in restaurants, typically ordering one big bottle for each friend at the table. When they buy beer to consume at home, they want to save money, as no one else is around for them to impress. Furthermore, using glass bottles that could be readily sourced locally and recycled pleased the Chinese government. These practices raised the local labor involved in bottling the contents of the products and eliminated solid-waste disposal problems brought into being by aluminum cans. For the company, this strategy whittled down the beer product cost by doing away with the need to import expensive aluminum cans from the United States.

Whirlpool markets the White Magic washing machine. The company runs a global factory network that makes basic models with 70 percent common parts; the remaining parts are readily modifiable to suit local needs. For the Indian market, it conceived a TV-based advertisement program to associate Indian housewives' belief that white means hygiene and purity with Whirlpool washing machines designed to be capable specifically of washing white fabrics in local water. The company offered incentives for local retailers to stock washing machines and hired contractors conversant in eighteen local languages to deliver products and collect cash payments. Annual sales of Whirlpool washing machines went up from $110 million in 1996 to $200 million in 2001—an impressive 80 percent gain.

Kodak has had tough competition in analog film sales and photo processing services from Fuji in Japan, Agfa in Germany, and other global players. In China, Kodak markets its franchise business, the chain of Kodak Express photo supply and development shops, to small entrepreneurs by (1) supporting the franchisee by offering Kodak equipment as collateral to secure local bank loans, and (2) supplying monthly training services to transfer know-how. Kodak was able to establish about 10,000 Kodak Express shops in China by the end of 2001. Its Chinese market share increased from 30 percent in 1995 to 60 percent in 2001.

Not adapting to local conditions could result in business failure. Walsin-CarTech, a joint venture of CarTech with Walsin-Lihwa in Taiwan, planned to build a steel mill in South Taiwan to produce 200,000 tons per year of stainless steel and carbon bar, rod, and wire products for the world markets. Unexpectedly, the local farmers around the intended plant site delayed the installation of electric power lines until they were financially compensated. The Taiwan government also complicated the plant's permitting process. Meanwhile, competitors added their production capacity for stainless steel. The plant needed more investment capital to build than originally expected. The two-year delay in the plant startup caused the joint venture to miss the window of opportunity. Subsequently, CarTech abandoned the joint venture in 1998 and moved on to form a steel joint venture in India.

Adapting to local markets will likely allow companies to increase their market share, augment their gross margin due to enhanced value to customers, and neutralize local competition. However, the cost increase associated with local adaptation must be commensurate with the value added to customers, inducing them to pay for the higher price charged. TGI Friday's incorporated a variety of local dishes (e.g., kimchi) into its menu when it entered the Korean market. This strategy backfired because Korean customers wanted to visit TGI Friday's to taste American, not Korean, food.

The degree of local adaptation may shift over time as the result of the global media, international travel variables, and a steady reduction in income disparity. Companies must constantly adjust their local adaptation strategies.

14.4.2 Economies of Global Scale

Companies may realize economies of global scale by taking a number of steps, such as (1) spreading fixed costs—R&D, operations, and advertising, (2) reducing capital and operating costs per unit when capacity is increased, (3) pooling purchase power—volumetric discounts and lower unit transactions costs by sourcing from a few large suppliers, and (4) creating a critical mass of talent—centers of excellence for specific products and technologies.

Autobytel refined a global baseline architecture that consisted of software modules that can be snapped together in various combinations, depending on the local needs. There are hooks for adding customer software when required. New features invented for a specific country may be incorporated back into the baseline if it seems likely that they will be used elsewhere.

There are a number of counterbalancing factors to consider. Too much centralization in product manufacturing can mandate higher costs of distribution. Concentrated production can also isolate the company from the targeted marketplace. Procurement from a few suppliers generates dependency and constraint, insofar as supply disruptions related to labor unrest, access to world-class technologies, and utilization of existing competencies are concerned.

14.4.3 Economies of Global Scope

Globalization allows products and services that do not require local adaptation to be marketed to multiple regions and countries. Companies can benefit by:

- Providing coordinated marketing approaches for standard products (e.g., PCs, software products, ketchup used in McDonald's) to achieve greater consistency in quality, faster or smoother coordination, and lower unit transaction costs.
- Leveraging market power and customer-specific insights, as a global supplier understands a global customer's value chain better and hence is better prepared to serve. For example, FedEx, as a multilocation logistics service provider, better understands the needs of Laura Ashley, a multilocation global customer.
- Specifying the same hardware platform design for all global locations. GM uses Unigraphics as its common computer-aided design and manufacturing tool and design environment, making it easy for global engineers to collaborate and do design work twenty-four hours a day.

In 2003, IBM entered an eight-year contract worth $1.2 billion to take over the North American and European information technology operations of the French tire company Michelin.

However, there is a challenge facing the management of centrally coordinated marketing programs: How should businesses reconcile the tension between the needs of headquarters and those of the regional units in the actual delivery of products and services?

14.4.4 Location-Based Optimization

This is another globalization-specific opportunity for companies to add value. Certainly, the intercountry differences in location-based cost structure and services must be considered. By optimally selecting the location for each activity in the value chain (e.g., R&D, procurement, component manufacturing, product assembly, marketing, sales, distribution, and service), global companies can secure advantages in several areas.

Performance Enhancement. To build and sustain world-class excellence conferred by talents, speed of learning, and the quality of external and internal coordination, Fiat chose Brazil, not Italy, as the place to design and launch its "World Car," the Palio. Microsoft established a corporate research laboratory in Cambridge, the United Kingdom, rather than in the United States.

Cost Reduction. Cost is, of course, a major concern to any company. Cost considerations relate to factors such as local manpower and other resources, transportation and logistics, government incentives, and local tax structures. For example, Texas Instruments set up a software development unit in India, and Nike sources the manufacture of athletic shoes from Asian countries (China, Vietnam, Indonesia, and others).

Risk Reduction. Beside economic and political risks, there are also currency risks associated with devaluation. A company might need to spread the manufacturing operations across a few locations to minimize such risks.

For instance, Texas Instruments has been designing integrated circuits in India since 1986, Sun Microsystems has hired Russian scientists for software and microprocessor research, and CrossComm Corp has its communications software written by Poles at the University of Gdansk.

To capture location-based opportunities to add value, companies need to have the right management skills with the flexibility and the ability to foster coordination.

Ford relocated some manufacturing operations to Mexico to become more selective in hiring, to achieve a reduction in turnover, and to realize better productivity by training. Doing so allowed Ford to achieve lower wage rates as well as higher productivity than it would have been able to do in the United States.

Location-specific conditions do evolve with time. Companies must be flexible in shifting production should the location-based conditions no longer justify a continuation of production at a given site.

Coordination is of critical importance for companies to maximize the value generated by location-based opportunities. Texas Instruments conceived the product concept of its TCM9055 (high-speed telecommunications chip) in collaboration with engineers in Sweden. It designed the product in France with the use of software tools advanced in Houston, manufactured the product in Japan and Dallas, and tested the product in Taiwan.

14.4.5 Knowledge Transfer across Locations

The global company may add value by actively transferring knowledge across locations. Knowledge about product or process innovations and about risk management options are of particular value.

Product and Service Innovations. Sharing new ideas among subsidiaries eliminates the "reinvention of the wheel" and speeds up product and process innovation.

Procter & Gamble used ideas conceived at different centers to design Liquid Tide in 1980: P&G built on technologies developed in Cincinnati (resulting in a new ingredient to help suspend dirt in wash water), Japan (cleaning agents), and Brussels (ingredients that fight the mineral salts present in hard water).

Procter & Gamble applied an efficient stock list-based distribution system from India to Indonesia and China and thus significantly minimized its cost of innovation.

In 1997, ABB, a $23 billion industrial-product company headquartered in Zurich, Switzerland, shifted 1,000-plus manufacturing jobs from Western Europe to emerging economies over a five-year period for the purposes of increasing efficiency, exploiting lower wages, and becoming more responsive to customers in growth markets. ABB set up a system that propels local ideas for new products and projects around the world in just three weeks. On the basis of key words contained in the proposal, principal global players comment and sign off within an allocated period; this minimizes the time from idea to approval.

Reduced Risks of Competitive Preemption. By rapidly transferring new innovations to all global locations, the global company can lessen the danger of losing ideas to competitors for replication in other markets.

Generally speaking, there are two types of knowledge that are important to a company. On the one hand, *codified knowledge* is typically embodied in chemical formulas and engineering blueprints and is documented in operations manuals. Such knowledge is readily transferable. On the other hand, the *tacit knowledge* embedded in people's minds, in behavior patterns, and in the skills of individuals or teams may be difficult to transfer. Examples of such tacit knowledge include the vision of a roadmap of new technologies or competency in managing global customer accounts. Managers in global companies need to find efficacious ways to transfer tacit knowledge across subsidiaries.

It is a natural tendency for people to want to preserve specific competencies (e.g., manufacturing superiority) for survival and competitive reasons. Global companies need to systematically recognize unique know-how that is worth transferring and encourage knowledge sharing across locations. All subsidiaries must also be encouraged to learn from peer units instead of being handicapped by the "not-invented-here" syndrome that some locations develop.

Global companies are blessed with location-based, value-added opportunities not readily available to companies that are domestically focused. Global leadership is needed to take advantage of these unique opportunities to confer competitive advantages.

14.5 PREPARATION FOR GLOBALIZATION

Pursuing globalization successfully requires that global companies understand the success factors gleaned from the experiences of others and that they are properly prepared. This section addresses the issues associated with personal readiness, success factors in globalization, global virtual teams, management styles, globalization pathways and avoidance of globalization mistakes.

14.5.1 Personal Preparation

Benjamin Franklin said: "An empty bag cannot stand upright." * It is indeed important for global managers to prepare themselves well. Global businesses require managers to be patient with, tolerant of, and open-minded toward divergent cultures, customs, and business practices. They should be dedicated to the mission at hand and assume a flexible negotiation style to win. They should possess stamina to endure personal hardships, the personality to properly handle uncertainties and ambiguity, and the conviction that what is different is not necessarily dangerous (Dalton, Ernst, Deal, and Leslie 2002; Marguardt and Berger 2000; McCall and Hollenbeck 2001).

The manager needs to recognize that, for global businesses to succeed, the new model of responsiveness, partnership, teamwork, and decentralization must replace the early management model of efficiency, hierarchy, control, and centralization.

Not everyone has the desire to become a global leader. Those who want to be global leaders need to become proactive in seeking opportunities for leadership development. Personal preparation can assist in refining these desirable traits.

Global management is demanding, indeed. According to Lamberton (2002), global managers must possess certain characteristics and savvy to be successful in a global environment.

---

* *Source*: Barlett, John. 2002. "*Barlett's Familiar Quotations*," 17th eds. Little Brown & company, Boston. Page 316.

Inquisitive Mind. Global business is highly complex and uncertain, due to variations in cultural, linguistic, political, social, and economic conditions. Global managers must constantly learn in order to succeed. Constant learning requires an inquisitive mind. Successful global leaders are adventuresome, curious, and open-minded. There are business models, strategic initiatives, operational tactics and customer service practices that are worth learning from various global companies. Taking a "head in the sand" approach or having a bunker mentality will not bestow business with advantages.

Inquisitiveness strengthens personal character growth, characterized by emotional connection to people and uncompromising integrity and duality, the capacity to handle uncertainty and the ability to balance tensions. Personal integrity inspires staff trust and commitment, which, in turn, affect the implementation results of any global strategy.

Global Mindset. "Thinking globally and acting locally" is regarded as a best practice for a global company to keep things in perspective while achieving practical results. Having a global orientation is no longer opulence, but a necessity for economic survival in many industries. Without an international perspective, global managers have a disadvantage in the global economy of the twenty-first century (Garten 1999). To be effective in global business, managers need to have a global mindset, which will enable them to do the following (Jeannet 2000):

- *Extend* concepts and modes from one-to-one relationships to holding multiple realities and relationships in one's mind simultaneously. Then act skillfully on this more complex reality (global think).
- *Change* management orientation from taking individual initiatives to adopting team and group initiatives.
- *Focus* simultaneously on hard issues (low-cost producers, bottom lines, budgets, manufacturing, marketing, distribution, head count, and finances) and soft issues (value, culture, vision, leadership style, innovative behavior, and risk taking).
- *Balance* the pressures of global integration (product standardization) and local responsiveness (adjusting to the needs of local markets). Recognize the interdependence of the global economy and view the world from a broad perspective. Seek trends that affect company business, balance contradictory forces, rethink boundaries, and build and maintain organizational networks at the global level.
- *Serve* as a catalyst within the company, being sensitive to, and capable of, managing cultural diversity. Become more tolerant of other people and cultures. Consider culture diversity an asset. Connect emotionally with people and the worldwide organization. It is worth noting that European managers are said to be more accustomed to exposure to cultural diversity than American managers.
- *Recognize* complex patterns in the global environment and thrive on ambiguity. Become proficient at managing uncertainty and dealing with conditions that change constantly and are inherently complex.
- *Preserve* a unique time and space perspective. Take a long-term view, extending personal space in geography and relationships.
- *Exhibit* business and organizational savvy. Recognize opportunities, grow in knowledge of available resources, and be capable of mobilizing them to take advantage of opportunities.

Additional useful advice on how to manage across cultures and conduct business with a global mindset are offered by Solomon and Schell (2009).

Example 14.3 "Think globally and act locally" has been the general guideline offered to managers at the headquarters of global companies that seek to achieve success in the global markets. The logic is rather compelling.

For those managers of global companies that operate in local regions or markets, perhaps the guideline should be "Think locally and act globally." What is your opinion on these guidelines?

ANSWER 14.3 The principal objective of requiring headquarters managers of global companies to "think globally and act locally" is to make sure that the company's products and services are sufficiently adjusted to the needs of local markets, while enjoying the economies of scale advantages of being global.

Numerous local managers of global companies have implemented innovative strategies and achieved remarkable success because of their understanding of the culture and customs in local markets. Oftentimes, the same insight and innovative strategies can be applied to other regions, with only minor modifications, for the headquarters managers of the global companies to realize the economies-of-scope advantages on a global basis.

Das (1993) spoke in favor of the concept of "thinking locally and acting globally," as a result of his personal experience as a local manager of Procter & Gamble in India.

Knowledge and Skills. Global managers must possess specific knowledge and capabilities to succeed. They should have a mastery over technology (information systems, telecommunications, and operations) and use it effectually. They need to be aware of the social and political conditions in different countries. They should be familiar with the specific culture and cross-cultural issues that affect management. Of great importance is their understanding of the global competitive practices in manufacturing and communications, such as total quality management, just-in-time delivery, factory automation, employee involvement, and outsourcing. Also helpful are some general knowledge about business and industry practices and the skills required to put knowledge into action, to become acculturated, to envisage a vision, and to motivate a diversified workforce.

Example 14.4 Continuous improvement is a critical requirement for all companies, domestic or global. How can the process of continuous improvement be implemented in a global company?

ANSWER 14.4 Continuous improvement helps companies to remain competitive in the marketplace. To manage the continuous improvement process, management should do the following:

- Specify objectives in consultation with top management. Assign the responsibility of managing the global continuous improvement process to someone with visibility. Commit resources and establish a central office to coordinate the global continuous improvement efforts. The mission of this office must be communicated to all global employees.

- Define specific goals in a number of areas, on the basis of inputs from various divisions, using standards derived by gleaning the available best practices in the industry.
- Form a number of task forces by interviewing and selecting capable and devoted people who have expertise in diversified disciplines and who are from various operational units in different regions. Each task force should be empowered to pursue betterment ideas in a specified domain.
- Hold teamwork training sessions for members of diversified cultural and technical backgrounds. Visit team members in various locations to establish contact, assure understanding, and build trust.
- Set up a communications system (e.g., an intranet, videoconferences, regional meetings, and phone, fax, multimedia technologies) to enable members to interact constantly. Apply Web-based tools to foster close collaboration.
- Establish "suggestion box systems" at each site for members to solicit and obtain inputs from knowledgeable employees. All employees are encouraged to contribute new ideas for refinement.
- Empower the task force teams to implement improvement ideas deemed useful and to apply resources made available from the central continuous improvement office.
- Reward employees who suggested those creative ideas that produced positive results after implementation. Present awards in well-advertised meetings to promote the continuous improvement effort. Publish awards and the positive results in company newsletters and other suitable media to practice positive reinforcement.
- Encourage the cross-pollination of ideas from global employees at various locations to take advantage of their diversified experience and viewpoints.
- Summarize and publicize results on a regular basis.

Global Business Savvy. Of critical importance to global companies is the business savvy of their global leaders to size up business opportunities and to have a vision of doing business worldwide.

Recognize global market opportunities. The ability to recognize new opportunities is a key leadership quality of global managers. They need to be able to do the following:

- Assess the cost and quality differences in production outputs and inputs, and exploit cost differentials for land, energy, labor, raw materials, and people talents.
- Identify market needs for goods and services from a deep and broad knowledge base, having mastered finance, accounting, marketing, human resources, operations, international relations, economics, industry conditions, and strategy disciplines.
- Size up opportunities for efficiency gains by (1) eliminating redundancies to wring out costs, (2) using economies of scale in procurement, (3) pursuing standardized outputs, and (4) selling to multiple markets.
- Generate competitive advantages by forming supply chain networks involving strategically selected, local and global partners with complementary resources and expertise.

Envision doing business worldwide to ultimately make money. The manager should also have a good overall perspective of what the company's business has to offer—what the core is, why the core is what it is, and what drives the core. This understanding is combined with the fundamental good business goal of making money for the company.

Demonstrate Global Organizational Savvy. Global managers are required to demonstrate organizational capabilities. These capabilities are built on specific qualities:

- *Know your company.* Global managers must have an intimate knowledge of their own companies with respect to subsidiaries' product lines, cost structures, and overall competitiveness. They should know the location and quality of technological resources available, including physical assets and managerial and employee talents. Global managers must be known to the company's key decision makers by having served on key committees, participated in task forces, and attended critical meetings.
- *Mobilize resources.* Global managers need to be able to mobilize resources to take advantage of global opportunities. Establishing trust with top management and key decision makers will assure their favorable response to these mobilization efforts.
- *Acquire insight.* Global managers must be able to identify critical knowledge and capabilities beyond merely understanding policies and programs.
- *Keep current.* Global managers must keep themselves constantly informed of what is going on at the headquarters.

Example 14.5 Companies in the future will rely more and more on temporary workers, specialized vendors, and consultants to flexibly satisfy unique needs and contingencies. Employer–employee relations will become peer-to-peer relations rather than hierarchical ones. Engineers will find that their careers are less stable than in previous generations.

In your opinion, what is the most important preparation for engineers under this volatile, uncertain, and dynamic scenario?

ANSWER 14.5 Engineers need to proactively manage their own careers more closely than ever before, while keeping their skills, knowledge, and experience marketable. A good way to focus is to pursue lifetime learning about the changes in technologies, tools, industry, and business. Engineers who do not take care of their own careers will rapidly become redundant.

The employer's position is well stated by Bahrami (1992): "You own your own career, we provide you with opportunity!"

14.5.2 Success Factors for Globalization

In order to attain long-term profitability, global companies must (1) build customer relationships supported by superior, worldwide, uniform service, (2) possess wide and deep knowledge about customers, (3) have strong and easily recognizable brands;, (4) hire and retain talented people, and (5) organize global virtual teams to effectually implement global strategies.

Other key traits of companies that have successfully entered the global market include:

- Home market strength, which provides a solid revenue basis for global expansion.
- A global business model that is easily replicated and scaled up to multiple markets.
- A powerful vision that motivates employees and communicates core values.
- Strong leaders who can articulate and carry the message globally.

General Electric (GE) has been recognized as a master of globalization. It has moved a large number of plants to countries of lower labor cost. Specifically, GE locates and relocates manufacturing plants to locations where the GE quality standards can be met at the lowest cost. GE pursues the business tenet of *continuous mobility*. As the emerging economies continuously upgrade their production skills and standards, GE relocates plants from one location to another to keep costs under control. General Electric Medical Systems, a division of GE, is known to have moved production plants from Paris to Budapest, from Milwaukee to Mexico City, and from Japan to Shanghai and Bangalore. According to Jack Welch, "Ideally, every plant you own would be on a barge." In Fig. 14.5, quality standards are plotted against cost of production. It is generally expected that in a given economy, products/services of high quality demand larger efforts and incur higher cost, thus the inclined line therein. For example, Line I could be for a developed country (US, Europe, and Japan) and Line II for a developing country (China, India, and Mexico). Since some developing countries are capable of generating product/services at levels acceptable to some global customers, it is thus only logical that selected companies in the developed economies will want to relocate their production plants from A to B, in order to gain a cost advantage without sacrificing quality. In fact, they would continuously move their production plants to seek out lower cost and quality production sites. This is so-called *continuous mobility strategy* as practiced by GE and other multinationals.

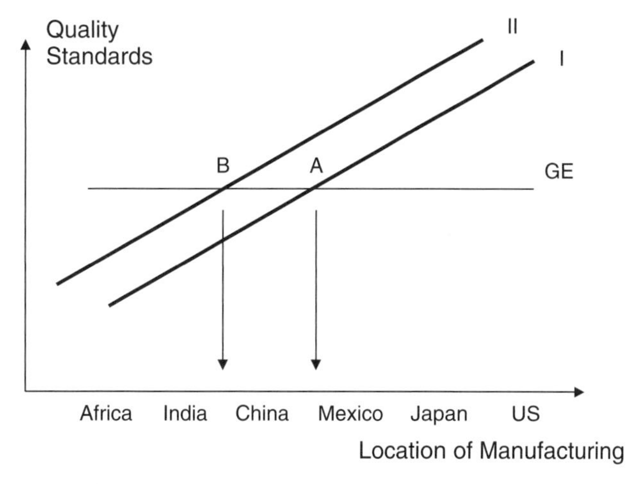

Figure 14.5 The concept of continuous mobility.

14.5.3 Global Virtual Team

To achieve global success, global virtual teams must function well, as the pace of change has increasingly forced global organizations to be more outward looking, market oriented, customer focused, and knowledge driven (Lipnack and Stamps 2000). Global virtual teams would typically be composed of members who are geographically dispersed, each with specific technical or business competencies, cultural and language backgrounds, working habits, and variable comfort levels with technology accessibility and utilization. Global teams may be ineffective because of cultural values, cultural and language differences, and other factors, such as the leader's approach, lack of organizational support, or individual rewards overshadowing the team's success (Dreo, Kunkel and Mitchell 2002; Hoefling 2001). The cultural barriers just mentioned could come from three factors:

1. *Function*—due to differences in reasoning styles, reactions, and getting motivated by people in various professions, such as engineers and marketing personnel.
2. *Organization*—due to different value perceptions and behaviors.
3. *Nationality*—due to different styles of human interaction because of national origin (e.g., in the United States the emphasis is on the individual, whereas in East Asia the emphasis is on the group and on reaching out for a consensus).

Cultural issues must be overcome. Metrics should also be set for goals and performance, and these should be focused on deliverables so that all team members fully understand their respective accountabilities. Managers also should convince all team members of the value of change made possible by the team activities.

Selecting the right team members is a crucial first step for the leader of any global virtual team. Preference should be given to people who can act on their own when needed. One model of membership selection is to find regional alliance partners who have the necessary core competencies to execute a global, centralized strategy. The best kind of alliance partners are those who think ahead, demonstrate commitment with investment, understand the company's requirements, share the same vision, and operate on behalf of the company.

When building global virtual teams, leaders need to pay attention to factors known to have a direct impact on team success. These factors include clearly articulating common goals, being aware of overlapping competencies and skills, acknowledging each other's contributions and needs, formulating clear procedures and guidelines for working together, and establishing common rules and technologies for sharing information and data.

In order to operate global virtual teams productively, proactive attention and preventive maintenance are needed. Members may need to be trained properly. A constant monitoring of the team progress is advisable. Roles and responsibilities must be clearly communicated and emphasized. To get team feedback, frequent communications are needed.

Communication is the key to keeping teams together and on track. Poor communication can bring forth stumbling blocks affecting team success. Some of these stumbling blocks are (1) commitment of team members may be misdirected from global activities to local priorities (the "out-of-sight, out-of-mind" syndrome), (2) trust between global team members may not be strong enough, (3) time-zone differences, resulting in work-time overlaps, can discourage frequent communications, (4) language barriers, as English is not universally spoken, can make verbal and written communication in other countries uneasy and time consuming for some, (5) culture differences produce different work habits and value perceptions, which can cause communication to be less free and open. Team leaders need to proactively schedule conference calls, in addition to e-mail, and have quarterly face-to-face meetings if budgets and schedules so allow.

Brett, Behfar, and Kern (2006) suggest four specific strategies to overcome conflicts in global teams:

1. *Accommodation*—acknowledging cultural gaps openly and working around them.
2. *Structural intervention*—changing the organization of the team.
3. *Managerial intervention*—setting norms earlier and bringing in a higher-up manager.
4. *Exit*—removing a team member when other options have failed.

Example 14.6 Global teams are often deployed to handle important tasks, such as devising specific implementation plans, proposing market entry strategies, and designing global products. How do you make global teams effective?

ANSWER 14.6 The following generalized steps could help make a global team effective in achieving its intended objectives:

- Select a team leader who is well recognized for leadership quality, interpersonal skills, and managerial capabilities. Above all, the team leader must enjoy the strong support of the company's top management. The president or another suitable top executive of the company should announce the appointment of the team leader to demonstrate company commitment.
- The team's objectives must be specified, including standards to measure progress and the expected outcomes. The potential impact of the team efforts on the company's profitability is clearly understood.
- The team leader specifies the qualifications of the team members on the basis of the expertise and level of experience needed to achieve the team objectives.
- The team leader solicits suggestions from various regional management centers and receives assurance that all required local support (time off, secretarial work, analysis, use of local facilities and engineering, marketing, and other resources) will be offered to team members.
- The team leader interviews specific candidates and selects the team members with the concurrence of local management. Team members with different skills and expertise may be needed as the team progresses through various stages, resulting in a constant flow of people moving into and out of the team. By organizing the team properly, the team leader ensures that a good working atmosphere prevails at all times.
- With the support of local management, the team leader should compile a comprehensive roster of corporate talents (who is specialized in doing what,

for how long, with what accomplishments, etc.). This roster is constantly maintained and could be used by all team members.

- The team leader makes sure that all members receive proper teamwork training regarding five functions: (a) Working in teams, (b) Communications, including the use of associated equipment and tools, (c) Problem solving, (d) Available support functions and resources that can be tapped, and (e) Group goals and expectations.
- The team leader specifies guidelines for fostering communications between team members (e.g., intranet, videoconferencing).
- The team leader enhances personal interactions and cooperation, and builds trust and confidence among team members by inviting all to be physically present at the first team meeting.
- The team leader plans out and implements subsequent team meetings to focus on achieving the team objectives. He or she encourages all to communicate, prevents the dominance of meetings by a few, and takes into account the diverse cultural backgrounds of the respective members.
- The team leader assigns members specific tasks to carry out (e.g., analysis, focus group inputs, activity-based costing, etc.).
- The team leader conducts field trips, on-site visits, and other activities needed to collect data and to exercise judgment.
- The team leader engages outside consultants and other resources to provide benchmarks, suggest alternatives, or overcome bottlenecks if needed.
- The team leader solves problems, resolves conflicts, and secures support functions needed for all team members.
- The team leader provides regular reports to all regional managers concerning the team's progress and team member performance.
- The team leader strives to achieve a consensus on major issues. He or she makes sure that the outcome serves as a valid solution to the problem under consideration.
- The team leader reports the final outcome of the team to the company president or to other upper management. He or she gathers the whole team to make a formal presentation of the results and to celebrate the successful completion of the team effort.
- The team leader documents the experience and preserves the learning gained by the team efforts, by collecting inputs from all team members.

14.5.4 Management Style

Management style plays a critically important role in globalization, because globalization makes communication and personal interaction necessary between people of different cultural, business, and personal backgrounds. For example, American managers may grow up in hierarchical and command and control systems. They typically perform thorough competitive analyses, using strong analytical tools and strategic audits. They focus on short-term profit objectives. They value being goal and achievement oriented. A great number of them are competitive, aggressive, ambitious, and intolerant of poor performance. Their management style is also influenced by American education, politics, and internal and external reward systems. In contrast, the Japanese management style is characterized by teamwork, market-share objectives, commitment to quality, and a philosophy that says, "The nail that stands up gets pounded back down."

It is thus important for managers of global companies to recognize and accept extreme differences in management styles practiced by people in different geographical regions. In fact, there is no style—American, Chinese, Japanese, German, French, or British—that must be rejected. To be successful in a global economy, a manager must respect the cultural norms of others by stressing shared goals and a common outlook; remaining open minded; adapting to the local customs, business practices, and value systems; and avoiding both cultural and intellectual arrogance (Schneider and Barsoux 2003; Brett 2001).

Companies involved in global businesses typically change their management styles over time. Initially, some global companies direct worldwide activities according to home-country standards, adopting central decision-making and control paradigms. As the global companies increase their foreign investments, they pursue a model of decentralization and autonomous global operations, using the home-country standards as a reference basis for managing worldwide operations. Gradually, the global companies build a global network and follow a transnational strategy that is integrated and interdependent.

Example 14.7 In pursuing global business, one commonly practiced strategy is to elicit the maximum possible collaboration with the right business partners in the host country. This is because their local knowledge is of tremendous value in facilitating the adaptation of foreign-made products or services to local market needs and for problem solving. Creating trust among the partners will naturally be a critical first step toward implementing such a strategy.

What are some American management practices that can be counterproductive in winning the collaboration of foreign partners?

ANSWER 14.7 As background information, Table 14.1 contrasts the typical American management practices with those of the Japanese.

Table 14.1 American and Japanese Management Practices

	United States	Japan
Employment	Short-term	Lifetime
Decision making	Individual	Group consensus
Responsibility	Individual	Collective
Evaluation	Rapid	Slow
Control	Explicit formal	Implicit informal
Career Path	Specialized	Nonspecialized
Concern	Segmented	Holistic

The following list describes some of the American practices that could be counterproductive in winning the collaboration of foreign business partners:

- *Exhibiting* a highly competitive and arrogant personal demeanor and an ignorance of local culture, customs, and other differences in business practice, alienates the foreign partners. Here are two examples of well-known language blunders: (1) An "escrow account" in English means a "gyp account" in French. (2) When an issue is "tabled" in America it means that it is not to be brought up again; but in England, it means the exact opposite—that it *will* be brought up.
- *Being* highly impatient and pushing aggressively for instant decisions fails to allow time for the foreign partners to achieve group consensus for the decisions at hand.
- *Emphasizing* short-term profitability is a barrier to recognizing the business goals of the foreign partners of seeking long-term, broad-based collaboration.
- *Becoming* proud of risk taking and exhibiting decisiveness, a command-and-control rationale and an excessive profit motive fails to recognize that the values favored by the foreign partners might be different.
- *Adopting* the "ugly American syndrome" makes American managers insist that foreign partners do exactly what Americans do.

14.5.5 Strategic Pathways to Globalization

Companies pursue globalization along any or all of geography-based, product-based, customer-focused, and Internet-based pathways.

The *geography-based pathway* is a pathway in which companies pursue globalization in geographical areas that have common cultural and linguistic ties—Canada and England for U.S. companies, China for Taiwanese companies, Southeast Asian countries for Chinese companies, and African countries for French companies.

The *product-based pathway* is one according to which companies conceive and perfect specific products that do not require local customization. The companies then distribute the products globally wherever there is a demand for them.

The *customer-focused pathway* suggests that global companies follow their major clients to foreign markets with a basket of products to serve the needs of the local customers of these clients more efficiently. Examples of such baskets of products include the combination of insurance, banking, and securities offered by Citicorp, and that of logistics and inventory management used by FedEx.

The *Internet-based pathway* prescribes that companies devise a Web presence and leapfrog over other competitors to reach end users in numerous global markets.

14.5.6 Avoidance of Globalization Mistakes

There are a number of mistakes commonly made by companies pursuing globalization. Among them are a *lack of company commitment*, when companies do not make a firm and sufficient corporate commitment to people, capital, and time; and *low management attention*, when senior managers get involved only when there is a crisis affecting earnings. Oftentimes, companies assign *low priority*, viewing the international businesses as "incremental," and do *not engage foreign partners decisively* (take a minority position when entering a joint venture with local partners). Should the business relationship turn adversarial, these companies can get blocked out of the target markets; thus, one should always attempt to keep 50–50 ownership to stay even.

A *lack of cultural sensitivity and understanding* has been a major source of frustration for global managers. An American in Japan is described in the following example (Glover, Freidman, and Jones 2002):

> A young American manager was sent to Japan to work with the Fuji villagers on a forest project. During his first week in Fuji, he requested a local village chief to send "three men to do an eight-hour job clearing a field." Each of the three men was to be paid an hourly wage.
>
> The next morning, forty able-bodied men from the village showed up to do the work. The American manager asked the group to select three men, reasoning that, as he did not need all forty of them, he would send the remaining thirty-seven back to the village. The chief responded that if all forty of them cleared the field, they would complete the work in one to two hours and then could go back to the village to do other work. Furthermore, the chief requested that the men not be paid individually. He would take the money and put it into the village fund, a traditional communal means for distributing money equally.
>
> The American manager sent the chief and all of his forty men away and paid higher wages to three Fujian Indian contract workers he got from a nearby city a week later. He complained that the Fujian villagers were not motivated to be productive and they did not seem to have any individual initiative.

This was clearly a case of a cultural clash between the occidental approach to productivity based on "scientific management" and the oriental approach of getting the work done in a speedy manner by a collective work group. Their culturally conditioned views of productivity were different. The American manager was trapped in the "one best way" he believed in—namely, "three people to do an eight-hour job." He was unable to see possibilities of adapting to the concern of the chief, who needed to secure external funds to augment the overall village operations, while delivering work at a faster rate by using all of his able-bodied men. The chief thought it ought to be the same to the American manager, as the total cost remained the same, whether the work was done by three or forty people, as long as it would be at the same hourly wage rate. Thus, the chief was equally frustrated by this exchange and viewed any future interchanges with this inflexible American manager with suspicion from that point on.

Example 14.8 In the 1980s, a lot of multinational companies were eager to conquer foreign markets. In a hurry, they committed a large number of culturally insensitive blunders, contributing to major marketing and business failures at the time. Name a few of such embarrassing examples.

ANSWER 14.8 The examples of culturally induced mistakes are plenty. Reviewing them from time to time is useful only for the purpose of preserving the learning opportunities they offer (Risks 1983).

- Chevrolet introduced "Nova" in Puerto Rico and found out only later that *No va* means "doesn't go" in Spanish.
- Ford introduced a low-cost truck, the Fiera, into some developing countries without success. It turned out that *fiera* means "ugly old woman" in Spanish.
- Esso, the oil company, went to Japan. The phonetic pronunciation of *Esso* in Japanese means "stalled car," which was not helpful in promoting the sales of gasoline there.
- Cadbury Schweppes, an English food company, introduced its Rondo soft drinks into the United States. It failed badly, although it was a success in England. Later, they found out that the people in the United States thought Rondo was a dog food.
- Rolls Royce, before Mercedes acquired it, marketed a car called "Silver Mist." When that name was translated into German, the "mist" became "excrement." It forced Rolls Royce to change the name.
- McDonald's promoted its food products in Japan using white-faced clowns at one time. White face in Japan is a death symbol.

14.6 PAST PRACTICES RELATED TO GLOBALIZATION

When pursuing globalization, the management of global companies can benefit from the international business experience gained by other companies. This section addresses emerging issues and the ways some global companies conduct global businesses.

Global companies face two emerging issues. The first is *fairness*. Traditionally, in the market economy, major multinational profit-seeking companies have pursued globalization. Home-country governments of these companies tend to set the macroeconomic policy and rules of the game, and they do not always have the interests of developing countries in mind. Calls have been issued by some developing countries to seek global governance, with the participation of all developing countries in order to ensure fairness to all involved.

The second issue is *conflicts of interest*. In industrialized countries, workers in certain "old-economy" sectors (e.g., mining, textiles, agricultural, and other manufacturing enterprises) face unemployment when jobs are transferred to developing countries that offer lower labor rates and are more competitive. Workers in the *knowledge economy* sectors—electronics, computers, and high-tech export businesses—are gaining. Surveys indicate that people with low incomes are generally opposed to globalization, while those with high incomes favor it.

Protests staged against the World Trade Organization (WTO) by the joint forces of labor (stumping for work rules to protect U.S. jobs), environmental (promoting the reduction of pollution), and human rights groups (protesting for the elimination of political and religious suppression) at various international places in recent years have indicated clearly that not everyone is in favor of globalization.

Even inside developing countries (e.g., China), globalization is not welcome by all. Increased privatization, foreign investment, market opening (telecommunications, banking, financial services, and others), and increased foreign trade can facilitate the destruction of countless existing state enterprises and thus cause massive unemployment in the state sector. Hence, globalization brings gains to some sectors and losses to others, as discussed previously. Global companies need to devise long-range programs to address these issues in order to sustain the benefits they realize from globalization.

Dholakia (2010) discusses various issues related to global organizations. Global companies engaged in international business may be classified into one of the four groups listed next, according to their corporate behavior in conducting global business.

14.6.1 The Defender

These companies are internally focused. They have no global orientation and no international element in their business strategies. They are focused on domestic markets and make no effort to understand other markets and cultures. They have limited skills and knowledge to pursue foreign markets. They look for governments to provide protection against foreign intrusion (e.g., through trade barriers, quota, duties, laws, and special agreements). Their view is, "What is different is dangerous." Examples include (1) the U.S. steel industry, which sought quotas from 1960 to 1980 to restrict Japanese steel imports to the United States; (2) the U.S. footwear industry, which attempted in vain in 1980 to get import protection; and (3) the U.S. textile and machine tools industries, which got government relief against imports in 1975. Today, a host of these industries are under the protection of bankruptcy laws or import tariffs. Some of them have barely survived under governmental subsidization programs.

14.6.2 The Explorer

These companies are largely inwardly oriented, with dominance in the domestic markets. They are aware that opportunities may exist abroad. They move into foreign markets very cautiously after closely studying the opportunities available. They have some knowledge about the markets abroad and possess a restricted set of skills to pursue them. Overall, they have small international business revenues. The home-based headquarters controls their businesses. They may pursue some export and franchising activities, but with rather limited investment commitment. Companies in this category include Seiko and Lotus (which has been acquired by Microsoft).

14.6.3 The Controller

These companies are more externally oriented than the explorers. They want to control the market abroad. They have sufficient knowledge and skills to pursue foreign markets, but have a limited global mindset. They generate a significant amount of overseas sales revenues with major investment commitment. They impose their home culture and practices on overseas operations, although they do tailor some strategic decisions to suit the local cultures or to optimize the interests of their home office and the local markets. They maintain financial and strategic control at the home office, while allowing some independence in overseas activities. Examples of companies in this category include Coke, McDonald's, and Pizza Hut.

Example 14.9 In the new millennium, engineering managers will need to be prepared to lead and manage technology-intensive companies and industries. What specific technology management activities should engineering managers be concerned with?

ANSWER 14.9 Generally speaking, there are eight activities that engineering managers should pay attention to (National Research Council 1987):

1. Integrating technology into the overall strategic objectives of the firm (strategic planning).
2. Getting into and out of technologies effectually (gate-keeping).
3. Assessing and evaluating technology more effectively (making decisions and choices).
4. Conceiving better methods for transferring and assimilating new technology (managing technical knowledge).
5. Reducing new-product development time (creating supply chains and applying innovations).
6. Managing large, complex, and interdisciplinary or interorganizational projects, programs, and systems (leading cross-functional or global teams).
7. Managing the organization's internal use of technology (guiding knowledge management and open innovation efforts).
8. Leveraging the effectiveness of technical professionals (managing knowledge workers).

Engineering managers focus not only on R&D management or management of technical professionals, but also on managing manufacturing and process technology, new-product development, and other technology-intensive functions in an organization. In addition, they strive to implement projects and programs correctly the first time and dedicate their efforts to continuous improvement. They are capable of commercializing technology products because they possess the required background and training in activity-based costing (cost accounting), NPV analysis (financial management), and customer relations management (marketing management). The critical roles of engineering managers are well recognized, as technology is of strategic importance to the national economy.

14.6.4 The Integrator

These companies have a global perspective based on heightened awareness (knowledge) and strengthened abilities (skills). They form a worldwide web of relationships, partnerships, and alliances with suppliers, developers, designers, distributors, competitors, and customers. They reconfigure these relationships over time as new threats and opportunities arise. They coordinate, rather than control, these networks of business partners. They focus on overall organizational effectiveness in delivering products or services of value to customers. They understand, bridge, and resolve differences between people, companies, values, and cultures. Their core strategy is to win in the marketplace by leveraging, sharing, and nurturing complementary capabilities. In this group of companies, General Electric, Toyota, and Dell are known to have formed networks with primary, secondary, and tertiary suppliers and subcontractors. To be globally successful, companies need to walk, talk, and act like integrators.

14.7 DEVELOPING GLOBAL STRATEGIES FOR SERVICE BUSINESSES

Global strategies for service must address the special characteristics of services (Lovelock and Yip 1996): (1) service outputs are intangible, (2) customers may participate to varying degrees in production and consumption, (3) customer experience is strongly influenced by people interactions, (4) services have no or little inventories, (5) services are strongly influenced by the time factors, (6) service suppliers are less capable in quality control, (7) customer evaluation is more difficult, and (8) use of electronic distribution means is possible. Global strategies need to be properly aligned with services in accordance with the degree to which customers are directly involved:

- *Active participation*. Customers are involved in people-processing services, such as passenger transportation, healthcare, food service, and lodging services.
- *Some participation*. Customers are involved in information-based services such as accounting, banking, consulting, education, insurance, legal services, and news.
- *No participation*. Customers are not involved in possession-processing services, such as freight transportation, warehousing, equipment installation, car repair, laundry, and disposal.

Herman (2010) suggests ways for enterprises to enhance value in the service economy. In general, several options are available for companies to pursue global strategies for their service businesses:

- Advertise the brand names to promote services, which are typically intangible and difficult for vendors to quality-control and for customer to evaluate.
- Selectively standardize the core and the supplements (see Fig. 8.9), and customize the rest of service elements to adjust to local conditions. It is easier for companies to achieve a profitable balance between the competing goals of economies of scale and local customization in services than in products. Citibank introduced "Citicard Banking Centers" with automatic teller machines in twenty-eight countries, allowing a 24/7 access. Hewlett-Packard maintains a globally standard set of services that range from site design to system integration and remote diagnostics and seamless service at any hour and from anywhere.
- Strategically place back-office and other standardizable procedure work in locations that offer wage and other advantages. Some banks and insurance companies send checks and claims to East Asia or Ireland for processing. McKinsey & company sends some of its work to India. It may also be useful to bring best practices proactively to emerging economies so that the outsourced work can be performed with cost-effectiveness, speed and quality (Hexter and Woetzel 2007).
- Organize local supply chains to capture expertise and technologies.
- Hire local talents to assure access to local knowledge and relationships.
- Transfer marketing advantages realized in one country to others.
- Follow global customers around—Some customers who expand globally will need services that are standardized (e.g., Big Six accounting firms engaged by numerous multinational corporations). Other services in these categories are banking, insurance, business logistics, management consulting, and travel related services.

14.8 FUTURE TRENDS

Several major trends already noticeable in the knowledge economy are induced by the growth of Web-based enablers and by the increasing demands of customers for better, faster, and cheaper products and services. The common underlying threads among these trends appear to be *effectiveness* (customer and environment), *efficiency* (internal structure and operation activities), and *integration* (one-stop consolidation). Discussed next are specific trends related to customer focus, enterprise resource planning and application integration, supply strategy, knowledge management, changes in organizational settings, and population diversity.

14.8.1 Customer Focus

In the knowledge economy, customer service will be driven by several distinct characteristics (Hiebeler, Kelly and Ketterman 1998; Schmitt 2003).

- *Speed of customer service*. It is preferable to reduce the processing time between searching, selecting, and entering and fulfilling orders. There will be no more excessive handoffs. Companies are moving toward a seamless integration of steps to accept orders, trigger receivables, send orders to production, route requisitions to warehouses, activate shipments by logistics partners (e.g., UPS, FedEx), replenish inventory, update accounting, replenish stock with suppliers, and track delivery status to ensure on-time delivery (Buss 1999).
- *Customer self-service.* Companies involved in real estate, insurance, travel, car buying, auction, parts sourcing, and retailing are increasingly moving toward empowering customers to serve themselves by creating 24/7 (i.e., 24 hours a day, 7 days a week) systems and cutting out intermediaries. The following three companies are successful examples of the trend to encourage and empower customer involvement in self-service:

  1. Gateway. Customers define their own needs, configure systems, place orders, pay for new computers, and get limited support.
  2. E-Trade. Customers trade securities without broker involvement, using a 24/7 Web site.
  3. Microsoft Expedia. Customers make reservations online to book flights and receive confirmations at a lower cost, while enjoying faster service.

- *Integrated solutions for customers*. Customers have moved away from best-of-breed individual solutions toward integrated systems. A specific example is Microsoft Office Suites with integrated functionality.

  Customers are motivated by the desire to spend less time shopping, shop at one-stop stores, make fewer shopping trips, and face fewer choices that may be time consuming and difficult. The following are examples of organizations responding to these desires:

  1. Gap. A one-stop clothing and accessory provider that encourages convenience-based "package" purchases. Mannequins are outfitted with shirts,

blue jeans, belts, baseball caps, sunglasses, socks, shoes, gloves, and a knapsack, marketing a hip image.

2. Citigroup. In 2002, this company started to offer the combined on-line services of banking, credit cards, automobile insurance, brokerage and investing, mortgage and loans, and e-mail cash for money abroad.
3. Automatic Teller Machines (ATMs). These machines are being expanded by countless financial institutions to include Web-based services such as e-mails, online purchases, and transactions.

- *Customized service and sales*. Statistics indicate that diverse companies lose 50 percent of their customers every five years. Generally, it costs five to ten times more to obtain a new customer than to retain one. One approach taken by some companies is to train service people to cross-sell and up-sell.

  For example, The Home Depot emphasizes service to do-it-yourself customers and provides customers with easy access to information. The company starts offering service before sales and continues its customer service after sales.
- *Consistent and reliable service*. Customers prefer to have single points of contact, which requires the company's service calls to be coordinated with supply-chain business partners (i.e., those members of the extended enterprise family under outsourcing, alliance, or partnership agreements).
- *Flexible fulfillment and convenient service*. Gevalia Kaffe imports coffee and makes home deliveries. It performs 200,000 transactions per week and has built an e-business infrastructure to take orders, find the lowest-cost routing distance between the customer and the nearest warehouse, check inventories, issue shipping orders, and activate shipping by the networked partners who deliver coffee to home addresses.
- *Transparent sales process*. Customers typically want to know the order status, product information, pricing, and availability (Larson and Lundberg, 1998). UPS is known to have set up a 24/7 sophisticated information system that performs two functions:

  1. Tracks air and ground parcels at any time and from anywhere.
  2. Achieves flawless delivery, which is now becoming the norm rather than the exception.

  Another example is Solectron, which makes circuit boards and electric assemblies for such customers as IBM, HP, and Intel and has plants in California, Washington, Malaysia, France, and Scotland. The company has refined a shop-floor tracking and recording system (STARS) that uses bar codes to enable customers to track the status of their orders. The enhanced process transparency helps induce new demand while retaining satisfied customers.
- *Continuous improvement in customer service*. Constantly learning new ways to enhance customer service is one of the keys for companies to succeed in today's marketplace. For example, Nordstrom bends over backward to please customers

with gold-plated service and a no-questions-asked return policy. This company found that 90 percent of its business is from a loyal 10 percent of its shoppers. (This is an example of the Pareto principle, described in section 5.6). Nordstrom applies three methods to achieve a high degree of customer retention and sustainable innovations:

1. Motivate employees by paying them very high rates of commission.
2. Use undercover shoppers to evaluate service and give cash rewards to employees who achieve a perfect score.
3. Delegate authority downward to allow autonomy and local decision making.

Nordstrom's strategy is compatible with the *service profit chain model* (Heskett, Jones, Loveman, Sasser, and Schlesinger, 1994). This model is unique in that it originates from company leadership (e.g., vision, values, energy, concern, and discipline). Superior leadership secures a good place for employees to work (e.g., through workplace design, job description and latitude, selection and development, communication and information, and tools for serving customers), which in turn produces satisfied employees. Happy and productive employees serve customers better. Satisfied customers confer increased profitability to the company.

- *Technology-enabled services.* The trend is toward an integration of various means of access available to customers, such as the Web, direct dial-up, interactive voice response, and kiosks.

  For example, customers are projected to be in the driver's seat in influencing the development of IT technologies (Moschella 2003). The waves of IT technology move from a technology-centered paradigm to a customer-centered practice.

  The first, system-centric, wave focused on developing proprietary systems. The second, PC-centric, wave centered on working out the hardware and software standards. The third, network-centric, wave dealt with Internet standards. The fourth, customer-centric, wave addresses the information content and transaction standards. This last wave aligns more closely the values of IT technologies with the needs of customers.

14.8.2 Enterprise Resource Planning and Application Integration

In the past, companies concentrated on achieving optimum performance in various individual functional departments. Each of these departments had the tendency to pursue optimal operations within its own boundaries. Each department, similar to a silo, acted like a tightly controlled organizational unit with limited communications capabilities to the outside world across its boundaries, except through its top. In practice, the deficiencies originated from poor, uncoordinated transactions between the functional groups more than offset the benefits of the local optimization they achieved. Thus, over time, it has become evident that corporate competitive advantages can be achieved only through a proper integration of the individual functional departments of, for example, engineering, manufacturing, design, customer service, procurement, finance, and accounting. Today's trend is toward enterprise application integration (Myerson 2002).

An integration of disparate departmental functions permits a greater access to information, while it links employees, business partners, and customers more effectively.

There are various enterprise application software products on the market that facilitate such integration. The obvious benefits produced by this type of integration are speed, accuracy, and cost reduction, because of reduced manual handoffs (e.g., manpower), which allow business decisions to be made faster and at higher quality and lower cost.

- Integrated communications systems. The trend is toward an integration of networks composed of telephones, cable TV, wireless, and computer data. The "last mile" bandwidth problem (from telephone switching in the office to the home) will likely be solved using fiber-optic systems. AT&T, Sprint, and MCI/WorldCom are said to be working on the integration of voice and data services. Browsers and modems are used as customer home contact points today. WebTV could very well replace these in the future.
- *Wireless applications*. The use of airwaves for other services in addition to phone calls is expected to increase. Data transfer to mobile units allows managers and leaders to make important decisions from anywhere at any time. For example, Palm Pilot (3Com) is a two-way personal communication tool in text format that displays real-time flight schedules, news headlines, and on-line transactions such as movie ticket purchases, stock trading, etc.). The iPhone 3G, as well as its latest competition, the Droid smartphone, is a phone, Web browser, music player, global positioning system, and personal messaging and data-organizing system. With hundreds of applications (*apps*), it can become almost any tool you would like it to be (see www.apple.com/iphone). In January 2010, Apple debuted its newest gadget, iPad, which further expanded several wireless capabilities, allowing it to compete directly against the e-reader Kindle marketed by Amazon.
- *Leveraged legacy systems*. Middleware consists of connectivity products that link the existing legacy systems on mainframe computers with client and servers and the Internet. Middleware provides the important function of making existing data widely available to employees and customers.

14.8.3 Supply Strategy

Traditional enterprises set up vertically integrated organizational structures and amass heavy physical assets to achieve competitive advantages in the marketplace. Knowledge-economy companies form flexible supply chain partnerships to increase the speed to market and vary the product features to better satisfy the ever-changing needs of customers. Use is made of the technological and marketing expertise of the networked partners. Competitive advantages are thus derived from their capability to form such knowledge-intensive business networks, rather than from the value of capital assets piled up on the ground.

Outsourcing is a major trend that favors the formation of virtual enterprises. A single company working alone is no longer viable, as diverse competencies are needed to compete in today's marketplace. In these virtual enterprises, certain noncore business processes will be strategically outsourced to achieve higher earnings and more pronounced competitive advantages. Examples include:

- Niagara Mohawk, which outsources human resources and purchasing functions.
- United Technologies and American Express, which outsource their procurement functions to IBM.

Other companies have started outsourcing some of their noncore operations, concentrating instead on doing what they do best. For example, Sun Microsystems has focused on design, electing to contract out or purchase all workstation components. This strategy permits Sun to (1) introduce new products rapidly, (2) achieve better quality, (3) ensure dependability, (4) shorten speed to market, (5) gain flexibility, and (6) realize cost advantages.

Another example is Sara Lee, which focuses on building new products, managing brands, and building market shares, while outsourcing the production of L'eggs hosiery, frozen desserts, Wonderbras, Coach briefcases, and Kiwi shoe polish.

In essence, this strategy allows the company to jump on and slide down someone else's experience curve (e.g., Boston Consulting Group's 85 percent curve) to maximize benefits while maintaining flexibility in switching partners.

The current push for better asset utilization (return on assets, or ROA) helps move companies toward becoming knowledge intensive (through supply chain and marketing), rather than capital-asset intensive (through in-house production). Better asset utilization can be achieved by creating contract partnerships, setting up global production networks, keeping overhead costs low, changing products frequently, and innovating through open sources.

U.S. companies that market products to end users have also started to outsource such corporate functions as production, back-office work, logistics, after-sales service, procurement, inventory management, and new-product design. As this outsourcing trend continues, more demands are induced for such corporate functions. Economies of scale dictate that new, vertically integrated parts and service providers are likely to be brought into being for the following purposes:

- *Vertically integrated parts suppliers are becoming huge multinationals in their own right, and they are tightly integrated to derive efficiency (e.g., Delphi Automotive).* Their factories are designed to be quickly rearranged in order for the same shop to make different products for different client companies. The focus is to manufacture products and to operate at capacity almost all the time. They get cheap components by buying in quantity. Parts may come from various regions of the world. Their gross margins are relatively small (6 to 8 percent on sales), but they generate good return on equity (for example, 20 percent).
- *Integrated service organizations take orders from banks, automakers, and pharmaceutical companies to handle financial advisement, accounting, and other services.* Services offered to different industries are now bundled together. Certain design service providers may have teams of industrial, mechanical, and chip engineers scattered around the world.

Engineers in these product and service provider companies will need to work in an interdisciplinary environment, as dictated by the companies' mission to serve clients in different industries. Since these product and service providers conduct no R&D, perform no marketing, and design no products, product innovations will have to come from their client companies' own R&D departments or from third-party startups.

Managers in the client companies will need to learn how to supervise the interactions between diversified product and service providers, and to ensure that their breakthrough idea—now entrusted to an outside provider to practice—will not benefit the competition. This is an important issue related to knowledge management.

14.8.4 Knowledge Management

Knowledge management refers to activities related to the preservation and enterprise-wide application of corporate expertise and know-how to magnify competitive advantages.

Business success depends on innovation and the expertise of competent knowledge workers, whose insights need to be properly preserved by documentation, knowledge sharing, recruitment, and retention. The importance of knowledge management is becoming increasingly evident in the knowledge economy, wherein knowledge-intensive companies forcefully strive, as a corporate strategy, to attract talented and more mobile knowledge workers.

14.8.5 Changes in Organizational Settings

There will be two broad types of corporations in the near future. The first type generates and markets products and services to consumers (the end users). These companies retain core competencies and outsource everything else to selected service providers. A network of contract partners may produce the products or services, each being particularly efficient and superior in supplying specific components or elements. Outsourcing will succeed for these companies, as it has for Dell Computers, with products that have acceptable qualities to the customers and that are composed of current technologies available from the company's networked partners. Outsourcing requires a perfect understanding of what customers need and how the products are specified. Under these conditions, companies compete on the basis of speed, flexibility, and cost. Going virtual is thus becoming useful. This is a major change in organizational settings, moving toward the virtual and away from vertical integration.

The other type of corporation will supply specialized parts, designs, and services to client companies under contracts. Acquiring various manufacturing facilities and organizing them into vertically integrated enterprises are the functions of these companies. Their manufacturing plants are laid out in a flexible manner in order to respond to divergent needs of global clients. This is a second change in organizational settings moving toward additional vertical integration.

Some virtually organized companies are swinging back toward vertically integrated operations to perform their creative and developmental tasks. The reason for this is that, to achieve breakthroughs in products and technologies, innovative performance is needed. Usually, virtual companies do not have sufficient information or resources to enable their networked partners to make parts for innovative products. However, a vertically integrated company can readily amass all of the needed resources under one roof and thus can be effective in advancing new technologies and products. Cisco Systems has integrated its operations to innovate new optical networks. So the pendulum swings back a little, from virtual to vertical integration, for those companies that intend to come up with innovative products.

Not all companies should become virtual; neither should all functional elements of a company be vertical. The organizational form may have to be frequently adjusted according to the company's objectives, strategies, and changing environments.

14.8.6 Population Diversity

The U.S. Census Bureau predicts that there will be significant shifts in the American population by 2050 (see Table 14.2).

For example, the projected increase in Hispanics is due primarily to immigration. As a consequence, the U.S. workforce is likely to become more diverse (Livers and Caver 2003). Managers may need more diversity training, as more women and minorities are expected to attain key positions.

The trends just described have become noticeable in the marketplace since the turn of the millennium. All of these trends have important effects on how business priorities are set and how companies are run. In the sections that follow, some of the anticipated changes will be elucidated, along with the challenges facing engineering managers in the new millennium.

14.8.7 Global Business Trends

A recent McKinsey Global Survey identified fourteen business trends (Anonymous 2008). The top six of these trends are regarded by some respondents to be more important.

1. Growing number of consumers in emerging economies/changing consumer tastes.
2. Increasing availability of knowledge and the ability to exploit it.
3. A faster pace of technological innovation.
4. Shift of economic activity between and within regions (e.g., to Asia or within the EU).
5. Increasingly global labor and talent markets.
6. Increasing constraints on the supply or usage of natural resources (e.g., environmental regulations).
7. Development of technologies that empower consumers and communities (e.g., online connectivity and self-serve).
8. Growing consumer demand for corporate contributions to the broader public good.
9. Growth of the public sector.
10. Shifting industry structure/emerging forms of corporate organization.
11. Increasing sophistication of capital markets.

Table 14.2 Population Diversity in the United States

	1995 (% of total)	2050 (% of total)
White (non-Hispanics)	73.60	52.80
Black (non-Hispanics)	12	13.6
Hispanics	10.2	24.5
Asian/Pacific Islanders	3.3	8.2
Native Americans	0.7	0.9

Source: Livers and Cave 2003.

12. Adoption of increasingly scientific and data-driven management techniques.
13. Geopolitical instability (e.g., cross-border conflict and terrorism).
14. An aging population in developed economies.

In addition, there are eight business technology trends that are worth watching (Manyika, Robinson, and Spraque 2007) in conjunction with the management of relationships, capital and assets, and information:

- Managing relationships
  1. Distributing co-creation (open innovation)—decentralize innovation
  2. Using consumers as innovators.
  3. Tapping into a world of talent—contracting outside talents.
  4. Extracting more value from interactions
- Managing capital and assets
  5. Expanding the frontier of automation
  6. Unbundling production from delivery
- Leveraging information in new ways
  7. Putting more science into management (data mining, etc.)
  8. Making business from information

Global companies need to act on these trends by initiating the appropriate strategies.

14.9 THE GLOBAL CHALLENGES AHEAD

Globalization brings about challenges to systems engineers as individuals, to managers who are involved in pursuing globalization-based business strategies, and to companies which wish to sustain a long-term leadership in globalization. Described below are problems and potential solutions related to job migrations due to globalization, the needs of developing local talents for global managers to be successful, and the six dimensional challenges faced by global companies.

14.9.1 Job Migration Induced by Globalization

Globalization has a profound impact on jobs, which are migrating to and from the developed countries as the push for globalization intensifies. For example, when foreign products (such as textiles, shoes, toys, and cheap electronics) overwhelmed U.S. markets in the 1980s, some American workers suffered. When Japanese automakers set up shop in the southern part of the United States in the 1990s to compete against Detroit carmakers, their investments created new jobs for American workers. When Nokia expanded its cell phone business into the United States, American workers benefited.

To maintain the competitive edge, global companies are constantly looking for ways to offer products and services better, cheaper, and faster. One way of achieving cost competitiveness is to practice the "continuous mobility" strategy spearheaded by General Electric (see Figure 14.5), seeking and finding the least-cost production sites that achieve acceptable product quality. GE moves its production plants to new locations whenever doing so will result in cost advantages without incurring penalties in quality. Factors in favor of outsourcing include reduced costs of Internet-based telecommunications, increasingly more powerful computers, and rising levels of skills abroad.

Besides reducing the cost of doing businesses, companies outsource work to conserve investment capital; shorten time to market; build a variable cost structure; secure needed skills, technologies, and expertise; and realize a round-the-clock operation.

Over the years, some U.S.-based multinational companies have implemented strategies that cause engineering and other jobs to migrate to developing countries. Figure 14.6 exhibits the reduction in manufacturing jobs in the United States during the last several decades.

It is clear that engineering and blue-collar jobs related to manufacturing have been disappearing at an alarming rate. However, as presented in Fig. 14.7, American manufacturing productivity has risen constantly over the last decade due to automation and a shift to higher value work.

For the year 2002, U.S. productivity reached an annual growth rate of 9.2 percent, the highest worldwide. U.S. Bureau of Labor Statistics (2004 B) has determined that outsourcing, to both overseas and domestic partners, has contributed about 1.5 percent toward this 9.2 percent growth.

Innumerable U.S. firms are involved in outsourcing engineering and science related work to developing countries. Engardio (2003) offers a comprehensive list of American companies engaged in outsourcing engineering and science jobs, including General Electric, Flour, Intel, Oracle, Texas Instruments, Hewlett-Packard, Boeing, and others.

Other big technology companies, including IBM and Microsoft, are bringing in foreign workers to America on L-1 visas, which allow the companies to pay workers their home-country wage rates for as long as seven years. The federal government places no limits on the number of L-1 visas issued to a company (Bridger 2003). Ford announced its plan to purchase auto parts worth $1 billion from China in 2004. With the stroke of a pen, Ford shifted thousands of jobs to China.

A large number of companies are also engaged in outsourcing various nonengineering jobs overseas (e.g., customer service, microbiology research, tax return preparation, back-office support, IT applications, and CT scan interpretation work). In fact, McCarthy (2002) predicts that 3.3 million U.S. white-collar jobs valued at $136 billion

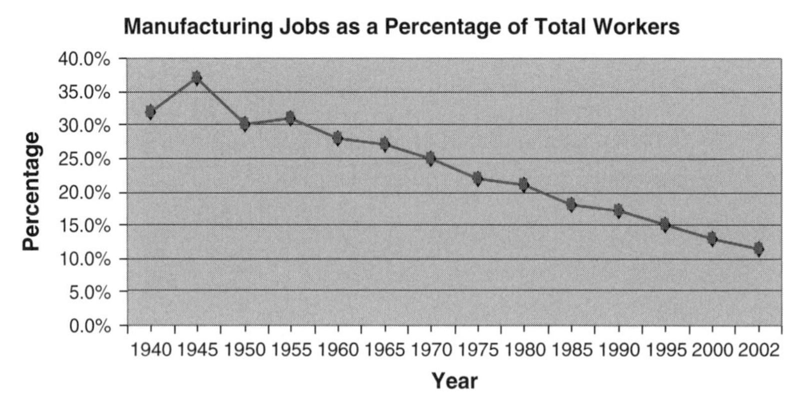

Figure 14.6 Loss of manufacturing jobs in the U.S. workforce. (U.S. Department of Labor.)

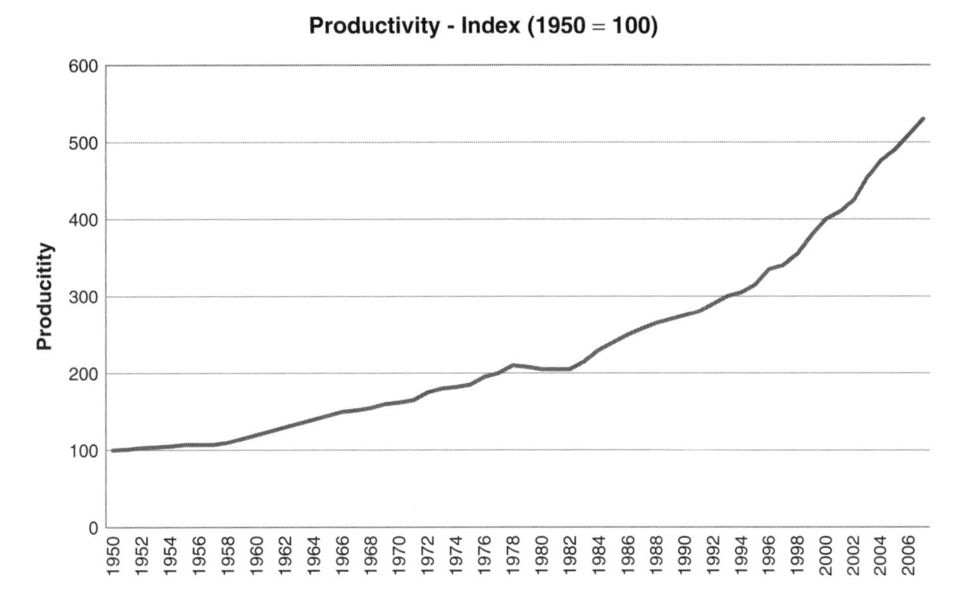

Figure 14.7 U.S. manufacturing productivity. (U.S. Bureau of Labor Statistics; Federal Research Board.) (Resource: Federal Reserve Bank of Chicago, October 2007.)

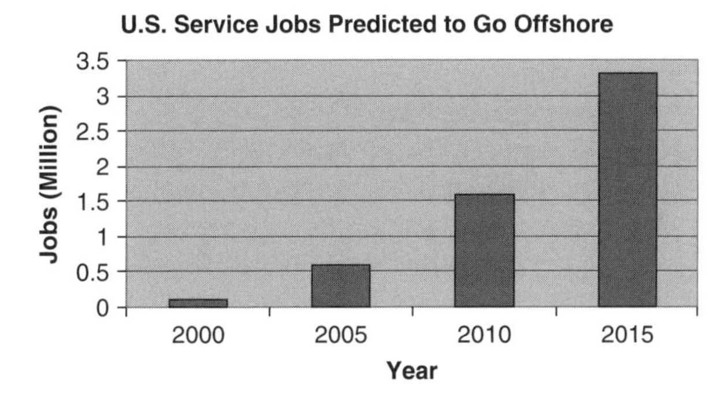

Figure 14.8 Service jobs. (Adapted from Schwartz, 2003)

in wages will migrate abroad by the year 2015 (see Fig. 14.8). More recently, Blinder (2009) estimated that 25 percent of all US jobs may be offshorable.

Service jobs are also being outsourced to low-wage countries from selected developed economies (Karmarkar 2004). Specifically, information-intensive services (e.g., data entry, credit card processing, software programming, market research, content management, publishing, engineering, management, financial services, and education) will be affected much more directly by outsourcing than physical services (e.g., nurses, construction workers, bricklayers, janitors, restaurant and hotel workers, mechanics, and others).

The white-collar job migration is a serious problem that must not be overlooked. It is a moving train viewed by many to be unstoppable.

Global Pie Concept. Is white-collar job migration all that bad? It is certain that some white-collar workers will be adversely affected. Some analysts believe, however, that the overall impact of offshoring may actually be positive for both the United States and the world at large. Agrawal and Farrell (2003) estimated that the benefit of about $1.45 may be derived from $1 spent by the United States in offshoring, suggesting that the global pie will be made bigger, not smaller, by the offshore investments. The key point is, of course, that the offshoring investment is not predicted as a zero-sum game, wherein the gain of one party is at the expense of the other. Rather, both parties are expected to benefit in the process.

How to Survive the White-Collar Migration. Agrawal and Farrell (2003) assume in their studies that the value of U.S. labor reemployed is $0.45 to $0.47 for each U.S. dollar invested in offshoring. This means that the investment funds freed by going offshore can be effectually deployed within the United States (e.g., hiring employees to do higher value work and initiating leading-edge R&D).

Engineers whose jobs may be outsourced unexpectedly need to post the question, "What will it take for them to become part of these 'reemployed resources' that companies will want to use the freed investment funds to engage?" There are a number of steps engineers may take to augment their relative competitiveness in this situation:

- *Get more education.* Currently, the white-collar jobs migrate primarily from the United States to four lower- wage countries: China, India, the Philippines, and Mexico. These recipient countries bring forth an overwhelmingly large number of engineering and science graduates at the B.S. level every year. In Fig. 14.9, the data points for the years 1989 and 1999 originated from the National Science Foundation (2004), whereas those for 2009 are predictions. In other words, the number of U.S. graduates at the B.S. level is predicted to become a smaller percentage of the five-country total—only 15 percent—by 2009. Each U.S. engineer or scientist graduating with a B.S. degree will have six other engineers or scientists as competitors in the job market. Similarly, the M.S. and Ph.D. graduates in the United States will make up about one-third of the five-country total by 2009. Each U.S. engineer or scientist with a M.S. or a Ph.D. degree will have only two others from the four low-wage countries to compete against.

These numbers suggest that it would be useful for American engineers to get advanced degrees as one of several steps to decrease the adverse effects of white-collar migration. Having an advanced degree is likely to be useful, but is still not a guarantee that an engineer will survive the threat imposed by the white-collar job migration.

- *Practice the "steady-ascent" strategy.* Engineers should constantly seek to upgrade the ways they perform their jobs, so that their employers can provide better, faster, and cheaper products and services to their customers. Innovations are desirable at the interface between engineering, design, and manufacturing to minimize labor, cut waste, enhance quality, and speed up time to market. The added value concept is quite ordinary in itself. However, it becomes critical in times of white-collar job migration (see Fig. 14.10).

Quality and value versus location is plotted in Fig. 14.10. The horizontal axis denotes production sites, from low cost on the left to high cost on the right. Production

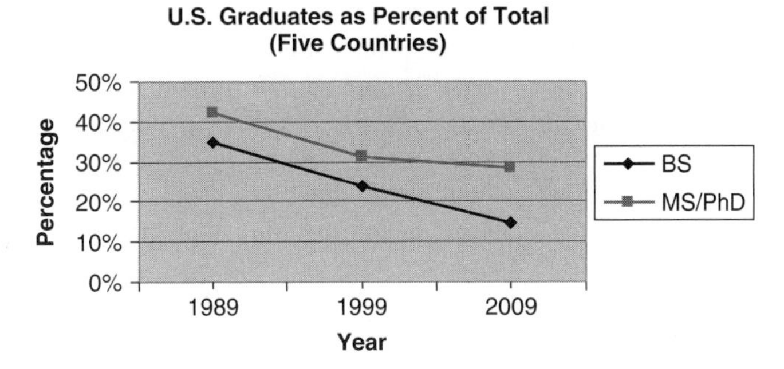

Figure 14.9 U.S. engineering and science graduates as percentage of the five-country total. (National Science Foundation, 2004)

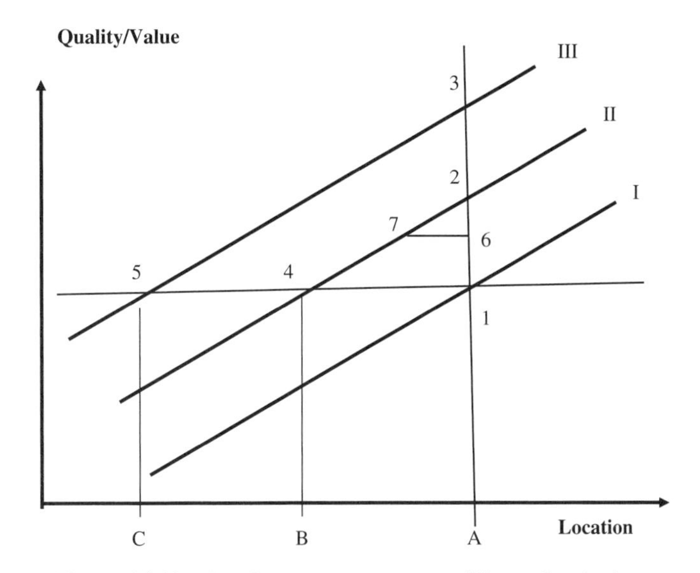

Figure 14.10 Steady-ascent strategy—adding value in time.

cost at A is higher than at B, which, in turn, is higher than at C. The straight lines I, II, and III symbolize the linear relationship assumed to exist between the quality–value combination and cost for high-, medium- and low-cost countries, respectively. It is reasonable to assume that the higher the product or service quality is, the higher its production cost will be thus, causing the inclined lines to rise from the lower-left to upper-right corners. In the course of time, such a straight line shifts from III to II to I, reflecting the constant strides made by developing countries in their capabilities of manufacturing products and supplying services at high quality, while allowing their wages to rise accordingly.

Let us assume that Point 1 is the current point of operation, with a quality–cost relationship depicted by the straight line I and its location selected at A (e.g., the United States). As the developing countries upgrade their engineering and production skills and become increasingly capable of supplying work of acceptable quality, the quality–cost curve available to U.S. firms. Under these circumstances, U.S. firms will logically shift production and other work from point 1 to 4 and relocate the associated operations from A to B (e.g., to India or China), while keeping the quality at an acceptable level. Should another country offer even lower wages in the future, the U.S. firms will not hesitate to move their facilities again from B to C to operate at point 5. This was the concept of *continuous mobility* discussed in section 14.5.2. The relocation of operations from A to B and from B to C will ensure that the firms remain cost competitive over a long period, without sacrificing the quality of their products or services.

For individual engineers to ensure that they do not get displaced by white-collar migration, Fig. 14.10 offers an obvious strategy. It is a given fact that the developing countries will constantly upgrade their engineering and production skills, as illustrated by the shifting of the straight line from III to II to I. Therefore, in order for engineers to survive the white-collar job migration, they need to periodically add enough value to ascend from point 1 to 2 and from 2 to 3 while employed in the home country (A in this example). In fact, should the individual fail to reach Point 2 his or her job may be outsourced due to the availability of line II. Specifically, if the value contributed by the engineer reaches only Point 6 rather than Point 2, then employers will seek to outsource the work to operate at Point 7, the lowest-cost operating point for the same value/quality.

The key message here is that engineers must pay attention to the rapid progress made by developing countries and use these external benchmarks as the correct yard-sticks to measure their own performance so that the incremental value they add is always on par with or higher than that which can be readily gained by their employers through off-shoring. Furthermore, this value must be added in time for them to remain competitive from the value–cost standpoint. Besides the magnitude of the value contributed, timing is critical. Engineers need to avoid being caught off guard in this dynamic environment. It is like rowing a boat upstream; resting means falling back.

Thus, the traditional mode of continuing education (e.g., taking a course or two from time to time) will probably not be sufficient to ensure job security. What is needed is fast learning guided by constant external benchmarking and adding value before equivalent value can be attained at lower cost (via off-shoring).

As global employers pursue continuous mobility, global engineers need to practice the *steady-ascent strategy* in a timely manner. Engineers are advised to *be selective* in choosing activities and functions that add recognizable value to their employers. Adhering to the following guidelines will make it more likely that engineers will be marketable at any given time:

1. Stay close to the employer's core competencies, which are usually preserved and nurtured in-house.
2. Learn to absorb complex information quickly and solve technical, business, and people problems creatively. Focus your own work on maximizing its value added. Practice the *steady ascent strategy*.
3. Become versed in interacting with workforce members of diversified backgrounds. Baby boomers who are managing the current operations of numerous companies are expected to retire in the next two decades. The gap left behind by them will need to be filled by capable people who can direct teams of diversified cultural and business backgrounds (Kaihla 2003).
4. Acquire the capabilities of designing and advancing the next generation of products and technologies and pursuing work that has high innovative content.

Apply new technologies to wring out product costs. Dell uses robots to assemble computers quickly and reliably in the United States, negating some of the cost advantages of moving plants offshore.

5. Cultivate visions with originality and a global orientation, and demonstrate leadership in strategic planning.
6. Maintain a broad business network that can be tapped into to add value for challenging situations.
7. Avoid attaching yourself for too long to functions or activities that are readily outsourced. Study the industry and stay away from labor-intensive "grunt" work—tasks that can be performed by following well-specified procedures and are more or less mechanical or operational in nature. Such work is easily learned and performed by other engineering graduates in low-wage countries. Examples include low-level engineering design, assembly operation, customer service, procurement, project management, troubleshooting equipment, engineering analysis using canned software programs, and laboratory tests.

14.9.2 Developing Local Talents

Developing local talents is a major challenge to a majority of global companies. As suggested by a McKinsey survey (Eddy, Hall, and Robinson, 2006), there is a general consensus among corporate leaders interviewed by McKinsey to integrate the nurturing of local leaders with broader localization efforts (e.g., promoting education, building a supplier base and improving the local business infrastructure). Several major reasons exist for the need to develop local talents:

- Requirement of some governments that 90 percent of workforce in natural resources companies be local
- Lower labor costs (local people tend to be less expensive)
- Better knowledge of local language and culture
- Deeper business relationships with government, industry and technology partners
- Corporate value –promoting a meritocratic global culture

Table 14.3 indicates the relative ranking of these key drivers in using local talents.

The techniques employed by some global companies to aggressively develop local talent are (1) offering short-duration overseas rotation program at headquarters (six to eighteen months), (2) support selected staff for MBA training abroad, and (3) assigning coaches to staff. They need to measure progress regularly by following the metrics such as the number of locals hired and promoted every year. Once the local talents are

Table 14.3 Key Drivers of Using Local Talents

#	Drivers	Percent of Responses
1	Operational effectiveness	52
2	Language, cultural issues	50
3	Company values leading to self-initiated localization	46
4	Legislation/commercial agreement with government	41
5	Cost reduction	32

developed, a subsequent challenge is to retain them beyond the obligatory duration of stay, which is typically required. There should be an attractive incentive program in place that is market-competitive.

A capability of attracting and retaining top talents is indeed a strong differentiator of success of global companies. According to Chandler (2009), it is wrong to assume that top local talents are inexpensive, as their packages are usually as expensive as those for their counterparts in the US and Europe. An out of the box hiring strategy is to hire from neighboring fields (consumer goods, hospitality) and take some risks with younger managers in the age group of 30s and early 40s. Local talents are starting to show sophistication with career planning. It is wise to assemble a team of mixed Western and local talents. Knowing where the top talents are is important. To retain local talents, the company environment must be right: (a) having a non-hierarchical environment with opportunity to advance to the top and (b) setting company values for which local talents will be proud to participate.

14.9.3 The Challenges Ahead

To confront the management challenges of the twenty-first century, engineering managers need to manage from the inside as well as from the outside, to lead from the present to the future, and to act locally and think globally. Table 14.4 contains descriptions of this alternative viewpoints.

On the *inside*, service systems engineers and managers plan, organize, lead, and control to implement projects and programs. They manage people, technologies, and other resources to add value to their service enterprises. They strengthen the company's core competencies and continue to modify services with features that customers want. They properly define (by activity-based costing and Monte Carlo simulations), monitor, and control costs. They appraise the enterprise's financial position and seize the right moments to initiate major service projects with high technological contents. These projects are supported by rigorous financial analyses in order to meet tough corporate evaluation criteria.

On the *outside*, service systems engineers and managers keep abreast of emerging technologies and screen new technologies that might affect the company's services.

Table 14.4 Management Challenges for Service Engineers and Leaders

Manage/Lead/Act/Think	Focuses
Inside	Core competencies, cost and quality control, production and engineering functions
Outside	Emerging technologies, supply chains, market orientation, customer relationship
Today	Organizing, controlling, planning, leading teams and projects; doing things right
Tomorrow	New projects, new core competencies, new products, new markets; doing the right things
Local	Implementation details, local adjustments
Global	Global resources, global scale and scope, global mindset and savvy

They proactively identify and introduce new tools related to service design, project management, plant operations, facility maintenance, and knowledge management to streamline the company's current operations. They define the best practices in the industry, emulate them as standards by which to evaluate their own in-house practices, and relentlessly strive to surpass these best practices. They look for potential supply chain partners whose alliances could produce competitive advantages for their enterprises in operations, service customization, and customer support. They pursue networking as a means to gain information and knowledge in a timely manner. They are sensitive to the enterprise's need to cultivate customer relationships. They pursue open innovation involving customers, suppliers and other external resources to expedite the generation of new service ideas and the rapid development of differentiable services to expand the customer base. They strive to add value to all stakeholders—customers, employees, suppliers, investors, and the communities in which the company operates.

For the *present*, service systems engineers and managers focus on keeping the company smoothly operating by "doing things right." They pay attention to details. They apply the best available tools (such as *Six Sigma, value stream mapping, failure mode* and *effect analysis*, and others) to promote operational excellence, which emphasizes cost-effectiveness and productivity. They aid in the important corporate function of knowledge management and contribute to the collection, transfer, and dissimilation of corporate know-how, expertise, and insights. They introduce a balanced scorecard to make sure that both financial and nonfinancial metrics are selected and deployed to monitor and evaluate the company's performance. They attempt to continuously augment current company operations. They take care of assignments (e.g., cost control, waste elimination, etc.) that should be accomplished for the company to achieve business success in the short term.

For the *future*, service systems engineers and managers strive to generate new ideas (both in-house and via open innovation) and manage the innovation process to bring about new values in both the core and supplement elements of differentiable new services. They design and introduce new services in a timely manner to ensure sustainable profitability for the company in the future. They seek emerging technologies to strengthen corporate competencies in the long term. They follow the future trends and capture new opportunities, with respect to new markets, new customer segments, new technologies, and new business alliances. They envisage a vision for the future, contribute to new company strategies related to technology, and assist company management in defining "the right things to do."

At the *local* level, services systems engineers and managers seek to best utilize the resources available (for example, people, technology, and business relationships) to achieve the company's objectives. They adjust to local conditions (e.g., language and culture, competition, market conditions, business practices, and governmental regulations) and take lawful, ethical, and proper actions to discharge their daily responsibilities. They help train local talents in order for the company to access knowledge and relationships at the local level. They maintain their local networks of professional talents and business relationships to heighten the company's productivity. They assist in customizing some supplemental service elements to better meet the needs of local customers. They communicate their experience and preserve lessons learned so that others at different sites within the company may benefit.

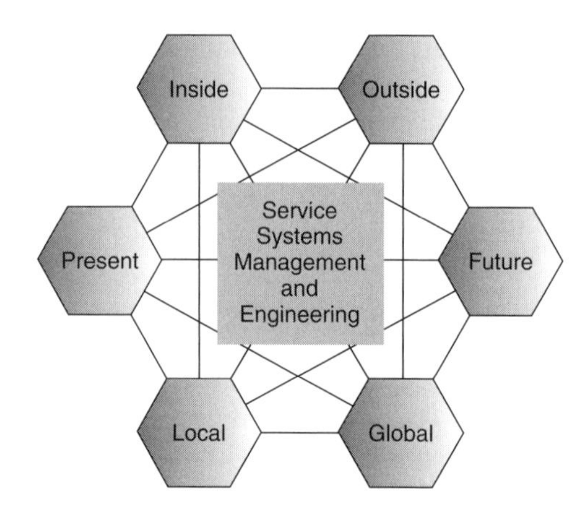

Figure 14.11 Six-dimensional challenges.

At the *global* level, service systems engineers and managers are sensitive in their pursuit of the optimal use of location-based resources to realize global economies of scale and scope, and to derive both cost and technology advantages for their enterprises. They help to manage global teams. They establish global networks of professional talents and business relationships and exploit innovative business opportunities. As companies pursue globalization over time, these professionals acquire a global mindset, become globally business savvy, and ready themselves to exercise leadership roles in international settings.

Service systems engineers and leaders need to be competent on the inside, well connected on the outside, productive at the present, visionary about the future, assertive locally and efficacious globally. Figure 14.11 elucidates the interrelationship between these six-dimensional challenges.

Because countless companies are affected by the rapid advancement of technology and the fast-paced evolution of globalization, the new millennium both offers ample opportunities for, and poses new challenges to, service systems engineers and managers (Kouzes and Poser 2002; Drucker 1999; Deep and Sussman 2000; Palus and Horth 2002; Jacobson with Setterholm and Vollum 2002). Those service systems engineers and managers who capture these new opportunities and meet these new challenges will be highly rewarded.

14.10 CONCLUSIONS

Globalization continues to be an inevitable business trend in the twenty-first-century economy. Three of the top five economies in the world are likely to shift to Asia by 2020.

Globalization creates ample value-added opportunities to those engineering managers who are properly prepared and equipped with the required global mindset, knowledge, and savvy.

Adjustments are needed for service systems engineers and managers to become leaders in the global arena. Their traditional American business attitude may need to be modified. They need to become more tolerant of, and adaptive to, that which is foreign and different. Past experience can serve as a useful basis for guiding future progress.

Service systems engineers and managers are encouraged to continuously follow the globalization process, to become sensitized to all issues involved in globalization, to make useful contributions to high-value work, and to prepare to lead in the globalization of the economy.

14.11 SUMMARY REMARKS FOR THE TEXT

This book is aimed at assisting service systems engineers and managers in assuming leadership positions in enterprises of the twenty-first century. A majority of these enterprises are affected by rapid changes in technology and the fast-paced advancement of globalization.

The first part of the book consists of the basic functions of engineering management, such as planning (Chapter 2), organizing (Chapter 3), leading (Chapter 4), and controlling (Chapter 5). These functions provide service engineers and managers with foundation skills to manage themselves, staff, teams, projects, technologies, and global issues of importance.

Best practices are emphasized as the pertinent standards for goal setting and performance measurement. Systems engineers and managers solve problems and minimize conflicts to achieve the company's objectives. They use the Kepner–Tregoe method to make rational decisions and take lawful and ethical actions. They apply Monte Carlo simulation methods to assess projects involving risks and uncertainties. They engage emerging technologies, motivate a professional workforce of diversified backgrounds, design new generations of services in a timely manner, and constantly surpass the best practices in industry.

The roles of systems engineers and managers in strategic planning, employee selection, team building, delegating, decision making, and managing creativity and innovation are explained. The augmentation of managerial competencies is emphasized.

The second part of the book covers the fundamentals of business management, including cost accounting and control (Chapter 6), financial accounting and management (Chapter 7), and marketing management (Chapter 8). This section is written to enable service systems engineers and managers to acquire a broadened perspective of the company business and its stakeholders and to facilitate their interactions with peer groups and units.

These chapters also prepare services systems engineers and managers to make decisions related to cost, finance, service, and capital budgets. Discount cash flow and internal rate of return analyses are reviewed. These discussions are of critical importance, as decisions made during the service design phase typically determine up to 85 percent of the final costs of the services offered. Additional applications are presented of activity-based costing (ABC), to define indirect costs related to products and services, and of economic value added (EVA), to determine the real profitability of an enterprise above and beyond the cost of capital deployed.

Also presented is capital formation through equity and debt financing. Resource allocation concepts based on adjusted present value (APV) for assets in place and option pricing for capital investment opportunities (e.g., for R&D and marketing projects) are addressed as well. By understanding the project evaluation criteria and the tools of financial analyses, service systems engineers and managers will be in a better position to secure project approvals. A critical step to refining technological projects is the acquisition and incorporation of customer feedback. For service systems engineers and managers to exercise leadership, they must meet the major challenges that arise from the initiation, development, and implementation of emerging technological projects that contribute to the long-term profitability of the enterprise.

The important roles and responsibilities of marketing in any profit-seeking enterprise are then introduced, along with the supporting contributions expected of service systems engineers and managers. Various progressive enterprises are increasingly concentrating on customer relationship management to grow their businesses. This customer orientation is expected to continue to serve as a key driving force for service design, project management, back-office operation, customer service, and countless other engineering-centered activities.

The third part of the book addresses six major topics: service systems engineers as managers and leaders (Chapter 9), ethics (Chapter 10), knowledge management (Chapter 11), innovation in services (Chapter 12), operational excellence (Chapter 13) and globalization (Chapter 14). These discussions provide additional building blocks to enhance the preparation that service systems engineers and managers must undertake to assume technology leadership positions in service enterprises and to meet the challenges in the new millennium. Figure 14.12 illustrates our "Three-Decker Leadership-Building Architecture" and the notion that service systems management and engineering leadership can be effectively fostered by a combination of the course materials contained in these chapters.

Engineers are known to possess strong skill sets that enable them to do extraordinarily well in certain types of managerial work. However, some of them may also exhibit weaknesses that prevent them from becoming leaders in engineering/technology organizations or even from being able to survive as engineers in the industry. The expected norms of effective leaders are described. Steps enabling service systems engineers and managers to augment their leadership qualities and attune themselves to the

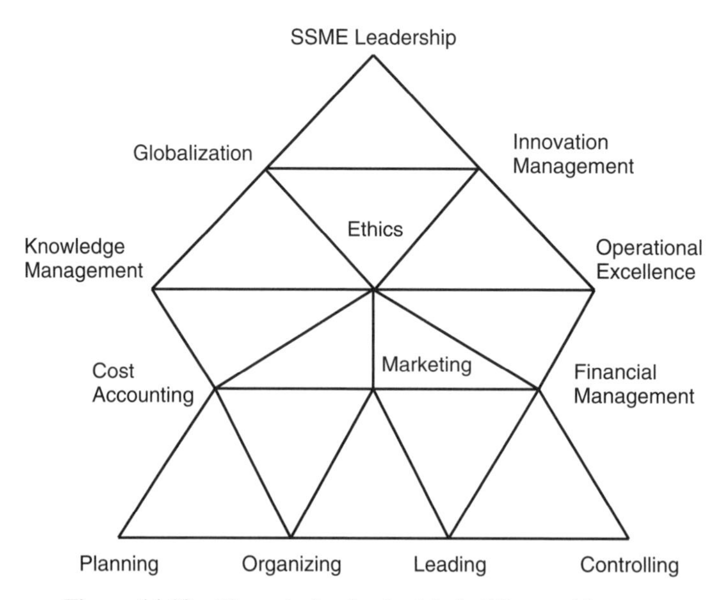

Figure 14.12 Three-decker leadership-building architecture.

value-centered business acumen are elaborated. Certain outlined steps should be of great value to those service systems engineers and managers who want to become better prepared to advance new services based on technology, integrate technology into their organizations, and lead technology-based service organizations.

A number of tried and true rules are included to serve as suitable guidelines for service systems engineers and managers to follow in becoming excellent leaders. Above all, service systems engineers and managers are expected to lead with a vision of how to apply company core competencies to add value, insights into how to seize opportunities offered by emerging technologies, and innovations in making services better, faster, and cheaper, so that they constantly excel in customer satisfaction. The concepts of value addition, customer focus, time to market, mass customization, supply chains, enterprise resources integration, and others are also elucidated.

Although engineers are ranked high in trustworthiness and integrity (ahead of businessmen, bankers, certified public accountants, lawyers, and others), it is important for all service systems engineers and managers to remain vigilant in observing a code of ethics. Other topics related to ethics are discussed, with examples of several difficult ethical dilemmas.

The changes wrought by the Internet are transforming most aspects of company business, including information dissemination, service distribution, and customer support. As processor design, software development, and transmission hardware technologies continue to advance, their roles in service business will surely grow steadily and affect various functions of engineering/technology management in the future. Progressive service systems engineers and managers engineering managers need to know which emerging technologies can be applied effectually to promote both economies of scale and service customization, expedite new services to market, align supply chains, foster team creativity and innovation, and enhance customer satisfaction.

Globalization expands the perspectives of service systems engineers and managers further with respect to divergence in culture, business practices, and value. Globalization is a major business trend that will affect innumerable enterprises in the coming decades. Service systems engineers and managers must become sensitized to the issues involved and prepare themselves to contribute to service enterprises wishing to cash in on the new business opportunities offered in the emerging global markets. They need to be aware of the potential effects of job migration caused by globalization and take steps to prepare themselves to meet such challenges. A major hurdle for service systems engineers and managers to overcome is failing to form global technical alliances to take advantage of new technological and business opportunities.

Service systems management and engineering will face external challenges in the new millennium. What these specific challenges are, how service systems engineers and managers need to prepare to meet these challenges, and how to optimally make use of location-specific opportunities to foster competitive advantages have been examined. Progressive service companies change organizational structures, engage in open innovations, set up supply alliances, and implement advanced tools to serve customers better, cheaper, and faster.

Globalization is also expected to constantly evolve. The United Nations has predicted that, by the year 2020, three of the five biggest national economies will be located in Asia. There will certainly be winners and losers as businesses become more and more global. It is important for future engineering managers to explore prudent corporate strategies for engineering enterprises in the pursuit of globalization, while

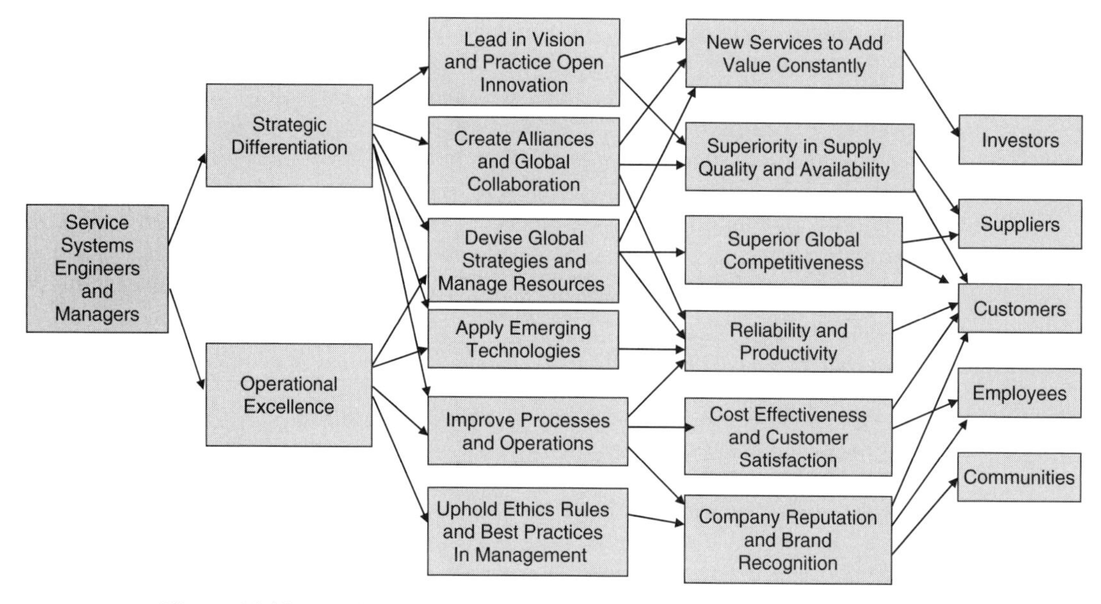

Figure 14.13 Emphasis on strategic differentiation and operational excellence.

minimizing any detrimental impact on the environment and maintaining above-standard human rights and labor conditions.

How can service systems engineers and leaders fulfill their roles to add value to the stakeholders of a service enterprise? The model depicted in Fig. 14.13 illustrates a possibility. From left to right, service systems engineers and managers add value by contributing to further the strategic differentiation and operational excellence of the firms (displayed in the second column from the left), take specific steps (lined up in the third column from the left), achieve the desirable end results (registered in the fourth column from the left), and eventually benefit all stakeholders (enumerated in the right-most column) of a service enterprise.

To enhance the strategic differentiation of a service enterprise, service systems engineers and managers provide vision toward the future, practice open innovation (involving customers, suppliers, and external resources) to expedite the development of differentiable new services, collaborate and align with both domestic and global partners to reduce time to market and service customers better, and manage global resources optimally to sustain profitability.

To enhance the operational excellence of a service enterprise, service systems engineers and managers apply emerging technologies (Web-based enterprise resource planning, project management, collaboration and other tools) to assure cost-effectiveness, implement best practices (e.g., Six Sigma, value stream mapping, failure mode and effect analysis, etc.) to promote productivity of existing operations, and follow ethics and other management practices to uphold company reputation.

David Starr Jordan said: "Wisdom is knowing what to do next; virtue is doing it." * The longest journey starts with a single step. Service systems engineers and

* *Source*: www.brainyquote.com/quotes/quotes/d/davidstarr106080.html

managers are advised to pursue these value-adding responsibilities by meeting the related challenges in six dimensions: inside, outside, present, future, local, and global.

14.12 REFERENCES

Agrawel, V., and D. Farrell. 2003. "Who Wins in Offshoring?" *The McKinsey Quarterly,* (4) (Global Directions).

Anonymous. 2008. "How Companies Act on Global Trends: A McKinsey Global Survey." *McKinsey Quarterly* (April).

Asheghian P., and B. Ebrahimi. 1990. *International Business: Economics, Environment and Strategies*. New York: Harper & Row.

Bahrami, H. 1992. "The Emerging Flexible Organization: Perspectives from Silicon Valley." *California Management Review*, Summer, pp. 32–52.

Begley, Thomas M., and David P. Boyd. 2003. "The Need for a Corporate Global Mind-Set." *MIT Sloan Management Review*, Vol. 44, No. 2.

Blinder, Alan S. 2009. "How Many US Jobs Might be Offshorable?" World Economics, 10(2), 41–78.

Brett, J. M. 2001. *Negotiating Globally: How to Negotiate Deals, Resolve Disputes, and Make Decisions Across Cultural Boundaries*. San Francisco: Jossey-Bass.

Brett, Jeanne, Kristin Behfar, and Mary C. Kern. 2006. "Managing Multicultural Teams." *Harvard Business Review* 84 (11).

Buss, D. D. 1999. Embracing Speed, *Nation's Business* 89 (June): 12.

Bridger, C. 2003. "Some Firms Are Sending Their White-Collar Work Overseas." *The Buffalo News*, August 30.

Brysk, A. (ed.). 2002. *Globalization and Human Rights*. Berkeley: University of California Press.

Calpan, R. 2002. *Global Business Alliances: Theory and Practices*. Westport, CT: Quorum Books.

Chandler, Clay. 2009. "Winning the Talent War in China," McKinsey Quarterly (November).

Collier P., and D. Dollar. 2002. *Globalization, Growth and Poverty: Building an Inclusive World Economy*. Washington, DC: The World Bank.

Condon, B. J. 2002. *NAFTA, WTO and Global Business Strategy: How AIDS, Trade and Terrorism Affect our Economic Future*. Westport, CT: Quorum Books.

Cundiff, E. W., and M. T. Hilger. 1988. *Marketing in the International Environment*. Englewood Cliffs, NJ: Prentice-Hall.

Dalton, M., C. Ernst, J. Deal, and J. Leslie. 2002. *Success for New Global Managers: What You Need to Know to Work across Distances. Countries and Cultures*. San Francisco: Jossey-Bass.

Das, Gurcharan. 1993. "Local Memoirs of a General Manager." Harvard Business Review (March–April).

Davis, Ian. 2009. "The New Normal." *McKinsey Quarterly* (2) (March).

Deep, S. D. and L. Sussman. 2000. *Act on It: Solving 101 of the Toughest Management Challenges. New York*: Perseus Books Group.

Dholakia, R. R. 2010. "Global Organization," Chapter 25 in Salvendy Gavriel and Waldemar Karwowski (Eds), *Introduction to Service Engineering*, John Wiley (January).

Dreo, H., P. Kunkel, and T. Mitchell. 2002. *Virtual Teams Guidebooks for Managers*. Milwaukee: ASQ Quality Press.

Drucker, P. F. 1999. *Management Challenges for the Twenty-first Century*. New York: Harper Collins Books.

Eddy James, Stephen J. D. Hall, and Stephen R. Robinson. 2006. "How Global Organizations Develop Local Talent." *McKinsey Quarterly* (3).

Engardio, P. et al. 2003. "The New Global Job Shift." *Business Week*, February 3.

Garten, J. E. 1999. *World View: Global Strategies for the New Economy*. Boston: Harvard Business School Press.

Ghenmawat, Pankaj. 2007. "Managing Differences." *Harvard Business Review* 85 (3).

Glover, J., H. Friedman, and G. Jones. 2002. "Adaptive Leadership: When Change Is Not Enough." *Organization Development Journal*, Part 1, Summer.

Govindarajan, V., and A. K. Gupta. 2001. *The Quest for Global Dominance: Transferring Global Presence into Global Competitive Advantage*. San Francisco: Jossey-Bass.

Gumbel, Peter. 2007. "Mapping Globalization's Winners and Losers." *Fortune* (October, 29).

Gupta, A., and V. Govindarajan. 2001. "Converting Global Presence into Global Competitive Advantage." *The Academy of Management Executive*, May.

Herman, A. 2010. "Enterprise Value Creation in the Global Service Economy," Chapter 5 in Salvendy, Gavriel and Waldemar Karwowski (Eds), *Introduction to Service Engineering*, John Wiley (January).

Heskett, J. L., T. O. Jones, G. Loveman, W. E. Sasser Jr. and L. A. Schlesinger. 1994. "Putting the Service-Profit Chain to Work." *Harvard Business Review* (March).

Hexter, Jimmy, and Jonathan R. Woetzel. 2007. "Bring Best Practice to China." *McKinsey Quarterly* (4).

Hiebeler, R., T. B. Kelly, and C. Ketterman. 1998. *Best Practices: Building Your Business with Customer-Focused Solutions.* New York: Arthur Andersen/Simon & Schuster.

Hoefling, T. 2001. *Working Virtually: Managing People for Successful Virtual Teams and Organizations*. Sterling, VA: Stylus Publishing.

Jacobson, R., with K. Setterholm and John Vollum. 2000. *Leading for a Change: How to Master Five Challenges Faced by Every Leader.* Woburn, MA: Butterworth-Heinemann.

Jeannet, J.-P. 2000. *Managing with a Global Mindset*. Upper Saddle River, NJ: Financial Times/Prentice Hall.

Johansson, J. K. 2000. *Global Marketing: Foreign Entry, Local Marketing and Global Management*. 2nd ed. Boston: Irwin/McGraw Hill.

Kaihla, P. 2003. "The Coming Job Boom." *Business 2.0 Magazine*, September.

Karmarkar, Uday. 2004. "Will You Survive the Services Revolution?" *Harvard Business Review* (June).

Kouzes J. M., and B. Z. Poser. 2002. *The Leadership Challenge*, 3rd ed. San Francisco: Jossey-Bass.

Kugler, R. L., and E. L. Frost (eds.). 2001. *The Global Century: Globalization and National Security*. Washington, DC: National Defense University Press.

Lamberton, D. 2002. *Managing the Global: Globalization, Employment and Quality of Life*. New York: St. Martin's Press.

Langhorne, R. 2001. The Coming of Globalization: Its Evolution and Contemporary Consequences. New York: Palgrave.

Larson, M., and D. Lundberg. 1998. The Transparent Market: Management Challenges in the Electronic Age. New York: St. Martin's Press.

Lipnack, J., and J. Stamps. 2000. *Virtual Teams: People Working Across Boundaries with Technology*. 2nd ed. New York: John Wiley & Sons, Inc.

Livers, A. B. and K. A. Caver. 2003. *Leading in Black and White: Working Across the Racial Divide in Corporate America.* San Francisco: Jossey-Bass.

Lovelock, Christopher H., and George S. Yip. 1996. "Developing Global Strategies for Service Businesses." *California Management Review* 38 (2).

Manyika, James M., Roger P. Roberts, and Kara L. Sprague. 2007. "Eight Business Technology Trends to Watch." *McKinsey Quarterly* (December).

McCarthy, J. 2002. "3.3 Million U.S. Services Job to Go Offshore." *Forrester Research* (Tech-Strategy Brief), November.

Marguardt, M. J., and N. O. Berger. 2000. *Global Leaders for the Twenty First Century*. Albany: State University of New York Press.

McCall, M. W., Jr. and G. P. Hollenbeck. 2001. *Developing Global Executives: The Lessons of International Experience*. Boston: Harvard Business School Press.

Moschella, D. 2003. *Customer Driven IT: How Users are Shaping Technology Industry Growth.* Boston: Harvard Business School Press.

Myerson, J. M. (ed.) 2002. *Enterprise Systems Integration.* Boca Roton, Florida: Auerbach.

National Research Council. 1987. Management of Technology: The Hidden Advantage. Washington D.C. National Academy Press.

National Science Foundation. 2004. *Science and Engineering Indicators 2004*. Washington, DC: U.S. Government Printing Office.

Nelson, James. 2006. "Daniel D. Fillippo: What is Driving Globalisation?", *New Zealand Management, 53, 5. p. 44*

Oppenheim, Jeremy M. 2004. "Corporations as Global Citizens." *McKinsey Quarterly*

Rogers, Kimberly B. and Gary Gereffi. 2007. "US and Japanese Lead Firm's Production Strategies and Labor in the North America Automotive Industry," ASA Conference Paper (January 17). www.allacademic.com//meta/p_mla_apa_research_citation/1/8/4/7/6/pages184767/p184767-1.php.

Sassen, S. 1998. *Globalization and Its Discontents*. New York: New Press.

Schmitt, B. 2003. Customer Experience Management: A Revolutionary Approach to Connecting with Your Customers. Hoboken, NJ: John Wiley & Sons, Inc.

Schneider, S. C. and J.-L. Barsoux. 2003. *Managing Across Cultures*. 2nd ed. Upper Saddle River, NJ: Financial Times/Prentice Hall.

Solomon, Charlene M., and Michael S. Schell. 2009. *Managing Across Cultures: The Seven Keys to Doing Business with a Global Mindset*. New York: McGraw Hill.

Steger, M. B. 2002. *Globalism: The New Market Ideology*. Lanham, MD: Rowman & Littlefield Publishers.

Sullivan, J. J. 2002. *The Future of Corporate Globalization: From the Extended Order to the Global Village*. Westport, CT: Quorum Books.

Thurow, L. C. 1997. "New Rules: The American Economy in the Next Century." *Harvard International Review* (Winter).

U.S. Bureau of Labor Statistics. 2004A. U.S. Had Largest Manufacturing Productivity Increase in 2002. Washington, DC: U.S. Government Printing Office.

U.S. Bureau of Labor Statistics. 2004B. *The Effect of Outsourcing and Offshoring on BLS Productivity Measures.* Washington, DC: U.S. Government Printing Office.

Wilson, D., and R Purushothaman. 2003. *Dreaming with BRICs: The Path to 2050*. Global Economics Paper #99, Goldman Sachs, New York (October 1).

World Bank. 1992, *Global Economic Prospects and the Developing Countries*. Washington, DC: World Bank.

World Bank. 2001. *Global Economic Prospects and the Developing Countries*. Washington, DC: World Bank.

World Bank. 2008, *Global Output Totals $59 Trillion in 2006*, Press Release (April 11), Washington, D.C: World Bank

World Bank. 2009. *Global Economic Forecast Update, Press Release (March31).* Washington D.C.: World Bank

14.13 QUESTIONS

14.1 Strategic planning on a global scale is a leadership function of great importance to any organization. Review the Harvard Business School case "General Electric Medical Systems, 2002," # 9-702-428 (rev. February 26, 2003) and answer the following questions. The case materials may be purchased by contacting its publisher at 1-800-545-7685 or at www.custserve@hbsp.harvard.edu.

A. What is the underlying logic behind the Global Product Company (GPC) idea?
B. What are the pros and cons of the Chinese recommendations to change GPC?
C. What are the pros and cons of slashing product prices in China?
D. What are the pros and cons of pursuing other options (1) genomics, (2) healthcare IT, and (3) others?
E. Which corporate strategy do you recommend for GE to consider?
   1. Modify PPC to fit China market
   2. Slash Price to gain market share
   3. Focus on genomics, healthcare IT or others
F. What are the lessons you have learned from having studied this case?

14.2 Study the case "Metsco Paper: Globalization of Finnish Metal Workshops," Harvard Business School Case # 9-805-057 (Revised December 13, 2004) and answer the following questions. The case materials may be purchased by contacting its publisher at 1-800-545-7685 or at www.custserve@hbsp.harvard.edu.

A. What is this case about?
B. Transforming an asset-intensive company into a global service and Knowledge-intensive company takes some major efforts. What are the major challenges in such a transformation? How did Metsco Paper handle these challenges?
C. Globalization increased the complexity of the company's business. Name some examples of this complexity? How did Metsco Paper successfully handle such complexity?
D. The roles of IT were critical. How does one measure the impact of IT at Metsco Paper? What is the IT-induced innovation? Explain the importance and nature of IT adoption.
E. What are the lessons you have learned from having studied this case?

14.3 It can be argued that democracy and capitalism are concepts that are fundamentally incompatible with each other. Democracy is built on the principle of equality—one person, one vote—regardless of the individual's intelligence, wealth, work ethic, or any other features that distinguish one individual from another. Capitalism, by contrast, fosters inequality. It uses incentive structures to encourage hard work and wise investment to realize differences in economic returns. Because future income from investments (in human or physical assets) depends on current income, wealth tends to generate wealth, and poverty tends to constrain the individual's economic growth. The cycle is self-reinforcing: Success breeds success, and failure compounds failure. "The economically fit are expected to drive the economically unfit out of existence. Thus, there are no equalizing feedback mechanisms in capitalism" (Thurow 1997).

A. What are some of the remedies capitalistic countries have introduced to mitigate such inequality?
B. Would globalization compound this condition in a capitalistic and democratic country? Why, or why not?

14.4 Globalization, which causes the countries involved to become more interconnected, clearly has tremendous social and political implications. It also has a cultural dimension to it, due to worldwide communications that facilitate the global connections. Cultural globalization may lead to a more civic global society with a greater consensus on civic values. It may also diminish the rich diversity of human civilization, as the Asian, Islamic, South American, and other non-Western values become increasingly generic. For many, the preservation of distinct cultural traditions is a very serious matter.

A. Is globalization a form of Western imperialism that may homogenize non-Western values? Why or Why not?
B. Can homogenization be avoided or mitigated?

14.5 During the new century, increased flows of products, services, technologies, capital, and workers across national borders will affect the economical, social, and political life of everyone involved. The United Nations is expected to play a critical role in this increasingly dynamic environment. In your opinion, what should be the major missions of the United Nations in addressing these issues?

14.6 Sustainable development refers to work that simultaneously satisfies economical, social, and environmental requirements (United Nations 2002). It is self-evident that work must be economically viable so that customers are willing to pay for the work supplied. Work must also be safe and otherwise socially compatible. Furthermore, work needs to be environmentally acceptable in that harmful discharges are minimized, wastes are decreased, material and energy resources are conserved, and any other detrimental impact on the environment is minimized.

Some academicians suggest that it is the engineer's responsibility to attain the ideals of sustainable development. They view it as the major challenge facing engineers in the future (Cruickshank 2003). Do you agree with this notion? Why or why not?

14.7 Leading technological innovation will be a major challenge for engineering managers in the new millennium. What are some of the success factors for technological innovation?

14.8 The new millennium is expected to see continued changes in communications technologies, business practices, worker diversity, customer empowerment, and marketplace conditions. Name a few leadership qualities that are deemed essential for engineering managers to achieve success in the new millennium.

14.9 U.S. productivity has improved noticeably in recent years, averaging 4 to 5 percent per year, while the U.S. economy grew by only 3.5 percent. The gain in productivity was due, in large part, to the use of technology, in addition to longer working hours by those who are lucky enough to have jobs. According to the Economic Policy Institute, the average U.S. worker has added 199 hours to a year since 1973. The United States achieved the per-hour productivity of $32, compared with $38 for Norway and $34 for Belgium. In other words, U.S. workers are simply working longer, not necessarily better or smarter.

They take less annual vacation time (only 10.2 days, on average), compared with 30 days in France and in Germany.

At the same time, a large number of U.S. companies are aggressively outsourcing work to low-wage countries, such as China, India, the Philippines, and Mexico. A 2003 study released by the University of California at Berkeley indicates that as many as 14 million U.S. service jobs are in danger of being shipped overseas.

A. Who is responsible for this peculiar position that U.S. workers are being forced into?

B. How can U.S. companies meet the new challenge of improving the quality of life for their workers without sacrificing the companies' relative competitiveness in the marketplace?

14.10 In the new millennium, speed to market will be a major harbinger of competitiveness. "The early bird gets the worm," as the saying goes. How can engineering managers meet this challenge by leading their companies to be the first movers?

14.11 Expertise and talents in various disciplines are critical to the success of any company. This is true, especially in the rapidly developing global economy of today, wherein knowledge workers are highly mobile and talents are dispersed everywhere. Companies will be better prepared to meet the future challenges in the marketplace if they can establish and maintain well-organized global supply chains of talents so that they may deploy them in the right place, at the right time, and for the right purpose.

Assume that you are in charge of establishing and maintaining such a global supply chain of talents in engineering, how are you going to plan, organize, lead, and control them?

14.12 Globalization is already upon us. In spite of the current recession in the United States, whose recovery is expected by various economists to begin before or around the end of 2009, economic activities are expected to increase dramatically (e.g., with a GDP at 5 percent to 9 percent) in Asia, South America and Africa, in the next fifteen to twenty years. The economic growth rates in North America, Europe and Japan are forecast to grow only marginally (e.g., with a GDP at 1 percent to 3 percent) during the same period. Engineering management leaders must prepare themselves in order to capitalize on the growth opportunities accorded by globalization.

Explain what types of preparations are critical and useful to ready service systems engineers and leaders for this global economy in the near future.

14.13 Globalization is generally accepted as a dominant business trend in the twenty-first century. The center of economic activities is expected to gradually shift toward Asia in the next twenty years, according to forecasts of the United Nations. Explain how companies in private sectors may seize opportunities offered by globalization.

14.14 For products intended for global markets, customers' wants and needs are different from one market to another. How can a centralized global team build up a product to serve as a "platform" for the global market?

14.15 Japanese companies face challenges similar to those faced by U.S. companies in that low-cost manufacturing capabilities are readily available in such countries as China, India, the Philippines, and Mexico. How can the Japanese companies plan to deal with these challenges?

Appendix

Selected Cases Relevant to Service Systems Management and Engineering

The case method is well established and widely practiced in management schools. Service systems engineers and leaders are advised to consult with a recent publication regarding the advantages of learning from cases (Ellet 2007).

Samples of SSME cases are selected to assist service systems engineers and leaders in gaining management and business perspectives. These cases, illustrating the complex service systems management and engineering issues involved, may be studied to augment the topics covered in this book. Therefore, these cases should be considered as a useful extension to the exercise problems listed at the end of each chapter. Service systems engineers and leaders are strongly encouraged to study these and other cases to glean useful lessons from them.

Class surveys indicate that students generally like case discussions, as they enhance critical thinking and stimulate active participation. They like the fact that well-organized case studies are the next-best alternatives, besides working in the right industrial settings, for them to acquire useful and broad-based experience in dealing with the real-world engineering and management issues. They understand also that the principle of *more in and more out* applies here, in that the more effort they put into preparing themselves for the case discussion, the more valuable insights they will get out of the cases discussed.

It is generally advisable for the case teacher to specify a list of open-ended questions ahead of the time, so that students can focus on them when preparing themselves for the case studies. By immersing themselves in the company's affairs and acting like external consultants, students are to recommend a strategy, solve a problem, devise a new procedure, capture a new opportunity in the marketplace, apply a new technology to improve customers' satisfaction, and more, for the benefit of the company.

It is equally important that a few students in the class do not dominate the discussions, that most students are encouraged to speak up, and that the discussions are driven by the thought-provoking questions. Each case takes, on average, about eighty to ninety minutes of class time. Usually, students are in favor of receiving from the case teacher a written case summary that captures the salient points of the case discussion. Past experience indicates that asking students to summarize in writing what lessons they have learned from each case, without reiterating the case details, helps them to internalize the "take-away" messages.

Table A.1 Classification of SSME Cases

Case	Name	Service Sector	SSME Issues	Chapter
1	Strategic Planning at United Parcel Service	Logistics	Strategic planning	2
2	Delta Airlines (A) The Low Cost Carrier Threat (most popular)	Airlines	Strategic planning	2
3	Microsoft in 2005 (#9-705-505 Rev.1-9-2006)	Software	Strategic planning	2
4	Zappos.com: Developing a Supply Chain to Deliver WOW!	Internet retailer	Supply chains	2
5	BMG Entertainment (most popular)	Music/entertainment	Strategy making	2
6	Equity Bank (A)	Financial services	Business strategy	2
7	Exel plc—Supply Chain Management at Haus Mart	Logistics provider	Supply management	2
8	Cityside Financial (2006)	Financial services	Diversity	3
9	The Wheels Group: Evolution of a Third party Logistics Service Provider	Logistics	Decision making	4
10	Time Share Exchange Fair (A)	Tourism/hotels	Decision analysis	4
11	Proctor & Gamble Global Business Services	Service operations	Decision making	4
12	Personal Shoppers at Sears: The Elf Initiative	Retailers	Performance metrics for services	5
13	Quality Wireless (A)—Call Center Performance	Service firms	Data statistical validation of data	5
14	Survey Masters, LLC	Service firms	Activity-based accounting	6
15	Critical Mass—IT Innovations	Marketing services	Innovations	6
16	Globalizing Cost of Capital and Capital Budgeting at AES	Service operations	Cost of capital	7
17	Outsourcing at Office Supply	Retails	Financial analysis of IT outsourcing	7
18	Santa Fee Relocation Services (Regional Brand Management)	Logistics	Brand management	8
19	Air France Internet Marketing	Travel industry	Online sales	8
20	Cabo San Viejo (Most Popular)	Service firm	Customer relationship management	8
21	Rosewood Hotels & Resorts (most popular)	Hotels	Marketing	8
22	Zipcar- influencing customer behavior (most popular)	Rental Cars Services	Control- Marketing	8
23	HSBC Credit Card Rewards Program	Financial Services	Marketing	8
24	Nector- Making Loyalty Pay (#505031)-12-5-2005 (most popular)	Retails	Customer relationship	8
25	Charles Schwab (#507005—Rev. 1-11-200) (most popular)	Financial services	Advertising	8
26	InvestorSoft	Software	Core marketing issues	8
27	Fidelity Investment's Charitable Gift Fund (A)	Financial Services	Patents, IPR	9
28	Excel Logistic Services	Logistics	SPC applications	9

Table A.1 (*continued*)

Case	Name	Service Sector	SSME Issues	Chapter
29	Premier Inc. (A) (award-winning)	Conflicts of interests in healthcare products	Ethics	10
30	Knowledge Management at Katzenbach Partners	Consultancy	Knowledge management	11
31	Eli Lilly and Company: Innovations in diabetic Care	Drug company	Innovation management	12
32	Pharmacy Service Improvement at CVS (A) (most popular)	Pharmacy	IT roles in business	12
33	Bank of America (A) (most popular)	Financial services	Innovation in Products/services	12
34	Southwest Airlines 2002: An Industry Under Siege (most popular)	Airlines	Low cost but different services	12
35	Google Inc. :	InterNet search solutions	New service strategies	12
36	University Health Service 2006	Healthcare	Operational excellence	13
37	McKinsey—Globalization of Consultancy Services	Consultancy	Globalization	14
38	Globalization of Wyeth	IT enabler	Globalization	14

Service systems engineers and leaders may purchase the selected case documents directly from Harvard Business School Publishing Company (www.hbsp.harvard.edu), even though my list includes some cases from other universities such as Stanford, Northwestern, Virginia, and Ontario.

Academic teachers may register at the same Web site of Harvard Business Online for Educators free of charge. Once registered, they may be allowed to download a free copy of any of these cases for evaluation purposes. To use these cases in class, students need to purchase the original case documents, possibly at an academic discount of about 50 percent. For information regarding order processing, contact the publisher at www.custserve@hbsp.havard.edu. or at 1-800-545-7685.

A total of thirty-eight (38) SSME cases have been carefully selected to cover the service sectors of interests and the SSME issues of relevance. They are also the most current ones available at the time of this writing. Ten of these cases have been designated by Harvard Business School Publishing as the "most popular." Table A.1 classifies these cases in accordance to the "Service Sector," "SSME issues," and "Chapter" to which they may belong.

REFERENCES

Ellet. 2007. The Case Study Book: How to Read, Discuss and Write Persuasively About Cases. Cambridge, MA: Harvard Business Press.

1. David Garvin, Lynne C. Levesque (2006), **"Strategic Planning at United Parcel Service,"** Harvard Business Case # 9-306-002 (Rev. June 19).

2. Jan W. Rivkin, Laurent Therivel (2005), **"Delta Airlines (A): The Low Cost Carrier Threat,"** Harvard Business Case# 9-704-403 (Rev. January 25).
3. Yoffie, David B., Dharmesh M. Mehta, and Rudina I. Seseri (2006), **"Microsoft in 2005,"** Harvard Business Case # 9-705-505 (Rev. January 9).
4. Michael Marks, Hay Lee, David W. Hoyt (2009), **"Zappos.com: Developing a Supply Chain to Deliver WOW!"** Stanford University Business Case #GS-65 (February13).
5. Jan W. Rivkin and Gerrit Meier (2005), **"BMG Entertainment,"** Harvard Business Case # 9-701-003 (Rev. September 22).
6. Garth Saloner, Bethany Coates (2007), **"Equity Bank (A),"** Stanford Business Case #E260 (July 20).
7. Zeynep Ton and Steven C. Wheelwright (2005), **"Exel plc—Supply Chain Management at Haus Mart,"** Harvard Business Case #9-605-080 (Rev. May 16).
8. Ely, Robin (2006), **"Managing Diversity at Cityside Financial Services,"** Harvard Business Case # 9-405-047 (Rev. April 17).
9. P. Fraser Johnson and Michael Sartor (2004), **"The Wheels Group: Evolution of a Third Party Logistics Service Provider,"** Ivey Business Case #904D04 (March 9).
10. Anto Ovchinnikov, Dmitry Krass, Scott Sampson (2007), **"Time Share Exchange Fair (A)"** University of Virginia Business Case #UV0901 (January 31),
11. Thomas J. Delong, Warren Brackin, Alex Cabana, Phil Shellhammer, David L. Ager (2005), **"Proctor & Gamble: Global Business Services,"** Harvard Business Case # 9-404-124 (Rev. July 12).
12. Kyle Murray, RamasastryChandrasekar (2008), **"Personal Shoppers at Sears: The Elf Initiative,"** Ivey Business Case # 907A19 (Version A—January 10).
13. Sunil Chopra (2006), **"Quality Wireless (A)—Call Center Performance,"** Kellogg Business Case # KEL153 (April 1).
14. William T. Brunn Jr. (2007), **"Survey Masters, LLC,"** Harvard Business Case #9-107-061(March 8).
15. Malcolm Munro and Sid L. Huff (2008), **"Critical Mass: The IT Creativity Challenges,"** Ivey Business Case # 908E10 (October 27).
16. Mihir A. Desai, Doug Schillinger (2003), **"Globalizing Cost of Capital and Capital Budgeting at AES,"** Harvard Business Case #9-204-109 (December 12).
17. Jeffrey, Mark and James Anfield (2009), **"Outsourcing at Office Supply, Inc."** Kellogg Business Case # KEL308.
18. Niraj Dawar, Nigel Goodwin (2006), **"Santa Fee Relocation Services (Regional Brand Management),"** Ivey Business Case # 905A29 (January 13).
19. Jeffery, Mark (2009), **"Air France Internet Marketing: Optimizing Google, Yahoo! And Kayad Sponsored Search,"** Kellogg Management Case # KEL319.
20. Youngme Moon, Seth Schulman, Gail McGovern (2006), **"Cabo San Viejo: Rewarding Loyalty,"** Harvard Business Case # 9-506-060 (March 10).
21. Chekitan S. Dev, Laure Mougeot Strook (2007), **"Rosewood Hotels & Resorts Expanding to Increase Customer Profitability and Lifetime Value"** Harvard Business Case # 2087 (June 15).
22. Frei, Frances X. (2005), **"Zipcar- Influencing Customer Behavior,"** Harvard Business Case #9-605-054 (Rev. June 30).
23. Fisher, Robert (2008), **"HSBC Credit Card Rewards Program,"** Ivey Business Case # 908A17 (September 17).

24. John Deighton (2005), **"Nector- Making Loyalty Pay,"** Harvard Business Case #9-505-031 (Rev. December 5).

25. Hohn A. Quelch and Laura Winig (2008), **"Charles Schwab & Co., Inc.: The "Talk to Chuck" Advertising Campaign,"** Harvard Business Case # 9-507-005, (Rev. January 11).

26. Robin Ritchie, Alim Merali (2005), **"Investor Soft,"** Ivey Business Case #905A30 (November 29).

27. Robert C. Poze (2008), **"Fidelity Investment's Charitable Gift Fund (A),"** Harvard Business Case # 9-309-002 (July 30).

28. Sunil Chopra (2004), **"Excel Logistic Services,"** Kellogg Business Case # KEL019.

29. Lawrence, Anne T (2005), **"Premier Inc. (A),"** Babson College Case # BAB117 (Award-winning), Harvard Business School Publishing (Rev. February 15).

30. Robert Burgelman and Brooke Blumenstein (2007), **"Knowledge Management at Katzenbach Partners, LLC"** Stanford Business Case # SM162 (June 25).

31. Clayton M. Christensen (2004), **"Eli Lilly and Company: Innovations in Diabetic Care,"** Harvard Business Case #9-696-077 (Rev. April 15).

32. Andrew McAfee (2006), **"Pharmacy Service Improvement at CVS (A)"** Harvard Business Case #9-606-015 (Rev. October 20).

33. Thomke, Stefan and Ashok Nimgade (2002) **"Bank of America (A),"** Harvard Business Case #603022 (Rev. October 25).

34. James Heskett (2003), **"Southwest Airlines 2002: An Industry under Siege,"** Harvard Business Case # 9-803-133, (Rev. March 11).

35. Herman, Kerry and Thomas R. Eisenmann (2006), "Google, Inc." Harvard Business School Case #806105 (Rev. November 9).

36. David M. Maister (2006), **"University Health Services: Walk-In Clinic,"** Harvard Business Case # 9-681-061 (Rev. June16).

37. Geoffrey G. Jones and Alexis Lefort (2008), **"McKinsey and the Globalization of Consultancy,"** Harvard Business Case # 9-806-035 (Rev. July 3).

38. Munir Mandviwalla, Jonathan W. Palmer (2008), **"Globalization of Wyeth,"** Ivey Business Case # 908M17 (April 10).

Index

F

G

P

S